FLORA OF SOUTHEASTERN WASHINGTON AND OF ADJACENT IDAHO

By

HAROLD ST. JOHN

Emeritus Professor of Botany, University of Hawaii
Professor of Botany, Washington State University, 1920-1929.
Botanist, Bishop Museum, Honolulu

THIRD EDITION

OUTDOOR PICTURES

ESCONDIDO, CALIFORNIA

1963

First Edition June 19, 1937
Revised Edition March 1, 1956
Third Edition, November, 1963

Third Edition, October, 1963

Printed and bound in U.S.A.
by Edwards Brothers, Inc.

TABLE OF CONTENTS

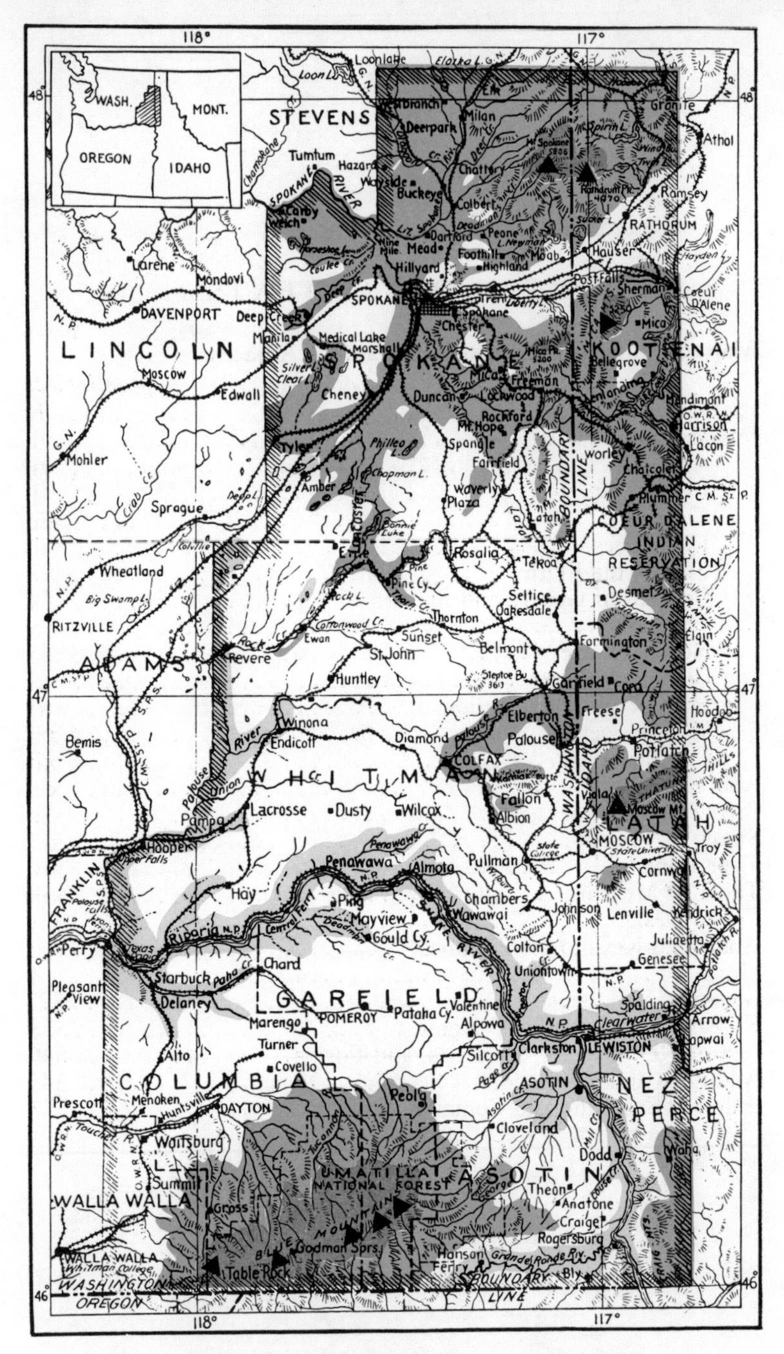

MAP OF SOUTHEASTERN WASHINGTON AND OF ADJACENT IDAHO.

Life Zones

Yellow. Upper Sonoran *(U. Son.)*.
White. Arid Transition, Timberless *(A.T.)*.
Green. Arid Transition, Timbered *(A.T.T.)*.
Blue. Canadian *(Can.)*.
Black Triangles. Hudsonian *(Huds.)*.
Scale: 1 inch = 23 miles.

PREFACE

The Pacific Northwest has a large and varied flora, still imperfectly known. There is a need for new and more accurate accounts of its classification and distribution.

Successive botanists at Washington State University have published floras of the adjoining region. In 1901 the Flora of the Palouse Region was published by Profs. C. V. Piper and R. K. Beattie. Later, in 1914, they published a much larger work, the Flora of Southeastern Washington and Adjacent Idaho. The writer acknowledges his indebtedness to the detailed and excellent work of these earlier botanists. His own first edition was the result of studies made during nine years of residence in Pullman, Wash., and seven years in Honolulu. The second edition, published in 1956, contained much additional material. The present or third edition has been thoroughly revised after studies by the author in the herbaria in Cambridge, Mass., and Pullman, Wash. It is presented in the hope that it may be of use to botanists, and that it may introduce students to the beauty, the value, and the family kinship of this northwestern flora.

Numerous specialists, students, nature-lovers, librarians, and others have aided the author. The officials of Washington State University have been encouraging and helpful for many years. Former students who have actively helped in the exploration and study of the flora include: C. S. Parker, W. D. Courtney, R. Sprague, C. S. Spiegelberg, F. A. Warren, Edith Hardin, Gladys Weitman, Rocelia C. Palmer, J. A. Moore, C. S. English, Jr., and G. N. Jones. The Spokane botanists, T. A. Bonser, and Miss Nettie M. Cook coöperated actively. My wife has worked untold hours on the manuscript. To all these, and to the many others who also helped, the author offers his hearty thanks.

Previous books on the flora of this and of adjacent regions have been freely consulted and used. Acknowledgments are hereby made to them.

All available reports on, and collections of, the flora have been consulted when possible. Nearly all of the collections are now deposited in the Herbarium of Washington State University, Pullman, Washington.

INTRODUCTION

Area

This flora covers southeastern Washington and adjacent Idaho. As shown on the map, the frontispiece, this area includes in Washington: Spokane, Whitman, Asotin, Garfield, Columbia Counties and part of Walla Walla County; in Idaho: a strip about 15 miles wide adjacent to the Washington area. This includes the regions around Spokane, Cheney, Colfax, Pullman, and Walla Walla in Washington; and Coeur d'Alene, Moscow, and Lewiston, Idaho.

Physiography

The region is one of intermontane plateaus and low mountains. These are mostly of old crystalline or metamorphic rocks, granite, or quartzite. The plains below, called the "Palouse Country," are formed by many successive lava flows forming great sheets, each some 50 feet thick, and totaling about 2000 feet in thickness. On top of these lavas is a layer of deep, rich soil, the Palouse siltloam, about forty feet thick. This soil is eroded by water and wind, and somewhat rearranged by wind into the endless rolling hills, 100 to 200 feet high. These are often crescent shaped, oriented to the prevailing southwesterly winds, and exposing the underlying basalt as "rim-rock" only on their windward slopes.

The plateau is deeply trenched by the major streams. Most spectacular of all is the canyon of the Snake River, which is cut to a depth of from 1000 to 3000 feet, and at Granite Point exposes the underlying granite rock.

The elevations vary from 523 feet at Perry to 6370 feet at Table Rock in the Blue Mts.

The drainage is in general westerly, the smaller streams all being tributaries of the Snake River or the Spokane River.

The Pleistocene glaciation invaded the area from the north. The ice sheet advanced as far south as Spangle, and profoundly changed the surface. Great deposits of gravel and sand partly fill the valleys to the north or are spread about as dry, rather sterile sand plains. These sands and gravels have a somewhat special flora.

From the wavy edge of the ice front came enormous torrents of water, in some cases ten miles wide. These swept off the Palouse soil, making bare, barren areas, the "Scablands," and cutting deep, spectacular canyons, the "Coulees." This area extends in an interlacing pattern from Spangle and Medical Lake southward to the mouth of the Palouse River and to our western boundary. Within the glaciated area and the Scabland area are many lakes or ponds. In the more arid regions some of these are alkaline.

Mountains occur in several areas. The southernmost part is the Blue Mts., a great dome of basalt, its highest peak in Washington being Table Rock, 6370 feet high. Across the Snake to the west is a lower, narrow basaltic ridge, the Craig Mts. Northward along the Idaho border is a series of low, wooded mountains, culminating in Moscow Mt. 5000 ft., Mica Peak 5250 ft., Rathdrum Peak 4970 ft., and Mt. Spokane 5208 ft. high. Outlying peaks to the eastward are Kamiak Butte 3650 fet. and Steptoe Butte 3614 ft. high.

Climate

The climate is rather severe, and of a continental type.

The rainfall tends to increase from the low regions near the Snake River to the higher areas to the northeast and the south. At Hooper the annual rainfall is 13 inches, as it is at Lewiston. At Dayton it is 24 inches, at Pullman 22 inches, and doubtless it is much higher on the forested mountain peaks to the south and east. A distinct dry season prevails in the summer, most of the precipitation coming during the remainder of the year. The winters are cold and the summers hot. The extremes (quoted from Washington Geological Survey, Bulletin 17) are for Pullman, minimum — 18° F., maximum 104° F., and later records would show greater extremes.

Life Zones

To any observer the division of the vegetation into distinct zones is obvious. The most useful classification is that into life zones made by Dr. C. H. Merriam. This concept was used by

Prof. C. V. Piper in his book, the Flora of Washington (1906), and well presented graphically on his map.

The *Upper Sonoran* is the zone of the arid regions. The vegetation is sparse and mainly shrubby. Dominant species and indicators of this zone are: scabland sagebrush, *Artemisia rigida;* the prickly pear, *Opuntia polyacantha;* rabbit brush, *Chrysothamnus nauseosus* var. *albicaulis; Chaenactis Douglasii* var. *Douglasii; Penstemon triphyllus* subsp. *triphyllus;* sand dock, *Rumex venosus;* salmon globe mallow, *Sphaeralcea Munroana,* etc.

The *Arid Transition Timberless* is the zone of grasslands or prairies, known locally as the "Bunch Grass Prairies." They are nearly treeless, except for the thickets along streams. Indicator species are: bunch grass, *Agropyron spicatum* var. *inerme; Poa Sandbergii; Geranium viscosissimum* forma *viscosissimum;* balsamroot, *Balsamorhiza sagittata;* pik, *Wyethia amplexicaulis; Iris missouriensis* forma *missouriensis;* purple Trillium, *Trillium petiolatum,* etc.

The *Arid Transition Timbered* is the zone of the open yellow pine forests. Indicator species are: yellow pine, *Pinus ponderosa;* buckbrush, *Ceanothus sanguineus;* huckleberry, *Vaccinium membranaceum;* mallow ninebark, *Physocarpus malvaceus.*

The *Canadian* is the zone of moist, dense woods on the middle slopes of the mountains. Indicator species are: white fir, *Abies grandis;* tamarack, *Larix occidentalis;* Engelmann spruce, *Picea Engelmanni;* giant cedar, *Thuja plicata;* fool's huckleberry, *Menziesia ferruginea;* mountain lover, *Paxistima myrsinites;* and bunchberry, *Cornus canadensis.*

The *Hudsonian* is the zone at the upper limit of trees on the mountains. It is not well developed in our area, but is recognizable on the peaks of our highest mountains. Indicator species are: subalpine fir, *Abies lasiocarpa;* pink heather, *Phyllodoce empetriformis; Senecio triangularis* var. *triangularis;* heliotrope, *Valeriana sitchensis* subsp. *sitchensis.*

Detailed studies have shown that few, even of the indicator, plants are absolutely constant to one life zone. If their entire geographic range is considered it is often found that they occur in two or more life zones. Also if a number of indicator species of one zone are studied and their ranges mapped in detail, it will be seen that their ranges do not exactly coinside. Hence, the life zones do not seem to be scientific concepts capable of precise definition. On the other hand they are generalizations of the mass association of plants characteristic of the great physiographic and climatic areas. They have a meaning and a use. To the naturalist, the name Upper Sonoran brings an indelible picture of hot, arid plains or canyons with sagebrush, cactus, jack-rabbits, and horned toads. The name Canadian brings an image of deep, moist woods in the mountains, with the shade and fragrance of spruce, fir, and cedar. It is because of this value as a generalization that the life zone concept is used in this book. Whenever known the life zone is stated for each species. It is based on its occurrence within this area, and disregards its zonal occurrence elsewhere, which may or may not be identical.

Size of the Flora

As here presented the flora consists of 459 genera, 1,187 species and 286 subdivisions of species. Of this total 1,266 are indigenous, and 207 are adventive. These species and infraspecific taxa are divided among the plant groups as follows: 27 Pteridophyta; 13 Gymnospermae; 337 Monocotyledones; and 1,096 Dicotyledones. Other kinds of plants, such as those present only as cultivated crops or as ornamentals in gardens, are excluded from this book.

Concept of Genera and Species

The writer does not subscribe to the recognition of minute genera and species. In weighing the evidence on such questions, differences should of course be considered, but the resemblances should also be considered and be given equal weight. Throughout this work the writer has endeavored to accept as genera and species the so-called Linnaean genera and species, and to reduce to subdivisions of species those minor elements so frequently announced as species by recent American botanists. The writer's concept of genera and species is not materially different from that of Piper.

Taxonomy

As a system of major classification that of A. Engler and K. Prantl is followed, largely that presented in the first edition of their Natürlichen Pflanzenfamilien, though the few volumes published of the second edition have also been used. In a few cases the writer has deviated from the Engler and Prantl system, as in the acceptance of the *Lobeliaceae,* the *Fumariaceae,* and the *Zannichelliaceae.*

The author is fully aware of the systems of plant taxonomy of Bessey, and of Hutchinson, but up to the present that of Engler and Prantl seems the best.

Nomenclature

The acceptance of the scientific names of plants is determined by the laws of nomenclature. In this flora the International Code of Botanical Nomenclature, 1956, is followed consistently, except where modified by the subsequent International Botanical Congress at Montreal, 1959.

Common Names

Common names are given when known. These are either genuine folk names actually in current use in this area, or names of widely distributed species that are almost universally accepted. The folk names are in some instances at variance with accepted usage in other regions or that of many books, as tansy for *Achillea millefolium,* var. *lanulosa,* yet by local usage it is in the truest sense a common name, as is heliotrope for *Valeriana sitchensis.* Neither priority nor botanical rule can determine the acceptance of a common name. Their use is subject to growth or modification just like other parts of a language. No attempt has been made to include book-made English names, such as the translations of the Latin scientific name. These are not known to the people, and in most cases never become real common names. A large proportion of our flora, even of the large and conspicuous flowers, have no common names. It is hoped that our plants will become better known to the people and that this will be evidenced by the acceptance or creation of more common names.

Habitat and Distribution

The habitat is given for each species, when known. This is a description of the kind of place in which the plant grows.

The distribution is also stated. When the plant is known from only a very few localities, these are cited exactly, together with the name of the collector and the number of the collection. When the species is more abundant, its occurrence is referred to the one or more life zones in which it is found.

Measurements

Statements of altitude and distance are given in English feet or miles. This is done for practical reasons, as the only good and detailed maps, the official Geological Survey, topographic quadrangles, are constructed on this system. Measurements of the dimensions of the plants or their parts are given in the metric system, in millimeters, centimeters, meters, etc. This is almost universally used in science, and is rapidly becoming more familiar in our country, due to its use in sport, in radio, in technical work, and foreign trade. For those not familiar with the metric system, a table of equivalents is given.

Conclusion

The writer is now on the staff of the B. P. Bishop Museum, in Honolulu, but he has a continued interest in the flora of the Pacific Northwest. He would be glad to hear of additions to or need changes in this flora.

Faculty of Science,
Cairo University,
Giza,
United Arab Republic.

ABBREVIATIONS

apm. = apomict.
auth. = authors.
Calif. = California.
cm. = centimeter.
Co. = County.
Cr. =Creek.
e. = east.
Eur. = Europe.
f. = filius; the son.
f. = forma (used only after binomials in the keys).
Ga. = Georgia.
Gr. = Greek.
Ida. = Idaho.
Jct. = Junction.
L. = Lake.
Lat. = Latin.
Ldg. = Landing.
m. = meter.
mm. = millimeter.
μ = mu, one micron; 1/1,000 of a millimeter.
Mts. = Mountains.
n. = north.
Natl. = National.
Ore. = Oregon.
Pa. = Pennsylvania.
R. = Range; or River.
R.R. = Railroad.
s. = south.
S. Am. = South America.
Sprs. = Springs.
ssp. = subspecies (used in the keys).
subsp. = subspecies.
T. = Township.
U. S. = United States.
var. = variety.
w. = west.
Wash. = Washington.

LIFE ZONES

	Legend on map.
U. Son. = Upper Sonoran	Yellow.
A. T. = Arid Transition Timberless	White.
A. T. T. = Arid Transition Timbered	Green.
Can. = Canadian	Blue.
Huds. = Hudsonian	Black Triangles.

SIGNS

— dash connects figures of the extremes of variation; as 6–8 mm. = from 6 to 8 mm.
× crossed with; the sign of a hybrid.

TYPE FONTS

Native or indigenous plants in Black Face.
Introduced or adventive plants in Roman.
Excluded species placed at the end of the genus, in Roman.
Common names in italics.
Synonyms in italics.

TABLE OF MEASURES

1 μ, one micron = 1/1,000 mm. = 1/25,400 in.
1 mm., one millimeter = 1/10 cm. = 1/25 in.
1 cm., one centimenter = 1/10 dm. = 2/5 in.
1 dm., one decimeter = 1/10 m. = 4 in.
1 m., one meter = 1/1,000 km. = 39.4 in.

1/25 in. = 1 millimeter (mm.)
1 in. = 25 millimeters (mm.) = 2.5 centimeters (cm.).
1 ft. = 30 centimeters (cm.) = 3 decimeters (dm.).
1 yd. = 91.4 centimeters (cm.) = almost a meter, 1 m. less 8.6 cm.

COVER DESIGN

The cover design is a silhouette of one of the commonest and most beautiful spring flowers of this region, the Lambs' Tongue (*Erythronium grandiflorum*).

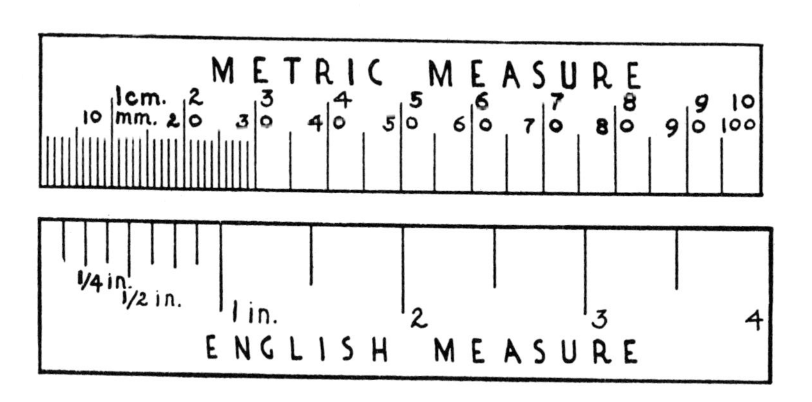
METRIC MEASURE
1cm.
mm.
ENGLISH MEASURE
1/4 in.
1/2 in.
1 in.

ANALYTICAL KEY TO THE FAMILIES

Plants with woody bundles, reproducing by spores. (See Fig. 1).
Subdivision PTERIDOPHYTA, xvii

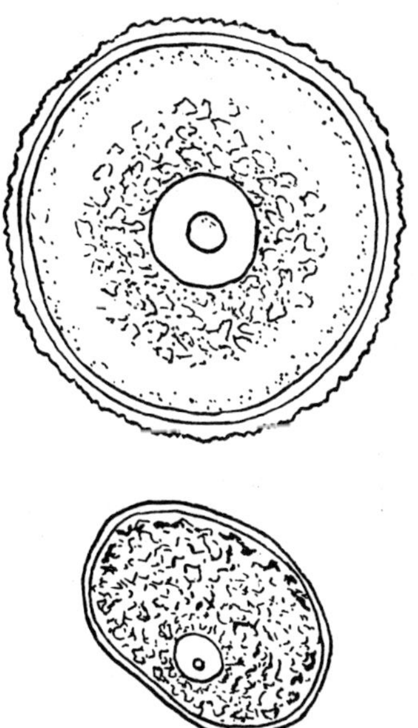

Fig. 1 Spores, highly magnified.

Plants with woody bundles, reproducing by seeds. (See Fig. 2).
Division EMBRYOPHYTA SIPHONOGAMA
(SPERMATOPHYTA), xix

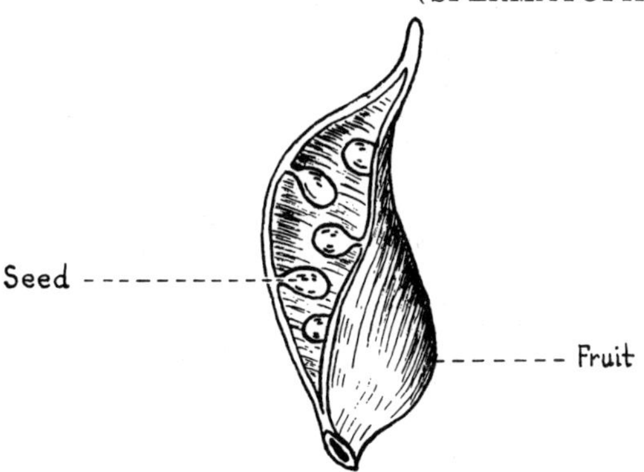

Fig. 2. Opened fruit showing seeds.

Subdivision PTERIDOPHYTA

Leaves few, large; stems mostly underground,
 Leaves 4-foliolate, clover-like; spore cases in closed pod-like sporocarps. — MARSILEACEAE, 10
 Leaves not 4-foliolate; spore cases not in sporocarps,
  Spore cases in the tissue of a prominent and distinct fertile lobe of the leaf. — OPHIOGLOSSACEAE, 2
  Spore cases formed of outgrowths from the surface of the leaf,
   Rhizome protostelic or solenostelic, a single vascular strand entering the stipe,
    Rhizome usually clothed with articulate hairs or bristles, or lindsayoid scales,
     Sori marginal; false indusium opening introrsely. — PTERIDACEAE, 8
     Sori dorsal, exindusiate, on the veins of the under side. — GYMNOGRAMMACEAE, 2
    Rhizome and often aerial parts clothed in narrow, flat scales,
     Sorus globose, on the distal vein end, but becoming confluent in an intramarginal line; indusium linear, continuous or interrupted; spores warty or spiny. — SINOPTERIDACEAE, 8
     Sorus linear, exindusiate, spores smooth,
      Sorus on the apex of the vein or sometimes between the veins under a false indusium. — ADIANTACEAE, 3
      Sorus generally along most of the vein's length; frond margin plane. — GYMNOGRAMMACEAE, 7
   Rhizome dictyostelic, with 2-7 vascular strands entering the stipe,
    Indusium inferior, calyciform or fringed. — WOODSIACEAE, 9
    Indusium, if any, superior,
     Rhizome simply dictyostelic, with 2 broad vascular strands entering the stipe, remaining distinct or united above; fronds not setulose with unicellular simple or forked, or articulate hairs; sori linear or oblong. — ASPLENIACEAE, 5
     Rhizome with a complicated dictyostele often mixed with parenchyma and dark sclerenchyma, with 3-7 strands, roundish and distinct in the stipe,
      Sori round or oblong, without paraphyses; indusium (when present) round-reniform with deep sinus or round and peltate; scales broad, copious; midrib and costa often with ctenitis-hairs. — ASPIDIACEAE, 3
      Sori elliptic-oblong; indusium entire, elliptic-oblong; scales of rhizome sparse or none; ctenitis-hairs none. — POLYPODICEAE, 7

Leaves numerous, small; stems aerial and underground,
  Leaves minute, whorled; stems jointed, hollow. — EQUISETACEAE, 10
  Leaves small, not whorled; stems not jointed, solid,
    Spore cases in the enlarged bases of the leaves; stems short, corm-like. — ISOETACEAE, 12
    Spore cases in the axils of the leaves; stems elongated. — SELAGINELLACEAE, 2

Division EMBRYOPHYTA SIPHONOGAMA (SPERMATOPHYTA)

Ovules and seeds not in a closed cavity; usually on the face of an open scale-leaf (see Fig. 3); stigmas none. — Subdivision GYMNOSPERMAE, xx

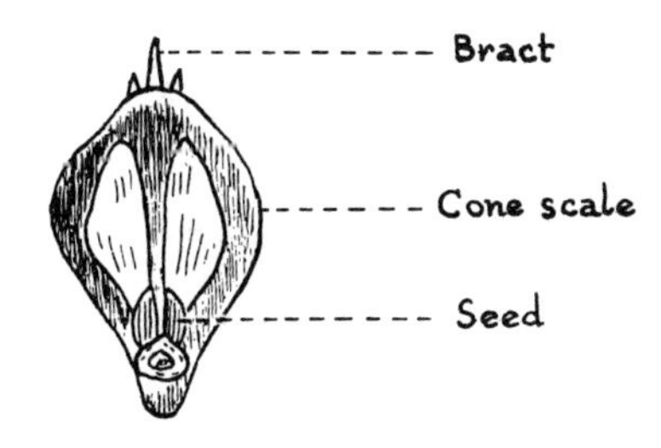

Fig. 3. Naked seeds.

Ovules and seeds contained in a closed cavity surrounded by one or more closed and modified leaves forming an ovary (see Fig. 4); stigmas present. — Subdivision ANGIOSPERMAE, xx

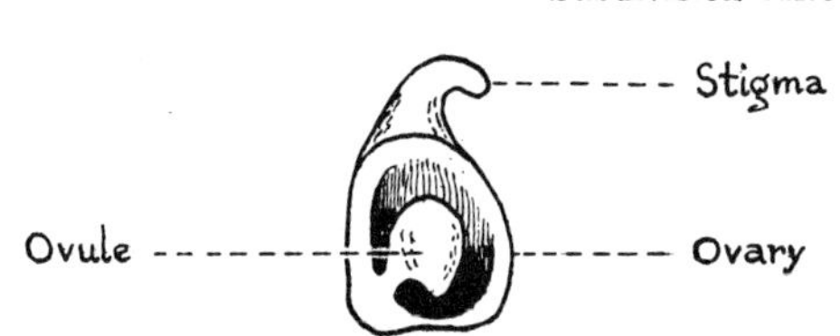

Fig. 4. Ovary containing ovule.

Cotyledon one; stem with no distinction into bark, wood and pith (*endogenous*); leaves usually parallel-veined; parts of the flowers nearly always in threes. — Class MONOCOTYLEDONES, xx
Cotyledons two; stem (in woody species) of bark, wood and pith (*exogenous*); leaves usually net-veined; parts of the flower in fours, fives or sixes, never in threes. — Class DICOTYLEDONES, xxi

Subdivision GYMNOSPERMAE

Fruit a cone or berry-like,
  Leaves and cone scales spirally arranged; ovuliferious scales free or slightly adnate to the bract. PINACEAE, 14
  Leaves and cone scales opposite or decussate or in whorls of 3-8; ovuliferious scale completely adnate to the bract. CUPRESSACEAE, 18
Fruit drupe-like. TAXACEAE, 14

Subdivision ANGIOSPERMAE

CLASS MONOCOTYLEDONES

Plants small, floating, with no distinction of stem and leaves. LEMNACEAE, 86
Plants with normal foliage,
  Inflorescence a fleshy spadix. (See Fig. 5). ARACEAE, 86

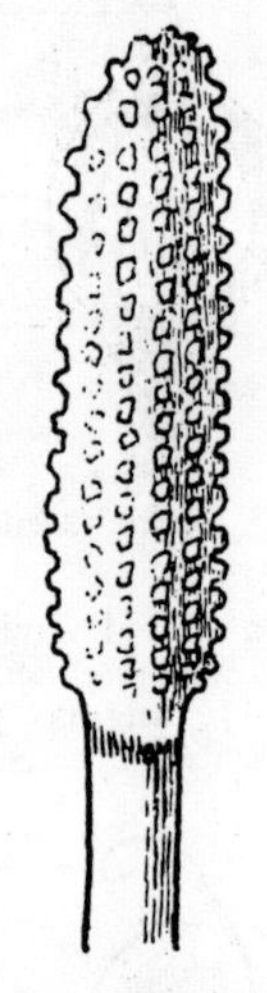

Fig. 5. Spadix.

Inflorescence not a fleshy spadix,
  Perianth none, or of bristles, chaffy scales, or a hyaline envelope,
    Flowers not in the axils of chaffy bracts,
      Perianth herbaceous or none,
        Flowers solitary, axillary. NAJADACEAE, 25
        Flowers in axillary or terminal clusters,
          Carpels 3 or 6, united into a compound ovary. JUNCAGINACEAE, 25
          Carpels 2–5, separate,
            Flowers perfect, in spikes; sepals 4; stamens 4. POTAMOGETONACEAE, 21
            Flowers unisexual, axillary; staminate flowers with 1–2 stamens but no perianth; pistillate flowers with a funnelform, undivided perianth. ZANNICHELLIACEAE, 24
      Perianth of bristles or chaffy scales,
        Flowers in terminal cylindrical spikes. TYPHACEAE, 20

Flowers in axillary globular heads. Sparganiaceae, 20
Flowers in the axils of chaffy bracts,
Stems mostly hollow, jointed; leaves 2-ranked. Gramineae, 28
Stems solid; leaves 3-ranked. Cyperaceae, 69
Perianth present, the parts glume- or petal-like,
Perianth of glume-like segments. Juncaceae, 88
Perianth at least in part petal-like,
Carpels distinct. Alismataceae, 26
Carpels united,
Ovary superior,
Flowers more or less irregular, surrounded by a spathe. Pontederiaceae, 87
Flowers regular, without a spathe. Liliaceae, 92
Ovary inferior,
Terrestrial; flowers perfect,
Flowers regular. Iridaceae, 104
Flowers irregular. Orchidaceae, 106
Aquatic; flowers dioecious or polygamous. Hydrocharitaceae, 27

Class DICOTYLEDONES

I. Petals Distinct to the Base or Wanting (see Fig. 6)

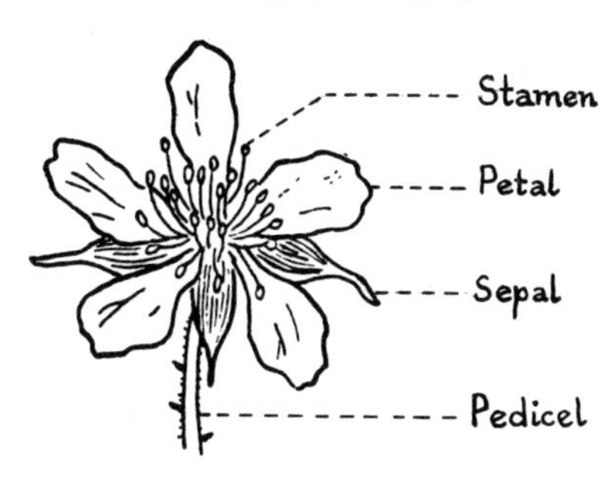

Fig. 6. Petals distinct to the base.

ARCHICHLAMYDEAE

A. Petals None

Plant parasitic on other plants and without chlorophyll. Loranthaceae, 125
Plant not parasitic, chlorophyll-bearing,
Trees or shrubs,
Leaves opposite. Aceraceae, 275
Leaves alternate,
Pistillate flowers not in aments,
Ovary 1-celled,
Plant without milky juice; flowers in sessile heads or racemes. Ulmaceae, 121
Plant with milky juice; flowers in pedunculate spikes. Moraceae, 122
Ovary 2–4-celled. *Rhamnus*, 278
Pistillate flowers in aments,
Calyx present. Betulaceae, 119

Calyx not present. SALICACEAE, 112
Herbs (sometimes somewhat woody at base),
Leaves opposite or whorled,
Leaves whorled,
Leaves dichotomously thrice forked. CERATOPHYLLACEAE, 160
Leaves simple or pinnatifid. HALORAGACEAE, 291
Leaves opposite,
Flowers perfect,
Style 1. *Ludwigia*, 301
Styles 2 or more,
Styles 2–5,
Fruit a many seeded capsule. CARYOPHYLLACEAE, 149
Fruit an achene. *Eriogonum*, 127
Styles many. *Clematis*, 165
Flowers monoecious,
Ovary 1-celled; stamens 2–5. URTICACEAE, 123
Ovary more than 1-celled; stamen 1,
Ovary 4-celled. CALLITRICHACEAE, 271
Ovary 3-celled. EUPHORBIACEAE, 270
Leaves not opposite,
Flowers monoecious,
Ovary 3-celled, 3-ovuled. EUPHORBIACEAE, 270
Ovary 1-celled, 1-ovuled,
Flowers with scarious bracts. AMARANTHACEAE, 143
Flowers bractless or, if bracted, the bracts not scarious. CHENOPODIACEAE, 137
Flowers perfect or dioecious,
Pistils more than one,
Stamens perigynous. ROSACEAE, 213
Stamens hypogynous. RANUNCULACEAE, 162
Pistil one,
Ovary more than 1-celled,
Ovary 2-celled,
Stamens 6, rarely 2 or 4; placentae parietal. CRUCIFERAE, 178
Stamens 2; placentae axile. *Besseya*, 396
Ovary more than 2-celled,
Ovary 6-celled. ARISTOLOCHIACEAE, 126
Ovary 3–5-celled. MOLLUGINACEAE, 145
Ovary 1-celled,
Ovary superior,
Fruit an achene. POLYGONACEAE, 127
Fruit not an achene. CHENOPODIACEAE, 137
Ovary partly inferior,
Flowers on a scape. *Heuchera*, 202
Flowers on a leafy stem. SANTALACEAE, 124

B. Petals Present

1. Stamens numerous, at least more than ten and more than twice the sepals or calyx-lobes.

Calyx free and separate from the ovary,
Pistils more than one,

Ovaries cohering in a ring around a central axis. MALVACEAE, 279
Ovaries separate or, if united, not cohering in a ring around the central axis,
Stamens perigynous. ROSACEAE, 213
Stamens hypogynous,
Aquatic plants; leaves not dissected. NYMPHAEACEAE, 161
Terrestrial plants or, if aquatic, the submersed leaves dissected. RANUNCULACEAE, 162
Pistil one, with one to several styles and stigmas,
Leaves minutely punctate with pellucid dots. GUTTIFERAE, 281
Leaves not punctate with pellucid dots,
Ovary simple,
Stamens hypogynous. RANUNCULACEAE, 162
Stamens perigynous. ROSACEAE, 213
Ovary compound,
Aquatics. NYMPHAEACEAE, 161
Terrestrial plants,
Ovary with parietal placentae. PAPAVERACEAE, 176
Ovary with a basal placenta. PORTULACACEAE, 145

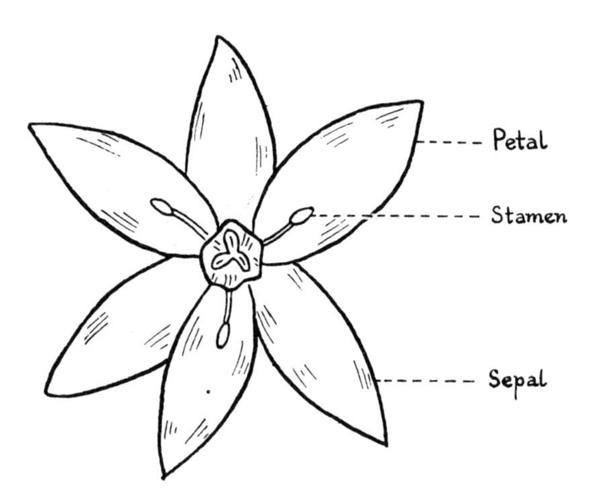

Fig. 7. Stamens opposite the petals.

Calyx more or less coherent with the surface of the compound ovary,
Ovary more than 1-celled,
Leaves alternate, with stipules. ROSACEAE, 213
Leaves opposite, without stipules. *Philadelphus*, 205
Ovary 1-celled,
Placenta basal, PORTULACACEAE, 145
Placenta parietal,
Plants fleshy, leafless or with minute leaves; petals many. CACTACEAE, 290

Plants not fleshy, with broad leaves,
Herbs; fruit a capsule. — LOASACEAE, 288
Shrubs or trees; fruit a pome. — *Crataegus,* 217

2. Stamens not more than twice as many as the petals.
Stamens opposite the petals or only partly so (see Fig. 7),
Ovary 2–4-celled. — RHAMNACEAE, 277
Ovary 1-celled,
Anthers opening by uplifted valves. — BERBERIDACEAE, 175
Anthers not opening by uplifted valves. — PORTULACACEAE, 145

Stamens not opposite the petals (see Fig. 8),

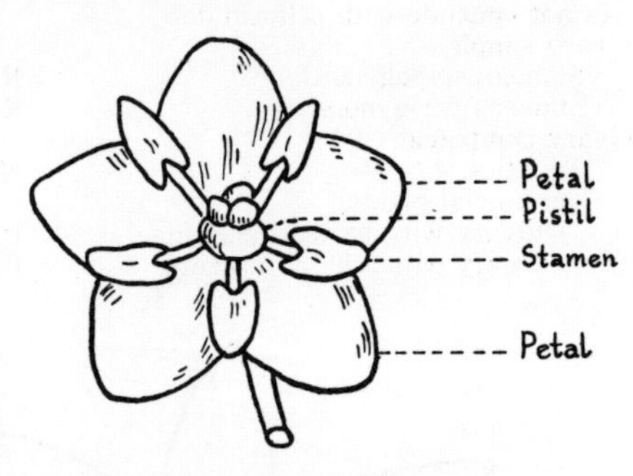

Fig. 8. Stamens not opposite petals.

Ovary wholly superior,
Ovaries two or more,
Ovaries somewhat united at the base, separate above,
Trees,
Fruit a single samara. — SIMAROUBACEAE, 269
Fruit a double samara. — ACERACEAE, 275
Herbs. — SAXIFRAGACEAE, 201
Ovaries entirely separate,
Stamens united with each other and with the stigma. — ASCLEPIADACEAE, 343
Stamens free from each other and from the stigma,
Stamens hypogynous,
Herbs,
Leaves not fleshy. — RANUNCULACEAE, 162
Leaves thick and fleshy. — CRASSULACEAE, 200
Shrubs or trees. — SIMAROUBACEAE, 269
Stamens perigynous,
Stamens just twice as many as the pistils. — CRASSULACEAE, 200
Stamens not just twice as many as the pistils,
Leaves without stipules. — SAXIFRAGACEAE, 201
Leaves with stipules. — ROSACEAE, 213
Ovary only one,
Ovary 3–5-lobed and beaked with a united style,

Leaves not pinnately compound. GERANIACEAE, 264
Leaves pinnately compound,
Cells of the ovary twice as many as the sepals. ZYGOPHYLLACEAE, 268
Cells of the ovary as many as the sepals (2–3) LIMNANTHACEAE, 272
Ovary not lobed and beaked,
Ovary simple with 1 parietal placenta. LEGUMINOSAE, 237
Ovary compound, as shown by the number of cells, placentae, styles or stigmas,
Ovary 1-celled,
Corolla irregular,
Stamens 6; petals 4. FUMARIACEAE, 177
Stamens and petals 5. VIOLACEAE, 283
Corolla regular or nearly so,
Ovule 1,
Shrubs or trees. ANACARDIACEAE, 272
Herbs. CRUCIFERAE, 178
Ovules more than 1,
Placenta central or basal,
Herbs. CARYOPHYLLACEAE, 149
Shrubs. *Glossopetalon*, 274
Placenta parietal,
Leaves punctate with pellucid dots. GUTTIFERAE, 281
Leaves not punctate,
Petals 4. CAPPARACEAE, 199
Petals 5. SAXIFRAGACEAE, 201
Ovary 2–several-celled,
Stamens neither just as many nor twice as many as the petals,
Stamens 6, tetradynamous, (or 2). CRUCIFERAE, 178
Stamens 5, regular. BALSAMINACEAE, 276
Stamens either just as many or twice as many as the petals,
Ovules 1 or 2 in each cell of the ovary,
Herbs,
Cells of the fruit 5. GERANIACEAE, 264
Cells of the fruit 10. LINACEAE, 267
Shrubs or trees,
Low shrub with pinnately veined leaves. CELASTRACEAE, 273
Trees with palmately veined leaves. ACERACEAE, 275.
Ovules several to many in each cell of the ovary,
Leaves opposite, with stipules. ELATINACEAE, 282
Leaves, when opposite without stipules,

Stamens on the calyx,
Style 1. LYTHRACEAE, 290
Styles 2–3. SAXIFRAGACEAE, 201
Stamens free from the calyx,
Style 1. ERICACEAE, 327
Styles 2–5.
Stamens and calyx free from the ovary. CARYOPHYLLACEAE, 149
Stamens and calyx united to the ovary. ARALIACEAE, 304

Ovary at least half inferior,
Ovules and seeds more than one in each cell of the ovary,
Perianth parts 2 or 4. ONAGRACEAE, 292
Perianth parts 5 (rarely 4). SAXIFRAGACEAE, 201
Ovules and seeds but one in each cell of the ovary,
Petals 2 or 4,
Styles or stigmas 4. HALORAGACEAE, 291
Style 1,
Stamens, 2 or 8; fruit indehiscent and nut- or bur-like. ONAGRACEAE, 292
Stamens 4; fruit a drupe. CORNACEAE, 326
Petals 5,
Trees or shrubs; flowers in corymbs. *Crataegus*, 217
Herbs; flowers in umbels or heads. UMBELLIFERAE, 305

II. Petals More or Less United into One Piece (see Fig. 9).

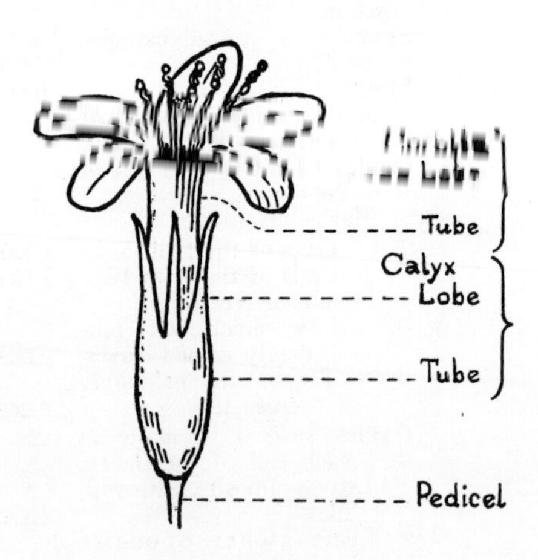

Fig. 9. Petals united.

METACHLAMYDEAE

Stamens more numerous than the corolla lobes,
  Ovary 1-celled,
    Ovary with 1 parietal placenta. LEGUMINOSAE, 237
    Ovary with 2 parietal placentae. FUMARIACEAE, 177
  Ovary 3–many-celled,
    Stamens free from the corolla. ERICACEAE, 327
    Stamens united with the base of the corolla. MALVACEAE, 279
Stamens as many as the corolla lobes or fewer,
  Stamens opposite the corolla lobes (see Fig. 10). PRIMULACEAE, 335

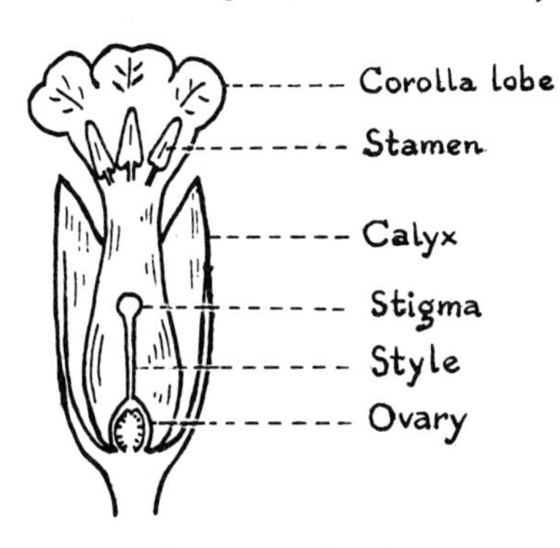

Fig. 10 Stamens opposite the corolla lobes.

  Stamens alternate with corolla-lobes or fewer,
    Ovary superior,
      Corolla more or less irregular,
        Fertile stamens 5. *Verbascum*, 414
        Fertile stamens 4 or 2,
          Ovules solitary in the cells of the ovary,
            Ovary 4-lobed, the style rising from between the lobes (see Fig. 11). LABIATAE, 381
            Ovary not lobed, the style rising from its apex. VERBENACEAE, 380
          Ovules 2 or more, usually numerous, in each cell,
            Placentae axile,
              Ovary 2-celled; land plants. SCROPHULARIACEAE, 395
              Ovary 1-celled; submerged aquatics. LENTIBULARIACEAE, 417
            Placentae parietal; parasitic herbs. OROBANCHACEAE, 418
      Corolla regular,
        Stamens fewer than the corolla-lobes,
          Corolla scarious. PLANTAGINACEAE, 419
          Corolla not scarious,
            Style 2-lobed. *Lycopus*, 383
            Style single. *Veronica*, 415

Stamens as many as the corolla-lobes,
  Ovaries 2, separate,
    Filaments distinct. — APOCYNACEAE, 342
    Filaments monadelphous. — ASCLEPIADACEAE, 343
  Ovary 1,
    Ovary deeply 4-lobed around the style (see Fig. 11),

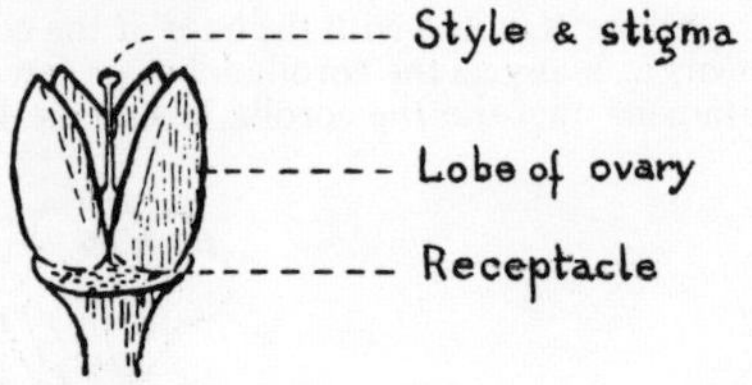

Style & stigma
Lobe of ovary
Receptacle

Fig. 11. Ovary deeply 4-lobed around the style.

      Leaves alternate. — BORAGINACEAE, 365
      Leaves opposite. — *Mentha,* 385
    Ovary not deeply lobed,
      Ovary 1-celled,
        Leaves entire, opposite or whorled. — GENTIANACEAE, 338
        Leaves, if entire, alternate or basal,
          Corolla conspicuously bearded on the upper surface,
            Leaves trifoliolate; flowers in racemes. — *Menyanthes,* 340
            Leaves simple; flowers solitary. — *Hesperochiron,* 359
          Corolla not conspicuously bearded. — HYDROPHYLLACEAE, 350
      Ovary 2 or more celled,
        Stamens free from the corolla. — ERICACEAE, 327
        Stamens on the corolla tube.
          Stamens 4. — PLANTAGINACEAE, 419
          Stamens 5,
            Fruit a many-seeded pod or berry. — SOLANACEAE, 391
            Fruit a few-seeded pod,
              Style 3-lobed. — POLEMONIACEAE, 347
              Style undivided or 2-cleft. — CONVOLVULACEAE, 345

Ovary inferior,
  Plants with tendrils. — CUCURBITACEAE, 431
  Plants without tendrils,

Anthers united into a ring or tube *(syngenesious)*,
- Flowers in an involucrate head on a common receptacle. — COMPOSITAE, 436
- Flowers separate, not involucrate. — LOBELIACEAE, 434

Anthers not united,
- Stamens on the ovary. — CAMPANULACEAE, 432
- Stamens on the corolla,
  - Stamens 1–3. — VALERIANACEAE, 430
  - Stamens 4–5,
    - Ovary 1-celled; flowers in a dense involucrate head. — DIPSACACEAE, 430
    - Ovary 2–5-celled; flowers not in an involucrate head,
      - Leaves alternate. — CAMPANULACEAE, 432
      - Leaves opposite or whorled,
        - Leaves opposite or whorled, when opposite with stipules. — RUBIACEAE, 420
        - Leaves opposite, without stipules. — CAPRIFOLIACEAE, 423

Flora of Southeastern Washington and of Adjacent Idaho

SUBDIVISION PTERIDOPHYTA. Ferns

Plant containing woody tissue and vessels in the stem and producing spores asexually which, on germination, develop very small structures called prothallia, on which are borne the sexual reproductive organs from which the asexual plant is developed. The sexual plant is rarely collected, and classification is based mainly on the characters of the asexual plant. (Name Gr., *pteris*, fern; *phuton*, plant.)

Class I. *LYCOPODINEAE*

Plant moss-like; stems branched, solid, with numerous small leaves; sporangia solitary in the axils of the leaves or on their upper surface. (Named from the genus *Lycopodium*.)

LYCOPODIACEAE

Low trailing or erect plants; stems mostly elongate, branched and leafy; leaves mostly evergreen, small, subulate to oblong; sporangia on the upper base of the sporophylls which are leafy or reduced. (Named from its genus *Lycopodium*.)

LYCOPODIUM

Perennials; leaves imbricate, 1-nerved; sporangia plump or flattened, 1-locular, 2-valved; spores powdery, inflammable. (Name from Gr., *lycos,* wolf; *pous,* foot.)

Lycopodium annotinum L., var. **annotinum.** Prostrate stems 0.5–3 m. long, forking, with few leaves; erect branches 0.6–3 dm. tall, simple or forking; leaves 6–8-ranked, 2.5–11 mm. long, linear to lance oblong, subulate-tipped, spreading or reflexed; cones single, sessile, 0.6–4.5 cm. long, 4.5–7.5 mm. thick; sporophylls about 3 mm. long, peltate, deltoid-ovate, the margin hyaline and erose. Alder Creek, Benewah Co., Ida., *G. N. Jones* 723, recorded by Flowers (1950).

SELAGINELLACEAE

Terrestrial, annual or perennial moss-like plants with branching stems and scale-like leaves, which are many-ranked and uniform, or four-ranked and of two kinds spreading in two planes; sporangia 1-celled, in the axils of leaves which are so arranged as to form more or less quadrangular spikes; spores of two kinds, the larger sporangia (*megasporangia*) containing four megaspores, the smaller (*microsporangia*) containing numerous orange microspores. (Named from the genus *Selaginella.*)

SELAGINELLA

Sporangia in the axils of leaves forming terminal cone-like spikes; sporangia minute, subglobose, opening transversely; megaspores globose, four in each megasporangium; microspores small, numerous. (Name Lat., a diminutive of *Selago,* the name of a species of *Lycopodium.*)

Selaginella Wallacei Hieron. Stems densely tufted, 5–10 cm. long, prostrate or ascending, much branched; leaves closely imbricate, narrowly lanceolate, ciliate margined, channeled on the back, tipped with a slender white awn; spikes quadrangular, 1–3 cm. long; bracts lanceolate to lance-ovate, the awns shorter and the margin more ciliate than on the leaves. Abundant on the ledges from the Snake River to the higher peaks, *U. Son.* to *Huds.* *S. scopulorum* Maxon, at least as to local collections.

The sheet from Mica Peak, *Suksdorf* 8,834, cited by Maxon in his original publication, and other similar ones, some seen by him, the writer cannot separate from *S. Wallacei.* All of the contrasting characters used by Maxon, the dorso-ventral branches, the leaves with 4–8 cilia, and the sporophylls broadly ovate, fail to separate our plants.

Class II. *FILICINEAE*

Plant highly organized, vascular, with green, usually large, leaves; spores borne within the tissue of, or in modified hairs on, modified or unmodified foliage leaves; stem solid, underground (in ours). (Named from the genus *Filix.*)

OPHIOGLOSSACEAE

Plant consisting of an underground stem bearing one or more leaves which rise above ground and are divided usually into two parts, a fertile portion and a sterile portion, the latter being the foliage part of the frond; frequently the fertile portion lacking in some of the fronds; sporangia borne within the tissue of the fertile portion, ringless, opening by a transverse slit. (Named for the genus *Ophioglossum,* from Gr., *ophis,* serpent; *glossa,* tongue.)

BOTRYCHIUM. Grape-Fern

Rootstock very short, with clustered fleshy roots; sterile part of the leaf ternately or pinnately divided or compound; veins free; fertile segment 1–3-pinnate, each pinnule bearing a double row of sessile sporangia; spores numerous, sulphur-yellow. (Name Gr., *botrus,* a bunch of grapes.)

Botrychium multifidum (Gmel.) Rupr., subsp. **silaifolium** (Presl) Clausen. *Leathery Grape-fern.* Stout, rather fleshy, 10–35 cm. high, stem very short and stout, swollen with the contained bud of the succeeding season; sterile blades 8–20 cm. wide, scarcely as long, ternate, one or two, their petioles stout, 2–12 cm. long, the primary divisions 3-pinnate or 4-pinnatifid; ultimate segments obliquely ovate, 1–1.5 cm. long, thick, entire or wavy, the veins few, obscure; sporophyll erect, the petiole stout, the fruiting portion 4-pinnate below, gradually simpler above; sporangia numerous, crowded, bright yellow. *B. silaifolium* Presl. Rare, in wet meadows, near Moscow, Idaho, *A. T.*

ADIANTACEAE

Rhizome erect, oblique, or creeping, with blackish brown, thick, entire scales; fronds uniform, with usually brown or ebony-colored, polished wiry stipes, not articulate; fronds 1–4-pinnate or pedate, soft herbaceous to chartaceous; pinnules stipitate, either unilateral and trapeziform or cuneate-flabellate; veins free flabellate; sori submarginal, borne along or sometimes between the apices of veins, covered by the reflexed false indusium which is from oblong to orbicular; sporangia globose pyriform; spores tetrahedral, smooth. (Named for the genus *Adiantum.*)

ADIANTUM

Graceful plants, ours of moist woods or rocks; sori appearing marginal, covered by the reflexed portion of the more or less altered margin of the leaf which bears sporangia on its under side from the tips of the free forking veins. (The name, Gr., *a,* not; *diaino,* to wet.)

Adiantum pedatum L. var. **pedatum.** *Maidenhair.* Rhizome thick, short, scaly; stipes slender shiny dark brown to black, 1–5 dm. tall; fronds 2-forked arching, each recurved branch bearing on one side long arching pinnate divisions; pinnules close, oblong to deltoid-oblong, the lower margin entire, the upper cleft, bearing the oblong to lunate-oblong sori. *A. pedatum aleuticum* of Piper. Mossy woods in Blue Mts., *Can.*

ASPIDIACEAE

Terrestrial or rarely climbing or epiphytic; rhizome usually short thick and erect, the scales broad and dentate or fimbriate, these often extending to the stipe and blade; fronds from simply pinnate to decompound, uniform, or somewhat dimorphic, not articulate, often hairy with brown or red, soft intestiniform hairs of 2–6 rod-like cells (known as ctenitis-hairs); the veins free or anastomosing; sori superficial, dorsal or apical on the veins, round or rarely oblong; indusium absent or present and centrally affixed, round-reniform or orbicular-peltate; spores bilateral, usually opaque. (Named from its genus *Aspidium.*)

Indusium peltate; fronds coriaceous, evergreen. *Polystichum.*
Indusium laterally attached and reniform, or absent; fronds herbaceous or evergreen. *Dryopteris.*

DRYOPTERIS

Rhizomes stout, erect, or short creeping, the scales entire or toothed, glabrous; stipes stout; blades pinnate-pinnatifid to tripinnate, mostly coriaceous, the ultimate segments mostly toothed; veins free or forked, ending in marginal hydathodes; sori dorsal on the veins; indusium reniform, persistent; spores bilateral. (Name from Gr., *drus,* oak; *pteris,* fern.)

Dryopteris spinulosa (O.F. Muell.) Kuntze var. **dilatata** (Hoffm.) Underw. *Spreading Wood-fern.* Rhizome stout, creeping, or ascending; stipe stout, brown scaly, especially below, 10–45 cm. tall; blades triangular-ovate, 10–90 cm. long, nearly or quite 3-pinnate, pinnae oblong-lanceolate, pinnules oblong to lanceolate acute pinnatifid or pinnate; toothed; sori round at end of vein near a sinus; indusia glabrous or glandular. *D. austriaca (Jacq.) Woynar* var. *dilatata* (Hoffm.) Fiori; *Thelypteris spinulosa* (O. F. Muell.) Nieuwl. var. *dilatata* (Hoffm.) St. John; *Aspidium dilatatum* (Hoffm.) Swartz. In moist shady woods, Lake Coeur d'Alene, *Can.*

POLYSTICHUM

Large or medium sized ferns, mostly with firm evergreen fronds, pinnate, bipinnate, or bipinnatifid, the pinnae serrate and usually auricled at the base on the upper side; veins free; indusium orbicular and peltate, depressed in the center and attached by a stalk to the middle of the sorus; sori round. (Name Gr., *polus,* many; *stichos,* row.)

Stipe 0.5–6 cm. long; middle pinnae 2–3 cm. long, falcate and obliquely lanceolate, the spinulose teeth spreading. *P. Lonchitis.*
Stipe 12–20 cm. long; middle pinnae 4–9 cm. long, linear or lance-linear, acuminate, the margin serrate with incurved, spinulose teeth. *P. munitum.*

Polystichum Lonchitis (L.) Roth. Rhizome woody, erect or decumbent about 1 cm. thick, scaly; fronds 15–60 cm. long, several, ascending in a close crown; stipes covered with rusty brown ovate, denticulate scales; blades 2–7 cm.

broad, linear-oblanceolate, pinnate, the rhachis stout and scaly; pinnae 25–40 on a side, alternate, coriaceous, evergreen, the lower pinnae much reduced and obliquely deltoid; sori mostly on pinnae of the upper half of the frond in 2 single or 2 double rows, later confluent; indusia 1–2 mm. wide. Moscow Mt., Ida., *Thomas,* reported by Flowers in 1950.

Polystichum munitum (Kaulf.) Presl. *Sword-fern.* Fronds simply pinnate, 60–120 cm. long, forming a crown; stipes stout, chaffy with numerous brown scales; rhachis also chaffy; pinnae linear or lanceolate-linear, acuminate, very sharply and often doubly serrate, sometimes chaffy on the midvein beneath, 3–10 cm. long; sori abundant, arranged in a row on each side of the midrib halfway to the margin. In deep woods, *Can.*

A large, beautiful, evergreen fern. There is an outlying station at 1,000 ft. alt. on Almota Creek, *Moore* 685.

ASPLENIACEAE

Rhizome dictyostelic, erect, creeping or scandent, scaly; fronds uniform, not articulate (except in a few species of *Asplenium*), usually stipitate, from simple to decompound pinnate, membranaceous to subcoriaceous, usually somewhat paleaceous, rarely pubescent; venation usually free, sometimes sparsely reticulate near the margin, or united by an intramarginal vein; sori superficial, usually oblong to linear, on one or both sides of the fertile veins, usually oblique to the midrib; indusia usually present, membranous, oval to linear; spores dark, bilateral, smooth, warty, or winged. (Named for the genus *Asplenium.*)

Indusium attached at a broad base under the sorus and covering one side of it like a flap. — *Cystopteris.*
Indusium reniform or absent,
  Stipe bundles united below the base of the blade; indusium none or attached on a line and horseshoe-shaped, the tip crossing the vein. — *Athyrium.*
  Stipe bundles free up into the blade; indusium none. — *Gymnocarpium.*

ATHYRIUM

Large or medium sized ferns with 1–3-pinnate or pinnatifid fronds; sori oblong or linear (nearly round when young), oblique, separate; indusium more or less curved, lunate, or sometimes horseshoe-shaped, often crossing to the outer or lower side of the vein; veins free. (The name Gr., *a,* without; *thurion,* a door.)

Scales fuscous; indusia ciliate; spores sparsely papillate, yellowish. — *A. Filix-femina* var. *Filix-femina.*
Scales black; indusia dentate or ciliolate; spores reticulate, blackish. — Var. *californicum.*

Athyrium Filix-femina (L.) Roth var. **Filix-femina** *Lady-fern.* Rootstock ascending, short, densely covered by the bases of the petioles; stipes tufted, 20–30 cm. long, straw-colored or brownish; blades delicate, glabrous, broadly lanceolate or acuminate at the apex, 3–9 dm. long, bipinnate to tripinnatifid; pinnae oblong-lanceolate, 5–20 cm. long; pinnules oblong, obtuse, obscurely 9–13-lobed, the lobes serrate; terminal pinnules confluent; sori short, straight or curved. Common in moist woods, *A. T. T., Can.*

Var. californicum Butters. A western variety. Reported from Moscow Mt. and Latah Co., Ida. by Flowers in 1950.

CYSTOPTERIS

Delicate rock-ferns; leaves 2–3 pinnate or pinnatifid; stipes slender; sori round, borne on the backs of the veins; indusium hood-like, attached by a broad base on the inner side partly under the sorus, early opening and withering away; in age the sorus appearing naked. (Name Gr., *kustis,* a bladder; *pteris,* a fern.)

Cystopteris Filix-fragilis (L.) Borbas. *Brittle Fern.* Rootstock short; stipes 2–20 cm. long; blades thin, oblong-lanceolate, only slightly tapering below, 10–25 cm. long, 3–7 cm. wide, 2–3-pinnatifid or pinnate; pinnae lanceolate-ovate, irregularly pinnatifid with bluntly or sharply-toothed segments along the margined or winged rhachis; texture membranous. *Filix fragilis* (L.) Underw.; *C. fragilis* (L.) Bernh. In shady woods, mostly on rocks, *U. Son.* to *Can.*

The species is widely distributed in the northern hemisphere.

Though Linnaeus changed his binomial, or basonym, to *Polypodium fragile* in 1755, and 1763 (and the identical 1764 reprint), it is not clear, as Merrill maintained (1935, that Linnaeus was correcting his error. His original name was *P. F. fragile* (=*P. Filix-fragile*), just as he had *P. F. mas* (*P. Filix-mas*), and *P. F. femina* (*P. Filix-femina*). Linnaeus retained his original *P. F. fragile* (1753) in his books in 1759, 1760, 1767, and 1770, so he was more consistent in using this double epithet. It seems correct to reestablish it as he originally published it, and more commonly employed it.

GYMNOCARPIUM

Delicate ferns with long creeping, branched rhizomes, the scales [illegible], entire, glabrous; fronds remote, deciduous, membranous; stipes longer than the blades, glandular or at base sparsely scaly; blades deltoid or pentagonal, compound, the lower pinnae long petiolulate, inaequilateral, glabrous or glandular; veins free, forked, reaching the margin; sori dorsal on the anterior vein branch, round; indusium none; spores bilateral, (Name from Gr., *gumnos,* naked; *karpos,* fruit, in reference to the naked sori.)

Gymnocarpium Dryopteris (L.) Newm. *Oak-fern.* Rootstock slender, horizontally creeping; stipes single, 15–40 cm. tall, pale straw-colored, shiny, bearing a few brownish scales toward the base; blades broadly triangular in outline, 10–20 cm. wide, ternate, the lateral primary divisions bipinnate, the terminal usually tripinnate, all naked at the base; pinnae oblong, 2–5 cm. long, glabrous, pinnately-cleft or divided into 15–25 obtuse lobes; sori near the margin, on the ends of free veins. *Phegopteris dryopteris* (L.) Fée; *Thelypteris Dryopteris* (L.) Slosson; *Dryopteris disjuncta* (Rupr.) Morton.

Common in rich woods, near or in the mountains, *Can.*

GYMNOGRAMMACEAE

Rhizome erect, oblique or short creeping with hairs or scales; fronds uniform or sub-dimorphic, pinnate or rarely simple or decompound, stipitate, not articulate, the frond with veins free or sparsely reticulate; sori superficial, linear, following the veins; indusium none; spores tetrahedral, translucent or opaque and warty. (Named from its genus *Gymnogramme.*)

PITYROGRAMMA

Small ferns; fronds uniform, 1–3 pinnate, covered beneath with a white or yellow powder; sori linear, following the veins; indusia wanting. (Name Gr., *pituron,* scurf; *gramme,* a line.)

Pityrogramma triangularis (Kaulf.) Maxon. *Silver Fern.* Rhizome short; fronds tufted; stipes slender, shiny brown, naked above, 5–15 cm. tall; blades triangular or pentagonal, dark green and glabrous above, 4–18 cm. long; lowest pair of pinnae much the largest and 2-pinnate; upper pinnae oblong, obtuse, 1-pinnatifid; sporangia confluent with age. *Gymnogramme triangularis* Kaulf.; *Ceropteris triangularis* (Kaulf.) Underw. Moist shaded crevices of Granite Point in Snake River Canyon, *St. John & Pickett* 2,999, *U. Son.* Very rare e. of the Cascade Mts.

POLYPODIACEAE

Terrestrial or epiphytic; rhizome creeping, the scales clathrate, dentate; fronds articulate, uniform or dimorphic, simple to pinnate, subcoriaceous; venation free or anastomosing; sori superficial or immersed, globose or oblong, borne on the vein ends or on a plexus of veins or fused into coenosori; spores bilateral, oblong-reniform, pale, smooth or warty. (Named from its genus *Polypodium.)*

POLYPODIUM

Mostly shade-loving ferns, many of them epiphytic; rhizomes scaly creeping; fronds uniform; blades simple, ours 1-pinnate; sori roundish, in rows parallel to the midrib. (Name Gr., *polus,* many; *pous,* foot.)

Polypodium vulgare L., var. **columbianum** Gilbert. *Licorice-root Fern.* Rhizome stout, horizontal scaly; stipes greenish, smooth, 1–10 cm. tall; blades 1-pinnate or pinnatifid, oblong-lanceolate, leathery, 2–15 cm. long; pinnae oblong-elliptic, obtuse, leathery, entire, 7–25 mm. long; sori brown, large, and becoming confluent. *P. glycyrrhiza* D. C. Eaton; *P. hesperium* Maxon. Shaded rocks, Kamiak, and Cheney, *A. T. T.*

PTERIDACEAE

Rhizome creeping, erect, or even arborescent, with articulate hairs or narrow scales; fronds uniform or nearly so, entire, ternate, pinnate or decompound, not articulate; venation various; sori marginal or intramarginal, as a rule coenosori, borne on the vascular commissure connecting the vein ends; indusium linear, formed by the modified, reflexed leaf margin, or double; spores tetrahedral or bilateral, smooth or verrucose. (Named from its genus *Pteris.)*

PTERIDIUM

Large, mostly coarse ferns, with pinnately divided leaves; sori marginal, linear, continuous on a slender thread-like receptacle which connects the tips of free veins; false indusium membranous, formed of the reflexed margin of the leaf. (Name Gr., *Pteris,* a fern.)

Pteridium aquilinum (L.) Kuhn, var. **pubescens** Underw. ex Heller. *Bracken* or *Brake.* Rootstock stout, black, subterranean, horizontally-creeping; stipes 30–90 cm. high, erect, pale-green or straw-color; blades 60–120 cm. long, 30–90 cm. wide, short pilose or glabrous above, pubescent beneath, ternate, the three branches bipinnate; pinnules oblong, acutish, mostly entire, the uppermost coalescent, the lower more or less lobed. Indusia narrow, villous and sparsely ciliate. *Pteris aquilina* L. var. *lanuginosa* Bong. Common in woods, occasional but infrequent elsewhere, *U. Son. to Can.*

Texture rough and mechanically harmful. If forming 20% of hay, it will kill horses.

SINOPTERIDACEAE

Rhizome short, ascending or creeping, solenostelic, the scales linear- or ovate-lanceolate, thick, entire, brown; fronds uniform or dimorphic 1–5-pinnate, not articulate; veins of blades free; sori round or oblong, on the vein ends, covered by a false indusium; spores globose-tetrahedral, opaque, warty or spiny or rarely smooth. (Named from the genus *Sinopteris.)*

CHEILANTHES

Mostly pubescent or tomentose rock-loving and small ferns with much divided fronds; sori on or near the ends of the veins; at first small and distinct, afterwards crowded; indusium the thinnish revolute margin of the frond; sporangia often concealed in the scales or hairs which in many species cover the segments. (Name Gr., *cheilos,* margin; *anthos,* flower.)

Blades glabrous; pinnules linear. *C. siliquosa*
Blades hairy, pinnules short, ovate to suborbicular,

Blades scaly and lanate; scapes sparsely scaly. *C. gracillima.*
Blades only lanate; scapes pilose. *C. Feei.*

Cheilanthes Feei Moore. *Slender Lip-fern.* Stipes densely tufted, 4–6 cm. high, brown, when young covered with long hair-like scales, at length glabrate; blades 4–8 cm. long, oblong-ovate; the lowest pinnae usually remote; ultimate segments orbicular or oblong, entire or crenate, crowded; upper surface with a few long hairs, the lower densely matted with whitish or pale-brown long hairs. Bluffs of Snake River near Almota, reported by Constance in 1936 above Bishop, *U. Son.*

Cheilanthes gracillima D. C. Eaton. *Lace Fern.* Stipes densely tufted, shining brown, 4–8 cm. high, bearing a few scattered lanceolate scales; blades 2–10 cm. long, oblong-lanceolate, bipinnate; pinnae numerous, crowded, pinnately divided into 5–9 oval mostly entire pinnules, glabrate above, pubescent beneath with rusty matted wool. Common in rock crevices in the mountains, and open woods, *A. T. T. to Huds.*

Cheilanthes siliquosa Maxon. *Oregon Cliff-Brake.* Densely tufted, 10–20 cm. high; stipes dark brown, longer than the blades; blades 3–6 cm. long, ovate or ovate-oblong, 3-pinnate; leaflets crowded, linear-lanceolate, 6–12 mm. long, mucronate, entire on the fertile leaves, serrate on the sterile ones. *Pellaea densa* (Brack.) Hook.; *Cheilanthes densa* (Brack.) St. John. On cliffs and among boulders, Blue Mountains, *Can.*

Usually found on magnesium rocks.

WOODSIACEAE

Rhizomes short, erect, bearing scales; fronds uniform, 1–2-pinnate or pinnate-bipinnatifid, herbaceous, often with chaffy scales and articulate hairs; veins free; sori superficial, dorsal, round, borne on a minute circular receptacle near the vein ends; indusium surrounding the receptacle; spores bilateral and dark or tetrahedral and translucent. (Named from its genus *Woodsia.*)

WOODSIA

Small or medium sized ferns, growing in rocky places; fronds once or twice pinnate or pinnatifid; sori round, borne on the backs of simply forked free veins; indusium attached under the sorus, round but cleft into filiform divisions, or star-shaped, delicate, early withering. (Named for *Joseph Woods,* English botanist.)

Fronds glabrous or nearly so; lobes of the indusium hair-like. *W. oregana.*
Fronds viscid-puberulent; lobes of the indusium broader at base. *W. scopulina.*

Woodsia oregana D. C. Eaton. *Oregon Woodsia.* Rootstock short; stipes glabrous, not jointed, brownish below; blades glabrous or slightly roughened, 5–28 cm. long, elliptic-lanceolate, pinnate, the sterile shorter than the fertile; pinnae triangular-oblong, obtuse, pinnatifid; lower pinnae reduced in size and somewhat remote from the others; rhachis straw-colored; segments oblong or ovate, dentate or crenate, the teeth often reflexed and covering the sori; indusium deeply cleft into hair-like segments. Common in crevices in rocks usually in shady places, *U. Son. to A. T.*

Woodsia scopulina D. C. Eaton. *Rocky-Mountain Woodsia.* Similar to *W. oregana,* but the blades puberulent with minute white jointed hairs and with stalked glands; indusium deeply divided into segments that are broader at the bases. In rock crevices, near Spokane and about Lake Coeur d'Alene, Idaho, *A. T.*

MARSILEACEAE

Plant perennial, herbaceous, rooting in the mud, with slender creeping rootstock and 4-foliolate or filiform leaves; sporangia borne within closed receptacles (*sporocarps*) which arise from the rootstock near the leafstalks or are consolidated with them; spores of two kinds, large ones (*megaspores*) and small ones (*microspores*), both contained in the same sporocarp. (Named from the genus *Marsilea.*)

MARSILEA

Marsh or aquatic plants, leaves slender petioled, 4-foliolate, commonly floating on the surface of shallow water; sporocarps ovoid or bean-shaped, peduncled and rising from the petiole or from the rootstock at the base of the petiole, composed of two vertical valves having several transverse compartments (*sori*) in each valve; also provided inside with a ring which at the opening of the valves swells and tears the sori from their positions; sori composed of both megasporangia and microsporangia. (Named for *Giovanni Marsigli,* Italian botanist.)

Marsilea vesita Hook. & Grev. Clover Fern. Rootstock slender, creeping; leaves more or less white pubescent; petioles slender, 4–15 cm. long; leaflets deltoid-obovate, 4–20 mm. long, mostly entire; sporocarps solitary on the stalks, [illegible] mm. long, [illegible] mm. wide, with a short raphe, a short blunt lower tooth, and an acute upper one, [illegible] with [illegible] hair-like scales; sori [illegible] in each valve. Common on shores, frequently [illegible], the leaves floating on the surface, in the spring [illegible]

Class III. *EQUISETINEAE*

Plants rush-like with hollow jointed stems rising from subterranean rootstocks; sterile leaves reduced to minute scales, whorled, forming sheaths at the joints; fertile leaves forming a short spike terminating the stem. (Named from the genus *Equisetum,* the horsetail.)

EQUISETACEAE

Branches, when present, whorled; sporangia 1-celled, clustered under the scales of the terminal cone-like spikes; spores of but one kind furnished with narrow ribbon-like appendages (*elaters*) attached at the middle, coiling around them when moist and spreading in the form of a cross when dry and mature; epidermis impregnated with silica, often rough. (Named from the genus *Equisetum,* the horsetail.)

EQUISETUM. Horsetail

Perennial plants with extensively creeping rootstocks; stems simple or with whorled branches, furrowed lengthwise, hollow or solid; sporangia adhering on the under side of shield-shaped scales of the spike, 1-celled, opening down the inner side; spores of one kind, with elaters. (Name Lat., *equus,* horse; *seta, bristle.*)

Aerial stems annual; spikes blunt or barely acute,	
Fertile stems brown or pale, soon withering,	
Fertile stems soon withering, not becoming green.	*E. arvense,* var. *arvense.*
Fertile stems at length green and branched.	*E. sylvaticum.*
Fertile stems green, persistent, 6–15 dm. tall,	
Stems unbranched, or with a few scattered branches.	*E. fluviatile, f. fluvatile*
Stems with definite whorls of 4–16 slender sterile branches.	*E. fluviatile, f. verticillatum.*
Aerial stems evergreen perennial; spikes often rigidly apiculate,	
Teeth partially persistent, similar to the sheath; stems 10 or less grooved, the central cavity ½ the diameter, or less, or wanting.	*E. variegatum.*
Teeth soon deciduous, differing from the sheath; stems 10–48 grooved, the central cavity larger,	
Sheaths green, even in age, but sometimes black-margined,	
Stems simple or nearly so.	*E. laevigatum,* f. *laevigatum.*
Stems with numerous weak sterile or fertile axillary branches.	*E. laevigatum,* f. *proliferum*
Sheaths in age ashy or blackened.	*E. hyemale,* var. *hyemale.*

Equisetum arvense L. var. **arvense.** *Field Horsetail.* Aerial stems annual, of two kinds; the fertile pale-brown or whitish and short lived, appearing in early spring before the sterile; fertile stems 10–25 cm. tall, simple, terete, bearing loose scarious distant sheaths, these whitish with 8–12 brownish lanceolate teeth; sterile stems pale green, 10–60 cm. tall, marked with 6–14 furrows, with numerous whorls of mostly simple, solid branches, these 4-angled or rarely 3-angled; cavity of the main stem small; spike 2–3 cm. long. Common in moist places, *U. Son.* and *A. T.*

Equisetum fluaviatile L. forma **fluviatile.** *Swamp Horsetail.* Aerial stems erect, simple, smooth, 60–150 cm. tall, 4–7.5 mm. thick, with 10–30 shallow grooves; central cavity 4/5 diameter of stem; sheaths usually close, nearly as broad as long, the teeth rigid slender and dark; spike short pedunculate cylindric, 1–2 cm. long. *E. limosum* L.; *E. fluviatile* forma *Linnaeanum* (Doell.)

Broun. Nez Perce Co., Spalding, *Rust,* reported by Flowers in 1950. Undoubtedly present in the area with the following occasional forma.

Forma **verticillatum** (Doell.) A. A. Eaton. Similar to the species, but the main stem producing weak sterile lateral branches. Shore of Lake Chatcolet, Idaho, *St. John* 9,060, *Can.*

Equisetum hyemale L., var. **hyemale.** *Scouring-rush.* Aerial stems evergreen, all alike, 40–120 cm. tall, 5–15 mm. thick, marked with 20–36 furrows; ridges roughened usually with a single series of transverse siliceous tubercles; sheath short, commonly marked with a black girdle at the base and another at the base of the early-falling teeth; spike nearly sessile in the uppermost sheath, 1–3 cm. long. Common in moist places, especially along Snake River, *U. Son.* and *A. T.*

The stems are usually simple but under certain conditions branches may be produced.

Equisetum laevigatum A. Br., forma **laevigatum.** *Braun's Scouring-rush.* Aerial stems 15–120 cm. tall, 2–8 mm. thick, marked with 14–30 furrows, the ridges nearly smooth; each sheath marked with a black girdle at the base of the deciduous white-margined teeth, and sometimes with another at its base; wall of the stem thin, the cavity large; spikes 1–2 cm. long, the stalk exceeding the sheath. *E. kansanum* Schaffner. Common along streams, *A. T.*

Forma **proliferum** Haberer. Occasional, differing only by the development of numerous weak axillary sterile or fertile branchlets. Pullman, *Piper* in 1899.

Equisetum sylvaticum L. Rhizomes creeping, branched; stems dimorphic, annual; sterile shoots 1–6 dm. tall, the central cavity more than half the diameter of the stem, 10–18 ridged, the ridges with 2 spinulose rows; sheaths 1–2 cm. long, green below, brown above, the teeth persistent cohering in 2's or 3's; branches whorled, numerous, often branched, the filiform branchlets 3-angled; fertile stems with shorter branches; spikes long-peduncled, 1.3 cm. long, 5–8 mm. thick, blunt. Alder Creek, Benewah Co., Ida., *G. N. Jones* 504; 696.

Equisetum variegatum Schleich. *Variegated Scouring-rush.* Aerial stems ascending, often tufted, 1.5–6 dm. tall, 2–8 mm. thick, 5–10 angled usually unbranched; sheaths longer than broad, slightly funnel-shaped, black or black rimmed or circled; the teeth blending into the sheath, with a hyaline margin and bristle which soon falls; spikes sessile, 5–10 mm. long. Open places, near streams, *A. T.*

ISOETACEAE

Plant aquatic, usually submersed or sometimes growing on moist soil, consisting of a short, 2–3-lobed fleshy stem with a dense tuft of fibrous roots and a compact cluster of rush- or grass-like leaves; sporangia in small lobes enclosed in the bases of the leaves; spores of two kinds, large (*megaspores*) and small (*microspores*). (Named for the genus *Isoetes.*)

ISOETES. Quillwort

Stem a fleshy corm rooting just above the base, surrounded above by the swollen bases of the awl-shaped linear leaves; sporangia large, enclosed in the bases of the leaves; those of the outer leaves with megaspores, those of the inner with microspores; the sides of the sporangia more or less covered with a fold of the inner side of the leaf base (the *velum*). (Name Gr., *isos,* equal; *etos,* year.)

Megaspores with short tubercles or crests,
  Megaspores more than 420 μ in diameter. — *I. Howellii,* var. *Howellii.*
  Megaspores smaller. — *I. Howellii,* var. *minima.*
Megaspores with short crests which, at least on basal face, branch to form a network. — *I. occidentalis.*

Isoetes Howellii Engelm. var. **Howellii.** Leaves 10–50, rather slender, 5–25 cm. long, erect or nearly so, semi-lunate or helmet-shaped in cross section, striate, with abundant stomata above; megasporangia dark-brown, covered ⅓ or less by the velum, the megaspores bright white, rough, with low, more or less confluent tubercles; microsporangia olivaceous, elliptic or oblong, much pitted, 6–8 mm. long, partly covered by the narrow wings of the velum, the microspores asymmetrical, spinulose on the ridges. *I. Underwoodii* Henders.; *I. melanopoda* J. Gay. Borders of ponds, *A. T.* and *Can.*

Var. **minima** (A. A. Eaton) Pfeiffer. Usually smaller than the species, and occasionally with a 3-lobed corm, but differing constantly only in the megaspores but 320–420 μ in diameter. Ponds or streams, *A. T.*

Isoetes occidentalis Henders. Corm bilobed; leaves 9–30, spreading, rigid, quadrangular, sharp-pointed, dark-green, 5–20 cm. long; velum narrow, covering about one third of the sporangium; megaspores covered with scattered irregular crests, 525–640 μ in diameter; microspores smoothish to spinulose. *I. paupercula* (Engelm.) A. A. Eaton. Submersed in 1–2 meters of water, Lake Coeur d'Alene, Idaho, *Can.*

DIVISION EMBRYOPHYTA SIPHONOGAMA (SPERMATOPHYTA). Seed Plants

Highly organized plants without distinct alternation of generations, mostly producing flowers and always producing seeds, each of which contains a young plant (the *embryo*) usually composed of a stem-like structure (the *caulicle* or *hypocotyl*), one or more rudimentary leaves (the *cotyledons*) and a terminal bud (the *plumule* or *epicotyl*); megasporangia (*ovules*) usually borne on the side or face of an open or closed modified leaf (the *carpel*); microsporangia (*anther-sacs*) on the end or side of a modified leaf (the *filament*) and bearing numerous microspores (*pollen grains*). (Name Gr., *sperma,* seed; *phuton,* plant.)

Subdivision GYMNOSPERMAE

Trees or shrubs, mostly evergreen.

Ovules (*megasporangia*) naked, not enclosed in an ovary, usually on the face of an open scale but sometimes on the axis, in which case the scale is rudimentary or wanting; stigmas none; cotyledons mostly several in a whorl, occasionally only two; perianth none. (Name Gr., *gumnos*, naked; *sperma*, seed.)

TAXACEAE. Yew Family

Trees or shrubs with linear leaves; flowers dioecious; ovule solitary, terminal on a short axillary branch. (Named from the genus *Taxus,* the yew.)

TAXUS

Evergreen trees or shrubs with spirally arranged, short-petioled linear flat leaves, spreading so as to appear 2-ranked; aments very small, axillary and solitary, sessile or nearly so; staminate aments consisting of a few scaly bracts and 5–8 stamens; ovules solitary, axillary, erect, subtended by a fleshy ring-shaped disk; fruit consisting of a fleshy disk which becomes cup-shaped, red and nearly encloses the bony seed. Wood strong, resilient, aromatic and but slightly resinous. (Ancient name from Gr., *taxus,* yew tree.)

Taxus brevifolia Nutt. *Western Yew.* Small tree, 4–25 m. high, the bark loose and reddish; branches slender, horizontal or drooping; leaves horizontal, 1–2 cm. long, 1–2 mm. wide, linear, acuminate, cuspidate, with revolute margins, shiny green above, yellowish-green beneath, abruptly narrowed at the base into a short petiole; staminate aments globose, 3 mm. broad; fruit bright red, insipid in taste; stone broadly ovoid, acute, somewhat flattened, 3–4 mm. long. Along streams in the mountains, Can.

The European *T. baccata* L., now cultivated as an ornamental, formerly furnished the best bows. Now our species is sought to supply them for the modern sport of archery.

PINACEAE. Pine Family

Trees or shrubs with resinous juice; leaves mostly entire, awl-shaped, needle-shaped, or scale-like; flowers dioecious; staminate catkins (*aments*) elongate, short-lived, yellow to purplish; pistillate catkins of few to many scales; ovules one to several on the base of the scales; fruit a hard, woody cone, with dry, often winged, seeds. (Named from the genus *Pinus,* the pine.)

Leaves in clusters or fascicles,
- Leaves evergreen, 2–5 in a fascicle, sheathed at base. *Pinus.*
- Leaves deciduous, 12–30 in a cluster, on short spurs, not sheathed. *Larix.*

Leaves single, opposite, or whorled,
Cones erect, their scales deciduous. *Abies.*
Cones pendulous, their scales persistent,
Branchlets rough with persistent leaf-bases; leaves deciduous when dried; scales longer than the bracts,
Leaves sessile, 4-sided, pungent. *Picea.*
Leaves petioled, flattened, blunt. *Tsuga.*
Branchlets not so; leaves persistent when dried; bracts 2-lobed and with aristate midrib, exceeding the scales. *Pseudotsuga.*

ABIES

Evergreen trees with linear flat scattered sessile leaves, spreading so as to appear 2-ranked but in reality spirally arranged, commonthly quite persistent in drying; staminate aments axillary; ovule-bearing aments dorsal, erect; ovules two on the base of each scale, reflexed; the scale shorter than or exceeding the thin papery bract; cones erect, subcylindrical or ovoid, their orbicular or broader scales deciduous from the persistent axis the first autumn. (Name Lat., *abies,* a fir.)

Leaves notched at the apex, usually spreading horizontally on the branches, white-lined with stomata only beneath. *A. grandis.*
Leaves not notched at the apex, not horizontally spreading; white-lined with stomata on both sides. *A. lasiocarpa.*

Abies grandis (Dougl. ex G. Don in Lamb.) Lindl. *White Fir.* Large tree, sometimes 100 m. tall and 2 m. in diameter, with thin, dark gray, rather smooth bark; branches horizontal or the lower drooping; leaves linear, obtuse or emarginate, shining green above, marked beneath by two white lines, 18–50 mm. long, usually arranged in two ranks, giving the foliage a flattened appearance; cones cylindric-oblong, 7–10 cm. long, dark green, more or less covered with drops of resin; scales broader than long, entire; bracts small. Wood soft and weak, light brown. In the mountain woods; especially abundant in the Blue Mountains, *A. T. T., Can.*

Abies lasiocarpa (Hook.) Nutt. *Subalpine Fir.* Narrowly conical, usually densely branched trees, 20–30 m. high and 20–150 cm. in diameter; bark pale, rather smooth but with large resin blisters; leaves 1–4 cm. long, acute, usually sharp-pointed, those of upper branches not flattened beneath; cones narrowly barrel-shaped, usually dark-purple, puberulent, 5–10 cm. long; bracts usually not exserted. Wood soft and weak, pale brown or whitish. Blue Mountains, at 5,000–6,000 feet altitude, *Huds.*

LARIX. Larch

Tall trees with horizontal or ascending branches and small linear deciduous leaves, without sheaths, in clusters on short lateral scaly bud-like branchlets; aments short, lateral; the staminate from leafless buds; ovule bearing buds commonly leafy at the base and the aments red; mature cones ovoid or cylindrical, small, erect; scales thin, spirally arranged, obtuse, persistent; ovules two on the base of each scale, ripening into two reflexed somewhat winged seeds. (Name Lat. *larix,* the larch.)

Larix occidentalis Nutt. *Tamarack.* A large tree, 30–70 m. high, 1–2 m. in diameter; bark thick, reddish, scaly; branches short, horizontal; buds spherical; leaves keeled, needle-like, 2–4 cm. long, in clusters of 12–30, deciduous; cones ovate-cylindric, reddish when young, brown when mature, 2–4 cm. long; scales broadly oblong, truncate, ciliate-fringed when young; bracts scarious, dilated at the base, the narrow terminal leaf-like part exceeding the scale. Wood heavy, the sapwood whitish, the heartwood reddish. A common tree, *A. T. T., Can.*

PICEA. Spruce

Evergreen conical trees, with linear short 4-sided leaves, spreading in all directions, falling away from the twig in drying, leaving it covered with small woody projections; leaf-buds scaly; staminate aments axillary, nearly sessile; ovule-bearing aments terminal, ovoid or oblong; ovules two on the base of each scale, reflexed, ripening into two more or less winged seeds; cones ovoid or oblong, obtuse, pendulous; their scales numerous, spirally arranged, thin, obtuse, persistent. (Name Lat., *pix,* pitch.)

Picea glauca (Moench) Voss, subsp. **Engelmanni** (Parry ex Engelm.) T. M. C. Taylor. *Englemann Spruce.* Slender pyramidal tree, 10–25 m. tall; bark light colored, thin, flaky; branchlets slender, 2–3 mm. thick, glandular puberulent; leaves bluish green, slender, 4-sided, rigid, sharp-pointed, 1.5–2.5 cm. long; unopened cones narrowly cylindric, 4–7 cm. long, about 1.5 cm. thick, opened cones oblong-elliptic, 3–4 cm. thick; cone scales rounded, erose, thin and papery, pale brown. Wood light and soft. *P. Engelmanni* Parry ex Engelm.; *P. columbiana* Lemmon. Common in upland woods, *Can.*

PINUS. Pine

Evergreen trees with two kinds of leaves; the primary ones scale-like with deciduous tips; the secondary ones forming the ordinary foliage, needle-like, arising from the axils of the former in clusters of 2–5; ovule-bearing aments solitary or clustered, each composed of numerous minute bracts, each with an ovule-bearing scale in its axil; ament, upon maturing, becoming a cone; the scales elongating and becoming woody; seeds two on the base of each scale. (Name Lat., *pinus,* the pine.)

Cone scales without prominent thickenings; leaves 5 in a fascicle, the sheath deciduous. — *P. monticola.*
Cone scales with promient thickenings; sheath persistent,
  Leaves 2 in a fascicle, 4–8 cm. long. — *P. contorta,* var. *latifolia.*
  Leaves 3 in a fascicle, 15–25 cm. long. — *P. ponderosa.*

Pinus contorta Dougl. ex Loud. var. **latifolia** Engelm. *Lodgepole Pine.* Small slender tree, 5–50 m. tall, 1–2 m. in diameter, the thin dark bark usually deeply checked; leaves 4–8 cm. long, dark green; cones small, ovoid, reflexed, often asymmetrical, 4–5 cm. long; scales thickened at the apex and armed with a stout point. *P. contorta* subsp. *latifolia* (Engelm.) Critchfield. Wood light, soft and weak, the sapwood whitish, the heartwood yellowish. In the mountains, often forming dense pure growth of nearly equal-sized trees, *A. T. T.* to *Huds.*

It is a fire-weed, springing up in dense stands on burns.

Pinus monticola Dougl. ex Lamb. *Western White Pine.* Tree 30–50 m. high, 1–2.4 m. in diameter; bark gray, rather smooth, longitudinally cracked; leaves pale green, in fascicles of five, 4–11 cm. long; cones narrowly cylindrical, 15–30 cm. long, about 4 cm. thick when unopened. Wood soft straight uniform, and of great value. In the mountains of Idaho at low altitudes, *Can.*

Pinus ponderosa Dougl. ex P. & C. Lawson. *Yellow* or *Bull Pine.* Large tree, 30–80 m. tall, 1–3 m. in diameter, the reddish bark thick and deeply checked; leaves in fascicles of three, 15–25 cm. long, minutely serrulate; staminate aments cylindric, somewhat flexuous, 4–6 cm. long, crowded at the base of young shoots; pistillate aments 1–6, greenish or purplish, borne near the apex of the shoots of the season; cones brown, ovoid, 7–15 cm. long, frequently in clusters of 3–5; scales much thickened near the apex and bearing a stout sharp point. Wood hard, strong, and pitchy, the sapwood whitish, the heartwood yellow or reddish. Throughout our limits where the soil is gravelly, forming open forests on the lower foothills, *U. Son.* to *Can.*

PSEUDOTSUGA

Very large trees, at first pyramidal and spruce-like, often at last more spreading; leaves linear, flat, somewhat 2-ranked by a twist at the base; aments from the axils of the leaves of the preceding year; staminate aments in an oblong or cylindrical column, surrounded or partly enclosed by numerous, conspicuous, round bud-scales; ovule-bearing aments with the scales much shorter than the broadly linear, acutely 2-lobed bracts with a linear projecting midrib; cones maturing the first year; scales persistent. (Name Gr., *pseudos,* false; Japanese, *tsuga,* hemlock.)

Pseudotsuga taxifolia (Lamb. ex Poir.) Rehder. *Douglas Fir,* or *Red Fir.* Very large tree, 50–100 m. high, 1–4 m. in diameter; bark thick, black, in age reddish within, deeply cracked longitudinally; branches usually short and horizontal; foliage usually reddish-green; leaves linear, obtusish, 20–30 mm. long, narrowed at the base, dark green above, paler beneath; staminate aments light brown, oblong-cylindric, 1 cm. long, half enclosed in the large bracts; pistillate aments green or purplish; cones pendent, cylindric-ovate, 3–10 cm. long, the tridentate bracts conspicuously exceeding the scale. Wood light, straight, strong and valuable; sapwood whitish; heartwood yellow or red, *P. mucronata* (Raf.) Sudw.; *P. Merrilli* Flous; *P. Menziesii* (Mirb.) Franco; *P. Douglasii* Carr., var. *caesia* Schwer. A very common tree, foothills and mountains, *A. T.* to *Can.*

Henry and Flood, Proc. Roy. Irish Acad. **35:** B 69, 1920, separate our inland tree from the coast tree. In the field and herbarium their characters have been used without success and a single isolated tree will produce both the large and the small cones.

There has been much dispute and distress concerning the name of this important tree. However, it is now clear that all of the earlier publications of *Abies taxifolia* were illegitimate, hence, that by Lambert ex Poiret is valid and with priority.

TSUGA. Hemlock

Trees with a pyramidal crown, horizontal branches, and a nodding leading shoot; leaves flat or angular, short petioled, twisted at base and appearing 2-ranked; staminate aments in subglobose clusters in axils of leaves of previous season; ovule-bearing aments terminal on 1-year old branchlets; cones pendulous ovoid to cylindric, maturing in 1 year; bracts minute; scales rounded thin, entire; seeds ovoid compressed, nearly surrounded by the short wing. (Name Japanese, *tsuga,* the hemlock.)

Tsuga heterophylla (Raf.) Sarg. *Western Hemlock.* Forest tree 30–70 m. tall, 0.5–3 m. in diameter; crown pyramidal; trunk straight; branchlets drooping; bark brown, becoming reddish-brown, thick, fissured and with flat longitudinal connected ridges; leaves unequal 6–20 mm. long, linear obtuse, whitened beneath, flat; cones ellipsoid, 15–25 mm. long; scales ovate. puberulent. Bark valuable in tanning, wood light and tough, sapwood whitish, heartwood pale brown. Benewah, Idaho, *St. John & Jones,* 5,409, *Can.*

CUPRESSACEAE

Shrubs or trees; leaves opposite or in whorls; juvenile leaves needle-like; leaves of mature growth scale-like or in a few species needle-like; plants monoecious or dioecious; cones axillary or terminal; filament short; sporophyll with an expanded tip bearing 3–6 sporangia; pistillate cone with 1–3 terminal ovules, the scales forming a leathery or woody cone, or even becoming fleshy and berry-like. (Named from its genus *Cupressus,* the cedar.)

Pistillate cone at maturity fleshy and berry-like, the scales coalescing. *Juniperus.*
Pistillate cone at maturity leathery or woody, the scales imbricate, distinct. *Thuja.*

JUNIPERUS Juniper

Evergreen shrubs or small trees with thin bark; leaves sessile, scale-like or needle-like, opposite or in whorls of three; flowers dioecious or monoecious, small and lateral; anther-cells 3–6, attached to the lower edge of the shield-shaped scale; ovule-bearing aments ovoid, of 3–6 fleshy coalescent scales, each 1-ovuled, in fruit red or bluish-black and berry-like. (Name Lat., *juniperus,* the juniper.)

Leaves mostly in 3's, resinous; fruit ripening in 1 season. *J. occidentalis.*
Leaves opposite, not resinous; fruit ripening in 2 (or 1) seasons. *J. virginiana* var. *scopulorum.*

Juniperus occidentalis Hook. *Western Juniper.* A tree 3–20 m. high, trunk stout, 7–10 dm. in diameter, bark thin, scaly, red-brown; crown thick, low and often spreading; foliage yellowish-green; leaves mostly in 3's, resinous from the active dorsal gland; berries glaucous blue, 6–8 mm. long; seeds 2 or 3, grooved or pitted. *Sabina occidentalis* (Hook.) Heller. Wood light and soft, the sapwood whitish, the heartwood red or brown. On ridges of the Blue Mts., occasional in the valleys of the Grand Ronde, and the Snake Rivers. *U. Son.* and *Huds.* Reaching its northern limit at Wawawai.

Juniperus virginiana L. var. **scopulorum** (Sarg.) Lemmon. *Rocky Mountain Juniper.* A shrub or scraggly tree, 1–15 m. high, much branched, the branches often drooping; foliage often glaucous; leaves small, acute, each with a linear indistinct gland on the back; berries blue-black with a thick whitish bloom, maturing the second or the first year, 4–6 mm. long; seeds usually 2, grooved longitudinally. Wood soft and durable, the sapwood white, the heartwood dull red or rose. *J. scopulorum* Sarg.; *Sabina scopulorum* (Sarg.) Rydb. Spokane, *A. T.*

THUJA. Arbor Vitae

Evergreen trees or shrubs; leaves small or minute, scale-like, appressed, opposite, 4-ranked; flowers monoecious, both kinds terminal, the staminate globose, the ovule-bearing ovoid or oblong, small, their scales opposite, each bearing 2, rarely 2–5 erect ovules; cones ovoid or oblong, mostly spreading or recurved, their scales 6–10, coriaceous, opposite, dry, spreading when mature. (Name Gr., *thuia,* an evergreen tree.)

Thuja plicata Donn ex D. Don in Lamb. *Giant Cedar.* Handsome pyramidal tree, 30–50 or even 80 m. high, 1–5 m. in diameter, the trunk rapidly tapering from the large buttressed base; branches usually somewhat drooping; bark reddish-brown, fibrous, longitudinally fissured; leaves oblong-ovate, bright green, rapidly tapering to an acuminate cuspidate apex; staminate aments minute, dark purple; pistillate aments usually crowded near the tips of the branchlets; cones oblong, 1–1.5 cm. long, light colored, consisting of about 6 pairs of scales, these elliptical, mucronate on the back near the apex. Wood soft, durable, straight grained, easily splitting, the heart-wood reddish, odorous. In moist places, Kamiak Butte, and the foothills of the Coeur d'Alenes, absent from the Blue Mountains, *A. T. T., Can.*

SUBDIVISION ANGIOSPERMAE

Ovules (*megasporangia*) enclosed in a cavity (the *ovary*) formed by the infolding and uniting of the margins of a modified rudimentary leaf (*carpel*), or of several such leaves joined together, in which the seeds are ripened; stigmas present; cotyledons one or two, very rarely wanting; perianth present or wanting. (Name Gr., *aggeion,* vessel; *sperma,* seed.)

Class I. MONOCOTYLEDONES

Embryo of the seed with but a single cotyledon, that is with the first leaves of the seedling alternate; stem composed of a mass of soft tissue (*parenchyma*) in which the woody bundles appear to be irregularly imbedded; no distinction into bark, wood, and pith; leaves usually parallel-veined, mostly alternate and entire, commonly sheathing the stem at the base and often with no distinction of blade and petiole; parts of the flowers mostly in threes. (Name Gr., *monos,* one; *kotuledon,* a cup-like hollow.)

TYPHACEAE. Cat-Tail Family

Marsh or aquatic herbs with creeping rootstocks and linear flat sheathing leaves; stems erect, terete; flowers monoecious, densely crowded in terminal spikes which are subtended by spathaceous bracts; ovary one, stipulate, 1–2-celled, with as many persistent styles; fruit a minute achene, endosperm copious. (Named for the genus *Typha.*)

TYPHA

Flowers in a dense cylindrical spike; staminate portion of the spike above but contiguous to the pistillate; stamens with very short connate filaments, mixed with numerous long hairs; ovary long-stalked, 1-celled, surrounded by a perianth of numerous bristles; fruit minute, usually splitting on one side. (Name Gr., *tuphe,* the cat-tail.)

Typha latifolia L. *Cat-tail.* Stout, 1–3 m. tall; leaves flat, sheathing at base, very long, 1–2 cm. wide; pistillate and staminate portions of the spike close together, each 8–10 cm. long, the pistillate dark brown; stigmas rhombic-spatulate; pollen grains in fours. In shallow water, not common, [illegible]

The dry leaves were used for matting and thatching, and the pollen and starchy rootstocks for food.

SPARGANIACEAE. Bur-reed Family

Marsh or pond herbs with creeping rootstocks and erect or floating stems; leaves linear, alternate, 2-ranked, sessile, sheathing at the base; flowers monoecious, densely crowded into globose heads on the upper parts of the stem and branches, the staminate above; spathes linear; perianth of a few chaffy scales; stamens usually 5; fruit mostly 1-celled, nutlike, embryo nearly straight, in copious endosperm. (Named for the genus *Sparganium.*)

SPARGANIUM. Bur-Reed

Characters of the family. (Name probably from Gr., *sparganon,* a band.)

Stigmas 2 on some or all of the carpels; mature carpels sessile, obpyramidal below, conical above; sepals nearly equaling the body of fruit. — *S. eurycarpum,* var. *Greenei.*

Stigmas 1; mature carpels fusiform; sepals not over ⅔ length of body of fruit. — *S. chlorocarpum.*

Sparganium chlorocarpum Rydb. Stems stout erect, 2–4 dm. tall; leaves coarse, broadly linear, exceeding the spike, 3–15 mm. wide; inflorescence a loose spike, but one or more of lowest pistillate heads long peduncled and attached above the node; achenes 5–10 mm. long, slender beaked. Along streams, *A. T.*

Sparganium eurycarpum Engelm., var. **Greenei** (Morong) Graebn. Stems stout, erect, 6–18 dm. high; leaves flat, slightly keeled; inflorescence branched; staminate heads many; pistillate heads 2–4, sessile or peduncled, 15–30 mm. in diameter; fruit 4–5-angled, 4–7 mm. long, abruptly tipped in the center. *S. Greenei* Morong. In marshes and on lake margins, *A. T.*

POTAMOGETONACEAE. Pondweed Family

Perennial herbs, submersed or floating in fresh or brackish water; leaves mostly 2-ranked; flowers small, regular, in spikes, perfect; perianth of 4 green sepals; stamens 4; carpels 4, 1-ovuled; fruit of small achenes, endosperm wanting. (Named for the genus *Potamogeton.*)

POTAMOGETON. Pondweed

Leaves alternate or the uppermost opposite, often of two kinds, the submersed mostly delicate and pellucid, the floating lanceolate, ovate or oval, coriaceous; spathes enclosing the young buds usually perishing soon after expanding; flowers small, perfect, spicate, green or red; parts of flower in fours; ovaries 4, sessile, distinct, 1-celled, 1-ovuled. (Name Gr., *potamos,* river; *geiton,* neighbor.)

Leaves all submersed.
- Blades ovate to ovate lanceolate. — *P. Richardsonii.*
- Blades linear,
  - Stipules adnate to the leaf base; spikes usually terminal, borne at the water surface,
    - Leaves serrulate, auricled; beak of achene central. — *P. Robbinsii.*
    - Leaves entire, not auricled; beak of achene at proximal edge. — *P. pectinatus.*
  - Stipules free; spikes from the middle axils, usually submerged,
    - Leaves 9–35-nerved; achene 4–5 mm. long. — *P. zosteriformis.*
    - Leaves 1–7-nerved; achene 1.8–3 mm. long,

Stipules connate into a tube, at least below the middle,
- Peduncles 4–30 mm. long, clavate; leaves usually without basal glands,
  - Stipules delicately fibrous, not coarsely ciliate, tardily splitting. — *P. foliosus.*
  - Stipules rigid, coarsely fibrous, ciliate with stiff bristles 1–3 mm. long, promptly splitting into fibers. — *P. fibrillosus.*
- Peduncles 1.5–8 cm. long, filiform; leaves often with a pair of basal glands,
  - Primary leaves 1–3 mm. wide. — *P. panormitanus,* var. *major.*
  - Primary leaves 0.3–1 mm. wide. — Var. *minor.*

Stipules flat or convolute, but not connate. — *P. pusillus.,* var. *pusillus.*

Floating leaves present,
- Submersed leaves linear,
  - Submersed leaves merely terete bladeless petioles, less than 2 mm. wide; floating blades subcrodate, 2.5–5 cm. wide; stipules acute, 5–15 cm. long. — *P. natans.*
  - Submersed leaves flat, ribbon-like, 2–10 mm. wide; floating blades cuneate, 0.4–3.5 cm. wide; stipules obtuse, 2–4 cm. long,
    - Submersed leaves 7–13-nerved; achenes 3–3.6 mm. wide. — *P. epihydrus,* var. *epihydrus.*
    - Submersed leaves 5–7-nerved; achenes 2–3 mm. wide. — Var. *Nuttallii.*
- Submersed leaves lanceolate,
  - Floating blades 30–50-nerved; peduncles 5–18 cm. long, enlarged upwards, over twice as large as the petioles. — *P. amplifolius.*
  - Floating blades [illegible]-nerved; peduncles [illegible] more slender,
    - Stipules acuminate; submerged leaves all petioled. — *P. nodosus.*
    - Stipules obtuse,
      - Submerged leaves 0.2–1 cm. wide, all or mostly sessile. — *P. gramineus,* var. *graminifolius.*
      - Submerged leaves 1.5–4 cm. wide, with petioles as much as 4 cm. long. — *P. illinoensis.*

Potamogeton amplifolius Tuckerm. Stems stout, simple or branched; floating leaf-blades oblong-oval, acute, rounded at the base, 5–10 cm. long, on petioles one to two times as long; submersed leaves mostly linear-lanceolate to lanceolate, the uppermost frequently oval or oblong; stipules large, 5–10 cm. long, spike stout, dense, 2–5 cm. long, on stout peduncles; fruit large, 4–5 mm. long, with a broad beak. In streams, *A. T.*

Potamogeton epihydrus Raf., var. **epihydrus.** Stems 1–20 dm. long, compressed, simple or branched; floating blades 3–8 cm. long, elliptic to broadly oblong-oblanceolate, bluntly cuspidate or merely rounded at tip, 19–41-nerved, tapering into the flattened petioles; submersed leaves 5–10 mm. broad, flaccid, ribbon-like, the broad space within the inner nerves loosely reticulate; their stipules up to 4 cm. long, hyaline, obtuse, convolute or flat; peduncles 1.5–16 cm. long, thickish; fruiting spikes 0.8–4 cm. long; fruits 3–4.5 mm. long, with a dorsal keel. *P. epihydrus,* var. *typicus* Fern. In streams, *U. Son., A. T.*

Var. **Nuttallii** (C. & S.) Fern. Floating blades 2–7.5 cm. long, narrowly oblong or oblong-lanceolate, rounded or rarely subacute at tip; submersed leaves 2–8 mm. broad; fruiting spikes 0.8–3 cm. long; fruits 2.5–3.5 mm. long. *P. pensylvanicus* Willd. Aquatic, Lewiston, *Piper* 2,662; Touchet R., Waitsburg, *Horner, U. Son., A. T.*

Potamogeton fibrillosus Fern. Stems compressed, branching; leaves submerged, the primary 1.2–2 mm. wide, 3–5-nerved, obtuse or subacute, the midrib bordered below the middle with 2–3 reticulate rows; peduncles 4–7 mm. long; spikes 4–5 mm. long, mostly 4-flowered, dense; fruits 1.8–2.2 mm. long, obliquely obovoid, compressed, shining, the back rounded and undulate keeled. Walla Walla Co., *Brandegee* 1127 (according to Fernald).

Potamogeton foliosus Raf. Stems filiform, compressed, mostly branched; leaves submersed, deep green to bronze, the primary ones 1.4–2.7 mm. broad, 3–5-nerved, acute or subacute, the midrib divided and reticulate below the middle; stipules 7–18 mm. long; peduncles 3–10 mm. long; spikes 2–5 mm. long, usually of 2–3 approximate whorls of 2 flowers; fruits 2–2.5 mm. long, the body obliquely suborbicular to obovoid, compressed, the keel thin, undulate or dentate. *P. foliosus,* var. *genuinus* Fern. Submerged aquatic, Moscow, *Henderson* 2,716; Pullman, *Piper* 1802; Waitsburg, *Horner* 645, *A. T.*

Potamogeton gramineus L., var. **graminifolius** Fries. Stem slender, branching, 3–40 dm. long; floating leaves thin, variable in shape, elliptic to lanceolate or ovate, mucronate at apex, 9–17-nerved, 1.5–7 cm. long; submersed leaves also variable, lanceolate to linear; stipules appressed-ascending, 15–25 mm. long; peduncles elongate, scarcely swollen, spikes 2–4 cm. long; fruit roundish, slightly compressed, scarcely keeled, 2.5–3 mm. long. *P. heterophyllus* Schreb. of Am. authors. Horseshoe Lake, *St. John et al.* 4,891, *A. T.*

Potamogeton illinoensis Morong. Stems simple or branched 1.5–5 mm. in diameter; submersed leaves with blades 5–20 cm. long, thin, elliptic, oblong-elliptic to lanceolate (rarely reduced to linear), 9–17-nerved, lacunae of 2–5 rows along midrib and larger nerves; stipules 2.5–8 cm. long, 2-keeled, conspicuous; floating leaves often absent, when present coriaceous, 4–13 cm. long, elliptic to ovate-elliptic, obtuse, mucronate, the base cuneate or rounded, the petioles 2–9 cm. long; leaves showing gradual transitional stages from submersed to floating common; peduncles 4–15 cm. long; spikes 3–6 cm. long, of 8–15 whorls in fruit crowded; fruits 2.7–3.5 mm. long excluding the beak, the sides flat keels prominent. Spokane Co., Turnbull Slough. *Sperry & Martin* 703.

Potamogeton natans L. Stems often branched, 6–15 dm. long; floating leaves coriaceous, ovate to elliptic, subcordate, 21–29-nerved, 2.5–10 cm. long; submersed leaves thick; spikes 3–6 cm. long, on peduncles as stout as the stem; fruit narrowly obovoid, scarcely keeled, 4–5 mm. long. Subalpine lakes, w. of Spirit Lake, *Sandberg et al.* 697; Paradise Cr., Moscow, *Henderson* 2,717, *A. T., Huds.*

Potamogeton nodosus Poir. Stems slender, flaccid, branched, 5–20 dm. or more long; floating leaves narrowly elliptic, pointed at each end, 5–10 cm. long, 17–27-nerved, narrowed at the base into a petiole about as long; submersed leaves linear-lanceolate, 10–30 cm. long; stipules narrow, 3–8 cm. long; spike dense, 2–4 cm. long, on rather stout peduncles; fruit about 3–4 mm. long. *P. americanus* C. & S.; *P. lonchitis* Tuckerm. In ponds and pools, not common *U. Son., A. T.*

Potamogeton panormitanus Biv., var. **major** G. Fischer. Stems capillary, much branched; leaves submersed, linear, firm, acute to obtuse, 3–5-nerved, midrib prominent, usually lacking any reticular border; stipules 6–17 mm. long, scarious, finally lacerate; spikes 6–12 mm. long, of 3–5 whorls, interrupted; fruits 1.9–2.8 mm. long, pale olive-colored, obliquely obovoid, smooth, hollowed on the sides, the rounded back with a low keel. In water, Lake Coeur d'Alene, *Epling & Houck* 10,055, 10,359; Moscow, *Henderson.*

Var. **minor** *Biv.* Differs only in its narrower leaves, 0.3–1 mm. wide. Lake Coeur d'Alene, *Henderson; Rust* 382.

Potamogeton pectinatus L. Stems filiform, branched, 3–20 dm. long; leaves narrowly linear, attentuate to the apex, acute, 1-nerved, 2–20 cm. long; peduncles slender; flowers in whorls; fruit obliquely obovoid, compressed, turgid, 3–4.5 mm. long; style or beak straight or nearly so. Brackish or fresh waters of the *U. Son.* and *A. T.*

Below the surface of the mud, the rootstocks produce starchy tubers, which are a favorite food of the wild ducks. To attract the ducks, this pondweed is often planted in the ponds in hunting preserves.

Potamogeton pusillus L., var. **pusillus.** Stems capillary, usually branching; leaves submersed, 0.5–1.5 mm. wide, flaccid, acute, with 1–2 rows of reticulations bordering the midrib below the middle, the base with a pair of translucent glands; stipules 3–14 mm. long, obtuse, flat or inrolled, somewhat hyaline; peduncles 3–30 mm. long, filiform; spikes 2–8 mm. long, of 1–3 few-flowered whorls; fruits 2–2.5 mm. long, dark olive-colored, obliquely obovoid, more or less rugulose, when dried the rounded back with a low keel. Aquatic, *A. T.*

Potamogeton Richardsonii (Ar. Benn.) Rydb. Stems usually branched; leaves all submersed, the lowest ovate to ovate-lanceolate, the upper 1.5–10 cm. long, 0.5–2 cm. wide, lanceolate, with 3–5 strong nerves and 4–25 weaker ones; stipules 1–2 cm. long, ovate to lanceolate, obtuse, whitish, eroding to fibers; peduncles 15–25 cm. long, often thickened upwards; inflorescence of 6–12 whorls, moniliform; fruits 2.5–3.5 mm. long, obovate, the beak 1 mm. long. Submerged aquatic, Cherry Cr., Ewan, Whitman Co., July 13, 1948, *Yocom.*

Potamogeton Robbinsii Oakes. Stems up to 1 m. long, commonly branched; blades all submersed, 3–10 cm. long, 3–8 mm. wide, ligulate, acute, 20–60-nerved, the margins cartilaginous and serrulate; stipules with the free portion 5–20 [illegible] long, [illegible], pale, eroding to fibers; inflorescence branching; spikes [illegible]–20 cm. long, of [illegible]; the [illegible] fruits [illegible] mm. long, obliquely obovoid, flattened, prominently keeled; plant propagating [illegible] tips and indurated buds. Submerged aquatic, first collected in 1948 by *Yokom,* [illegible] *T.*

Potamogeton [illegible] [illegible] much branched; principal [illegible] mm. broad, linear, obtuse or [illegible]; stipules 15–35 mm. long, firm fibrous, the lower obtuse, the upper [illegible] peduncles 15–55 mm. long; spikes 1.5–3 cm long, the flowers in 7–11 somewhat remote whorls; fruits quadrate-oblong or -suborbicular, the back with a more or less dentate keel. Submerged aquatic, channeled scablands, Spokane and Whitman Cos., *A. T., U. Son.*

ZANNICHELLIACEAE

Submersed perennial herbs; leaves all linear; flowers unisexual, in axillary clusters; staminate flowers single, without perianth and with with 1–2 stamens; pollen spherical; pistillate flowers 2–5 in each axillary cluster, with funnel-shaped perianth; ovary arcuate; style 1; stigma shield-like; ovule solitary, straight and pendulous. (Named for the genus *Zannichellia.*)

ZANNICHELLIA

Submersed herbs; leaves small, linear; stems, flowers, and leaf buds at first enclosed in a hyaline envelope; flowers unisexual, in axillary clusters, each composed of one staminate and 2–5 pistillate flowers; staminate flower with no perianth; pistillate with a funnel-shaped undivided perianth; ovary flask-shaped with a short style. (Named for *J. J. Zannichelli* of Venice.)

Zannichellia palustris L. *Horned Pondweed.* Stems slender, branching, leafy, 10–60 cm. long; leaves thin, filiform, 1-nerved, 5–8 cm. long; fruit nearly sessile, flattened, somewhat incurved, often more or less toothed on the back, the body 2–3.5 mm. long, about twice as long as the style. Fresh and sub-alkaline waters, *U. Son., A. T.*

NAJADACEAE

Submersed aquatic herbs with slender branching stems; leaves flat or filiform, spiny toothed, opposite, alternate, or whorled; flowers monoecious or dioecious, axillary, solitary; staminate enclosed in a membranous sheath; stamen 1, sessile or stalked with a 1–4-celled anther; ovary 1 with a short style, 2–4 stigmas, 1-celled, 1-ovuled; endosperm none. (Named for the genus *Najas.*)

NAJAS

Characters of the family. (Name Gr., *Naias,* a water nymph.)

Najas flexilis (Willd.) Rostk. & Schmidt. Stems slender; leaves numerous, linear, acute or acuminate, 1–2 cm. long, minutely serrulate, the sheath broadly-oblong; fruit 3 mm. long, lance-cylindric, tipped with a persistent style; seed pale-brown shining. Near Viola, Idaho, *Henderson*; Philleo L., Spokane Co., Aug. 14, 1889, *Suksdorf, A. T.*

JUNCAGINACEAE. Arrow Grass Family

Swamp plants with narrow leaves and terminal racemose or spike-like inflorescences; flowers perfect or unisexual, naked or with a usually evanescent bract-like perianth; stamens 2 or 3; carpels 3–6, each 1–2 ovuled, more or less united till maturity, dehiscent or indehiscent. (Name from Lat., *juncus,* the rush, alluding to the similarity to the rush.)

TRIGLOCHIN. Arrow Grass

Flowers small, perfect, in terminal scapose, spike-like racemes; perianth segments 3–6, greenish, evanescent; stamens 3–6; anthers 2-celled, on very short filaments; carpels 3–6, united into a compound pistil; ovules solitary; capsule, when ripe, splitting into 3–6 carpels with a persistent central axis. (Name Gr., *tri,* thrice; *glochin,* point.)

Triglochin maritima *L.* Scapes subterete, 20–40 cm. high, exceeding the leaves; leaves fleshy, narrow; ligules entire or emarginate; leaf bases persisting on the rootstock, blades obcompressed; raceme elongate, 15–40 cm. long; fruit ovoid, angled, 5–6 mm. long; carpels usually 6, sometimes 3. Restricted to alkaline marshes. Medical Lake, *Henderson;* Cheney, May 27, 1889, *Suksdorf, A. T., U. Son.*

The fresh tissues contain hydrocyanic acid, though when dried this deadly poison largely disappears. On alkaline flats, where other forage is scarce, cattle are killed by eating this plant.

ALISMATACEAE. Water-plantain Family

Aquatic or marsh herbs with fibrous roots, scape-like stems and basal long-petioled leaves; inflorescence a raceme or panicle; flowers regular, perfect, monoecious or dioecious, pedicelled; the pedicels whorled and subtended by bracts; sepals 3, persistent; petals 3, deciduous; stamens 6 or more; ovaries numerous or rarely few, 1-celled, usually 1-ovuled; carpels becoming achenes in fruit; endosperm none. (Named for the genus *Alisma.*)

Flowers perfect; carpels in a ring upon a small flat receptacle; stamens 6–9. *Alisma.*

Flowers monoecious or dioecious; carpels in many series upon a large convex receptacle; leaves sagittate (in ours); stamens many. *Sagittaria.*

ALISMA. Water-plantain

Perennial or rarely annual herbs with erect or floating leaves; inflorescence a panicle or umbel-like panicle; flowers perfect, small, numerous, on unequal 3-bracted pedicels; petals small; stamens 6–9; ovaries few or many, in one whorl on a small flat receptacle. (The ancient Gr. name.)

Achenes as wide as long, (ridged on the back); peduncles and pedicels recurved. *A. gramineum,* var. *gramineum.*

Achenes longer than wide (grooved on the back); peduncles and pedicels straight, ascending. *A. Plantago-aquatica,* var. *Plantago-aquatica.*

Alisma gramineum Gmel., var. **gramineum.** Plants diffuse; leaf blades lanceolate to linear-lanceolate entire, glabrous, 3–9 cm. long; petioles slender, exceeding the blades; scapes 2–4, pedicels 1–2 cm. long; petals pink, 2–4 mm. long, exceeding the sepals; achenes suborbicular, flat, 2–2.5 mm. long. Rare, Stoner, *St. John et al.* 7,160; s. end of Rock L., *Pickett et al.* 1,622, *U. Son.*

Alisma Plantago-aquatica L., var. **Plantago-aquatica.** *Water-Plantain.* Scapes stout, 30–100 cm. tall; leaves all radical, erect or floating, the petioles usually long; the blades ovate to oblong-lanceolate, acute, rounded or subcordate at the base, 5–15 cm. long, 5–7-nerved; flowers in a large panicle composed of 3–6 whorls of branches, these again branched once or twice; flowers on pedicels 1–5 cm. long; sepals suborbicular, 3–4 mm. long, exceeded by the white petals; achenes flat, semicordate, emarginate at base, 2–3 mm. long, compressed. Com-

mon in wet places, *A. Plantago-aquatica* subsp. *brevipes* (Greene) Sam.; also var. *Michaletii* (Asch. & Graebn.) Buch. *U. Son., A. T.*
The herbage is acrid and indigestible.

SAGITTARIA. Arrowhead

Perennial aquatic or marsh herbs with basal long-petioled leaves; flowers monoecious or dioecious, borne near the summits of the scapes in whorls of 3, the staminate uppermost; petals usually conspicuous; stamens usually numerous; carpels numerous, crowded in globose heads. (Name Lat., *sagitta,* arrow.)

Achenes with a long horizontal beak. *S. latifolia,* var. *latifolia.*
Achenes with a short vertical beak. *S. cuneata.*

Sagittaria cuneata *Sheldon.* Terrestrial or aquatic, 10–50 cm. tall; petioles stout, ascending, 10–30 cm. long; blades 6–18 cm. long, sagittate, acute, the basal lobes diverging and usually much smaller than the terminal one; sepals becoming reflexed; petals white, 1–2 cm. long; fruiting head globose, 8–15 mm. in diameter; achenes 2–2.5 mm. long, obovate-cuneate, much flattened, with a minute erect beak. Common on the margins of ponds and streams, *A. T.*

Sagittaria latifolia Willd., var. **latifolia.** *Wapato.* Wholly or partly emersed, 2–14 dm. tall; leaves very variable, sagittate, the lobes from ovate to linear, blades 15–40 cm. long; petals white 1.5–2 cm. long; achenes obovate, about 3 mm. long. Wet shores, *A. T.*

At the ends of the rootstocks are produced white starchy tubers. These can be freed from the muddy bottoms of the ponds or streams by poking around with sticks, or with toes as did the squaws. They were a favorite staple food of the Indians, and could be stored fresh for 2–3 months.

HYDROCHARITACEAE Frog's Bit Family

Aquatic herbs with dioecious or polygamous regular flowers; stamens 3–12, distinct or monadelphous; anthers 2-celled; stigmas 3 or 6; ovary 1–3-celled, ripening under water, indehiscent, many seeded. (Named for the genus *Hydrocharis,* from Gr., *hudor,* water; *charis,* ornament.)

ELODEA. Water-weed

Submerged aquatics, stems elongate, branching; leaves opposite or whorled, 1-nerved, pellucid; flowers perfect or dioecious; pistillate flowers and perfect flowers borne to the surface by the elongating stipe-like hypanthium, staminate similar, or sessile and at anthesis set free, rising to, then expanding on the surface of the water; stamens 3–9; stigmas 3, entire or 2-cleft; style often elongate; fruit cylindric, several seeded. (Name Gr., *helodes,* marshy.)

Leaves linear, flaccid, not imbricated; staminate spathes sessile. ***E. Nuttallii.***
Leaves ovate-oblong, firm, imbricated above; staminate spathes stipitate. *E. canadensis.*

Elodea canadensis Michx. *Canadian Water-weed.* Middle and upper leaves in whorls of 3, usually ovate-oblong, dark green, 6–12 mm. long, 1–5 mm. broad; staminate flowers stipitate; pistillate spathe cylindric, bidentate; pistillate flowers exserted by the long slender hypanthium; petals white, 2–6 mm. long, staminodia 3, acicular. *Anacharis canadensis* (Michx.) Planch.; *Philotria canadensis* (Michx.) Britton. Little Spokane River, *St. John* 3,423; 3,438; Fish L., Spokane Co.; Kamiak Butte, *D. Preston, A. T.*

This native N. Am. species was introduced into Europe in 1836. It soon spread over most of Europe and Asia, multiplying so as to obstruct rivers and canals.

Elodea Nuttallii (Planch.) St. John. Lowermost leaves opposite, ovate, reduced, other leaves in whorls of 3, 6–13 mm. long, 0.3–1.5 mm. wide; staminate spathes ovoid, apiculate, 2 mm. long; petals usually wanting, stamens 9; pistillate spathes cylindric-ovoid, bifid, 10–15 mm. long, long stalked, petals white, 1–3 mm. long. *E. occidentalis* (Pursh) St. John. Lakes Coeur d'Alene and Chatcolet, *Can.*

These stations are the known w. limit of the species.

GRAMINEAE. Grass Family

Annual or perennial herbs, rarely shrubs or trees; stems (*culms*) generally hollow; nodes solid; leaves 2-ranked, sheathing, the sheaths usually split to the base on the side opposite the blade; a scarious or membranous appendage (the *ligule*) borne at the opening of the sheath, rarely obsolete; inflorescence a spike, a raceme, or a panicle, consisting of spikelets composed of two to many 2-ranked imbricated bracts; the lowest two (*glumes*) without flowers or rarely wanting; one or more of the upper (*lemmas*) containing in its axil a flower, which is usually enclosed by a bract-like, generally 2-keeled, awnless organ (*palea*) opposite the lemma and with its back away from the axis (*rhachilla*) of the spikelet; lemma sometimes bearing an indurated thickening (*callus*) at the base; flowers perfect or sometimes monoecious or dioecious, subtended by 1–3 minute hyaline scales (*lodicules*); stamens 1–6, usually 3; ovary 1-celled, 1-ovuled; styles 1–3, commonly 2 and lateral; stigmas hairy or plumose; fruit a seed-like grain (*caryopsis*) or rarely a utricle; endosperm starchy. *Poaceae.* (Name Lat., *gramen*, grass.)

Spikelets with 1 perfect terminal floret (except in staminate or neuter spikelets) and 1 staminate or sterile floret or rudiment below; glumes 2, 1, or 0; articulation below the spikelet which falls entire or in clusters; spikelets or fruits dorsally compressed. Tribe 8. *Paniceae.*

Spikelets 1 to many flowered, the reduced florets, if any, above the perfect florets (except in *Phalaris*); articulation usually above the glumes; spikelets laterally compressed,

Staminate, neuter, or rudimentary lemmas below the fertile one. Tribe 6. *Phalarideae.*

Staminate, neuter, or rudimentary lemmas not below the fertile ones,

Spikelets unisexual, 1-flowered, terete or nearly so, articulate below the glumes. Tribe 7. *Zizaneae.*
Spikelets usually perfect and articulate above the glumes,
Spikelets pedicellate in panicles which are open, contracted, or even spike-like,
Spikelets 2–many flowered,
Glumes shorter than the lowest floret; lemmas awnless or awned from the tip or bifid apex. Tribe 1. *Festuceae.*
Glumes as long as or longer than the lowest floret; lemmas awned from the back or awnless. Tribe 3. *Aveneae.*
Spikelets 1-flowered. Tribe 4. *Agrostideae.*
Spikelets sessile in a spike or some sessile and some pedicillate in compact racemes,
Spikelets in 2 rows on opposite sides of the single rhachis. Tribe 2. *Hordeae.*
Spikelets on 1 side of each rhachis; rhaches several, digitate or spicate. Tribe 5. *Chlorideae.*

Tribe 1. FESTUCEAE

Lemma 3-nerved or rarely 1-nerved,
Rhachilla with hairs as long as the lemmas. *Phragmites.*
Rhachilla glabrous or with shorter hairs,
Glumes obtuse, very unlike. *Sphenopholis.*
Glumes acute, subequal,
Panicle close; spikelets 2–4-flowered. *Koeleria.*
Panicle loose; spikelets usually many-flowered. *Eragrostis.*
Lemma 5- or more-nerved,
Upper florets sterile, broad, folded about each other. *Melica.*
Upper florets not so,
Plants dioecious; lemma glabrous. *Distichlis.*
Plants not dioecious or if so the lemmas villous,
Inflorescence a dense, spike-like raceme. *Sclerochloa.*
Inflorescence a panicle,
Lemmas keeled on the back,
Spikelets strongly compressed, crowded in 1-sided clusters at the ends of the stiff, naked panicle branches. *Dactylis.*
Spikelets not so,
Lemmas awnless,
Spikelets 15–25 mm. long. *Bromus brizaeformis.*
Spikelets shorter. *Poa.*
Lemmas awned from the minutely bifid apex. *Bromus.*
Lemmas rounded on the back (or slightly keeled towards the tip),
Lemmas awnless, mostly obtuse, their nerves parallel, not or slightly converging at summit,
Lemma prominently 5–9-nerved; style present. *Glyceria.*

Lemma faintly 5-nerved; style none. *Puccinellia.*

Lemmas awned or pointed, their nerves converging at the summit,

Lemmas entire, awned from the tip or pointed.

Perennials; florets opening, not distended in anthesis; the 3 anthers and plumose stigma exserted; grain ellipsoid or ovoid. *Festuca.*

Annuals; florets cleistogamous, in anthesis distended above; the 1 (rarely 3) short stamen and small stigma included; grain linear-cylindric. *Vulpia.*

Lemmas awned or awn-tipped from a minutely bifid apex. *Bromus.*

Tribe 2. HORDEAE

Spikelets solitary at each node of the rhachis,

Lemmas with their backs towards the rhachis,

Inner glume wanting. *Lolium.*

Both glumes present. (Festuceae) *Sclerochloa.*

Lemmas with their sides towards the rhachis,

Nerves connivent at tip of glumes. *Agropyron.*

Nerves not connivent at tip of glumes. *Aegilops.*

Spikelets 2–6 at the nodes of the rhachis or at least at the middle nodes,

Spikelets 3 at each node, 1-flowered, the lateral ones pedicelled and commonly sterile. *Hordeum.*

Spikelets usually 2 at each node, all alike, 2–6-flowered,

Rhachis disjointing at maturity; glumes often deeply parted. *Sitanion.*

Rhachis of spike or panicle persistent; glumes entire. *Elymus.*

Tribe 3. AVENEAE

Perfect floret 1, the other staminate,

Lower floret staminate; awn twisted, geniculate, exserted. *Arrhenatherum.*

Lower floret perfect, awnless; awn of upper floret hooked. *Holcus.*

Perfect florets 2 or more,

Awn from between the teeth of the bifid apex, flattened, twisted. *Danthonia.*

Awn dorsal, not flattened,

Spikelets more than 15 mm. long. *Avena.*

Spikelets less than 10 mm. long,

Lemmas keeled, bidentate, their awns from above the middle. *Trisetum.*

Lemmas convex, truncate and erose-dentate at tip, their awns from below the middle. *Deschampsia.*

TRIBE 4. AGROSTIDEAE

Lemma indurate or firm, closely enveloping the grain, long awned,
- Lemma firm; callus not well developed. — *Muhlenbergia.*
- Lemma indurate; callus well developed, oblique, bearded,
  - Awn of lemma 3-branched. — *Aristida.*
  - Awn of lemma simple, twisted and bent. — *Stipa.*

Lemma membranous, loosely enveloping the grain, awnless or short awned,
- Inflorescence a dense spike-like raceme or panicle,
  - Lemma awnless. — *Phleum.*
  - Lemma awned,
    - Glumes awnless. — *Alopecurus.*
    - Glumes awned. — *Polypogon.*
- Inflorescence a loose panicle,
  - Rhachilla articulate below the glumes; stamen 1. — *Cinna.*
  - Rhachilla articulate above the glumes; stamens 3,
    - Lemma 1-nerved, (rarely 3-nerved); fruit a utricle. — *Sporobolus.*
    - Lemma 3–5-nerved, fruit a grain,
      - Callus with a tuft of long hairs at base. — *Calamagrostis.*
      - Callus naked. — *Agrostis.*

TRIBE 5. CHLORIDEAE

Perfect florets 2 or more. — *Leptochloa.*

Perfect florets 1, often with imperfect ones above,
- Glumes unequal, narrow. — *Spartina.*
- Glumes equal, broad and boat-shaped. — *Beckmannia*

TRIBE 6. PHALARIDEAE

Lower florets staminate; spikelets brown, shining. — *Hierochloë.*

Lower florets neuter; spikelets green or yellowish. — *Phalaris.*

TRIBE 7. ZIZANEAE

Only one genus represented. — *Zizania.*

TRIBE 8. PANICEAE

Spikelets subtended by 1 to several stiff bristles. — *Setaria.*

Spikelets not subtended by bristles,
- Glumes or sterile lemma awned. — *Echinochloa.*
- Glumes and sterile lemma awnless,
  - Spikelets in slender, 1-sided, more or less digitate racemes,
    - Racemes 3–12; lemma margins not inrolled. — *Digitaria.*
    - Racemes only one pair; lemma margins inrolled. — *Paspalum.*
  - Spikelets in a loose panicle. — *Panicum.*

AEGILOPS. Goat Grass

Tufted annuals; leaves flat; spikelets sessile at each joint of the rhachis, 3–several-flowered, parallel to the rhachis; terminal spikelet, or the upper 2–3, sterile, more slender than the others; glumes leathery, convex, not keeled, many nerved on the back, truncate, entire, or toothed, or long awned; lemmas papery, convex, not keeled, many nerved, 1–3-toothed or awned. (Name Gr. *aigos,* goat; *ops,* eye.)

Aegilops cylindrica Host. Plant 3–6 dm. tall; stems erect or ascending; leaves stiff, somewhat hispid; spike 4–12 cm. long, linear cylindric, at maturity disjointing at each node; spikelets 5–10, oblong cylindric, sunken in the axis, the upper spikelets with 3 awns, 2–8 cm. long; glumes 6–7 mm. long. Introduced from Eurasia, now established as a weed in Pullman, grassy places, *A. T.*

AGROPYRON. Wheatgrass

Annuals or perennials (like all ours) with simple stems and terminal spikes; spikelets 3–many-flowered, sessile, usually single and alternate at each notch of the usually continuous rhachis, the side of the spikelet, that is the edge of the glumes, toward the rhachis; glumes 2, equal, usually firm, several-nerved, usually shorter than the lemmas; lemmas rigid, convex on the back, 5–7-nerved, usually acute or awned at the apex; palea often with 2 hairy keels; grain pubescent at the apex. (Name Gr. *agros,* field; *puros,* wheat.)

Rhachis disjointing at maturity; glumes 2-nerved, with 1–2 awns much longer than the body. — *A. saxicola.*

Rhachis not disjointing; glumes 3–5-nerved, awnless or short awned,

- Lemmas pilosulous. — *A. dasystachyum.*
- Lemmas glabrous or merely scabrous towards the tip,
  - Plants spreading by rhizomes, not forming large tufts. — *A. repens.*
  - Plants without rhizomes, forming large, dense erect tufts,
    - Awn of lemma divergent, usually bent; blades puberulent above. — *A. spicatum.*
    - Awn of lemma ascending, straight, or wanting,
      - Blades 1–3 mm. wide, soon involute, puberulent above. — Var. *inerme.*
      - Blades mostly wider, flat,
        - Awns of lemma wanting or very short or not more than half its length,
          - Spikelets scarcely imbricated; rhachilla usually scabrous or strigose, tightly imbricate. — *A. trachycaulum,* var. *trachycaulum.*

Spikelets well imbricated; rhachilla usually villous, free. Var. *noviae-angliae.*
Awns from nearly equaling to much longer than the body of the lemma. Var. *unilaterale.*

Agropyron dasystachyum (Hook.) Scribn. Culms erect, 4–11 dm. tall, smooth, from elongated creeping rootstocks; stem leaves 4; sheaths smooth, loose, exceeding the internodes; blades 2–6 mm. wide, becoming involute, ascending, strigose above, smooth beneath; spikes erect, 5–30 cm. long; spikelets not crowded, 1–2.5 cm. long, 4–10-flowered; glumes 7–10 mm. long, somewhat pilose, lanceolate, the second broader; lemmas 8–12 mm. long, lanceolate. *A. lanceolatum* Scribn. & Smith; *A. Elmeri* Scribn. & Smith; Common on the dry sandy bars of Snake River, *U. Son.*

Agropyron repens (L.) Beauv. *Couch* or *Quack Grass.* Culms erect, smooth, 3–15 dm. high; leaf sheaths hairy or smooth; blades 3–15 mm. broad, flat, pilose above, the nerves not prominent; spike 5–15 cm. long; spikelets 3–7-flowered, 8–14 mm. long; glumes 5–7-nerved, 8–11 mm. long, acuminate or awn-pointed; lemmas strongly 5-nerved, 8–11 mm. long, acuminate or awned. European grass, introduced and now a noxious weed here. Meadows and fields, *A. T., A. T. T.*

Agropyron saxicola (Scribn. & Smith) Piper. Tufted plants, without rhizomes; culms 5–10 dm. tall, slender, wiry, pale striate; sheaths close, glabrous or pilose; blades 1.5–4 mm. wide glabrous or pilose, flat or involute; spike 7–12 cm. long, slender, or occasionally with 2 spikelets at some nodes; spikelets 2–5-flowered; glumes with the body 3–6 mm. long, linear, entire or 2–3-lobed, scabrous with a terminal, divergent, slender, scabrous awn, 1.5–2.5 cm. long; lemma with the body 8–10 mm. long, scabrous towards the tip, the 4 lateral nerves obscure, the mid-nerve prominent, projecting between the 2 membranous teeth into a scabrous awn, 2–5 cm. long. Following Hitchcock we list in synonomy *A. flexuosum* Piper. Wawawai, *Piper* 3,965; 3,967, *A. T.?*

This species with its disarticulating rhachises and slender long awned glumes is intermediate to *Elymus and to Sitanion.*

Agropyron spicatum (Pursh) Scribn. & Smith. *Blue Bunch Wheatgrass.* Culms 4.5–12 dm. tall, glabrous and glaucous; stem leaves 3; sheaths close, smooth; blades 1–4 mm. wide, becoming strongly involute, scabrous beneath; spikes erect, 5–20 cm. long, slender, often loose; spikelets 3–6-flowered, 10–20 mm. long; glumes 5–10 mm. long, acute or awned; lemmas 8–10 mm. long, tipped with a stout scabrous awn, 1–2.5 cm. long. The second authority must remain Scribn. & Smith since they included *Festuca spicata* Pursh as the first synonym, even though they designated as type, *Geyer,* Upper Missouri, and despite the fact that for that reason Hultén (1942) credits the combination to Rydberg (1900). If in wet places it forms stolons and a sod and is usually sterile. *Elymus spicatus* (Pursh) Gould. An important native forage grass, abundant on dry hills, *U. Son.,* to *A. T. T.*

Var. **inerme** (Scribn. & Smith) Heller. *Bunch Grass.* Culms 3–10 dm. tall, smooth; cauline leaves 3, the sheaths glabrous; spike 7–30 cm. long, loose; spikelets 10–15 mm. long, 6–10-flowered; glumes 5–10 mm. long, acute, or obtuse; lemmas 8–12 mm. long, acute or even awn-tipped. *A. inerme* (Scribn. & Smith) Rydb. Dry rocky soil or deep silt loam, *U. Son.* to *A. T. T.,* still very abundant in *A. T.*

A few individuals with short awns on some of the spikelets are transitional to the less common *A. spicatum.* Ours was the dominant plant of the Bunch Grass Prairies of the Palouse Country, now largely in wheat. It is a perennial forage of high quality.

Agropyron trachycaulum (Link) Malte, var. **trachycaulum** *Slender Wheatgrass.* Culms erect, 3–12 dm. high, rather slender, glabrous; stem leaves

3 or 4, the sheaths loose, glabrous; blades 2–6 mm. wide, flat or becoming involute, scabrous; spike 4–25 cm. long, narrow, slender; glumes 10–12 mm. long, pointed, the hyaline margins 0.4–0.6 mm. wide; lemmas acuminate. *A. tenerum* Vasey; *Elymus pauciflorus* (Schwein.) Gould, *A. pauciflorum* (Schwein.) Hitchc. Common in low ground, *U. Son.* to *A. T. T.*

A choice native forage, now cultivated with success as a hay crop.

Var. **noviae-angliae** (Scribn.) Fern. Differing from var. *trachycaulum* by having the glumes herbaceous to subcoriaceous, the hyaline margins 0.1–0.4 mm. broad. *A. trachycaulum,* var. *tenerum,* f. *ciliatum* (Scribn. & Smith) Pease & Moore. Meadows and prairies, *A. T.*

Var. **unilaterale** (Cassidy) Malte. *Bearded Wheatgrass.* Culms erect 3–10 dm. tall; leaf-sheaths smooth; blades 2–6 mm. wide, flat, scaberulous; spikes 7–22 cm. long, rather dense, somewhat nodding; spikelets 4–5-flowered; glumes 12–18 mm. long, lanceolate, with 5 scabrous nerves, awn-tipped; lemmas lanceolate, 3–5-nerved, glabrous or finely scabrous, smooth and rounded below, with a scabrous terminal awn, 7–40 mm. long. *A caninum* of Pease & Moore; *A. subsecundum* (Link) Hitchc.; *A. trachycaulum,* var. *ciliatum,* f. *pubescens* (Scribn. & Smith) Malte. Deer Park, *Bonser* 29, *A. T.*

AGROSTIS. Bent Grass

Annual or usually perennial grasses with small 1-flowered spikelets, in diffuse or compact panicles; glumes membranous, keeled, acute; lemma shorter, obtuse, hyaline, sometimes bearing a dorsal awn; palea shorter than the lemma, sometimes minute or wanting; grain free, enclosed in the lemma. (Name Gr. *agrostis,* a grass.)

Rhachilla prolonged behind the palea; lemma with an awn at least three times its length. — *A. interrupta.*
Rhachilla not prolonged; lemma awnless or short awned
 Palea at least half as long as the lemma,
  Stolons scaly, subterranean, or lacking; panicle pyramidal, loose in anthesis. — *A. alba, var. alba*
  Stolons leafy, creeping on the surface; panicle oblong-linear in anthesis and afterwards. — Var. *palustris.*
 Palea minute or wanting,
  Plants with rhizomes — *A. pallens,* var. [illegible]
  Plants tufted, without rhizomes,
   Panicle narrow, the branches ascending-appressed, often spikelet-bearing from near the base,
    Glumes 1.5–2 mm. long; lemmas 1.5 mm. long. — *A. variabilis.*
    Glumes 2.5–3.5 mm. long; lemmas about 2 mm. long. — *A. exarata.*
   Panicle broad and open, the branches mostly divergent at maturity, spikelet-bearing only near the tips,
    Panicle very diffuse, the lower branches much exceeding the upper; herbage scabrous. — *A. scabra.*
    Panicle not diffuse, the lower branches slightly exceeding the upper; herbage smooth. — *A. idahoensis.*

Agrostis alba L., var. alba. **Redtop.** Perennial, tufted, or matted with subterranean stolons; culms smooth, decumbent or erect, up to 1 m. tall; sheaths ridged; blades scabrous, 4–10 mm. wide; panicle exserted pyramidal to oblong, generally purplish; glumes 2–3.5 mm. long; lemma 1.5–2.2 mm. long, usually awnless. *A. vulgaris* With.; *A. stolonifera* L., var. *major* (Gaud.) Farw.; *A. palustris* of Piper and Hitchc. Naturalised from Europe. Moist meadows, common, *U. Son., A. T.*

Var. **palustris** (Huds.) Pers. Densely matted; culms decumbent, the tips ascending; blades 2–3 mm. wide; spikelets straw-colored. *A. maritima* Lam. Often in saline habitats. *A. stolonifera* L., var. *compacta* Hartm. Wawawai, *Piper* 3,531.

Agrostis exarata Trin. *Western Bent Grass.* Perennial, tufted, the stems 2–12 dm. tall; leaves erect, the blades flat, 2–8 mm. broad, 5–10 cm. long, scabrous, the upper distant from the panicle; panicle strict, pale green, spike-like or interrupted, the short branches appressed and spikelet-bearing to the very base; glumes 2.5–3.5 mm. long, usually scabrous on the back. Common in moist soil, *A. T.*

Agrostis idahoensis Nash. Delicate, loosely-tufted, glabrous, perennial, 10–30 cm. high; blades flat, 1–6 cm. long, less than 1 mm. wide; panicle loose, green or purple, 5–10 cm. long; rays capillary; spikelets about 1.5 mm. long; lower glume scabrous on the keel, slightly larger than the upper; lemma truncate, awnless, 1 mm. long; palea minute. In wet meadows, Craig Mountains, *Can.*

Agrostis interrupta L. Annual; culm smooth, 2–5 dm. tall; leaves short, flat, scabrous, 1–2 mm. wide; panicle exserted, narrow, contracted, often interrupted below, 6–18 cm. long; branches spikelet-bearing nearly to the base; glumes unequal, lanceolate, acuminate, broadly hyaline-margined, 2–2.5 mm. long; lemma brown, scabrous above, the slender awn attached in the sinus of the notched apex. Native of Eurasia, recently introduced, Spokane, *Bonser* in 1922; Ewan, Whitman Co., 1934, *Pickett* 1,612.

Mrs. Chase (1950) has classed this as in a different genus, *Apera,* but the resemblances outweigh the differences, so it seems best kept in *Agrostis.*

Agrostis pallens Trin., var. **Vaseyi** St. John. Culms erect, stiff, 5–12 dm. tall; blades, slender, scabrous, 1–3 mm. wide; panicle green or purplish, narrow, the branches mostly appressed, the shorter spikelet-bearing from the base; glumes equal, scabrous, 2–3 mm. long; lemma as long, or nearly so, smooth. *A. foliosa* Vasey, not R. & S. (1817); *A. pallens* Trin., var. *foliosa* (Vasey) Hitchc.; *A. diegoensis* Vasey. Meadows, *A. T.*

As this var. differs only in its larger and looser panicle, from the species, which is restricted to the sea coast, there seems little justification for maintaining it as a species.

Agrostis scabra Willd. *Tickle Grass.* Annual, tufted, 15–90 cm. tall; leaves mostly basal, flat or involute, narrow, very scabrous, 0.5–3 mm. wide; panicle 15–30 cm. long; branches in whorls of 2–12, capillary, 5–15 cm. long, branched above the middle and spikelet-bearing only near the tips. Common in both dry and moist places, *A. T., Can.*

Agrostis variabilis Rydb. *Alpine Redtop.* Tufted, 10–20 cm. high; leaves scabrous, 2–5 cm. long, 1–2 mm. broad; flat or folded; panicle narrow, erect, 3–6 cm. long, or occasionally longer, green or purple. *A. Rossae* sensu Piper, and of St. John, not of Vasey. Meadows on the highest ridges of the Blue Mts., *Huds.*

ALOPECURUS. Foxtail

Annuals or perennials; panicles dense and spike-like; glumes equal awnless, usually somewhat united at base; lemma often equaling the glumes, with a slender dorsal straight or geniculate awn; palea none. (Name Gr., *alopex,* fox; *oura,* tail.)

Glumes united for a third, or half of their length. — *A. myosuroides*
Glumes united one quarter or less of their length,
  Glumes 5 mm. long. — *A. pratensis.*
  Glumes 2–3 mm. long,
    Glumes 2–2.4 mm. long; awn attached near middle of lemma, not geniculate, included or nearly so. — *A. aequalis.*
    Glumes 3 mm. long; awn attached near base of lemma, geniculate, exserted. — *A. pallescens.*

Alopecurus aequalis Sobol. At first often floating, later tufted and erect, 15-50 cm. tall; culms often decumbent, smooth; blades usually short, 2–6 mm. wide; inflorescence usually pale green, slender, 2.5–6 cm. long, 4–5 mm. thick; keel of lemma hispid ciliate. *A. aristulatus* Michx. Wet meadows, abundant, *U. Son., A. T.*

Alopecurus myosuroides Huds. Smooth or scabrous, 3–7 dm. tall; sheaths shorter than the internodes; blades 2–6 mm. wide; inflorescence 4–10 cm. long, slender, 4–8 mm. thick; glumes shining, smooth above, the midrib hispidulous; lemma slightly exceeding the glumes, smooth; awn nearly basal, geniculate, twice the length of the lemma. *A. agrestis* L. Weed, introduced from Europe, not common, *A. T.*

Alopecurus pallescens Piper. Perennial, tufted, erect, 15–50 cm. tall; sheaths somewhat inflated; blades scabrous, 2–5 mm. wide; inflorescence pale green, 2.5–8 cm. long, 4–8 mm. thick; spikelets 3–3.3 mm. long; glumes pilose; lemma smooth, about 3 mm. long, near the base bearing a geniculate awn which is exserted 3–5 mm. The holotype is from Pullman, *Piper* 1,743. Wet places, *A. T.*

Alopecurus pratenis L. *Meadow Foxtail.* Nearly or quite glabrous, 3–8 dm. tall, sheaths loose, shorter than the internodes; blades 2–6 mm. wide; inflorescence stout, 4–7 cm. long, 8–12 mm. thick; glumes strongly ciliate on the keel, pale, purplish banded along the nerves; lemma equaling the glumes, smooth; awn attached near the base, geniculate, exserted 4–5 mm. An European species, cultivated, occasionally established. Meadows, *A. T.*

ARISTIDA. Three-awn

Annuals or perennials, spikelets narrow, 1 flowered; glumes 2, narrow, slightly keeled, lemma 1, convolute, bearing a 3-branched awn; palea short; grain free, tightly enclosed in the lemma. (Name Lat., *arista,* awn.)

Aristida longiseta Steud., var. **robusta** Merrill. Perennial, tufted, 15–40 cm. high; radical leaves numerous; stem leaves about 3, all strongly involute and filiform, scabrous; panicle loose, purplish, 10–20 cm. long; second glume twice as long as the first, awns of the slender lemma 4–6 cm. long, the central slightly longer than the others. *A. purpurea robusta* (Merrill) Piper. Maturing in early summer, dry soils, *U. Son., A. T.*

The tops break loose and blow. By means of their retrosely scabrous awns they bore into things, even into the flesh of animals and sometimes cause their death.

ARRHENATHERUM. Oatgrass

Tall perennials with long narrow panicles; spikelets 2-flowered, lower flower staminate, upper perfect; rhachilla extending beyond the flowers; glumes thin-membranous, keeled, very acute or awn-pointed, unequal, persistent; lemmas 2, rigid, 5–7-nerved, deciduous, the first bearing a long bent and twisted dorsal awn inserted below the middle, the second unawned; palea hyaline, 2-keeled; grain ovoid, free. (Name Gr., *arren,* masculine; *ather,* awn.)

Arrthenatherum elatius (L.) Presl. *Tall Meadow Oatgrass.* Glabrous perennial, 1–1.5 m. high; stem leaves 3, the blades flat, minutely scabrous, the sheaths shorter than the internodes, 5–10 mm. wide; panicle shining, 15–25 cm. long, rather narrow, the branches suberect; glumes scabrous, the second equaling the floret; lemmas 6–8 mm. long, the lower bearing a long dorsal geniculate awn. European grass, sparingly escaped from cultivation, *A. T.*

The second authority is Presl, not Mcrt. & Koch as cited by Hitchcock, Man. Grasses U. S. 298, 802, 1935.

AVENA. Oats

Annuals or perennials with large spikelets; inflorescence a panicle; spikelets 2–many-flowered; lower flowers perfect, the upper often staminate; glumes somewhat unequal, membranous, persistent; lemmas rounded on the back, indurate below, acute, generally bearing a dorsal awn; apex often 2-toothed; palea narrow, 2-toothed; grain oblong, deeply furrowed, enclosed in the lemma and palea, free or sometimes adherent to the latter. (The ancient Lat. name.)

Avena fatua L., var. glabrata Peterm. *Smooth Wild Oat.* Stems stout, pale, smooth, 50–100 cm. tall, erect; leaf blades flat, 5–14 mm. wide; the sheaths about equaling the internodes; panicle loose 15–30 cm. long; spikelets erect or drooping, on slender branches; glumes 18–28 mm. long; lemma smooth, except the scabrous apex and the ring of stiff white hairs at the base, bearing on the back a geniculate stout, twisted awn below the 2-cleft apex. A troublesome weed, introduced from Europe. In cultivated land, *A. T.*

BECKMANNIA.

Tall erect annuals or perennials; inflorescence a terminal panicle of erect spikes; spikelets 1–2-flowered, globose, compressed; glumes membranous, saccate, obtuse or abruptly acute; lemmas 1 or 2, narrow, thin-membranous; palea hyaline; grain oblong, free, enclosed in the lemma. (Named for *Prof. Johann Beckmann* of Germany.)

Beckmannia Syzigache (Steud.) Fern. *Slough Grass.* Annual, the stems stout, 3–10 dm. tall, glabrous throughout; blades 10–30 cm. long, 4–13 mm. wide, scabrous, the loose sheaths exceeding the internodes; panicle narrow, 10–30 cm. long, the densely-flowered branches mostly solitary and erect; spikelets nearly orbicular, flattened, 2 mm. long, 1-flowered or with an aborted second one; glumes 2.5 mm. long; lemma smooth, acuminate, 3 mm. long. *B. erucaeformis* (L.) Host. In shallow water or very moist places, frequent, *A. T., A. T. T.*

BROMUS. Bromegrass

Annuals, biennials, or perennials with flat leaves and terminal panicles of large spikelets; peduncles thickened at the summit; spikelets few–many-flowered; glumes unequal, acute; lemmas rounded on the back, or compressed-keeled, 5–9-nerved, the apex usually 2-toothed, generally bearing an awn just below the summit; palea shorter than the lemma; grain adherent to the palea. (Name Gr., for oats, from *broma,* food.)

Spikelets much flattened, the lemmas keeled,
Lemma with awn less than 7 mm. long,
Sheaths pilose; lemma pilose. *B. marginatus.*
Sheaths glabrous; lemmas scabrous or somewhat scabrous-puberulent. *B. marginatus,* var. *seminudus.*
Lemma with awn more than 7 mm. long,
Lemma puberulent or pilose. *B. carinatus.*
Lemma scabrous,
Sheaths pilose; spikelet 3–4 cm. long. *B. carinatus,* var. *Hookerianus.*
Sheaths glabrous; spikelet 2.5–3 cm. long. *B. carinatus,* var. *californicus.*
Spikelets not so,
Creeping rhizomes present; lemma awnless or mucronate. *B. inermis.*
Creeping rhizomes wanting; lemmas awned,
Perennials,
Panicle narrow, the branches erect; lemma 10–12 mm. long. *B. erectus.*
Panicle open, the branches spreading or drooping; lemma 8–10 mm. long. *B. vulgaris.*
Annuals,
Awn much longer than the body of the lemma,
Second glume usually less than 1 cm. long; pedicels capillary, flexuous,
Lemma villous or pilose. *B. tectorum.*
Lemma glabrous. *B. tectorum,* f. *glabratus*
Second glume 1.1–3 cm. long; pedicels stouter,
Awn about 2 cm. long; first glume about 8 mm. long. *B. sterilis.*
Awn 3–5 cm. long; first glume about 15 mm. long. *B. rigidus.*
Awn shorter than the lemma or none,
Awn wanting or nearly so; lemma as broad as long or nearly so, becoming inflated. *B. brizaeformis.*
Awn 3–9 mm. long; lemma not more than two thirds as broad as long, not inflated,
Sheaths glabrous; florets becoming distant, their bases and the rhachilla visible. *B. secalinus.*

Sheaths pilose; florets close, overlapping, their bases and the rhachilla concealed,
Panicle loose; pedicels slender, exceeding the spikelet, usually much exceeding it. *B. commutatus.*
Panicle dense; pedicels short, stout, usually not equaling the spikelet,
Lemmas glabrous. *B. racemosus.*
Lemmas pilose,
Spikelets pilose. *B. mollis.*
Spikelets glabrous or scabrous. *B. mollis,* f. *leiostachys.*

Bromus brizaeformis Fisch. & Mey. *Rattlesnake Brome.* Culms erect 1–6 dm. tall; sheaths densely pilose; blades pilose; panicle 5–25 cm. long, loose, one-sided, nodding; spikelets 15–25 mm. long, elliptical, compressed; lemmas 8–10 mm. long, smooth, white scarious margined, awnless, or the uppermost with short awns, 1–2 cm. long. Introduced from Europe and now abundant, *U. Son., A. T.*

Bromus carinatus H. & A. *California Bromegrass.* Short lived perennial or sometimes annual; awns of the lemmas 7–10 mm. long; otherwise similar to *B. marginatus.* Common, *U. Son.* to *A. T. T.*

Var. **californicus** (Nutt.) Shear. Leaves nearly glabrous. This is accepted as a variety though Shear does not definitely so state it. However, in all the adjacent comparable trinomials, he does clearly call them varieties. It seems fair, then, to interpret this one likewise. Pullman, *Hunt* 44.

Var. **Hookerianus** (Thurb.) Shear. Robust; panicle large, spreading. Occasional *A. T.*

Bromus commutatus Schrad. Very similar to *B. secalinus,* panicle more drooping; lemma broadly elliptic, not inrolled in fruit, tipped with a stout straight awn 7–8 mm. long. *B. pratensis* Ehrh., not of Lam. Introduced from Europe, Pullman, *Piper, A. T.*

Bromus erectus Huds. Culms smooth, 6–9 dm. tall; sheaths sparsely pilose or glabrous; blades sparsely pubescent; panicle 10–20 cm. long; spikelets 5–10-flowered; glumes acuminate, the first 6–8 mm. long, the second 8–10 mm. long; lemma glabrous or scabrous-pubescent, with an awn 5–6 mm. long. Introduced from Europe. Reported at Steptoe, *Vasey, A. T.*

Bromus inermis Leyss. *Smooth Brome.* Culms smooth, 5–10 dm. tall; sheaths smooth; blades smooth, flat 5–10 mm. wide; panicle loose, 10–20 cm. long, the slender branches usually ascending; spikelets narrow, often purplish, 4–9-flowered; glumes smooth, the first 4–5 mm. long, the second 6–8 mm. long; lemma smooth or scabrous, the awn up to 2 mm. in length. Introduced from Europe, becoming established, *A. T.*

A valuable pasture grass in regions of low rainfall, but moderate summer temperatures.

Bromus marginatus Nees. *Mountain Bromegrass.* Perennial, tufted; the stout stems 4–12 dm. high; leaf blades coarse, sparsely pilose; panicle 10–25 cm. long, rather narrow, the branches in whorls of 2–4, about 7 cm. long, and bearing two spikelets; spikelets 2.5–4 cm. long, 5–7 mm wide, compressed and keeled, 7–9-flowered; lemma rough-puberulent to pilose, 11–14 mm. long, 2-toothed at apex, and bearing a stout awn 4–7 mm. long. *Bromus marginatus latior* Shear. Abundant, *A. T., A. T. T.*

B. marginatus with its varieties together with *B. carinatus* and its varieties form a continuous series of variations. Their division into two species is artificial, not natural. A thorough revision would probably result in their reduction to one species with several variations.

Var. **seminudus** Shear. Differs in being pubescent, or glabrous. Occasional, *A. T., A. T. T.*

Bromus mollis L. *Soft Cheat.* Culm puberulent, erect, 2–8 dm. tall; sheaths retrorse pilose; blades pilose or smooth; panicle narrow, 1–4 cm. wide; spikelets erect, 12–15 mm. long, 5–12-flowered; lemma bearing a stout straight or somewhat twisted awn 6–9 mm. long. *B. hordeaceus* of Am. authors, not of L. Weed, introduced from Europe, common, *U. Son. to A. T. T.*

Forma leiostachys (Hartm.) Fern. Differs only in being less hairy. *B. hordeaceus,* f. *leptostachys* (Pers.) Wieg. Weed, introduced from Europe, common, *U. Son., A. T.*

Bromus racemosus L. Plant 3–9 dm. tall; culms glabrous or nearly so; sheaths finely villous with retrorse hairs; blades 3–6 mm. wide, villous on both sides; panicle mostly 1–2 dm. long, erect, then nodding, racemiform, the branches solitary or in pairs, mostly shorter than their 1 (-2) spikelets; spikelets 12–22 mm. long excluding the awns, 5–10-flowered, glabrous or minutely scaberulous; first glume 4.5–6.5 mm. long, narrowly lanceolate, 3-nerved; lemmas 6.5–10 mm. long, elliptic, 7–9-nerved, bearing awns 3–10 mm. long; anthers 2–2.5 mm. long. Weed, introduced from Europe, to grasslands, *A. T.*

Bromus rigidus Roth. *Ripgut.* Culms puberulent above, decumbent at base, 2–7 dm. tall; sheaths pilose; blades flat, pilose, 2–8 mm. wide; panicle becoming broad, the spikelets mostly single, 5–7-flowered; glumes smooth or scabrous on the ribs, acuminate; lemma narrow, tapering to both ends, scabrous or puberulent, 2.5–3 cm. long. *B. maximus* Desf. Weed, introduced from Europe, *U. Son., A. T.*

Bromus secalinus L. *Chess.* Erect, culm smooth, 3–10 dm. tall; leaf-blades sparsely pilose above, smooth beneath, 1–8 mm. wide; panicle 8–18 cm. long, pyramidal, erect at first, in fruit somewhat drooping; spikelets oblong-lanceolate, turgid in fruit, 10–18 mm. long; lemma elliptic, glabrous or nearly so, 6–8 mm. long, its margins involute in fruit; awn 2–5 mm. long, palea nearly equal to the lemma. Weed introduced from Europe, common in fields, *A. T., A. T. T.*

Bromus sterilis L. Stems smooth, 5–10 dm. high, usually decumbent at base; sheaths short pilose and sparsely villous; blades more or less pilose, flat, 1–4 mm. wide; panicle 10–20 cm. long, loose and drooping, its branches in whorls of 2–6, long and slender, usually bearing but one spikelet; spikelets drooping, 2.5–3.5 cm. long, 6–10-flowered; lemma narrow, lanceolate, 17–20 mm. long, bearing a stout rough awn, 20–30 mm. long. Weed from Europe, in waste places, infrequent, *U. Son., A. T.*

Bromus tectorum L. *Cheat.* Culms 20–60 cm. tall, mostly erect, puberulent above; sheaths pilose; blades pilose, flat, short, 1–5 mm. wide; panicle one-sided, drooping, the slender branches in threes and fours, and bearing several spikelets above the middle; spikelets nodding, 15–20 mm. long, narrow, 4–7-flowered; lemma lanceolate, 10–13 mm. long, the awn 13–15 mm. long. An European weed, now one of the most abundant weeds in arid eastern Washington and adjacent states. Dry places, *U. Son.* to *Huds.*

It has taken complete possession of all over-grazed areas. It provides no forage after flowering, and its rough awns are harmful to stock.

Forma glabratus (Spenner) St. John. Differs by having glabrous spikelets. *B. tectorum* f. *nudus* (Klett & Richt.) St. John.

The occasional plants differing only in pubescence seem properly classified as a forma.

Bromus vulgaris (Hook.) Shear. *Narrow-leaved Brome.* Culms 4–12 dm. tall, the nodes pilose; sheaths pilose; ligule 3–5 mm. long; blades flat, more or less pilose, 5–10 mm. wide, 10–20 cm. long; spikelets narrow, 4–7-flowered;

glumes sparsely hispidulous, the first acute, 1-nerved, 5–8 mm. long; lemma sparsely hispidulous, the awn 6–8 mm. long. *B. eximius umbraticus* Piper. A valuable native forage grass. *A. T.* to *Can.*

CALAMAGROSTIS. Reed Bent Grass

Tall perennial grasses with small spikelets in many-flowered terminal panicles; spikelets 1-flowered, the pubescent rhachilla prolonged behind the palea; glumes subequal, longer than the lemma which is hyaline and obtuse, and bears a dorsal awn; callus copiously hairy; palea shorter; grain free, enclosed in the lemma. (Name Gr., *kalamos,* reed; *agrostis,* grass.)

Awn strongly geniculate; callus hairs sparse, much shorter than the lemma; panicle dense; collar of sheath pubescent. — *C. rubescens.*
Awn straight; callus hairs not much shorter than the lemma; collar of sheath glabrous,
  Panicle broad; leaf-blades soft,
    Glumes equal or nearly so; awn inserted on lower two-thirds of lemma,
      Spikelets 2.8–5.5 mm. long; glumes exceeding the lemma. — *C. canadensis,* var. *canadensis.*
      Spikelets 2.2–2.8 mm. long; glumes nearly or quite equaling the lemma. — Var. *Macouniana.*
    Glumes unequal; awn inserted on upper fourth of lemma. — Var. *pallida.*
  Panicle narrow, rather close,
    Leaf blades soft, not stiff; plant scarcely caespitose; glumes scabrous on the veins. — *C. neglecta.*
    Leaf blades stiff; plant densely caespitose; glumes very scabrous. — *C. inexpansa,* var. *brevior.*

Calamagrostis canadensis (Michx.) Beauv., var. **canadensis.** *Bluejoint.* Plants with creeping rhizomes; culms smooth, 6–12 dm. tall; leaves scattered; sheaths long, smooth; ligules hyaline, 3–5 mm. long; blades flat, lax, scabrous, 4–10 mm. wide; panicle long exserted, rather loose; glumes narrowly lanceolate acuminate, scabrous or rough puberulent, at first purplish; lemma shorter, smooth; awn slender, straight, attached at the middle of the lemma, about equaling it; callus hairs abundant soft, equaling the lemma. *C. canadensis acuminata* Vasey; *C. Langsdorfii* of Am. auth.; *C. canadensis,* var. *robusta* Vasey. These have been separated as species or varieties solely on the size of the spikelet. As the gradation from one extreme to the other is continuous, there seems no good reason to divide the series at any one or several points. Wet meadows, common, *A. T., Can.*

Var. **Macouniana** (Vasey) Stebbins. Stems often branched below; leaf-blades long attenuate, 4–8 mm. wide; panicle 10–12 cm. long, purple, narrow or pyramidal, densely flowered, the branches in fives, slender, naked below; lemma 2-lobed at apex, bearing a very short, straight awn from above the middle of the back. Wet banks of Palouse River, not common, *A. T.*

Var. **pallida** (Vasey & Scribn.) Stebbins. Erect, 12–15 dm. tall; blades 5–8 mm. wide; panicle whitish green, pyramidal, 10–15 cm. long; glumes membranous and translucent at tip and margins, the upper more acuminate, 3 mm. long; lemma 2.3 mm. long, bifid, with a projecting awn. Along streams, Blue Mts., *Horner* R495B536.

Calamagrostis inexpansa Gray, var. **brevior** (Vasey) Stebbins. Caespitose perennial with marcescent sheaths at base; culm 3–6 dm. tall, scabrous above; sheaths scabrous between the prominent ribs; stipules puberulent fimbriate, 1–3 mm. long; blades scabrous, becoming involute, 4–5 mm. wide; panicle 7–15 mm. long, the hispidulous branches spikelet-bearing nearly to the base; glumes lanceolate, 3–4 mm. long; lemma 2.5–3.5 mm. long, acute, scabrous; awn straight, included, attached below the middle of the lemma, callus hairs numerous, shorter than the lemma. *C. hyperborea* Lange. Wet meadows, not common, *A. T.*

Calamagrostis neglecta (Ehrh.) G. M. & S. Rootstock slender; stems slender, 3–10 dm. high; leaves soft, 1–3 cm. long, smooth; panicle narrow, glomerate and lobed, 5–15 cm. long; spikelets 3–5 mm. long; glumes acute; callus-hairs a little shorter than the floret and as long as those of the rudiment; awn from the middle of the thin lemma or lower, scarcely exceeding it. Spokane County, *Suksdorf.*

Calamagrostis rubescens Buckl. *Pine Grass* or *Nigger Hair.* Stems tufted, from creeping rhizomes, 60–100 cm. tall; stem leaves 3 or 4, 2–4 mm. wide, scabrous, somewhat involute; panicle usually narrow and dense, or interrupted below, 8–17 cm. long, erect, pale green, rarely purple, the branches short and spikelet-bearing to the base; spikelets 3–5 mm. long; lemma shorter than the glumes, obtuse, 4-toothed at apex, bearing a stout, nearly basal, geniculate awn, as long or longer than itself. *C. Suksdorfii* Scribn. The most abundant grass in pine forests, somewhat rough, but a valuable forage plant, especially early in the season, *A. T.*

CINNA

Tall perennial grasses with flat leaves, conspicuous hyaline ligules and usually many-flowered nodding panicles; spikelets 1-flowered; rhachilla articulated below the glumes and prolonged behind the palea into a minute bristle; glumes narrow; lemma 3–5-nerved, bifid; palea 1–2-nerved; stamen 1. (Name Gr., *kinna,* a kind of grass.)

Cinna latifolia (Trev.) Griseb. *Wood Reed Grass.* Stems solitary or few, [illegible] panicle pale green, drooping, 10–30 cm. long; glumes subequal, scabrous, 4 mm. long; lemma shorter, scabrous, bearing a short straight awn, or awnless. In moist places in woods, *Can.*

DACTYLIS

Perennial grasses with flat leaves; inflorescence a densely clustered or interrupted panicle; spikelets 2–5-flowered, short-pediceled, in small one-sided fascicles, the flowers all perfect or the upper staminate; glumes thin, membranous, keeled, unequal, mucronate; lemmas larger than the glumes, rigid, 5-nerved, keeled, the midnerve extending into a point or short awn; palea shorter than the lemma; grain free, enclosed in the lemma and palea. (Ancient name for a grass, from Gr., *daktulos,* finger.)

Dactylis glomerata L. *Orchardgrass.* Tufted, the stout stems 6–15 dm. tall; sheaths nearly smooth; blades flat, scabrous, 6–8 mm. broad; panicle 3–15 cm. long, pyramidal-ovate, greenish or purplish; branches solitary, ascending, spikelet-bearing above; lemmas conspicuously ciliate on the keels, 7 mm. long. Escaped from cultivation, naturalized from Europe, *A. T.*

DANTHONIA. Wild Oat-Grass

Tufted perennials; spikelets solitary or few in a small raceme or panicle; spikelets several-flowered, the flowers all perfect or the upper staminate; rhachilla pubescent, extending beyond the flowers; glumes keeled, acute, subequal, persistent, generally extending beyond the uppermost lemma; lemmas rounded on the back, 2-toothed, deciduous, the bent awn flat and twisted at the base, arising from between the acute or awned teeth; palea hyaline, 2-keeled near the margins, obtuse or 2-toothed; grain free, enclosed in the lemma. (Named for *Étienne Danthoine* of Marseille.)

Spikelets mostly solitary; blades villous. *D. unispicata.*
Spikelets in small racemes,
  Blades pilose. *D. californica,* var. *americana.*
  Blades merely scabrous. Var. *Piperi.*

Danthonia californica Boland., var. **americana** (Scribn.) Hitchc. Erect 6–8 dm. tall; culm glabrous; leaf blades flat, the basal 15–30 cm. long, the cauline 6–13 cm. long, 2–4 mm. wide, pilose on both sides, scabrous above, linear, tapering; sheaths uniformly soft pilose from pustulate bases, not hairy tufted at the summit; spikelets 2–4 in a pilosulous raceme; peduncles 13–30 mm. long; glumes 14–21 mm. long, subequal, lanceolate, acuminate, glabrous; florets 4–6; lemma body 8–9.5 mm. long, glabrous except for the villous lower margins, the awn 7–9 mm. long; palea pilosulous on the ribs. Var. *palousensis* St. John.

Idaho: Potlatch, June 16, 1911, *R. K. Beattie* 4,061.

D. californica americana (Scribn.) Hitchc., a trinomial, Biol. Soc. Wash., 41: 160, 1928, later as a var. *americana* Hitchc., Man. Grasses U. S. 305, 831, 1935.

Var. **Piperi** St. John. Differs from *D. californica* Boland. by having the sheaths pilose throughout.

Washington: low marshes, Pullman, July 20, 1893, *C. V. Piper* 1,744 (type in Herb. Washington State University); also *D. A. Brodie* in 1898; *R. M. Horner* 879; and *A. D. E. Elmer* 1,011; Steptoe, Whitman Co., *G. R. Vasey,* June 1900; Kitsap Co., *Piper* 821.

Danthonia unispicata (Thurb.) Monro ex Macoun. Culms slender, glabrous below, appressed puberulent above, 15–30 cm. tall; sheaths glabrous below, papillose-villous above, with a conspicuous long villous ring at summit; blades flat, ribbed, slender, long acuminate, 1–2 mm. wide; spikelets solitary, seldom two, purplish tinged; glumes lanceolate acuminate, unequal, 12–22 mm. long; lemmas oblong-elliptic, nearly smooth, pilose-ciliate on the lower margins, the body 5–6 mm. long, the two terminal teeth linear-lanceolate, acuminate, scaberulous, half to two-thirds as long as the body, awn blackish below, 5–10 mm. long. Dry soils, *A. T.*

DESCHAMPSIA. Hairgrass

Annuals or perennials; inflorescence a contracted or open panicle; spikelets 2-flowered, both flowers perfect, the hairy rhachilla extending beyond the flowers or rarely terminated by a staminate one; glumes keeled, acute, membranous, shining, persistent; lemma of about the same texture, each bearing a dorsal awn from or below the middle, the apex erose-truncate; palea narrow; grain oblong, free, enclosed in the lemma. (Named for *Dr. Deschamps* of St. Omer, France.)

Glumes scarcely equaling the upper lemma; perennial. *D. caespitosa.*
Glumes exceeding the upper lemma,
  Panicle narrow, the branches appressed, some spikelet-bearing to near the base; awn straight; perennial. *D. elongata.*
  Panicle loose, the branches becoming divergent, mostly spikelets bearing only at the tip, awn geniculate; annual. *D. danthonioides.*

Deschampsia caespitosa (L.) Beauv. *Tufted Hairgrass.* Densely tufted; stems 6–15 dm. high, much exceeding the numerous basal leaves; stem-leaves 3, the blades flat or involute, 5–12 cm. long, 1.5–3 mm. wide, scabrous, the smooth sheaths shorter than the internodes; panicle 10–45 cm. long, usually open, the capillary branches in whorls of 2–5, spikelet-bearing above the middle; spikelets shining, greenish or purplish, 3–6 mm. long; lemma hairy at the base, erose-toothed at the apex, bearing near the base a slender, slightly exserted awn. Plentiful in wet ground, *A. T., Can.*

Though Linnaeus (1753) published this as *Aira cespitosa,* an acceptable epithet, he corrected it himself to *caespitosa* in his Sp. Pl. ed. 2, 96, 1762.

Deschampsia danthonioides (Trin.) Munro ex Benth. *Annual Hairgrass.* The slender stems erect, 15–60 cm. tall; leaves short, involute, scabrous, the blades 2–6 cm. long, less than 1 mm. wide; panicle very loose, 7–20 cm. long, with subequal branches mostly in twos, these capillary, branched above; spikelets pale green, sometimes purple-tinged; glumes unequal, 4–8 mm. long; lemma long-hairy at the base, 2–3 mm. long, minutely toothed at the apex, bearing a geniculate awn three or four times as long. Very abundant in dry soil, *A. T., Can.,* also on high ridges of the Blue Mts., *Umat.*

Deschampsia elongata (Hook.) Munro ex Benth. *Slender Hairgrass.* Stems densely tufted, smooth, 30–120 cm. tall; leaves filiform, commonly involute, scabrous; panicle narrow, 15–45 cm. long, somewhat nodding; branches very unequal, ascending or appressed, very slender; spikelets pale green, usually purple-tinged near the apex; glumes equal, 3–6 mm. long; lemmas hairy at base, shining, irregularly toothed at apex, 1.5–2 mm. long, bearing a slender awn about twice as long. Dry hillsides or meadows, common, *A. T.* to *Can.*

DIGITARIA

Annual or perennial; racemes slender, digitate or aggregate at top of culm; spikelets lanceolate or elliptic, sessile or short pediceled, solitary or in twos or threes in rows, 1 on each side of a continuous narrow or winged rhachis; glumes 1–3-nerved, 1st glume minute or wanting, 2nd glume often equaling the sterile lemma; fertile lemma leathery, with pale hyaline margins. (Name Lat., *digitus,* finger.)

Digitaria sanguinalis (L.) Scop. *Crab-grass.* Annual, branched at base; culms often decumbent or creeping, 3–12 dm. long; sheaths more or less papillose-hirsute; blades flat, 5–12 cm. long, 4–10 mm. wide, often pilose; racemes 3–12, slender, 5–12 cm. long; spikelets in twos, one subsessile, the other pediceled, 3–3.5 mm. long, with raised nerves, smooth or hispidulous; fruit lead-colored. *Syntherisma sanguinalis* (L.) Dulac. Introduced from Eur., an occasional weed, *U. Son., A. T.*

DISTICHLIS

Low dioecious perennials of the salt or alkaline marshes; rootstock scaly, creeping; pistillate spikelets rigid; glumes unequal, broad, acute, keeled; rhachilla disarticulating above the glumes and between the mostly 3-nerved lemmas; staminate spikelets dissimilar, elongate, purplish; lemmas membranous. (Name Gr., *distichos,* two-ranked.)

Distichlis stricta (Torr.) Rydb. *Salt Grass.* Rhizomes numerous, long, rigid; plant forming tufts or sod; culms 5–35 cm. tall; leaves coarse, rigid, acute and serrulate at tip, flat or involute, 0.5–2 mm. wide; pistillate spikelets smooth, pale, 8–15 mm. long; lemma subacute, 5–6 mm. long; staminate spikelets thinner, loose, 15–25 mm. long; lemma lanceolate, 6–7 mm. long. Alkaline shores and flats, in the scabland along the western border, *U. Son.*

ECHINOCHLOA

Coarse annuals with long leaves; spikelets 1-flowered, with sometimes a staminate flower below, nearly sessile in 1-sided spikes or panicle branches; glumes unequal, hispid, mucronate; sterile lemma similar and usually awned from the apex; fertile lemma smooth, shining, with inrolled margins. (Name Gr., *echinos,* hedgehog; *chloa,* grass.)

Spikelets or some of them awned, the awns mostly 5–10 mm. long. *E. crusgalli,* var. *crusgalli.*
Spikelets nearly awnless or the awns shorter than 3 mm. *E. crusgalli,* var. *mitis.*

Echinochloa crusgalli (L.) Beauv., var. crusgalli. *Barnyard Grass.* Glabrous; stems stout, branching at the base, 3–15 dm. tall; panicle dense, 10–20 cm. long, composed of many ascending or spreading racemes; spikelets green or purple, 3 mm. long, densely crowded in 3 or 4 rows; sterile lemma awned. Introduced from Europe, occasional weed in moist ground, *U. Son.*

Var. **mitis** (Pursh) Peterm. Differs from the species by having the spikelets awnless or nearly so. Common weedy plant, *U. Son, A. T.*

ELYMUS. Ryegrass

Perennial or annual grasses with spikes or rarely spike-like panicles which do not readily break up into segments; spikelets 1–7-flowered, usually 2–6 at each joint of the rhachis or rarely solitary at some of the lower nodes; glumes entire, lanceolate or narrower and rigid, awned or awnless, 1–several-nerved; lemmas acute, acuminate or awned, entire, rounded on the back, obscurely 5-nerved; palea shorter than the lemma; grain hairy at tip. (Name Gr. *elumos*, millet.)

Florets 1 (-2); awn of lemma 5-11 cm. long; annuals with fibrous roots. — *E. caput-medusae.*
Florets 2-7; awn of lemma shorter or none; perennials with stouter, long-lived roots,
- Glumes subulate,
  - Blades 8-20 mm. wide; plant stout, tufted, 1-3.3 m. tall,
    - Sheaths or blades or both short pilose. — *E. cinereus.*
    - Sheaths and blades smooth or scabrous. — Forma *laevis.*
  - Blades 2-8 mm. wide; plant slender, not tufted, with rhizomes, 4-14 dm. tall. — *E. triticoides.*
- Glumes lanceolate or the body lanceolate,
  - Glumes shorter than their awns; awns of lemmas bent and divergent. — *E. canadensis,* f. *glaucifolius.*
  - Glumes awnless; awns of lemmas straight and ascending, or none,
    - Lemmas pilose, acute or merely awn-pointed; inflorescence usually paniculate. — *E. arenicola.*
    - Lemmas merely scabrous at tip, tipped by an awn 7-25 mm. long; inflorescence spicate,
      - Blades scabrous; sheaths glabrous or merely ciliate. — *E. glaucus.*
      - Blades and sheaths pilose. — *E. glaucus,* f. *Jepsoni.*

Elymus arenicola Scribn. & Smith. Plants with long rhizomes; culms 6-15 dm. tall, erect, smooth; leaf sheaths smooth; blades 2-4.5 mm. wide, glaucous, scabrous above and somewhat pilose towards the base, glabrous beneath, flat or becoming involute; panicle narrow and subspicate; rhachis pilose on the angles; spikelets in pairs or single, 4-6-flowered; glumes 7-15 mm. long, linear-lanceolate, acuminate, glabrous, with an evident midnerve and 1 obscure lateral nerve; lemmas [illegible] mm. long, [illegible]-nerved. A native "sand-binder" grass. Sand dunes, Ilia, *St. John et al.* 9,260, *U. Son.*

Elymus canadensis L., forma **glaucifolius** (Muhl.) Farw. Stems 8-20 dm. tall, stout, pale and somewhat glaucous throughout; leaf-sheaths glabrous; blades 5-15 mm. wide, flat or involute, nearly smooth; spike 8-25 cm. long, dense, with 2-4 spikelets at the nodes, pale green; spikelets 2-5-flowered, somewhat divergent; glumes with their awns 15-35 mm. long, the body linear-lanceolate, 3-nerved, tapering into the scabrous awn; lemma 9-16 mm. long, 5-nerved, hirsute, the awn 2-4 cm. long. *E. philadelphicus* L., var. *robustus* (Scribn. & Smith) Farw. Low valleys, *U. Son., A. T.*

Elymus caput-medusae L. Stems 2-12 dm. long, erect or decumbent at base; sheaths pubescent or glabrous; blades 1-2 mm. broad, becoming involute, ciliate at base; spike stout, green or becoming purplish, 1-4 cm. long without the awns; spikelets paired; glumes 15-37 mm. long, subulate, rigid, scabrous; lemma with the body 6-8 mm. long, 5-nerved, narrow, subterete, smooth or scabrous, tipped with a stout scabrous awn. A weed from the Mediterranean region, introduced near Steptoe, *G. R. Vasey* in 1901.

Elymus cinereus Scribn. & Merr. Differing from the forma only in having the sheaths or blades or both short pilose. *E. condensatus* Presl, var. *pubens* (Piper ex Jepson) St. John. River bottoms, Colfax, *Parker* 674, *A. T.*

Forma **laevis** St. John forma nova. *Ryegrass.* Leaf-sheaths striate, smooth; blades flat, glaucous, smooth, scabrous especially on the margin, spikes 1.5–4 dm. long, erect; dense with 3 or more spikelets at the median nodes; spikelets 3–6-flowered, usually several at each node, commonly densely crowded; glumes 7–14 mm. long, 1-nerved, scabrous-puberulent; lemmas 8–11 mm. long, 5-nerved, scabrous or puberulent at least above, mucronate-pointed. Abundant and often in pure stands in stream bottoms or coulees, especially if the soil is alkaline, *U. Son., A. T* .

Laminis vaginisque laevibus.

A good forage at all times, but valuable in winter as the self-cured hay stands erect and in easy reach.

Holotype: Washington, Clark's Springs, 10 miles n. of Spokane, July 7, 1902, *F. O. Kreager* 107 (UC).

Elymus glaucus Buckl. *Smooth Wild Rye.* Plants 5–12 dm. tall, erect, usually tufted; leaf blades 6–16 mm. wide, spreading or drooping, scabrous; spike narrow, erect or rarely nodding above, 5–20 cm. long, greenish or purplish, glaucescent; spikelets usually appressed, mostly in pairs, 3–6-flowered; glumes 6–12 mm. long, 3–5-nerved; lemmas 7–10 mm. long, 5-nerved, scabrous toward the tip. Open places, *U. Son., A. T.*

A good and persistent native forage grass.

Forma **Jepsoni** (Davy) St. John. Leaf sheaths or blades or both pilose. Originally made a variety, but as it differs from the species only in the pubescence of the foliage, it is worthy only of the rank of a form. *E. glaucus* Buckl., var. *Jepsoni* Davy in Jepson; *E. glaucus* subsp. *Jepsonii* (Davy) Gould. Wooded valleys, *A. T.*

Elymus triticoides Buckl. Stems erect; leaf-sheaths smooth, scabrous, or the lower pilose; blades green, flat or becoming involute, sometimes scabrous; inflorescence 10–21 cm. long, a spike or paniculate below the middle, usually purplish, rather loose; glumes 6–15 mm. long, 1–3-nerved, scabrous towards the tip; lemmas 6–14 mm. long, 7–9-nerved, glabrous, turning brownish, acuminately short-awned. Moist bottomlands and alkali flats, Walla Walla and Snake Rivers, *U. Son.*

ERAGROSTIS. Love Grass

Annual or perennial grasses; inflorescence a contracted or open panicle; spikelets many-flowered, much flattened laterally; rhachilla jointed; glumes unequal, shorter than the lemmas, keeled, 1-nerved; lemmas membranous, 3-nerved, awnless; palea shorter than the lemmas, prominently 2-nerved or 2-keeled, usually persisting on the rhachilla after the lemma has fallen; grain free, loosely enclosed in the lemma and palea. (Name Gr. *eros,* love; *agrostis,* grass.)

Spikelets unisexual or polygamous; stems creeping; leaves 15–30 mm. long. *E. hypnoides.*
Spikelets perfect; stems erect or slightly decumbent; leaves 2–12 cm. long,
Plants not glandular on branches or lemmas. *E. pectinacea.*

Plants glandular there,
  Spikelets 2.5–4 mm. wide; glands prominent on keel of lemma. — *E. megastachya.*
  Spikelets 1.5–2 mm. wide; glands present on the branches. — *E. lutescens.*

Eragrostis hypnoides (Lam.) B.S.P. Annual; stems prostrate, creeping, much branched from the base, forming dense circular mats, partially buried in the sand; leaf blades 2–3 cm. long, somewhat pubescent; sheaths loose, inflated, hairy; panicles very numerous, pale green, dense, 2–4 cm. long, usually with their bases included in the uppermost sheaths; spikelets oblong, 4–7 mm. long; lemmas lanceolate, membranous, scabrous on the 3 green prominent ribs, 1.5 mm. long. Sandy banks of Snake River, common, *U. Son.*

Eragrostis lutescens Scribn. Annual, stems tufted, erect, 10–30 cm. high, branched near the base; herbage pock-marked with glandular pits; leaf blades 2–12 cm. long, 2–4 mm. wide, scabrous above; panicle silvery green, 5–10 cm. long, the branches naked below; spikelets oblong, 4–7 mm. long, usually 10–12-flowered; lemmas obtuse, 2–2.2 mm. long. Very rare, discovered on the sandy banks of the Snake River near Almota, *Piper* 2,624, *U. Son.*

Eragrostis megastachya (Koel.) Link. *Stink-grass.* Annual, malodorous, branched from base; culms 10–90 cm. tall, with glandular pits below nodes, and on sheaths; blades 2–8 mm. wide, the margins with glandular pits; panicles 5–20 cm. long, greenish lead-colored, ovoid, rather dense; pedicels 0.5–2 mm. long; spikelets 10–40-flowered, 5–17 mm. long, lance-oblong, flat, the florets closely imbricate; lemmas 2–2.5 mm. long, papery, strongly 3-nerved; palea nearly as long, its two nerves strong and ciliolate; grain 0.7 mm. long, ovoid. *E. cilianensis* sensu Lutati, not All. Adventive European weed, established in 1939 on sandy shores of Snake River at Wawawai, *U. Son.*

Eragrostis pectinacea (Michx.) Nees. Annual, tufted, freely branching at the base, and tardily above; culms slender, smooth, 15–40 cm. tall; sheaths smooth, inflated; blades flat, very scabrous, 2–10 cm. long, 1–3 mm. wide, the auricles long pilose, spikelets ovate-lanceolate when young, becoming linear, 5–18-flowered, 1.5 mm. wide; glumes with scabrous keels, the lower 1 mm. long; lemmas ovate, scabrous on the ribs, 1.5 mm. long; palea persistent. *E. caroliniana* (Spreng.) Scribn. *E. Purshii* Schrad. Common on sandy banks of Snake River, *U. Son.*

FESTUCA. Fescue

Perennials with terminal panicles; spikelets 3-many-flowered, glumes more or less unequal, acute, keeled, the lower 1-, rarely 3-nerved, the upper 3-, rarely 5-nerved; lemmas lanceolate, firm in texture at least near the base, narrow, rounded on the back or slightly keeled, always 5-nerved, acute or usually awned; palea usually about equaling the lemma; stamens 3; grain elongated, often adherent to the palea. (Name Lat., *festuca,* straw.)

Blades broad, flat,
  Lemma awnless or nearly so, indurate, not keeled. — *F. elatior.*
  Lemma awned, membranous, keeled. — *F. subulata.*
Blades narrow, involute,
  Culms red fibrillose and decumbent at base, somewhat stoloniferous,
    Lemma glabrous or scabrous. — *F. rubra.*
    Lemma villous. — *F. rubra,* var. *lanuginosa.*

Culms not so; stolons wanting,
  Culms scabrous; lemma scabrous puberulent throughout. *F. scabrella.*
  Culms smooth, or if slightly scabrous, the lemma not puberulent,
    Lemma awnless or nearly so. *F. viridula.*
    Lemma awned,
      Awn shorter than or sightly exceeding the lemma; ovary glabrous,
        Blades scabrous. *F. ingrata.*
        Blades smooth. *F. idahoensis.*
      Awn in upper florets well exceeding the lemma; ovary hispidulous at apex. *F. occidentalis.*

Festuca elatior L. *Meadow Fescue.* Stem stout 5–15 dm. tall, smooth and glabrous throughout; leaves dark green, 7–15 cm. long, 3–8 mm. wide; panicle narrow, erect, 10–25 cm. long, the branches solitary or in twos; spikelets rather crowded, lanceolate-oblong, 12–18 mm. long, 5–10-flowered; glumes oblong-lanceolate, the upper 4 mm. long; lemma scarious-margined, oblong-lanceolate, smooth, somewhat scabrous at tip, 5–7 mm. long. Native of Europe, escaped from cultivation. *A. T.*

Festuca idahoensis Elmer. Culms single or few, smooth, 6–10 dm. tall; panicle narrow, open, 7–15 cm. long; spikelets 3–5-flowered; glumes smooth, the lower narrow, the upper lanceolate, a third longer, 4–5 mm. long, lemma nearly smooth, 5–6 mm. long, the awn 1–3 mm. long. Meadows and wooded hills, *A. T., A. T. T.*

Festuca ingrata (Hack.) Rydb. *Blue Bunchgrass.* Densely tufted, the whole plant pale or glaucescent; stems 2–10 dm. tall, smooth or slightly scabrous; leaf blades very numerous, setaceous, mostly basal, firm and harsh; panicle 3–27 cm. long, narrow, one-sided, the branches erect; spikelets 3–5-flowered; glumes linear, the upper 5 mm. long; lemma 5–7 mm. long, tipped by an awn of from half to equal length. *F. ovina ingrata* Hack. Abundant on the prairies, *A. T., A. T. T.*

Though Mrs. Chase (1950) reduced this to the synonymy of *F. idahoensis,* it seems different.

Festuca occidentalis Hook. Densely tufted, the whole plant smooth and bright green; stems 3–8 dm. high; leaf blades smooth, becoming longitudinally grooved when dry; panicle narrow, loose, 8–20 cm. long, often drooping at the summit; spikelets green, 3–5-flowered; glumes lanceolate, puberulent at tip, the upper 4 mm. long; lemmas scabrous near the apex or smooth, 5–6 mm. long, bearing a slender awn of equal length. In open coniferous woods, common, *U. Son., A. T. T.*

Festuca rubra L. *Red Fescue.* Plant smooth, stems 4–10 dm. tall, slender; blades 10–20 cm. long; panicle narrow, erect, 3–20 cm. long, the short erect branches in twos, spikelet-bearing to the base; spikelets frequently purplish, 10–12 cm. long, 4–6-flowered; lemma smooth, 5–7 mm. long, tipped with an awn 3–4 mm. long. Native of n. N. Am., Europe, cultivated, sparingly established, wet places, *A. T.*

Var. lanuginosa Mert. & Koch. Differs only in its pubescent spikelets. Introduced from Europe. Pullman, *Brodie.*

Festuca scabrella Torr. Densely tufted, 3–10 dm. tall, scabrous throughout; leaf blades very scabrous, strongly involute, elongate, breaking away early from the sheaths which remain for several seasons at the base; panicle 7–15 cm. long, narrow, the ascending branches mostly in pairs, spikelet-bearing above the middle; spikelet 8–12 cm. long, 3–5-flowered; glumes with broad hyaline margins, the upper oblong-lanceolate, 6–9 mm. long, the lower narrower, two thirds as long; lemmas 6–7 mm. long, very short awned. *F. hallii* (Vasey) Piper. *F. altaica* Trin. in Ledeb., subsp. *scabrella* (Torr.) Hultén. Steptoe, *G. R. Vasey* 5.

Festuca subulata Trin. in Bong. Loosely tufted, glabrous or nearly so; stems 4–12 dm. high; leaf blades scabrous, thin, 3–10 mm. broad; panicle very loose and somewhat drooping, 15–40 cm. long; rays in 3–5 sets; spikelets pale green, 3–5-flowered, 7–12 mm. long; glumes lanceolate, the lower a third shorter, the upper 3–4 mm. long; lemma narrowly lanceolate, keeled its entire length, 5–7 mm. long, with an awn 5–20 mm. long. *F. subulata* Trin., var. *Jonesii* (Vasey) St..-Yves. In wet places in woods, common *A. T. T., Can.*

Festuca viridula Vasey. *Mountain Bunch Grass.* Densely tufted, dark green, smooth; stems 5–10 dm. high; leaves erect, soft, 7-nerved, the basal ones involute, panicle loose and open, suberect, 10–15 cm. long; spikelets 3–6-flowered; glumes 4 and 6 mm. long; lemma firm, keeled toward the apex, 5-nerved, acute or sometimes mucronate, smooth or nearly so, 6–8 mm. long. *F. viridula* Vasey, var. *Vaseyana* St.-Yves. Mountain meadows, Mt. Spokane, and the Blue Mts., *Huds.*

GLYCERIA. Mannagrass

Tall marsh perennial grasses with terminal panicles; spikelets few–many-flowered, terete, or somewhat flattened; glumes unequal, obtuse or acute, 1–3-nerved; lemmas membranous, obtuse, rounded on the back, with 5–9 prominent raised nerves; palea scarcely shorter than the lemma, rarely longer; grain smooth, enclosed in the lemma and palea, free, or when dry slightly adhering to the latter. (Name Gr., *glukeros,* sweet.)

Spikelet 10–18 mm. long, linear. *G. borealis.*
Spikelet 2–8 mm. long, ovate or oblong,
  Lemma with 5 prominent nerves. *G. pauciflora.*
  Lemma with 7 prominent nerves,
    Spikelet 4–6 mm. long; lemma 2–3 mm. long. *G. grandis.*
    Spikelet 2–4 mm. long; lemma 1.5–2 mm. long,
      First glume 1–1.4 mm. long; leaves [illegible] mm. wide. *G.* [illegible]
      First glume [illegible] mm. long; leaves [illegible] mm. wide. *G. striata,* var. *stricta.*

Glyceria borealis (Nash) Batch. Culms weak, erect, glabrous, 3–10 dm. tall; leaf blades pale green, nearly smooth, 8–20 cm. long, 3–8 mm. wide; sheaths loose, longer than the internodes; panicle lax, narrow, 15–30 cm. long, the branches single or in twos, usually short and erect; glumes hyaline, the second 2.5–3.5 mm. long, twice as long as the first; lemmas thin, 3.5–4 mm. long, 7-nerved, the nerves minutely hispid, hyaline on margins and tip. *Panicularia borealis* Nash. Common in shallow ponds, *A. T.*

Glyceria elata (Nash) Hitchc. Culm smooth, 6–20 dm. tall; blades scabrous; panicle loose, the branches ascending, or at length reflexed, naked below; spikelets green or purplish, 5–8-flowered; glumes broad, thin; lemma scarious on the margin and tip. *Panicularia nervata elata* (Nash) Piper. Wet woods, *A. T. T.*

Glyceria grandis Wats. *Reed Meadow Grass.* Culm stout, 1–2 m. high; blades 6–15 mm. wide; panicle large and loose, 20–40 cm. long, nodding at the top; spikelets mostly purplish, 4–7-flowered; glumes whitish, the second 2–2.5 mm. long. *Panicularia americana* (Torr.) MacM. In wet places, *A. T.*

Glyceria pauciflora Presl. Culm 3–12 dm. high, smooth; blades acute, scabrous, 10–20 cm. long, 6–14 mm. wide; sheaths nearly equaling the internodes; panicle loose, 10–20 cm. long, green or purplish, branches slender, spreading, 2–5 at a node, spikelet-bearing above the middle; spikelets 4–6-flowered, 4–5 mm. long; glumes obtuse, 1 and 1.5 mm. long; lemmas prominently 5-nerved, 1.5–2.5 mm. long. *Panicularia pauciflora* (Presl) Kuntze; *Torreyochloa pauciflora* (Presl) Church; *Puccinellia pauciflora* (Presl) Munz. In moist places, common, *A. T.* to *Can.*

Glyceria striata (Lam.) Hitchc., var. **stricta** (Scribn.) Fern. Culms slender 2–9 dm. high; blades flat or folded; panicle loose, 5–20 cm. long, the branches ascending, in age rarely reflexed; spikelets purplish, 5–6-flowered; lemma with a blunt scarious tip. *Panicularia nervata* (Willd.) Kuntze. Meadows, *A. T.*

HIEROCHLOË

Perennial, erect, sweet smelling grasses; panicles terminal; spikelets with 1 terminal and 2 staminate florets, disarticulating above the glumes; glumes equal, 3-nerved, thin, papery, smooth, acute; staminate lemmas nearly as long as the glumes, boat-shaped, hispidulous; fertile lemma somewhat indurate, nearly smooth; palea 3-nerved, rounded on back. (Name from Gr., *hieros,* sacred; *chloe,* grass, because the fragrant European species was strewn on church steps on saint's days.)

Hierochloë odorata (L.) Beauv. *Sweet Grass.* Rhizomes slender; basal leaves reduced, soon withering; the 2–3 cauline leaves having sheaths glabrous; blades 2–5 mm. wide, mostly reduced; sterile shoots with several blades 20–80 cm. long, culm 3–9 dm. tall; panicle 3–14 cm. long, ovoid, loose; spikelets 4–8 mm. long, bronze or purplish; glumes 4–5.5 mm. long, ovate; staminate lemmas 3–3.5 mm. long, ovate, brown hairy; lemma of fertile floret hairy tufted at apex. Introduced from Eurasia. Rocky meadow, head of Rock L., Whitman Co., *Beattie & Lawrence, 2,418.*

HOLCUS

Perennial grasses with densely-flowered terminal panicles; spikelets deciduous, 2-flowered, the lower flower perfect, the upper staminate; glumes membranous, keeled, the lower 1-nerved, the upper 3-nerved and often short-awned; lemmas 2, papery, that of the upper flower bearing a bent awn; palea narrow; grain oblong, free, enclosed in the lemma. (Name from Gr., *holkos,* attractive.)

Holcus lanatus L. *Velvet Grass or Mesquite.* The whole plant downy with pale pubescence; stems erect, 30–90 cm. high; leaf blades flat, rather short, 3–10 mm. broad; sheath loose, shorter than the internodes; panicle pale purplish, 5–10 cm. long, rather densely-flowered, the branches in twos or threes; spikelets 4–6 mm. long; upper glume short-awned near the apex; lemma of the staminate flower with a hook-like awn. *Ginannia lanata* (L.) Hubb. Sparingly introduced from Eurasia or Africa, *A. T.*

HORDEUM. Wild Barley

Caespitose annuals or perennials with terminal cylindrical spikes; spikelets 1-flowered, usually in threes at each joint of the rhachis, the lateral generally short-stalked and imperfect; the central one sessile or short-stalked, perfect; rhachilla produced as a bristle, beyond the floret, sometimes itself with a reduced floret; glumes often reduced to awns and forming an apparent involucre around the spikelets, rigid; attached in front of the central spikelet, fused to the joint of the rhachis and the spikelets falling with the joint when it disjoints at maturity; lemmas rounded on the back, 5-nerved at the apex, awned; palea scarcely shorter than the lemma; grain usually adherent to the lemma, hairy at the summit. (The ancient Lat. name for the barley.)

Floret of central spikelet stalked; some of glumes ciliate towards the base. — *H. leporinum.*
Floret of central spikelet sessile; glumes not ciliate,
  Glumes 2–6 cm. long,
    Glumes 4–6 cm. long, — *H. jubatum.*
    Glumes 2–3 cm. long. — *H. caespitosum.*
  Glumes 8–20 mm. long,
    Lemma of central spikelet 10 mm. long; glumes 15–20 mm. long. — *H. brachyantherum.*
    Lemma of central spikelet 6–8 mm. long. — *H. Hystrix.*

Hordeum brachyantherum Nevskii. Tufted perennial; culms 3–10 dm. tall, smooth; lower sheaths often pilose; blades 3–8 mm. wide, flat, scabrous or pilose; spikes 3–10 cm. long; lateral spikelets perfect, staminate, or neuter, with a lemma 6 mm. long. *H. boreale* Scribn. & Smith, non Gdgr. (1881). Marshy creek bank, Buffalo Rock, Snake River Canyon, *St. John et al.* 8,227, *U. Son.*

Hordeum caespitosum Scribn. Tufted perennial; culms 3–6 dm. tall; sheaths somewhat scabrous; blades 2–5 cm. wide, flat, scabrous; spike 3–8 cm. long; lemma of central spikelet 6 mm. long, scabrous at tip; lemma of lateral spikelets 2–4 mm. long; glumes 2–3 cm. long, setose, scabrous. *H. jubatum* L. var. *caespitosum* (Scribn.) Hitchc. Stream or pond shores, infrequent, *U. Son., A. T.*

Hordeum Hystrix Roth. Tufted annual from fibrous roots; culms 1–7 dm. tall, decumbent at base, smooth; sheaths smooth or pilose; blades 1–2.5 mm. wide, flat, pilose; spikes 2–3.5 cm. long, stout, stiff; glumes rigid, glabrous or scabrous, the awn 8–15 mm. long; floret of lateral spikes reduced, the awn 4–5 mm. long. *H. depressum* (Scribn. & Smith) Rydb. Dry places, *U. Son., A. T.*

Hordeum jubatum L. *Squirrel-tail Barley.* Perennial, tufted; culms 2–6 dm. tall, erect, smooth; sheaths usually covering the stem and the base of the spike, the lower ones pilose; blades 2–4 mm. wide, flat, becoming involute, scabrous-puberulent; spike 5–10 cm. long, pale green or yellowish, central spikelet with the lemma 7–8 mm. long, scabrous at tip, with an awn 3–6 cm. long; the lateral spikelets merely awns; glumes all subulate, 4–6 cm. long. Hillsides, meadows, or alkali flats, *U. Son., A. T.*

The sharp-pointed, bearded spikelets when detached are injurious to stock. Due to their backward pointed barbs, they work their way into the eyes, mouth parts, or even the skin of the animals, and may in extreme cases cause death.

Hordeum leporinum Link. *Wall Barley*. Annual, with fibrous roots; culms 1.5–4 (–17) dm. tall, several, decumbent at base, smooth; sheaths smooth, the upper inflated; blades 2–8 mm. wide, flat, sparsely pilose; spikes 5–10 cm. long, thick, often partly included; both glumes of the central spikelet and the inner glume of the lateral spikelets ciliate and flattened, bearing awns 20–30 mm. long; outer glumes of lateral spikelets neither flattened nor ciliate; lemma 8–14 mm. long, scabrous at the apex, with a scabrous awn 15–40 mm. long. A weed, introduced from Eurasia or Africa, roadsides and fields, *U. Son.*, occasional in *A. T.*

KOELERIA

Annual or tufted perennial grasses with narrow leaves and densely flowered spike-like panicles; spikelets 2–4-flowered; rhachilla articulated between the lemmas; glumes acute, subequal, keeled, scarious on the margins; lemmas 5-nerved, keeled, acute; palea hyaline, acute; grain free, enclosed in the lemma and palea. According to Hitchcock, *Koeleria* should be placed in the *Aveneae*. (Named for *Georg Ludwig Koeler* of Germany.)

Koeleria cristata (L.) Pers. *June Grass*. Perennial, the erect stems densely tufted, 30–90 cm. tall, glabrous below, puberulent above; leaves narrow, commonly involute; panicle dense, silvery, greenish or purplish, shining, 5–15 cm. long, often more or less interrupted; spikelets 4–5 mm. long, 2–4-flowered; glumes and lemmas scabrous. Doubtfully *K. macrantha* (Ledeb.) Spreng. Common on dry hillsides, *U. Son., A .T.*

A valuable native forage grass.

LEPTOCHLOA

Annuals or perennials; blades flat; spikes numerous, borne on a common axis; spikelets two–several-flowered, approximate along one side of a slender rhachis; glumes 1-nerved; lemmas 3-nerved, sometimes pubescent or 3-toothed. (Name Gr. *leptos,* slender; *chloa,* grass.)

Leptochloa fascicularis (Lam.) Gray. Annual, culms erect, 3–6 dm. tall, or spreading in a small rosette; sheaths smooth, inflated, the upper partially or wholly enclosing the inflorescence; spikes numerous, 7–12 cm. long; spikelets short pedicelled, 7–11-flowered, the florets much longer than the lanceolate glumes; lemma hairy margined at base, short awned from the cleft apex. Sandy bank of Snake River, Wawawai, *St. John* 6,734, *U. Son.*

LOLIUM. Darnel

Annuals or perennials with flat leaf blades; spikes terminal, 2-sided; spikelets several-flowered, solitary, set in alternate notches of the rhachis, the spikelets placed sidewise to the rhachis, the glume of the side next the rhachis wanting; outer glume 3–7-nerved, rigid, as long as or longer than the lowest floret; lemmas 5–7-nerved, broad, rounded on the back; palea 2-keeled, shorter than the lemma. (The ancient Lat. name.)

Lolium perenne L. *English or Perennial Rye-grass.* Perennial with rhizomes; culms 3–7 dm. tall, erect, slender, smooth; sheaths smooth; blades 2–4 mm. wide, smooth, folded; spikes 7–20 cm. long, slender; spikelets 8–12 mm. long, 5–10-flowered; lemmas 6–7 mm. long, blunt or pointed but awnless. A valuable forage, introduced from Europe or northern Africa, persisting as a weed in lawns, *A. T.*

MELICA. Melic Grass

Perennial grasses with simple stems, often cormose at base; inflorescence a contracted or open panicle; spikelets 2–several-flowered; rhachilla extending beyond the flowers and usually bearing 2–3 empty club-shaped or hooded lemmas twisted around each other; glumes membranous, the lower 3–5-nerved, the upper 5–9-nerved; lemmas larger, rounded on the back, 5–7-nerved, the margins more or less scarious; palea broad, shorter than the lemma; grain free, enclosed in the lemma and palea. (Old Italian name for sorghum, from Lat., *mel,* honey.)

Lemma notched at the apex; panicle broad, the branches spreading,
- Lemma 10 mm. long, awned. — *M. Smithii.*
- Lemma 4–6 mm. long, awnless. — *M. fugax.*

Lemma not notched at the apex, awnless; panicle narrow, the branches ascending or appressed,
- Lemma lanceolate, long acuminate. — *M. subulata.*
- Lemma elliptic, obtuse,
  - Plants tufted; first glume 7–8 mm. long. — *M. bulbosa.*
  - Culms separate; first glume 4.5–5 mm. long. — *M. spectabilis.*

Melica bulbosa Geyer ex Porter & Coult. *Purple Melic Grass.* Stems 30–90 cm. tall, erect, smooth, with bulbous bases about 1 cm. in diameter, these appearing on the apices of stout rootstocks; stem leaves 3, rarely 4; sheaths strongly veined, scabrous; blades flat short pilose above, 2–5 mm. wide, panicle erect, narrow, 5–16 cm. long; spikelets ovate lanceolate, 10–15 cm. long; 5–9-flowered, usually purple; lower glume 6 mm., the upper 8–10 mm. long; lemmas about as long, hispidulous. Open places, *A. T., A. T. T.*

The corms are edible, sweet and nut-like.

Melica fugax Boland. Tufted perennial from white, sweet, bulbous bases; culms 2–5 dm. tall, hispidulous-scabrous; sheaths hispidulous-scabrous; blades 1–4 mm. wide, flat, hispidulous scabrous beneath, remotely pilose above; inflorescence loose, scabrous; spikelets 7–13 mm. long, 2–5-flowered, purplish; first glume 3–4 mm. long, 3-nerved; second glume 5–6 mm. long, 5-nerved, oval, scarious margined; fertile lemmas 7-nerved, obovate, scaberulous towards the emarginate tip; palea slightly shorter, the nerves and tip ciliate. *M. Macbridei* Rowland. Gravel ridges, Blue Mts.., *Huds.*

Melica Smithii (Porter) Vasey. Glabrous; stems tufted, usually few, not bulbous at base, 6–12 dm. tall; sheaths scabrous; blades flat, scabrous, 6–12 mm. wide; panicle loose, the branches at length reflexed; spikelets 3–6-flowered; glumes unequal; lemmas scabrous above, 10 mm. long, the straight awns half as long or less. Rich open woodlands, Blue Mountains, *Lake & Hull* 117, *Can.*

Melica spectabilis Scribn. Stalk bulbous at base, single, not tufted, 3–10 dm. tall; sheaths glabrous or pilosulous between the nerves; blades 2–4 mm. wide, elongate, glabrous; inflorescence 10–15 cm. long, strict, narrow, the branches appressed ascending; pedicels slender, scabrous; spikelets 4–7-

flowered, elliptic, 10–15 mm. long; glumes purplish below, bronze above; lemmas 7–8 mm. long, purplish where exposed, scabrous, strongly 7–nerved. Open places, Big Butte, T7N, R44E., Asotin Co., *Cronquist 5,852.*

Melica subulata (Griseb.) Scribn. Glabrous; stems tufted, usually few, bulbous at base, 4–12 dm. tall; sheaths nearly smooth; blades flat, shining, scabrous above, 3–12 mm. wide; panicle rather loose, suberect, 10–20 cm. long; spikelets 3–6-flowered; glumes unequal, usually purplish; lemmas pubescent, scabrous to hispidulous, 7–12 mm. long. In open woods and copses, especially abundant in the Blue Mts., *A. T. T., Can.*

MUHLENBERGIA

Annual or perennial grasses with small 1-flowered spikelets; glumes usually unequal and shorter than the lemma, acute or sometimes awned; lemma narrow, 3–5-nerved, with a straight awn from the apex or from between the teeth, or merely mucronate. (Named for *Rev. Henry Muhlenberg* of Pa.)

Lemmas awnless or mucronate,
  Plants with scaly rhizomes,
    Glumes 1–1.5 mm. long; panicle narrow. *M. Richardsonis.*
    Glumes 0.7–1 mm. long; panicle diffuse. *M. asperifolia.*
  Plants tufted; glumes 0.6–0.9 mm. long. *M. idahoensis.*
Lemmas awned; blades 6–14 cm. long, soft, flat. *M. mexicana,* f. *ambigua.*

Muhlenbergia asperifolia (Nees & Meyen) Parodi. Perennial, freely branching; culms decumbent at base, 5–60 cm. tall; sheaths smooth, keeled; blades rigid flat, scabrous, 2–5 cm. long, about 2 mm. wide; panicles diffuse, ovoid, partially or wholly exserted, 4–15 cm. long, the scabrous branches bearing spikelets only at their tips; glumes subequal, nearly as long as the lemma; lemma dark, about 1.5 mm. long. *Sporobolus asperifolius* Nees & Meyen. Rare, in alkali flats, Rock Lake, *St. John & Warren* 6785, *U. Son.*

Muhlenbergia idahoensis St. John. Tufted perennial from numerous slender, fibrous roots; basal leaves numerous; the sheaths about 1 cm. long, loose, expanded towards the base, with 5–7 prominent nerves, hyaline-margined; the blades 3–7 cm. long, 0.1–0.3 mm. wide, filiform involute, striate nerved, scabrous; cauline leaves usually 4, with close sheaths and the blades shorter than those of the basal leaves; culms 3–10 cm. tall, slender, nearly smooth; panicle 15–35 mm. long, strict, 1–2 mm. wide, the branches erect and appressed; spikelets 1-flowered, 1.6–2 mm. long, on short, straight, scaberulous pedicels; glumes similar, 1-nerved, the outer 0.6–0.8 mm. long, obovate, scarious, scabrous on the mid-nerve, obtuse or erose at tip; second glume 0.7–0.9 mm. long, broader as well as dentate at tip; lemma 1.5–1.9 mm. long, lanceolate, membranous, dark or purplish tinged, 5-nerved, appressed pilose in lines at base, scabrous towards the tip, the strong midrib running into an acute or mucronate tip; palea 1.5–1.6 mm. long, scabrous at tip; anthers oblong, 0.5 mm. long; grain 1.5 mm. long. Along creek, Zaza, Craig Mts., *St. John* 9,085 (type), *Can.*

Chase (1950) reduced this to the different appearing annual, *M. filiformis* (Thurb.) Rydb.

Muhlenbergia mexicana (L.) Trin., forma **ambigua** (Torr.) Fern. Perennial with short scaly creeping rhizomes; culms simple below, more or less branched above, scabrous below the nodes, 6–10 dm. tall; sheaths scabrous; blades scabrous, 2–5 mm. wide; panicle narrow, spike-like, interrupted below; glumes narrow, acuminate, 3 mm. long; lemmas pilose below, 3 mm. long, the terminal awn 2–5 mm. long, hairs of callus not over 1.5 mm. long. *M. sylvatica setiglumis* Wats.; *M. foliosa* (R. &. S.) Trin., forma *ambigua* (Torr.) Wieg. Moist rocky places, uncommon, *A. T.*

Muhlenbergia Richardsonis (Trin.) Rydb. Perennial, tufted or creeping from slender scaly rootstocks; culms erect or decumbent, 5–60 cm. tall; leaves less than 2 mm. wide; panicles narrow, often interrupted 2–15 cm. long; glumes 1–1.5 mm. long; lemmas lanceolate, 2 mm. long. *Sporobolus Richardsonis* (Trin.) Merrill; *M. squarrosa* (Trin.) Rydb. and *S. depauperatus* (Torr.) Scribn. Stream banks and flats, common. *U. Son., A. T.*

PANICUM. Panic Grass

Spikelets with one perfect flower, often with a staminate one below it; glumes 2, membranous; lemmas 2, the lower empty or including the staminate flower, the upper papery, shining, enclosing a similar palea and the perfect flower; awns none (in ours); grains free, enclosed in the hardened lemma and palea. (Name ancient Lat. for millet.)

Spikelets acuminate; annuals,
- Panicles usually purplish, partly included; spikelets long pediceled. — *P. capillare.*
- Panicles not so; spikelets mostly short pediceled. — *P. capillare,* var. *occidentale.*

Spikelets blunt; perennial,
- Spikelets 3.2 mm. or more long. — *P. Scribnerianum.*
- Spikelets less than 2 mm. long,
  - Spikelets turgid, blunt, strongly nerved. — *P. occidentale.*
  - Spikelets not so. — *P. Lindheimeri,* var. *fasciculatum.*

Panicum capillare L. *Old Witch Grass.* Annual; stems solitary or several, erect or decumbent at the base, 15–60 cm. tall, mostly simple; leaf blades 2–30 cm. long, pilose; sheaths villous; panicle usually purplish, 10–35 cm. long, during anthesis the lower branches crowded, ascending, and included in the sheath, the rhachis scabrous to pilose, the capillary branches solitary or in twos, ascending, branched and spikelet-bearing above the middle; spikelets 2–3 mm. long, [illegible] Nash; [illegible]; *P. barbipulvinatum* Nash var. [illegible] Suksd. Dry, usually sandy places, *A. T.*, often a weed.

At maturity the stem breaks freeing the whole panicle, which acts like a tumble-weed.

Var. **occidentale** Rydb. *Western Witch Grass.* Similar to the species but the panicle seldom purplish, soon exserted and the branches quickly divaricate, spikelets short pediceled. Alkali Lake, *St. John et al.* 4,882, *U. Son.*

Panicum Lindheimeri Nash, var. **fasciculatum** (Torr.) Fern. Vernal phase light green, 3–6 dm. tall, more or less pilose throughout; blades 5–10 cm. long, 5–8 mm. wide, acuminate; axis of panicle pilose; spikelets 1.6–2.1 mm. long, obtuse. Autumnal phase prostrate, forming large winter rosettes of short thick dark leaves; upper nodes forming axillary branches with reduced, less pilose leaves, and small panicles. Pond and lake shores, *A. T.*

Panicum occidentale Scribn. Tufted perennial; vernal phase with several erect, unbranched culms from the tufted base; culms 28–48 cm. tall, slender, pilose; internodes 3–6 cm. long, mostly included; nodes nearly glabrous, but subtending a pilose ring; sheaths nearly equaling or even exceeding the internodes, striate-nerved, spreading pilose; ligule hairs 5–7 mm. long; cauline blades 4–10 cm. long, 6–10 mm. wide, linear-lanceolate, ascending, firm, plane, the dark green upper surface remotely pilosulous but soon glabrate except for the remote long pilosity near the base, the margin scabrous, at base long pilose-ciliate, the lower surface pilosulous; the terminal panicle 6–8 cm. long, 2–3 cm. wide, the lower branches ascending, the middle and upper ones divergent, the axis pilose below, the branches sinuous, forking; spikelets 1.5–1.8 mm. long, 0.8–1 mm. wide, broadly ellipsoid, crisply white pilosulous; first glume 1/6 the length of the spikelet or less, with a deltoid tip; second glume and sterile lemma equaling the fruit, 1.6–1.8 mm. long, obovate, strongly parallel ribbed, pilose; fertile lemma 1.5–1.7 mm. long. *P. Brodiei* St. John. Wawawai, Snake River. *U. Son.*

Panicum Scribnerianum Nash. Vernal phase erect, 20–50 cm. high, culms smooth or pubescent; sheaths loose, ciliate, striate, pilose or nearly glabrate; cauline leaves 5–10 cm. long, 6–12 mm. wide, glabrous or sparsely hairy only beneath; panicles compact, 4–8 cm. long; spikelets 3.2–3.3 mm. long, obovate, turgid, blunt, usually short pilose. Autumnal phase branching above, these axillary branches with reduced blades and small partly included panicles. Common along Snake River and its larger tributaries, *U. Son.*

PASPALUM

Mostly perennial grasses; spikes paired, these pairs single or racemose along axis; spikelets plano-convex, subsessile, solitary or in pairs, in two rows on one side of a rhachis, the back of the fertile lemma towards the rhachis; 1st. glume usually wanting; sterile lemma about as long as the glume; fertile lemma indurate, the margins inrolled.

Paspalum distichum L. Culms 30–60 cm. tall, decumbent and rooting at base, glabrous, but nodes villous; blades 3–9 mm. wide, mostly sparsely villous; racemes two, 1.5–7 cm. long, the rhachis flattened, the margins puberulous; spikelets 2.5–4 mm. long, ovate, acute, sparsely pilosulous; sterile lemma glabrous.

Probably introduced from the eastern U. S., bank of Snake R., below Clarkston, 1941, *Ownbey 2,515.*

PHALARIS

Annuals or perennials with spike-like or narrow panicles; spikelets crowded, 1-flowered; glumes 2, about equal, compressed; lemmas 3, the first two much reduced and sterile, the third coriaceous, enclosing a palea and a perfect flower; stamens 3. (Name Gr., *phalaris,* coot, or some grass.)

Phalaris arundinacea L. *Reed Canary Grass.* Perennial with creeping rhizomes, forming large tufts; culms smooth, stout, 6–15 dm. tall; leaves large coarse, scabrous beneath, 6–20 mm. wide; inflorescence looser at anthesis, otherwise spike-like, pale, 7–17 cm. long; glumes acuminate, ribbed, 5 mm. long; fertile lemma shining, sparsely villous, 3 mm. long. Locally abundant, forming pure stands in marshes, *A. T.*

Often cultivated, producing good forage on lands too wet for other crops.

PHLEUM. Timothy

Annuals or perennials; inflorescence a spike-like panicle; spikelets 1-flowered; glumes membranous, compressed, keeled, the apex obliquely truncate, the mid-nerve produced into an awn; lemma much shorter; grain ovoid, free, enclosed in the lemma and palea. (Name Gr., *phleos,* reed.)

Spike-like racemes slender cylindric; awns of glumes 1 mm. long. *P. pratense.*
Spike-like racemes ovate-oblong; awn of glume 2 mm. long. *P. alpinum.*

Phleum alpinum L. *Mountain Timothy.* Stems 10–50 cm. high; leaf blades flat, 2–15 cm. long, 3–7 mm. wide; inflorescence 1–4 cm. long, usually purple, cylindric; body of glume 2 mm. long. In meadows, Craig Mountains, Hunter, *Huds.*

Phleum pratense L. *Timothy.* Perennïal, tall, erect, 50–150 cm. tall; leaf blades 7–20 cm. long, 4–8 mm. wide, smooth or scabrous; inflorescence 3–20 cm. long; body of glumes 2.5 mm. long; ciliate on the keels. Abundantly introduced, meadows, *A. T.*

PHRAGMITES. Reed

Very large perennials, leaves flat; panicles large, terminal; spikelets loosely 3–7-flowered, the rhachilla disarticulating above the glumes and between the florets, long silky hairy; lowest floret staminate or neuter; glumes unequal carinate, acute, the lower 3-nerved, the upper 3- or 5-nerved and shorter than the florets; lemmas long acuminate, glabrous, ciliate on the keel, 3-nerved. (Name Gr., *phragmites,* growing in hedges.)

Phragmites communis Trin., var. **Berlandieri** (Fourn.) Fern. *Common Reed.* From a stout creeping rootstock; culms smooth, 1.5–4 m. tall; sheaths smooth, overlapping; blades pale green, somewhat scabrous beneath, linear-lanceolate, 1–5 cm. wide; panicle dense, 15–40 cm. long, the branches nearly smooth, ascending; first glume 4–6 mm. long; second glume 6–8.5 mm. long; lemma linear-lanceolate, acuminate, smooth below, 10–15 mm. long. Springy places, or stream banks, [illegible], *A. T.*

POA. Bluegrass

Annual or perennial grasses with simple stems and narrow leaves; inflorescence a contracted or open panicle; spikelets 2–6-flowered, compressed upper floret reduced or rudimentary, the rhachilla usually glabrous; flowers perfect or rarely dioecious; glumes membranous, keeled, 1–3-nerved; lemmas membranous, keeled, awnless, longer than the glumes, often with a tuft of cobwebby hairs at the base (the *web*), 5–7-nerved, marginal nerves and the midrib often pubescent; palea a little shorter than the lemma, 2-nerved or 2-keeled; grain free, or sometimes adherent to the palea. (Name Gr., *poa*, grass.)

Florets altered to bulblets, normal lemmas rare. *P. bulbosa*, var. *vivipara*.

Florets normal, bulblets wanting,
  Creeping rhizomes present,
    Lemma with a silky web at base,
      Lemma 2.5–3 mm. long; panicle open, the branches spikelet-bearing above the middle. *P. pratensis*.
      Lemma 2–2.5 mm. long; panicle narrow, compact, the branches spikelet-bearing to near the base. *P. compressa*.
    Lemma without a web. *P. nervosa*.
  Rhizomes wanting,
    Lemma with a silky web at base,
      Lemma glabrous or scabrous, 3–4 mm. long. *P. Horneri*.
      Lemma pilose on the midrib and marginal nerves, 2–3 mm. long,
        Panicle erect, the branches ascending,
          Ligule 0.5 mm. or less long; empty glumes awl-shaped. *P. nemoralis*, var. *nemoralis*.
          Ligule 1 mm. or more long; empty glumes broader, short acuminate to acute. Var. *interior*.
        Panicle drooping, the branches spreading; ligule 2–3 mm. long. *P. palustris*.
    Lemma not webbed,
      Lemma villous on the nerves below; annual. *P. annua*.
      Lemma glabrous or generally puberulent or short pilose on the lower half; perennial,
        Lemma puberulent or short pilose on the lower half,
          Lemma flattened, keeled, acute. *P. gracillima*.
          Lemma rounded on the back, scarcely keeled, obtuse. *P. Sandbergii*.
        Lemma glabrous or somewhat scabrous,
          Panicle narrow, mostly over 10 cm. long,
            Blades flat. *P. ampla*.
            Blades involute. *P. juncifolia*.
          Panicle open or somewhat so, 2–8 cm. long,
            Cauline blades flat; lemma 3–4 mm. long. *P. vaseyochloa*.
            Blades all filiform, involute; lemma 4–5 mm. long,
              Inflorescence dense. *P. Cusickii*.
              Inflorescence open, loose, the branches spikelet-bearing only toward the tip. *P. filifolia*.

Poa ampla Merrill. Tufted, pale and glaucous throughout; the stout stems 6–9 dm. tall; basal leaves numerous, 30–40 cm. long, flat or involute, smooth or scabrous, about 1 mm. wide; stem leaves 2; panicle erect, 10–25 cm. long, narrow, dense, the appressed branches in half-whorls of 3–5, the shorter ones spikelet-bearing to the base; spikelets 8–12 mm. long, 4–8 flowered; lemmas minutely scabrous, 4–5 mm. long. *P. brachyglossa* Piper. A common bunch grass, *U. Son., A. T.*

Poa annua L. *Annual Bluegrass*. The compressed stems 5–30 cm. tall; leaves glabrous, the blades short and soft, 0.5–3 mm. wide; panicle pyramidal, sometimes 1-sided; spikelets 3–7-flowered, crowded, 4 mm. long; lemma 2–2.5 mm. long. A weed, introduced from Europe. In moist places becoming common, *U. Son, A. T.*

Poa bulbosa L., var. vivipara Koeler. Perennial; bulbous at base; culms smooth, erect, 1–5 dm. tall; leaf sheaths smooth; blades flat or channeled, somewhat scabrous, 0.5–2 mm. wide; panicle narrow, 3–10 cm. long, dense, and at maturity plumed by the slender green leafy tips of the bulblets; glumes ovate-lanceolate, acuminate, firm, green with scarious margins, scabrous on the midnerve, 2–3 mm. long. Introduced from Europe, now abundant, spreading from towns to open grasslands, *A. T.*

Poa compressa L. *Canada Bluegrass.* Smooth perennial; spreading and forming dense mats; the much compressed stems 15–45 cm. tall; decumbent at base; blades flat, 1–3 mm. wide; panicle 3–7 cm. long, contracted; spikelets 3–9-flowered; lemmas obscurely nerved, the midrib and marginal nerves short pilose below. Weed, introduced from Europe, well established about towns, *A. T.*

Poa Cusickii Vasey. Pale green, in dense tufts; stems 30–70 cm. high, scabrous above; leaves scabrous, 15–30 cm. long; ligule very short; panicle narrow, the rays in threes or fours, scabrous; spikelets 5–9 mm. long, loosely 4–7-flowered; lemma scabrous. *P. capillarifolia* Scribn. & Williams. Rocky slopes, Snake River Canyon, *U. Son.*

Poa filifolia Vasey. Tufted perennial; culms 3–6 dm. tall; basal leaves 10–25 mm. long, numerous, involute, flaccid, minutely scaberulous; cauline leaves 2–3, similar, the blades 2–5 cm. long, the ligule 2–3 mm. long, acute, panicle 5–10 cm. long, loose, the branches flexuous, spikelets 6–10 mm. long, loosely 2–7-flowered, purplish or green; glumes 2–4 mm. long, ovate-lanceolate, subequal, the first 1-nerved; lemmas 4–5 mm. long, oval, thin-margined, 5-nerved, scaberulous. *P. idahoensis Beal.* Meadows, near Lewiston and Anatone, *U. Son., A. T. T.*

Poa gracillima Vasey. Densely tufted 20–60 cm. tall; basal leaves numerous, green, the blades 2–6 cm. long, flat, slightly scabrous, about 2 mm. broad; stem leaves 2; ligules well developed; sheaths loose, becoming scarious; panicles pale or purple, erect, rather loose, 4–10 cm. long, the rays mostly in twos; spikelets lanceolate, 4–6 mm. long, loosely 3–5-flowered; glumes subacute, smooth, subequal, about 4 mm. long; lemma obscurely 5-nerved, oblong, 4–4.5 mm. long. *P. saxatilis* Scribn. & Williams. In rocky places, high ridges of the Blue Mts., *Piper* 2555, *Huds.*

Poa Horneri St. John. Tufted, glabrous perennial; roots fibrous; basal leaves several, ascending; the sheaths loose, striate-nerved, about 1 cm. long; the blades 15–45 mm. long, 1–4 mm. wide, flat, linear, acute; cauline leaves 2–3; the striate sheaths [illegible] cm. long, often equaling the internodes; the blades [illegible] long, [illegible] wide, ascending, flat; culms [illegible] cm. tall, striate; panicles 8–10 cm. long, seldom over 1 cm. in width, the branches strict, erect, scabrous, the longer ones spikelet-bearing in their upper halves; pedicels 0.5–8 mm. long; spikelets 3–5-flowered, 5.5–7.5 mm. long, green, the florets loose, exposing both their base and the rhachis; first glume 1.8–2.2 mm. long, ovate-lanceolate; second glume 2.5–3 mm. long, broadly oval-lanceolate, scabrous on the hyaline margin and on the 3 nerves; lemmas 7-nerved, with the intermediate nerves distinct, hyaline margins, 3–4.5 mm. long, with a silky web at base, scabrous, especially above; palea 2.5 mm. long, hyaline, with 2 firm, scabrous keels; anthers 1 mm. long, oblong, notched at each end, purple; grain not seen.

This species is most similar to *P. Bolanderi* Vasey and it was included in that species in Piper's Flora of Washington, and by other authors, and now by Chase (1950). The Californian species differs in being an annual; and in having culms 15–60 cm. tall; the panicle open with spreading or reflexed branches; the spikelets 1–3-flowered, about 4 mm. long; the florets close, overlapping; and the lemmas 2.5–3 mm. long, faintly 5-nerved or the intermediate ones obsolete.

The species is named for Prof. Robert M. Horner of Waitsburg, Wash., in recognition of his extensive explorations of the Blue Mts. Moist woods, Blue Mts., *Can.*

The holotype is from Indian Corral Spring, Columbia Co., *Darlington.*

Poa juncifolia Scribn. *Alkali Bluegrass.* Perennial, forming dense tufts 5–10 dm. tall; culms smooth; leaves mostly basal; sheath smooth; blades involute, appearing filiform, smooth; inflorescence 10–20 cm. long, 4–15 cm. wide, the branches erect, strict, scabrous; spikelets 3–6-flowered; 7–10 mm. long, lanceolate, only slightly flattened; glumes subequal, the outer 4–4.5 mm. long; lemmas 3.5–4.5 mm. long, lanceolate, the midrib and lower half scabrous. *P. brachyglossa* Piper. Alkaline flats, Ewan, [in scablands], 1934, *Pickett* 1,554 & 1,614, *U. Son.*

Poa nemoralis L. var. nemoralis. Similar to *P. palustris;* stems tufted, slender, 30–70 cm. high; blades scabrous margined, 2 mm. wide; panicle open in flower, pyramidal, 4–10 cm. long; spikelets 2–5-flowered. Rare, probably introduced, Spokane *Piper* 2596, *A. T.*

Var. **interior** (Rydb.) Butters & Abbe. Panicle 5–10 cm. long, lowest node with 1–2 branches; lemmas 3–3.5 mm. long; lemmas below middle with midrib and laterals long silky. *P. interior* Rydb. Spokane Co., *Suksdorf.*

Poa nervosa (Hook.) Vasey. Glabrous tufted perennial, the stems 3–8 dm. high; leaf blades numerous, flat or folded, 1–3 mm. wide; ligule small, usually 1–1.5 mm. long; panicle green or purple, narrow, loose, more or less drooping; spikelets 4–6-flowered, 7–10 mm. long; lemma smooth, or scabrous, 3–4 mm. long. *P. Olneyae* Piper. In open pine woods, *A. T. T.*

Poa palustris L. *Fowl Meadow Grass.* Tufted, the stems 3–15 dm. high; leaves smooth, 2–4 mm. wide; ligule conspicuous; panicle yellowish to purple, 10–30 cm. long, the branches mostly in fives, naked below; spikelets 2–5-flowered; lemma obscurely nerved. *P. triflora* Gilib. In wet meadows, *A. T., A. T. T.*

Poa pratensis L. *Kentucky Bluegrass.* Perennial, with conspicuous running rootstocks, the terete stems 3–10 dm. tall; blades flat or folded, often elongate, 1–5 mm. wide; panicle pyramidal, open, 6–20 cm. long, the ascending or spreading branches in whorls of 3–5; spikelets 3–5-flowered; lemmas 3–4 mm. long with prominent intermediate nerves, pilose on the midnerve and margins. Introduced from Europe, thoroughly established, wet or dry places, *U. Son., A. T.*

Poa Sandbergii Vasey. Glabrous, not glaucous, tufted; the stems 15–60 cm. tall; basal leaves numerous, the blades 5–10 cm. long, linear, 1.5 mm. or less in width; stem leaves 2, low down, the blades 1–2 cm. long, the sheaths often purple; panicle 4–17 cm. long, narrow, the ascending branches in whorls of 2–5, spikelet-bearing near the top; spikelets 2–4-flowered, usually purplish; lemmas oblong, 3.5 mm. long, sparsely pubescent above, villous or puberulent near the base. *P. Helleri* Rydb.; *P. secunda* sensu Scribn., Hitchc., and Marsh, not of Presl. Open places, *A. T.*

The commonest native species, flowering early.

Poa vaseyochloa Scribn. Tufted, culms slender, smooth, 10–28 cm. tall; leaves of the basal shoots involute; cauline sheaths smooth, the blades short, 1–1.5 mm. wide; panicle open, ovate, 2–5 cm. long, the branches bearing 1–2 spikelets; spikelets purplish, 3–6-flowered. Blue Mts., *Horner* 222, *Huds.*

POLYPOGON

Annual grasses; inflorescence a spike-like panicle; spikelets 1-flowered; glumes each extended into an awn; lemma smaller, generally hyaline, short-awned, with a palea which is shorter than the lemma; grain free, enclosed in the lemma and palea. (Name Gr., *polus,* many; *pogon,* beard.)

Polypogon monspeliensis (L.) Desf. *Beard Grass.* Annual, decumbent or erect, 8–60 cm. high; leaf blades flat, 3–6 mm. wide, more or less scabrous; sheaths loose, shorter than the internodes; spike-like panicle densely flowered, oblong, rarely interrupted, 2–15 cm. long; glumes hispidulous elliptic, notched at the apex, about 2 mm. long, bearing terminal awns 3 or 4 times as long; lemma truncate, 1 mm. long, bearing a terminal awn somewhat longer. Appearing native, but said to be introduced from Europe. Moist places, especially where somewhat alkaline, *U. Son., A. T.*

PUCCINELLIA

Halophytes, usually perennials; inflorescence a panicle; spikelets several-flowered, the rhachilla disarticulating above the glumes and between the florets; glumes unequal, shorter than the first lemma; lemmas firm, rounded on the back, usually obtuse and scarious-erose at tip, 5-nerved; palea shorter than or equaling the lemma. (Named for *Prof. Bendetto Puccinelli* of Italy.)

Midrib of lemma prominent to its tip, often excurrent; branches spikelet-bearing to the base. *P. rupestris.*
Midrib not so; lower half of branches naked. *P. Suksdorfii.*

Puccinellia rupestris (With.) Fern. & Weath. Perennial; culms usually decumbent, smooth, 1–4.2 dm. long; blades flat, 2.5–6 mm. wide; panicle ellipsoid, glaucous, dense, 2–7 cm. long, the branches approximate, ascending; spikelets glaucous 3–5-flowered, 5–8 mm. long; glumes 3-nerved, the second obtuse, 2–2.5 mm. long; lemma ovate, obtuse, thick with a narrow sub-entire hyaline tip, 3–3.5 mm. long; anther 0.8–1 mm. long; grain 2 mm. long. Native of Europe, sparingly introduced, Johnson, *Purnell* in 1926, *A. T.*

Puccinellia Suksdorfii St. John. Caespitose annual; culms erect or decumbent, smooth, 1–5 dm. long; sheaths ribbed, smooth; basal blades smooth, 0.5–1.5 mm. wide; cauline blades 3–4 mm. wide; aestival panicles loose, 12–20 cm. long; the numerous autumnal panicles much smaller; panicle branches or some of them reflexed; spikelets 3–8-flowered, 3–6 mm. long; second glume 1–1.3 mm. long; lemma elliptic obtuse, green, [illegible] [illegible] [illegible] mm. long. [illegible] and [illegible]. Alkali Lake, *St. John et al.* [illegible]; Medical Lake, *St. John et al.* 6755, *U. Son.*

Hitchcock (1935), and Chase (1950) have reduced this species to *P. distans* (L.) Parl., a different species which is only adventive in North America.

SCLEROCHLOA

Low, tufted annual; upper sheaths broad; blades folded, boat-shaped at tip; racemes dense, spike-like; spikelets subsessile, imbricate in 2 rows on one side of the broad rhachis; spikelets 3-flowered, the upper floret sterile; rhachilla thick, the spikelet falling entire; glumes firm, obtuse, the margins hyaline, the first 3-nerved, the second 7-nerved; lemmas rounded on the back, 5-nerved, obtuse; palea hyaline, keeled. (Name from Gr. *skleros,* hard; *chloa,* grass.)

Sclerochloa dura (L.) Beauv. Culms 2–7 cm. long, erect or spreading; sheaths glabrous, the upper 2.5–3 mm. wide from keel to margin; blades 7–18 mm. long, 1–3 mm. wide, glabrous but for the scabrous keel and margins; ligule obtuse or truncate, 2 mm. or less in length; racemes 1–2 cm. long; spikelets 6–7 mm. long; first glume 2.5–3 mm. long; second glume 3.5–4 mm. long; lower lemma 5 mm. long. *Crassipes annuus* Swallen. Weed, introduced from Europe, established at Moscow, and Pullman, *A. T.*

SETARIA

Annual or perennial grasses with erect stems and flat leaves; spikelets with one perfect flower and rarely also a staminate one, in spike-like panicles; pedicels bearing bristles; glumes and lower lemma membranous, the latter often containing a palea and rarely a staminate flower; upper lemma papery with a similar palea and a perfect flower. (Name Lat., *seta,* bristle.)

Spikelet subtended by 5 or more bristles. *S. glauca.*
Spikelet subtended by 1–3 bristles. *S. viridis.*

Setaria glauca (L.) Beauv. *Yellow Foxtail.* Annual, 1–12 dm. tall, the herbage glabrous, branching at base, erect or decumbent at base, the culms compressed, the basal nodes geniculate; blades 3–10 mm. wide; panicle 5–10 cm. long, about 1 cm. in diameter, yellowish, dense, ascending, spike-like; bristles 4–8 mm. long, upward barbed; spikelets 3–4 mm. long, elliptic, compressed; fertile lemma strongly rugose, cartilaginous. *S. lutescens* (Weigel) F. D. Hubbard; *Chaetochloa lutescens* (Weigel) Stuntz.

Weed, introduced from Eurasia. Lewiston-Pomeroy Highway, 1913, *Dillon* 913.

Setaria viridis (L.) Beauv. *Green Foxtail.* Annual, usually tufted, green; stems 3–9 dm. high; sheaths glabrous; leaf-blades flat, 4–10 mm. wide, scabrous on the margins; spikes green, 3–9 cm. long, the rhachis villous; bristles 1–3, upwardly barbed, 6–12 mm. long; spikelets 2–2.5 mm. long; fertile lemma faintly wrinkled. *Chaetochloa viridis* (L.) Scribn. Introduced in fields and waste places, *U. Son., A. T.*

SITANION. Squirrel-Tail

Caespitose perennials, with spike-like panicles, which readily break up into segments; spikelets 2–3 at each joint of the rhachis, rarely solitary, 1–5-flowered; glumes subulate and entire, or lanceolate and bifid, or parted into several long-awned lobes; lemmas 5-nerved, lanceolate and acute, or those of the lowest floret sterile and subulate, entire with a single awn, or trifid and three-awned; palea 2-keeled. This genus is scarcely distinct from *Elymus.* (Name Gr., *sitos,* grain or food.)

Glumes broadened at base into a lanceolate body, all or some 3-nerved. *S. Hanseni.*
Glumes not broadened, cleft to or nearly to the base into rigid awns, 2-nerved,
Glumes cleft almost to base into 3–several awns. *S. jubatum.*

Glumes entire or 2-cleft,
  Lemmas smooth at base, scabrous above. *S. Hystrix.*
  Lemmas puberulent. *S. velutinum.*

Sitanion Hanseni (Scribn.) J. G. Smith. Culms 7–11 dm. tall; leaf-sheaths pilose; blades 2–7 mm. wide, flat or involute, pilose; spike 10–20 cm. long, slender, usually erect, long-exserted; glumes 2, 1.5–4 cm. long, including the scabrous awns, equal, entire, or cleft; lemma with the body 8–11 mm. long, smooth at base, scabrous above, bifid at apex and tipped with a straight awn, 3–4 cm. long. *S. anomalum* J. G. Smith; *Elymus Leckenbyi* Piper. Bars of Snake River at Wawawai, *U. Son.*

Sitanion Hystrix (Nutt.) J. G. Smith. Culms 1–5 dm. high; sheaths smooth to pilose; blades 1–5 mm. wide, flat or at length involute; somewhat hairy, usually puberulent; spike 2–13 cm. long, erect or nearly so, bushy; glumes or divisions 2–3.5 cm. long, all awned; lemma 2–7 cm. long, 3-awned, *S. rigidum* J. G. Smith; *S. hordeoides* Suksd. In dry soil, *U. Son.,* to *A. T. T., Huds.*

Sitanion jubatum J. G. Smith. Culm 2–6 dm. tall, minutely puberulent; sheaths pilose or scabrous; blades 2–4 mm. wide, usually involute, puberulent and more or less strigose; spike 2.5–10 cm. long, the base enclosed in the elongated upper leaf sheath; glumes 3–8-parted, each lobe bearing a slender awn 3–8 cm. long; lowest floret sterile, its lemma resembling the glumes; lemma of perfect florets lanceolate, 3-awned, the middle awn stout, 4.5–11 cm. long, the lateral ones slender and usually shorter. *S. villosum* J. G. Smith. Common in dry ground, *U. Son. to A. T. T.*

Sitanion velutinum Piper. Culms 3–6 dm. tall; whole plant densely puberulent with white soft hairs; blades 3–8 mm. wide, flat; spikes 4–11 cm. long, erect, bushy; glumes 3–5 cm. long, puberulent, entire or rarely cleft or divided, slightly dilated at base, all awned; lemma 3-awned, the awns 3–7 cm. long. Type locality, Steptoe, *A. T.*

This was reduced to *S. jubatum* by Chase (1950), but is seems distinct.

SPARTINA. Cord Grass

Coarse perennial grasses with strong creeping rootstocks, rigid simple stems and long tough leaves; [illegible] alternate spikes, spikelets 1-flowered, narrow, deciduous, borne in two rows on the rhachis, articulated with the [illegible] below the [illegible]; glumes keeled, [illegible] equal; lemma keeled, equaling or shorter than the second glume; palea often longer than its lemma; grain free. (Name Gr., *spartine,* cord.)

Second glume awned; spikelets 12–14 mm. long. *S. pectinata.*
Second glume acute; plant slender; spikelets 6–9 mm. long. *S. gracilis.*

Spartina gracilis Trin. *Alkali Cord Grass.* Culm 3–10 dm. high, erect, stiff, leaf-blades pale green, flat or slightly involute, 15–30 cm. long, 2–4 mm. wide, somewhat scabrous; spikes mostly 5–7, appressed, 1–5 cm. long; glumes acute, scabrous on the keels; lemma entire, acute, membranous, scabrous on the keels, 7 mm. long. In alkaline soil in scablands along w. border, *U. Son.*

Spartina pectinata Link. *Cord Grass.* Culm 7–20 dm. tall, simple, smooth; blades flat, keeled, long-acuminate, involute in age, scabrous on the margins; spikes 5–20, 5–12 cm. long, ascending, sometimes peduncled; spikelets closely imbricated; glumes very scabrous on the keels, awn-pointed; lemma scabrous on the midrib, which terminates below the 2-toothed apex, 7–9 mm. long. *S. Michauxiana* Hitchc. Stream banks, *U. Son.*

SPHENOPHOLIS

Slender perennials; panicles contracted; spikelets 2–3-flowered; the rhachilla extending beyond the flowers; glumes shorter than the spikelet; the lower linear, acute, 1-nerved; the upper much broader, 3–5-nerved, obtuse or rounded at the apex, or sometimes acute, the margins scarious; lemmas narrower, generally obtuse, 1–3-nerved, awnless; palea narrow, 2-nerved; grain free, loosely enclosed in the scale and palea. According to Hitchcock, *Sphenopholis* belongs in the *Aveneae*. (Name Gr., *sphen,* wedge; *pholis,* scale.)

Panicle dense, erect; second glume truncate. *S. obtusata,* var. *lobata.*
Panicle loose, usually nodding; second glume subacute. *S. pallens.*

Sphenopholis obtusata (Michx.) Scribn., var. **lobata** (Trin.) Scribn. ex. Robins. Somewhat tufted, perennial, 30–100 cm. tall, the stems smooth; leaf-blades flat, scabrous, 10–15 cm. long, 3–5 mm. wide; panicle erect, narrow, 5–18 cm. long, pale green, the branches short and densely flowered; lemma narrow, obtuse, scabrous, about 2 mm. long. *S. obtusata lobata* (Trin.) Scribn., Rhodora 8: 144, 1906. Scribner indicated that he was making this a subspecies. Its differences are so slight that the writer considers that it should be reduced to a variety. Rare on sandy bars of Snake River, *U. Son,*

Sphenopholis pallens (Biehler) Scribn. Panicle slender, 8–15 cm. long, usually loose and nodding, the branches 3–7 cm. long; lemma narrow, acutish. *S. pallens major* (Torr.) Scribn. Infrequent, *A. T.*

The name of the first author, Biehler, is incorrectly spelled by Chase (1950).

SPOROBOLUS. DROP-SEED

Annuals or perennials with small 1-flowered spikelets; inflorescence an open or contracted panicle; glumes membranous; lemma membranous, 1-nerved, awnless, equaling or longer than the glumes; palea equaling or exceeding the lemma; grain free and early deciduous. (Name Gr., *spora,* seed; *ballein,* to throw.)

Glumes nearly equal, panicle usually enclosed in the sheath. *S. neglectus.*
Outer glume not over half as long as the lemma; panicle large, partly or wholly exserted. *S. cryptandrus.*

Sporobolus cryptandrus (Torr.) Gray. Perennial, tufted; stems 50–100 cm. tall, erect, usually simple, smooth; leaf-blades flat, becoming involute, 10–15 cm. long, 2–5 mm. wide; ligule minute; sheaths long, bearded at the throat; panicle erect, narrowly pyramidal, 10–25 cm. long, its base enclosed in the uppermost sheath; branches mostly in pairs, ascending; spikelets crowded, leaden-colored, 1.5 mm. long. Sandy bars of Snake River, *U. Son.*

Sporobolus neglectus Nash. Annual, culms decumbent or erect, 2–30 cm. tall; sheaths smooth, usually inflated; blades flat or involute, scabrous on the margin, 1–2 mm. wide; panicles spike-like, 2–3 cm. long; glumes ovate, white scarious with strong green mid-nerve, equal, nearly as long as the lemma;

lemma whitish, acute, 2–2.5 mm. long. Woods and river banks near Spokane *A. T.*

STIPA. Needle Grass

Tufted perennial grasses; inflorescence a panicle; spikelets 1-flowered, narrow; glumes membranous or papery; lemma 1, convolute, indurate, bearing a twisted or bent awn which is spiral and articulated at the base; grain narrow, free, tightly closed in the lemma. The twice geniculate awn is flattened on one side in its first segment. It is hygroscopic, loosening with moisture, spiraling with dryness. These coilings and the hairs or scabrous surface, tend to push the lemma and grain forward. It may penetrate the ground, or even the flesh of animals. (Name Gr., *stupe*, tow.)

Awn plumose below,
  Sheaths pilose. — *S. Elmeri.*
Sheaths glabrous or scaberulous,
    Ligule 3–6 mm. long; awn 4–5 cm. long. — *S. Thurberiana.*
    Ligule 1–2 mm. long; awn 2.5–3.5 cm long. — *S. californica.*
Awn scabrous or appressed hispidulous,
  Lemma 8–12 mm. long, callus 3 mm. long; glumes 15–30 mm. long,
    Third segment of awn 10–12 cm. long; leaves smooth or scaberulous. — *S. comata,* var. *comata.*
    Third segment of awn 3–5 cm. long,
      Leaves smooth or scaberulous; third segment of awn 3 cm. long. — *S. comata,* var. *intermedia.*
      Leaves pilose; third segment of awn 5 cm. long. — *S. comata,* var. [illegible]
  Lemma 6–7 mm. long; callus 0.5–1 mm. long; glumes 8–10 mm. long,
    Glumes broad, abruptly acuminate, rather firm, the first 5-nerved. — *S. Lemmoni.*
    Glumes narrow, gradually acuminate, usually hyaline, the first 3-nerved,
      Sheaths glabrous, naked at the throat, awns 2–2.5 cm. long. — *S. columbiana,* var. *columbiana.*
      Sheaths somewhat pilose at the throat; awns 3.5–5.5 cm. long. — Var. *Nelsoni.*

Stipa californica Merr. & Davy. Culms smooth or puberulent below, 3–12 dm. tall; sheaths glabrous or the lower puberulent, villous at the throat; leaves becoming involute, scabrous above; panicle narrow, 10–30 cm. long; glumes equal, hyaline, 3-nerved, about 1 cm. long; lemma 6–7 mm. long, villous, the hairs at the summit much longer than those below. Not common, rocky slopes, *A. T.*

Stipa columbiana Macoun, var. **columbiana.** Perennial, tufted, 30–100 cm. tall, dark green; leaf-blades involute, long-attenuate, 10–30 cm. long, scabrous; sheath shorter than the internodes, loose, smooth; ligule 1–2 mm. long; panicle erect, rather dense, 8–20 cm. long, narrow, its base included in the uppermost sheath, the branches mostly in pairs; spikelets purple-tinged; lemma appressed villous. *Stipa minor* (Vasey) Scribn. Grassy hillsides and pine woods, *A. T.*

Var. **Nelsoni** (Scribn.) St. John. Differs from the species by having the sheaths somewhat pilose at the throat, and the awns 3.5–5.5 cm. long. *S. nelsoni* Scribn.; *S. columbiana nelsoni* (Scribn.) Hitchc. Grassy prairies, *A. T.*

According to Dr. Hitchcock it is a larger plant than the species, but such is not the case with our northwestern material. Since the two plants grow in the same areas and differ only in characters of secondary importance, they seem properly classified as a species and a variety.

Stipa comata Trin. & Rupr., var. **comata.** Tufted, pale green, 4–10 dm. high; sheaths glabrous or scabrous; blades involute; ligule 3–5 mm. long, conspicuous; panicle narrow, loose, 10–30 cm. long, often included at base, the branches spreading and few-flowered; glumes 5-nerved, subequal; lemma short appressed villous or glabrate above. Dry soils, Spokane; common along Snake River, *U. Son., A. T.*

Var. **intermedia** Scribn. & Tweedy. Differs from the species by its shorter awn, the glumes and lemmas averaging a little longer, and the panicle usually exserted. Creek bank, head of Rock Lake, *Beattie & Lawrence* 2363, *A. T.*

Var. **Suksdorfii** St. John. Differs from the species in having the leaf-sheaths and blades pilose, and the third segment of the awn 5 cm. long.

Washington: in a grove s.w. of Philleo Lake, Spokane Co., Aug. 3, 1916, *W. Suksdorf* 8990 (type).

This variety was reduced to the species by Chase (1950) without discussion of its characters.

Stipa Elmeri Piper & Brodie ex Scribn. Tufted, pale green, 60–100 cm. high, the whole plant pubescent; blades mostly involute; ligules very short; panicle erect, rather dense, 5–20 cm. long, its base usually included; glumes subequal, about 12 mm. long, the lower 5-nerved, the upper 3-nerved; lemma 6–7 mm. long, pubescent; awn 20–50 mm. long, pubescent to the second bend. Near Spokane, *A. T.*

Stipa Lemmoni (Vasey) Scribn. Stems slender, rigid, glabrous, 40–80 cm. high; leaves rather numerous; sheaths glabrous, shorter than the internodes; ligule 1–3 mm. long; blades 5–20 cm. long, 1–3 mm. wide, glabrous beneath, striate and pubescent above, strongly involute; panicle narrow, compact, erect, 5–12 cm. long; spikelets pale green or purplish, shiny; glumes nearly equal; lemmas oblong, sparsely pilose; callus obtuse, short-bearded; awns 20–35 mm. long, sparsely pilose to the second bend. In rocky soil, Spokane, and common in the Blue Mts., *A. T.*

Stipa Thurberiana Piper. Tufted, pale green; stem slender, 30–75 cm. high, pubescent at the nodes; blades involute, rather rigid; panicle erect, 8–15 cm. long, rather dense, often included at base in the upper sheath; lower glume 5-nerved, longer than the 3-nerved upper glume; lemma 6–7 mm. long, pubescent; awn plumose to the second bend. Dry soils, Spokane, Rock Lake, *A. T.*

TRISETUM

Annuals or caespitose perennials; inflorescence a spike-like or open panicle; spikelets 2–12-(usually 3-)flowered, the flowers all perfect or the uppermost staminate; rhachilla glabrous or with long soft hairs, extending beyond the flowers; glumes unequal (in ours), acute, persistent; lemmas usually shorter than the glumes,

2-toothed, bearing a dorsal awn below the apex or sometimes awnless; palea narrow, hyaline, 2-toothed; grain free, enclosed in the lemma. (Name Lat., *tres,* three; *seta,* bristle.)

Lemmas awnless or nearly so; panicle narrow. *T. Wolfii.*
Lemmas awned,

  Panicle dense and spike-like; ovary smooth. *T. spicatum,* var. *molle.*

  Panicle loose; ovary hairy at apex,
    Panicle open; second glume 3–5 mm. long; lower lemma 5 mm. long. *T. cernuum.*
    Panicle narrow, the branches appressed; second glume 6–7 mm. long; lower lemma 7 mm. long. *T. canescens.*

Trisetum canescens Buckl. Perennial; culms 4–12 dm. tall; sheaths pilose; blades flat, more or less pilose, 2–11 mm. wide; panicle erect, 7–20 cm. long, the branches spikelet-bearing to near the base; lemma with a bent awn nearly twice its length. *T. cernuum,* var. *canescens* (Buckl.) Beal. Coniferous woods, *A. T. T., Can.*

Trisetum cernuum Trin. Perennial; stems 60–100 cm. tall, erect; sheaths glabrous; leaf-blades flat, merely scabrous, 15–20 cm. long, 4–10 mm. wide; panicle 10–30 cm. long, loose, nodding; branches in distant whorls, capillary, cernuous, flower-bearing above the middle; spikelets 2–4-flowered, 6–7 mm. long without the awn; lemma bearing an awn about twice its length. Moist woods and copses, *A. T.* to *Can.*

Trisetum spicatum (L.) Richt., var. **molle** Beal. Densely tufted, perennial, 10–80 cm. high, pubescent to glabrous with the sheaths pilose; panicle elliptic to cylindric, becoming interrupted below, 2.5–11 cm. long, 5–20 mm. thick, shining and often brownish-purple; glumes scabrous on the keels, the upper 4.5–6.5 mm. long; lemma 5–6 mm. long, its awn divergent and about as long. *Avena mollis* Michx., Fl. Bor.-Am. 1:72, 1803, not *A. mollis* (L.) Salisb. (1796) or Koeler (1802); *T. spicatum molle* (Michx.) Piper. *T. spicatum* var. *Michauxii* St. John. Upper ridges in the Blue Mountains, *Huds.*

Trisetum Wolfii Vasey. Perennial; stems erect, 30–75 cm. high; blades flat, scabrous or sometimes pubescent; panicles narrow, erect, rather dense; spikelets 2-flowered; glumes subequal, acute; lemmas smooth, [illegible] short awned. *T. muticum* (Boland.) Scribn. A rare grass only [illegible] limits [illegible]

VULPIA

Annuals; first glume very small; lemmas compressed subulate, slender-tipped or long awned; filaments very short, the usually single anther appressed to the palea or the stigma; grain attenuate at both ends. (Named from Lat. *vulpes,* fox, from the brush of long awns in the panicle.)

Spikelets densely 5–13-flowered. *V. octoflora* var. *tenella.*
Spikelets loosely 2–5–(6)-flowered,

  Spikelets reflexed. *V. reflexa.*
  Spikelets ascending or somewhat divergent,
    First glume less than half the length of the

second; lemma hispid ciliate above. *V. Myuros,* var. ***hirsuta.***
First glume nearly equaling the second; lemma not ciliate. *V. pacifica.*

Vulpia Myuros (L.) Gmel., var. **hirsuta.** Hack. Tufted, 10–65 cm. tall, smooth; sheaths smooth; blades pilose above, becoming involute; panicle spike-like, elongated, flexuous, 5–25 cm. long, pale-green; spikelets 4–5-flowered, 2–2.5 cm. long, including the awns; lemmas lanceolate, scabrous, the body 5 mm., the awn 10–15 mm. long. *Festuca megalura* Nutt. Dry places, *U. Son., A. T.*

Vulpia octoflora (Walt.) Rydb., var. **tenella** (Willd.) Fern. Tufted, 8–40 cm. tall: sheaths ribbed, short pilose; blades sparse, short, involute, scabrous; panicle 2–13 cm. long, rather dense, often one-sided; spikelets 6–10 mm. long; glumes subulate-lanceolate; lemmas 3–5 mm. long, scabrous, attentuate into an awn 2–4 mm. long. *F. tenella* Willd. Dry places, *U. Son., A. T.*

Vulpia pacifica (Piper) Rydb. Stem single or a few in a loose tuft, 5–60 cm. tall; sheaths glabrous or puberulent, striate; blades glabrous, involute; panicle 4–15 cm. long; branches mostly solitary, longest below, all erect at first but becoming spreading or reflexed by means of a prominent pulvinus at the base; spikelets 2–6-flowered; lemma 6–7 mm. long, the awn two to three times as long; upper florets somewhat scabrous. *Festuca pacifica* Piper. Dry places, common, *U. Son., A. T.*

Vulpia reflexa (Buckl.) Rydb. Culms single or few, erect, 5–50 cm. tall; sheaths smooth or pubescent; blades smooth, involute; panicle open, 5–12 cm. long, the branches divergent or reflexed; spikelets 1–3-flowered; glumes lanceolate, glabrous, the second 4.5–5 mm. long, the first half as long; lemma scabrous above, 4.5–8 mm. long, the awn 5–12 mm. long. *Festuca reflexa* Buckl. Dry rocky places, Palouse Falls, *St. John & Pickett* 6,190, *U. Son.*

ZIZANIA. Indian Rice

Tall aquatic annuals or perennials with monoecious flowers; upper part of the inflorescence pistillate, the lower staminate; pistillate spikelets lacking glumes, the lemmas long awned, embracing the palea and grain; staminate spikelets with first glume 5-nerved, the second 3-nerved, the short awned lemma and the palea subequal; stamens 6; grain long cylindrical, black. (Name Gr., *zizanion,* a wild grain.)

Zizania aquatica L., var. interior Fassett. *Wild Rice.* Plants without creeping rootstocks, stout, 0.9–3 m. tall; culms and sheaths smooth; blades large, 1–3 cm. broad; ligules 1–1.5 cm. long; pistillate lemmas firm, shining, glabrous except on the margins, nerves, and awns, 1.5–2 cm. long. A native of e. N. Am., introduced to attract waterfowl. *Z. interior* (Fassett) Rydb. Well established on Lake Chatcolet, *St. John* 9,056.

CYPERACEAE. Sedge Family

Grass-like or rush-like herbs; stems slender, solid (rarely hollow), triangular, quadrangular, terete, or compressed; leaves narrow with closed sheaths; flowers perfect or unisexual in spikelets, one (rarely two) in the axil of each scale (glume or bract); spikelets solitary or clustered, 1- to many-flowered; scales 2-ranked or in a spiral, persistent or deciduous; perianth hypogynous, of bristles, or interior scales, or wanting; stamens 1–3, rarely more; ovary 1-celled, 1-ovuled; style 2–3 cleft; fruit a lenticular or three-sided achene; endosperm mealy; embryo minute. (Named from the genus *Cyperus*.)

Flowers monoecious or dioecious; perigynium present. *Carex.*
Flowers of the spikelet all, or at least one of them perfect; spikelets all similar; perigynium wanting,
  Spikelets laterally compressed; scales 2-ranked. *Cyperus.*
  Spikelets not compressed; scales in a spiral,
    Base of style swollen, persistent as a tubercle on the achene. *Eleocharis.*
    Base of style not swollen, deciduous or persistent as a subulate tip,
      Bristles present; hyaline scales none. *Scirpus.*
      Bristles none; perianth of a single minute posterior scale. *Hemicarpha.*

CAREX. Sedge

Grass-like sedges, perennial by rootstocks; stems mostly triangular; leaves 3-ranked, the upper (bracts) elongated or very short, and subtending the spikes of flowers, or wanting; flowers solitary in the axils of bracts (scales), monoecious or dioecious; spikes either wholly staminate, or pistillate, or sometimes with both staminate and pistillate flowers; perianth none; staminate flowers of 3 stamens; pistillate flower of a single pistil with a style and 2–4 stigmas borne in the axil of a second bract (the perigynium) which completely encloses the achene; achene lenticular or planoconvex, 3-angled, or 4-angled. (The ancient Lat. name, perhaps from the Gr. *kerein*, to cut.)

Culms bearing a single spike or apparently so; stigmas 3; achenes trigonous,
  [illegible] [illegible]
  [illegible]
    [illegible] filiform. *C. filifolia.*
  Perigynia glabrous; leaves broader,
    Achenes rounded at apex; spikes apparently single (the lateral ones on much shorter filiform basal peduncles). *C. saximontana.*
    Achenes pointed at apex; spikes single, on long, equal scapes. *C. Geyeri.*
Culms bearing several (often closely crowded) spikes,
  Stigmas 4; achenes tetragonous. *C. concinnoides.*
  Stigmas 2–3; achenes 2–3-sided,
    Stigmas 2; achenes lenticular,
      A. Lateral spikes sessile, short; terminal spike with both staminate and pistillate flowers,
        Rootstocks long; culms remote,
          Rootstocks slender; beak as long as the lanceolate body of the perigynium. *C. Douglasii.*
          Rootstocks stout; beak ½ as long as the

ovate body of the perigynium. *C. praegracilis.*
Rootstocks short or the plant caespitose,
Upper flowers of spikes staminate,
Perigynia ovoid-elliptic, abruptly contracted to a minute beak. *C. disperma.*
Perigynia lanceolate with a prominent beak,
Perigynia rather abruptly contracted into the beak,
Inflorescence a head. *C. Hoodii.*
Inflorescence looser, at least the lower spikes distinct,
Blades 1–2.5 mm. wide; perigynia 2-2.7 mm. long, shining, not concealed. *C. diandra.*
Blades 2–6 mm. wide; perigynia 3–4 mm. long, dull, concealed by the scales. *C. Cusickii.*
Perigynia long tapering into the beak. *C. stipata.*
Upper flowers of spikes pistillate,
Perigynia at most thin edged,
Perigynia at maturity spreading or ascending,
Spikes aggregated in a head. *C. arcta.*
Spikes all or at least the lower remote,
Beak of perigynium strongly serrulate the body broadest near the base,
Beak of perigynium bluntly bidentate, the ventral false suture none inconspicuous. *C. interior.*
Beak of perigynium sharply bidentate, the ventral false suture conspicuous. *C. cephalantha.*
Beak of perigynium with few weak serrulations, the body broadest near the middle. *C. laeviculmis.*
Perigynia appressed,
Beak ⅓ as long as the body of the perigynium. *C. leptopoda.*
Beak ½ as long as the body of the perigynium. *C. Bolanderi.*
Perigynia wing-margined,
Lower bracts leafy, much longer than the head. *C. athrostachya.*
Lower bracts not conspicuously leafy, not or but slightly exceeding the head,
Beak of perigynium flat and serrulate,
Perigynia with the body narrowly ovate, brown. *C. Bebbii.*
Perigynia greenish,
Scales brown to whitish;

body or perigynium suborbicular. — *C. brevior.*

Scales reddish brown; body or perigynium ovate. — *C. multicostata.*

Beak of perigynium terete towards the apex, smooth or nearly so,

Perigynia lanceolate, 6–7 mm. long. — *C. petasata.*

Perigynia ovate or broader, or if lanceolate less than 5.5 mm. long,

Perigynia with thinnish, submembranous walls,

Perigynia 3.5–5 mm. long,

Perigynia lanceolate, narrowly margined, spreading. — *C. microptera.*

Perigynia ovate, widely margined, appressed. — *C. festivella.*

Perigynia 4.5–6 mm. long. — *C. Haydeniana.*

Perigynia with thick firm walls,

Spikes densely capitate; beak obliquely cut, dark tipped,

Perigynia 1.5–2.2 mm. wide, nearly nerveless ventrally. — *C macloviana,* var. *pachystachya.*

Perigynia 1–1.5 mm. wide, nerved ventrally. — *C. subfusca*

[illegible] beak bidentate, reddish tipped. — *C. Preslii.*

A. Lateral spikes elongate, often peduncled; terminal spike usually staminate,

Lowest bract long sheathing; pistillate spikes not very many flowered. — *C. Garberi,* var. *bifaria.*

Lowest bracts sheathless or, if somewhat sheathing, the pistillate spikes very many flowered,

Culms not leafy at base. — *C. prionophylla.*

Culms leafy at base,

Beak markedly bidentate. — *C. nebrascensis.*

Beak not deeply bidentate,

Perigynia turgid, spreading. — *C. aperta.*

Perigynia not turgid, appressed or

ascending. *C. Kelloggii.*
Stigmas 3; achenes trigonous,
Perigynia pubescent,
Pistillate spikes with from few to 25 flowers. *C. Rossii.*
Pistillate spikes with more numerous flowers,
Leaf-sheaths and blades glabrous,
Spikelets hispidulous, oval; leaves 2–4 mm. wide. *C. lanuginosa.*
Spikelets puberulent, suborbicular; leaves 2 mm. or less wide. *C. lasiocarpa.*
Leaf-sheaths and lower surface of blades hairy; perigynia sparsely pilose. *C. Sheldonii.*
Perigynia glabrous,
Style jointed, finally withering and deciduous. *C. amplifolia.*
Style continuous, indurated and persisting on the achene,
Leaves pilose. *C. atherodes.*
Leaves glabrous or merely scabrous,
Perigynia coarsely ribbed,
Lower perigynia reflexed; bracts many times exceeding the spikes. *C. retrorsa.*
Lower perigynia not reflexed; bracts moderately exceeding spikes,
Perigynia ascending; lower sheaths filamentose,
Perigynia 5–8 mm. long. abruptly contracted into the beak. *C. vesicaria.*
Perigynia 7–10 mm. long, tapering into the beak. *C. exsiccata.*
Perigynia spreading; lower sheaths not filamentose. *C. rostrata.*
Perigynia finely and closely ribbed. *C. comosa.*

Carex amplifolia Boott. Stems 5–10 dm. tall, scabrous at tip; sheaths brownish tinged; blades flat, smooth but for the scabrous margins, 8–18 mm. broad, longer than the stem; bracts very large, the lower overtopping the stem; bracts long and leafy; spikes 5–7, the uppermost staminate, 4–8 cm. long; pistillate spikes narrowly cylindrical, straight or curved, 2–10 cm. long, densely flowered, dark olivaceous, the lower ones remote and long-peduncled; scales lanceolate, brown margined, cuspidate from the notched tip, shorter than the perigynia; perigynium triangular subglobose, glabrous, the curved beak with an oblique entire orifice. Along streams in woods, *A. T. T.*

Carex aperta Boott. Tufted, with short stolons; stems 5–10 dm. high, scabrous above; leaves flat but keeled, 2.5–6 mm. broad, rather shorter than the stems, scabrous near the tip; spikes 3–4, the terminal staminate, mostly short peduncled; pistillate spikes 10–48 mm. long, densely flowered, 5 mm. thick: bracts leafy, scales equaling the perigynia, acuminate, purple with a green midrib; perigynia 3 mm. long, turgid, broadest at the middle, tapering to each end, brown when mature, the two or three angles pale; beak short and entire. On overflowed river bottoms, *A. T.*

In some regions very abundant and cut for hay.

Carex arcta Boott. Loosely tufted, pale green; culms 1.5–8 dm. high, scabrous at tip; leaves pale green, scabrous, 2–4 mm. wide, usually longer than the stems; head ovate-oblong, green or brownish, of 5–15 crowded spikes; 1–2 bracts often leafy; spikes oblong or ovoid, 5–10 mm. long; scales hyaline, often brownish, acute, shorter than the perigynia; perigynia spreading, ovate somewhat cordate, gradually tapering into the serrulate beak, strongly nerved on the outer face, 2–3 mm. long. In moist meadows, Spokane and Whitman Cos., *A. T.*

Carex atherodes Spreng. Stolons stout; culms scabrous, 6–12 dm. high; leaves numerous, 4–12 mm. wide, flat, nodulose, scabrous, exceeding the stems, the lower surface and sheaths pilose; spikes 2–8, scattered, short peduncled, the upper 1–4 staminate; staminate spikes pale, slender, 2–10 cm. long; pistillate spikes 2–5 cm. long, rather loosely flowered; bracts leafy, the lower longer than the inflorescence; scales oblong-lanceolate, pale, aristate, ciliate, shorter than the perigynia; perigynia ovoid-lanceolate, strongly nerved, the body 5 mm. long, the stout recurved teeth of the beak 2 mm. long. In wet places, infrequent, *A. T.*

Carex athrostachya Olney. Stems tufted, scabrous at the tip, 1–9 dm. tall; leaves 1.5–2.5 mm. wide, shorter than the stems; inflorescence a dense ovoid straw-colored head composed of 5–20 crowded spikes, these staminate below; lower bracts 2–5, exceeding the head; scales ovate or narrower, acuminate, brownish, about equaling the perigynia; perigynia 3–4 mm. long, lanceolate, spongy at base, the long beak obliquely cut, its margins serrulate. Common in wet places, *A. T.*

Carex Bebbii Olney. Stems slender, erect, 2–8 dm. high, tufted, rough at tip: leaves not stiff, 2–4.5 mm. wide; shorter than the stems; bracts setaceous; spikes 3–12, brown, ellipsoid, 5–8 mm. long, closely crowded in an ovoid cluster; scales oblong, acuminate; perigynia ascending, firm, narrowly ovate, 3–4 mm. long, exceeding the scales, nerveless or faintly nerved. In low meadows, rare, northern part of the *A. T. T.*

Carex Bolanderi Olney. Tufted, pale green; stems 2–10 dm. long, rather slender, weak and spreading; leaves soft, smooth, 2–5 mm. wide, shorter than the stems; inflorescence of 4–10 scattered spikes; spikes oblong, 10–15 mm. long, sessile or nearly so; scales white, scarious, with a broad green midvein, ovate-lanceolate, acuminate-cuspidate, as long as the perigynia; perigynia lanceolate, faintly nerved, 4 mm. long, the body gradually tapering into the nearly equal serrulate deeply 2-toothed beak. In moist woods in the Blue Mts., *A. T. T.*

Carex brevior (Dewey) Mack. Stems tufted, slender, erect, often scabrous at tip, 3–10 dm. high; leaves 1.5–3 mm. wide, flat, stiff, shorter than the stem; sheaths with a pale marginal band; spikes 3–6, straw-colored, approximate, broadly ovoid, obtuse, 8–12 mm. long, staminate at base; scales ovate, obtuse or acute, thin, brown or whitish, with a heavy green midrib, shorter than the perigynia; perigynia spreading, firm, broadly ovate to suborbicular, strongly 7–15 nerved on the outer face, 4–5.5 mm. long, the [illegible] minutely bidentate. *C. festucacea* Schk., var. brevior (Dewey) Fern. In rather dry soil, not common, *U. Son., A. T.*

Carex cephalantha (Bailey) Bicknell. Plant 2.5–7.5 dm. tall, densely caespitose; rootstocks not prolonged; blades 1–2 dm. long, 1.5–2.5 mm. wide, flat or canaliculate, stiff, light green; inflorescence of 3–7 spikes, more or less separate, 2.5–5 cm. long, gynecandrous; spikes 4–10 mm. long, 5–25-flowered, subglobose or short oblong; scales ovate, acute or somewhat cuspidate, equaling the body of perigynia, yellowish or chestnut brown tinged, with sharply keeled green midvein extending to apex and ill defined opaque margins; perigynia 3.5–4 mm. long, 1.5–2 mm. wide, ovate, plano-convex, subcoriaceous, strongly many nerved dorsally, finely nerved ventrally, round truncate at base, serrulate above, contracted into a serrulate beak half the length of body, the apex strongly bidentate, reddish brown tipped; stigmas 2; achenes 1.5 mm. long, lenticular, golden yellow, substipitate, apiculate, jointed with the deciduous style. Newman L., Spokane Co., July 9, 1916, *Suksdorf* 8,776.

Carex comosa Boott. Tufted, the stem stout, 5–15 dm. high, rough or smooth, sharply angled; leaves broad, 6–15 mm. wide, scabrous on the margins, nodulose, scabrous, the sheaths nodulose; bracts leafy, exceeding the culm; spikes 4–6, drooping, on slender peduncles, the uppermost staminate, linear, brownish, 3–9 cm. long, often partly pistillate; pistillate spikes pale green, cylindric, about 1.5 cm. thick, densely flowered; scales brownish, lance-linear awned, about as long as the perigynia; perigynia spreading, firm, lance-ovate, 5–7 mm. long, strongly many-nerved, attenuate into a long 2-toothed beak, the teeth 1–2 mm. long and spreading. Usually in shallow water, *A. T.*

Carex concinnoides Mack. Stoloniferous; stems 1–4 dm. long, erect, later arching; leaves numerous, firm, pale, strongly striate, scabrous, shorter than the stems, 2–4 mm. broad; stem leaves bladeless or with very small blades: bract short; pistillate spikes 1 or 2, 5–10 mm. long, nearly sessile, few-flowered; staminate 8–20 mm. long, with purple ovate obtuse scales; scales ovate-lanceolate, acute, purple, hyaline-margined, shorter than the perigynia; perigynia oval, shortly stipitate, abruptly short-beaked, loosely pubescent, 2.5–3 mm. long. Common in open woods, *A. T. T.*

The only known species with 4 stigmas and a 4-sided achene.

Carex Cusickii Mack. Densely tufted, the slender stems 7–12 dm. high; leaves narrow, 2–6 mm. broad, mostly shorter than the stems, rough margined; spikes 6–20, brown, ovoid, 3–6 mm. long, in elongated often nodding usually branched inflorescences, 5–12 cm. long; terminal florets staminate; scales brown, ovate, acuminate-cuspidate, with a thick, scabrous midrib, as long as the perigynia; perigynia spreading, ovoid, stipitate, nerved near the base on both sides, 3–4 mm. long, the serrulate beak as long as the thick body. *C. diandra ampla* (Bailey) Piper. In boggy places, *A. T.*

Carex diandra Schrank. Plant caespitose, 3–7 dm. tall; rootstocks short, slender, blackish, fibrillose; sheaths tight, striate, ventrally thin and reddish-brown-dotted, prolonged and truncate or convex at the mouth; blades up to 3 dm. in length, 1–2.5 mm. wide, flat or canaliculate, light green, roughened on margins and toward apex; inflorescence 2.5–5 cm. long, 10 mm. wide, somewhat compound; spikes numerous, small, androgynous or pistillate, the lower more or less separated; scales nearly equalling perigynia, acute or cuspidate, brownish with hyaline margins and lighter midrib; perigynia 2–2.7 mm. long, 1 mm. wide, ascending or spreading, serrulate above, biconvex, brown, shining, few nerved dorsally, sharp edged, stipitate; stigmas 2, short, reddish brown; achenes 1 mm. long, lenticular, stipitate, apiculate, jointed with the short style. Spokane Co., Cheney, May 27, 1889, *Suksdorf* 3,046.

Carex disperma Dewey. Growing in clumps but producing long slender stolons; culms slender and weak, 1.5–6 dm. tall, scabrous at tip; leaves flat, 1–1.5 mm. wide, weak but scabrous; all or at least the lower spikes well separated; spikes with 1–5 perigynia below and 1–2 staminate flowers above; bracts none or minute; scales ovate, pointed, white hyaline nearly throughout, shorter than the perigynia; perigynia 2 mm. long, ovoid-elliptic, slightly flattened, with many fine nerves, abruptly beaked, the beak 0.2 mm. long. *C. tenella* Schk. Shady creek banks, Craig Mts., *St. John & Mullen* 8,632, *Can.*

Carex Douglasii Boott. Dioecious; rootstocks creeping, slender, glabrous, but scabrous on the margins; stems erect, 15–30 cm. high, smooth; leaves pale, 1–2.5 mm. wide, tapering, shorter than the stems; staminate spikes narrow, yellow to brownish; bracts slender; pistillate heads ovate-oblong, pale brown, 2–3 cm. long, composed of many crowded spikes; scales lanceolate, acute or cuspidate, much longer than the perigynia; perigynia 4 mm. long, lanceolate, narrowly margined, stipitate, acuminate, obscurely nerved, the serrulate beak as long as the body; stigmas 2, very long. Arid or alkaline spots, *U. Son.;* occasional, *A. T.*

Carex exsiccata Bailey. Rootstock short; culms stout, scabrous at tip, 3–10 dm. tall, purplish at base; leaf-sheaths nodulose and filamentose; blades

3–6 mm. wide, flat, scabrous; lower bracts leafy, longer than the inflorescence; spikes 4–7, the upper 2–4 staminate, narrow; pistillate spikes sessile or short stalked, the lower remote, 2–7 cm. long, 10–14 mm. wide, cylindric; scales lance-ovate, ½ as long as the perigynia; perigynia lanceolate, 7–10 mm. long, the body tapering into the smooth, bidentate beak. Wet places, *A. T.*

Carex festivella Mack. Tufted, the culms 3–10 dm. tall; blades flat, 2–6 mm. wide, shorter than the culm, smooth; spikes ovoid, 5–12 mm. long, 5–20 massed in an ovoid or broader head; bracts minute; scales ovate, obtuse or acute, dark chestnut-brown, shorter than the body of the perigynium; perigynia appressed, 3.7–5 mm. long, pale to brownish, ovate, tapering into a serrulate, shallowly bidentate beak, ⅓ as long as the body. Wet or dry slopes, *U. Son.* to *Can.*

Carex filifolia Nutt. *Wool grass.* Densely matted in extensive tufts; stems wiry, 5–30 cm. tall; leaves numerous, scabrous, filiform, about equaling the stem; inflorescence a solitary terminal spike, the upper portion staminate; scales broadly obovate, obtuse, longer than the perigynia, brownish with thin white margins; perigynium triangular-ovoid, 3 mm. long, pale below, darker at apex and very minutely puberulent, the short beak with an entire orifice. In dry soil, sometimes very abundant, *U. Son.,* occasional in *A. T.*

Exceedingly difficult to plow up. It is very palatable forage.

Carex Garberi Fern., var. **bifaria** Fern. Stems loosely tufted, from slender horizontal root-stocks; culm slender, nearly smooth, 3–40 cm. high; leaves flat, pale green, 2–4 mm. wide, usually overtopping the stems; spikes 3–6, all stalked, the uppermost staminate and slender, or rarely pistillate at base, the rest pistillate, narrowly cylindrical, loosely flowered, 4–20 cm. long; bracts mostly exceeding the stem; scales ovate brownish, shorter than the perigynia; perigynia nearly globose, 2–3 mm. long, very minutely beaked, smooth, yellow or reddish when mature, strongly nerved. In low meadows, rare, *A. T.*

Carex Geyeri Boott. Densely tufted, culms rough, 1–4.5 dm. high; leaves flat, rigid, very scabrous, 2–3.5 mm. wide, shorter or longer than the stems; inflorescence a single terminal straw-colored spike, the upper and larger portion of which is staminate; pistillate flowers several, usually but one maturing; scales pale or rusty, elongated, acuminate, exceeding the perigynia; perigynium smooth, pale, 6 mm. long, oblong, with a very short entire beak, 1-nerved on each side. Very abundant on dry hillsides, *A. T., A. T. T.*

Carex Haydeniana Olney. Densely tufted, the culms 1–6 dm. tall, nearly smooth; leaves 2–3 mm. wide, flat, soft, much shorter than the culms; spikes 4–7, ovoid to globose, 5–9 mm. long, massed in an ovoid or globose head; bracts minute; scales ovate-lanceolate, brown, as long as the body of the perigynium; perigynia ovate-lanceolate, 4.5–5.5 mm. long, dark tinged, especially on the bidentate beak, ½ as long as the body. *C. nubicola* Mack. Wet woods in the Blue Mts., *Huds.*

Carex Hoodii Boott. Slender, erect, tufted, smooth below, 3–6 dm. high; leaves many, pale, 1.5–3.5 mm. broad, shorter than the stems; spikes about 6, densely crowded into an ovoid or oblong head, 1–2 cm. long; scales lance-ovate, acuminate, scarious margined, brownish, as long as the perigynia; perigynia spreading, lanceolate, rather thick, 4–5 mm. long, dark when mature, obscurely nerved, narrowly wing-margined, somewhat stipitate at base, narrowed into a rough-margined beak half as long as the body. In moist places, not common, *A. T. T.*

Carex interior Bailey. Plant 1.5–5 dm. tall, densely caespitose; sheaths tight, hyaline ventrally, thin, concave at mouth; blades 1–3 dm. long, 1–3 mm. wide, flat or canaliculate, thin, yellowish green, roughened on margins and toward apex; inflorescence of 2–4 (–6) spikes, the terminal gynecandrous 5–10 mm. long, 4 mm. wide, the staminate 5–10 mm. long, 5 mm. wide, the lower pistillate and about 4 mm. long and wide; scales obtuse, yellowish brown with white hyaline margins and 3-nerved green center, the staminate more acute; perigynia 2.2–3.2 mm. long, 1.5–2 mm. wide, 1–10 on a spike, radiating, concave-convex, subcoriaceous, olive green, stramineous, or light brownish, nerved dorsally, sharp edged, the lower third spongy, substipitate; stigmas 2; achenes 1.5 mm. long, 1.2 mm. wide, lenticular, in upper part of perigynium, stipitate, apiculate, joined with the deciduous style. Spokane Co., Spangle, June 27, 1884. *Suksdorf* 3,017.

Carex Kelloggii W. Boott. Stems tufted, slender, 1.5–7 dm. tall, erect, scabrous at tip; leaves numerous, flat, 2–3 mm. wide, commonly as long as the stem, sometimes longer; spikes 4–15, the bracts commonly exceeding the stem; staminate spike usually single, slender, purplish; pistillate spikes 3 to many, sessile or the lower stalked, greenish, dense, 1.5–4 cm. long, 3 mm. thick; scales green with purple margins, elliptic, shorter than the perigynia; perigynia promptly deciduous, 2.5 mm. long, smooth, flattened, 3- or 4-nerved on the outer face, stipitate at the base, short-beaked, the beak emarginate. On wet stream banks and lake shores. *A. T.* to *Can.*

Carex laeviculmis Meinsh. Glabrous throughout, tufted; stems very slender, 2–6 dm. high; leaves numerous, flat, soft, 1–2 mm. broad, shorter than the culms; spikes 3–8, sessile, 3–6 mm. long, the lower ones scattered; bract solitary, usually shorter than the inflorescence; scales ovate, hyaline except the midrib, shorter than the perigynia; perigynia spreading when mature, shortly stipitate, strongly curved, flat above, convex beneath, pale green, 2.5–4 mm. long, the body elliptic-ovate, the beak ⅓ as long or less, faintly 7-nerved on each face. Moist places in woods, *A. T. T., Can.*

Carex lanuginosa Michx. Stems erect, slender, 30–60 cm. tall, somewhat tufted from long, stout rootstocks; basal sheaths reddish, often filamentose; leaves flat, 2–4 mm. wide, as long or nearly as long as the stem, nodulose, scabrous; bracts long and leafy; staminate spikes 1–3, sometimes pistillate at base; pistillate spikes 1–3, mostly long-peduncled, cylindric, 1–5.5 cm. long; scales lanceolate, acuminate, reddish brown, shorter or longer than the perigynia; perigynia 2.5–3.5 mm. long, oval, densely hispidulous, with a short 2-toothed beak. Swamps, not rare, *U. Son., A. T.*

Carex lasiocarpa Ehrh. Plant 3–12 dm. tall, loosely caespitose and stoloniferous; rootstocks long; lower sheaths long, filamentose; blades 7 dm. long, 2 mm. or less in width, flattish at base, involute above, thinnish, septate-[illegible] and soon drying; sheaths [illegible] thick, 1–3, cylindric; scales acute to cuspidate or awned, the staminate light reddish brown with dull hyaline margins and lighter center, the pistillate purplish brown with dull hyaline margins and 3-nerved green center; perigynia 15–50 to a spike, 3–5 mm. long, 1.7 mm. wide, ascending, suborbicular, inflated, coriaceous, dull brownish green, densely puberulent, obscurely ribbed; stigmas 3, blackish; beak 1 mm. long, bidentate; achenes 1.7–2 mm. long, 1.5 mm. wide, triangular with concave sides and blunt angles, loosely enveloped, yellowish brown, punctate, bent-apiculate, jointed with the straight or flexuous style. Spokane Co., Cheney, May 27, 1889, *Suksdorf* 2,887.

Carex leptalea Wahl. Very slender, tufted plant, the rootstocks short and slender; culms nearly smooth, filiform, 2–6 dm. tall, usually longer than the leaves; leaves flat or folded, slender, 0.5–1.2 mm. wide; scapes with single, terminal spike; bract none; spike narrowly cylindric, staminate at tip, 4–15 mm. long; scales ovate, pointed, brown, with a median green stripe, ½ as long as the perigynia; perigynia elliptic, 2.5–4.2 mm. long, green, finely many nerved, rounded and beakless. Wet shaded stream bank, Dartford, *Turesson* 68, *A. T. T.*

Carex leptopoda Mack. Rootstocks short, plant often tufted; stems erect, slender, 3–7.5 dm. tall, scabrous at tip; leaves shorter than the stems, flat, 2.5–5 mm. wide, smooth but scabrous margined; spikes staminate at base, 4–7, the lower scattered, oblong, 5-10 mm. long; scales ovate, acute, thin margined, shorter than the perigynia; perigynia ovate-lanceolate, 3.5–4 mm. long, light green, stipitate at base, obscurely nerved, the shallowly bidentate beak about ⅓ as long as the body. Meadows or wet woods, *Can.*

Carex Macloviana d'Urv., var. **pachystachya** (Cham.) Kükenth. Stems tufted, 3–7 dm. tall; leaves flat, flaccid, 1–4 mm. broad, smooth, shorter than the stems; inflorescence a dense oblong head 10–24 mm. long, composed of 3–12 crowded sessile brownish spikes; scales ovate, acutish, reddish brown, shinning, equaling the body of the perigynia; perigynia spreading, ovate-lanceolate, flat, 3.5–5 mm. long, the beak bidentate, serrulate on the thin margins. *C. pachystachya* Cham. Common in wet meadows, *A. T., A. T. T.*

Carex microptera Mack. Stems tufted, 5–10 dm. tall, smooth; leaf-blades flat, 2–3.5 mm. wide, smooth, shorter than the stem; spikes 5–10, closely aggregated, ovoid, 5–8 mm. long, staminate at base; scales ovate-lanceolate, acute, shorter than the perigynia, brown with a light midvein; perigynia 3.5–4 mm. long, broadly lance-shaped, tapering into a serrulate beak ⅓ as long as the body. Marshy places, *Can.;* and Latah Cr., Spokane Co., *A .T. T.*

Carex multicostata Mack. Caespitose, the culms 3–9 dm. tall, rather stout, many striate, slightly roughened at apex, much exceeding the leaves; lowest leaves bladeless; well developed leaves 3–4 to a culm, on its lower fourth, but not crowded, blades 1–3 dm. long, 2.5–6 mm. wide, flat; spikes 5–10, closely aggregate, but distinguishable; spikes 6–16 mm. long, 6–9 mm. wide, oblong-ovoid, with 10–30 perigynia appressed in many rows, but their tips not conspicuous; staminate flowers inconspicuous; lowest 1–few bracts prolonged, the others scale-like, ovate, obtuse or acute, light reddish brown, with light, 3-nerved center, and hyaline margins; perigynia 3.5–5.5 mm. long, plano-convex, the body ovate, green to straw-colored, thickish, winged to the base, serrulate to below the middle, the nerves many dorsally, few ventrally, abruptly contracted into a broad, flat, bidentate beak; achenes lenticular substipitate. Indian Corral Spring, Blue Mts., *Darlington* 65.

Carex nebrascensis Dewey. Stems 2.5–10 dm. tall, from stout horizontal rootstocks; leaves pale, flat, scabrous, 3–8 mm. broad, shorter than the stems, the sheaths nodulose; inflorescence of from 3–6 spikes, [illegible] [illegible] lanceolate, purple, with a green midrib, usually shorter than the perigynia; perigynia oval or obovate, flattened, 3–3.5 mm. long, brownish, resinous-dotted, short-beaked, the beak bidentate. Wet places, frequent, *A. T.*

Carex petasata Dewey. Culms erect, tufted, 3–8 dm. tall, scabrous above; leaves flat, 2–4 mm. wide, shorter than the stems; spikes 3–6, ovoid or oblong, pointed at each end, 2 cm. long, all pistillate at tip, pale, sessile, distinct but usually close together; scales brownish, ovate, acuminate, about as long as the perigynia; perigynia smooth, broadly lanceolate, 6–7 mm. long, the broad beak exceeding the body in length. *C. Liddoni* Boott. Grassy slopes, *U. Son., A. T.*

Carex praegracilis W. Boott. Stems crowded, slender, erect, scabrous, 2–6 dm. high, from stout scaly black rootstocks; leaves pale, 1–3 mm. broad, scabrous margined, shorter than the stems; inflorescence narrow, 2–4 cm. long, of about 6–12 crowded spikes; spikes brown, ovate, 4–7 mm. long, scales ovate, acute to cuspidate brown, hyaline-margined, as long as the perigynia; perigynia stipitate, broadly ovate, beaked, bidentate, narrowly margined, minutely serrulate, 3–4 mm. long. This species tends to be dioecious and is sometimes truly so. *C. marcida* Boott; *C. camporum* Mack. Usually on alkaline flats, scablands near the western border, *U. Son.*

Carex Preslii Steud. Densely tufted; culms erect, 2.5–7 dm. tall, scabrous at tip; leaves flat, 1.5–4 mm. wide, soft; spikes 3–8, the upper aggregated, the lower usually separated; spikes 5–8 mm. long, ovoid or broader; bracts minute; scales ovate, pointed, reddish brown with a green median stripe, shorter than the perigynia; perigynia ovate, 3.5 mm. long, the brownish, bidentate, serrulate beak nearly ⅓ as long as the body. *C. multimoda* Bailey. Open ridges in the Blue Mts., *Huds.*

Carex prionophylla Holm. Stoloniferous, in loose clumps; stems slender, 60–100 cm. high, wing-angled and rough near the top; sheaths reddish, the lower bladeless; leaves flat, scabrous, bright green, 4–5 mm. wide, as long as the stems; bract 5–8 cm. long, not equaling the inflorescence; spikes 3–4, sessile, the terminal staminate, 15–25 mm. long; pistillate spikes oblong, 1–2 cm. long; scales blackish purple, oblong-ovate, nearly as long as the perigynia; perigynia 2.5 mm. long, obovate, narrowed at base, abruptly short, black, emarginate, beaked. In moist woods, Mt. Spokane [Mt. Carlton] *Kreager* 264, *A. T. T.*

Carex retrorsa Schwein. Tufted; stolons short; stems stout, smooth, 4–10 dm. high; leaves 4–10 mm. broad, longer than the stem, nodulose, scabrous; the sheaths pale, nodulose; bracts leaf-like; staminate spikes usually 1–4, slender, 2–5 cm. long, sometimes only part of the terminal spike staminate, or the upper 1–3 staminate, linear; pistillate spikes 3–8, usually close together, sessile or nearly so, oblong cylindric, 2–5 cm. long; scales lanceolate, as long as the body of the perigynium; perigynia, thin, much inflated, 8–10 mm. long, ovoid, attenuate into a long beak, strongly nerved, reflexed. Wet places, *A. T., A. T. T.*

Carex Rossii Boott. Densely tufted, the stems 5–40 cm. tall; leaves pale, flat, scabrous, 1–3 mm. wide, commonly equaling the stems; bracts leafy, overtopping the spikes; inflorescence of 1–4 spikes, the uppermost staminate; staminate spike slender, 2–3 mm. long, pale, long-stalked; pistillate spikes 1–3, distinct, loosely few-flowered; scales purple with hyaline margins, or greenish, ovate, sharply acuminate, shorter than the perigynia; perigynia 3–4.5 mm. long, puberulent, the body globose, slender stipitate, with a 2-toothed beak. In stony soil, in the mountains, *Can.,* also occasional in *A. T.*

Carex rostrata Stokes. Stems stout, spongy at base, scabrous, 3–10 dm. [illegible] nodose [illegible] stem; staminate spikes [illegible] top; pistillate spikes 2–6, peduncled, cylindric, green or brownish, 4–10 cm. long, often staminate at the top; scales lance-oblong, purple, mostly pointed, shorter than the perigynia; perigynia 5–6 mm. long, ovoid-conic, inflated, shining, gradually contracted into the beak, strongly nerved. *C. utriculata* Boott. In wet places *A. T.*

Carex saximontana Mack. Caespitose; culms 1–2.5 dm. tall, weak, sharply 3-angled, winged towards the tip; leaves rather stiff, basal, 2–5 mm. wide, somewhat scabrous, much longer than the culms; lateral spikes on thread-like basal peduncles, hence all the stalks appearing to have but 1 spike; terminal spike on a stout peduncle 8–30 cm. long; bracts leafy; spikes staminate at tip, the rhachis stout and winged; pistillate flowers 2–5; lower scales leafy, broadened at base; perigynia 5 mm. long, stipitate, short beaked, the body obovoid to suborbicular; achene brown, tapering to the base, rounded at tip. Moist thickets, Spangle, *Suksdorf* 8625, *A. T.*

Carex Sheldonii Mack. Strongly stoloniferous; culms nearly smooth, 6–9 dm. tall; leaves 5–6 mm. wide, scabrous, flat, shorter than the culm, sparingly pilose beneath, and the sheaths pilose; staminate spikes 2–4; pistillate spikes usually 2, widely separated, nearly sessile, cylindric, 2–5 cm. long; lowest bracts sheathing, longer than the inflorescence; scales ovate-lanceolate, thin, aristate; perigynia lanceoloid, 5-6 mm. long, ribbed, sparsely pilose, gradually tapering into a stout, bidentate beak, ⅓ as long as the body. Wet places, Pullman, *Hunt* 283, *A. T.*

Carex stipata Muhl. in Willd. Stems numerous, tufted, 3-10 dm. tall, sharp-angled; leaves flat, 4–8 mm. wide, shorter than the stem; sheaths somewhat rugulose on the inner side; inflorescence a dense ovate head of many crowded spikes, 3–10 cm. long, straw-colored or olivaceous, commonly exceeded by the slender lowest bract; scales ovate-triangular, thin margined, as long as the body of the achene; perigynia spreading, 4–5 mm. long, lance-shaped, truncate, then stipitate at base, many-nerved, tapering into a long stout 2-toothed beak, whose margins are serrulate. Moist woods and meadows, *A. T.*, *A. T. T.*

Carex subfusca W. Boott. Caespitose; rootstocks short, blackish, fibrillose; culms 2–6.5 dm. tall, slender, nearly smooth; 3–5 blades on a fertile culm; blades 7–15 cm. long, 1.5–3.5 mm. wide, roughened near the attenuate apex; the 4–12 spikes 4–10 mm. long, ovoid to oblong, with 8–24 appressed ascending perigynia; lowest bract a cusp shorter than the head; scales ovate, acute, with light midvein and narrow hyaline margins; perigynia 3–3.5 mm. long, plano-convex, thickish, the body ovate, the serrulate beak half as long, bidentate; achenes 1.2 mm. long, short oval, apiculate. Lake Chatcolet, and Plummer, Ida., *Daubenmire* 38207.

Carex vesicaria L. Stolons short; culms 3–10 dm. tall, rather slender, sharp angled and very rough; leaves pale, flat, 3–6 mm. wide, shorter than the stem or little exceeding it, more or less nodose, reticulate, scabrous, lower sheaths fibrose-margined; inflorescence of from 3–6 spikes, overtopped by the long bracts; staminate spikes 1–4, 2–4 cm. long, sometimes pistillate at top; pistillate spikes 1–3, yellow or brownish, thick cylindric, 2.5–7 cm. long, short-stalked or sessile; scales pale, ovate, acuminate, much shorter than the perigynia; perigynia 5–8 mm. long, ovoid, turgid, strongly nerved, with a stout 2-toothed beak. *C. monile pacifica* Bailey. Wet places, common, *A. T.*, *A. T. T.*

CYPERUS

Annuals or perennials; stems simple, triangular, leafy near the base, and with one or more leaves at the summit which form an involucre for the simple or compound umbellate or capitate inflorescence, rays of the umbel sheathed at the base, usually very unequal, one or more of the heads or spikes commonly sessile; spikelets flat or roundish, few- to many-flowered; scales concave, 2-ranked, all but the lower one flower-bearing; flowers perfect; perianth none; stamens 1–3; stigmas 2–3; achene lenticular or triangular. (*kupeiros,* the ancient Gr. name.)

Rhachilla of spikelet not winged,
- Scales 2–3 mm. long, with recurved aristate tips; annuals. *C. aristatus.*
- Scales without recurved tips,
  - Scales 3-nerved, 0.8 mm. long; annuals from fibrous roots. *C. acuminatus.*
  - Scales 11-nerved, 2–2.5 mm. long; perennials from corms. *C. filiculmis.*

Rhachilla of spikelet winged,
  Wing of rhachilla separating into scale-like pieces; annual. *C. erythrorhizos.*
  Wing of rhachilla entire and persistent; perennials,
    Spikelets except the 2 lower scales deciduous; stems tuberous at base. *C. strigosus.*
    Spikelets with deciduous scales but a persistent rhachilla; scaly rootstocks present. *C. esculentus.*

Cyperus acuminatus Torr. & Hook. Smooth annual; culms tufted, slender, 7–40 cm. tall; leaves flat or folded, ribbed, shorter than the culm, less than 2 mm. wide; inflorescence of one or more sessile and a few short rayed heads; spikelets flat, ovate-oblong, 4–8 mm. long; scales ascending, greenish, oblong-lanceolate, achenes gray, pointed ellipsoid, trigonous. Rare on bank of the Snake River, *U. Son.*

Cyperus aristatus Rottb. Annual, smooth; stems 1–15 cm. tall, ascending; leaves flat, 0.5–2 mm. wide, often curved, about equaling the stems; bracts much exceeding the inflorescence; spikelets ovate-lanceolate, 3–6 mm. long, in dense heads, terminating the branches of an unequally rayed umbel, or rarely all aggregated into a single compound head; scales green, becoming brown, 7–11-nerved, all with strongly recurved awn-like tips; achenes gray, oblong-oblanceoloid, trigonous, 0.8 mm. long. On drying the plant develops a strong, pleasant, aromatic odor. *C. inflexus* Muhl. Abundant on sandy shores of the Snake River, *U. Son.;* occasional by streams, *A. T.*

Cyperus erythorhizos Muhl. Annual; stems usually 1.5–6 dm. tall sometimes much smaller; leaves flat, 2–8 mm. broad, shorter than the stem; involucral leaves 4–8, broad at base, far exceeding the inflorescence; spikelets bright chestnut, linear, 3–10 mm. long, densely crowded into flattened spikes, 1–3 cm. long, the latter umbelled on the unequal branches of the primary umbel; scales oblong, keeled, 1.5–2 mm. long, the green midrib prolonged into a short, mucronate tip, the brown membranous margin nerveless, wings of the rhachis separating to the base, forming pairs of small scales; achenes shining, white, oblong-ellipsoid, trigonous, 0.9 mm. long. Banks of Snake River, frequent, *U. Son.*

Cyperus esculentus L. *Nut Grass, Chufa.* Perennial, with horizontal, scaly, tuber-bearing rootstocks; stems 3–8 dm. tall; leaves flat, 4–8 mm. wide, exceeding the stem; involucral leaves as long as the others, far exceeding the [illegible] unequal, the [illegible] long, strongly 7-nerved, scarious margined, [illegible] short points; wing of rhachis narrow, not becoming divided into scale-like parts; achene obovoid, obtuse, trigonous. Banks of Snake River, Almota, *Piper* 2,651; and *Ownbey* 2,509, *U. Son.*

Cyperus filiculmis Vahl. Perennial from small, woody corms; scapes smooth, 1–6 dm. tall, longer than the leaves; leaves 3–4 mm. wide; umbel capitate or with 1–5 rays; spikelets flat, linear, 8–16 mm. long; scales firm, shining, oblong, mucronate, 3 mm. long; achene oblong, trigonous. *C. Bushii* Britton; *C. Houghtonii* Torr., var. *Bushii* (Britton) Kükenth. Sandy bank of Grande Ronde River, *St. John & Brown* 3833, *U.Son.*

Cyperus strigosus L. Perennial from a globose tuber; stems erect, 1–9 dm. high; leaves rough margined, flat, 4–6 mm. wide; umbel simple or compound, the longest rays 8–12 cm. long, their sheaths terminating in 2 bristles; spikelets linear, 8–19 mm. long, 7–15-flowered; scales pale, oblong-lanceolate, appressed, subacute, 2–4 mm. long, 5–7 nerved, green down the middle; achene linear-oblong, acute, trigonous, pale whitish brown, papillose, 1.6 mm. long. Sandy bars of Clearwater River, *Piper* 2661, *U. Son.*

ELEOCHARIS. Spike Rush

Annual or perennial; stems simple, triangular, quadrangular, terete, flattened or grooved, the leaves reduced to sheaths or the lowest very rarely blade-bearing; spikelets solitary, terminal, erect, several–many-flowered, not subtended by an involucre; scales concave, in a spiral; perianth of 1–12 bristles usually barbed, or none; stamens 2 or 3; achene 3-angled or biconvex; base of the style persistent on the summit of the achene forming a terminal tubercle. (Name Gr., *elos,* marsh; *charis,* grace.)

Tubercle confluent with the achene. *E. parvula.*
Tubercle not confluent with, instead, clearly delimited from the achene,
  Achenes obscurely trigonous to subterete; stigmas 3,
    Spikelets 3–6 mm. long; anthers 1 mm. long. *E. acicularis,* var. *occidentalis.*
    Spikelets 2–3 mm. long; anthers 0.4 mm. long. *E. bella.*
  Achenes lenticular; stigmas 2,
    Tubercle flattened; caespitose annuals,
      Tubercle narrower than the rounded top of the achene. *E. obtusa.*
      Tubercle as broad as the truncate top of the achene,
        Bristles about equaling the achene. *E. Engelmanni,* var. *monticola.*
        Bristles rudimentary or none. *E. Engelmanni,* var. *monticola,* forma *leviseta.*
    Tubercle not flattened, usually conical; stoloniferous perennials,
      Spikelet with 2–3 basal sterile scales,
        Culms 1–4 dm. tall; achenes 1.2–1.7 mm. long. *E. macrostachya.*
        Culms 5–19 dm. tall; achenes 1.4–2.1 mm. long. *E. palustris,* var. *major.*
      Spikelet with 1 clasping basal sterile scale,
        Median scales 3–5 mm. long. *E. uniglumis.*
        Median scales 1.8–3 mm. long. *E. calva.*

Eleocharis acicularis (L.) R. & S., var. **occidentalis** Svens. Perennial by creeping rootstocks, usually forming extensive mats; stems very slender, striate, 4–15 cm. tall, erect; sheaths reddish at base, loose and scarious at tip; scales rigid, brown, brown keeled; spikes 6–20-flowered, 3–6 mm. long; bristles none; achene oblong or obovoid, somewhat 3-angled, marked with 9–12 longitudinal ribs, with 40–50 cross-lines between them; tubercle broad, contracted at its junction with the achene. Wet places, *A. T.*

Eleocharis bella (Piper) Svens. Densely tufted annual; culms 2–6 cm. tall, light green, soft, capillary, sheaths loose, pale, acute, inflated at tip; spikelets 2–3 mm. long, ovoid, 8–many-flowered; scales ovate-lanceolate, 1–1.5 mm. long, green with purple striped sides; style 3-parted; achene yellowish or white 0.7–0.8 mm. long, linear obovate, obtusely 3-angled, with 3 primary and many secondary longitudinal ridges, the cross-lines about 30; bristles none.

E. acicularis bella Piper. First discovered by Prof. Piper in Pullman. Wet places, *A. T.*

Eleocharis calva Torr. Stoloniferous to slightly caespitose; culms 1–6.5 dm. tall, nearly filiform, terete; spikelet linear-lanceoloid to lance-ovoid, densely 40 or more flowered, 9–17 mm. long; basal sterile scale suborbicular, encircling the culm; fertile scales oblong to ovate, thin, reddish to pale brown; bristles 1–4, delicate, or none; achenes pyriform, 1–1.4 mm. long; tubercle conical, 0.2–0.4 mm. wide. *Heleocharis calva.* Palouse, *Beattie* 4,181, *A. T.*

Eleocharis Engelmanni Steud., var. **monticola** (Fern.) Svens. Tufted annual; stems usually erect, 1–4 dm. tall; sheaths thin and oblique at tip; spikelets pale brown, ovate-lanceoloid to oblong-lanceoloid, 5–16 mm. long; scales lance-ovate, 2–2.5 mm. long, the midrib green, bordered with brown, the margins white scarious; bristles retrorsely toothed, about as long as the achene; achene broadly obovate, lenticular, narrowed at base, brown, smooth, the body 1.1 mm. long; tubercle dark brown, deltoid, less than ¼ as long as the achene. Banks of Snake River, Almota, *Piper,* July 30, 1897, *U. Son.*

Eleocharis Engelmanni Steud., var. **monticola** (Fern.) Svens., forma **leviseta** (Fern.) Svens. Differs from var. *monticola* by having the bristles rudimentary or none. Wet places, mouth of the St. Joe River, Idaho, *Humphrey, Can.*

Eleocharis macrostachya Britton. Rootstocks extensively creeping; stems terete, firm; spikelet oblong, lanceoloid to ovoid, 5–20 mm. long; basal sterile scales 2–3, ovate, firm; fertile scales oblong ovate, reddish brown, with pale, scarious margins, the median 3–4 mm. long; bristles 4, usually exceeding the obovate smooth biconvex achene; tubercle lanceoloid to conic-ovoid, taller than broad, obtuse, contracted at its junction with the achene. Abundant at the edges of ponds or in shallow water, *U. Son.,* common in *A. T.*

Eleocharis obtusa (Willd.) Schult. Tufted annual; stems erect or ascending, 3–50 cm. high; sheaths purplish at base, oblique and firm at tip, apiculate; spikelets brownish, ovoid, obtuse, 2–13 mm. long; scales ovate, thin margined; bristles retrorsely toothed, 6–7, exceeding the achene; achene 1–1.5 mm. long, turbinate-obovoid, narrowed at base, brown, smooth; tubercle deltoid, acute, [illegible] nearly as wide as the achene. Wet shores, *U. Son., A. T.*

spikelets [illegible]
of streams or lakes, *A. T.*

Eleocharis parvula (R. & S.) Link. Densely tufted; roots fibrous, often with tuberous stolons; the stems capillary, becoming grooved, 2–7 cm. high; spikelet ovoid, greenish, bearing 2–9 florets; scales ovate; green to yellowish or brown acutish, the lowest usually larger, ½ the length of the spikelet; bristles usually 6, about as long as the achene; achenes obovoid trigonous, 1–1.4 mm. long, straw-colored, shining; style-base small, triangular, greenish. *Scirpus nanus* Spreng. In wet places, near Viola, Idaho, *Sandberg, MacDougal & Heller* 1052.

Eleocharis uniglumis (Link) Schult. Loosely stoloniferous or caespitose; culms 3–70 cm. tall, terete; spikelets lanceoloid to lance-ovoid, 3–17 mm. long, loosely 5–30-flowered; basal sterile scale suborbicular, castaneous, encircling the culm; fertile scales oblong-ovate, castaneous to purplish, lustrous; bristles none or delicate; achene ellipsoid to narrowly obovoid, 1.2–1.8 mm. long, yellow to brown, tubercle low conical, ½–¾ as wide as the achene. Marshy place, Pullman, *Hunter,* July 20, 1899, *A. T.*

HEMICARPHA

Annual, low tufted plants; stems and leaves erect or spreading, almost filiform; spikelets terete, terminal, clustered or solitary, subtended by a 1–3 leaved involucre; scales in a spiral, deciduous, all subtending perfect flowers; perianth of a single hyaline bract between the flower and the rhachilla; bristles none; stamens 1–2; style 2-cleft, deciduous, not swollen at the base; achene oblong, turgid or lenticular, the surface papillate. (Name Gr., *hemi,* half; *karphos,* straw.)

Hemicarpha micrantha (Vahl) Pax, var. **aristulata** Cov. Tufted, glabrous, the numerous stems 5–8 cm. tall, erect or spreading; leaves narrow, involute, the recurved blades 1–3 cm. long, as long as or longer than the sheaths; spikelets ovoid, obtuse, solitary or two in a cluster, 2–4 mm. long; involucral bracts 3, the uppermost 1–3 cm. long; the others much shorter; scales 1–1.5 mm. long, broadly obovate, the apex abruptly acuminate recurved; achene oblong, ovate or obovate, 0.5 mm. long, with a short beak, the surface gray, very minutely and regularly hexagonal-reticulate and rugose; styles shorter than the achene, 2-cleft half-way to the base; hyaline bract triangular-ovate. *H. aristulata* (Cov.) Smyth. Very inconspicuous plant of the sandy banks of Snake River, *U. Son.*

SCIRPUS

Annual or perennial; stems leafy or the leaves reduced to basal sheaths; spikelets terete (in ours), solitary, or in a terminal cluster, when it is subtended by a 1–several-leaved involucre; scales in a spiral, usually all fertile, or 1–3 of the lowest sometimes empty; flowers perfect; perianth of 1–6 often barbed bristles, or sometimes none; stamens 2–3; style 2–3-cleft, not swollen at the base; achene triangular, lenticular, planoconvex, or trigonous. (The Lat. name of the bulrush.)

Involucral bract 1, erect, appearing to be a continuation of the stem,
  Stem sharply triangular; spikelets 1–7, sessile. — *S. americanus,* var. *polyphyllus.*
  Stem terete; spikelets more numerous in a compound umbel,
    Achenes lenticular or plano-convex, brownish. — *S. validus.*
    Achenes trigonous, fuscous. — *S. heterochaetus.*
Involucral bracts several, spreading or ascending, leaf-like,
  Spikelets 15 mm. or more in length; achene lenticular. — *S. paludosus.*
  Spikelets 8 mm. or less in length,
    Spikelets numerous, massed in a few heads 6–10 mm. in diameter; achenes trigonous. — *S. pallidus.*
    Spikelets separate or few in the smaller heads of the freely compound inflorescence; achenes lenticular. — *S. microcarpus.*

Scirpus americanus Pers. var. **polyphyllus** (Boeckl.) Beetle. Perennial with rootstocks; stems sharply 3-angled, 1–11 dm. high; leaves 1–4, smooth, shorter than the stem, channeled; bract 2–10 cm. long, pointed; spikelets 1–7, ovoid, closely crowded, 8–12 mm. long; scales brown, ovate, 2-cleft at apex and often awned; bristles 2–6, shorter than the achene; style 3-fid; achene smooth, brown, planoconvex, oblanceolate, 3.5 mm. long. In wet places especially where somewhat alkaline, *U. Son., A. T.*

Scirpus heterochaetus Chase. Perennial by stout rootstocks; leaves reduced to basal sheaths; involucral leaf shorter than the inflorescence; culms terete, pithy, 2 m. or less in height; inflorescence a loose compound umbel, the primary rays as much as 1 dm. long; spikelets solitary, ovoid to ellipsoid, 8–15 mm. long; scales ovate, glabrous, brownish, and reddish spotted, 4 mm. long, short awned from the notched tip; bristles 2–4, unequal, shorter than the achene, downwardly barbed; achene fuscous obovoid, cuneate trigonous, 2.6–3 mm. long. Lake shores, n. Idaho, *A. T .T.*

Scirpus microcarpus Presl. Perennial, with scaly rootstocks; the stout stems 9–16 dm. tall; leaves 3–15 mm. broad, smooth beneath, rough on the margins and upper surface of the midvein, the upper leaf usually exceeding the stem; inflorescence a two to three times compound umbel, the primary rays 3–10 cm. long, unequal; bracts leaf-like, about equaling the inflorescence; spikelets ovoid, numerous, dark green, 3–5 mm. long, in heads of 3–12 or more; scales 2 mm. long, broadly ovate, acute; stamens 2; styles bifid; bristles 4, backwardly barbed, longer than the achene; achene 1.2 mm. long, white, oblong-obovate, very short-beaked. Common on the margins of ponds and streams, *A. T., A. T. T.*

Scirpus pallidus (Britton) Fern. Perennial from a short rootstock; culms stout, triangular, 9–13 dm. tall; leaves pale, flat, smooth, 6–15 mm. wide; spikelets oblong, in dense sessile or short rayed heads; scales ovate, greenish or pale, 1.5 mm. long, tipped with an awn half as long; bristles 6, downwardly barbed, about as long as the achene; achene straw-colored, ovate, 1 mm. long, short beaked. Banks of Touchet River, Waitsburg, *A. T.*

Scirpus paludosus A. Nels. Perennial with tuber-forming rootstocks; stem smooth, triangular, 3–7 dm. tall; leaves pale, smooth, 3–9 mm. wide; involucral leaves 2–3, elongate; spikelets ovoid to cylindric, drab to chestnut-brown, 15–25 mm. long, 3–10 in a dense head or with a few short rays; scales 6 mm. [illegible] from the notched tip; [illegible]

The name *S. campestris* Britton is a later homonym. As *S. paludosus* (18[illegible]) is the earliest name in the specific category for the plant with lenticular achenes, it must be maintained as the species, and the other variants made varieties.

Scirpus validus Vahl. *Tule.* Perennial from stout rootstocks, creeping beneath the mud; stems stout, terete, smooth, 1–3 m. tall; sheathed at base with bladeless leaf-sheaths; involucral leaf shorter than the umbel; rays 0–6 cm. long; spikelets solitary or in small clusters, 5–18 mm. long, ovoid to cylindric; scales ovate to orbicular, spotted, hairy along the middle, awned, equaling or longer than the achenes; bristles 4–6, equaling the achene, backwardly barbed; achenes plano-convex obovoid, brownish, 2–3 mm. long. *S. occidentalis* Chase; *S. acutus* Muhl.; *S. validus,* var. *creber* Fern. Fresh or alkaline shores, often in several feet of water, *U. Son, A .T.*

The strong pithy stems dry without rotting. They were used in basketry and matting by the Indians.

ARACEAE. Arum Family

Plants with acrid or pungent juice and simple or compound leaves; flowers crowded on a spadix, perfect, monoecious or dioecious; spathe present or none; perianth of 4–6 sepals or none; fruit a berry or utricle, 1–several-celled, each 1–several-seeded. (Named for the genus *Arum*).

Spadix lateral; spathe leaflike, continuing the scape; leaves linear. *Acorus.*
Spadix terminal, with an elliptic, sheathing spathe; leaves petioled and with elliptic or oblanceolate blades. *Lysichiton.*

ACORUS

Perennial herbs from stout rhizomes; leaves basal, linear, erect; flowers perfect, in a cylindric spike; perianth of 6 segments; stamens 6; ovary 2–3-celled; fruit obpyramidal, hard, dry, but within gelatinous, 1–3-seeded. (Name Lat., *acorus,* an aromatic plant.

Acorus Calamus L. *Sweet Flag.* Rhizome 2–3 cm. thick, strongly and pleasantly aromatic; leaves 0.5–1.5 m. long, 7–20 mm. wide, ligulate, firm, thick and pithy, the midrib offcenter; the spathe 2–6 dm. long; spadix 5–10 cm. long, 1.2–2 cm. thick; flowers green or yellowish green, 3 mm. long, obovate; filaments membranous winged; fruit a reddish berry, rarely produced.
Eurasian, doubtless introduced, Newman L., July 21, 1948, *Yocom.*

LYSICHITON

Acaulescent swamp herbs with large leaves from a thick vertical rootstock; spathe sheathing at base, with or without a broad colored lamina, at first enveloping the cylindrical spadix which later becomes long-exserted on a stout peduncle; flowers perfect, crowded, covering the spadix, perianth 4-lobed; stamens 4, opposite the perianth-segments; ovary 2-celled, 2-ovuled, fruit fleshly, somewhat immersed in the rhachis and coalescent. (Name Gr., *lusis,* release; *chiton,* tunic).

Lysichiton americanum Hultén & St. John. *Yellow Skunk Cabbage.* Leaves large, 3–15 dm. long, oblanceolate or elliptic, acutish, narrowed at base into a short margined petiole; spathe golden yellow, the blade oblong, acute, narrowed into a sheathing petiole; peduncle stout, 20–30 cm. long; spadix cylindric, in fruit 5–14 cm. long. *Lysichitum americanum* Hultén & St. John. Open swamps or wet woods, *A. T.* to *Can.* Only in northern part of area.
The brilliant malodorous flowers appear before the leaves.

LEMNACEAE. Duckweed Family

Very small thallose plants floating free on the water, propagating by the division of the thallus; flowers 1–3, monoecious, from the edge or upper surface, rare; fruit a 1–7-seeded utricle.

(Named for the genus *Lemna.*)

Thallus 1–5-nerved, with a single rootlet. *Lemna.*
Thallus 5–12-nerved, with several rootlets. *Spirodela.*

LEMNA. Duckweed

Thallus flattened; flowers produced from a cleft in the margin of the thallus, usually 3 together, surrounded by a spathe; two staminate, each of a single stamen, the other pistillate of a simple pistil; anthers 2-celled, didymous, dehiscing transversely; ovary 1-celled; utricle ovoid. (Name an ancient Gr. one).

Thalli rounded, separate or soon separating. *L. minor.*
Thalli oblong-lanceolate, stipitate, remaining connected. *L. trisulca.*

Lemma minor L. Thalli round to elliptic-ovate, 2–5 mm. long, thick, green, very obscurely 3-nerved; seeds oblong-obovate, amphitropous, with prominent operculum. Common floating in ponds and slow streams, very rarely fruiting, *A. T.*

Lemma trisulca L. Thalli thin, 3-nerved, 6–13 mm. long, usually several remaining connected, seeds ovate, amphitropous, with a small round operculum. Submerged near Potlatch, Idaho., *A. T. T.*

SPIRODELA

Very similar to *Lemna,* but rootlets several, with axile vascular tissue; anther-cells divided by a vertical partition and dehiscing longitudinally. (Name Gr. *speira,* cord; *delos,* evident.)

Spirodela polyrhiza (L.) Schleid. Thallus round-obovate, 3–8 mm. long, thick, purple and rather convex beneath, dark green above, palmately, mostly 7-nerved. *Lemna polyrhiza* L. Floating on still water, *A. T.*

from the 3-celled ovary; stamens 3 or 6, unequal or dissimilar, on the throat of the perianth; style 1; fruit a 1–3-celled capsule or 1-celled utricle. (Named for the genus *Pontederia.*)

HETERANTHERA

Low herbs living in mud or shallow water, with a 1–few-flowered spathe bursting from the sheathing side or base of a petiole; perianth-lobes subequal; stigma 3-lobed; capsule 1- or incompletely 3-celled. (Name Gr. *hetera,* different; *anthera,* flowering.)

Heteranthera dubia (Jacq.) MacM. *Water Stargrass.* Stems slender, branched, leafy, 30–100 cm. long; leaves sessile, linear, acute; spathe terminal,

1-flowered; flowers small, pale yellow, the tube very slender, 3–6 cm. long; the lobes linear 8–12 mm. long; capsule 1-celled, 6–8 mm. long. Rare, aquatic, Marshall Junction, *Suksdorf, A. T.*

JUNCACEAE. Rush Family

Perennial, grass-like, usually tufted herbs; inflorescence a compound panicle, corymb, or umbel, with the flowers singly or loosely clustered or aggregated into spikes or heads (rarely reduced to a single flower); flowers small, regular, with or without bractlets; perianth 6-parted, the parts glumaceous; stamens 3 or 6; pistil superior, tricarpellary; ovary 3-celled, or 1-celled with basal, axial, or 3 parietal placentae; ovules 3–many; stigmas 3; fruit a loculicidal capsule; seeds 3–many, small. (Named for the genus *Juncus.*)

Leaf-sheaths open; capsule 1- or 3-celled, many-seeded; placentae parietal or axial. *Juncus.*
Leaf-sheaths closed; capsule 1-celled, 3-seeded; placentae basal. *Luzula.*

JUNCUS. Rush

Annual or perennial plants; stems leaf-bearing or scapose; leaves glabrous, terete, grass-like, or channeled; inflorescence a panicle or corymb, often one-sided, bearing its flowers either singly and with two bractlets or in heads and without bractlets, but each head in the axil of a bract; stamens 6, rarely 3; ovary 1-celled or 3-celled; placentae parietal or axial; seeds several to many. (Name from Lat., *jungere,* to join.)

Lowest bract of inflorescence terete, appearing like a continuation of the stem; the inflorescence appearing lateral,
  Flowers 1–3,
    Inner sheaths bristle-tipped; capsule retuse. *J. Drummondii,* var. *longifructus.*
    Inner sheaths with blades; capsule acuminate. *J. Parryi.*
  Flowers more numerous, in panicles,
    Plants forming large tussocks; stamens 3. *J. effusus,* var. *caeruleimontanus.*
    Plants forming running rootstocks; stamens 6,
      Anthers longer than the filaments; main bract usually much shorter than the stem. *J. balticus,* var. *montanus.*
      Anthers equaling the filaments; main bract as long as the stem. *J. filiformis.*
Lowest bract not so; inflorescence terminal,
  Leaves equitant, the blades gladiate,
    Heads few flowered, semiglobose or smaller, usually numerous; flowers pale (or occasionally brown). *J. xiphioides,* var. *montanus.*
    Heads many flowered, globose or nearly so, few,

dark,
Heads brown. Var. *macranthus.*
Heads brownish black. Var. *triandrus.*
Leaves not gladiate, not truly equitant,
Annuals: roots fibrous,
Capsule oblong-ovoid, 3–5 mm. long. *J. bufonius.*
Capsule subglobose to ovoid, 2–3 mm. long. *J. sphaerocarpus.*
Perennials: rootstocks clustered or creeping,
Blades flat and thin, or channeled above.
Flowers attached singly, each 2-bracted; plants tufted,
Auricles cartilaginous, opaque, yellow. *J. Dudleyi.*
Auricles membranous, whitish,
Capsules oblong, 3-celled, about equaling the erect perianth,
Perianth segments straw-colored, 4–5 mm. long, the outer scarious only at base. *J. brachyphyllus.*
Perianth segments fuscous, 3.5–4 mm. long, scarious margined to the apex. *J. confusus.*
Capsules ovoid, 1-celled, not over ¾ length of the spreading perianth. *J. macer.*
Flowers 1-bracted, in heads; plants stoloniferous,
Auricles present on sheaths; perianth smooth. *J. longistylis.*

Auricles wanting; perianth rough,
Perianth greenish or straw-colored. *J. orthophyllus,* f. *orthophyllus.*
Perianth dark brown. Forma *congestus.*
Blades terete,
Filaments much longer than the anthers, cross partitions not prominent in blades. *J. Mertensianus.*
Filaments shorter than the anthers; cross

shorter than the inflorescence,
Stamens 3; anthers shorter than the filaments. *J. acuminatus.*
Stamens 6; anthers longer than the filaments,
Flowers dark brown; panicle loose. *J. nevadensis,* var. *nevadensis.*
Flowers pale brown; heads aggregated into several clusters. Var. *columbianus.*

Juncus acuminatus Michx. Rootstocks tufted; stems 2–10 dm. tall; leaves basal and cauline; sheaths striate, ribbed above; blades terete with the cross ribs prominent, 10–20 cm. long; panicle loose with 2–50 dense, often spherical heads; perianth segments lanceolate, acuminate, green to brownish, 3–5 mm. long; capsules lance-ovoid apiculate. Sandy beach, Newman Lake, *Turesson* 49, *A. T.*

Juncus balticus Willd., var. **montanus** Engelm. Stems, terete, naked, 2–8 dm. tall, from stout horizontal rootstocks; leaves basal, the outer bladeless, the inner with a bristle; leaf-sheaths dull or shining brown or yellow, the inner 3–11 cm. long; panicle apparently lateral, the elongated bract appearing like a continuation of the stem; panicle simple or compound, 1–10 cm. long; perianth segments lanceolate, acute or obtusish, green, brown margined, 3–5 mm. long, usually brown; capsule acutely angled and short-beaked; seeds oblong, the surface reticulated. *J. balticus* sensu Am. auth., not of Willd. Abundant in wet places, *U. Son., A. T.*

Juncus brachyphyllus Wieg. Stems tufted, stiff, erect, 2–5 dm. tall; leaves flat, or channeled, 1–2 mm. broad, commonly one-fourth to one-third the length of the culms; inflorescence many-flowered, short and rather crowded, usually exceeded by the bracts; perianth-segments subulate, acute; capsule oblong, the triangular apex usually obtuse. Blue Mountains, and Craig Mountains, Idaho, *A. T. T.*

Juncus bufonius L. *Toad Rush.* Annual, branching from the base, 2–35 cm. tall; leaf-blades flat or involute, 1 mm. or less in width; flowers loosely scattered, usually on but one side of the long branches of the panicle; perianth segments pale, scarious-margined, lanceolate, acuminate, 4–7 mm. long; stamens 6; capsule with a short blunt beak; seeds very finely reticulate. Forming pure stands, wet or exsiccated mud, *U. Son.* to *A. T. T.*

Juncus confusus Cov. Pale green, tufted, erect, the culms 40–50 cm. high; leaves very narrow, flat or involute, one-half to two-thirds as long as the culms; inflorescence dense, 5–20 mm. long, exceeded by the bracts; perianth segments straw-colored with 2 brown stripes, lanceolate, acutish, 3.5–4 mm. long, capsule oblong, the triangular apex retuse. Open moist place, *A. T., A. T. T.*

Juncus Drummondii E. Mey., var. **longifructus** St. John. Tufted; stems slender 1–3 dm. high; bract as long as or longer than the inflorescence; perianth segments acute or acuminate, brownish, the outer longer, 4–6 mm. long; capsule brown oblong retuse, 5.5–7 mm. long; seeds tailed. Blue Mountains, *Piper 2274, Huds.*

Juncus Dudleyi Wieg. Tufted, stiffly erect, 2–12 dm. tall; leaves basal, the sheaths usually less than half as long as the stems; blades usually channeled, very narrow, 4–17 cm. long; panicle rather close, 2–7 cm. long; perianth segments spreading, pale green, lanceolate, 4–5 mm. long, slightly exceeding the 1-celled ovoid capsule; stamens 6. Common in low ground, *U. Son* to *A. T. T.*

Juncus effusus L., var. **caeruleimontanus** St. John. Perennial; stems terete pithy strong, 1.3–2.2 mm. thick, 6–7.5 dm. tall; leaves all basal, the blades reduced to bristles, the sheaths dull brown, firm, close, the inner 4.5–7.5 cm. long; inflorescence loose, 1-sided, up to 5 cm. long; perianth segments linear-lanceolate, rigid, the middle green, the margins brown scarious, 2–2.8 mm. long, equaling the obovoid apiculate capsule. Known only from Tucannon River, *H. T. Darlington* 336. Swampy places, *A. T., A. T. T.*

Juncus filiformis L. Rootstocks slender; stems very slender, 15–60 cm. high, most of them usually sterile; sheaths obtuse, the blades small and bristle-like; inflorescence few-flowered, cymose, nearly simple; segments of the perianth green, sepals lanceolate, acute, 4–5 mm. long, petals broader, acute, 3–4 mm. long; capsule green, obovoid, shorter than the perianth; seeds thick, short-pointed at each end. Moist places, Lake Tesemini, [Spirit L.] Idaho, *Sandberg et al* 880, *Can.*

Juncus longistylis Torr. Stems 30–60 cm. tall; basal leaves flat, 5–25 cm. long, 1–2 mm. wide; cauline leaves 1–3, bract short; panicle consisting of 2–8 heads, these 3–12-flowered; bracts scarious, whitish; perianth segments broadly lanceolate, acute or acuminate, green flanked with brown, white scarious at edge, 5–6 mm. long; stamens 6; capsule oblong, 3-angled above, much shorter than the segments. Wet places, Marshall Jct. to Rock Lake, *A. T.*

Juncus macer S. F. Gray. Densely tufted, 1.5–6 dm. tall; leaves flat or somewhat involute, narrow, shorter than the stems; auricles whitish, 1–1.5 mm. long; panicle loose, seldom exceeded by the slender bract; perianth segments lanceolate, green with white scarious margins, 3–4.5 mm. long; capsule ovoid, thin-walled, rounded at the apex, 1-celled with 3 parietal placentae; seeds minutely reticulate. Open prairies, *A. T.*

Juncus Mertensianus Bong. Rootstocks short, branching; stems terete, weak, crowded, 10–30 cm. high; basal leaves reduced; cauline leaves nearly terete or somewhat flattened, 5–15 cm. long, about 2 mm. wide; ligules scarious; heads solitary, globose, dark brown, about 1 cm. broad; perianth segments lanceolate, acuminate 2.5–3.5 mm. long, exceeding the obovoid retuse apiculate capsule. Moist meadows, Blue Mountains, *Huds.*

Juncus nevadensis Wats., var. **nevadensis.** Stems terete, leafy, 15–85 cm. tall from running rootstocks; leaves slightly flattened, with few partitions, 2–3 mm. wide, 15–30 cm. long; ligules present; panicle loose, composed of numerous small dark brown hemispherical heads, these few-flowered; perianth segments lanceolate, acuminate, 3.5–4.5 mm. long, smooth; stamens 6; capsule oblong, abruptly acute, apiculate, usually shorter than the perianth; seeds infrequent. *J. Suksdorfii* Rydb. Common along streams, *A. T.*

Juncus nevadensis Wats., var. **columbianus** (Cov.) St. John. Differs only in having the flowers pale brown, and the heads commonly aggregated into several clusters. *J. columbianus* Cov. Growing with the species along streams, *A. T.* Though commonly kept as a species its characters are few and mere matters of degree. There are numerous intermediates. For instance, the sheet, Chatcolet, Idaho, June 25, 1904, *C. Crozier* has the pale flowers, but a loose inflorescence. Another, Pullman, Washington, *C. V. Piper* 3,026 has dark brown flowers, but the heads are congested. Other intermediates could be cited, but these substantiate the reduction of this plant to a variety.

Juncus orthophyllus Cov., forma **orthophyllus.** Perennial; stems leafy, 2–7 dm. tall; leaves flat, 2–5 mm. broad, pale green, shorter than the stems; ligules absent, flowers usually aggregated into 8–10-flowered heads, these panicled; perianth segments greenish or straw-colored, lanceolate, acuminate, scarious-margined, rough, the inner the longer, 5–6 mm. long; stamens 6; capsule 3-angled, oblong-ovoid, short-beaked. *J. latifolius* (Torr.) Buch., not Wulf; and var. *paniculatus* Buch. Common along streams, *A. T.* to *Can.*

[illegible] St. John. Differs only in having the perianth

[illegible] with terete blades 4–10 cm. long; bract much exceeding the inflorescence; perianth segments brown, acute, the outer 7 mm. long, slightly exceeding the inner; capsule slender, exceeding the perianth; seeds tailed. Open summit of Mt. Spokane, *Huds.*

Juncus sphaerocarpus Nees. Very similar to *J. bufonius* but the perianth 3–4 mm. long; the capsule rounder. Dried up ditches, *A. T.*

Juncus Torreyi Cov. Rootstock slender, with tubers; stems single, stout, 3–10 dm. tall; leaves mostly cauline; sheaths elongate, loose; auricles firm, 2–4 mm. long; blades divergent, stout, terete, 2–5 mm. thick; panicle often congested; heads 30–80-flowered, dense, spherical; perianth segments lanceolate, acuminate, the outer longer, 4–5 mm. long; stamens 6, 2.5 mm. long; capsule 1-celled, exceeding the sepals. Stream banks, Wawawai, *St. John* 9267; Pampa, *St. John et al.* 9697, *U. Son.*

Juncus xiphioides E. Mey., var. **macranthus** Engelm. Heads many flowered, globose or nearly so, brown. Wet places, Blue Mts., *Can.*

Var. **montanus** Engelm. Rootstocks stout, creeping; stems flattened, 2-edged, 1–6 dm. tall; leaves equitant, gladiate, the basal ones reduced, the cauline 5–20 cm. long, 2–6 mm. wide, the cross ribs partial, irregular; heads 2–12, panicled; perianth segments 2.5–3 mm. long; stamens 6; capsule oblong, mucronate. *J. ensifolius* Wiks., var. *major* Hook.; *J. saximontanus* A. Nels. Wet places, *U. Son.* to *Huds.*

Var. **triandrus** Engelm. Similar to var. *macranthus,* but heads brownish black; stamens usually three. *J. ensifolius* Wiks. Wet places, *A. T.*

LUZULA. Wood Rush

Perennials (as to our species), with glabrous herbage or commonly with the leaf base and the summit of the sheath long ciliate; stems leafy, leaf sheaths with united margins; blades grass-like; inflorescence umbellate, paniculate or congested into head-like clusters; flowers bracteolate; ovary 1-celled with 3 basal ovules. (Name Lat., a diminutive of *lux,* light, hence, little shining ones.)

Flowers congested into head-like clusters. *L. multiflora,* var. *comosa.*
Flowers solitary or in clusters of 2–3,
  Plant glabrous; perianth 3–3.5 mm. long. *L. glabrata.*
  Leaves ciliate at base; perianth 2–2.5 mm. long. *L. parviflora.*

Luzula multiflora (Retz.) Lejeune, var. **comosa** (E. Mey.) St. John. Plants tufted, 6–47 cm. tall; leaves flat, 2–5 mm. broad, 5–15 cm. long, villous ciliate; cauline leaves 1–3; spikes globose to cylindric, the longer 10–30 mm. long, borne in cymes; longest bract usually exceeding the inflorescence; peduncles 1–10 cm. long, erect or nodding; flower bracts pale, membranous, lacerate-ciliate; perianth segments straw-colored or brownish, lanceolate, acuminate, 2.5–4.5 mm. long; capsules triangular-subspherical, short-beaked, often equaling the perianth. *L. comosa* E. Mey., Syn. Luzula 21, 1823. Dry woods, *A. T. T.*

Luzula glabrata (Hoppe) Desv. Glabrous throughout; stems 2–5 dm. high, from creeping rootstocks; leaves dark green, flat, 4–20 cm. long, 5–10 mm. broad, acute; panicle loose, erect, 5–7 cm. long; bracts ciliate; perianth segments dark brown, lance-ovate, acute, shorter than the apiculate nearly black capsule. Mt. Carlton [Mt. Spokane], *Kreager,* Blue Mts., *Darlington* 75, *Huds.* A species greedily eaten by horses.

Luzula parviflora (Ehrh.) Desv. Panicle nodding, diffuse, perianth green or brownish; capsule ovoid, green or brownish. *Juncoides parviflorum* (Ehrh.) Cov. Moist woods, *Can.*

LILIACEAE. Lily Family

Terrestrial herbs or sometimes climbers or woody plants; stems usually from bulbs, corms or more or less thickened rootstocks; flowers mostly perfect, regular; perianth not scale-like, of 6 separate or united segments free from the ovary; stamens 6, opposite the perianth-segments, hypogynous or perigynous; ovary 3-celled; fruit a few- to many-seeded, 3-celled capsule or berry. (Named for the genus *Lilium.*)

Fruit a berry; rootstocks present,

- Leaves 3, whorled. *Trillium.*
- Leaves several, alternate,
  - Leaves basal; berry blue. *Clintonia.*
  - Leaves cauline; berry reddish,
    - Flowers axillary, 1 (or 2) as a node. *Streptopus.*
    - Flowers in a terminal inflorescence,
      - Flowers in racemes or panicles. *Smilacina.*
      - Flowers solitary or in few-flowered umbels. *Disporum.*

Fruit a capsule,

- Flowers in umbels on naked scapes; from bulbs or corms,
  - Perianth parts distinct or nearly so, the stamens on their bases. *Allium.*
  - Perianth parts united, stamens on the throat. *Brodiaea.*
- Flowers not in umbels; stems leafy,
  - Rootstocks present; styles distinct,
    - Flowers polygamous; leaves lanceolate or broader, pleated. *Veratrum.*
    - Flowers perfect, leaves linear, flat or channeled. *Xerophyllum.*
  - Bulbs or corms present; styles united, 3-parted, or wanting,
    - Sepals mostly scarious, narrower than the petals; corms present. *Calochortus.*
    - Sepals petaloid, similar to the petals,
      - Leaves 2, opposite, cauline; corms present. *Erythronium.*
      - Leaves not so, bulbs present,
        - Perianth parts with a prominent gland; capsules septicidal. *Zigadenus.*
        - Perianth parts without a prominent gland; capsules loculicidal,
          - Racemes scarious bracted; leaves basal. *Camassia.*
          - Flowers leafy bracted; cauline leaves present,
            - Anthers versatile; perianth parts cuneate. *Lilium.*
            - Anthers basifixed; perianth parts broad based. *Fritillaria.*

the leaves and the usually erect stem from a coated bulb; leaves narrowly linear, or rarely lanceolate or oblong, sheathing; inflorescence a terminal simple umbel, in the axils of 2 or 3 membranous separate or united bracts; pedicels slender, not jointed; flowers white or pink; perianth-segments 6, separate or slightly united at base; stamens 6; ovary sessile or nearly so, obovoid-globose, rounded at tip or with 3–6 subterminal ridges or crests, completely or incompletely 3-celled; ovules 1–6 in each cell. (Name Lat. for the garlic.)

Key to Flower Characters

Petals serrulate,

- Sepals entire, 8–12 mm. long. *A. acuminatum.*
- Sepals serrate, 12–15 mm. long. *A. dictuon.*

Petals entire,
  Stamens exserted or equaling the perianth,
    Ovary not crested; leaves 10–25 mm. wide. *A. Douglasii.*
    Ovary 6-crested; leaves 2–4 mm. wide. *A. Nevii.*
  Stamens well included,
    Scapes decumbent, 5–20 cm. long,
      Perianth white, the stamens ½ as long. *A. fibrillum.*
      Perianth pink, the stamens ⅔ as long. *A. Tolmiei,* var. *Tolmiei.*
    Scapes erect, 1–5 dm. tall,
      Perianth 8–9 mm. long; bulb coats fibrous. *A. Geyeri.*
      Perianth 4–6 mm. long; bulb coats not fibrous or reticulate. *A. macrum*

KEY TO BULB CHARACTERS

Outer bulb-coat fibrous. *A. Geyeri.*
Outer bulb-coat membranous,
  Outer bulb-coat not reticulate or very obscurely so,
    Leaves flat 10–25 mm. wide. *A. Douglasii.*
    Leaves flat or channeled, less than 7 mm. wide,
      Leaves flat, 2–7 mm. wide. *A. Tolmiei,* var. *Tolmiei.*
      Leaves channeled, 2–4 mm. wide. *A. macrum.*
  Outer bulb coats reticulate,
    Reticulations fine, intricately sinuous. *A. fibrillum.*
    Reticulations not sinuous,
      Reticulations heavy walled, regular, squarish, in longitudinal rows,
        Reticulations prominent, the meshes forming distinct pits. *A. acuminatum.*
        Reticulations fine, not forming pits. *A. Nevii.*
      Reticulations thinner, irregular, rhomboidal or polygonal, not in regular rows. *A. dictuon.*

Allium acuminatum Hook. Bulbs globose, the outer coats straw colored, finely reticulate, the units heavy walled, squarish, in regular longitudinal rows; scape terete, 10–35 cm. tall; leaves 2–4, linear, shorter than the stem, 2–3 mm. wide, withered by flowering time; perianth segments pink to whitish, gibbous at base, ovate, acuminate, 8–12 mm. long; petals serrulate; stamens ⅔ the length of the perianth; ovary only obscurely 3-crested; *A. acuminatum,* var. *cuspidatum* Fern. Abundant, dry rocky places, *U. Son.* to *Can.*

Allium dictuon St. John. Bulbs obliquely lanceoloid, the inner coat sinuous reticulate; leaves channeled or flat, 1.5 mm. wide, about 10 cm. long; scape terete, 20–27 cm. tall; bracts 2, with 3 nerves, elliptic-lanceolate, long acuminate; perianth parts magenta-pink, lanceolate, serrulate, acuminate, 8–14 mm. long, more than twice as long as the stamens; capsule globose, not crested. Known only from the type, from dry soil, Weller's Butte, Blue Mts., *Darlington* 114, *Can.*

Allium Douglasii Hook. Bulb ovoid, 2 cm. long, the coats thin, not reticulated; scapes stout, terete, 15–40 cm. tall; leaves two, flat, falcate, 10–25 cm. long, 4–14 mm. wide, flowers very numerous, in a dense often globose umbel, rose-purplish; pedicels 15–20 mm. long; perianth segments broadly lanceolate, acuminate, gibbous at base, strongly keeled, 6–9 mm. long; stamens equaling the perianth; ovary not crested. In gravelly places, local, *A. T.*

Discovered by David Douglas in 1826 in the Blue Mts., behind Walla Walla. The bulbs are very mild, sweet, and edible.

Allium fibrillum M. E. Jones. Usually caespitose; bulbs multiplying, ovoid, the outer coats marked with fine sinuous reticulations; scapes 5–20 cm. high; leaves narrowly filiform or linear 0.5–6 mm. wide, often exceeding the scape; perianth segments lanceolate, white (or pink tinged), 7–8 mm. long; stamens ½ length of perianth; capsule globose, obscurely ridged at summit. *A. collinum* Dougl., not Guss. Bare rocky ridges, Mica Peak, Craig Mts., and very abundant on the summits of the Blue Mts., *A. T. T.* to *Huds.*

Discovered by David Douglas "on the Blue Mountains."

Allium Geyeri Wats. Bulbs lance-ovoid, the outer coats of loose gray fibers, the inner membranous, strongly rhomboidal reticulate with similar veins; leaves 2–4, blunt, 1–3 mm. wide, 1–3.5 dm. long; scapes slender, 1–5 dm. tall; perianth segments rose-colored, ovate, acute or acuminate; stamens ¾ the length of the perianth; ovary deltoid-obovoid with 6 low terminal crests. Meadows and scab-rock areas, locally abundant, *A. T.*

Discovered by C. A. Geyer on "the Kooskooskie" [Clearwater River], Idaho.

Allium macrum Wats. Bulb about 15 mm. long, ovoid, the coats lacking reticulations; leaves mostly exceeding the scape; scapes 10–15 cm. tall, slightly angled; the two bracts ovate, acute; pedicels 5–10 mm. long; perianth segments 5–6 mm. long, lanceolate, acuminate, white or pale rose, veined with purple, scarcely longer than the stamens; capsule ridged along the upper margin. Dry soil, Little Almota Cr. Divide, April 20, 1941, *Ownbey & Aase.*

Allium Nevii Wats. Bulb ovoid, about 2 cm. long, the coats usually purplish and the reticulations very obscure, transversely polygonal; scape terete or slightly 2-edged, 10–35 cm. tall; leaves 2, channeled, shorter than the scape, 2–4 mm. broad; pedicels slender 5–18 mm. long; flowers pink; perianth segments ovate, cuspidate, gibbous at base, keeled, 6–7 mm. long; stamens exserted; ovary 6-crested. Scablands, Spokane to Rock Lake, *A. T.*

Allium Tolmiei Baker var. **Tolmiei.** Bulb ovoid, 1–2 cm. long, the outer coats black, not reticulated; scape slightly flattened, 2-edged, curved, usually prostrate, 6–15 cm. long; leaves two, falcate, flat, 2–7 mm. broad, exceeding the scape; pedicels slender, 15–20 mm. long; perianth segments lanceolate, acuminate, keeled, pink to whitish, at least the sepals gibbous at base, 7–10 mm. long; the stamens ⅔ as long as the perianth; capsule obovoid, 3-crested. *A. Cusickii* Wats. Thin rocky soil, Craig Mts., Blue Mts., and along Snake River down to [illegible]

[illegible] a membranous or fibrous coated corm; leaves linear; inflorescence a several-bracted umbel of few to many flowers, on jointed pedicels; perianth funnelform to open campanulate, blue, purple, yellow, or white, the segments united; stamens 3 on the throat opposite the inner lobes and alternate with three sterile stamens, or 6 in one or two rows; ovary stalked or sessile, 3-celled, each cavity containing 3–8 seeds. *Hookera.* (Named for *James Brodie* of Scotland.)

Flowers white, the stamens in one row. *B. hyacinthina,* var. *hyacinthina.*
Flowers blue, the stamens in two rows. *B. Douglasii,* var. *Douglasii.*

Brodiaea Douglasii Wats. var. **Douglasii.** *Wild Hyacinth.* Corm ovoid, the outer coats coarsely brown fibrous; leaves 2–3 shorter than the stem, 2–4 mm. broad; scape 3–8 dm. tall; pedicels 2–50 mm. long; perianth tubular-campanulate, blue, 2–3 cm. long, the tube longer than the lobes; stamens in two rows, the upper and longer ones on the petals; filaments linear-lanceolate; capsule oblong-ovoid, stipitate. *Hookera douglasii* (Wats.) Piper; *Triteleia grandiflora* Lindl.; *B. grandiflora* (Lindl.) Macbr. Rocky slopes, grass-lands, or open woods, *U. Son, to A. T. T.;* very abundant, showy spring flower on prairies of the *A. T.*

Daubenmire (1939) reported a collection with albino flowers.

Brodiaea hyacinthina (Lindl.) Baker, var. **hyacinthina.** The new corm forms a root, 5–13 cm. long, which contracts, draws the corm down, planting it deeper in the soil; leaves linear shorter than the stems, 3–8 mm. broad, scapes 3–8 dm. tall; perianth open-campanulate, white, 10–15 mm. long, each lobe with a green midvein, the tube about half as long as the lobes; stamens in one row, the filaments broadly deltoid, dilated, equal; capsule stipitate, subglobose. *Hookera hyacinthina* (Lindl.) Kuntze; *B. lactea* (Lindl.) Wats.; *Hesperochordum hyacinthinum* Lindl. Low meadows, common in summer, *A. T.*

CALOCHORTUS. Sego Lily

Branched or simple herbs with coated corms; leaves linear or linear-lanceolate; flowers large, showy, peduncled; perianth segments separate, spreading, yellow, whitish, purplish, or variegated; the three outer sepal-like, narrow, the three inner petal-like, gland-bearing, and bearded or spotted within; stamens 6, on the base of the perianth segments; ovules numerous; capsule ellipsoid to linear, angled or winged. Name Gr. *kalos,* beautiful; *chortos,* grass.)

Ovary almost linear, not sharply angled or winged; cauline leaves several, involute, linear,
- Gland cordate; petals lilac-purple; capsule 4–6 cm. long. — *C. macrocarpus.*
- Gland elliptic; petals white; capsule 7–9 cm. long. — *C. maculosus.*

Ovary elliptic, sharply angled or winged; cauline leaves usually 1, flat, tape-like,
- Petals yellowish; gland purple-black. — *C. apiculatus.*
- Petals white to purplish; gland not purple-black,
  - Capsules nodding; petals densely hairy throughout. — *C. elegans,* var. *elegans.*
  - Capsules erect; petals sparsely hairy, glabrous at tip.
    - Gland oblong; stems bulbiferous. — *C. longibarbatus.*
    - Gland round or cordate; stems not bulbiferous,
      - Petals not ciliate, with a rounded spot. — *C. eurycarpus.*
      - Petals ciliate, with a crescent-shaped spot. — *C. nitidus.*

Calochortus apiculatus Baker. Stem erect, 2–4 dm. high, bearing 1–3 flowers; leaf solitary, 1–2 cm. long, 6–18 mm. broad; bracts narrowly lanceolate, acuminate; sepals greenish, purple dotted, acute, lanceolate, 1–2 cm. long; petals yellowish, broadly obovate, 2–3 cm. long, ciliate, sparingly yellowish pilose, gland small, circular, bounded by purple-black hairs; anthers lanceolate, caudate, as long as the filaments; capsule lance-elliptic, sharply 3-angled, 2–3 cm. long. Woods and meadows, Spokane and Kootenai Cos., *A. T. T.*

Calochortus elegans Pursh, var. **elegans.** *Cat's Ear.* Corms ovoid, dark coated, 1–2.5 cm. long; stems slender, flexuous, 5–15 cm. tall, with a narrowly lanceolate leaf, 8–20 cm. long, 2–12 mm. wide; flowers erect, solitary, or 2–5 in a cyme, the pedicels ascending, 2–5 cm. long, each subtended by a lanceolate-acuminate bract; sepals broadly lanceolate, acuminate, greenish or purplish, 12–18 mm. long, shorter than the petals; petals obovate, obtuse, greenish-white with a purple-violet blotch at base, ciliate, very hairy on the upper surface; gland lunate, covered by a fringed scale; anthers lanceolate, caudate; capsule sharply 3-angled, elliptic, 1.5–2 cm. long. Occasional in prairies, *A. T.;* common, dry woods, *A. T. T., Can.*

First collected by Capt. M. Lewis on the Kooskoosky [Clearwater] River. He said that the corms were eaten by the Indians.

Calochortus eurycarpus Wats. Stems erect, 2.5–4.5 dm. tall, with 1–4 erect flowers; leaf grass-like, 2–12 mm. wide, shorter than the stem; sepals lanceolate, acuminate, 2–3 cm. long; petals white, fading lavender, with a rose-purple median, purplish spot, broadly cuneate, rounded at tip, sparsely villous, 25–45 mm. long; anthers oblong; gland round to subcordate, with a narrow scale, but concealed by a mass of yellow hairs; capsules elliptic, wing-angled, 2–3 cm. long. *C. nitidus* sensu Ownbey (1940), non Dougl. Blue Mts., Asotin Co., *A. T., A. T. T.*

M. Ownbey (in Herb. W.S.) recorded in 1948 a change in interpretation from that of his monograph (1940).

Calochortus longibarbatus Wats. Stem slender, 2–3.5 dm. tall, bearing a small lateral bulblet near the ground surface; leaf 1, grass-like, 1.5–3.5 dm. long, 2–9 mm. wide; flowers 1–2, the pedicels erect, stout; sepals lanceolate acute, 15–22 mm. long; petals lilac, yellow at base, obovate, rounded but erose at tip, 20–34 mm. long, nearly smooth, but long hairy above the gland; gland transverse, oblong, covered with a fringed scale; anthers obtuse, oblong, half the length of the filaments; capsule elliptic, acute, angled, 20–25 mm. long. Oakesdale, *Suksdorf* 8,856, *A. T.*

Calochortus macrocarpus Dougl. *Sego Lily.* Stems stout, 2–6 dm. tall; leaves 3–5, involute, linear, 5–25 cm. long; flowers 1–4; sepals greenish or purplish, linear-lanceolate, acuminate, scarious-margined, 3–5 cm. long, equaling the petals; petals broadly obovate, abruptly acute or acuminate, lilac-purple, with a deeper blotch above the gland, long-hairy only on the paler base, midrib green; gland covered by a densely ciliate scale; anthers narrowly lanceolate, sagittate at base; capsule lance-linear, not winged. Early summer, rocky or dry places, from Spokane down the scablands to the Snake and Touchet

[illegible] cm. long; petals white, fading lilac, midrib green, oblanceolate, acute, 4–6 cm. long, the base pilose, yellow; gland hairy, with a marginal fimbriate scale; anthers narrowly lanceolate, sagittate at base; capsules linear. The type is from Lewiston, *Henderson* 2,727. In midsummer, *U. Son., A. T.* *C. macrocarpus,* var. *maculosus* (Nels. & Macbr.) Nels. & Macbr.

Calochortus nitidus Dougl. Stems slender, 2.5–6 dm. tall; cauline leaf linear, nearly as long as the stem; flowers usually 2–4 in an umbel-like cyme; sepals ovate, acuminate, early deciduous, 2–3 cm. long, scarious-margined, purple tinged; petals cuneate-obovate, truncate to acute, erose ciliate, lilac-purple, sparsely pilose, 4–5.5 cm. long; gland cordate, covered with a small densely hairy scale; anthers oblong-lanceolate; capsule oval, acuminate, prominently winged, 2–3 cm. long. *C. Douglasianus* Schultes f.; *C. pavonaceus* Fern. Common in low meadows, *A. T.*

M. Ownbey in Herb. W.S. in 1948 retracted his 1940 interpretation, and in 1958 recorded examining the Douglas type which must have come from the Craig Mts., and bore the earliest and valid name for *C. pavonaceus* Fern.

CAMASSIA

Scapose herbs from membranous, coated, edible bulbs; leaves linear, all basal; inflorescence a terminal, bracted raceme; flowers showy, blue or white; pedicels jointed at base of flower; perianth segments distinct, equal, spreading, persistent, 3–7-nerved; stamens 6, on bases of segments; style filiform, stigma 3-lobed; capsule 3-angled, loculicidal. (Named from its Indian name, *camas.*)

Perianth segments not or scarcely narrowed below,
  Perianth segments withering separately,
    Flowers dark blue. — *C. Quamash,* subsp. *Quamash.*
    Flowers white. — *Forma pallida.*
  Perianth segments withering connivent over the capsule. — Subsp. *breviflora.*
Perianth segments clawed at base, pale blue. — Subsp. *Teapeae.*

Camassia Quamash (Pursh) Greene, subsp. **Quamash.** *Camas.* Bulbs black coated, ovoid, 12–38 mm. long; scapes stout, 2–9 dm. tall; leaves numerous, dark green, mostly shorter than the scape, 3–20 mm. wide; raceme 7–37-flowered, 5–28 cm. long; pedicels short; bracts narrowly lanceolate, about equaling the flowers; flowers dark or purplish blue, the perianth parts lanceolate, 15–25 mm. long; capsule ovoid, 3-lobed, 10–15 mm. long. *Quamasia quamash* (Pursh) Cov.; *C. esculenta* sensu Shinners, not of (Nutt.) Robins. Abundant in low, wet meadows, *A. T.*

A staple food root of the Indians. Discovered by Lewis and Clark at Weippe, Ida.

Forma **pallida** St. John. It differs by having white flowers. It is very rare, in spite of the abundance of the blue-flowered species. The type was from meadow, Moscow, *St. John* 10,558, *A. T.*

Subsp. **breviflora** Gould. Leaves 1.5–3 dm. long, 1–17 cm. wide; pedicels 0.5–1.5 cm. long; at least some of the flowers irregular, all bright blue to deep blue-violet, 1.5–2 cm. long, the perianth segments 3–5 mm. broad, 3– (or 5–)-nerved; anther 0.4–5 mm. long, bright yellow. In meadows, Blue Mts., *A. T.*

Gould has recorded his subsp. *breviflora* from the vicinity of Pullman, based upon the collections: in 1896, *Climer;* and *Elmer 223.* These were misdetermined by him, and are really subsp. *Quamash.* His statement that the two subspecies overlap in the vicinity of Pullman, should now be changed. Only the subsp. *Quamash* occurs there. The northern limit of subsp. *breviflora* is on the northern slope of the Blue Mts., Wash.

Subsp. **Teapeae** (St. John) St. John. Bulbs 12–25 mm. long, ovoid, brown coated; roots from a short vertical rhizome; scapes 26–67 cm. tall, smooth; leaves 15–38 cm. long, 3–9 mm. wide, basal, linear, flat or folded, scaberulous beneath, withering before fruiting time; raceme 7–22-flowered, in flower 5–13 cm. long, in fruit as much as 20 cm. long; bracts 12–50 mm. long, narrowly lanceolate, white or bluish, dark veined; pedicels ascending, in flower 6–12 mm., in fruit 6–20 mm. long; buds 16–20 mm. long, pale blue, boat-shaped, axially flattened or concave; perianth segments 12–30 mm. long, 3–6 mm. wide, narrowly lanceolate, obtuse, 3–7-nerved, contracted to a narrow claw, lower segment descending, the other five equally spaced in the upper 180°; stamens diverging, nearly equaling the perianth; filaments subulate; anthers 3–5 mm. long, oblong, yellow; style sinuous, ascending, 12–16 mm. long; stigma shallowly 3-parted; capsule 10–15 mm. long, 3-lobed, ovoid, with 2–3 convex faces, the base of the style persisting; seeds 3–5 mm. long, numerous, ovoid or lanceoloid, usually asymmetric, black and shining, wrinkled. *C. Teapeae* St. John. Meadows and pine woods, From Spokane Co., and Kootenai Co. northward, *A. T. T.*

Named in honor of Mrs. Nancy Miller Teape, of Spirit L., Ida.

CLINTONIA

Somewhat pubescent herbs, with slender rootstocks and erect simple scapes; leaves few, broad petioled, sheathing, basal; flowers solitary or umbelled on the erect scapes, bractless; perianth-segments distinct, equal or nearly so, erect or spreading; stamens 6, inserted at the base of the perianth segments; ovary 2- or 3-celled; ovules 2–several in each cavity. (Named for *De Witt Clinton,* naturalist, early Governor of N. Y.)

Clintonia uniflora (Schult.) Kunth. *Queen Cup.* Rootstock creeping; scapes slender, pilose, naked or with one or two small linear bracts, 6–15 cm. tall, bearing a solitary white flower or rarely 2–3; proper stem short; leaves 2–5, oblanceolate, acuminate, pilose beneath, 8–25 cm. long, perianth parts white, oblanceolate, obtuse, 18–25 mm. long; filaments attenuate, pilose below; berry globose or pyriform, blue, 8–12 mm. thick. Common in mountain woods, *Can.*

DISPORUM

More or less pubescent herbs with slender rootstocks; stems branching, scaly below, leafy above; leaves alternate, somewhat asymmetrical, sessile or clasping; flowers drooping, whitish or greenish-yellow, perianth segments 6, narrow, equal, separate; stamens 6, hypogynous; ovary 3-celled; ovules 2 or sometimes several in each cavity. (Name Gr., *dis,* double; *spora,* seed.)

Stigma 3-cleft; berry globose, papillose. *D. trachycarpum.*

[illegible]

ganum (Wats.) W. Miller. Moist woods, *A. T. T., Can.*

Disporum trachycarpum (Wats.) B. & H. *Fairy Bells.* Rootstocks short, compact, with a mass of strong roots; stems dichotomously branching above, 3–10 dm. tall, the branches crinkled pilose when young; leaves ovate or lanceolate, acuminate, pilose or glabrate, 3–11 cm. long; flowers solitary or in pairs, yellowish-white, campanulate on more or less curved pedicels, 5–8 mm. long; perianth segments oblong or oblanceolate, 8–16 mm. long; berry orange-yellow, 6–10 mm. in diameter, roughened with minute shallow pits. *D. majus* (Hook.) Britton. In woods, *A. T. T.*

ERYTHRONIUM. Dog-tooth Violet, Adder-tongue

Low herbs, from deep-seated, membranous-coated corms; stem simple; leaves but one pair, broad or narrow, usually below the middle of the stem where it rises from the ground and thus appearing basal; flowers large, nodding, bractless, one to several; perianth segments separate, lanceolate, oblong or oblanceolate; stamens 6, hypogynous, shorter than the perianth; ovary sessile; style elongate; ovules numerous or several in each cavity. (Name Gr., *eruthros,* red.)

Perianth bright yellow,
  Anthers reddish-purple. *E. grandiflorum,* var. *grandiflorum.*
  Anthers white. Var. *pallidum.*
Perianth whitish, with yellow eye;
  anthers white. *E. idahoense.*

Erythronium grandiflorum Pursh, var. **grandiflorum.** *Lamb's Tongue.* Scape 1.5–6 dm. tall; corm elongate; leaves dark green, oblong-lanceolate, acute, 7–23 cm. long; petioles short, grooved above; flowers 1–2 or rarely 3–10 in a raceme; perianth segments yellow, pale at base, recurved; anthers reddish-purple; capsule oblanceoloid; attenuate at base, 2–4.5 cm. long. Abundant on north hillsides, *A. T., A. T. T.,* sporadic from *Can.* to *Huds.*

Var. **pallidum** St. John. *Yellow Avalanche Lily.* Flowers 1–2, in length 2–4.5 cm. *E. parviflorum* of Piper not *E. grandiflorum,* var. *parviflorum* Wats. Open slopes or woods, *Can., Huds.,* and sporadic in *A. T.*

Erythronium idahoense St. John & G. N. Jones. Corms ellipsoid, with a slender rhizome up one side; plants 8–25 cm. tall, 1–3-flowered, leaves green, unequal, oblanceolate, with a cucullate tip, the larger 7–13 cm. long, 2.5–4.5 cm. wide; perianth segments greenish or creamy white, yellow marked at base, recurved, linear-lanceolate, 3–4.5 cm. long; the petals 4-saccate at base; anthers white. *E. grandiflorum candidum* Piper; *E. grandiflorum,* var. *idahoense* (St. John & Jones) R. J. Davis. Abundant, yellow pine woods, Harvard and northward, *A. T. T.*

Discovered in 1926 at Worley. The type was *St. John et al.* 3,719.

FRITILLARIA

Herbs from scaly bulbs; stems simple, leafy; inflorescence of rather large nodding solitary or racemed flowers; perianth mostly campanulate, of 6 separate and nearly equal oblong or ovate segments, each with a nectar-pit or spot at the base; stamens 6, hypogynous; ovary nearly or quite sessile; ovules numerous in each cavity. (Name Lat., *fritillus,* a dice-box.)

Flowers yellow; styles united; capsule not winged. *F. pudica.*
Flowers brownish spotted; styles partly cleft; capsule broadly
  6-winged. *F. lanceolata.*

Fritillaria lanceolata Pursh. *Rice-root.* Stems 1.5–6 dm. tall, glaucous; leaves in two or three whorls of 3–6, or the upper opposite or alternate, lanceolate, obtuse, sessile, 3–12 cm. long; flowers ill smelling, subglobose, 1–6, brown-purple with greenish-yellow spots; segments oblong-lanceolate, 2–3.5 cm. long; nectaries conspicuous; capsule short stalked, ovoid, truncate at both ends, the 6 angles broadly winged, 2–3 cm. long. Woods, from Palouse northward and northeastward, *A. T. T.*

Fritillaria pudica (Pursh) Spreng. *Yellow Bell.* Bulb-scales round, numerous, stems 1–4 dm. tall; leaves several, oblong-lanceolate, 4–15 cm. long, 2–15 mm. wide, in a single whorl or scattered near middle of stem; flowers commonly single, sometimes 2–6, deep cup-shaped, on drying shrinking at base; perianth-segments oblong, obtuse, 1–3 cm. long, yellow, each developing a reddish mark at the base; stamens about equaling the united styles; capsule obovoid, truncate, 2–5 cm. long. *Ambilirion pudicus* (Pursh) Raf.; *Ochrocodon pudicus* (Pursh) Rydb. Common in spring, open places, *U. Son.* to *A. T. T.*

LILIUM. Lily

Tall perennial herbs from scaly bulbs; stems simple; leaves alternate or whorled, sessile; flowers one to several, large and showy, campanulate; the 6 segments usually recurved; stamens on the base of the perianth; capsule subcylindric, 3-celled, the seeds numerous, in 2 rows in each cell. (Name Lat., *lilium,* a lily.)

Lilium columbianum Hanson ex Baker. *Wild Tiger-lily.* Bulbs white, 2–5 cm. thick; plant smooth; stems 3–16 dm. tall; middle cauline leaves in whorls of 5–9, lanceolate, sessile, 5–10 cm. long; flowers one to several, nodding; perianth segments similar, reflexed, lanceolate, orange with purple-brown spots, 1–6 cm. long; capsule obovoid, 3–4 cm. long. Open woods, *A. T. T.*, Newman Lake, 5/15/28, *Cook.*

SMILACINA. False Solomon's Seal

Herbs with slender or short and thick rootstocks; stem simple, scaly below, leafy above; leaves alternate, short-petioled or sessile, ovate, lanceolate, or oblong; inflorescence a terminal raceme or panicle; flowers white or greenish, small; perianth of 6 separate spreading equal segments; stamens 6, on the base of the perianth segments; ovary 3-celled, sessile; ovules 2 in each cavity; fruit a globose berry. (Name a diminutive of the Gr., *Smilax,* hence, a little *Smilax.*)

Flowers shorter than their pedicels, in a raceme. *S. stellata.*

[illegible]

panicle pyramidal or narrower, 3–16 cm. long; pedicels short; perianth spreading, the segments oblong-lanceolate, about 2 mm. long; filaments subulate, longer than the perianth; berry purple-specked, turning clear red, 4–6 mm. thick. *Vagnera amplexicaulis* (Nutt.) Greene. Thickets or woods, *U. Son.* to *Can.*

Smilacina stellata (L.) Desf. *Spikenard.* Rootstocks slender; stems arching, 2–5 dm. tall; leaves lanceolate, puberulent, or glabrate and somewhat glaucous beneath, 4–16 cm. long; flowers 3–14, white; the segments lanceolate, 4–6 mm. long; filaments almost filiform; stamens about ⅔ as long as the perianth; berry green, black-striped, turning purplish, 6–10 mm. thick. *Vagnera stellata* (L.) Morong. *V. sessilifolia* (Baker) Greene was accepted for the plants with flat, spreading leaves. Several other segregates have been accepted in the West. There are many intermediates and a complete intergradation. They seem to be ecological responses to the habitats. After comparative study, the writer is unable to distinguish or accept any of them.

Moist places, *U. Son.* to *Can.*

STREPTOPUS

Herbs with stout or slender rootstocks; leaves thin, sessile or clasping, alternate; flowers slender-peduncled, white to greenish or purplish, small, nodding; peduncles usually bent or twisted at about the middle; perianth campanulate, its segments 6, recurved or spreading, the outer flat, the inner keeled; stamens 6, hypogenous; ovary 3-celled; ovules numerous, in two rows in each cavity, fruit a juicy berry. (Name Gr., *streptos,* twisted; *pous,* foot.)

Streptopus amplexifolius (L.) DC., var. **americanus** Schultes. *Twisted Stalk.* Stems branched above, 3–10 dm. tall, from compact rootstocks with a mass of roots, glabrous except at the base; leaves ovate to lanceolate, acuminate, cordate and clasping at base, 5–12 cm. long, glaucous beneath; flowers solitary in the upper axils; pedicels bent at the base and geniculate in the middle, so that the flower is concealed by the leaf; perianth narrowly campanulate, the segments whitish, lanceolate, acuminate, the tips strongly recurved, 8–12 mm. long; anthers acuminate; berry ellipsoid, clear red, about 1 cm. thick. *S. amplexifolius,* var. *chalazatus* Fassett. Moist woods, in the mountains, *Can.*

TRILLIUM. Trillium

Glabrous erect unbranched herbs, with short rootstocks; leaves netted veined, 3 in a whorl at the summit of the stem just under the sessile or peduncled, solitary, bractless flower; the sepals green, the petals white, pink, purple, or greenish; stamens 6, hypogynous; ovary sessile, 3–6-angled or lobed; ovules several or numerous in each cavity; fruit a large fleshy berry. (Name Lat., *tres,* three.)

Leaves sessile; flowers peduncled. *T. ovatum.*
Leaves petioled; flowers sessile. *T. petiolatum.*

Trillium ovatum Pursh. *Western Trillium.* Stems 1.5–5 dm. tall, from a stout horizontal rootstock, 2–5 cm. long; leaves broadly ovate or rhombic, acuminate or acute, 5–22 cm. long; flowers fragrant; peduncles erect, 2–8 cm. long; petals lanceolate to ovate, 2.5-6.5 cm. long, acute, white, in age changing through various shades of pink to dark red, exceeding the lanceolate green sepals; anthers yellow. In this species the shape and size of leaves, sepals, petals, and anthers are notably variable. Common, open or dense woods, *A. T. T., Can.*

Trillium petiolatum Pursh. *Purple Trillium.* Stems buried, 4–15 cm. long, usually enclosed in sheath-like bracts, arising from a stout oblong upright rootstock; blades brownish, oval or orbicular, 5–15 cm. long; petals linear-oblanceolate, dark purple, 3–5.5 cm. long, scarcely longer than the sepals; anthers dark purple. Rich hillsides and copses, *A. T.*

VERATRUM. False Hellebore

Tall perennial herbs, from short thick erect poisonous rootstocks; stem pubescent; leaves mostly broad, clasping, strongly veined; inflorescence pubescent, of terminal panicles; flowers greenish, white, or purple, rather large, on short stout pedicels; perianth segments 6, glandless or nearly so, adnate to the base of the ovary; stamens opposite the perianth segments and free from them, short, mostly curved; ovary ovoid, 3-celled, the cavities several-ovuled. (Name Lat., *vere,* truly; *ater,* black.)

Perianth segments green, subacute, 2–4 mm. wide. *V. Eschscholtzii.*
Perianth segments white, obtuse, 4–7 mm. wide. *V. speciosum.*

Veratrum Eschscholtzii Gray. *Green Hellebore.* Stems stout, 1–2 m. tall; leaves from broadly round-oval to lanceolate in the upper, white pilose beneath, 15–40 cm. long; panicle open, slender, 3–6 dm. long, attenuate, the lower branches more or less drooping, usually simple; perianth segments obovate, pilose without, 8–10 mm. long; principal bracts foliaceous, like the upper leaves; capsule ovoid. *V. Eschscholtzianum* (R. & S.) Rydb. Wet places in mountain woods, *Can.*

Veratrum speciosum Rydb. *White Hellebore.* Stem 1–3 m. tall; leaves puberulent or glabrate above; panicle large, compact; perianth segments 9–17 mm. long; capsules lanceoloid, 3-lobed, 2–3 cm. long. Wet meadows, *A. T.*, and occasional in *Can.*

XEROPHYLLUM

Perennials with a short thick woody rootstock and numerous radical stiff linear leaves; perianth white, of six separate spreading petal-like segments; stamens 6, at the base of the perianth segments; styles distinct, linear, reflexed; ovary sessile, ovate, 3-lobed. (Name Gr., *xeros,* arid; *phullon,* leaf.)

Xerophyllum tenax (Pursh) Nutt. *Pine Lily* or *Bear Grass.* Stems stout 1–2 m. tall, stiff and erect; basal leaves very numerous harsh and stiff, linear, keeled, 3–6 mm. broad, 6–9 dm. long; cauline similar but smaller and dilated

baskets used for cooking.

ZIGADENUS

Glabrous erect perennial herbs from coated bulbs each crowning a very short rhizome, and with leafy stems; leaves linear; inflorescence a terminal panicle or raceme; flowers perfect or polygamous, greenish, yellowish, or white; perianth segments lanceolate or ovate, separate or united below, sometimes adnate to the lower part of the ovary, with one or two glands above the narrowed base; stamens free from the perianth segments, about equaling them; ovary 3-celled; ovules several or numerous in each cavity. (Name Gr., *zugon,* a yoke; *aden,* gland.)

Upper cauline leaves without sheaths; petals and sepals equally long clawed. *Z. venenosus.*
Cauline leaves with scarious sheaths; sepals subsessile or shorter clawed. *Z. gramineus.*

Zigadenus gramineus Rydb. *Grassy Death Camas.* Bulb ovoid, up to 5 cm. in length, the outer coat blackish; basal leaves several, prominent, linear, channeled, shorter than the stem; cauline leaves usually 2, the blades reduced, the base, white, hyaline, long sheathing; stems 2–8 dm. tall; inflorescence a raceme, or on large plants paniculate at base; sepals with a short claw; petals oblong or ovate-lanceolate, blunt or subacute, with longer claws; gland with its upper margin indefinite; anthers exserted; capsule cylindric, 5–15 mm. long, the styles and carpel tips divergent. Rocky slopes, or open woods, *U. Son.* to *Huds.*

The plants and particularly the bulbs are poisonous to all animals, but especially to sheep. Half a pound is fatal for a 100 lb. sheep. This is the most virulent of all the species and causes the death of many sheep.

Zigadenus venenosus Wats. *Death Camas.* Stems 3–6 dm. tall, from ovate, dark-coated bulbs; leaves several, linear, keeled, 3–5 mm. broad, shorter than the stems; raceme many-flowered, 5–10 cm. long, much longer in fruit; flowers yellowish, on pedicels 5–10 mm. long; bracts awl-shaped; perianth segments clawed, elliptical, the blade obtuse; nectaries with thick margins. In meadows, rare, *A. T.*

Plant poisonous, especially the bulbs and seeds; pollen poisonous to honey bees.

IRIDACEAE. Iris Family

Perennial herbs from rootstocks or bulbs; leaves narrow, equitant, 2-ranked; flowers perfect, regular or irregular, mostly clustered, subtended by bracts; perianth of 6 segments or 6-lobed, its tube adnate to the ovary, the segments or lobes in 2 series; stamens 3, inserted on the perianth opposite its outer series of segments or lobes, distinct or monadelphous; ovary inferior, mostly 3-celled; ovules mostly numerous in each cell; style 3-cleft, its branches sometimes divided; capsule loculicidal. (Named for the genus *Iris.*)

Styles petaloid; sepals and petals unlike. *Iris.*
Styles filiform; sepals and petals alike, rotate. *Sisyrinchium.*

IRIS. Iris, Flag, Fleur-de-Lis

Herbs with creeping or horizontal, poisonous, often woody and sometimes tuber-bearing rootstocks; stems erect; leaves erect or ascending; flowers large, regular, terminal, solitary or clustered; perianth of 6 clawed segments united below into a tube, the three outer dilated, spreading or reflexed, the three inner narrower, smaller, usually erect; ovary 3-celled; divisions of the style arching over the stamens, bearing the stigmas immediately under their mostly 2-lobed tips; style-base adnate to the perianth tube. (Name from Gr., *iris,* rainbow.)

Flowers violet-blue. *I. missouriensis,* f. *missouriensis.*
Flowers white. Forma *alba.*

Iris missouriensis Nutt., forma **missouriensis.** *Western Blue Flag; Iris.* Rootstocks stout, short, branched, brown-scaly; stems 2–7 dm. tall, naked or with one or two leaves; leaves pale or glaucous, usually shorter than the stem, 3–10 mm. wide; flowers violet-blue, especially along the veins, 2–4 in each umbel; bracts scarious, 2–7 cm. long, acute; sepals 4–7 cm. long, oblanceolate, the honey guides yellowish; petals erect, shorter and narrower; capsule 3–8 cm. long, oblong-cylindric, bluntly 6-angled. *I. missouriensis,* forma *angustispatha* Foster. After studying the type and the other specimens of this forma in the Gray Herbarium, it is confidently reduced to synonymy. The width of the spathe is variable and lacks significance. Common in wet meadows, forming dense patches, *A. T.*

The poisonous rootstocks were used as medicine by the Indians.

Forma **alba** St. John. Like the species, except for its white flowers. Rare, in meadows and bunchgrass prairies, *A. T.*

The holotype is: between Pullman and Wawawai, June 3, 1916, *F. L. Pickett.*

SISYRINCHIUM. Blue-eyed Grass

Perennial, tufted, slender herbs, with short rootstocks; stems simple or branched, 2-edged, 2-winged, or terete; leaves linear, grass-like; flowers bluish to purplish, red, yellow, or white, terminal, umbellate, from a pair of erect green bracts; perianth tube short or none, the 6 spreading segments oblong or obovate, equal; filaments somewhat united; style branches filiform, undivided, alternate with the anthers. (Name Gr., used by Theophrastus for a bulbous plant like *Iris.*)

Filaments united nearly to the top; perianth blue, evanescent. *S. idahoense.*
Filaments united about half-way; perianth not evanescent

[illegible], compressed and winged, the edges smooth or serrulate; leaves shorter than the stems, linear, acute, 1–3.5 mm. wide, all radical, or occasionally one on the stem; outer bract 2–6 cm. long, narrow, attenuate; inner bract 1.5–3.5 cm. long, broad, hyaline-margined; flowers 3–9, dark blue, with a small yellow center; pedicels 1.5–3 cm. long; perianth segments 12–18 mm. long, mucronate; capsules 4–6 mm. long, brown, ovoid. Low meadows, common, *U. Son., A. T.*

The type specimen was from Nez Perces Co., Idaho, collected by J. H. Sandberg.

Sisyrinchium inflatum (Suksd.) St. John, forma **inflatum.** *Grass Widow.* Roots fascicled, stout; stems 1.5–5 dm. tall, usually clustered, terete, striate; cauline blades 5–10 cm. long, erect, shorter than the sheaths, thick, terete, the basal leaves reduced to scales; lower bract of the spathe 4–7 cm. long, the upper 2–3 cm. long; flowers 1–8, the pedicels shorter than the largest bract; perianth segments 14–22 mm. long, oblanceolate, obtuse to acuminate; filament tube with a slight gradual enlargement near the base. *Olsynium inflatum* Suksd.

Meadows and hillsides, *U. Son.* to *A. T. T.*, very abundant in early spring, characteristic of the *A. T.*

Forma **album** St. John. An albino with pure white flowers. Spokane, *Sampson* in 1922, *A. T.*

ORCHIDACEAE. Orchid Family

Perennial herbs, with corms, bulbs, or tubers; leaves entire, sheathing, or sometimes reduced to scales; flowers perfect, irregular, bracted, solitary, spiked, or racemed; perianth of 6 segments, the three outer (*sepals*) alike or nearly so; two of the inner ones (*petals*) lateral, alike; the third inner one (*lip*) dissimilar, often markedly so, usually larger, often spurred, usually inferior by the twisting of the ovary or pedicel; stamens variously united with the style into an asymmetrical *column;* anther one (sometimes two); pollen more or less coherent into waxy, stalked masses (*pollinia*); ovary inferior, usually long and twisted, 3-angled, 1-celled; ovules numerous on three parietal placentae; seeds numerous; seed-coat loose, hyaline, reticulate; endosperm none. (Named from the genus *Orchis.*)

Perfect anthers 2; lip an inflated sac, the upper margins infolded, closed. — *Cypripedium.*
Perfect anther 1; lip not saccate, or if saccate with an open gaping throat,
- Green leaves none; scaly saprophytes with yellowish or purplish stems,
  - Flowers often short spurred; stems from a mass of white coralloid rhizomes. — *Corallorrhiza.*
  - Flowers not spurred; rhizome single, creeping. — *Cephalanthera.*
- Green leaves present,
  - Flower and leaf single; stem bulbous above the ground. — *Calypso.*
  - Flowers and leaves several; stem not bulbous above the ground,
    - Leaves 2, opposite, near the middle of the stem. — *Listera.*
    - Leaves not so,
      - Flowers spurred. — *Habenaria.*
      - Flowers not spurred,
        - Anther terminal; cauline leaves broad, numerous. — *Epipactis.*
        - Anther dorsal; cauline leaves linear, few or reduced to bracts,
          - Lip saccate, without callosities at base. — *Goodyera.*
          - Lip not saccate, with 2 horn-like callosities at base. — *Spiranthes.*

CALYPSO

Herbs with mostly coral-like roots; leaf petioled; scape low, sheathed by two or three loose scales; flower large, terminal, showy, bracted; sepals and petals similar, nearly equal; lip large, saccate or swollen, 2-lobed below; column dilated, petal-like, bearing the lid-like anther just below the summit; pollinia 2, waxy, each 2-parted. (Named for the Gr. nymph, *Kalupso,* daughter of *Atlas.*)

Calypso bulbosa (L.) Oakes, forma **occidentalis** Holz. *Calypso; Lady Slipper.* Stems 10–23 cm. tall, enwrapped toward the blade with 3–4 scarious sheaths, the uppermost prolonged into a narrow bract; leaf solitary, radical, ovate, acute, 3–8 cm. long, on a petiole about as long; sepals and petals ascending, lanceolate, rose-purple, 14–25 cm. long; lip 17–24 mm. long, brown-lined, rose-purple at the edge of the cavity, or slightly throughout, sac-like, two-lobed at the apex, with 3 lines of white hairs within; the apex of the slipper prolonged into two tooth-like projections; column half as long as the petals. Abundant, mossy woods, *A. T. T., Can.*

Replacing the yellow bearded *C. bulbosa* in our region and westward to the Pacific.

CEPHALANTHERA

Saprophytic herbs with leaves reduced to scarious bracts; flowers erect, white in a terminal raceme, lateral sepals spreading, strongly keeled; upper sepal and petals erect, somewhat connivent; lip free, shorter, the saccate base with broad winglike margins, articulate at the middle, with 2 basal callosities; stigma beakless; anther short-stalked. (Name from Gr., *cephale,* head; *anthera,* anther.)

Cephalanthera Austinae (Gray) Heller. *Phantom Orchid.* Whole plant waxy white; stem 2–5 dm. high, slender, erect; bracts linear-lanceolate, the lower with dilated sheaths; raceme 5–15 cm. long; flowers 3–20, nearly sessile; sepals and petals oblong-lanceolate, subequal, 10–15 mm. long; lip shorter, saccate at base. *C. oregana* Reichenb.; *Eburophyton austinae* (Gray) Heller; *Serapias Austinae* (Gray) A. A. Eaton. In deep woods, rare, *A. T. T., Can.*

base with the foot of the column, forming a short spur or protuberance, the other one free, the spur adnate to the summit of the ovary, or wanting; petals about as long as the sepals, 1–3-nerved; lip 1–3-ridged; column nearly free, slightly incurved, somewhat 2-winged; anther terminal; pollinia 4, in two pairs, oblique, free, soft-waxy. (Name from Gr., *korallion,* coral; *riza,* root.)

Lip entire, purple striate; spur none. — *C. striata.*
Lip 3-lobed, not striate; spur present,
  Sepals and petals 1-nerved; lip white. — *C. trifida,* var. *verna.*
  Sepals and petals 3-nerved,
    Lip spotted; spur wholly attached to ovary. — *C. maculata.*
    Lip dark red; apical half of spur free. — *C. Mertensiana.*

Corallorrhiza maculata Raf. *Spotted Coral-root.* Whole plant reddish or sometimes green, 2–5 dm. tall; raceme 10–35-flowered, the buds ascending, the old flowers and fruits drooping; sepals and petals 6–10 mm. long, 3-nerved, mostly brownish purple; lip ovate, white, mottled with purple, 3-lobed, with prominent ridges; middle lobe obtuse or notched, the lateral ones acute; column nearly as long as the petals. *C. multiflora* Nutt. Moist woods, *A. T. T., Can.*

Corallorrhiza Mertensiana Bong. *Mertens' Coral-root.* Scape 2–5.5 dm. tall, glabrous, purplish; raceme 10–40-flowered; sepals and petals 6–10 mm. long, lanceolate, dull purplish; lip broadly oblong, narrowed at base, thin, concave, reddish purple; column slender, nearly equaling the petals. Meadows and woods, *A. T. T.*

Corallorrhiza striata Lindl. *Striped Coral-root.* Whole plant reddish-purple, 2–5 dm. high, raceme 15–30-flowered; sepals and petals 10–15 mm. long, purplish, each with 3 conspicuous, dark purple nerves; lip entire, darker purple, somewhat fleshy, ovate, narrowed below, concave and bearing 2 short prominent ridges near the base. Moist woods, Blue Mountains and Thatuna Hills, *A. T. T.*

Corallorrhiza trifida Chat., var. **verna** (Nutt.) Fern. *Early Coral-root.* Scapes 1–3 dm. tall, pale yellowish; raceme 3–15-flowered; sepals 3–5 mm. long, lanceolate, somewhat acute, greenish yellow; petals like the sepals but broader, obtuse; lip oblong, obtuse or notched, somewhat 3-lobed; column shorter than the petals. Damp woods, *Can.*

CYPRIPEDIUM. Lady's Slipper

Glandular-pubescent herbs with leafy stems or scapes and thick fibrous roots; leaves large, broad, many-nerved; flowers one or several, drooping, large, showy; sepals spreading, separate or two of them united under the lip; column bearing a sessile or stalked anther on each side and a dilated petal-like sterile stamen above, covering the summit of the style; pollinia granular; stigma terminal, broad, obscurely 3-lobed. (Name from Gr., *Kupris,* Venus; *pedilon,* shoe.)

Lip yellow; sterile stamen triangular. — *C. Calceolus,* var. *parviflorum.*
Lip white, purple-veined; sterile stamen ovate or obovate. — *C. montanum.*

Cypripedium Calceolus L., var. **parviflorum** (Salisb.) Fern. *Yellow Lady's Slipper.* Plant viscid puberulent; stems leafy 2–6 dm. high; leaves 5–15 cm. long, elliptic to lanceolate; flowers 1–3; sepals 2–5 cm. long, ovate-lanceolate; upper petals 3–4.5 cm. long, linear, twisted; lip 2–3 cm. long, yellow, more or less marked with purple; sterile anther triangular, yellow and purple spotted. *C. parviflorum* Salisb. In springy places about Spokane, *A. T. T.*

Cypripedium montanum Dougl. ex Lindl. *Mountain Lady's Slipper.* Whole plant glandular-puberulent; stems 3–7 dm. tall, erect; leaves 4–6 oval, acuminate, 6–12 cm. long, with many strong ribs; flowers 1–3, pedicelled, leafy bracted; sepals brown, usually dark, narrowly lanceolate, 3–6 cm. long; upper petals like the sepals, but narrower, wavy-twisted, 4–6 cm. long; lip white, veined with purple, 2–3 cm. long; sterile anther yellow with purple spots, about 1 cm. long. Mossy woods, *A. T. T., Can.*

EPIPACTIS

Tall stout leafy herbs with creeping rootstocks; leaves green, clasping the stem; flowers in terminal leafy-bracted racemes;

sepals and petals separate; lip free, sessile, broad, concave below, the upper portion dilated and petal-like; column erect, short; anther 1, erect; capsule oblong beakless. (The ancient Gr. name for *Helleborus.*)

Epipactis gigantea Dougl. ex Hook. *Stream Orchis.* Stout and leafy, 3–10 dm. high, nearly smooth; leaves ovate below, reduced upwards and narrowly lanceolate, 5–20 cm. long, acute or acuminate, somewhat scabrous on the veins beneath; raceme puberulent, loose; flowers 3–15, greenish, strongly veined with purple, with large foliaceous bracts, on slender pedicels, 4–6 mm. long; sepals ovate-lanceolate, 12–16 mm. long, the upper concave; petals a little smaller; the lip 15–18 mm. long. *Amesia gigantea* (Dougl.) Nels. & Macbr. Spokane Co., *Suksdorf* 240; Almota, *Dillon* 188.

GOODYERA. Rattlesnake Plantain

Herbs with bracted erect scapes and thick fleshy-fibrous roots; leaves basal, tufted, often blotched with white; flowers in bracted spikes; lateral sepals free, the upper ones united with the petals into a hood-like structure (*galea*); lip sessile, entire, roundish, ovate, concave or saccate, without protuberances, its apex reflexed; anther without a lid, erect or incumbent, attached to the column by a short stalk; pollinia composed of angular grains, one in each sac, attached to a small disk which coheres with the top of the stigma. (Dedicated to *J. Goodyer* of England.)

Goodyera oblongifolia Raf. *Rattlesnake Plantain.* Scape 15–40 cm. tall, erect, glandular-pilose, bearing several short bracts; leaves 3–15 cm. long, ovate-lanceolate, acute, short petioled, with a broken white stripe down the midvein, dull green or white blotched; spike bracteate, densely many-flowered, 6–10 cm. long, somewhat 1-sided; perianth glandular-pilosulous, white, the sepals and petals 6–8 mm. long, about as long as the glandular pilose ovary; [illegible]

HABENARIA. Rein Orchis

Stems erect, leafy at least at base, from fleshy tuberous roots; inflorescence a terminal spike or raceme; flowers white, yellow, green, pink, or purple; sepals equal, the lateral spreading; petals erect, connivent with the upper sepal; lip spreading or drooping, entire or variously fringed or divided, produced at base into a slender spur; column short; stigma sometimes appendaged; anther-sacs divergent. (Name Lat., *habena,* rein.)

Stem leafy (or the leaves linear if basal),
  Flowers white or whitish,
    Spur about equaling the lip. *H. dilatata,* var. *dilatata.*
    Spur longer than the lip. Var. *leucostachys.*

Flowers greenish,
Spur slender, not saccate. *H. hyperborea.*
Spur short, saccate at tip. *H. saccata.*
Leaves basal, rarely more than 2,
Lateral sepals free; leaves less than twice as long as broad. *H. orbiculata,* var. *Menziesii.*
Lateral sepals adnate to the base of the lip; leaves more than twice as long as broad,
Spur shorter than the ovary; flowers greenish. *H. unalascensis.*
Spur longer than the ovary; flowers whitish. *H. elegans.*

Habenaria dilatata (Pursh) Hook., var. **dilatata.** *Tall White Bog Orchis.* Tubers thick, forking; stem 3–9 dm. tall, leafy; leaves 7–20 cm. long, lanceolate, the lower obtuse, the upper acute; spike 5–27 cm. long, loose or dense; perianth 5–10 mm. long; lip rhombic lanceolate, usually dilated at base. *Limnorchis dilatata* (Pursh) Rydb. Meadows and swampy woods, *A. T.* to *Can.*

Var. **leucostachys** (Lindl.) Ames. *White-flowered Bog Orchis.* Similar to the species and blending into it, but the spike 5–30 cm. long, and the spur longer than the lip. *Limnorchis leucostachys* (Lindl.) Rydb.; *L. leucostachys,* var. *robusta* Rydb. Swampy woods, Moscow Mt., and Blue Mts., *Can.*

Habenaria elegans (Lindl.) Boland. Stem 4–7 dm. tall; basal leaves 2–4, oblong or lanceolate, 10–15 cm. long, 2–6 cm. wide, obtuse or acute, soon withering; cauline bracts oblong, acuminate, 1–3 cm. long; spikes slender or dense, 8–28 cm. long; bracts ovate, acuminate, about equaling the flowers or the mature ovary; sepals and petals similar, 3–5 mm. long, obscurely 3-nerved; lip lanceolate to ovate, 3–6 mm. long; the spur slender, 7–18 mm. long; capsule oblong, nearly sessile. *Piperia elegans* (Lindl.) Rydb.; *P. leptopetala* Rydb.; *P. michaeli* (Greene) Rydb.; *P. multiflora* Rydb. Pullman, very rare, *A. T.*

Habenaria hyperborea (L.) R.Br. *Green Bog Orchis.* Stems 2–5 dm. high from fusiform tubers; basal leaves 5–29 cm. long, oblanceolate, obtuse, the upper lanceolate, acute; flowers 10–12 mm. long; upper sepal ovate, broader than the spreading lanceolate lateral ones; lip lanceolate, obtuse, 4–5 mm. long, equaling the clavate spur. *Limnorchis viridiflora* (Cham.) Rydb. Spokane County, Suksdorf 452; Craig Mt., *Hunter* 107.

Habenaria orbiculata (Pursh) Torr., var. **Menziesii** (Lindl.) Fern. *Large round-leaved Orchis.* Stem 1–2 dm. tall; leaves usually 2, elliptic to orbicular, prostrate on the ground, bright green and shining above, glaucous beneath, thick, usually 10–17 cm. long; scape bearing 1–3 lanceolate bracts; spike loose, 10–20-flowered; flowers greenish white; lateral sepals ovate, the upper orbicular; lip linear-oblong, 8–15 mm. long; spur slender, 16–22 mm. long. *Lysias orbiculata* (Pursh) Rydb. In deep mossy woods, usually in colonies, Cedar Mt., *Piper* in 1902, *Can.*

Habenaria saccata Greene. *Slender Bog Orchis.* Roots tuberous; stems stout, 2.5–10 dm. tall; leaves lanceolate, acuminate, 5–18 cm. long, gradually diminishing upward into the bracts; spike slender, 5–45 cm. long, loose, the lower flowers remote; bracts narrowly lanceolate, exceeding the lower flowers, equaling the upper ones; sepals 3-nerved, 3–6 mm. long; lip 5–7 mm. long, linear, thick, longer than the spur, which is inflated and very obtuse at the tip. *Limnorchis stricta* (Lindl.) Rydb.; *L. laxiflora* Rydb. Springy places in the mountains, *Can.*

Habenaria unalascensis (Spreng.) Wats. Tubers rounded; stems 3–9 dm. tall, leafy at base, bracteate above; leaves 2–3, oblanceolate, obtuse, 1–2 dm. long, soon withering; spike narrow rather dense, 1–4 dm. long; bracts triangular, acute or acuminate, shorter than the ovary, a few along the stem below the spike; sepals and petals lanceolate, 1-nerved, 2–4 mm. long; lip oblong, scarcely exceeding the petals; capsule oblong, sessile. *Piperia unalaschensis* (Spreng.) Rydb. Usually in moist woods, *A. T.* to *Can.*

LISTERA. Twayblade

Small herbs, with fibrous or sometimes rather fleshy roots; flowers in terminal racemes, spurless; sepals and petals nearly alike, spreading or reflexed, free; lip notched or cleft at apex; anther without a lid, erect, jointed to the column; pollinia 2, powdery. (Named for *Martin Lister,* an English naturalist.)

Lip 4–7 mm. long; ovary glabrous,
- Lip broadly wedge-shaped, emarginate. *L. Banksiana.*
- Lip lance-linear, cut halfway into two linear lobes. *L. cordata.*

Lip 9–13 mm. long; ovary glandular on the angles. *L. convallarioides.*

Listera Banksiana Lindl. *Northwestern Twayblade.* Stems slender, 10–35 cm. high, glabrous below the leaves; leaves sessile, ovate or elliptic, obtuse or acutish, glabrous, 3–6 cm. long; inflorescence glandular-puberulent; the slender pedicels longer than the bract or the ovary; sepals and petals lanceolate, spreading; lip cuneate, obovate, emarginate, with a slender tooth on each side near the base; capsule ovoid, 5–6 mm. long. *L. caurina Piper.* Deep woods in the mountains, *Can.*

Listera convallarioides (Sw.) Torr. *Broad-lipped Twayblade.* Stem 10–25 cm. tall, glabrous below the leaves; leaves 3–6.5 cm. long, glabrous, oval or ovate, obtuse; inflorescence glandular-puberulent; the pedicels scarcely longer than the bracts or the ovary; sepals and petals lanceolate, reflexed; lip clawed, cuneate, retuse, with a short triangular tooth on each side near the base. In deep moist woods, *Can.*

Listera cordata (L.) R.Br. Plant very slender, glabrous except near the leaves, height 0.5–3 dm.; leaves 1–3 cm. long, sessile, ovate, cordate, mucronate, much shorter than the peduncle; racemes loosely 4–20-flowered; pedicels 2 mm.

SPIRANTHES. Ladies' Tresses

Roots tuberous, clustered; stem bracted above, leaf-bearing below or at the base; flowers small, white, yellowish- or greenish-white, spirally-twisted in a raceme; lateral sepals lanceolate, the upper united with the oblong petals or free; perianth segments more or less connivent; lip short-stalked with a callus on each side of the base, the summit somewhat dilated; column short, bearing the ovate stigma on the front and the 2-celled erect anther sessile or nearly so on the back. (Name from Gr., *speira,* coil; *anthos,* flower.)

Spiranthes Romanzoffiana Cham. *Hooded Ladies' Tresses.* Plants 8–47 cm. tall from roots 5–8 mm. thick, leafy below and leafy-bracted above; leaves 3–13 mm. long, about half as long as the stem, linear to linear-lanceolate; scape glandular-pubescent above; cauline bracts 2–3; spike 2–19 cm. long, dense, cylindrical, in 3 spiral ranks; perianth yellowish or whitish, 6–12 mm. long; bracts of the raceme often much longer than the fragrant flowers; sepals and petals united. Blue Mountains, *Horner* 471.

Class II. *DICOTYLEDONES*

Embryo with two cotyledons; stems if woody differentiated into bark, wood and pith; wood cells forming annual rings; leaves mostly net-veined; parts of flowers in fours or fives, rarely in twos, threes, or sixes. (Name Gr., *di,* two; *kotuledon,* a cup-like hollow.)

SALICACEAE. Willow Family

Trees or shrubs with light wood, bitter bark and brittle twigs; leaves alternate; stipules minute or well developed; flowers solitary in the axil of each bract, dioecious, both staminate and pistillate in aments which expand with or before the leaves; staminate aments often pendulous; pistillate pendulous, erect, or spreading; staminate flowers of 1–many hypogynous stamens, subtended by a gland-like or cup-shaped disk; pistillate flowers of a sessile or short-stipitate 1-celled ovary subtended by a minute disk; ovules usually numerous; fruit an ovoid or oblong or conic 2–4-valved capsule; seeds minute, with abundant coma; endosperm none. (Named for the genus *Salix.*)

Ament bracts entire; stamens 1–10; winter buds with 1-scale. *Salix.*
Ament bracts fimbriate or incised; stamens more numerous; winter buds with decussate scales. *Populus.*

POPULUS

Trees with soft, usually whitish wood, scaly, resinous buds and rounded or angled twigs; leaves long-petioled, with minute, fugacious stipules; disk cup-shaped, oblique, lobed or entire; staminate aments dense, pendulous; stamens 12–60, the filaments distinct; pistillate aments sometimes raceme-like by the elongation of the pedicels, pendulous, erect or spreading; ovary sessile; tuft of hair on the seed very conspicuous. (Name is the ancient *Lat.* for poplar tree.)

Bark smooth; petioles flattened; buds slightly resinous. *P. tremuloides,* var. *aurea.*
Bark rough; petioles terete; buds sticky resinous,

Capsule 3-valved.	*P. hastata.*
Capsule 2-valved.	*P. balsamifera.*

Populus balsamifera L. *Balsam Poplar.* Tree, sometimes 33 m. tall, up to 2 m. in diameter; bark reddish brown, becoming rough on old trunks; leaves petioled, the blades ovate-lanceolate, acuminate, rounded or cordate at base, finely serrate, nearly glabrous, dark shiny green above, pale, glaucous, or rusty beneath, resinous when young; staminate aments stalked, the disk oblique, stamens 20–30, red; pistillate aments becoming 10–13 cm. long; fruits oblong-ovoid, about 7 mm. long. Wood pale brown, with white sapwood, soft, light, but of value for boxwood and pulpwood. *P. tacamahacca* Mill. Moscow Mt. Ida., *W. B. Wise* in 1927. *Can.*

Populus hastata Dode. *Cottonwood.* Tree, 10–20 m. tall, 60 cm. in diameter; trunk gray; blades 6–13 cm. long, ovate to subcordate, acuminate, rounded or cordate at base, crenate or subentire, dark green above, whitened beneath; staminate aments 2–7 cm. long, the disk pale, cupulate; stamens about 30, purplish; pistillate aments 5–20 cm. long, pilosulous, the capsules 6–7 mm. long, ovoid, warty, glabrate. Wood soft, pale, of secondary value. Common along streams, *U. Son., A. T., A. T. T.*

Populus tremuloides Michx., var. **aurea** (Tidestr.) Daniels. *Aspen.* Small tree, 10–20 m. tall, with smooth light bark; branchlets brown or gray; blades 3–8 cm. long, broadly ovate to semiorbicular, subcordate or obtuse or cuneate at base, abruptly acuminate, entire or crenate-serrate, ciliate when young, glabrate; staminate aments dense, 3–5 cm. long; stamens 7–10; bracts long-ciliate; pistillate aments 5–10 cm. long. Often treated as a species distinct from the eastern *P. tremuloides* Michx., because of the alleged golden yellow color of the autumn foliage, the larger and more irregularly cleft bracts. After field and herbarium study, the author cannot support these. However the trees from the Rocky Mts. and Pacific Coast regions have the anthers 0.7–1.25 mm. long (instead of 0.5). Hence the plant is maintained as a variety. Common, forming groves on north hillsides and along streams, *A. T., A. T. T.*

SALIX. Willow

Trees or shrubs with single-scaled buds, the scales with an [illegible] pistillate aments usually erect or spreading; staminate flowers with 1–10, mostly 2 or 5, stamens; ovary sessile or short stipitate. (The classic Lat. name)

Key to Staminate Plants

Stamens 5–9; floral scales yellowish; aments appearing with the leaves,	
Blades green on both sides,	
Shoots pilosulous.	*S. caudata,* var. *caudata.*
Shoots glabrous.	Var. *Bryantiana.*
Blades glaucous beneath,	
Petioles glabrous, glandless.	*S. amygdaloides.*
Petioles more or less pilose and glandular at apex.	*S. lasiandra.*

Stamens 1–2,
  Stamen 1; filament glabrous; floral scales dark brown to blackish; aments appearing before the leaves. *S. sitchensis.*
  Stamens 2,
    Filaments pilose at base,
      Floral scales yellowish; filaments distinct; blades subsessile, not glaucous; aments appearing with the leaves,
        Blades linear-oblanceolate or -lanceolate, soon glabrate. *S. melanopsis.*
        Blades linear or lance-linear, white sericeous, or tardily more or less glabrate,
          Foliage permanently silky villous. *S. argophylla.*
          Mature foliage sparsely sericeous or more or less glabrate. *S. exigua.*
      Floral scales, blackish; filaments united at base; blades petioled, aments appearing before the leaves,
        Blades glaucous beneath. *S. lasiolepis,* var. *lasiolepis.*
        Blades silvery tomentose beneath. Var. *Sandbergii.*
    Filaments glabrous; aments appearing before the leaves,
      Filaments united at the base; floral scales dark brown to blackish,
        Blades glaucous beneath. *S. mackenzieana.*
        Blades scarcely paler green beneath. *S. monochroma.*
      Filaments distinct,
        Floral scales yellowish, narrowly elliptic. *S. Bebbiana.*
        Floral scales dark brown to blackish, oval,
          Twigs glabrous,
            Twigs not pruinose. *S. Drummondiana.*
            Twigs pruinose,
              Stipules 3–6 mm. long. Var. *bella.*
              Stipules smaller or none. Var. *subcoerulea.*
          Twigs pilosulous to glabrous; blades oblanceolate to narrowly obovate, beneath pilosulous to nearly glabrate. *S. Scouleriana.*

Key to Pistillate Plants

Floral scales yellowish, caducous; aments appearing with the leaves; capsule glabrous,
  Leaves subsessile; pedicels 0.5–1 mm. long. *S. melanopsis.*
  Leaves petioled; pedicels 1–2 mm. long,

Blades green on both sides,
Shoots pilosulous. — *S. caudata, var. caudata.*
Shoots glabrous. — Var. *Bryantiana.*
Blades glaucous beneath,
Petioles glabrous, glandless. — *S. amygdaloides.*
Petioles more or less pilose and glandular at apex. — *S. lasiandra.*
Floral scales persistent or in part tardily deciduous,
Capsule glabrous,
Floral scales at least in part deciduous, yellowish to pale brown; blades linear or lance-linear,
Aments appearing with the leaves,
Capsule pilose; foliage permanently silky villous. — *S. argophylla.*
Capsule pilosulous to glabrate; foliage sparsely sericeous or more or less glabrate. — *S. exigua.*
Aments appearing before the leaves; blades lanceolate to ovate-lanceolate. — *S. mackenzieana.*
Floral scales persistent, brown to blackish; aments appearing before the leaves,
Blades silvery tomentose beneath. — *S. lasiolepis,* var. *Sandbergii.*
Blades glabrous or glabrate beneath,
Blades glaucous beneath. — Var. *lasiolepsis.*
Blades scarcely paler beneath. — *S. monochroma.*
Capsules pubescent; aments appearing before the leaves; leaves white hairy beneath,
Pedicels 0.2–0.3 mm. long,
Twigs not pruinose. — *S. Drummondiana,* var. *Drummondiana.*
Twigs pruinose,
Stipules 3–6 mm. long. — Var. *bella.*
Stipules smaller or none. — Var. *subcoerulea.*

Floral scales brown, tardily deciduous; pedicels 0.5–1 mm. long; capsule 4–6 mm. long. — *S. sitchensis.*
Floral scales blackish, persistent; pedicels 1–1.5 mm. long; capsule 7–9 mm. long. — *S. Scouleriana.*

Salix amygdaloides Anderss. Tree, 3–12 m. tall, young shoots glabrous or glabrate; young twigs brownish or yellowish; trunk 2–4.5 dm. in diameter, the bark reddish brown, longitudinally furrowed; stipules 3–15 mm. long, semicordate, acute or obtuse, serrulate; petioles 3–30 mm. long, glabrous, slender; blades 3–15 mm. long, serrulate, glaucous beneath, those of the flowering shoots mostly lanceolate to oblanceolate, later leaves mostly ovate-lanceolate, acuminate; aments terminal on short lateral branches, appearing with the leaves; staminate aments 3–5 cm. long; bracts 2.5 mm. long, yellowish, elliptic, pilose below the tip; stamens 5–9; filaments pilose at base, distinct;

pistillate aments 4–13 cm. long, loose; bracts 3–4 mm. long, yellowish, linear, pilose near the base and within, caducous; pedicels 1–2 mm. long; style 0.1–0.2 mm. long; stigmas 2-cleft; capsule 3–5 mm. long, lanceoloid, acuminate, glabrous. Wood pale yellowish brown, soft, brittle. *S. amygdaloides,* f. *pilosiuscula* Schneid. Banks of Snake River, *U. Son.*

Salix argophylla Nutt. Differs from *S. exigua* only in having the foliage permanently silky villous, and the capsules being permanently pilose. Banks of Snake River and its tributaries, *U. Son., A. T.*

Salix Bebbiana Sarg. Shrub 1–3 m. tall, or rarely a tree up to 8 m. tall and 2 dm. in diameter; bark thin, gray green; young shoots whitened, densely appressed pilose; branches reddish brown, glabrate; stipules 3–5 mm. long, semicordate, dentate; petioles 3–12 mm. long, appressed pilosulous; blades 2–8 cm. long, elliptic to oval-obovate, entire or nearly so, firm chartaceous, above dull green, appressed pilosulous or nearly glabrate, below white and densely to sparsely appressed pilose; aments from lateral buds on the old wood, appearing before the leaves, short peduncled and finally more or less leafy bracted at base; staminate aments 1–2 cm. long, dense; scales 2–3 mm. long, yellowish, narrowly elliptic, densely villous throughout; stamens 2; filaments glabrous, distinct; pistillate aments 2–6 cm. long, loose; scales 1.5–2 mm. long, pale brownish, narrowly elliptic, persistent, sparsely villous; pedicels 2–5 mm. long; stigmas subsessile, 2-parted; capsule 6–8 mm. long, white appressed pilosulous, long acuminate from an ovoid base. By streams, and in moist thickets, *U. Son., A. T.*

Salix caudata (Nutt.) Heller, var. **caudata.** Shrub, 2–5 m. tall or tree up to 10 m. tall; youngest shoots pilosulous, glabrate; branches chestnut, glabrous, shining; old bark brown, smooth; stipules 1–5 mm. long, semicordate, glandular serrate; petioles 1–12 mm. long, short at anthesis, pilosulous or glabrate, glandular at apex; blades 2–18 cm. long, paler green beneath, linear-lanceolate, often acuminate, glandular serrulate, glabrate or nearly so; aments terminal on short leafy, lateral branches, appearing with the leaves; staminate aments 1.5–4 cm. long; bracts 3–4 mm. long, yellowish, elliptic, the tip entire or erose, sparsely pilosulous below the tip; stamens 5–7; filaments pilose at base, distinct; pistillate aments 2–6.5 cm. long, loose; scales 3–5 mm. long, yellowish, caducous, sparsely pilosulous only at base, erose at apex; pedicels 1–2 mm. long; style 0.2–0.3 mm. long; the stigmas lobed; capsule 4–5.5 mm. long, lanceoloid, glabrous. By streams, *U. Son.,* common in *A. T.*

Var. **Bryantiana** Ball & Bracelin. Differs from the species only in having the vegetative organs glabrous. Common by streams, *U. Son.,* A. T.

Salix Drummondiana Barratt. Shrubs 1–4.5 m. tall; twigs yellowish brown, puberulent, later dark- to purplish-brown, glabrate, shining; stipules on summer twigs 5–9 mm. long, obliquely lanceolate, on fruiting twigs minute or none; petioles 5–10 mm. long; blades 3–10 cm. long, 1–4 cm. wide, narrowly to broadly elliptic-lanceolate or -oval, entire or minutely serrulate, dark green, above puberulent, below densely silvery tomentose and the veins almost hidden; stamens 2; aments mostly before the leaves, subsessile, the pistillate 1–5 cm. long; scales 1–2 mm. long, elliptic to ovate, mostly acutish; pedicels 0.5–1.5 mm. long, silky puberulent; styles 0.8–1.2 mm. long; stigmas 0.3–0.6 mm. long, entire or divided; capsules 4.5–6 mm. long, pyriform, silvery pubescent. Spokane Co., *Piper* 3517.

Var. **bella** (Piper) Ball. Shrub 2–4 m. tall; new twigs glabrous; young leaves appressed pilosulous; branches yellowish brown to reddish or blackish, glabrous, pruinose; stipules 3–6 mm. long, lanceolate, subentire, revolute; petioles 3–15 mm. long, glabrous; blades 3–11 cm. long, oblong-lanceolate, acute, mostly entire, firm chartaceous, above green, remotely appressed pilosulous, finally nearly glabrate; below densely white close pilose velvety; aments from axillary buds on the old wood, sessile, naked, appearing before the leaves;

staminate aments 2–2.5 cm. long, dense; scales 2–2.5 mm. long, oval, dark brown to blackish, villous; stamens 2; filaments glabrous, free; pistillate aments 2–4 cm. long, dense; scales 1.5–2 mm. long, ovate, dark brown, densely villous, persistent; pedicels 0.2–0.3 mm. long, pilosulous; style 0.5–1.3 mm. long; stigmas linear, entire or bifid; capsule 4–5.5 mm. long, lanceoloid, densely appressed pilosulous. *S. bella* Piper. By streams, *A. T.*, *A. T. T.*

The type was collected by Prof. *L. F. Henderson,* at Garrison, Whitman Co., in 1895–96.

Var. **subcoerulea** (Piper) Ball. Leaves thinner, mostly narrowly oblong-oblanceolate and acute on fruiting twigs, tomentum beneath less dense than in the species. Spokane Co., Latah Creek, Spangle, Suksdorf 8638; 8639; 8640, *A. T. T.*

Salix exigua Nutt. Shrub or small tree, 2–7 m. tall; young shoots sericeous; branches brown, glabrous, shining; bark rough, fissured; stipules wanting, rudimentary or rarely 1–2 mm. long, ovate, denticulate; leaves subsessile 2–11 cm. long, finally sparsely sericeous or more or less glabrate, linear or lance-linear, entire or remotely denticulate; aments terminal on lateral leafy shoots, appearing with the leaves; staminate aments 2–4 cm. long; scales 1.5–2 mm. long, elliptic, yellowish or pale brown, densely pilose below the tip; filaments pilose at base, distinct; pistillate aments 2–5 cm. long, dense; scales 1.5–2.5 mm. long, yellowish to pale brown, pilose towards the base, the lower caducous, the others tardily so or persistent; pedicels 0.5–1 mm. long, pilose; stigmas sessile, bifid; capsule 4–6 mm. long, pilosulous to glabrate, lanceoloid. Banks of Snake River and its tributaries, *U. Son.*, *A. T.*

Salix lasiandra Benth. Shrub or tree up to 15 m. tall and 65 cm. in diameter; young shoots pilosulous to glabrate; branches brown to yellow, glabrous, shining, brittle; old bark brown, fissured; stipules 3–12 mm. long, semicordate, glandular serrate; petioles 3–25 mm. long, pilose or glabrate, more or less glandular at apex; blades 3–16 cm. long, pale and somewhat glaucous beneath, elliptic-lanceolate to oblanceolate, glandular serrate, the later acuminate; aments terminal on lateral, leafy branchlets, appearing with the leaves; staminate aments 2–6 cm. long; scales 2–3 mm. long, yellowish, elliptic to obovate sometimes dentate, pilosulous below; stamens 5–9; filaments pilose below, distinct; pistillate aments 3–10 cm. long; scales 2 mm. long, pale, elliptic oblanceolate, pilose, caducous; pedicels 1–2 mm. long; stigmas subsessile,

brate; bark smooth; stipules 2–15 mm. long, semicordate, serrulate; petioles 2–17 mm. long, pilosulous or glabrate; blades 3–10 cm. long, narrowly to broadly oblanceolate, entire or subentire, beneath glaucous and pilosulous or glabrate; aments subsessile from lateral buds on older branches, appearing before the leaves; staminate aments 1.5–4 cm. long; scales 1.5–2 mm. long, blackish, oval, densely villous; stamens 2; filaments pilose below, united at base; pistillate aments 2–5 cm. long; scales 2–3 mm. long, brown to blackish, elliptic, densely villous; pedicels 1–3 mm. long; style 0.5 mm. long; stigmas bifid; capsules 4–5 mm. long, lanceoloid, glabrous. Creek banks, *A. T.*

Var. **Sandbergii** (Rydb.) Ball. Blades oblanceolate to obovate, silvery tomentose beneath. *S. Sandbergii* Rydb.

The type collection was from Nez Perce Co., Hatwai Cr., *Sandberg, Mac Dougall & Heller* 71.

Salix mackenzieana (Hook.) Barratt. Shrub or tree, 2–6 m. tall, 7–13 cm. in diameter; young shoots sparsely pilosulous, soon glabrate, branches brown or yellow, glabrate; bark gray, smooth, stipules 3–10 mm. long, prominent, semicordate, glandular crenate; petioles 2–25 mm. long, pilosulous to glabrate; blades 3–11 cm. long, lanceolate to ovate-lanceolate, serrate, glabrous, beneath glaucous; aments from lateral buds on the old wood, appearing before the leaves, naked, or sometimes finally with a few leafy bracts; staminate aments 2–3.5 cm. long; scales 1.5 mm. long, dark brown or blackish, villous below the tip; stamens 2; filaments glabrous, united at base; pistillate aments 3–6, rather dense; scales 1–1.5 mm. long, dark brown, villous at base, the lower deciduous, the upper persistent; pedicels 2–3 mm. long, style 0.2–0.5 mm. long; stigmas 2-parted; capsules 3–4.5 mm. long, glabrous, lanceoloid, acuminate. Stream banks, *U. Son., A. T.*

Salix melanopsis Nutt. Shrub or small tree, 3–5 m. tall; young shoots somewhat pilose, early glabrate; branches brown or gray, smooth; stipules 3–7 mm. long, obliquely broad lanceolate serrate; leaves subsessile. 4–13 mm. long, early glabrate, remotely serrulate, linear-oblanceolate or -lanceolate, paler green beneath, aments terminal on leafy lateral shoots, appearing with the leaves; staminate aments 2–5 cm. long; scales 1.5–2 mm. long, yellowish, oval, pilosulous towards the base; stamens 2; filaments pilose at base, distinct; pistillate aments 2–5 cm. long, dense; scales 1.5–2 mm. long, oblong-ovate, more or less erose, yellowish, pilosulous below; pedicels 0.5–1 mm. long; stigmas sessile, bifid; capsule 4–5 mm. long, glabrous, ovate-lanceoloid. Banks of Snake and Spokane Rivers, *U. Son., A. T.*

Salix monochroma Ball. Shrub 1–3 m. tall, young shoots pilosulous or glabrous; branches dark or reddish brown, shining; bark reddish brown with pale flakes; stipules 2–14 mm long, prominent, reniform, glandular crenate; petioles 4–10 mm. long, pilosulous above; blades 3–9 cm. long, lanceolate to ovate-lanceolate, glandular serrate, scarcely paler green beneath; aments from lateral buds on the old wood, subsessile, appearing before the leaves, naked or finally with a few leafy bracts at base; staminate aments 2–6 cm. long; scales 1.5–2 mm. long, dark brown to blackish, villous below the tip; stamens 2; filaments glabrous, united at base; pistillate aments 2–6 cm. long, dense; scales 1–1.5 mm. long, dark brown, persistent, villous below the middle; pedicel 2.5–4 mm. long; style 0.3–0.5 mm. long; stigmas bifid; capsule 4–5 mm. long, lanceoloid, acuminate, glabrous. Stream banks, *U. Son., A. T.*

Ball (in Herb. WS.) has changed his determinations of part of these collections to *S. lutea* Nutt., var. *platyphylla* Ball. The writer sees no basis for this change, so retains all these specimens in *S. monochroma*.

Salix Scouleriana Barratt. Shrub or tree, 4–10 m. tall, up to 25 cm. in diameter; young shoots densely to sparsely pilosulous; branches yellow to reddish brown, densely pilosulous to glabrous; bark gray to brown, smooth or fissured; stipules 3–12 mm. long, semicordate, glandular serrate; petioles 3–10 mm. long, more or less pilosulous; blades 3–11 cm. long, oblanceolate to narrowly obovate, entire or crenulate, firm chartaceous, above dark green and nearly glabrate, beneath whitish, at first densely pilosulous, later less so or nearly glabrate, and whitish or rusty; aments from lateral buds on the old wood, subsessile, naked, appearing before the leaves; staminate aments 1–3.5 cm. long, dense; scales 2–3 mm. long, oval, blackish, densely villous; stamens 2; filaments glabrous, distinct; pistillate aments 1.5–7 cm. long, dense; scales 3–4.5 mm. long, elliptic, blackish, villous, persistent; pedicels 1–1.5 mm. long pilosulous; style 0.2–0.4 mm. long; stigmas 2-parted; capsule 7–9 mm. long lanceoloid, long acuminate, appressed pilosulous. Streams banks or moist thickets, *U. Son., A. T., A. T. T.*

Salix sitchensis Sanson. Shrub 2–9 m. tall; young shoots appressed pilosulous; branches dark brown to blackish, pilosulous, later glabrate; stipules 3–13 mm. long, semicordate, glandular denticulate; petioles 3–13 mm. long, pilosulous; blades 3–8 cm. long, oblong-oblanceolate to oblanceolate, entire, above green and finally only sparsely pilosulous; beneath densely white pilose sericeous; aments from lateral buds on the old wood, appearing before the leaves, naked, or finally with a few leafy bracts at base; staminate aments 2–7 cm. long; scales 1–1.5 mm. long, dark brown to blackish, oval, densely villous; stamen 1; filaments glabrous; pistillate aments 3–9 cm. long, dense; scales 1–1.5 mm. long, brown or dark brown, oval to lance-ovate, sparsely pilose, tardily deciduous; pedicels 0.5–1 mm. long, pilosulous; styles 0.2–0.5 mm. long; stigmas entire or short bifid; capsule 4–6 mm. long, white appressed pilosulous, the lower half ovoid, the upper acuminate. Moist thickets, *A. T. T., Can.*

BETULACEAE. Birch Family

Trees or shrubs, without terminal buds; leaves alternate, petioled, simple, with deciduous stipules; flowers small, monoecious, in aments; staminate aments pendulous, linear-cylindric; pistillate aments erect, spreading or drooping, spike-like or capitate; staminate flowers 1–3 in the axil of each bract, with a membranous 2–4-parted calyx, or calyx none, and 2–10 hypogynous stamens; pistillate flowers with or without a calyx adnate to the solitary 1–2-celled ovary; ovules 1–2 in each cavity; fruit a small compressed or ovoid-globose, mostly 1-celled and 1-seeded nut or samara; endosperm none. (Named for the genus *Corylus,* the hazelnut.)

Fruit of winged nutlets borne in a cone; staminate flowers usually 3 in scale axils,
  Pistillate aments solitary, their scales thin, 3-lobed, deciduous. *Betula.*
  Pistillate aments racemose, their scales erose, persistent, becoming woody. *Alnus.*
Fruit a nut in a leafy involucre; staminate flowers solitary in scale axils. *Corylus.*

ALNUS. Alder

mostly 4-parted perianth and 1–4 stamens, and subtended by 1–4 bractlets; filaments short, simple; anther sacs adnate; pistillate flowers 2 or 3 in the axil of each bract, without a perianth, but subtended by 2–4 minute bractlets; ovary sessile, 2-celled; styles 2; bracts woody, 5-toothed or erose, persistent, forming a woody, ovoid cone; nut small, compressed, wingless or winged. (The ancient Lat. name.)

Pistillate peduncles slender, at least the lower longer than their cones, borne on new leafy shoots. *A. sinuata.*
Pistillate peduncles stout, shorter than their cones, borne from last year's twigs,
  Leaves lobed, doubly serrate; stamens 4. *A. tenuifolia.*
  Leaves not lobed, serrate or slightly doubly serrate; stamens 1–3, usually 2. *A.rhombifolia.*

Alnus rhombifolia Nutt. *White Alder.* Tree 10–35 m. tall, 3–10 dm. in diameter, with whitish bark which on older trees becomes broken into rectangular blocks; twigs glabrate, reddish, glutinous; petioles glandular, pilosulous, 5–17 mm. long; blades 5–8 cm. long, ovate, elliptic, or obovate, mostly cuneate at the base and obtuse at the apex, glandular-denticulate, at first finely puberulent on each surface, somewhat glandular beneath; fruiting aments oblong-ovate, 1–1.8 cm. long, nutlets 2.5 mm. long, broadly obovate, the margins thickened. Wood light brown, the sapwood nearly white, soft and brittle. Along streams, Snake River valley and tributaries, *U. Son.*

Alnus sinuata (Regel) Rydb. Sitka Alder. Shrub or small tree, even 15 m. tall, up to 20 cm. in diameter; erect or ascending; dark bluish gray; twigs glutinous at first, chestnut brown; buds smooth, gummy; petioles 1–3 cm. long, slender, glutinous; blades 5–11 cm. long, ovate, acuminate, obtuse or cuneate at the base, bright green, doubly dentate, glabrous above, nearly so beneath, except on midrib and vein axils, thin, very gummy when young; fruiting aments 9–17 mm. long; nutlet obovate, as wide as the delicate wings. The branches are decumbent at base, pointing down hill. Abundant, forming impenetrable thickets along mountain streams, *Can., Huds.*

Alnus tenuifolia Nutt. *Mountain Alder.* Shrub or small tree, 5–10 m. tall; bark whitish; young branches reddish brown, glutinous puberulent, becoming gray, glabrate; buds puberulent; petioles 1–3 cm. long, pilosulous; blades 3–10 cm. long, ovate or somewhat obovate, mostly obtuse, rounded or subcordate at the base, with a small lobe at the end of each vein, doubly dentate-serrate, dull-green, sparsely pilosulous on both sides, becoming glabrous above; fruiting aments 10–18 mm. long, ovoid, resinous; in our region frequently parasitized; nutlets obovate, with narrow, thinner margins. Along streams, *A. T., A. T. T.*

BETULA. Birch

Trees and shrubs; bark resinous, with long horizontal lenticels, often separating into thin layers; leaves dentate or serrate; buds scaly; flowers all in aments, expanding before and with the leaves; the pistillate aments erect or spreading; staminate flowers with a membranous usually 4-toothed perianth and 2 stamens, and subtended by 2 bractlets, filament forked, each fork bearing an anther sac; pistillate flowers 2 or 3 (rarely 1) in the axil of each bract; perianth none; ovary sessile, 2-celled; nut small, a samara, shorter than the bracts. (The ancient Lat. name.)

Bark of papery layers; trees,
- Leaves broadly ovate, cuneate, or rounded at base, thin, sharply doubly serrate; twigs glandular. — *B. papyrifera,* var. *commutata.*
- Leaves ovate, subcordate, thick, coarsely and saliently dentate or serrate; twigs mostly smooth. — Var. *subcordata.*

Bark not of papery layers; small bushy tree. — *B. occidentalis,* var. *fecunda.*

Betula occidentalis Hook., var. **fecunda** Fern. *Spring Birch.* Small tree, 5–12 m. high, 4 dm. in diameter; the bark dark bronze; branchlets drooping, grayish, resinous warty; petioles 7–23 mm. long; blades 2–7.5 cm. long, broadly ovate, obtuse or acute, cuneate or subcordate at base, coarsely serrate-dentate, glandular and sparsely hairy on each surface; pistillate aments solitary, cylindric, 2–5 cm. long, drooping with the branchlets, sessile or on short peduncles;

fruiting bracts pilosulous and ciliolate, the central lobe longest and narrowest; wings as broad as the obovate nutlet. Wood yellowish, sapwood whitish. *B. fontinalis* Sarg.; *B. Piperi* Britton emend. Piper; *B. microphylla* sensu St. John, not of Bunge. Common along streams, *U. Son., A. T., A. T. T.*

Betula papyrifera Marsh., var. **commutata** (Regel) Fern. *Western Paper Birch.* Tree, up to 40 m. tall and 13 dm. in diameter; bark white, or orange-brown especially in colder or more exposed habitats; twigs brown, somewhat pilose; petioles 10–25 mm. long; blades 4–12 cm. long, glabrate except in vein axils; fruiting aments 2–4 cm. long, 10–12 mm. thick, cylindric; nutlet oval, as wide as the wings. *B. papyrifera,* var. *occidentalis* sensu Sarg., not Hook. Moist woods, *A. T. T., Can.*

Var. **subcordata** (Rydb.) Sarg. *Heart-leaved Paper Birch.* Tree, 10–24 m. tall, very similar to var. *commutata,* and differing principally in the shape and toothing of the leaves. *B. subcordata* Rydb. Moist woods, *A. T. T., Can.*

CORYLUS. Hazelnut

Shrubs with alternate, simple leaves; flowers in aments appearing before the leaves; staminate aments drooping from last year's twigs; stamens 4–8; calyx none; pistillate flowers several in rounded, scaly, mostly terminal buds; calyx adherent to the ovary; style short; stigmas 2, elongate, red. (Name Gr. from *korus,* helmet.)

Corylus cornuta Marsh., var. **californica** (A. DC.) Sharp. Spreading shrub 2–10 m. tall; bark becoming smooth; branchlets brown, at first hispid; petioles 4–18 mm. long, hispid; blades 3–11 cm. long, obovate, acute, sharply doubly serrate or shallowly lobed, cordate at base, hispidulous or partly glabrate, dark-green above, pale beneath; fruit often clustered; involucres closely enwrapping the nut at base, hispid especially below, prolonged into a narrow cylindric tube 12–25 mm. long. *C. californica* (A. DC.) Rose. Woods, Fan Lake, Deer Park, *Nettie M. Cook* in 1928, *A. T. T.*

ULMACEAE. Elm Family

Trees or shrubs; leaves alternate, simple, serrate or entire, petioled, pinnately veined, with usually fugacious stipules; flowers small, monoecious, dioecious, perfect, or polygamous; sepals 3–9 free or united; petals none; stamens erect in bud as many as the sepals and opposite them; ovary 1–2-celled; styles 2; ovule pendulous, one; fruit a samara, drupe, or nut; endosperm none. (Named for the genus *Ulmus.*)

CELTIS

Trees or shrubs; leaves serrate or entire, pinnately veined; flowers polygamous or monoecious, borne in the axils of the season's leaves, the staminate clustered, the fertile solitary or in 2–3-flowered clusters; calyx 4–5 parted; ovary sessile; stigmas 2; fruit an ovoid or globose drupe, with a large, thick-walled nutlet. (Name Lat. *celtis,* the lotus.)

Celtis Douglasii Planch. *Hackberry.* Round-topped tree, 3–10 m. tall; bark rough, fine-checked, grayish brown; twigs green, pilose, becoming glabrate and brown; petioles 3–17 mm. long, pilose; blades 4–14 cm. long, oblique-ovate, acuminate, cuneate to cordate at base, sharply serrate, hispidulous when young, glabrate and shiny above when mature, somewhat hispidulous on the veins beneath, scabrous, reticulated; sepals 3 mm. long, yellowish, ciliate; fruit cinnamon-brown, globose, smooth, 5–7 mm. in diameter, on slender peduncles, 1–2 cm. long; the thin sweet pulp edible; stone with raised, reticulate ridges. Wood pale, very hard. Rocky slopes, valley of Snake River and its principal branches, *U. Son.*

A handsome tree when grown in good soil, but under natural conditions very scraggly; leaves often distorted as a result of insect attacks.

MORACEAE. Mulberry Family

Trees, shrubs, or rarely herbs, with milky juice; leaves usually alternate; stipules 2, caducous; flowers in heads, disks, or hollow receptacles, monoecious or dioecious, regular; staminate flowers in cylindric spikes; calyx-lobes usually 4, stamens 4, opposite the sepals; petals none; anthers 2-celled, dehiscent lengthwise; ovary rudimentary or none; pistillate flowers sessile, the ovary 1–2-celled; ovule 1, usually pendulous from the apex; fruit of achenes, nuts, or drupes. (Named for the genus *Morus.*)

Staminate and pistillate flowers in spikes; syncarp cylindric, juicy. *Morus.*
Staminate flowers in racemes, the pistillate in heads; syncarp globose, dry. *Maclura.*

MACLURA. Osage Orange

Trees with slightly acrid juice; leaves petioled, pinnately veined; stout axillary spines present; flowers dioecious; stamens inflexed in bud; style filiform, exserted; pistillate calyces enlarging and becoming fleshy in fruit. (Named for the American geologist, *William Maclure.*)

Maclura pomifera (Raf.) Schneid. *Osage Orange.* Tree 10–30 m. tall, 1 m. in diameter; bark brown, rough; branches spreading; branchlets green, pilose, but soon glabrate and brown; petioles 1–5 cm. long; blades 7–15 cm. long, ovate to oblong-lanceolate, entire, acuminate, cuneate to subcordate at base, pilose at least on the veins; spines sometimes 7 cm. long; staminate raceme 10–25 mm. long; flowers 1 mm. across, slender pedicelled; pistillate head peduncled, pendulous, about 2.5 cm. in diameter, ripening into a hard, yellowish, tubercled syncarp, 5–15 cm. in diameter. Wood orange turning brown, heavy, very strong and durable. Escaped from cultivation, along the Snake River, Couse Creek, *St. John et al.* 8,256, *U. Son.*

The bark of the roots furnishes a yellow dye.

MORUS. Mulberry

Trees or shrubs with 3-nerved leaves and fugacious stipules; flowers monoecious or dioecious; ovary sessile; stigmas 2, linear, spreading; fruiting juicy perianth enclosing the achene; exocarp succulent; endocarp crustaceous; endosperm scanty; embryo curved. (The ancient Lat. name of the mulberry.)

Morus alba L. *White Mulberry.* Small tree, up to 13 m. tall and 8 cm. in diameter; bark gray, rough; branchlets green, turning brownish, glabrous; petioles ciliate, shorter than the blades; blades 5–15 cm. long, ovate, varying from serrate to variously lobed, pilose on the veins beneath, acute or acuminate, rounded to cordate at base; staminate spikes about 2.5 cm. long, drooping; pistillate spikes at maturity 10–15 mm. long, white or purplish, edible. Introduced from Europe as food for silkworms. Established along Snake River, Couse Cr., *St. John* 8,257; 9,628, *U. Son.*

URTICACEAE. Nettle Family

Herbs (in ours), shrubs, or trees; leaves alternate or opposite, simple, with or without stipules, the epidermal cells with prominent cystoliths; flowers small, monoecious, dioecious, or polygamous; calyx of 2–5 nearly separate or cup-shaped sepals; calyx of pistillate flowers similar, but often enlarging in fruit; petals none; stamens as many as the calyx-lobes and opposite them; ovary 1-celled, mostly superior; ovule solitary, erect; fruit an achene; endosperm scanty or none. (Named for the genus *Urtica.*)

Leaves opposite, with stinging hairs and stipules. *Urtica.*
Leaves alternate, without stinging hairs or stipules. *Parietaria.*

PARIETARIA

Annual or perennial, diffuse or erect herbs; leaves alternate, entire, 3-nerved, petioled; flowers axillary, in compact clusters, polygamous, subtended by an involucre of leafy bracts; calyx of the staminate flowers deeply 4-parted, that of the fertile flowers tubular or campanulate, 4-lobed, surrounding the ovary and the achene; stigma spatulate, recurved. (The ancient Lat. name from *paries,* a wall.)

Parietaria occidentalis Rydb. Annual, 1–4 dm. tall, the stems pellucid, succulent, villous; petioles 3–25 mm. long, villous; blades 5–60 mm. long, elliptic-lanceolate to oval, obtuse, thin, sparsely villous or hispid; bracts of the involucre linear, about 2.5 mm. long; flowers 1.5 mm. long, green; achene about 1 mm. long, ellipsoid, brownish, polished and shining. Shady or moist places, abundant in Snake River Canyon, *U. Son.*

URTICA. Nettle

Annual or perennial simple or branching herbs with stinging hairs; leaves 3–7-nerved, petioled, dentate or incised, with free stipules; flowers greenish, very small and numerous, dioecious or monoecious; staminate flowers with a deeply 4-parted calyx and 4 stamens; pistillate flowers with 4 sepals, the two inner larger and in fruit enclosing the ovary; the two outer smaller and spreading; achenes flattened; stigma sessile, tufted. (The ancient Lat. name, from *urere,* to burn.)

Stipules herbaceous, greenish to pale brown, narrowly lanceolate, attenuate. *U. gracilis.*
Stipules subcoriaceous, turning deep brown, usually obtuse,
  Stem nearly glabrous; leaves glabrous or nearly glabrate. *U. Lyallii.*
  Stem and leaves densely pubescent,
    Stem and lower leaf surface velvety. *U. holosericea.*
    Stem and lower leaf surface hispidulous. *U. serra.*

Urtica gracilis Ait. Slender, 3–10 dm. tall, stem glabrous above, or somewhat setulose and sparingly pilose; leaves lanceolate to ovate, those subtending the lowest inflorescences 5–15 cm. long, rounded to cordate at base, glabrous on both sides or sparingly pilose beneath, coarsely serrate; inflorescence slender, usually forking, mostly moniliform or interrupted. *U. cardiophylla* Rydb. The commonest species. Wet or shady places, *A. T., A. T. T.*

Urtica holosericea Nutt. Tall, 1–3 m. high, usually somewhat bristly, densely soft-pubescent throughout; blades 5–16 cm. long, ovate-lanceolate, or the lower cordate, acuminate, subcordate, coarsely and evenly dentate, whitened and more densely, soft-pubescent beneath; staminate flowers in loose narrow panicles shorter than the leaves; pistillate panicles denser, in the axils of the uppermost leaves. Common in moist places, especially in the warmer valleys, *U. Son., A. T.*

Urtica Lyallii Wats. Stem 1–2 m. tall; blades 3–15 cm. long, ovate or cordate, acuminate, thin, coarsely dentate; inflorescences shorter than the leaves, the pistillate ones scarcely exceeding the petioles; sepals ovate, usually shorter than the achenes. Grassy hillside, Anatone, *V. Gessell* in 1926, *A. T. T.*

Urtica serra Blume. Stems 1–3 m. tall, finely strigose; blades 6–20 cm. long, cordate or lanceolate, with very large, coarse teeth often 1–1.5 cm. wide, cinereous puberulent or below densely hispidulous; staminate panicles shorter than the leaves; pistillate panicles exceeding the petioles. Meadows, Pullman and Union Flat, *A. T.*

SANTALACEAE. Sandalwood Family

Herbs or shrubs or trees; leaves alternate or opposite, entire, without stipules; flowers clustered or solitary, axillary or terminal, perfect, monoecious, or dioecious; calyx adnate to the base of the ovary, 4–5 cleft; petals none; stamens as many as the calyx-lobes and opposite them; ovary 1-celled; ovules 2–4; fruit a drupe or nut with only one seed. (Named for the genus *Santalum.*)

COMANDRA

Glabrous erect perennial herbs, sometimes parasitic on the roots of other plants; leaves alternate, oblong, oval, lanceolate or linear, entire, pinnately veined; flowers perfect, in terminal panicles or corymbs of small umbels, each subtended by a foliaceous involucel; calyx campanulate, the base of its tube adnate to the ovary; limb 5-lobed; stamens 5, rarely 4, at the bases of the calyx-lobes and between the lobes of the disk; fruit a dry nut crowned by the persistent calyx. (Name Gr., *kome,* hair; *aner,* man.)

Comandra pallida A. DC. *Bastard Toadflax.* Plant glabrous and glaucous throughout; rootstocks long running, horizontal, with indigo blue cortex; stems erect, 15–40 cm. high, simple or branched above; leaves 1–4 cm. long, sessile, lanceolate or linear, acuminate, umbels clustered; flowers 4–5 mm. long, the whitish calyx-lobes little spreading, acute, puberulent above, scarcely equaling the green tube; fruit nut-like, sweet and edible, globose or ovoid, 7–10 mm. long, glaucous, reticulated, tipped by the persistent calyx. Gravelly soil, common in *U. Son.,* occasional in *A. T.*

LORANTHACEAE. Mistletoe Family

Parasitic shrubs or herbs, yellow or yellowish-green, growing on woody plants and absorbing food from their sap through specialized roots (*haustoria*); leaves opposite, foliaceous or reduced to scales; flowers regular, terminal or axillary, clustered or solitary, dioecious (in ours); calyx-tube adnate to the ovary petals none; stamens 2–6; ovary solitary, inferior, 1-celled; style simple or none; stigma terminal; fruit a berry; seed solitary; endosperm usually copious and fleshy. (Named for the genus *Loranthus.*)

ARCEUTHOBIUM. Dwarf Mistletoe

Small fleshy glabrous plants, parasitic on the branches of coniferous trees; branches 4-angled; leaves reduced to opposite connate scales; flowers not bracted, solitary or several together in the axils of the scales; staminate flowers with 2–5-(usually 3-) parted calyx and usually equal number of stamens; berry fleshy, ovoid, more or less flattened, circumscissile, near the base, when fully ripe explosively dehiscent; the mucilaginous pulp glues the seed onto any branch it may touch, making possible seed dispersal. (Name from Gr., *arkeuthos,* juniper; *bios,* life.)

Staminate flowers paniculate, nearly all terminal on distinct peduncle-like joints. — *A. americanum.*
Staminate flowers mostly axillary, in simple or clustered spikes,
- Staminate stems 5–12 cm. long. — *A. campylopodum.*
- Staminate stems shorter,
  - Stems 3–5 cm. long, branched. — *A. laricis.*
  - Stems 1–3 cm. long, nearly simple. — *A. Douglasii.*

Arceuthobium americanum Nutt., ex Engelm. in Gray. Stems greenish yellow, slender, much branched, the branches in pairs or whorls; staminate plants 6–10 cm. long, 1–2 mm. thick at base, the pistillate 1.5–3 cm. long; staminate flowers 2 mm. wide, with ovate acutish lobes; pistillate flowers smaller; fruit bluish-green, 2 mm. long. *Razoumofskya americana* (Nutt) Kuntze. On lodgepole pine, *Pinus contorta,* var. *latifolia* (and related species elsewhere); often abundant and causing swellings and distortions, *A. T. T.*

Arceuthobium campylopodum Engelm. in Gray. Stems stout, 5–12 cm. long, branched, the staminate deep yellow, 4 mm. thick at base, the pistillate more slender, olivaceous; staminate flowers in dense spikes, that are 5–20 mm. long and 2.5–3 mm. thick, the calyx-lobes 3 or 4, oblong-ovate, acutish; fruit 4–5 mm. long, obovate, acute, bluish-green, drooping on curved pedicels. *Razoumofskya campylopoda* (Engelm.) Piper. Abundant on yellow pine, *Pinus pondersosa, A. T. T.*

Arceuthobium Douglasii Engelm. Stems olivaceous, scattered, usually solitary, the branches never in whorls; flowers in short few-flowered spikes, the staminate with broadly ovate, acutish, often purplish lobes; fruit 5 mm. long. *Razoumofskya Douglasii* (Engelm.) Kuntze. On douglas fir, *Pseudotsuga taxifolia,* and reported elsewhere on white fir, *Abies grandis,* on various other species of *Abies* and *Picea.* Probably not uncommon but the minute plants are hidden by the needles of the host. Collected only on *Pseudotsuga* at Kamiak Butte, and at the Touchet River, *A. T. T.*

Arceuthobium laricis (Piper) St. John. Staminate plants swollen, yellow, about 2 mm. thick at base, the flowers in short 3–7-flowered spikes; lobes ovate, acute; pistillate plants olivaceous clustered, 5–8 cm. long, branched; joints 1.5–2 mm. thick, sharply 4-angled; fruit oblong, acutish, bluish or green, 4 mm. long. *Razoumofskya laricis* Piper. Common on tamarack, *Larix occidentalis, A. T. T.,* and as shown by J. R. Weir, Bot. Gaz. **66**: 15–19, 1918, very rare on *Pinus ponderosa, P. contorta,* and *Abies grandis.* It readily grows on other species of *Larix.* The parasite seems to hold its characters even when transferred to different hosts. This would justify its right to specific rank, so I made the necessary transfer to *Arceuthobium,* a genus conserved by the International Rules.

Reduced to synonymy is *A. campylopodum* Englem., *forma laricis* (Piper) L. S. Gill, a name representing a classification based solely upon the identity of the host.

ARISTOLOCHIACEAE. Birthwort Family

Low herbs or twining shrubs, with bitter tonic, aromatic, or stimulating properties; leaves alternate or basal, petioled, mostly cordate or reniform, without stipules; flowers axillary or terminal, solitary or clustered, perfect; calyx-tube at least at the base adnate to the ovary, its limb 3-lobed, 6-lobed, or irregular; petals none; stamens 6–many, united with the style; ovary partly or wholly inferior, mostly 6-celled; ovules numerous in each cavity; fruit usually a 6-celled capsule or berry; seeds many, ovoid or oblong, angled or compressed. (Named from the genus *Aristolochia.*)

ASARUM. Wild Ginger

Stemless perennial herbs, with slender aromatic branched rootstocks; roots thick, fibrous-fleshy; leaves long-petioled, reniform or cordate, entire; flowers solitary, peduncled, brown-purple or mottled, borne near the ground; calyx bell-shaped or hemispheric, regularly 3-lobed; stamens 12, inserted on the ovary; ovary partly or wholly inferior, 6-celled, the parietal placentae intruded; capsule coriaceous. (An ancient name, of obscure derivation.)

Asarum caudatum Lindl. *Long-tailed Wild Ginger.* Rootstocks 10–30 cm. long; blades 5–14 cm. wide, reniform-cordate, obtuse or acutish, hispid; petioles 3–23 cm. long, crinkled villous; peduncles 1–4 cm. long, crinkled villous; flowers terminal, reclining; calyx 2.5–8.5 cm. long, crinkled villous, brownish purple without, maroon within; calyx-lobes oblong-lanceolate, attenuate. Deep mountain woods, *A. T. T., Can.*

POLYGONACEAE. Buckwheat Family

Herbs or shrubs, or in tropics vines or trees; stems jointed; leaves alternate or sometimes opposite or whorled, simple, mostly entire, with usually sheathing united stipules (*ocreae*); flowers small, regular, perfect, dioecious, monoecious, or polygamous, in spikes, racemes, corymbs, umbels, or panicles; petals none; calyx free from the ovary, 2–6-cleft or -parted, the segments sometimes petal-like; stamens 2–9, inserted near the base of the calyx or in staminate flowers crowded toward the center; pistil 1; ovary superior, 1-celled; style 2–3-cleft; ovule 1; fruit a lenticular or 3-angled, rarely 4-angled achene; endosperm mealy; embryo straight or curved. (Named for the genus *Polygonum.*)

Flowers subtended by involucres; leaves without stipules. *Eriogonum.*
Flowers not subtended by involucres; leaves with sheathing stipules,
  Perianth 6-parted; stigmas 3, tufted. *Rumex.*
  Perianth 5- (or 4–6-) parted; stigmas 2 or 3, capitate. *Polygonum.*

ERIOGONUM

Annual or perennial acaulescent or leafy-stemmed herbs or shrubs; stems simple or branched, often tufted; leaves entire, alternate or whorled; flowers small, fascicled, cymose, umbellate or capitate, subtended by 5–8-toothed or -cleft campanulate or cylindric involucres; calyx 6-cleft or -parted, usually colored; segments equal or the outer ones larger; stamens 9; styles 3; achenes pyramidal, 3-angled, more or less swollen near the base. (Name from Gr., *erion,* wool; *gonu,* knee.)

Joint apparently near middle of pedicel, due to the narrow stipe-like base of the perianth,
  Perianth very hairy,
    Tube of involucre 2–4 mm. long, with reflexed linear or oblong lobes. *E. Douglasii.*
    Tube of involucre 6–8 mm. long, with minute, erect lobes. *E. Piperi.*
  Perianth glabrous or nearly so,
    Flowering stem with a whorl of leaves near the middle; leaves linear to oblanceolate. *E. heracleoides.*
    Flowering stem scape-like; leaves broader,
      Leaves all or some of them cordate at base. *E. compositum,* var. *compositum.*
      Leaves not cordate,

Umbel compound,
Flowers white; leaves lanceolate. Var. *lancifolium.*
Flowers bright yellow; leaves oblanceolate to broadly oval. *E. stellatum.*
Umbel simple. *E. subalpinum.*
Joint at tip of pedicel beneath the flower which is not stipe-like at base,
Shrubs; leaf margins revolute. *E. microthecum.*
Herbs; leaf margins not revolute,
Outer perianth segments subcordate at base; tube of involucre woolly. *E. niveum.*
Outer perianth segments not cordate; tube of involucre glabrous,
Flowers 1.5–2.5 mm. long, numerous on wand-like branches, in the axils of bracts. *E. vimineum.*
Flowers 2.5–3 mm. long, few, terminal or in the axils of the 2–3-forking cymes. *E. strictum.*

Eriogonum compositum Dougl. ex. Benth., var. **compositum.** Stems 20–43 cm. tall, stout, simple, glabrous or nearly so; petioles 3–10 cm. long, somewhat lanate; blades 2–8.5 cm. long, deltoid to ovate, acute or obtuse, densely white-woolly beneath, green and less so above, umbels 6–10-rayed, mostly compound, rather dense; principal bracts linear or oblanceolate; rays short or 2–10 cm. long; involucre body obconic, 3–5 mm. long, pilose at tip or glabrate, deeply 5-lobed, the deltoid lobes reflexed; flowers white or yellow, sparsely pilose or glabrous, 3–4, becoming as much as 7 mm. long; outer calyx-lobes ovate, the inner obovate, nearly equal in width, but becoming longer; achene pilose at tip. Including *E. compositum leianthum* Benth. Dry gravelly slopes, *U. Son., A. T.*

Var. **lancifolium** St. John & Warren. Leaves lanceolate, cuneate, or occasionally subcordate at base; scapes slightly pubescent; flowers white. Dry gravelly slopes, Wawawai, *Elmer 773, U. Son.*

Eriogonum Douglasii Benth. Plant woolly throughout; branches woody, prostrate or nearly so; leaves in tufts on woody caudex branches, 1–2 cm. long, oblanceolate or spatulate, cuneate, densely white woolly; peduncles 5–10 cm. tall, with a whorl of 5–6 linear-oblanceolate bracts near the middle; flowers yellow to purplish, in simple spherical, head-like umbels; perianth 3–6 mm. long, pilose, the lobes obovate, the inner ones slightly larger; achene 4 mm. long, trigonous-lanceoloid, pilose at tip. Open gravels, above 4,000 ft., Blue Mts., *A. T.* to *Huds.*

Discovered by D. Douglas in the Blue Mts.

Eriogonum heracleoides Nutt. Loosely tufted, thinly tomentose throughout; flowering stems erect, 1–4 dm. tall; leaves clustered at end of branches; petioles 2–30 mm. long; blades 1–6 cm. long, acute, attenuate at base, woolly or glabrate above, paler and more pubescent beneath; bracts similar to the leaves, but smaller; umbels 1–10-rayed, simple or compound; involucres woolly, the tube campanulate, 3–4 mm. long, the linear lobes as long, reflexed; flowers whitish, 3.5 mm., becoming as much as 7.5 mm. long; outer perianth lobes oval, the inner narrower, long clawed with an oval limb; achenes 3.5 mm. long, pilose on the 3-angled tip. Dry, often rocky places, *U. Son.* to *A. T. T.,* abundant in *A. T.*

Eriogonum microthecum Nutt. Shrub 2–10 dm. tall; young branches woolly, the older brown, smooth and shreddy; leaves alternate; petioles 2–5 mm. long, woolly; blades 5–27 mm. long, oblanceolate to almost linear, densely white woolly beneath, green and often glabrate above; inflorescence a cyme; involucres peduncled, narrowly funnelform, the tube 2 mm. long, glabrate, the lobes 0.5 mm. long, lanate, inflexed; flowers 1.5–3 mm. long, white or pink, the outer lobes oval, the inner elliptic, half as wide, united nearly half their length; achenes 2 mm. long, linear-lanceoloid, smooth. Dry hillside, Snake River Canyon, Wilma, *St. John* 6,797; Lime Pt., *St. John* 9,290, *U. Son.*

Eriogonum niveum Dougl. ex Benth. *Canyon Heather.* Densely white-tomentose throughout; plant tufted from a loose woody caudex, stems 2–5 dm. tall, several-times forked, the branches in twos or threes; basal leaves numerous; petioles 1–4 cm. long; blades 1–3 cm. long, elliptic or ovate, obtuse, densely white-woolly on each side, cauline narrower, nearly sessile, whorled at the forks, the upper reduced; involucres axillary and sessile or terminal, narrowly campanulate, the tube 3–3.5 mm. long, the stiff deltoid lobes erect, 1–2 mm. long; flowers white, pink, or rose, 4–5 mm. long; the outer lobes suborbicular, the inner oblanceolate; achenes 3.5 mm. long, glabrous, trigonous, lanceoloid, the upper part cellular reticulate. *E. niveum,* var. *Suksdorfii* Gdgr. is probably not distinct. Dry slopes, abundant, *U. Son.,* occasional in *A. T.*

Discovered by Douglas in the Blue Mts. The flowers are everlasting and the plant is a favorite for winter bouquets.

Eriogonum Piperi Greene. Densely tufted on a stout woody caudex; stems erect, leafless, tomentose, 1–3 dm. tall; leaves basal; petioles 1–4 cm. long, pilose; blades 2–4 cm. long, lanceolate or oblanceolate, acute or obtuse, white and densely hairy beneath, green and less hairy above; umbels with several rays 5–15 mm. long; bracts 3–8, oblanceolate, 1–3 cm. long; involucre short-toothed, villous; flowers yellow, very villous, 5–7.5 mm. long, the outer perianth lobes elliptic, the inner obovate, slightly broader; achene 3 mm. long, trigonous, the sides flattened, glabrous. Rocky places, *Huds.*

Named for Prof. C. V. Piper who discovered the plant in the Blue Mts.

Eriogonum stellatum Benth. In loose tufts, woody at base, somewhat tomentose; stems 1–4 dm. tall, scape-like; leaves clustered at end of branches; petioles 4–40 mm. long; blades 8–35 mm. long, obtuse, densely white tomentose beneath; bracts leaf-like; umbel compound; involucre tube 2 mm. long, campanulate, lanate, the reflexed oblong lobes longer; flowers 4–5 mm., becoming 7 mm. long from tip to joint, glabrous; outer perianth-lobes oblanceolate, obtuse; inner lobes obovate, larger; achenes 5 mm. long, trigonous, lanceoloid, cuneate, pilose on the angles at tip. Rocky slopes and peaks, Blue Mts., *Huds.*

Eriogonum strictum Benth. Plant tufted from a slender woody base and caudex; stems 15–40 cm. tall, woolly below; leaves basal or nearly so; petioles 1–6 cm. long, woolly; blades 1–3 cm. long, white woolly beneath, sparsely so or glabrate above, slender, elliptical to oblanceolate, acute, mostly on short branches from the caudex; inflorescence glabrous, forking into several slender, ascending branches; flowers yellowish-white; involucres sessile, axillary or terminal, the tube 2–3 mm. long, narrowly campanulate, glabrous, the lobes 1 mm. or less in length, deltoid, erect, carinate, the sinus woolly; flowers white, the outer lobes broadly oval, the inner elliptic, about half as broad; ovary lanceoloid, glabrous. Rocky slopes, Blue Mts., *Huds.*

Eriogonum subalpinum Greene. *Sulphur Plant.* Perennial, 1–3 dm. tall, the short woody branches forming mats; flowering stems scape-like, woolly, with a ring of leafy bracts at tip; leaves basal or on sterile shoots; petioles 3–40 mm. long, woolly; blades 8–30 mm. long, elliptic, woolly beneath, above green and soon glabrous; inflorescence an umbel, the rays 1–3 cm. long; involucre woolly, the tube 2–3 mm. long, campanulate, the lobes 3.5–5 mm. long, oblong; flowers 4–7 mm. long, cream-colored, turning rose-colored, glabrous, the lobes oval, of about the same width, but the inner the longer; achenes 3–4 mm. long, linear-lanceoloid, pilose at tip. *E. umbellatum* Torr., var. *majus* Benth. Exposed summits, Mt. Spokane, *Kreager* 236; Mica Peak, *Suksdorf* 8,806, *Huds.*

Eriogonum vimineum Dougl. ex Benth. Annual; stems wiry, erect, 15–40 cm. tall, much branched above the base, tomentose below, the branches long and slender; leaves commonly in a single rosette at or near the base; petioles 3–28 mm. long; blades 7–50 mm. long, broadly oval or orbicular, densely tomentose beneath, less so above; inflorescence soon glabrate, slender branched; involucres sessile in axils of branches or bracts, the tube 2.5–3.5 mm. long, narrowly cylindric, the deltoid lobes 0.3 mm. long, erect, carinate, ciliate; flowers pink, the outer segments broadly obovate, the inner narrower, oblanceolate, retuse; achene narrowly fusiform, trigonous, scabrous puberulent above the cuneate base. Dry places, *U. Son.*

E. heracleoides Nutt., var. micranthum Gdgr., Bull. Soc. Roy. Bot. Belg. 189, 1906, has not been available for study.

POLYGONUM. Knotweed

Annual or perennial, terrestrial or aquatic herbs, some species woody; stems erect, prostrate, climbing or floating; leaves alternate, sessile or petioled, continuous with or jointed to the cylindric funnelform or 2-lobed often lacerate or fringed ocreae; flowers small, perfect, green, white, pink, or purple, variously clustered, the clusters terminal or axillary; pedicels jointed, subtended by the sheaths (*ocreolae*); calyx 4–5-parted or -cleft, often petaloid, the outer segments larger than the inner; stamens 3–9; stigmas capitate; achenes lenticular or 3-angled (rarely 4-angled), invested by or exceeding the calyx. (Name Gr., *polus,* many; *gonu,* knee.)

Stems twining; leaves cordate. *P. Convolvulus.*
Stems and leaves not so,
  Leaves jointed on the petioles, small; ochreae 2-lobed, becoming lacerate; inner filaments dilated,
    Fruit reflexed,
      Perianth campanulate, the outer segments thickened, green or purplish on the back,
        Upper bracts subulate; leaves lanceolate. *P. Douglasii,* var. *Douglasii.*
        Upper bracts foliaceous; leaves oblanceolate. Var. *latifolium.*
      Perianth funnelform, the outer segments white, merely with a narrow green midrib. *P. majus.*
    Fruit erect,
      Flowers in rather dense, terminal, bracteate spikes,
        Bracts elliptic, with broad, white, scarious margins. *P. polygaloides.*
        Bracts linear-lanceolate, green,
          Achenes brown, smooth. *P. Kelloggii.*
          Achenes black, striate. *P. confertiflorum.*
      Flowers axillary, in loose or interrupted spikes,
        Stems striate with numerous prominent smooth ridges,
          Achenes exserted, about twice as long as the calyx. *P. exsertum.*

Achenes entirely or nearly enclosed by the calyx. *P. aviculare.*
Stems scaberulous striate, becoming terete and smooth at base. *P. minimum.*
Leaves not jointed at base, large; ocreae not 2-lobed; filaments slender,
Styles 3; roots large, tuberous,
Leaves puberulent beneath. *P. bistortoides,* var. *bistortoides.*
Leaves glabrous beneath. Var. *oblongifolium.*
Styles 2; roots slender or fibrous,
Racemes 1 or few, terminal; aquatic or marsh perennials,
Branches aquatic; leaves or most of them glabrous.
Peducle glabrous; raceme ovoid, 1–3 cm. long. *P. amphibium,* var. *stipulaceum,* f. *fluitans.*
Peduncle glandular puberulent, raceme narrowly cylindric, 3–10 cm. long. *P. coccineum,* f. *natans.*
Branches terrestrial; leaves hairy,
Leaf sheaths with spreading, foliaceous border. *P. amphibium,* var. *stipulaceum,* f. *simile.*
Leaf sheaths close, not foliaceous,
Racemes mostly 4–18 cm. long and petioles attached midway on the sheath. *P. coccineum,* var. *pratincola.*
Racemes mostly 4–8 cm. long; petioles attached near the base of the sheath. Var. *coccineum, forma terrestre.*
Racemes axillary and terminal, numerous; annuals, often of drier habitats,
Ocreae without marginal bristles,
Leaves green; achenes 1.5–2 mm. broad. *P. lapathifolium.*
Leaves white tomentose beneath; achenes 2.5–2.9 mm. broad. Var. *salicifolium.*
Ocreae bristle-fringed,
Fruiting raceme about 1 cm. thick; perianth not punctate. *P. Persicaria.*
Fruiting raceme about 5 mm. thick; perianth glandular punctate. *P. Hydropiper,* var. *projectum.*

Polygonum amphibium L., var. **stipulaceum** (Coleman) Fern., forma **fluitans** (Eaton) Fern. Perennial aquatic; stems floating or submersed, 5–7 mm. in diameter; petioles 1–6 cm. long, attached to upper part of the sheath;

blades 7–12 cm. long, elliptic or elliptic-oval, (or lanceolate if stranded) subcoriaceous, glabrous, the base rounded, apex acute, entire; sheaths cylindric, scarious; peduncle glabrous; raceme spike-like 1–3 cm. long, ovoid, dense; ocreolae 3–4 mm. long, triangular, membranous; calyx 3–4 mm. long, pink, parted to the middle into rounded lobes; stamens and styles dimorphic; achenes 2.5–2.7 mm. long, thick lenticular, dull and pitted. *P. natans* A. Eaton, forma *genuinum* Stanf. Ponds and streams, *A. T.*

Var. **stipulaceum** (Coleman) Fern., forma **simile** Fern. Differing in being more or less stranded; stems erect, more or less hirsute; petioles 1–5 mm. long; blades 10–15 cm. long, lanceolate, chartaceous, more or less hirsute, at least on the margin; sheaths cylindric, almost membranous, hirsute, with a spreading foliaceous border about 1 cm. in diameter; raceme rare and usually sterile. Exsiccated shores, *A. T.*

Polygonum aviculare L. *Knotweed.* Annual, glaucous, glabrous; stems 1–6 dm. long, slender, usually prostrate, much branched, forming dense mats, the branches leafy to the top; leaves 1–3 cm. long, oblong or lanceolate, acute, nearly sessile; sheaths scarious, cut into lobes; flowers 1 or few in the upper axils, pinkish margined, on very short pedicels which are included in the sheaths; achenes 2.5–3.5 mm. long, 3-angled, dull, minutely granular. A European weed, established in paths, lawns, and fields, *U. Son., A. T.*

Much relished by stock. Many recent Europeans recognize numerous subdivisions of this species. Lindman in Svensk Bot. Tidskr. **6**: 673–696, 1912, maintains three binomial subspecies and many varieties under the collectve species, *P. aviculare.*

Polygonum bistortoides Pursh, var. **bistortoides.** Perennial, from a thick tuber, stem erect, simple, 2–8.5 dm. tall; radical leaves oblong or elliptic lanceolate, acute or acuminate, 5–20 cm. long, 3–7 cm. broad, on slender usually shorter petioles; cauline 3–4, sessile, the uppermost much reduced, lanceolate; leaves all glaucous and puberulent beneath; spikes 2–6 cm. long, 1–2 cm. thick, cylindric dense; calyx 4–6 mm. long, white, deeply 5-cleft; achenes 4 mm. long, chestnut-brown, 3-angled, smooth, shiny. In the original description Pursh wrote that the leaves were glabrous on both sides. Despite this, the type specimen, now at the Philadelphia Academy of Natural Sciences, has the leaves crisp puberulent beneath. It was collected by Capt. M. Lewis at "Quamash-flats," that is at Weippe, Idaho. This puberulent plant is abundant and characteristic there, and in the low meadows of northern Idaho from Potlatch, the Thatuna Hills, Moscow, and the Craig Mts. It is also found in western Idaho, eastern Oregon and Colorado. There are isolated western stations at Winchester Mt., *St. John* 8,976; Horseshoe Basin, *St. John & Ridout,* 3,660; and Olympic Mts., *Elmer* 2,681, Washington; and Monmouth, *Coleman;* and Yeoman, *Nelson* 597, Oregon. In the local area, abundant in moist meadows near the mountains, eastern part, *A. T., A. T. T.*

Var. **oblongifolium** (Meisn.) St. John *P. Bistoria* L., var. *oblongifolium* Meisn.; *P. vulcanica* Greene. Differing from the species in having the leaves glabrous throughout. This is the beautiful *Mountain Dock* of the alpine meadows of the Cascade Mountains of Washington and Oregon, the Sierras of California, and the Blue Mountains of Washington. The writer has studied an isotype of the variety in the Gray Herbarium, and observed the leaves to be glabrous.

The type specimen of this variety is *Geyer* 405, Highlands of the Nez Percez. In Hooker's "Catalogue of Geyer's Collection," Journ. Bot. & Kew Miscell. **5**: 262, 1853, the habitat is more fully stated as "Moist deep grassy meadows, on the high and cold plains of the Nez Percez Indians, with Veratrum viride." C. A. Geyer's own, "Notes on the Vegetation of . . . Oregon Territories . . . ," London Journ. Bot. **4**: 483, 1845, and **5**: 520, 521, 1846, gives more details: "To [Rev. Henry Spalding] I owe the means of visiting another field, the

Highlands of the Nez-Percez Indians, where he accompanied me on my excursions, and also afforded facilities to investigate the flowery Koos Kooskee valley over again, where previous botanists had but cursorily passed." . . . "Following these rivulets to their source in the plains, we come to a vegetation of *Camass, Veratrum, Carices, Polygonum* (405), . . ." "Leaving the main ridge of the Blue Mountains to our extreme left, we descend again, at the junction of the Koos-Kooskee and Lewis river, . . . " To use the modern names, it is clear that Geyer and Spalding climbed from near Clarkston or Asotin, Washington, to the head of a fork of Asotin Creek. Here in the meadows at the edge of the woods he collected his *Polygonum* No. 405. This is in the immediate vicinity of Anatone, where the glabrous variety and it only is still abundant, and where a green flowered *Veratrum* similar to *Veratrum viride* occurs. There are five recent collections of the *Polygonum* from that region, and it is the only one in the Blue Mountains, which stretch westward from Anatone.

The range of this glabrous plant is largely distinct from the pubescent species. It might be considered a species, as did Dr. Greene, but the sole character of pubescence is a meager one. Then there are some intermediates. From the alpine meadows of Mt. Baker, is a collection, *St. John* 7,876, which has the leaves of young plants puberulent beneath, though the older ones are glabrous. There is a similar intermediate from South Baldy, *St. John & Smith* 8,828. The existence of such intermediate specimens convinces the writer that the glabrous plant is best classified as a variety.

Polygonum coccineum Muhl., var. **coccineum,** forma **terrestre** (Willd.) Stanf. Emersed or stranded aquatic plant; stem 10–15 dm. tall, striate, above hirsute or glandular hirsute and often branched; petioles 2–3.5 cm. long, attached near the base of the sheath; blades 10–18 cm. long, lanceolate to ovate-lanceolate, the base from rounded to cuneate, subcoriaceous, the margin bristly scabrous, the surfaces mostly more or less appressed hirsute; sheaths 2–3.5 cm. long, membranous, enlarged at base, hirsutulous, the margin entire or ciliate; peduncles 3–7 cm. long, hirsutulous or glandular hirsutulous; raceme 3–10 cm. long, spike-like, narrow; ocreolae 3–4 mm. long, lance-deltoid, hirsutulous; flowers dimorphic as to stamens and styles, on separate plants; calyx 3–3.5 mm. long, in fruit 4–5 mm. long, deep pink, parted nearly to the base into broadly oval segments; achenes 2.5–3.3 mm. long, thick lenticular, minutely roughened, dark brown. *P. emersum* (Michx.) Britton. Wet shores, *A. T.*

Forma **natans** (Wieg.) Stanf. Differs from f. *terrestre* in being submerged or floating; petioles 2–9 cm. long; blades 10–15 cm. long, ovate-lanceolate to lanceolate, the base rounded to cordate, glabrous or glabrate; sheaths glabrous; peduncles usually glandular puberulent. Known from the type locality, Moscow, Ida., *Muenscher* 129, *A. T.*

Var. **pratincola** (Greene) Stanf. Differs from f. *terrestre* in being more or less minutely canescent with weak simple, or with glandular hairs; petioles attached midway up the sheath; sheath mostly acute and densely hirsute; racemes usually longer. Gravel island in Grande Ronde River, Rays Ferry, Asotin Co., *St. John et al.* 9,691, *U. Son.*

Polygonum confertiflorum Nutt. ex Piper. Annual, glabrous; stems very slender, 3–15 cm. tall, with few branches; sheaths scarious, soon split nearly to base into slender lobes; leaves linear, 1–3 cm. long; spikes few, rather dense; bracts longer than the pink flowers, the margins revolute and keeled; perianth 2 mm. long; stamens 3–5; achenes 1.7–2 mm. long. *P. Watsoni* Small, in part. In thin soil, rare, *A. T.*

Polygonum Convolvulus L. *Black Bindweed.* Annual; stems prostrate or climbing, 1–12 dm. long, glabrous or pubescent, scurfy below; ocreae close, rough, oblique; petioles equaling or shorter than the blades; blades 2–6 cm. long, cordate or somewhat hastate, acuminate, minutely scurfy beneath; flowers in clusters or slender axillary leafy racemes; calyx 5-parted, 3.5–4 mm. long, green, scaberulous, closely investing the fruit; achene 3.5 mm. long, 3-angled, dull black, minutely roughened. *Bilderdykia Convolvulus* (L.) Dumort; *Tinaria Convolvulus* (L.) Webb and Moq. Weed, from Europe, introduced to cultivated or waste soil, *U. Son., A. T.*

Polygonum Douglasii Greene, var. **Douglasii.** Annual; stems 1–6 dm. tall, angled and scaberulous to glabrous, erect; ocreae scarious, the veins prolonged in the acuminate teeth; leaves 2–6 cm. long, acute at each end, sessile at the jointed base; flowering branches slender; flowers 1–3, from the axils of the scattered bracts, soon becoming deflexed; calyx purple or whitish margined, 2.5–4 mm. long; achenes 3–4 mm. long, 3-angled, black, shiny. Common on gravelly or dry soil, *U. Son.* to *A. T. T., Huds.*

Var. **latifolium** (Engelm.) Greene. A more leafy variety. *P. douglasii montanum* Small. Gravelly soil in pine woods in the mountains, uncommon, *A. T. T., Huds.*

Polygonum exsertum Small. Annual, 2–10 dm. tall, glabrous, sometimes slightly glaucous; tap-root 2–6 mm. in diameter, brown; stem usually erect, striate, much branched; leaves 1–6 cm. long, linear-lanceolate to oblanceolate, cuneate at base, subsessile; sheaths funnelform, hyaline and silvery, becoming brownish and lacerate; flowers 2–4 in the axils; pedicels 3–3.5 mm. long, erect; calyx 3–3.5 mm. long, 5–6-parted, the segments oval, obtuse, with hyaline white or pink margins; achenes 4–6 mm. long, 3-angled, ovoid-pyramidal, somewhat acuminate, smooth, greenish to chestnut-colored. Blue Mts., *Horner* 427.

Polygonum Hydropiper L., var. **projectum** Stanf. Annual, with peppery, acrid herbage; stem 2–6 dm. tall, erect or ascending, or the tips drooping, glabrous; ocreae scarious, loose, brown, ciliate with bristles usually 2 mm. long, swollen at base by the concealed panicles; blades 3–5 cm. long, ovate or ovate-lanceolate, subsessile, glandular-punctate; panicles numerous; pedicels exserted; perianth 2–2.3 mm. long, green or pink margined; achenes 2–2.5 mm. long, blackish, dull, lenticular or trigonous. Slough, near mouth of Potlatch River, Idaho, *St. John et al.* 9,746, *U. Son.*

Polygonum Kelloggii Greene. Delicate annual; stem 3–8 cm. tall, striate above; simple or branched at base; ocreae scarious, the acuminate teeth soon split; leaves 5–10 mm. long, linear, acute; flowers axillary; perianth 1.5–2 mm. long, white; achenes 1.2 mm. long, 3-angled. Bare ridges, Blue Mountains, *Huds.*

Polygonum lapathifolium L. Stems branched, 2–10 dm. high, erect or ascending; ocreae cylindric, ribbed or striate; leaves 5–20 cm. long, lanceolate, acuminate, scabrous on the midrib and margins; peduncles minutely glandular; spikes axillary and terminal, slender, 2–10 cm. long, erect or nodding; flowers 2–2.5 mm. long, white or pink; outer segments with raised anchor-like nerves; stamens 6; achenes 2 mm. long, brown, ovate, lenticular. *Persicaria lapathifolia* (L.) S. F. Gray. European weed, sparingly introduced, *U. Son., A. T.*

Var. salicifolium Sibth. Annual, simple or branched, 10–30 cm. high; leaves lanceolate, acute, 2–6 cm. long, green above, white woolly beneath; peduncles decidedly glandular; spikes shorter, 1–3 cm. long. On river banks, rare. A European weed. Recorded by Piper & Beattie, not verified.

Polygonum majus (Meisn.) Piper. Annual, 1–4 dm. tall, much branched from the base, erect or nearly so; stems wiry, terete, inconspicuously striate; ocreae scarious, with 2 acuminate teeth; leaves 2–6 cm. long, linear-lanceolate, spikes elongate, 5–12 cm. long; flowers remote, short-pedicelled, horizontal or spreading; bracts consisting of the scarious sheaths and short subulate blades, about as long as the buds; calyx 3–6 mm. long; styles separate for one half their length; fruit reflexed; achene black, shiny, smooth, 3-angled, acuminate, 3–3.5 mm. long. Common in stony soil, *U. Son., A. T.*

Polygonum minimum Wats. Annual, slightly scabrous; stems 5–20 cm. high, often branched from the base, wiry, red; ocreae scarious, turning brown, papillose, the margin with several acuminate teeth; leaves 5–12 mm. long, obovate to elliptic acute or apiculate; flowers axillary, crowded above, 1.5–2 mm. long, white- or pink-margined, erect on short pedicels; stamens 5–8; achenes 2–2.3 mm. long, smooth, shiny. In moist places at high altitudes, Blue Mountains, *Huds.*

Polygonum Persicaria L. *Lady's Thumb.* Annual, erect or ascending, glabrous or puberulent; stems 2–9 dm. long; ocreae 10–15 mm. long, tubular; leaves 2–25 cm. long, lanceolate, acute or acuminate at each end, short-petioled, frequently with a brown-purple spot near the middle, glabrous or sparsely puberulent below and on the midrib and margins above; spikes slender, 2–4 cm. long, slender peduncled, erect; calyx 2.5 mm. long, pink, 5-lobed; achene 2–3 mm. long, lenticular, or trigonous, black and shining. *Persicaria Persicaria* (L.) Small. A European weed, sparingly introduced, *U. Son.*

Polygonum polygaloides Meisn. Annual, glabrous; stems slender, 5–22 cm. tall, much branched from the base, ocreae scarious, soon cleft into long acuminate lobes; leaves 1–5 cm. long, linear, sessile, 1-nerved; spikes 5–10 mm. long, dense, numerous, terminal; bracts crenulate, obtuse or acuminate, exceeding the flowers; perianth 2 mm. long; stamens 8; achenes 1.5 mm. long, 3-angled, long acuminate, longitudinally striate. *P. natans* A. Eaton, forma *Hartwrightii* (Gray) Stanf. Moist meadows, *A. T.*

First collected by Rev. Henry Spalding, along the Clearwater River.

RUMEX. Dock

Perennial or annual leafy-stemmed herbs, some species slightly woody; stem grooved, mostly branched; leaves entire or lobed, undulate, flat or crisped; ocreae usually cylindric, brittle, soon falling away; inflorescence of simple or compound often panicled racemes; flowers green or reddish, perfect, dioecious, or polygamo-monoecious, whorled, on jointed pedicels; perianth 6-parted, the three outer ones unchanged in fruit, the three inner ones mostly developed into winged valves which are entire, dentate or fringed with bristle-like teeth; stamens 6; achenes 3-angled, the angles more or less margined. (The ancient Lat. name.)

Flowers dioecious; foliage acid,
- Inner perianth parts not longer than the granular achene; leaves hastate. — *R. Acetosella.*
- Inner perianth parts longer than the smooth achene; leaves lanceolate. — *R. paucifolius.*

Flowers not dioecious; foliage not acid,
- Inner perianth parts without tubercle-like thickening in fruit,
  - Fruiting valves cordate, 15–30 mm. long,
    - Fruiting valves 15–30 mm. long. — *R. venosus.*
    - Fruiting valves 5–7 mm. long. — *R. Patientia.*
  - Fruit valves ovate, 6–10 mm. long. — *R. occidentalis.*
- Inner perianth parts or some of them with tubercle-like thickenings in fruit,
  - Annuals; valves with bristle-like marginal teeth. — *R. maritimus,* var. *fueginus.*

Perennials; valves denticulate,
  Leaves crisped on margins; tubercles ovoid,
    Lower leaves cordate or rounded at base; tubercle mostly half as long as the valve. *R. crispus.*
    Lower leaves narrowed or acuminate at base; tubercle less than one-third as long as the valve. *R. Patientia.*
  Leaves flat; tubercle lance-ellipsoid,
    Leaves lanceolate; all inner perianth parts with a tubercle. *R. mexicanus.*
    Leaves broadly oval, cordate; one inner perianth part with a tubercle. *R. obtusifolius.*

Rumex Acetosella L. *Sheep Sorrel.* Spreading by creeping rootstocks; stems 1–6 dm. tall, glabrous, slender, simple or somewhat branched; leaves 1–15 cm. long, glabrous, mostly hastate, the basal lobes entire or toothed; panicle narrow, the branches ascending; bracts wanting; flowers clustered; sepals remaining small, much shorter than the 3-angled achene, 1 mm. long. A European weed, now abundant in lawns and pastures, *U. Son., A. T.* The sour leaves are edible.

Rumex crispus L. *Yellow Dock.* Stem stout, 3–16 dm. high, grooved, from an elongated fusiform root; leaves oblong-lanceolate, obtuse, truncate or rounded at the base, 15–30 cm. long; petiole short; panicle rather dense, 20–40 cm. long, greenish; pedicels 5–10 mm. long; valves 3–5 mm. long, broadly ovate or cordate; achenes 2–2.5 mm. long, shining, brown. A European weed established in our grain fields and along roadsides, *U. Son., A. T.*

Rumex maritimus L., var. **fueginus** (Phil.) Dusén. Minutely hispidulous; stems erect or procumbent, branched, 1–10 dm. high; leaves 3–25 cm. long, linear-lanceolate, truncate or slightly cordate at base, wavy-margined; panicle dense, the flowers short-pedicelled in numerous close whorls; valves ovate, each bearing a linear-lanceolate tubercle on the back and 2 or 3 slender bristles on the margins; achenes 1.2 mm. long, brown, shining. *R. fueginus* Phil. In wet places, especially where alkaline, Snake River, and the "Scablands," *U. Son.,* occasional in *A. T.*

Rumex mexicanus Meisn. Stems erect or decumbent, 2–10 dm. long, slightly grooved, often branched below; leaves 8–15 cm. long, oblong-lanceolate or linear-lanceolate, mostly acute, entire, rounded or cuneate at base; petioles rather short; panicle 10–25 cm. long, the branches short, leafy-bracted; valves 3.5–6 mm. long, triangular-ovate, slightly toothed, achenes 2 mm. long, red-brown, shining, margined. *R. triangulivalvis* (Danser) Rech. f.; *R. utahensis* Rech. f. Wet shores, *U. Son., A. T.*

Rumex obtusifolius L. *Bitter Dock.* Perennial from stout taproot, glabrous, dark green; stem 6–12 dm. tall, stout, erect, simple or sparingly branched, roughish; racemes in terminal panicle, upright, with slightly remote whorls of flowers; basal blades 10–35 cm. long, broadly to narrowly ovate, long petioled, membranous, cordate to subcordate at base, rounded or blunt at apex, usually red-veined, crenulate; upper leaves oblong to lanceolate, acute; pedicels longer than the fruit; valves 3–5 mm. long, triangular-ovate, submembranaceous, with prominent teeth, only one valve (rarely all) with an ovate tubercle; achene 2 mm. long, reddish brown, shining. Weed, introduced from Europe, in 1947, Forest, Craig Mts., Nez Perce Co., *Ownbey 3,142.*

Rumex occidentalis Wats. Stout, perennial, 5–20 dm. tall; leaves 1–3.5 dm. long, oblong-ovate to lanceolate, mostly obtuse, entire or undulate, cordate at the base; petioles of the lower leaves long and slender, of the upper stout; panicles dense, usually reddish; pedicels 5–15 mm. long, slender, clavate; valves obtuse, more or less toothed; achenes 3–4 mm. long, brown, shining. *R. fenestratus* Greene. Swampy places, *A. T., A. T. T.*

Rumex Patientia L. Perennial 6–15 dm. tall; stems erect, from the middle branching and flowering; basal leaves with petioles mostly glabrous, mostly a third shorter than the blades; blades 2–3 dm. long, ovate- or oblong-lanceolate, acute, like all the others somewhat fleshy, whitish green, the margin undulate and usually crisped and crenulate, beneath glabrous, but scaberulous on the veins; cauline leaves with shorter petioles; the blades oblong-lanceolate, usually acute, truncate at base; uppermost blades lanceolate to linear, subsessile; racemes erect, dense or somewhat interrupted in fruit, the glomerules 10–16-flowered; pedicels slender, 2–4-times as long as the valves, jointed below the middle; fruiting valves 6–8 mm. long, rounded cordate, rounded or subacuminate at apex, reddish brown, reticulate, membranous, entire, minutely crenate, or dentate, only one valve bearing a tubercle; tubercle about 1.5 mm. long, globose or ovoid; achene 3 mm. long, ovoid, acuminate, brown, shining, the 3 sides concave. *R. Patientia* L., subsp. *eu-Patientia* Rech. f. A weed, introduced from Eurasia. Spokane Co., *Turesson;* Waitsburg, 1897, *Horner* 184.

Rumex paucifolius Nutt. Perennial; stems 2–5 dm. tall, erect and slender, sparingly leafy; leaves 5–12 cm. long, narrowly to linear-lanceolate, acute to acutish, attenuate to a slender petiole, not very acid; panicle branches slender, erect; flowers reddish, small, in loose fascicles; valves 4 mm. long, rounded cordate, winged. Sandy prairies, Craig Mts., *Henderson* 2,704, *A. T. T.*

Rumex venosus Pursh. *Sand Dock.* Stems 3–6 dm. tall, commonly woody, strongly grooved, erect or decumbent, from long, woody creeping rootstocks; leaves 3–15 cm. long, elliptic or lanceolate, entire, acute at each end; petioles stout, rather dense, 1–15 cm. long; valves bright red, entire; achenes 6–8 mm. long, brown, sharply thin-angled. In sandy soil especially on sand dunes, where it is a natural sand-binder. Abundant, *U. Son.,* occasional, *A. T.*

CHENOPODIACEAE. Goosefoot Family

Mostly annual or perennial herbs; stems angled, striate or terete; leaves alternate or sometimes opposite, simple, entire, toothed or lobed, mostly petioled, without stipules; flowers perfect, polygamous, monoecious or dioecious, small, greenish, regular or slightly irregular, commonly in panicled spikes, with or without bracts; calyx persistent, 2–5-lobed or -parted, or rarely reduced to a single sepal or in some pistillate flowers wanting; petals none; stamens as many as the calyx lobes or fewer and opposite them; disk usually none; ovary mostly free from the calyx, 1-celled; styles and stigmas 1–3; ovule 1; fruit a utricle; endosperm mealy, fleshy or wanting; embryo in a ring, coil, or spiral. (Named from the genus *Chenopodium.*)

Leaves terete, spine tipped. — *Salsola.*
Leaves not so,
  Thorny shrubs; staminate flowers in aments. — *Sarcobatus.*
  Herbs or thornless shrubs; flowers not so,
    Embryo in a spiral; leaves linear, fleshy. — *Suaeda.*
    Embryo in a ring,
      Flowers perfect, all with perianth; fruit not enclosed by 2 bracts,
        Fruit laterally flattened, much larger than the calyx. — *Corispermum.*

Fruit enclosed in the calyx,
Sepal 1, bract-like; stamen 1. *Monolepis.*
Sepals 3–5; stamens 1–5,
Calyx in fruit not winged; pubescence, stellate, glandular, or none. *Chenopodium.*
Calyx in fruit with a horizontal wing; pubescence pilose. *Kochia.*
Flowers unisexual; fruit enclosed by 2 accrescent bracts. *Atriplex.*

ATRIPLEX. Saltbush

Annual or perennial herbs, usually mealy or scurfy; leaves alternate or the lower opposite; flowers in leafy spikes, monoecious or dioecious; staminate without bracts, with a 3–5-lobed calyx and 3–5 stamens; pistillate of a naked pistil between 2 appressed foliaceous bracts, which are enlarged and sometimes united in fruit; perianth usually none. (The ancient Lat. name.)

Herbs; bracts not 4-winged,
Foliage green, mostly glabrate; fruiting bracts united only near the base. *A. patula,* var. *patula.*
Foliage whitish, scurfy; bracts united ⅓ or more,
Bracts broadest below the middle, acute. *A. rosea.*
Bracts broadest at or above the middle, truncate. *A. truncata.*
Shrubs; bracts 4-winged. *A. canescens,* var. *canescens.*

Atriplex canescens (Pursh) Nutt., var. **canescens.** *Fourwing Saltbush.* Erect, 6–15 dm. tall, much branched, the branches pale, shreddy; leaves 1–5 cm. long, subsessile, linear to spatulate, pale scurfy; flowers usually dioecious; the staminate in yellow glomerules in spikes or panicles; the pistillate in axillary glomerules or in spikes; pedicels 2–15 mm. long; bracts 6–20 mm. long, united nearly to the apex, oval, the margins and the backs developing wings, the margins undulate or dentate; seeds 1.5–2.5 mm. broad, brown. *A. canescens* (Pursh) Nutt., subsp. *typica* H. & C. Spokane, *A. T.*

The herbage and especially the seeds are palatable and very nutritious to grazing animals.

Atriplex patula L., var. **patula.** Stems 6–45 cm. tall, erect; branches ascending; leaves 2–10 cm. long, lanceolate or linear, entire or denticulate, petioled; inflorescence often leafy at base; bracts 3–8 mm. long, rhomboid or roundish, narrowed to the base, the backs smooth, the margins denticulate or entire; seeds 1.5–2.5 mm. broad, brownish or black. *A. patula* L., subsp. *typica* H. & C. An uncommon weedy plant, Colfax, *Laney* in 1926, *A. T.*

Atriplex rosea L. Annual, 1–20 dm. tall, branched from the base, bushy; leaves 2–6 cm. long, sessile or petioled, ovate or rhombic-ovate to lanceolate, cuneate or rounded at base, remotely dentate; flowers monoecious; staminate glomerules in upper axils or dense terminal spikes, the perianth 4–5-lobed; pistillate without perianth; bracts 4–6 mm. long, rhombic or ovate, indurate united to the middle, the margins greenish, dentate, the backs tubercled; seeds 2–2.5 mm. long, brown, dull. A European weed, now well established, dry soils, *U. Son., A. T.*

Atriplex truncata (Torr.) Gray. *Wedgescale.* Annual, 2–10 dm. tall, usually with ascending, slender branches; leaves 1–4 cm. long, mostly sessile, deltoid to rounded-ovate, truncate or rounded at base, entire or merely undulate; flowers monoecious; upper glomerules mostly staminate, sepals 4–5; pistillate without perianth; bracts 2–3 mm. long and broad, broadly cuneate; scarcely compressed, united to the summit, backs smooth or with 1–2 minute tubercles; seeds 1–1.5 mm. wide, brown, shining. Alkaline soil, western part, *U. Son.*

CHENOPODIUM. Goosefoot

Annual or perennial, green and glabrous, white-mealy or glandular-pubescent herbs; leaves alternate, petioled, entire, sinuate-dentate or pinnately-lobed; flowers very small, sessile, bractless, clustered in axillary or terminal, often panicled or compound spikes; calyx 2–5-parted or lobed; stamens 1–5; utricle embraced or enclosed by the calyx, the segments of which are herbaceous or fleshy in fruit; endosperm mealy. (Named from Gr., *chen,* goose; *pous,* foot.)

Calyx red and very fleshy in fruit; seed vertical. — *C. capitatum.*
Calyx green or reddish, herbaceous,
  Plant glandular hairy,
    Stem glandular hirsutulous. — *C. pumilio.*
    Stem glandular puberulent; embryo horseshoe-shaped. — *C. Botrys.*
  Plant not glandular hairy; embryo forming a ring,
    Stamens 1–2; calyx reddish and slightly fleshy in fruit; seeds vertical,
      Stems usually 1–10 dm. tall, erect; blades sinuate dentate. — *C. rubrum.*
      Stems usually 2 dm. or less in height, prostrate; blades hastate, entire or subentire. — *C. chenopodioides.*
    Stamens 5; calyx green, herbaceous, seeds horizontal,
      Leaves with large, divaricate, acute lobes, seeds 1.5–2 mm. in diameter. — *C. gigantospermum.*
      Leaves entire or toothed; seeds 0.5–1.5 mm. in diameter,
        Leaves linear or lance-linear, entire or nearly so. — *C. pratericola,* var. *leptophylloides.*
        Leaves broader, toothed,
          Panicles axillary, mostly shorter than the leaves, leaves bright green, glabrate. — *C. murale.*
          Panicles terminal, not very leafy; leaves scurfy and pale at least beneath,
            Leaves often 3-lobed; seeds 1.3–2 mm. wide. — *C. album.*
            Leaves never 3-lobed; seeds 0.8–1 mm. wide. — *C. Berlandieri,* subsp. *Zschackei.*

Chenopodium album L. *Lamb's Quarters, Pigweed.* Annual, 3–20 dm. tall, erect, stout, usually simple below the inflorescence, more or less white-mealy throughout; leaves 2.5–8 cm. long, rhombic-ovate, sinuate or dentate, obtuse or acute, greener and glabrate above; petioles slender, nearly equaling the blade; uppermost leaves lanceolate and entire; panicle narrow, commonly 30 cm. long; spikes axillary or terminal, rather dense; fruiting calyx 1 mm. broad, the sepals keeled and arched over the lenticular fruit; seed black, minutely pitted, shining. A Eurasian weed. Common in waste or cultivated ground, *A. T.*

Chenopodium Berlandieri Moq., subsp. **Zschackei** (Murr) Zobel. Annual, 4–15 dm. tall, erect, ill-scented, simple or with ascending branches; petioles slender ½ as long to as long as the blades; blades 1.2–4 cm. long, oval-deltoid, rounded at both ends, subentire or with small teeth, scurfy when young, glabrate; flowers in glomerules in dense or interrupted panicles; calyx scurfy, enclosing the fruit; pericarp adherent; seeds nearly smooth, black, shining. Waste places, *U. Son., A. T.*

Chenopodium Botrys L. *Jerusalem Oak.* Stems 1–6 dm. tall, branched from the base, the branches erect or spreading, sweet-scented; leaves 1–5 cm. long, oblong, pinnately divided into 5–6 irregular toothed lobes; petioles short, or the uppermost leaves sessile; flowers abundant, in loose axillary racemes or panicles, these forming strict narrow panicles 10–30 cm. long; sepals dry, ovate, loosely enclosing the fruit; pericarp close, whitish; seeds 0.6 mm. wide, dark brown, dull, vertical or horizontal. A weed introduced from Eurasia or Africa, becoming abundant in the warmer valleys, *U. Son., A. T.*

Chenopodium capitatum (L.) Aschers. *Strawberry Blite.* Glabrous throughout; stems 1.5–8 dm. tall, erect or spreading, somewhat branched; leaves 2–9 cm. long, triangular or rhombic-oblong, coarsely and unevenly laciniate dentate, thin, often turning reddish, on petioles of about the same length; flower clusters 5–10 mm. in diameter, globose, dense, resembling strawberries, in the axils of the reduced upper leaves, and of the upper bractless nodes; calyx-lobes 3–5; stamens usually 1; seeds 1–1.5 mm. broad, brown, usually dull, subglobose. A European weed, introduced near settlements, *A. T.*

Chenopodium chenopodioides (L.) Aellen. Annual, simple or branched; petioles shorter than the blades; blades 7–25 mm. long, rhombic-ovate or obovate, cuneate at base, bright green, fleshy; flowers in dense sessile, axillary glomerules; sepal-lobes 3–5, shorter than the fruit; pericarp green; seeds 0.5–1 mm. broad, reddish brown, shining. Dried shores of alkaline ponds, in the "Scablands," *U. Son.*

Chenopodium gigantospermum Aellen. Annual, 2–15 dm. tall, green, glabrous; branches erect or spreading, angled; petioles ¼–½ as long as the blades; blades 4–17 cm. long, angular-ovate, sinuately 3–5-angled, acuminate, rounded or subcordate at base; flowers mostly in a loose, branching, leafless panicle; sepals partly covering the fruit; pericarp adherent; seeds black, pitted. *C. hybridum* L., var. *gigantospermum* (Aellen) Rouleau. A native plant, of weedy habit, *U. Son., A. T.*

Chenopodium murale L. Annual, 1–6 dm. tall, simple or commonly branched; petioles slender, often as long as the blades; blades 3–8 cm. long, rhombic-ovate, cuneate, irregularly sinuate-dentate; sepal-lobes carinate, partly enclosing the fruit; pericarp green, adherent; seeds 1.2–1.5 mm. broad, blackish, dull, pitted. A European weed, established in waste ground, *U. Son., A. T.*

Chenopodium pratericola Rydb., var. **leptophylloides** (Murr) Aellen. Erect, annual, 2–8 dm. tall, slender, simple or branched, more or less mealy throughout; leaves 1–6 cm. long, 3-nerved; inflorescence paniculate; flowers in dense clusters, these in spikes; calyx-lobes strongly keeled, enclosing the fruit; pericarp free; seeds 1 mm. wide, black, smooth, shining. *C. pratericola* sensu Wahl, in part. A native plant commoner in arid area to west, Waitsburg, *Horner* 1089.

Chenopodium pumilio R. Br. Annual; stem 2–8 dm. long, branched from base, glandular hirsutulous; petioles 3–15 mm. long; blades 5–25 mm. long, bright green, elliptic in outline, coarsely sinuate dentate, above appressed puberulent, below gland-dotted but the veins glandular hirsutulous; flowers in short axillary clusters; calyx 0.6–0.8 mm. long, gland-dotted, the lobes elliptic, acute; utricle 0.7–1 mm. long, dark brown; seed 0.5–0.6 mm. in diameter, brown, cochleate, reticulate, vertical. *C. carinatum* of Am. authors, not of R. Br. Weed, adventive from Australia. Wawawai, in 1949, *Gaines & Hafercamp 339.*

Chenopodium rubrum L. Glabrous; stems stout, angled, branching, often reddish; petioles slender, usually shorter than the blades; leaves 3–15 cm. long, triangular-ovate or lanceolate, acute, cuneate at base, the upper ones linear-lanceolate and usually entire; spikes axillary, leafy-bracted, the flowers in dense clusters; calyx-lobes 2–5, obtuse; stamens 1–2 in outer flowers, 5 in the inner; pericarp green; seed 0.8–1 mm. wide, dark brown, shining. Shores of the Snake River, *U. Son.*

CORISPERMUM. Bugseed

Annual herbs; leaves alternate, narrow entire, 1-nerved; flowers bractless, small, green, solitary in the upper axils, forming terminal narrow leafy spikes, with the upper leaves shorter and broader than the lower; calyx of 1–3 thin broad sepals; stamens 1–3, rarely more and one of them longer; utricle ellipsoid, mostly plano-convex; pericarp membranous, adherent; seed erect; endosperm fleshy. (Name Gr., from *koris,* bug; *sperma,* seed.)

Fruit 2–3 mm. long; lower bracts narrower than the fruit. *C. nitidum.*
Fruit 3–4.5 mm. long; lower bracts as broad or broader than the fruit. *C. hyssopifolium.*

Corispermum hyssopifolium L. Stems 1–6 dm. tall, spreading, much branched, glabrous or finely stellate, striate; leaves 1–7 cm. long, linear or linear-lanceolate, cuspidate, glabrous, sessile; spikes dense, 2–4 cm. long; bracts 4–10 mm. long, ovate-lanceolate, acuminate, scarious-margined, stellate or glabrate; achenes blackish, broadly ovate or orbicular, short-mucronate, narrowly pale winged. *C. marginale* Rydb. Introduced from Europe. Abundant, sandy or gravelly banks of Snake River, *U. Son.*

Corispermum nitidum Kit. Stems 2–5 dm. tall, much branched, striate, glabrate; leaves 1.5–6 cm. long, narrowly linear, cuspidate, finely stellate to glabrate; spikes slender usually lax, scarcely imbricate; upper bracts 2–8 mm. long, ovate-lanceolate, scarious margined, as broad as the fruit, stellate; lower bracts 8–20 mm. long, linear lanceolate; fruit greenish black, pale winged. Introduced from Europe. Bank of Snake River, Wawawai, *St. John* 6,738.

KOCHIA

Herbs or low shrubs; leaves entire, narrow, often terete; flowers perfect or pistillate, sessile, axillary; perianth subglobose, 5-lobed, the lobes incurved, mostly coriaceous in age, developing scarious horizontal wings, distinct or confluent; stamens 5, exserted, the filaments compressed; stigmas 2–3; utricle depressed-globose; seed horizontal. (Named for the German botanist *W. D. J. Koch.*)

Kochia Scoparia (L.) Schrad. *Summer Cypress.* Annual, 2–20 dm. tall; stems erect, striate, paniculately branched above, sparsely pilose above, but glabrate below; leaves 1–7 cm. long, 2–8 mm. wide, sessile, lance-linear, flat, narrowed to a petiolar base, sericeous; flowers pilose, in short leafy spikes; calyx in fruit 2 mm. wide, the wings flabellate, 0.6 mm. long; seeds 1.5 mm. wide. A sturdy weed, introduced from California, Pullman, in 1925, *Pickett, A .T.*

MONOLEPIS

Low annual branching herbs; leaves small, narrow, alternate, entire, toothed or lobed; flowers in small axillary clusters; calyx of one persistent herbaceous sepal; utricle flat; seed erect, compressed, endosperm mealy. (Name Gr., from *monos,* one; *lepis,* scale.)

Monolepis Nuttalliana (R. & S.) Greene. Stems 1–3 dm. tall, erect or decumbent at base, branched below, glabrous or very sparsely mealy throughout; leaves 1–7 cm. long, lanceolate, attenuate at each end, acute or obtuse, commonly with a single large tooth or lobe on each side near the middle, the upper sessile or nearly so, the lowermost slender-petioled; flowers clustered in the axils of the leaves on the elongate, erect, simple branches; sepal oblanceolate or spatulate, acutish; pericarp pitted; seed 1 mm. wide, brownish. In bare soil, especially where it is somewhat saline. *U. Son.,* occasional in *A. T.*

SALSOLA

Annual or perennial, bushy-branched herbs or shrubs; leaves rigid, subulate; flowers sessile, perfect, 2-bracteolate, solitary in the axils, or sometimes several together; calyx 5-parted; stamens usually 5; utricle flattened, enclosed by the calyx, the segments of which are appendaged by a broad membranous horizontal wing in fruit; pericarp free from the seed; embryo spiral, usually green; endosperm none. (Name Lat., from *salsolus,* a little bit salty.)

Salsola pestifer A. Nels. Russian Thistle. Much branched from the base, forming hemispherical plants 3–10 cm. tall, glabrous or hispidulous, the scales not imbricate; branches striate, often reddish below; early leaves 3–9 cm. long, linear, bluish green, sessile, fleshy; later leaves shorter, broad based, becoming dry and stiff; calyx 3–10 mm. wide, the segments sharp pointed; wings of the calyx persistent, membranous, whitish or pinkish, making the fruit 3–8 mm. in diameter. *S. ruthenica* Iljin. A European weed, introduced about 1898, and now an abundant and troublesome weed of arid plains and grain fields, *U. Son., A. T.*

When young the plant makes palatable forage or silage. When mature it is hard and spiny, and it blows and spreads its seed as a "tumble weed."

SARCOBATUS

Leaves alternate or opposite, sessile, narrow, fleshy; flowers monoecious or dioecious; staminate aments terminal the flowers without calyx, the stamens 1–3 underneath a peltate, stipulate, scarious scale; pistillate flowers sessile, 1–2 in an axil, the perianth turbinate, compressed, partly fused with the ovary, forming a membranous, horizontal wing; ovary thin; seed erect, orbicular; embryo spiral; endosperm none. (Name Gr., from *sarx,* flesh; *batos,* a prickly bush.)

Sarcobatus vermiculatus (Hook.) Torr., var. **vermiculatus.** *Greasewood.* Much branched, 6–30 dm. tall; older stems smooth, nearly white; young stems glabrate; lateral branches stout, forming thorns; leaves 5–30 mm. long, linear or somehwat spatulate, mostly glabrous; staminate aments 7–30 mm. long; fruit body 4–5 mm. long, the wings 5–13 mm. broad, often reddish. Alkaline flats, in the "Scablands," western part of our area, *U. Son.,* also at Colfax, *Parker* 673.

Salty to taste, but a useful forage. If hungry sheep are allowed to graze on it to excess, it will poison or kill them, due to the content of sodium and potassium oxalate.

SUAEDA. Sea Blite

Annuals or perennials; leaves alternate, entire; flowers perfect or polygamous, axillary, bracteate; sepal-lobes 5, fleshy; stamens 5; utricle compressed; pericarp usually free from the seed; endosperm scanty or none. (An Arabic name.)

Suaeda occidentalis Wats. Annual, 1–6.5 dm. tall, erect or spreading, green or glaucous, glabrous; leaves 10–25 mm. long, acute, those of the inflorescence little reduced; flowers 1–3 in the axils; calyx-lobes transversely winged in age; seed horizontal 0.8–1 mm. wide, black, shining. *Dondia occidentalis* (Wats.) Heller. Alkaline pond shore, Revere, *St. John et al.* 7,137.

AMARANTHACEAE. Amaranth Family

Weedy herbs, shrubs, or rarely trees; leaves thin, simple, mostly entire, without stipules, alternate or opposite; flowers small, green or white, perfect, monoecious, polygamous, or dioecious, bracteolate, usually in terminal spikes or axillary heads; calyx herbaceous or membranaceous, 1–5-parted, the segments distinct or united at the base, equal, or the inner ones smaller; petals none; stamens 1–5, mostly oposite the calyx-segments, hypogynous, or perigynous; ovary 1-celled; ovule 1 (in ours); fruit a utricle, 1-seeded (in ours); endosperm mealy, usually copious; embryo in a ring or horseshoe. (Named from the genus *Amaranthus.*)

AMARANTHUS. Amaranth

Annual branched erect or diffusely spreading, glabrous or pubescent herbs; leaves alternate, petioled, pinnately veined, entire, undulate or crisped; flowers small, green or purplish, mostly 3-bracteolate, in dense terminal spikes or axillary clusters; calyx of 1–5 distinct sepals; stamens 1–5; filaments filiform or subulate, distinct; anthers oblong, 4-loculed; fruit an ovoid or oblong utricle,

2–3-beaked, with the persistent style; seed erect. Mostly pernicious weeds. (Name Lat. *amarantus,* amaranth, from the Gr. *amarantos,* unfading.)

Flowers in dense terminal spikes,
  Sepals sharp pointed, mostly shorter than the utricle,
    Bracts reddish or purplish; utricle longer than the calyx. — *A. cruentus.*
    Bracts green or pinkish; utricle not longer than the calyx. — *A. hybridus.*
  Sepals mostly blunt and longer than the utricle. — *A. retroflexus.*
Flowers in axillary clusters,
  Plant erect; sepals 3; seeds 0.8 mm. wide. — *A. albus.*
  Plant prostrate; sepals 4 or 5; seeds 1.3–1.5 mm. wide. — *A. graecizans.*

Amaranthus albus L. *Tumbleweed.* Erect, 1.5–12 dm. tall, pale green, much branched, glabrous or nearly so; petioles 0.3–5 cm. long; blades 0.5–7 cm. long, oblong or narrowly obovate, obtuse or emarginate, pale green, glabrous; bracts 2–4 mm. long, oblong lanceolate, spiny-pointed; staminate sepals oblong, scarious; pistillate sepals oblong to linear, thin, green along the nerve; stamens 3; utricle circumscissile, rugose; seed discoid, polished, mahogany-colored; fruit roughened. *A graecizans* sensu Uline & Bray, not L. A common native weed in grain fields, *U. Son., A. T.*

The ascending branches give the mature plant a rounded form. High winds break it loose, and drive it across the fields, dropping seeds as it bounds.

Amaranthus cruentus L. Stems 5–20 dm. tall, erect, simple or branched, usually villous above; petioles 2–20 cm. long; blades 3.5–30 cm. long, elliptic to rhombic-ovate, green or purplish, sparsely pubescent or glabrate; bracts lanceolate, pungent-tipped, reddish or purplish, often longer than sepals; staminate sepals oblong-ovate, acute, with an excurrent nerve; pistillate sepals oblong, obtuse or erose, scarious, purplish to green; stamens 5; utricle globose, circumscissile; seed 1 mm. wide, rotund, black or brownish, shining. *A. paniculatus L.* A Chinese ornamental, occasionally escaped, Clark's Springs, 10 miles n. of Spokane, *Kreager* 569, *A. T. T.*

Amaranthus graecizans L. Stems 1.5–6 dm. long, usually branched, glabrous or nearly so; petioles 2–20 mm. long; blades 0.8–4 cm. long, broadly spatulate, obovate, or oblanceolate, pale green, as least the smaller ones white-margined; bracts 2–3 mm. long, short acuminate, aristate; staminate sepals 4–5, scarious, acute; pistillate sepals 2.5–3 mm. long, oblong, greenish; stamens 3; utricle circumscissile; seed 1.3–1.5 mm. wide, discoid, black. *A. blitoides* sensu Wats., not of L. Probably a native, now a troublesome weed in fields, *U. Son., A. T.*

Amaranthus hybridus L. *Spleen Amaranth.* Coarse annual, 3–15 dm. tall, freely branched, the stems often reddish below, sparsely hirsutulous to glabrate; petioles 1–3.5 cm. long; blades 1–8 cm. long, ovate-lanceolate to rhombic-ovate, entire, olive-green, firm chartaceous; flowers in terminal and axillary, cylindric spikes, monoecious; bracts ovate to lanceolate, acerose-tipped, twice as long as the calyx; sepals 5, oblong, mucronate, 1.5–2 mm. long; utricle 3-horned; seed 1.1–1.3 mm. in diameter, discoid, polished, black. Weed, introduced from Europe. In field, 1950, Wawawai, *Gaines et al. 471.*

Amaranthus retroflexus L. *Green Amaranth.* Stem 3–30 dm. tall, stout, erect, simple or branched, villous at least above; petioles 1.5–8 cm. long; blades 3–12 cm. long, ovate or rhombic-ovate, acute or obtuse, obscurely crenate or entire, minutely roughened, glabrate above, villous or puberulent beneath; spikes very dense, cylindric, erect, in large panicles; bracts 4–5 mm. long, ovate, green subulate-tipped; staminate sepals ovate to lanceolate, scarious; pistillate sepals 5, oblong, acute, obtuse or emarginate; stamens 5; utricle subglobose, circumscissile; seed 1 m. broad, rotund, polished mahogany-colored. A native weed in waste ground, *U. Son.,* to *A. T. T.*

MOLLUGINACEAE. Carpet Weed Family

Soft herbs, sometimes fleshy or succulent; leaves mostly whorled or opposite; stipules wanting, or scarious; flowers small, regular, perfect, solitary, cymose, or glomerate; calyx 4–5-cleft or -parted; petals and stamens sometimes numerous, but petals often wanting; ovary usually free from the calyx, 2–several-celled; ovules numerous in each cell (in ours); fruit a capsule; endosperm scanty or copious. (Named from the genus *Mollugo.*)

MOLLUGO. Carpet Weed.

Mostly annual, much branched herbs; leaves whorled, sometimes basal or alternate; stipules scarious, membranaceous, deciduous; flowers small, whitish, cymose or axillary; sepals 5, white inside, scarious-margined, persistent; petals none; stamens hypogynous, 3 and alternate with the 3 cells of the ovary or 5 and alternate with the sepals; ovary and capsule usually 3-celled, loculicidally dehiscent. (An ancient Lat. name for some soft plant.)

Mollugo verticillata L. *Carpet Weed.* Annual, glabrous throughout, prostrate; stem 5–30 cm. long, slender, freely branched; leaves 1–3 cm. long, spatulate or oblanceolate, acute or obtuse, entire, narrowed at the sessile base, 3–8 in a whorl; flowers axillary, on pedicels of the same length or longer; sepals 2–3 mm. long, oblong, capsules 4–5 mm. long, ovoid; seeds 0.6 mm. long, kidney-shaped, brown, shining, several ribbed. A weed from tropical America, common on the sandy banks of streams, *U. Son., A. T.*

PORTULACACEAE. Purslane Family

Annual or perennial usually succulent herbs; leaves entire, alternate or opposite; flowers regular, perfect; sepals 2 or 4–8; petals 4 or 5, hypogynous, early withering; stamens 2–5 or more and adherent to the bases of the petals; ovary 1-celled; style 2–3-cleft or divided; ovules few to many; endosperm mealy. (Named from the genus *Portulaca.*)

Ovary half inferior, circumscissile; sepals partly united. — *Portulaca.*
Ovary superior,
  Capsule circumscissile; sepals 2–8. — *Lewisia.*
  Capsule 3-valved; sepals 2,
    Petals equal or unequal, separate or slightly coherent on one side; stamens 5 or 3. — *Claytonia.*
    Petals unequal, united; stamens 3. — *Montia.*

CLAYTONIA

Low and glabrous succulent herbs, annual from fibrous roots or perennial from corms or thickened rootstocks; basal leaves petioled, the cauline opposite or alternate; flowers small, white, pinkish, or yellowish, in loose terminal racemes, lasting more than one day; sepals 2, ovate, herbaceous, persistent; petals 5; style 3-notched or cleft; seeds flattened. (Named for *Dr. John Clayton* of Virginia.)

Perennials with thick roots or corms; cauline leaves 2, opposite. *C. lanceolata,* var. *lanceolata.*
Perennials or annuals; roots not corm-like,
  Stems with more than two opposite leaves, or leaves alternate,
    Cauline leaves several pairs, opposite; stolons present. *C. Chamissoi.*
    Cauline leaves alternate; stolons not present,
      Seeds 2 mm. broad; sepals 4 mm. long. *C. linearis.*
      Seeds 0.7–0.9 mm. broad; sepals 2 mm. long. *C. dichotoma.*
  Stems with only two leaves, these opposite,
    Cauline leaves united, at least at the base,
      Calyx 4 mm. long; seeds 2 mm. broad. *C. perfoliata.*
      Calyx 2 mm. long; seeds 1–1.5 mm. broad,
        Plants erect; racemes elongate; basal leaves usually linear or nearly so. *C. parviflora,* var. *parviflora.*
        Plants low; racemes scarcely exceeding the leaves; basal leaves usually rhombic to reniform. Var. *depressa.*
    Cauline leaves not united,
      Leaves linear or narrowly linear-oblanceolate. *C. arenicola.*
      Leaves broader,
        Pedicels mostly bractless; petals 8–13 mm. long. *C. cordifolia.*
        Pedicels mostly subtended by bracts; petals 6–8 mm. long. *C. sibirica.*

Claytonia arenicola Henders. Annual; stem 5–25 cm. tall, erect or ascending, simple; leaves 2–10 cm. long, the basal slender petioled; raceme of 2–14 flowers, each from the axil of a small bract; pedicels up to 2 cm. in length, slender, spreading; sepals 2–3.5 mm. long, acute; petals 5–8 mm. long, pink, notched; capsule shorter than the calyx; seeds 1–1.2 mm. long, oblong, notched at the hilum, black, shiny. *Limnia arenicola* (Henders.) Rydb. Shady, moist places, in the warm valleys, abundant along the Snake River, *U. Son., A. T.*

Claytonia Chamissoi Ledeb. Perennial, with slender elongate stolons, which finally bear tubers at the apex; flower stems mostly simple, 5–40 cm. tall; leaves 3–10 pairs, broadly spatulate, or oblanceolate, 1–7 cm. long; raceme bracted at base, few-flowered; pedicels slender, up to 3 cm. in length; sepals 2 mm. long, orbicular; petals 6–8 mm. long, white or pink, obtuse; stamens 5; capsule equaling the calyx; seeds 1.5 mm. long, black, kidney-shaped, minutely roughened. *Montia Chamissoi* (Ledeb.) Dur. & Jacks.; *Crunocallis Chamissonis* (Ledeb.) Rydb. Moist copses, *A. T.*

Claytonia cordifolia Wats. Perennial, with short rootstocks, somewhat tufted; stems 15–52 cm. tall, bearing a single pair of leaves below the inflorescense; leaf blades 2–8 cm. long, rather fleshy, broadly ovate, somewhat cordate, obtuse, the cauline sessile, the radical often reniform, slender-petioled; raceme peduncled, 3–12-flowered; pedicels up to 3 cm. in length, slender; sepals 3–4.5

mm. long; petals white, pellucid; capsule longer than the sepals; seeds 1–1.5 mm. broad, discoid, black, reticulate, shiny. *Limnia asarifolia* (Bong.) Rydb.; *C. asarifolia* sensu T. & G., not Bong.; *Montia cordifolia* (Wats.) Pax & K. Hoffm. In cold woods along streams in the mountains, *Can.*

Claytonia dichotoma Nutt. ex T. & G. Annual, 2–8 cm. tall; leaves 5–30 mm. long, linear, terete, fleshy; raceme bractless except at base, 1-sided, 3–10-flowered; petals 2–2.5 mm. long, white; stamens 3; seeds black, shining, but the cell walls visible, depressed. *Montia dichotoma* (Nutt. ex T. & G.) Howell. Open places, *A. T.*

Claytonia lanceolata Pursh, var. **lanceolata.** *Spring Beauty.* Stems simple, 5–20 cm. tall, erect, 1–24 from a globose corm 1–4 cm. in diameter; radical leaves few, lanceolate, acute, long-petioled, usually withered before flowering; cauline 2–5.5 cm. long, sessile, lanceolate or ovate-lanceolate, acute; flowers 3–17 in a short raceme; pedicels up to 4 cm in length, all but the lowest bractless, slender; petals 8–11 mm. long, white with pink veins, oblong or ovate, emarginate; seeds black, shiny, 2 mm. broad. Mountain meadows and open mountain woods, descending to the plains at Spangle, *A. T. T.* to *Huds.*

The starchy corms were eaten by the Indians.

Claytonia linearis Dougl. ex Hook. Annual, branched below, 5–30 cm. tall; leaves 2–7 cm. long, linear, succulent, scarious-margined at base; raceme 4–10-flowered, 1-sided, the pedicels curving downward; sepals orbicular; stamens 3; capsule shorter than the sepals; petals 4–4.5 mm. long, white, notched; seeds shiny, black, discoid. *Montia linearis* (Dougl. ex Hook.) Greene; *Montiastrum lineare* (Dougl. ex Hook.) Rydb. Abundant in spring on moist, open ground, *A. T.*

This species and *C. dichotoma* are somewhat intermediate to the genus *Montia.*

Claytonia parviflora Dougl. ex Hook., var. **parviflora.** Annual, branched from the base; stems 15–45 cm. high, erect; radical leaves long-petioled; cauline pair united into an orbicular somewhat angled disk, 1–5 cm. in diameter; raceme usually loose, interrupted, somewhat 1-sided; petals 3–5 mm. long, pink or white; seeds shiny, black minutely muricate. *Montia perfoliata* (Donn) Howell, forma *parviflora* (Dougl.) J. T. Howell; *Limnia parviflora* (Dougl. ex Hook.) Rydb. In moist thickets, abundant, *U. Son.,* to *A. T .T.*

This plant is extremely variable in stature and shape of foliage, blending into the var. *depressa.* Several of the leaf variants have been described as species.

Var. **depressa** Gray. Smaller and depressed, more fleshy; whole plant commonly reddish; cauline leaves sometimes nearly separate. *Limnia depressa* (Gray) Rydb.; *C. rubra* (Howell) Tidestr; *C. perfoliata,* var. *rubra* (Howell) Poellnitz. Very common, *U. Son., A. T.*

Claytonia perfoliata Donn. *Miner's Lettuce.* Annual, branched from the base, 1–5 dm. high, erect; basal leaves long-petioled, the blades usually rhombic-ovate, but varying to spatulate-linear; cauline leaves united, forming an orbicular or somewhat angled disk, 3–5 cm. broad; racemes 1-sided, usually interrupted, with the flowers fascicled; petals 3–5 mm. long, white; seeds black, shiny, minutely cellular reticulate, kidney-shaped. *Montia perfoliata* (Donn) Howell; *Limnia perfoliata* (Donn) Haw. Occasional, moist shady places, *U. Son.,* and *A. T. T.*

Claytonia sibirica L. Annual or perennial with offsets; stems 1–4 dm. tall, simple, erect or ascending; basal leaves ovate to lanceolate, 2–6 cm. long, contracted into long margined petioles; cauline leaves 10–55 cm. long, acute, ovate, sessile; raceme loose, the flowers on slender pedicels up to 5 cm. in length; petals white or pink with red veins; seeds 1.7 mm. wide, discoid, black, the surface somewhat shiny, roughened by the projecting truncate cells. *Limnia* *sibirica* (L.) Haw.; *Montia siberica* (L.) Howell. In wet places, especially in woods, *A. T. T., Can.*

Rydberg (1932) restricted this species to Siberia-Alaska, and for ours adopted the name *Limnia alsinoides* (Sims) Haw., but von Poellnitz (1932) disagreed.

LEWISIA

Low acaulescent fleshy perennial herbs, with fleshy roots or a corm; flowers on short scapes, often showy; sepals 2–8; petals 3–16, white or reddish; stamens 5 to many; styles 3–8; capsule circumscissile; seeds many; black, shining. (Named for its first collector, *Capt. Merriwether Lewis.*)

Sepals 4–8; scape jointed and with an involucre of 5–9 bracts. *L. rediviva.*
Sepals 2; scape with only 2 bracts,
With fleshy conical root; stem leaves 2, bract-like. *L. nevadensis.*
With globose corm; stem leaves 2–4, linear. *L. triphylla.*

Lewisia nevadensis (Gray) Robins. in Gray. Root 2–4 cm. long; basal leaves 2–10 cm. long, linear, fleshy; scapes simple, 2–8 cm. high, decumbent in fruit, with a pair of linear-lanceolate, scarious bracts and 1–3 flowers; sepals 7–12 mm. long, broadly ovate, acute, subentire; petals 6–8, white, 9–15 mm. long; seeds 1.5 mm. long, ovoid-discoid. *Oreobroma nevadensis* (Gray) Howell; *L. pygmaea* (Gray) Robins. in Gray, var. *nevadensis* (Gray) Fosb. Gravelly ridges, in the Blue Mountains, *Huds.*

Lewisia rediviva Pursh. *Rock Rose; Bitter Root.* Roots thick, fusiform, often forked; leaves numerous in a basal cluster, linear, thick and fleshy, 2–5 cm. long, mostly withered before anthesis; scapes 2–8 cm. high, 1-flowered, jointed near the middle above a whorl of scarious bracts; sepals 6–8, imbricate, 10–25 mm. long, the inner somewhat petal-like; petals 12–16, oblong, 2–3 cm. long, rose-colored; seeds 2–2.2 mm. wide, discoid, black, smooth. Thin rocky soil, Rock Lake, *Beattie & Lawrence* 2333, and abundant northward in the "Scablands," *U. Son., A. T.*

The Indians gather the thick, bitter, starchy roots for food.

This beautiful flower was adopted as the state flower of Montana.

Lewisia triphylla (Wats.) Robins. Stems 2–10 cm. high, very slender, from a deep-seated small globose corm; cauline leaves usually 3 in a whorl, sessile, 1–7 cm. long; flowers few or many in a loose paniculate cluster; sepals 3–5 mm. long, rounded, entire; petals 3–10, oblong, 4–6 mm. long, white; seeds 1.2 mm. long, flattened ovoid, mahogany-colored. *Erocallis triphylla* (Wats.) Rydb. Moist gravels, Blue Mts., *Huds.*

MONTIA

Low branching glabrous succulent annual herbs; leaves mostly opposite; flowers small, white, nodding, axillary or racemose; sepals 2, ovate, herbaceous, persistent; petals 5, unlike, two larger and three smaller; stamens alternate with the smaller petals; style very short; stigmas 3; capsule 3-valved. (Named for *Prof. Giuseppe Monti* of Italy.)

Montia fontana L., var. **tenerrima** (Gray) Fern. & Wieg. Stems procumbent or ascending, rooting at the nodes, 3–10 cm. high or when aquatic often longer; leaves 4–14 mm. long, spatulate or obovate; raceme terminal or axillary, few-flowered; sepals 1 mm. long, orbicular; petals about 2 mm. long, white; seeds barely 1 mm. long, black, discoid to kidney-shaped, muriculate. In wet places and running water, *U. Son., A. T.*

PORTULACA

Low fleshy herbs; leaves alternate or partly opposite; stipules scarious or none, or reduced to hairy tufts; flowers terminal and sessile, expanding in direct sunshine before mid-day, soon closing; sepals 2, coherent at the base in a tube and adnate to the ovary, the upper part deciduous; stamens 7 to many; ovules numerous. (Old Lat. name.)

Portulaca oleracea L. *Purslane.* Annual, prostrate, fleshy, forming mats 1–5 dm. in diameter; leaves 1–2 cm. long, alternate below, clustered at the branch tips, narrowly obovate, obtuse or truncate, cuneate, fleshy, glabrous; calyx-lobes 3–5 mm. long, ovate, keeled; petals 4–6 mm. long, yellow, notched at the apex; stamens 7–12; capsule conical, acute, dehiscing below the middle; seeds 0.5–0.7 mm. wide, kidney-shaped, black, with stellate reticulations. Introduced from Europe, abundant on banks of Snake River, *U. Son.;* occasional *A. T.*

CARYOPHYLLACEAE. Pink Family

Annual or perennial herbs; stems often swollen at the nodes; leaves opposite, entire, with or without stipules; flowers regular and mostly perfect; sepals 4–5, imbricate and separate or united into a calyx-tube; petals as many as the sepals or none; stamens not more than twice as many as the sepals, hypogynous or perigynous; styles 2–5; ovary 1-celled, rarely imperfectly 3–5-celled; ovules several or many, basal or attached to a free, central column; fruit a capsule, or utricle; endosperm mealy. (Name from the generic name *Caryophyllus.*)

Sepals united, the calyx teeth short; stamens hypogynous; styles free,
  Calyx ribs at least twice as many as the teeth, running to the teeth and to the sinuses,
    Styles 5, alternate with the foliaceous calyx-teeth. *Agrostemma.*
    Styles 3, or if 5 opposite the short calyx-teeth,
      Styles 5; capsule 1-celled. *Lychnis.*
      Styles 3; capsule usually partly several-celled at base. *Silene.*
  Calyx 5-ribbed, or nerveless, or striate nerved,
    Petals with a crown; calyx not strongly angled. *Saponaria.*
    Petals without a crown,
      Capsule 5-toothed; calyx sharply 5-angled. *Vaccaria.*
      Capsule 4-parted to the middle or below; calyx not sharply angled. *Gypsophila.*
Sepals distinct or nearly so; stamens often perigynous; styles free or united,
  Stipules present, scarious,
    Styles 3; leaves opposite. *Spergularia.*

Styles 5; leaves apparently whorled. *Spergula.*
Stipules wanting,
Petals deeply 2-cleft or 2-parted,
Capsule cylindric, often curved at the 10-valved tip; styles usually 5. *Cerastium.*
Capsule ovoid or nearly so, straight, usually 6-valved at tip; styles usually 3. *Stellaria.*
Petals entire, erose, or shallowly notched,
Petals denticulate or jagged. *Holosteum.*
Petals entire or wanting,
Styles fewer than the sepals. *Arenaria.*
Styles as many as the sepals and alternate with them. *Sagina.*

AGROSTEMMA

Annual or biennial pubescent herbs; leaves linear or linear-lanceolate, acute or acuminate, sessile; flowers solitary at the ends of long axillary peduncles, large, erect, white to purple; calyx ovoid-oblong, not inflated, narrowed at the throat, 5-lobed, 10-nerved; petals 5, shorter than the calyx-lobes, not appendaged; stamens 10; capsule 1-celled; seeds numerous, black. (Name Gr., *agros,* field; *stemma,* wreath.)

Agrostemma Githago L. *Corn Cockle.* Erect, branched above, 3–10 dm. tall, the whole plant more or less appressed pilose; leaves 3–10 cm. long; flowers few; calyx-lobes 1.5–5 cm. long, linear, unequal; petals shorter than the sepal-lobes, obovate-cuneate, emarginate, purplish red; capsule ovoid, cartilaginous, longer than the calyx-tube; seeds 2.8–3.5 mm. long, suborbicular, flattened and wedge-shaped at one end, black with rows of tubercles. A weed from Europe, occasional, usually in wheat fields, *A. T.*

ARENARIA. Sandwort

Annual or perennial, mainly tufted herbs; leaves sessile, often subulate and more or less rigid; stipules none; flowers white, cymosely panicled or capitate, rarely solitary and axillary; sepals 5; petals 5, very rarely minute or wanting, entire or emarginate, white or nearly so; stamens 10, or often fewer; styles generally 3, rarely 2–5; capsule globose or oblong, dehiscent at the apex by as many or twice as many valves as there are styles. (Name from Lat., *arena,* sand.)

Leaves filiform, mostly basal or on basal shoots,
Flowers in compact heads,
Bracts narrowly lanceolate, wholly scarious except for the midrib. *A. congesta,* var. *cephaloidea.*
Bracts ovate, merely scarious margined. Var. *congesta.*
Flowers in loose (rarely in close) cymes. *A. capillaris,* subsp. *americana.*
Leaves flat, broader, the cauline ones numerous,
Plants with rhizomes; seeds appendaged at the hilum,

Sepals obtuse; leaves usually obtuse. *A. lateriflora,* var. *lateriflora.*
Sepals acuminate; leaves usually acute. *A. macrophylla.*
Plants with fibrous roots; seeds not appendaged,
Plant glabrous; valves of the capsule entire. *A. pusilla,* var. *pusilla.*
Plant scaberulous puberulent; valves of the capsule 2-cleft. *A. serpyllifolia,* var. *serpyllifolia.*

Arenaria capillaris Poir., subsp. **americana** Maguire. Perennial, tufted glabrous below, glandular-pubescent above; sterile branches numerous, very leafy; flowering stems 1–3 dm. tall; leaves 2–7 cm. long, narrowly linear, rather rigid, sharply cuspidate, chiefly grouped in fascicles at the bases of the erect stems, somewhat pungent, little spreading; cauline few, reduced; flowers in loose cymes; bracts lanceolate, more or less scarious margined; sepals 3–4.5 mm. long, ovate, scarious margined, obtuse or some acute; petals 6–9 mm. long, obovate. *A. capillaris* Poir. and *A. glabrescens* (Wats.) Piper. Exposed summits, Blue Mts., *Piper* in 1896; Mt. Carleton [Mt. Spokane], *Kreager* 238, *Huds.*

Arenaria congesta Nutt., ex T. & G., var. **congesta.** Nearly glabrous throughout, tufted; stems 1–3 dm. tall, erect, joints prominent; leaves narrowly linear, rather rigid, scabrous ciliate, sharply cuspidate, the basal numerous on short sterile branches, 2–6 cm. long, the cauline pairs shorter, rather distant; flowers congested into 1–3 heads on each stalk; bracts 4–5 mm. long; sepals 4–6 mm. long, ovate-lanceolate, acuminate, 3-nerved, with broad scarious margins; petals twice as long, white, spatulate; capsule 1.5 mm. long, ovoid. Rocky places, Blue Mts., and Moscow Mts., *A. T. T.* to *Huds.*

Var. **cephaloidea** (Rydb.) Maguire. Tufted glabrous perennial, from a woody base; stems 2–4 dm. tall, strict; leaves from the base and from the sterile branches 3–10 cm. long, filiform subulate, scabrous ciliate, cuspidate; cauline leaves shorter; heads 1–3; sepals 4–5 mm. long, lanceolate petals half again as long, white, narrowly spatulate; capsule ovoid, 5 mm. long; seeds 1 mm. long, auriculate, yellow. *A. cephaloidea* Rydb. Grassy or rocky slopes, common, *A. T., A. T. T.*

Arenaria lateriflora L., var. **lateriflora.** Perennial; stems 5–40 cm. tall, slender, retrorse puberulent; leaves 5–30 mm. long, elliptic or ovate-lanceolate, sessile, slightly connate, puberulent on the margins and midrib, occasionally throughout; cymes lateral or becoming so, 1–6-flowered; sepals 2–3 mm. long, ovate, obtuse, glabrous, green, hyaline margined; flowers dimorphic, one form having large anthers on stamens twice the length of the calyx and having petals thrice as long as the calyx, the other form having small or imperfect stamens or stamens shorter than or equaling the calyx and with petals less than twice as long as the calyx; petals 2–3-times as long, oval, entire, white; capsule nearly twice as long as the calyx, ovoid, 6-toothed on dehiscence. *Moehringia lateriflora* (L.) Fenzl; *A. lateriflora,* var. *typica* (Regel) St. John. Grassy slopes or thickets, *A. T., A. T. T.*

Arenaria macrophylla Hook. Perennial from slender rootstocks; stems 4–15 cm. tall, puberulent; leaves 1–5 cm. long, lanceolate or linear-lanceolate, puberulent; cymes terminal or becoming axillary, 1–5-flowered; flowers dimorphous and apparently functionally dioecious; male flowers with stamens about twice as long as the sepals, and petals well exceeding the sepals; pistillate flowers with stamens included, and with petals shorter than the calyx; sepals 2.5–4 mm. long, ovate, lanceolate; petals white, capsule 4–5 mm. long, ovoid; seeds 1.5 mm. long, reniform, reddish black, shining. *Moehringia macrophylla* (Hook.) Torr. Woods, *A. T., A. T. T.*

Arenaria pusilla Wats., var. **pusilla.** Stems 2–5 cm. tall, very slender, usually branched; leaves 2–3 mm. long, connate oblong or ovate, distant; sepals 2–3 mm. long, lanceolate, acuminate; petals wanting, or unequal and shorter than the calyx, ovoid. *Minuartia pusilla* (Wats.) Mattf. Stony soil, common but very inconspicuous, Pullman, *Piper* 1885; Waitsburg, *Horner* 128, *A. T.*

Arenaria serpyllifolia L., var. serpyllifolia. *Thyme-leaved Sandwort.* Annual or biennial, scabrous puberulent; stems branched, diffuse, 2–25 cm. tall; leaves 2–8 mm. long, ovate, strongly veined; inflorescence a loose, leafy bracted cyme; pedicels 4–8 mm. long; sepals 3–4 mm. long, ovate-lanceolate; petals shorter, ovate, white; capsule ovoid, equaling or exceeding the calyx; seeds 0.3–0.6 mm. wide, brown, reniform, warty. A weed, introduced from Europe, grassy places, *U. Son.,* to *A. T. T.*

CERASTIUM. Mouse-Ear Chickweed

Annual or perennial, generally pubescent or hirsute herbs; leaves flat, rarely subulate; flowers in terminal bracted dichotomous cymes; sepals 5; petals 5, retuse or bifid, very rarely subentire, white; stamens 10, sometimes fewer; styles 5, sometimes 4 or 3, opposite the sepals; capsule 1-celled, dehiscent by 10, rarely 8, apical teeth. (Name Gr., *keras,* horn.)

Herbage white tomentose throughout. *C. tomentosum.*
Herbage not tomentose,
  Petals not longer than the sepals,
    Pedicels longer than the calyx. *C. vulgatum,* var. *hirsutum.*
    Pedicels not longer than the calyx. *C. viscosum.*
  Petals decidedly longer than the sepals,
    Annual; pods nodding. *C. nutans.*
    Perennial; pods erect. *C. arvense.*

Cerastium arvense L. *Field Chickweed.* Perennial, tufted, with depressed basal branches bearing abundant axillary fascicles and marcescent leaves; plant glandular or glandless, pubescent throughout or sometimes nearly glabrous; stems weak, erect, 2–6 dm. tall, nearly naked above; leaves 1–6 cm. long, linear or lanceolate; flowers cymose, on slender pedicels; sepals 4.5–8.5 mm. long, oblong, acute; petals deeply notched, 2–3-times as long as the sepals. A very polymorphic species. Open slopes or thickets, common and showy, especially in the Snake R. Canyon, *U. Son., A. T.*

Cerastium nutans Raf. Stems erect, 15–40 cm. high, usually branched at the base; leaves oblong-lanceolate, or the lowest spatulate, acute, 1–3 cm. long; cyme open, rather many-flowered; pedicels elongated, nodding, especially in fruit; calyx 3–4 mm. long, exceeded by the petals; pods curved, three times as long as the calyx. Near Lewiston, Idaho, and about Lake Coeur d'Alene. The writer has been unable to verify it.

Cerastium tomentosum L. Perennial, 15–30 cm. tall, the stems diffuse and often creeping or decumbent; sterile, leafy shoots numerous; leaves 12–20 mm. long, linear to linear-lanceolate; the nodes commonly with axillary leafy shoots; cymes 7–15-flowered; pedicels 2–3-times longer than calyx; calyx 6–7 mm. long, tomentose; sepals ovate, the margins membranous, glabrous; petals 12–15 mm. long, bifid up to ⅓ way; capsules 10–12 mm. long. An ornamental, from Europe, often in rock gardens, escaped and established, Kamiak Butte, June 19, 1948, *Hafercamp 38.*

Cerastium viscosum L. *Mouse-ear Chickweed.* Annual, viscid-hirsute, erect or nearly so, 1–3 dm. high; leaves 5–25 mm. long, oval to ovate or obovate, very obtuse, the lowest narrowed into short-margined petioles; inflorescence rather close in flower, looser in fruit; bracts herbaceous; sepals 3–4 mm. long, lanceolate, acute; petals 2-cleft; capsule 5–8 mm. long; seeds 0.4 mm. broad, brown, warty. A weed introduced from Europe, grasslands, *U. Son., A. T.*

Cerastium vulgatum L., var. hirsutum Fries. *Common Mouse-ear Chickweed.* Perennial, hirsute throughout, tufted, 1–6.5 dm. high; leaves 5–40 cm. long, oblong, obtuse, or the upper ones acutish, inflorescence loose, the pedicels longer than the calyx; bracts scarious margined; sepals 4–7 mm. long, ovate-lanceolate, as long as the 2-cleft petals; capsule 7–11 mm. long; seeds 0.5–0.7 mm. wide, reddish, tuberculate, somewhat curved. A weed introduced from Europe, lawns and pastures, *A. T., A. T. T.*

GYPSOPHILA

Annual or perennial, branching or diffuse, mostly glabrous and glaucous herbs; leaves narrow; flowers perfect, axillary or panicled; calyx narrowly turbinate or campanulate, bractless, 5-nerved, 5-toothed; petals 5, with narrow claws, without a crown; stamens 10; styles 2; pod sessile, 1-celled. (Name Gr., *gupsos,* gypsum; *phylein,* to love.)

Gypsophila paniculata L. *Baby's Breath.* Perennial; stems 3–9 dm. tall, bushy branched, puberulent below, glabrous above; leaves 15–30 mm. long, connate, linear-lanceolate, somewhat fleshy, mostly 3-nerved, reduced to bracts in the inflorescence; panicle diffuse, at times 1000-flowered; pedicels 3–10 mm. long; calyx 2 mm. long, campanulate, deeply 5-lobed; petals 4–6 mm. long, white; capsule depressed spherical, exceeding the calyx; seeds 1.5–1.7 mm. long, flat, warty. A cultivated Eurasian plant, becoming established since 1929, in Spokane and Whitman Cos., *A. T.*

HOLOSTEUM

Annual or perennial herbs with cymose-umbellate, white flowers on long terminal peduncles; sepals 5; petals 5; stamens 3–5, hypogynous; styles 3; ovary 1-celled, many ovuled; capsule ovoid-cylindrical, dehiscing at the tip by 6 short teeth; seeds compressed, attached by the inner face. (Name Gr., *holosteon,* used by Dioscorides for a *Plantago.*)

Holosteum umbellatum L. *Jagged Chickweed.* Annual, more or less glandular puberulent; stems several, 3–25 cm. tall; basal leaves oblanceolate, petiolate; cauline leaves 6–25 mm. long, lance-ovate, umbel 3–8-flowered; sepals 3–4 mm. long, ovate, acute; petals 4–6 mm. long, jagged, white; pod 5–7 mm. long, ovoid; seeds 0.5–0.7 mm. wide, yellowish brown, papillose, shield-shaped, keeled on the inner side. A weed, recently introduced from Europe, collected in 1926. Fields, Pullman, *G. N. Jones* 285; 1223; opposite Clarkston, *Constance et al.* 990 (as *Arenaria serpyllifolia*); Lewiston, *Rodock* 15, *U. Son., A. T.*

LYCHNIS

Herbs, mostly erect; calyx 5-toothed, 10-nerved, tubular ovoid, or inflated; petals 5, narrowly clawed, the blade entire, 2-cleft, or laciniate, usually crowned; stamens 10; ovary 1-celled or incompletely several-celled, many ovuled; styles 5 (rarely 4 or 3), opposite the calyx-teeth; capsule dehiscing by 10 or less, apical teeth. (Name Gr., *luchnos,* lamp.)

Plant viscid pilose or puberulent; calyx-teeth straight. *L. alba.*
Plant white woolly; calyx-teeth twisted together. *L. Coronaria.*

Lychnis alba Mill. *White Campion.* Annual or biennial, freely branching, 4.5–12 dm. tall; leaves 2–10 cm. long, ovate to lanceolate, acute, the lower tapering into a margined petiole; flowers few, opening at dusk, dioecious, in a loose cyme; calyx 18–25 mm. long, whitish, with green ribs, cylindric, enlarging with the fruit, the teeth 3–5 mm. long, lanceolate; petals 2.5–4 cm. long, white, (or pinkish), deeply notched, 2-toothed on the side, with a lobed crown; capsule 1.6–2 cm. long, ovoid; seeds 1.3–1.5 mm. broad, dark brown, tubercled. A European weed first noticed in our area in 1923. *Melandrium album* (Mill.) Garcke. Dry roadsides, *A. T.*

Lychnis Coronaria (L.) Desr. Perennial, 3–10 dm. tall, stout, simple or branched; lower leaves 5–10 cm. long spatulate, narrowed into margined petioles; upper leaves shorter, sessile, lanceolate, acute; flowers few, long peduncled; calyx 15–18 mm. long, narrowly campanulate, 5-angled, the teeth 5–7 mm. long, lanceolate; petals reddish purple, the limb wedge-shaped, emarginate; capsule ovoid, dehiscing with 5 teeth; seeds 1.3 mm. wide, kidney-shaped, black tubercled. A European plant, often cultivated, spreading to roadsides, Mead, *Lackey* in 1923, *A. T.*

SAGINA. Pearlwort

Low matted annuals or perennials; leaves filiform or subulate; flowers axillary, long pediceled; sepals 4–5; petals 4–5, entire or notched, white, or none; stamens 4, 5, 8, or 10; capsule 4–5-valved, dehiscent to the base, the valves opposite the sepals; seeds numerous. (Name Lat., *sagina,* fattening, an ancient name of *Spergula.*)

Flowers 4-parted; sepals longer than the petals. *S. procumbens.*
Flowers 5-parted; sepals shorter than the petals. *S. saginoides,* var. *hesperia.*

Sagina procumbens L. *Poverty.* Annual or perennial, glabrous; stems 2–8 cm. long, depressed or spreading; leaves 5–12 mm. long, linear, or filiform, subulate, connate at base; pedicels longer than the leaves; sepals 2 mm. long, ovate; petals shorter or none; capsule exceeding the spreading calyx; seeds 0.3–0.5 mm. long, reniform-discoid, papillose, brown. Reported by Piper as a weed on golf greens, Spokane.

Sagina saginoides (L.) Dalla Torre, var. **hesperia** Fern. Perennial, glabrous; stems prostrate or ascending, filiform, 2–20 cm. long; leaves 5–20 mm. long, linear, mucronate, opposite, with short axillary branches with shorter leaves; pedicels 5–35 mm. long; sepals 1.3–2 mm. long, ovate; petals 1.5–3 mm. long, ovate; styles almost wanting; capsule 2.5–3 mm. long; seeds 0.3 mm. long, flattened ovoid, brown. *S. occidentalis* Wats. of Piper. Meadows in the Blue Mts., Indian Corral, *Darlington* 149; Asotin Co., *G. N. Jones* 1876, *Huds.*

SAPONARIA

Annual or perennial herbs; leaves broad; flowers large; calyx ovoid or tubular, 5-toothed; petals 5, entire or emarginate, long clawed, with forked appendages at base of the limb; stamens 10; ovary 1-celled or incompletely 2–4-celled; styles 2; capsule dehiscent by 4 apical teeth. (Name Lat., *sapo,* soap.)

Saponaria officinalis L. *Bouncing Bet.* Perennial, 3–8 dm. tall, sparingly branched; leaves 5–8 cm. long, lance-oval or ovate, acute, 3–5-ribbed, narrowed to a short petiole; flowers in dense, corymbiform leafy bracted cymes; calyx 1.5–2.3 cm. long, pale, short toothed; petals 3–4 cm. long, pink, the limb obcordate; capsule lance-ovoid, as long as the calyx; seeds 1.8 mm. long, rounded reniform, blackish, tubercled. A European plant, persisting after cultivation, Waitsburg, *Horner* 395, *A. T.*

If rubbed with water the leaves make a lather due to their content of saponin.

SILENE. Catchfly

Annual or perennial herbs; flowers solitary or in cymes, mainly pink, red, or white; calyx more or less inflated, tubular, ovoid, or campanulate, 5-toothed, 10–many-nerved; petals 5, narrow, clawed, mostly crowned with a scale at the base of the blade; stamens 10; style 3, (rarely 2, 4, or 5, if 5 opposite the calyx-teeth); ovary 1-celled or incompletely 2–4-celled; ovules many; capsule 3- or 6-toothed. (Name Lat., *saliva,* saliva.)

Upper internodes glabrous, each with a dark glutinous band. *S. antirrhina.*
Upper internodes pubescent, without a dark glutinous band,
  Inflorescence leafy; flowers single, axillary, 1 cm. or less in length,
    Lower stem with glandless hairs, the hairs thick walled and with peg-like thickenings. *S. Menziesii,* var. *Menziesii.*
    Lower stem with thin walled, glandular hairs. Var. *viscosa.*
  Inflorescence merely bracted cymose or spicate; flowers longer,
    Blade of petal 2-parted and cleft into 4–6 linear segments. *S. oregana.*
    Blade of petal 2-cleft, sometimes also with short lateral teeth,
      Petals not or scarcely exceeding the calyx. *S. Spaldingii.*
      Petals markedly exceeding the calyx,
        Cauline leaves broadly lanceolate,
          Fruiting calyx cylindric, less than 15 mm. long, the nerves prominent and hirsute. *S. dichotoma.*
          Fruiting calyx inflated, the nerves anastomosing,
            Fruiting calyx 25 mm. or more long, the nerves prominent, hirsute. *S. noctiflora.*
            Fruiting calyx 20 mm. long, glabrous, the nerves inconspicuous. *S. Cucubalus.*
        Cauline leaves linear to linear-oblanceolate,
          Upper parts viscid puberulent; calyx subcylindric, narrowed and contracting at base,
            Basal leaves 6–25 cm. long, 4–30 mm.

broad; calyx scarcely inflated in fruit. *S. Scouleri,* subsp. *Scouleri.*
Basal leaves 3–8 cm. long, 4–10 mm. wide; calyx inflated in fruit. *S. Parryi.*
Plant puberulent, but not viscid; calyx ovoid-campanulate, rounded at base. *S. Douglasii,* var. *Douglasii.*

Silene antirrhina L. *Sleepy Catchfly.* Annual, glabrous or puberulent; stem 1–9 dm. tall, slender, mostly simple; leaves 2–5 cm. long, linear, lanceolate, or oblanceolate, acute, the upper sessile; flowers in a forked compound cyme, long-pedicelled; calyx 4–8 mm. long, 10-nerved, ovoid, cylindrical in fruit; petals white or pinkish, notched, usually a little exceeding the calyx; capsule 5–8 mm. long, ovoid, 6-toothed on dehiscence; seeds 0.5 mm. wide, reniform, flattened, brown, papillose. *S. antirrhina,* var. *vaccarifolia* Rydb. Sandy or gravelly soil, *U. Son.* to *A. T. T.*

Silene Cucubalus Wibel. *Bladder Campion.* Robust perennial, 10–45 (–80) cm. tall, from a creeping rhizome, usually glabrous but occasionally hirsutulous; leaves 3–8 cm. long, 1–3 cm. wide, ovate-lanceolate to oblanceolate, thin, abruptly acuminate, sometimes ciliolate, the cauline often clasping; inflorescence open, 5–30-flowered, the upper bracts small; cyme long peduncled, open paniculate, with stiff branches, few–many-flowered, the upper bracts reduced; calyx 15–18 mm. long in fruit, ovoid, membranaceous, conspicuously bladdery-inflated, glabrous, with fine anastomosing veinlets connecting the 20 obscure nerves; petals 12–16 mm. long, white or rarely pinkish, 2-lobed, crownless; capsule stipitate, included; stipe 2–3 mm. long; capsule 3-celled; seeds 1–1.5 mm. long, subreniform, warty papillate. *S. latifolia* (Mill.) Britten & Rendle; *S. inflata* Sm. Eurasian weed. introduced, Spokane, 1952, *Gaines et al. 746.*

Silene dichotoma Ehrh. Annual, 3–8 dm. tall, sparingly branched, with strongly hirsute stems; leaves 3–8 cm. long, 3–35 mm. wide, lanceolate to oblanceolate, lower leaves usually with ciliate petioles, upper ones sessile; inflorescence usually once or several times dichotomous, the ultimate branches racemose; flowers mostly perfect; calyx 10–15 mm. long, narrowly tubular, the 10 green nerves stiffly hirsute; corolla white to reddish, the petals bifid with a short obtuse crown, auricles lacking, appendages truncate, about 0.2 mm. long, blades 5–9 mm. long, rhombic-cuneate, deeply 2-lobed; stamens usually exserted but sometimes vestigial; stipe 2–4 mm. long; capsule 3-celled; seeds 1–1.3 long, dark gray-chocolate, finely rugose. Weed, introduced from Europe, Pullman, in 1916, *Pickett 590.*

Silene Douglasii Hook., var. **Douglasii.** Perennial, tufted, puberulent throughout; flowering stems 2–6.3 dm. tall, several; leaves 2–8 cm. long, linear or linear-oblanceoate, acute or acuminate; cymes few-flowered, long-peduncled; calyx-tube 6–12 mm. long, somewhat inflated, pale green, the 10 nerves inconspicuous, the teeth 2–4 mm. long, obtuse, scarious-margined; petals 14–20 mm. long, white or pink, 2-lobed; scales oblong, obtuse; claw broad, with a short auricle on each side at the apex; gynophore 2–4 mm. long; capsule 9–11 mm. long, ellipsoidal, 6-toothed on dehiscence; seeds 1.5 mm. long, reniform, brown, reticulate, tubercled on the back. Grassy hillsides, common, *U. Son.* to *A. T. T.*

Silene Menziesii Hook., var. **Menziesii.** Perennial, glandular-puberulent; stems weak, erect or decumbent, usually branched, 1–3 dm. high, leafy; leaves elliptic-lanceolate, acuminate or acute at each end, 2–8 cm. long; calyx 5–8 mm. long, 5-toothed, narrowly campanulate, nerves not prominent, the teeth deltoid; petals 2-cleft, 6–10 mm. long, white, usually without a crown; gynophore 1 mm. long; capsule 6–8 mm. long, ovoid, 6-toothed; seeds 0.7 mm. wide, ellipsoid, dark mahogany colored, tuberculate. Grasslands, thickets, or woods, common, *U. Son.,* to *A. T. T.*

Var. **viscosa** (Greene) C. L. Hitchc. & Maguire. Differs by having the lower internodes rather densely villous to pilose, the hairs spreading and gland-tipped, thin walled, not sculptured on outer surface, and the septa very evident. *U. Son.* to *A. T. T.*

The type is *Horner* 2, Blue Mts., Columbia Co.

Silene noctiflora L. *Night-flowering Catchfly.* Annual, viscid pilose or puberulent; stems 3–10 dm. tall, stout, erect; lower leaves 5–12 cm. long, oblanceolate or spatulate, petioled; the upper 2.5–8 cm. long, acute, sessile; flowers white, opening at dusk, few and long-peduncled, in a loose cyme; fruiting calyx 20–25 mm. long, ovoid, white with green veins, the teeth 5–8 mm. long, subulate; petals 2–3 cm. long, the limb 2-forked; capsule 14–16 mm. long, 6-toothed; seeds 1–1.2 mm. wide, reddish brown, kidney shaped, tubercled. *Melandrium noctiflorum* (L.) Fries. A weed, introduced from Europe, grain fields, *A. T.*

Silene oregana Wats. Perennial; whole plant viscid pilosulous; stems several, erect, 1.5–7 dm. high; basal leaves 5–12 cm. long, oblanceolate, long-petioled; cauline 2–10 cm. long, usually lanceolate and sessile; inflorescence narrow, rather dense, often raceme-like; tube of calyx 8–12 mm. long, cylindric or narrowly ellipsoid, scarious with 10 green nerves; the teeth 2–3 mm. long, ovate-deltoid to lanceolate; petals 13–20 mm. long, white, the claws narrow and auricled; gynophore, or stipe of ovary, 2–6 mm. long; capsule 6–8 mm. long, ovoid, 6-toothed on dehiscence; seeds 1.5 mm. wide, compressed, angular, thin winged, reticulate. Open or wooded slopes, *A. T. T.* to *Huds.*

Silene Parryi (Wats.) C. L. Hitchc. & Maguire. Perennial 20–60 cm. tall, above glandular puberulent, below glandular or non-glandular puberulent; basal leaves 3–8 cm. long, linear-oblanceolate or spatulate, at base attenuate to long petioles; cauline leaves of 2–3 pairs, mostly narrower; inflorescence of one to several cymes; these 1–7-flowered, not glomerate; pedicels 5–40 mm. long; calyx 12–16 mm. long, tubular-campanulate, glandular pubescent, with 10 prominent green or purple ribs, the lobes 2–3 mm. long, the body deltoid but with thin apical acute broadly ovate expansion; petals 9–14 mm. long, the limb 3–7 mm. long, 2-lobed for nearly half their length (rarely 4-lobed), each lobe usually with a lateral tooth; gynophore 3 mm. long, puberulent; styles 3 (or 4–5), exserted; seeds 1–1.5 mm. long, brown, rugose-tessellate. Blue Mts., fide C. L. Hitchc. & Maguire, not seen by the author.

Silene Scouleri Hook., subsp. **Scouleri.** Perennial, puberulent throughout, glandular-viscid puberulent above; stems erect, simple 3–8.5 dm. tall, solitary or tufted; radical blades 5–12 cm. long, linear-lanceolate or oblanceolate, on slender petioles of equal length; cauline several pair, sessile at the swollen nodes, 2–9 cm. long; inflorescence narrow, elongated; flowers short-pediceled, one or several in the axils; calyx-tube 8–12 mm. long, scarious, with 10 dark nerves and scarious-margined teeth, 2–4 mm. long; petals 15–20 mm. long, white or purplish, 2-cleft, the lobes notched; auricles of the claw narrow, cleft into narrow lobes; gynophore 2–5 mm. long; capsule 6–9 mm. long, ovoid, 6-toothed on dehiscence; seeds 1–2 mm. long, reniform, flattened, brown, reticulate. Prairies and dry woods, *A. T., A. T. T.*

Silene Spaldingii Wats. Perennial, very viscid villous; stems erect, swollen at the nodes, 3–4 dm. high, very leafy; leaves 3–7 cm. long, sessile, lanceolate or oblong-lanceolate, the basal short petioled; inflorescence narrow and raceme-like; calyx-tube 14–21 mm. long, subcylindric, narrowed below, becoming scarious, with 10 green nerves; the lobes 3–6 mm. long, lanceolate; petals greenish white, appendages 4, claw with broad auricles; gynophore 3 mm. long; capsule 12–20 mm. long, 6-toothed on dehiscence; seeds 1.5–2.2 mm. long, reniform, flattened, thin, brown. Prairies, uncommon, *U. Son., A. T.*

First collected on the Clearwater R. (probably near Lapwai) by Mrs. Spalding, and named for the Rev. Henry Spalding.

SPERGULA

Annual branched herbs; leaves subulate, fascicled; flowers in a cymose panicle; sepals 5; petals, 5, white; stamens 10, rarely 5; ovary 1-celled; capsule-valves 5, opposite the sepals. (Name from Lat., *spargere,* to scatter.)

Spergula arvensis L. *Corn Spurry.* Minutely and sparsely glandular puberulent throughout; stems several, mostly simple below the inflorescence, 1–5 (–10) dm. long, decumbent or ascending; leaves 1–3 cm. long, narrowly linear, in fascicles of 6–15; flowers in a loose cyme, the pedicels becoming deflexed; sepals 3–4 mm. long, ovate; petals white, 4–5 mm. long, equaling or exceeding the sepals; capsule 5–7 mm. long, ovoid; seeds 1 mm. wide, black, discoid, acutely margined, minutely roughened. A weed from Europe, in grain fields, Pullman, *Hardwick* in 1895, *A. T.*

SPERGULARIA

Low annual biennial or perennial herbs; leaves fleshy, linear or setaceous, opposite, often with axillary fascicles of smaller leaves; flowers in terminal racemose cymes; sepals 5; petals 5 (rarely fewer or none), purplish or white; stamens 2–10, commonly 10; ovary 1-celled; capsule-valves 3, (or when 5, alternate with the sepals.) (Name derived from the generic name *Spergula.*)

Stipules lanceolate, 4–6 mm. long. *S. rubra.*
Stipules triangular, 1 mm. long. *S. diandra.*

Spergularia diandra (Guss.) Boiss. Much like *S. rubra* but densely glandular puberulent throughout; leaves, excepting the reduced upper ones, not cuspidate; petals pinkish, shorter than the sepals. Introduced from Eurasia or Africa. Banks of Snake River at Almota, *Piper,* July 30, 1897, *U. Son.*

Spergularia rubra (L.) J. & C. Presl. *Sand Spurry.* Annual, sometimes biennial or perennial, prostrate or decumbent; stems slender, 2–30 cm. long, smooth below, glandular puberulent above; leaves 5–12 mm. long, linear, cuspidate; stipules 4–6 mm. long, silvery attenuate-lanceolate, often lacerate; flowers in a small cyme; pedicels filiform, exceeding the leaf-like bracts; sepals 3–4 mm. long, ovate-lanceolate, scarious margined; petals pink, ovate, hardly longer than the sepals. A weed, from Europe, waste lands, *U. Son.,* to *A. T. T.*

STELLARIA. Chickweed, Stichwort

Annual or perennial, tufted generally diffuse herbs; leaves flat, rarely subulate; stipules none; flowers solitary or cymose, terminal or becoming lateral; sepals 5 or 4; petals 5 or 4, (rarely none), always more or less deeply bifid, often divided almost to the base, white; stamens 3–10; styles 3 or 4, rarely 5; capsules globose, ovoid or oblong, dehiscing by twice as many valves as there are styles. (Name from Lat., *stella,* star.)

Lowest leaves ovate, petiolate,
  Stems pubescent with a line of hairs; leaves ovate. *S. media.*
  Stems glabrous, except at base; leaves shiny, the upper linear-lanceolate. *S. nitens.*

Lowest leaves or all leaves sessile,
 Bracts of the inflorescence leafy, not much reduced,
  Leaves linear-lanceolate. *S. calycantha*, var. *Bongardiana*.
  Leaves broader, often ovate,
   Capsule longer than the calyx. *S. crispa*.
   Capsule not longer than the calyx. *S. viridula*.
 At least the upper bracts of the inflorescence small, scarious,
  Pedicels erect; cyme few-flowered. *S. longipes*.
  Pedicels spreading; cyme loosely many-flowered,
   Leaves linear, acute at each end; sepals 2–3 mm. long. *S. longifolia*.
   Leaves linear-lanceolate, broadest near the base; sepals 4–5 mm. long. *S. graminea*.

Stellaria calycantha (Ledeb.) Bong., var. **Bongardiana** (Fern.) Fern. Perennial, glabrous throughout, the weak stems ascending, branched, 15–60 cm. long; leaves lanceolate to lance-linear, the midnerve prominent, 15–45 mm. long; pedicels slender, often deflexed in fruit, 15–30 mm. long; sepals 4–5.5 mm. long, ovate-lanceolate, acute, scarious-margined; petals wanting, or if present shorter than the sepals; styles mostly 4; capsule 5–7 mm. long, ovoid; seeds 0.7–0.9 mm. long, brown, oval with one flattened side. *S. borealis* Bigel., var. *Bongardiana* Fern; *Alsine borealis* (Bigel.) Britton. In wet places in the mountains, *Can.*

Stellaria crispa C. & S. Perennial, glabrous: stems 15–70 cm. long, weak, decumbent or prostrate, simple or with but few branches; leaves 5–23 mm. long, thin, ovate, acuminate, entire or crisped on the margin, sessile or subsessile; pedicels slender, 6–20 mm. long; sepals 2–4 mm. long, lanceolate, 3-nerved, very acute, with a narrow scarious margin; petals deeply cleft, shorter than the sepals, or usually wanting; capsule 4–6 mm. long, ovoid; seeds 0.8–1 mm. wide, dull brown, discoid but slightly pointed near the hilum. *Alsine crispa* (C. & S.) Holz. By streams in woods, *Can.*

Stellaria graminea L. *Lesser Stichwort.* Whole plant, glabrous, shining; stems 3.5–6 dm. tall, 4-angled, ascending, branched above; leaves 2–5 cm. long, acute; flowers in a loose much-forked cyme; bracts scarious, somewhat ciliate; pedicels slender, spreading or reflexed; sepals lanceolate, acute, 3-nerved, ciliate; petals 2-cleft, as long as the sepals; capsule oblong, exceeding the sepals; seeds 1 mm. wide, reddish brown, rounded, minutely roughened. *Alsine graminea* (L.) Britton. Very sparingly introduced, Pullman, *Piper,* Sept. 6, 1899.

Stellaria longifolia Muhl. Erect or nearly so, 2–5 dm. tall; the stem sharply 4-angled, glabrous; leaves 2–6 cm. long, linear or oblong-linear, acute at each end, often ciliate near the base, 2–3 cm. long; cyme open, long-peduncled, becoming lateral; pedicels spreading or at length deflexed; sepals ovate-lanceolate, acute; petals exceeding the sepals, deeply parted; capsule 3.5–4 mm. long, ovoid; seeds 0.7 mm. long, oval, flattened, brown. *Alsine longifolia* (Muhl.) Britton. In wet meadows, Marshall Jct., *Piper* 2258, *A. T.*

Stellaria longipes Goldie. Somewhat tufted, with creeping rootstocks, smooth and shining throughout; stems 1–3 dm. tall, 4-angled, erect, branched above; leaves 1–3 cm. long, linear-lanceolate, tapering from the base to the acute apex, prominently 1-nerved; flowers solitary or in very loose cymes; bracts reduced and scarious or, when the flowers are few or solitary, foliaceous; pedicels 2–10 cm. long, slender, erect; sepals 3–5 mm. long, obtuse or acute; petals 2-cleft, exceeding the sepals; capsule blackish, longer than the calyx; seeds 1 mm. long oval-discoid, one side overlapping at the hilum, brown. *Alsine longipes* (Goldie) Cov.; *A. strictiflora* Rydb. is apparently not distinct. Moist places, not rare.

Stellaria media (L.) Cyrill. *Common Chickweed.* Annual or biennial, weak and spreading; stems weak, 2–40 cm. long, glabrous except for a line of hairs; the upper leaves sessile, all acute, 1–4 cm. long; flowers solitary in the axils or somewhat cymose on slender pedicels; sepals 3–5 mm. long, oblong, glandular pilose; petals deeply 2-parted, shorter than the sepals; stamens 3, 5, or 10; capsule ovoid, longer than the calyx; seeds 1–1.4 mm. wide, reddish brown, flattened, with rows of tubercles. *Alsine media* L. A weed, introduced from Europe, well established in lawns and by streams, *U. Son., A. T.*

The seeds are indigestible to lambs.

Stellaria nitens Nutt. ex T. & G. *Shining Chickweed.* Annual, somewhat tufted; stems 4–30 cm. tall, erect, slender, shining, branched above, sparsely pilose only at base; lowest pair of leaves acute, 4 mm. long, the others linear-lanceolate and sessile, 3–12 mm. long; cymes loose, few-flowered, the bracts scarious; pedicels thread-like, 1–5 cm. long; sepals 2.5–5 mm. long, 3-nerved, lanceolate, acute, scarious-margined; petals deeply 2-cleft, half as long as the sepals or wanting; capsules 3–4 mm. long, lanceoloid; seeds 0.5 mm. wide suborbicular, flattened, asymmetric at the hilum end, yellowish brown. *Alsine nitens* (Nutt. ex. T. & G.) Greene. Stony hillsides, abundant, *U. Son., A. T.*

Stellaria viridula (Piper) St. John. Perennial, forming prostrate mats, 10–20 cm. broad; herbage entirely glabrous except a few ciliate hairs at the bases of the leaves; stems branched; leaves 5–15 mm. long, ovate, short-acuminate, dull, thin but the veins obscure, each abruptly narrowed into a short petiole which bears a few long hairs; flowers solitary axillary; pedicels mostly shorter than the leaves; sepals 2–2.5 mm. long, oblong-ovate, acutish, green with a narrow white scarious margin, thin, 3-nerved; petals none; capsule ovoid, not longer than the calyx; seeds 0.5 mm. long, brown, discoid, pointed near the hilum. *Alsine viridula* Piper. Along rivulets in woods, altitude 1400 m., Blue Mts., Columbia Co., *Piper,* 2,328.

VACCARIA

Annual or perennial herbs; leaves ovate or ovate-lanceolate, sessile or petioled; flowers in corymbs or cymes; calyx 5-toothed, ovoid, oblong, or cylindric in flower, in fruit inflated; petals 5, longer or shorter than the calyx; stamens 10; styles 2; ovary 1-celled or incompletely 2–4-celled; capsule dehiscing with 4 teeth. (Name Lat., *vacca,* cow, alluding to use as fodder.)

Vaccaria vulgaris Host. *Cow Cockle.* Annual, glabrous and glaucous, 2–10 dm. tall, usually much branched above; leaves 2–8 cm. long, acute, somewhat connate at base; calyx 8–16 mm. long, with 5 salient angles, herbaceous and the intervals scarious; petals 2 cm. long, emarginate, rose-pink, without appendages. *V. vaccaria* (L.) Britton; *Saponaria Vaccaria* L. A weed from Europe, troublesome in wheat fields, *U. Son., A .T.*

CERATOPHYLLACEAE. Hornwort Family

Submerged aquatic herbs; leaves verticillate, thrice dissected dichotomously into filiform, stiff divisions; flowers monoecious, inconspicuous, sessile, axillary; sepals 6–12, herbaceous, valvate; petals none; stamens 10–24, filaments very short; pistil 1; ovary 1-celled; fruit nut-like, with a persistent style. (Named from the genus *Ceratophyllum.*)

CERATOPHYLLUM. Hornwort

The only genus. Characters of the family. (Name from Gr., *keras,* horn; *phullon,* leaf.)

Ceratophyllum demersum L. *Hornwort.* Plant glabrous; stem 1–10 dm. long, freely branching; leaves 7–27 mm. long, 5–12 at a node, linear, 2–3-times dichotomously forked, rigid, spiny denticulate; staminate flowers with 10–20 stamens nearly as long as their campanulate, deeply lacerate involucre; pistillate flowers with a similar involucre; fruits 4–6 mm. long, oval, with a straight or curved spine-like beak, and below tubercled, or winged, or 2-spurred. Submerged in slow streams or ponds, Dartford, *St. John & Warren* 6751, mouth of Potlatch R., *St. John et al.* 9739; Clear L., Spokane Co., in 1948, *Yocom, U. Son., A. T.*

NYMPHAEACEAE. Waterlily Family

Aquatic perennial herbs with horizontal rootstocks; leaves peltate or cordate, floating, submersed, or rarely emmersed; flowers perfect, solitary, axillary, on long peduncles; sepals 3, 4, 6 or more; petals 5–many, often grading into the sepals or stamens; stamens 5–many; carpels 3–many, indehiscent, free or immersed in a fleshy receptacle or more or less coalescent into a fleshy fruit; endosperm present or none. (Named for the genus *Nymphaea.*)

Leaves cordate,
  Flowers white or not yellow; stamens epigynous. *Nymphaea.*
  Flowers yellow; stamens hypogynous. *Nuphar.*
Leaves peltate; flowers purple; carpels separate. *Brasenia.*

BRASENIA. Watershield

Aquatic herbs; leaves alternate, long-petioled, centrally peltate, oval, floating; flowers axillary, small; sepals 3 or 4; petals 3 or 4, linear, sessile; stamens 12–18; the filaments filiform; pistils 4–18, forming club-shaped indehiscent pods, each with 1–2 seeds. (Name of uncertain derivation.)

Brasenia Schreberi J. F. Gmel. *Watershield.*. Leaves long-petioled, entire or obscurely crenate, 2–10 cm. long; flowers dull purple; sepals lance-ovate, 8–13 mm. long; petals oblong, 14–18 mm. long; carpels ellipsoid, greenish, 5–8 mm. long; seeds ovoid, yellow, 3–4 mm. long; submerged parts of the plant coated with a tough transparent jelly. In lakes, Spokane County, *A. T.*

NUPHAR. Yellow Pond Lily

Acaulescent perennials from stout rootstocks, commonly slightly milky; leaves with united stipules which are sometimes adnate to the base of the petioles; calyx more showy than the corolla; sepals 5–12, concave, roundish, mostly yellow, and petal-like; petals 10–20, hypogynous, small and thick, the innermost or sometimes all of them stamen-like; stamens numerous, hypogynous; stigmas radiate on the summit of the 10–25-celled ovary; ovules and seeds numerous. (The ancient Arabic name, *Nouphar,* used by *Dioscorides.*)

Nuphar polysepalum Engelm. *Tule Lily, Wokas.* Leaves orbicular or broadly oval, deeply cordate, 20–30 cm. long, 12–20 cm. broad; calyx subglobose, about 8 cm. broad; sepals 6–12, yellow, often red-tinged, 3–5 cm. long; petals 12–18, ovate-cuneate, truncate, 1–1.5 cm. long; stamens very numerous, reddish; fruit subglobose, truncate, 4–5 cm. long; seeds ellipsoid, yellowish, 3.5–4 mm. long. *Nymphaea polysepala* (Engelm.) Greene; *Nymphozanthus polysepalus* (Engelm.) Fern. In lakes and slow streams, *A. T.*
The Klamath Indians roasted and ate the seeds.

NYMPHAEA. Water Lily

Aquatic perennial herbs, with thick rhizomes from which rise the long petioles of the floating leaves; sepals 4, nearly free; petals numerous, in many series, the inner transitional to the stamens; stamens numerous, inserted on the ovary; ovary 12–35-celled, the apex concave with radiating stigmas; fruit depressed globular, maturing under water; seeds enclosed by a sac-like aril. (Ancient name from the water nymphs.)

Nymphaea odorata Ait. *Pond Lily, Water Lily.* Rhizome on pond bottom; petioles elongating to reach the surface; blades floating, 0.5-2.5 dm. across, nearly circular, with a narrow sinus, green above, usually purple below; flowers opening for 3-4 days from early morning to about noon; sepals 2.8-8 cm. long, ovate to ovate-lanceolate; petals 17–32, white, 1–2.2 cm. wide; stamens 36–100; seeds 1.5–2.3 mm. long, ellipsoid, exceeded by the aril. *Castalia odorata* (Ait.) Woodville & Wood. Planted, and now established in Newman Lake, Spokane Co.

RANUNCULACEAE. Buttercup Family

Annual or perennial herbs or sometimes woody plants with acrid sap; leaves usually alternate (opposite in *Clematis*), simple or compound; stipules none but the base of the petiole often clasping or sheathing; flowers regular or irregular; sepals 3–15, generally soon withering, often petal-like; petals 3–15, or more, or wanting; stamens numerous, hypogynous; carpels numerous or few or rarely solitary, separate, 1-celled, 1–many-ovuled; endosperm present. (Named for the genus *Ranunculus.*)

Carpels numerous, 1-ovuled; fruit an achene,
  Leaves opposite, *Clematis.*
  Leaves alternate or basal,
    Flowers subtended by a leaf-like involucre remote from the calyx, *Anemone.*
    Flowers not subtended by involucres,
      Petals none,
        Leaves ternately decompound. *Thalictrum.*

Leaves simple, palmately lobed. *Trautvetteria.*
Petals present,
Sepals spurred; achenes in a long slender spike. *Myosurus.*
Sepals not spurred; achenes in a head,
Petals with a nectariferous spot or scale at base; ovule erect. *Ranunculus.*
Petals not nectariferous; ovule pendent. *Adonis.*
Carpels few, 2–many-ovuled; fruit a follicle or berry,
Flowers irregular,
Upper sepal spurred; petals 4 (or 2). *Delphinium.*
Upper sepal hood-like; petals 2–5. *Aconitum.*
Flowers regular,
Petals spurred. *Aquilegia.*
Petals not spurred,
Petals 1–2 cm. long, reddish, inserted on a fleshy disk; flowers solitary. *Paeonia.*
Petals smaller, disk none, flowers solitary, or in racemes or umbels,
Carpels becoming berries; flowers in racemes. *Actaea.*
Carpels becoming follicles, flowers solitary or in umbels. *Coptis.*

ACONITUM. Aconite

Tall erect perennial herbs; leaves palmately-lobed or divided; flowers large, showy, in terminal racemes or panicles; sepals 5, petal-like, very irregular, the upper one hooded or helmet-shaped; petals 2–5, the upper two hooded, on long claws, concealed in the helmet; stamens numerous; pistils 3–5, many-ovuled, forming follicles at maturity. (The ancient Gr. and Lat. name.)

Flowers blue. *A. columbianum,* f. *columbianum.*
Flowers pale, yellowish, or greenish, to white. Forma *ochroleucum.*

Aconitum columbianum Nutt. ex T. & G. forma **columbianum.** *Aconite; Monkshood.* Stems erect, 5–17 cm. tall, somewhat pubescent or viscid pilose above; leaves glabrous or the upper puberulent, roundish, 5–15 cm. broad, palmately 5-lobed; lobes cuneate-obovate, incisely serrate or cleft; petioles mostly shorter than the blades; flowers in loose viscid pilose racemes or panicles; hood 2–3 cm. long, the helmet-shaped upper part higher than broad, 15–25 mm. long, strongly beaked; follicles oblong-linear, 1–2 cm. long, the slender beak usually recurved. Along streams especially in the mountains, *A. T. T., Can.* Not recorded from the Blue Mts.

All parts of the plant, but especially the roots and seeds are dangerously poisonous. They contain alkaloids similar to aconitin, which cause paralysis of the respiration.

Forma **ochroleucum** (A. Nels.) St. John. Differing from *A. columbianum* in having the flowers greenish, or yellowish, to white. *A. ochroleucum* A. Nels., *A. columbianum* Nutt., subsp. *pallidum* Piper. Common in woods in the Blue Mts., *Can.*

ACTAEA. Baneberry

Erect perennial herbs; leaves large, 2–3-ternately compound; flowers small, white, in a terminal raceme; sepals 3–5, petal-like;

petals 4–10, small, spatulate or narrow-clawed; stamens numerous; ovary 1–many-ovuled, in fruit forming a large somewhat poisonous berry. (Name Gr. from *aktea,* the elder, transferred to the baneberry by Linnaeus.)

Actaea rubra (Ait.) Willd., var. **arguta** (Nutt. ex T. & G.) Lawson. Stems 3–9 dm. tall erect; leaves cauline, triternately decompound; leaflets obliquely ovate or lanceolate, acuminate, coarsely incised-serrate or lobed, puberulent when young, 3–12 cm. long; racemes densely flowered, becoming loose in age and 4–10 cm. long; sepals oblong, 4–5 mm. long, deciduous; petals white, 3–6 mm. long; stamens white, showy, 3–7 mm. long; berries bright red, ellipsoid to globose, 9–12 mm. long; seeds pitted, brown, obliquely lanceoloid, 3–4 mm. long. *A. arguta* Nutt. ex T. & G.; *A. rubra,* subsp. *arguta* (Nutt. ex T. & G.) Hultén. In woods in the mountains, *A. T. T., Can.*

The berries of this species are probably poisonous, as are those of other species in the genus.

ADONIS

Herbs; stem leafy; leaves alternate, finely pinnatifid, the pinnules linear; sepals 5; petals 3–20; stamens and petals numerous. (In Greek mythology, the beautiful youth, *Adonis,* was the lover of Aphrodite.)

Adonis autumnalis L. Annual, 25–60 cm. tall; stem erect, simple or branched, usually glabrous; lower leaves petioled; blades 3–4-pinnatifid, the pinnules linear, entire or 3-toothed; flowers single, erect; sepals glabrous, becoming reflexed; petals 6-10, obovate, dark red, with a black blotch at base, concave connivent; receptacle fusiform; achenes ovoid, the body reticulate, the beak subulate, half as long as the body.

Introduced from southern Europe, a weed in wheat fields, 1932, *F. A. Warren.*

ANEMONE. Windflower

Erect perennial herbs; leaves compound or divided, all radical, except 2 or 3 cauline which form an involucre usually remote from the flower; peduncles 1-flowered, solitary or in umbels; sepals 4–20, petal-like; petals none; stamens numerous; pistils numerous; achenes pointed or tailed, flattened not ribbed. (Name Gr., *anemos,* wind, or fide Fernald, it commemorates *Na'man,* the Semitic equivalent of *Adonis,* the youthful lover of *Venus.*)

Anemone Piperi Britton. Rootstock usually erect; stems erect, 1-flowered, 1–3.5 dm. tall, smooth or sparingly sericeous; radical leaf and three of the involucre similar, ternate or often 5-foliolate, petiolate, the leaflets or divisions oblong-cuneate or ovate, acute or acuminate, incisely toothed or lobed, sparsely pilose, paler beneath, 3–5 cm. long; sepals 4–7, oval, white, or tinged with pink, 8–17 mm. long; achenes elliptic, short pilose, the style short. Moist woods, abundant, *A. T. T., Can.*

This species was named in honor of Prof. C. V. Piper from specimens he collected in 1893 in Latah County, Idaho.

The record by Prof. M. L. Fernald, in Rhodora **30**: 188, 1928, of *A. oregana* Gray, is based upon a duplicate type sheet of *A. Piperi* Britton. The writer can see no mixture or confusion in the specimens. They have the characters of genuine *A. Piperi,* and do not seem to be *A. oregana* of the Cascade Mountains.

AQUILEGIA. Columbine

Erect perennials with 2–3-ternately compound leaves; flowers regular, showy, on the ends of the branches; sepals 5, regular, petal-like; petals 5, all alike, each with a short lamina and produced backward into a large hollow spur much longer than the calyx; stamens numerous; pistils 1–10, with slender styles; pods erect, many-seeded. (Name perhaps from Lat., *aquila,* eagle.)

Spurs sturdy, much swollen and asymmetric at tip. *A. formosa,* var. *formosa.*
Spurs very slender, but slightly swollen at tip, not asymmetric. Var. *wawawensis.*

Aquilegia formosa Fisch. in DC. var. **formosa.** Usually sparingly pubescent; stems erect, 35–90 cm. high; leaflets 1–8 cm. long, firm, broadly cuneate, paler beneath, lobed; flowers nodding, 3.5–4 cm. long; sepals bright red, spreading, ovate-lanceolate, acute, 21–26 mm. long; laminae yellow, 2–5 mm. long; spurs red, nearly straight, about 2 cm. long; follicles pilose, beaked, 2–2.5 cm. long. Common in the Blue Mts., in woods or on wet rocks, *Can.*

Var. **wawawensis** (Pays.) St. John. Differing from *A. formosa* by having leaflets very thin; sepals and spurs pale red; spurs straight, slender. First collected by *Rex Hunt* in 1906. The only locality in our area, is his type locality, in the Snake River Canyon, Garfield County, opposite Truax. *A. wawawensis* Pays. Dripping ledges, *U. Son.*

CLEMATIS. Clematis, Virgin's Bower

Perennial herbs, low and erect, or more or less woody vines, climbing by the petioles; leaves opposite; flowers perfect or some or all of them unisexual; sepals 4, rarely 3, or more than 4, petal-like; petals none or minute; stamens numerous; pistils numerous, 1-ovuled; styles feathery in fruit. (An ancient Gr. name, used by Dioscorides.)

Tufted herbs; leaves 2–3 pinnate; leaflets lanceolate or linear, *C. hirsutissima.*
Half woody vines; leaves 1-pinnate; leaflets ovate,
Flowers solitary, blue. *C. columbiana.*
Flowers white, numerous in cymes. *C. ligusticifolia.*

Clematis columbiana (Nutt.) T. & G. *Purple Clematis.* Half-woody climber with slender stems; leaves ternate, long petioled; leaflets ovate, acuminate, often cordate, sparsely pilose beneath, entire or crenate; flowers on long naked peduncles; sepals ascending, oblong-lanceolate, acute or acuminate, ciliate, 3–5 cm. long; outer stamens with the filaments more or less dilated and petal-like, the anthers wanting; achenes pilose, the long plumose styles 4–5 cm. long. Moist woods in the mountains, *A. T. T., Can.*

Clematis hirsutissima Pursh. *Sugar Bowls.* Stems tufted, 3–7 dm. tall, from a tough woody caudex; leaves 3 or 4 pairs, twice or thrice pinnately compound, hirsute when young, becoming glabrous; flower solitary, at first nodding, then erect, usually long-peduncled; sepals madder purple within, lavender and hirsute without, thick, 2.5–4.5 cm. long, closely contiguous; achenes obliquely lanceolate, cuneate, hirsute, their tails very plumose 5–6 cm. long. *C. Wyethii* Nutt. Common in meadows and grass-lands, *A. T.*

C. A. Geyer reported that the Nez Perce Indians used this plant to revive horses that were nearly dead from exhaustion. When macerated roots were rubbed in the horse's nostrils, it sprang up in convulsions, then was bathed in the river, and soon seemed completely recovered. The leaves taste like strychnine. The tissues of this plant seem to contain a very powerful drug.

Clematis ligusticifolia Nutt., ex T. & G. Half-woody climber, the stem 2–10 m. long, pilose on the nodes and petioles; leaves pinnately compound, of 5–7 leaflets, or the lower pair ternate; leaflets ovate or lanceolate, sometimes 3-lobed, coarsely incised-dentate or subentire, 3–8 cm. long, sparsely pubescent, becoming glabrous; flowers dioecious, in large cymes, the staminate more showy; sepals oblong, soft hairy, about 1 cm. long; achenes pilose, ovate, the plumose styles 3–5 cm. long. Common in the warmer valleys, *U. Son., A. T.*

COPTIS. Goldthread

Low glabrous perennials with slender rootstocks; leaves all radical, ternately or pinnately divided or compound; flowers on scapes, solitary, or in few-flowered umbels; sepals 5–7; petal-like; petals 5–7, small, linear, or hooded at tip; stamens numerous; pistils 3–7, on slender stalks, in fruit forming a cluster of divergent follicles. (Name from Gr., *koptein,* to cut.)

Coptis occidentalis (Nutt.) T. & G. Scapes 2–3-flowered, 10–25 cm. tall; leaves trifoliolate, evergreen; leaflets long-petioled, suborbicular, 3–7 cm. long, shining, deeply 3-lobed, the lobes obtuse, dentate, or again lobed; sepals linear, 3-nerved, white, 1 cm. long; petals 5–6 mm. long, short-clawed at the base, broadened at the nectary, attenuate beyond, obtuse; stamens about 12, shorter than the carpels; mature carpels 1.2–2 cm. long, the fruiting portion spreading, longer than the erect stipe. *Chrysocoptis occidentalis* Nutt. Common in the evergreen woods, Idaho, *A. T. T., Can.*

DELPHINIUM. Larkspur

Annual or perennial erect branching herbs; leaves palmately lobed or divided; flowers showy; in a raceme or panicle; sepals 5, petal-like, the upper one prolonged backwards into a spur; petals 4, sometimes 2; the two posterior ones spurred; the lateral, when present, small; stamens numerous; pistils 1–5, sessile, many-ovuled, forming follicles at maturity. (Name Gr., *delphinion,* larkspur, from *delphin,* dolphin.)

Roots elongate, not tuber-like; stems about 1 m. tall,
  Axis glandular villous. — *D. occidentale,* var. *cucullatum.*
  Axis cinereous appressed puberulent. — Var. *griseum.*
Roots a cluster of tubers; stems shorter,
  Pedicels mostly shorter than the spurs; sepals erect or but little spreading,
    Flowers usually 10-30; pubescence villous, somewhat viscid, — *D. cyanoreios,* f. *idahoense.*

Flowers usually more numerous; pubescence of short appressed hairs,
Leaves puberulent. *D. Burkei.*
Leaves glabrous or nearly so. Var. *distichiflorum.*
Pedicels mostly longer than the spurs; sepals wide spreading,
Inflorescence glabrous or nearly so. *D. Nuttallii.*
Inflorescence hairy,
Stems weak, slender, tapering to a thin, disarticulating portion above the root,
Sepals 7.5–12 mm. long, 4.5–7 mm. wide. *D. Nuttallianum.*
Sepals 5–8 mm. long, 3.5–5 mm. wide. *D. depauperatum*
Stems strong, erect, not tapening to a basal disarticulating portion,
Sepals markedly unequal, the two lower the larger, the upper the smallest,
Rhachises of racemes felted with short curved hairs. *D. bicolor,* f. *McCallae.*
Rhachises of racemes sparsely hirsute. Forma *Helleri.*
Sepals not very unequal,
Racemes pyramidal; rhachis of racemes hoary pubescent or glandular villous. *D. Menziesii,* subsp. *pyramidale.*
Racemes oblong; rhachis of racemes sparsely puberulent or subglabrous. *D. Nelsonii.*

Delphinium bicolor Nutt., forma **Helleri** (Rydb.) Ewan. Perennial 20–30 cm. tall, from a cluster of slender woody fibrous roots; stem erect, glabrous or nearly so, rarely sparsely hirsute; leaves few, the blades 3–4.5 cm. wide, palmatifid nearly to the base into mostly undivided narrow, almost linear primary divisions; lower bracts of racemes leafy, ample; racemes interrupted, almost pramidal; pedicels ascending spreading; spur 15–20 mm. long; follicles 8–20 mm. long, glabrate; seeds 2 mm. long, ovoid, shining, black, narrowly white-angled. Grassy bank or open forests, *A. T., A. T. T.*

Forma **McCallae** Ewan. Azure blue sepals often contrasting strongly with the pale blue lower petals. Open woods, *A. T. T.*

Delphinium Burkei Greene. Roots short, thick; stems strict, erect, puberulent, 3–16 dm. tall, usually simple, rarely with a few erect branches; leaves puberulent, all divided into narrow lobes, linear in the upper leaves, broader in the lower ones; petioles shorter than the blades except in the lower leaves; raceme spike-like, the pedicels shorter than the pale dull blue flowers; sepals 8–10 mm. long, shorter than the spur; follicles puberulent, erect, 6–10 mm. long. *D. simplex* Dougl., non Salisb.; *D. strictum* A. Nels. Common in low meadows, *A. T.*

Var. **distichiflorum** (Hook.) St. John. Pubescence of minute mostly curved and appressed hairs, never villous; stems strictly erect, 3–9 dm. high, simple or rarely with a few erect branches; leaves rather numerous, thickish, deeply cleft or parted, the lower into cuneate rather broad segments, the upper into narrow segments, all glabrous or nearly so; raceme very dense, many-flowered, almost spicate; calyx puberulent externally, the spur usually longer than the sepals; follicles puberulent, not spreading, about 1 cm. long. *D. simplex,* var. *distichiflorum* Hook. In moist meadows especially in Spokane County, *A. T.*

Delphinium cyanoreios Piper, forma **idahoense** Ewan. More or less pubescent with villous somewhat viscid hairs; stems strictly erect, simple or rarely with a few branches, 3–10 dm. high; leaves rather few, mostly near the base, orbicular in outline, 2–5 cm. broad, thickish, the lower cleft into broad cuneate lobes, the upper into narrower or linear lobes or divisions; raceme moderately dense, usually 10–20-flowered; calyx villous, 1 cm. long, the spur about as long as the sepals; upper petals whitish, tinged with blue; follicles densely and finely villous, not spreading, 12–20 mm. long. In the mountains, usually at considerable altitudes. Dried swales, Craig Mts., *Henderson,* June 23, 1896; and Anatone, *Cronquist & Jones* 5,863, *A. T. T.*

Delphinium depauperatum Nutt. ex T. & G. Tubers fascicled; stems 2–8 dm. tall, villous, puberulent, or glabrate; leaves rounded, 2–10 cm. wide, thick, puberulent or glabrate, dissected into linear or narrowly oblong lobes; inflorescense a loose raceme, yellowish viscid villous, the pedicels ascending, the lower longer than the flowers; flowers dark blue, somewhat pilose; sepals elliptic, 1–1.7 mm. long, the spur half again as long; upper petals blue-veined, lower petals blue; follicles villous, divergent at tip, 6–20 mm. long. Open slopes or open wood, *U. Son.* to *A. T. T.*

Delphinium Menziesii DC., subsp. **pyramidale** Ewan. Perennial 20–75 cm. tall, from a cluster of shallow, globose tubers 1–3 cm. in diameter; stems soft pubescent, the upper part hoary or fuscous glandular villous, the buds, pedicels, and young follicles pilose; principal blades 5–7 cm. wide, pentagonal or orbicular, thick, palmatifid into approximate, narrow, cuneate division; racemes 12–20-flowered, pyramidal; lower pedicels 3.5–8 cm. long in flower; sepals 12–17 mm. long, broadly ovate. Asotin Co., Anatone, May 31, 1933, *R. Sprague.*

Delphinium Nelsonii Greene. Tubers fascicled; stems 1–8 dm. tall, appressed puberulent or glabrate; blades appressed puberulent or glabrous. dissected into linear lobes; inflorescence appressed puberulent; flowers like those of *D. depauperatum;* pods glabrous or strigose, about 15 mm. long. Rocky hillsides, *A. T., A. T. T.*

One of the poisonous species of *Low Larkspur.* It is less harmful to sheep than to cattle.

Delphinium Nuttallianum Pritzel ex Walp. Perennial 12–30 (–45) cm. tall, from a cluster of tubers or a short, thickened rootstock; stems glabrous or puberulent; leaves few, mostly cauline; principal blades 3–5 cm. wide, palmatifid into 3 or 5 aproximate primary divisions, almost simple, narrowly ovate, obtuse or closely palmately cleft into short oblong, obtuse ultimate segments, glabrous or nearly so; racemes hirsute, few-flowered; flowers nodding, bright dark- or purplish-blue; bracts leafy, linear; pedicels slender, ascending; sepals oblong-ovate, obtuse or acute; spur 10–12 mm. long, straight; upper petals whitish, shorter than the sepals; follicles 10–18 mm. long, sparsely hirsute; seeds 1–1.5 mm. long, quadrate-oblong, blackish, glabrous, narrowly white wing-angled. Grasslands or woods, *U. Son., A. T., A. T. T.*

Delphinium Nuttallii Gray. Habit and appearance of *D. depauperatum* but the herbage glabrous and usually a little glaucous; inflorescence glabrous or nearly so; follicles glabrous, spreading moderately at maturity. *D. columbianum* Greene. Open slopes, *A. T.*

Delphinium occidentale (Wats.) Wats., var. **cucullatum** (A. Nels.) Ewan. Tufted; stems 1–2 m. high, glabrous below; leaves orbicular in outline, 3–5-cleft or parted, the lobes again cleft, those of the lower leaves cuneate and rather broad, of the upper narrower or linear; larger leaves 8–14 cm. broad, the petioles as long or longer and scarcely dilated at the base; inflorescence paniculate, or on smaller stems racemose, the principal axis densely flowered and spike-like, often 30 cm. long, glandular villous; flowers dull blue; calyx glandular villous outside, the sepals erect or but little spreading, about 1 cm. long, as is also the spur; follicles 2–2.5 cm. long. Blue Mts., *Huds.*

Var. **griseum** St. John. Differs from var. *cucullatum* by having the inflorescence cinereous appressed puberulent, and the corolla subappressed pilose without; all the hairs being non-glandular. High ridges of the Blue Mts., *Huds.*

MYOSURUS. Mouse Tail

Very small annual herbs; leaves entire, linear or spatulate, in basal tuft; scapes simple, 1-flowered; sepals 5 (or 6–7), spurred at the base; petals 5–7, greenish-yellow, with long nectariferous claws, or none; stamens 5–25; pistils numerous, borne on a central axis, the receptacle, which becomes greatly elongated in fruit; ovule 1. (Name Gr., *mus,* mouse; *oura,* tail)

Beak more than ½ as long as the body of the achene; fruiting spike 3–8 mm. long. *M. aristatus.*
Beak very short; fruiting spike 1–4.5 cm. long. *M. minimus,* spp. *minimus.*

Myosurus aristatus Benth. Scapes 1–10 cm. high; leaves linear-spatulate, half the length of the scapes; sepals 1–1.7 mm. long, whitish; petals often none; achene body 1–3 mm. long, oblong, the back with a prominent keel which is prolonged into a stout beak. The plant usually becomes reddish when mature and often covers considerable areas. Rocky slopes, *U. Son., A. T.*

Myosurus minimus L., subsp. **minimus.** Plant 3–20 cm. tall; leaves linear, one-fourth as long as the scapes; sepals spreading, 2 mm. long; achenes scarcely keeled on the back, the body 1.5–2 mm. long. In dried-up ponds, infrequent, *U. Son., A. T.*

PAEONIA. Paeony

Robust perennial herbs with ternately or pinnately compound leaves and showy flowers; sepals 5, herbaceous, persistent; petals 5–10; stamens numerous, inserted on a fleshy disk; pistils 2–5; fruit of 2–5 leathery, several-seeded follicles. (The ancient Gr. and Lat. name.)

Paeonia Brownii Dougl. ex Hook. Roots massive, woody; plant glabrous and glaucous, 2–5 dm. high, at first erect or ascending, in fruit decumbent; leaves thick, once or twice ternately divided or parted, the ultimate segments from narrowly oblong to obovate; sepals green or reddish without, concave, unequal; petals dull brownish red, thick, 1–2 cm. long; follicles usually 5, oblong, smooth, about 3 cm. long; seeds brown, ellipsoid, 10–13 mm. long. On open hillsides in the Blue Mountains, *A. T.* to *Huds.*

First discovered by David Douglas in the Blue Mts.

RANUNCULUS. Buttercup, Crowfoot

Annual or perennial herbs; cauline leaves usually alternate; flowers solitary or corymbed; sepals usually 5, deciduous; petals as many or more, conspicuous or minute, with a nectariferous pit and with or without scale at the base of the blade; stamens numerous, occasionally few; pistils numerous, 1-ovuled; achenes capitate or spicate, generally flattened, each tipped with a minute or an elongated style. (Name Lat., a diminutive of *rana,* frog.)

Petals white or mostly so, with a naked pit at base; achenes transversely wrinkled; aquatics,
  Immersed leaves only present. *R. aquatilis*, var. *capillaceous.*
  Dilated floating leaves also present. Var. *hispidulus.*
Petals yellow, with a nectar pit and scale at base; achenes not transversely wrinkled,
  Plants aquatic or subaquatic; leaves finely dissected when submersed, less so when aerial,
    Achenes callous margined; petals 8–12 mm. long. *R. flabellaris.*
    Achenes without a thickened border; petals 4–8 mm. long. *R. Gmelini*, var. *terrestris.*
  Plants terrestrial, but often growing in very wet places; leaves never finely dissected,
    Achenes thin walled, the faces nerved; leaves crenate; plant spreading by runners. *R. Cymbalaria*, var. *saximontanus.*
    Achenes not thin walled or nerved,
      Leaves entire or nearly so,
        Stems creeping, rooting at the nodes. *R. Flammula*, var. *ovalis.*
        Stems erect, not rooting at the nodes,
          Plants 3–6 dm. high; leaves lanceolate or oblong. *R. alismaefolius*, var. *alismaefolius.*
          Plants 1–3 dm. high; leaves cordate or subcordate. *R. Populago.*
      Leaves or some of them lobed or divided,
        Faces of the achene scabrous, or the body much smaller than the beak,
          Achene lanate, with two lateral swellings, the body much smaller than the subulate beak. *R. testiculatus.*
          Achenes bristly or spiny,
            Petals 1–2 mm. long; achenes 1.5–2 mm. long. *R. hebecarpus.*
            Petals 5–8 mm. long; achenes 5–6 mm. long,
              Achenes 10–20, not spiny on margins; stems glabrous. *R. muricatus.*
              Achenes 5, the margins long spiny; stems thinly hirsute. *R. arvensis.*
        Faces of the achenes smooth or merely pilose; mostly perennials,
          Herbage glabrous or nearly so,
            Achenes 2.5–4 mm. long. *R. Macounii*, var. *oreganus.*
            Achenes 1–2 mm. long,
              Basal leaves 3-toothed or 3–5-lobed or rarely entire; cauline ones 3-cleft or -parted. *R. glaberrimus*, var. *glaberrimus.*
              Basal and cauline leaves 3–5-lobed or parted. *R. sceleratus*, var. *multifidus.*

Herbage pilose or hirsute; mostly tall and coarse species,
Beaks of achenes hooked at the tip,
Achenes hispid on the faces; stems pilose at base. *R. uncinatus*, var. *parviflorus.*
Achenes smooth; stems glabrous. Var. *uncinatus.*
Beaks of achenes not hooked at the tip,
Beak of achene 2–3 mm. long. *R. orthorhynchus*, var. *platyphyllus.*
Beak of achene shorter,
Head of achenes cylindric; petals not longer than sepals. *R. pensylvanicus.*
Head of achenes globose or ovoid; petals longer than sepals. *R. Macounii*, var. *Macounii.*

Ranunculus alismaefolius Geyer ex Benth., var. **alismaefolius.** Glabrous throughout, rather stout, branched above; radical leaves long-petioled, the blades lanceolate or ovate, obtuse, usually cuneate at the base, entire or obscurely denticulate, 5–10 cm. long; cauline leaves narrower, mostly sessile; peduncles elongated; sepals 3–7 mm. long; petals yellow, cuneate-obovate, strongly nerved, 6–10 mm. long; achenes turgid, smooth, short-beaked, 2.5 mm. long. Wet shores, common near Pullman, *A. T.*

Ranunculus aquatilis L., var. **capillaceus** (Thuill.) DC. *Water Crowfoot.* Stems 5–40 cm. long, slender, growing in water; stipules half fused to the petiole base; leaves only submersed, flaccid, finely divided, the segments 8–20 mm. long; petals 1–3-times longer than the sepals, white, yellow at base; achenes 1–2 mm. long, thick, transversely wrinkled, the style short; receptacle hairy. *R. trichophyllus* Chaix, var. *typicus* W. B. Drew. In ponds or slow streams, *A. T.*

Var. **hispidulus** E. R. Drew. Both immersed dissected and floating leaves present, the latter 3–5-lobed, acute or obtuse at tip. In streams, *A. T.*, *A. T. T.*

Ranunculus arvensis L. Plant erect; stems one or several, 1.5–5 dm. tall; basal leaves 1.5–3.5 cm. long, 1.5–4.5 cm. wide, deeply 3-parted, the lobes obovate-lanceolate, sometimes again shallowly lobed, glabrous or thinly appressed pilosulous; sepals 6–7 mm. long, lanceolate, pubescent; the 5 petals 5–8 mm. long, obovate, yellow; stamens 10–15; achenes 5 in a whorl, 5 mm. long, short-stalked, obovate, compressed, the faces with short spines, the thickened border produced into long spines, the beak stout, curving dorsally; receptacle not enlarged in fruit, slightly villous at the summit. Weed, introduced from Eurasia or Africa; first collected in 1936, in Latah Co., now occasional, *A. T.*, *A. T. T.*

Ranunculus Cymbalaria Pursh, var. **saximontanus** Fern. Perennial, glabrous; leaves clustered at the base and at the joints of the long slender stolons, ovate or ovate-cordate, thick, 5–35 mm. long; scapes 1–7-flowered, 3–20 cm. high; petals 5–8, pale yellow, 4–9 mm. long, exceeding the sepals; heads of achenes oblong, 6–12 mm. long. Moist soil especially where alkaline, by the Snake River and in the "Scablands," *U. Son.*

Ranunculus flabellaris Raf. in Bigel. Aquatic or in drying ponds subaquatic or terrestrial; submersed leaves ternately dissected into capillary lobes; emersed leaves roundish, 5–7-parted into cuneate lobes; peduncles stout; petals 5–8, bright yellow; achenes ovate, turgid, margined towards the base with a thick corky border, and tipped with a straight beak, the body 2 mm. long; terrestrial forms (forma *riparius* Fern.) have more or less pubescent leaves parted into cuneate lobes, and smaller flowers. *R. delphinifolius* Torr. In ponds and slow streams, infrequent, Spokane Co., *A. T.*

Ranunculus Flammula L., var. **ovalis** (Bigel.) Benson. Stems slender, ascending, then creeping, rooting at the joints, 1–7 dm. long; leaves narrowly-lanceolate to oblanceolate, short-petioled, acute at each end, 2–10 cm. long, shorter than the internodes; flowers yellow, mostly solitary on the ascending tips of the stems; sepals 2–2.8 mm. long; petals narrowly obovate, 2.5–5 mm. long; fruiting head globose; achene 2 mm. long, oval, flattened, smooth, short-beaked. *R. Flammula* L., var. *unalaschensis* (Bess.) Ledeb.; *R. reptans* L., var. *ovalis* (Bigel.) T. & G. Wet shores, *A. T., A. T. T.*

Ranunculus glaberrimus Hook., var. **glaberrimus.** *Buttercup.* Nearly glabrous throughout, the stems erect or ascending, 5–25 cm. high; roots fascicled, elongate, fleshy; radical blades 2–3 cm. long, ovate to obovate or orbicular, rarely entire, usually crenately 3-lobed at apex, 1–3.5 cm. long, on petioles as long or longer; cauline cuneate-obovate, deeply 3-lobed, usually sessile; petals 5–15, broadly obovate, 5–15 mm. long, bright yellow; achenes obovoid 2 mm. long, sparsely pilose, short-beaked, aggregated in a globose head about 1 cm. in diameter. Dry or moist places, or open woods, *U. Son., A. T., A. T. T.*

Flowers faintly, but pleasantly fragrant. Common, blooming in earliest spring.

The var. *ellipticus* Greene is a recognizable variation, but none of the specimens so called from our area are correctly determined, rather they all belong in var. *glaberrimus.*

Ranunculus Gmelini DC., var. **terrestris** (Ledeb.) Benson. Perennial aquatic, wholly or party submerged, or becoming stranded on muddy shores, the emersed form more or less hirsutulous; often rooting at the nodes; petioles 2–60 mm. long; submersed blades 2–3 cm. long, palmately cleft, and often 1–3-times further divided into oblong membranous divisions; emersed leaves smaller, firmer, and less divided; sepals 3–4 mm. long, suborbicular, thin-margined, early deciduous; petals oval, bright yellow; achenes 1–1.5 mm. long, obovoid, compressed, smooth, tipped by a style 0.5 mm. long. *R. Purshii* Richards. In spring water, Waitsburg, *Horner* B40, *A. T.*

Ranunculus hebecarpus H. & A. Annual; whole plant sparsely villous; stems slender, mostly erect, 15–35 cm. tall; leaves 1–3.5 cm. broad, 3-parted, the lobes incisely 3-lobed; the upper leaves with narrower, lance-shaped segments; petioles of the lower leaves much longer than the blades; petals pale yellow, 1.5–2 mm. long, about equaling the sepals; achenes few, rounded, flattened, 2 mm. long, with short hooked beaks, the sides roughened and covered with hooked hairs. Shaded creek banks, Wawawai, *U. Son.*

Ranunculus Macounii Britton, var. **Macounii.** Usually bristly hairy throughout; stems ascending or reclining, 3–6 dm. long, petioles hirsute or hispid; leaves ternately divided, the segments 2–8 cm. long, stalked, broadly ovate, 3-cleft or -parted, and incisely toothed; petals bright yellow, obovate, 5–7 mm. long; achenes obovate, flattened, 2.5 mm. long, with a stout straight flattened beak, 1 mm. long. In low wet meadows, *A. T., A. T. T.*

Var. **oreganus** (Gray) Davis. Plant nearly glabrous. Pullman, *Piper* 3,526, *A. T.*

Ranunculus muricatus L. Annual, glabrous; stems several, reclining 10–50 cm. long, glabrous or with a few stiff hairs; basal leaves with petioles longer than the blades, the blades 2–4 cm. long, in outline suborbicular to reniform, the base subtruncate, 3–5-lobed or -parted, the lobes deeply crenate; cauline leaves shorter petioled, the blades similar; pedicels 1–4 cm. long; sepals 4–5 mm. long, thin, more or less pubescent; petals 5, yellow, 5–8 mm. long, clawed; stamens 10–20; achenes 10–20, sessile, 5 mm. long, elongate ovate, the margin thick-keeled, the beak curved, horn-like, half as long as the body. Introduced from southern Europe; Pullman in 1932, *F. A. Warren* 1508.

Ranunculus orthorhynchus Hook., var. **platyphyllus** Gray. Roots fascicled, thick-fibrous; stem stout, 3–9 dm. high, usually hirsute; leaves pinnately 3–5-divided, the divisions stalked and again 3–5-cleft or -parted; ultimate segments 2–9 cm. long, cuneate-oblanceolate or obovate, incisely few-toothed; petioles of the radical leaves exceeding the blades; flowers in an open cyme, long-peduncled; sepals pilose; petals 5, bright yellow, obovate, 9–16 mm. long; achenes orbicular deltoid, flat, 3 mm. long, smooth, with stout straight beaks, fruiting heads globose. *R. platyphyllus* (Gray) A. Nels.; *R. maximus* Greene. Common in low meadows, *A. T.*

Ranunculus pensylvanicus L. f. Plant stout, hirsute with spreading hairs, erect, 3–10 dm. high, the root usually annual; leaves ternately compound, the leaflets 2–8 cm. long, ovate, acute, 3-cleft and sharply toothed; flowers small; petals pale yellow, 2–4 mm. long; achenes flat, obovoid, 3 mm. long, obscurely margined, each tipped with a short straight beak, crowded. In damp places, *A. T., A .T. T.*

The specific epithet was spelled by Linnaeus, the son, with a single ***n***, as William Pen often did his own name, and as Pensylvania was spelled with a single ***n*** on some contemporary maps.

Ranunculus Populago Greene. Flaccid and glabrous; stems solitary or sometimes two or three from a fascicle of fibrous roots, erect, leafy at tip; basal leaves 1–5 cm. long, thin, round-reniform to cordate-ovate, obtuse, entire or obscurely crenate, long-petioled; the cauline smaller, ovate or ovate-lanceolate, sessile; peduncles many, slender, in the axils of and longer than the cauline leaves; petals 5–6, yellow, elliptic, 3–7 mm. long; achenes obovate, flattened, 1.5–1.7 mm. long, with a short straight beak, in small nearly globose heads. Moist ground, Blue Mountains, *Huds.*

Ranunculus sceleratus L., var. **multifidus** Nutt. ex T. & G. *Cursed Crowfoot.* Annual, glabrous; stems erect, hollow, 1.5–6 dm. high; leaves thick, pale green, the basal ones 2–6 cm. wide, reniform orbicular, deeply 3-lobed, the cauline 3–5-lobed or -parted and the divisions cleft or lobed; sepals 3–4 mm. long, pilose; petals pale yellow, scarcely longer than the calyx; achenes obovoid, flattened, 1 mm. long, short-beaked, smooth, numerous in a cylindric head. Wet shores, in the "Scablands," *U. Son.,* also *A. T.*

The plant juice is acrid and poisonous. If rubbed on the skin it will cause blisters or ulcers. Professional beggars use it to produce sores on their skins.

Ranunculus testiculatus Crantz. Annual, 2–10 cm. tall, once- or commonly several-branched from the base, white arachnoid except for the few lowest leaves; leaves basal, 3–35 mm. long, 3-lobed, the lobes ligulate; scapes 1-flowered; flowers 5–10 mm. wide; sepals 2–3 mm. long, elliptic, green; petals 4–5 mm. long, elliptic, membranous, with several strong nerves, yellowish, fading to white; achenes 5–7 mm. long, the fertile base distended. Weed, introduced from Eurasia; Pullman, in 1947, *Ownbey 3,147.*

Ranunculus uncinatus D. Don. ex G. Don, var. **uncinatus.** Similar to var. *parviflorus* but less pubescent; stems glabrous; petioles nearly glabrous; achenes smooth. *R. tenellus* Nutt., not Viviani; *R. Bongardi* Greene, var. *Douglasii* (Howell) Davis; *R. Bongardi,* var. *tenellus* (Gray) Greene. Abundant, moist places, *A. T., A .T. T.*

Var. **parviflorus** (Torr.) Benson. Erect, 2–9 dm. tall; stems pilose at base; leaves deeply 3–5-cleft, the lower cuneate-obovate, incisely 2–5-toothed; the upper leaves divided into lanceolate, sharply toothed segments; petioles pilose, longer than the blades; flowers few, in open cymes, long-pedicelled; petals 5, pale yellow, 3–4 mm. long; achenes in globose heads, much flattened, the body

2–3 mm. long, each with a slender circinate beak, ⅔ as long. *R. Bongardi* Greene. Moist shady places, *A. T. T.*
All intergrades with var. *uncinatus* occur.

THALICTRUM. Meadow Rue

Erect perennial herbs; leaves 2–3-ternately compound, radical and cauline, the latter alternate; flowers perfect, polygamous, or dioecious, generally small, greenish-white, in corymbs, panicles, or racemes; sepals 4 or 5, petal-like or greenish, dull-colored; petals none; stamens numerous, the filaments often dilated; pistils 4–15, commonly few, 1-ovuled; stigmas elongate; achenes ribbed or nerved. (A Gr. name, used by Dioscorides, for some plant.)

Plant glandular puberulent,
- Achenes about 4 times as long as broad. — *T. occidentale*, var. *occidentale*.
- Achenes about 2–3 times as long as broad,
  - Achenes 6–8 mm. long. — Var. *megacarpum*.
  - Achenes 4.5–5 mm. long. — Var. *palousense*.

Plant glabrous or early glabrate. — *T. confine*, var. *columbianum*.

Thalictrum confine Fern., var. **columbianum** (Rydb.) Boivin. Dioecious perennial, 15–90 cm. tall, glabrous or glabrate; leaflets 1–2 cm. long, firm chartaceous, flabellate to suborbicular, mostly subequally 7–25-toothed; pistillate panicle compact, 2–4 cm. wide; achenes 4–5 mm. long, 1.5–2 mm. wide. *T. columbianum* Rydb. Common, meadows or woods, *A. T., A. T. T.*

Thalictrum occidentale Gray, var. **occidentale.** Plant dioecious, 3–10 dm. tall, glandular puberulent; leaves ternately decompound, the leaflets 1–4 cm. long, membranous, suborbicular to obovate-cuneate or -cordate, 3-lobed and coarsely crenate, green above, whitened beneath; staminate panicle 3–10 cm. long; sepals 3.5–4.5 mm. long, ovate, whitish or somewhat purplish tinged; filaments filiform; anthers 3.5–4 mm. long, linear, acute; pistillate panicles 8–30 cm. long; sepals 1.5–2 mm. long, ovate; achenes 6–9 mm. long, 1.5–2 mm. wide, fusiform, strongly ribbed, thin-walled. Moist woods, *A. T. T.*
The plant is avoided by grazing sheep.

Var. **megacarpum** (Torr. ex Rydb.) St. John. Differs from the species by having shorter, broader achenes, 6–8 mm. long, 2.5–3.5 mm. wide. *T. megacarpum* Torr. ex Rydb., Fl. Rocky Mts. 290, 1917. Trelease only mentioned it in synonymy. T. *occidentale* Gray, var. *megacarpum* (Torr.) St. John. Deep woods, Mt. Carleton [Mt. Spokane], *Kreager* 597; Spokane, *Henderson,* June 1, 1892, *A. T. T.*
Boivin (1944) published the opinion that this is based upon abnormal plants. Upon reexamination the writer still considers the fruits and seeds to be normal and fertile.

Var. **palousense** St. John. Differs by having the achenes 4.5–5 mm. long, 2.2–2.8 mm. wide. In pine woods, Kamiak Butte; Blue Mts. *A. T. T.*
The holotype is Blue Mts., Walla Walla Co., July 1896, *C. V. Piper.*

TRAUTVETTERIA. False Bugbane

Tall erect perennial herbs; leaves palmately-lobed, the radical large and long-petioled; the cauline few, short-petioled or sessile;

flowers white, in corymbs; sepals 3–5, broad, concave; petals none; stamens numerous, the filaments clavate; pistils numerous, 1-ovuled; achenes capitate, sharply-angled, inflated, tipped with minute styles. (Named for *Prof. Ernst Rudolph von Trautvetter* of Russia.)

Trautvetteria grandis Nutt. ex. T. & G. Stems 3–10 dm. tall, glabrous or nearly so; leaves broader than long 8–43 cm. across, 5–9-cleft, the lobes oblong or obovate, acute, incisely lobed and toothed; pedicels pilose; sepals oval, pilose, 3–4 mm. long, whitish, deciduous; stamens white and showy; achenes smooth, ovate, 4–6-angled, inflated, 4 mm. long, tipped with a slender recurved beak. Meadows or moist woods, *Can., Huds.*

BERBERIDACEAE. Barberry Family

Shrubs or herbs, often with rhizomes or tubers; leaves alternate, simple, compound, or divided, with stipules or dilated bases; flow ers perfect, the bracts, sepals, petals, and stamens all opposite; all the parts distinct and hypogynous; sepals and petals each usually in two rows of three, imbricate or the outer valvate; stamens 4–9, opposite the petals, anthers opening by two valves or lids hinged at the top; carpel single; style short or none; fruit a berry or pod; seeds few or several; endosperm present. (Named from the genus *Berberis,* the barberry.)

BERBERIS Barberry

Shrubs with yellow wood; leaves evergreen or deciduous, often spiny, rhachis jointed; flowers yellow, in clustered racemes; bractlets 2–6; sepals 6, petal-like; petals concave, 6, in two rows, each with two basal glands; stamens 6, short, filament irritable; stigma peltate; fruit an acid berry, used for jelly. (Name from *Berberys,* the Arabic name of the fruit.

Nodes unarmed; leaflets several,
  Leafllets palmately nerved; bud scales 1.5–4.5 cm. long, persistent. — *B. nervosa.*
  Leaflets pinnately nerved; bud scales smaller, deciduous,
    Leaflets 5–11, shining, strongly spinulose. — *B. Aquifolium.*
    Leaflets 3–7, dull, often glaucous, weakly spinulose. — *B. repens.*
Nodes spiny; leaves 1-foliolate. — *B. vulgaris.*

Berberis Aquifolium Pursh. *Oregon Grape.* Shrub often 1–2 m. high, erect or nearly so, glabrous; leaves 7–23 cm. long, leaflets evergreen, oblong or ovate, 3–10 cm. long, with numerous spiny teeth; bud scales 5–8 mm. long, deltoid; racemes 2.5–8 cm. long, usually clustered, subterminal; petals 5–6 mm. long, oval, yellow; berries 7–12 mm. long, black with a bloom, appearing blue, usually pear-shaped. *Odostemon nutkanus* (DC.) Rydb. Gravelly woods about Spokane; also Rock Lake, *Weitman* 105, *A. T. T.*

This is the state flower of Oregon.

Berberis nervosa Pursh. *Dull Oregon Grape.* Plants glabrous, 1–6 dm. tall, erect, simple; leaves 15–75 cm. long, with 11–19 leaflets, these 2–9 cm. long, ovate or lanceolate, acuminate, spinulose-dentate; bud scales lanceolate, acuminate, persistent, becoming dry and rigid; racemes terminal, one or several, 5–20 cm. long; pedicels equaling or shorter than the fruit; petals 4–5 mm. long, elliptic, yellow, often pink-tinged; berries 7–10 mm. long, globose, purple-black with a white bloom, appearing blue, very acid. *Odostemon nervosus* (Pursh) Rydb. Moist woods near Lake Coeur d'Alene, *A. T. T.*

Berberis repens Lindl. *Small Oregon Grape.* Low depressed shrub, 20–30 cm. high, often with subterranean stolon-like branches; leaves 6–27 cm. long, pinnately compound; leaflets evergreen, 3–9 cm. long, ovate, obtuse or acute, pale or glaucous, sinuately dentate with numerous spinulose teeth; bud scales 3–8 mm. long, deltoid; racemes 3–10 cm. long, terminal, clustered; petals 4–6 mm. long, oval, yellow; berries 7–9 mm. long, ellipsoid-globose, blackish with a bloom, appearing blue, sour. Common in gravelly or stony ground, *U. Son.* to *A. T. T.*

Berberis vulgaris L. *Common Barberry.* Shrub 1–3 m. tall, glabrous; branches upright or arching; nodes with 1–3 spines; leaves alternate or on short axillary shoots; rhachis 1–2 mm. long; petiole 2–10 mm. long; blades 2.5–5 cm. long, obovate or spatulate, obtuse, bristly serrate; many of the leaves on young shoots reduced to 3-pronged spines; racemes 2.5–7.5 cm. long, terminal on lateral shoots, drooping; flowers yellow, with a too sweet perfume; petals 3–5 mm. long, obovate; berries 6–10 mm. long, narrowly ellipsoid, scarlet, acid, edible. Planted as an ornamental or hedge plant, spreading, *A. T.*

It is the alternate host of the rust of wheat.

PAPAVERACEAE. Poppy Family

Herbs with colored juice, rarely shrubs or trees; leaves alternate, or the floral ones opposite or whorled, simple or compound, without stipules; pubescence of simple or barbellate hairs; flowers perfect, regular, mostly showy; sepals 2–3; petals 4–12, separate deciduous, imbricate, in 2 series, often crumpled; stamens numerous and distinct; ovary 1-celled, of 2 or more united carpels, with 2 or more parietal placentae, rarely falsely 2-celled; fruit a dry one-celled pod with numerous seeds. (Named from the genus *Papaver.*)

PAPAVER. Poppy

Annual to perennial herbs, or subshrubby, hispid or glaucous; with milky sap; leaves lobed or dissected; flower buds nodding; sepals 2, rarely 3, deciduous; petals 4–6, showy; stamens numerous; stigmas united into a radiate, persistent disc; capsule dehiscent by slits or pores near the tip; seeds pitted. (The classic Lat. name for the poppy, *papaver.*)

Capsule clavate, hispid; disc entire. *P. Argemone.*
Capsule turbinate, glabrous; disc lobed. *P. Rhoeas.*

Papaver Argemone L. *Headache.* Slender annual or biennial, 1.5–5 dm. tall, often branched; stems appressed hispid; leaves, except the uppermost, petioled; blades lanceolate, 1–2-times pinnately cut into linear or linear-lanceolate, acute segments; flowers single; sepals 2, greenish-yellow, hispid, early deciduous; petals 12–25 mm. long, 4, narrowly obovate, scarlet, black-marked at base; capsule 15–17 mm. long; stigmas 4–6; seeds semi-lunate, dark. A European weed, becoming established, *U. Son., A. T.*

Papaver Rhoeas L. *Corn Poppy.* Annual, 2.5–9 dm. tall, simple or branched, hispid throughout; lower leaves petioled, the upper smaller, sessile, all pinnatifid, the lanceolate, dentate segments divaricate; sepals 2, green, deciduous; petals 2–4 cm. long, full and imbricate, scarlet or purplish red, often dark spotted; stigmas 5–18, usually 10; pods 10–22 mm. long; seeds reniform, brown. A garden ornamental, introduced from Europe. Occasional as a weed, Pullman, *Bennett,* June 27, 1928, *A. T.*

FUMARIACEAE. Fumitory Family

Annual, biennial, or perennial herbs, with watery sap; leaves dissected, alternate or basal, without stipules; flowers perfect, irregular; sepals 2, small, scale-like; petals 4, the 2 outer spreading above and one or both saccate or spurred at base, the 2 inner narrower, thickened and united at tip over the stigma; stamens 6, diadelphous in sets of 3, hypogynous, middle anthers 2-celled, lateral ones 1-celled; carpels 2, united into a single, 1-celled pistil, with 2 parietal placentae and 2 deciduous valves. (Named from the genus *Fumaria,* the fumitory.)

Corolla 2-spurred at the base; pedicels 2-bracted above the middle. *Dicentra.*
Corolla 1-spurred at the base; pedicels not bearing bracts. *Corydalis.*

CORYDALIS

Erect pale or glaucous herbs; leaves radical and cauline, decompound; flowers in racemes, terminal or opposite the petioles; inner petals keeled; style entire, dilated or lobed; capsule oblong or linear. (Name from Gr., *korudallos,* the crested lark.)

Upper petal with a median, lacerate crest; flowers creamy yellow. *C. washingtoniana.*
Upper petal merely keeled; flowers golden yellow. *C. aurea.*

Corydalis aurea Willd. *Golden Corydalis.* Perennial; glabrous, the branched stems decumbent, 1–6 dm. long; leaves petioled, tripinnately compound, pale and glaucous; ultimate segments cuneate-obovate to linear-oblong, acute; racemes 1–5 cm. long; bracts linear-lanceolate, acuminate; flowers 12–16 mm. long; spur blunt, shorter than the body; capsule 2–3 cm. long, 2 mm. thick, torulose, long-beaked; seeds 1.9 mm. wide, reinform-discoid, black, very shiny. Spokane, *Piper,* May 8, 1898.

Corydalis washingtoniana Fedde. Perennial, glabrous, glaucous; branches 15–43 cm. long, ascending; leaves petioled, tripinnatifid, the ultimate segments cuneate, oblanceolate to linear, apiculate; racemes 1–6 cm. long; bracts oblanceolate, acuminate; flowers 12–13 mm. long, spur about ¼ the length of the petal; capsule 1.5–2.5 cm. long, 2 mm. thick, torulose, long beaked; seeds 2.2 mm. wide, discoid-reniform, black, shining. *Capnoides aureum* (Willd.) Kuntze. Thickets by streams, common in the Snake River Canyon, *U. Son., A. T.*

The type is *Elmer* 1,018, in Whitman Co.

DICENTRA

Erect or diffuse glabrous perennial herbs; leaves ternately compound or dissected; flowers in racemes or panicles, nodding; petals slightly united into a 2-spurred or swollen nectariferous often withering-persistent corolla; stamens opposite the outer petals; style slender; capsule oblong or linear. (Name from Gr., *dis,* twice; *kentron,* spur.)

Petals spurred at base, outer ones with tips merely spreading; flowers racemose. *D. Cucullaria.*
Petals cordate at base, outer ones revolute almost to the base; flowers single. *D. uniflora.*

Dicentra Cucullaria (L.) Bernh. *Dutchman's Breeches.* Glabrous throughout; rootstock short, covered with grain-like tubers from the axils of delicate sheaths; petioles 10–24 cm. long; leaves all basal; blades 4–15 cm. long, twice-ternately compound; ultimate divisions linear or oblanceolate, 1–5 mm. wide, acute; scapes exceeding the leaves, bearing a one-sided raceme of 4–14 nodding flowers; sepals 1.5–3 mm. long, cordate to deltoid; corolla 13–16 mm. long, and as wide, white or at first pinkish, yellow at the summit, the blunt spurs widely diverging; capsule 11–15 mm. long, fusiform, attenuate at each end; seeds 1.7–2 mm. long, reniform, black, polished and shining. *Bikukulla cucullaria* (L.) Millsp.; *B. occidentalis* Rydb. is identical, the duplicate type having the sheathing basal bracts. Moist copses, blooming in early spring, *U. Son., A. T.*

Though a lovely and favorite flower, the plant is poisonous. Either the tops or the bulbs will kill cattle, especially young stock.

Dicentra uniflora Kellogg. *Steer's Head.* Plants scapose, from a cluster of fusiform tubers and a tuber-bearing crown; leaves basal; petioles about as long as the blades; blades 2–3-ternate, 4–6 cm. long, the pinnules elliptic or spatulate, slightly pubescent, glaucous below; scapes exceeding the leaves; sepals 4–5 mm. long, ovate-lanceolate; flowers pink, 15 mm. long; outer petals with cordate base, the upper part abruptly narrowed, linear-spatulate; inner petals with a sagittate blade; scapes prostrate in fruit; capsule 12 mm. long, ovoid; seeds lunate-semiorbicular, black, shining. *Bikukula uniflora* (Kellogg) Howell. Rocky slope, top of Cedar Mt., Ida., 5,000 ft. alt., *F. A. Warren* 904, *Huds.*

CRUCIFERAE. Mustard Family

Herbs, rarely somewhat woody, with watery sap, glabrous or with simple, lepidote, or stellate hairs; leaves alternate (rarely opposite); stipules none; flowers perfect, regular, in usually bractless racemes, spikes, or corymbs; sepals 4, imbricate in 2 series, usually oblong; petals 4, rarely none, hypogynous, in the form of a cross, equal, generally clawed, alternate with the sepals; glands usually present on the receptacle; stamens 6, rarely fewer, hypogynous, of unequal length, the two shorter opposite the sepals, the four longer opposite the petals; pistil 1, of two united carpels; pod usually 2-celled, dehiscing by the separation of two valves from the central partition, or rarely indehiscent, either much longer than broad (a *silique*) or short (a *silicle*). (Name from Lat., *crux,* cross; *ferro,* to bear.)

Pod splitting transversely into numerous 1-seeded joints; flowers red. *Chorispora.*
Pods splitting longitudinally or indehiscent,
Pods indehiscent,
Fruit broadly winged, much flattened; pubescence of simple hairs. *Thysanocarpus.*
Fruit wingless, not or little flattened,
Pubescence of branched hairs; pods orbicular or oval. *Athysanus.*
Pubescence of simple hairs; pods not so,
Pod cordate-ovoid, somewhat flattened. *Lepidium.*
Pod subspherical, inflated. *Cardaria.*
Pods dehiscent,
Fruit a silique (usually more than 4-times as long as wide),
Pods compressed parallel to the broad partition,
Valves nerveless; leaves all petioled. *Cardamine.*
Valves 1-nerved; cauline leaves sessile,
Pods lanceolate; anthers subsagittate. *Phoenicaulis.*
Pods linear; anthers not subsagittate. *Arabis.*
Pods terete or slightly 4-angled,
Pods 4 cm. long or longer,
Pods stipitate from the calyx-scar. *Thelypodium.*
Pods non-stipitate (though pedicelled),
Flowers yellow,
Pods terete or nearly so. *Sisymbrium.*
Pods 4-angled,
Cauline leaves clasping by a cordate base. *Conringia.*
Cauline leaves not clasping. *Erysimum.*
Flowers purple. *Hesperis.*
Pods less than 4 cm. in length,
Pods produced beyond the valves into a conspicuous beak. *Brassica.*
Pods not beaked, or merely tipped with the persistent style,
Flowers white,
Plant glabrous; leaves pinnate. *Nasturtium.*
Plant with forked hairs; leaves simple or dentate. *Arabidopsis*
Flowers yellow,
Pods 4-angled,
Hairs of the stem lepidote. *Erysimum.*
Hairs of stem attached by the base. *Barbarea.*
Pods terete or nearly so,
Embryo accumbent. *Rorippa.*
Embryo incumbent. *Sisymbrium.*
Fruit a silicle (short, usually less than 4-times as long wide),
Pods compressed parallel to the partition,
Flowers solitary on scapes; seeds winged. *Idahoa.*
Flowers in racemes; seeds wingless,
Pod suborbicular, thinner at the margin. *Alyssum.*
Pod narrower, not thin margined. *Draba.*

Pods compressed at right angles to the partition or not compressed,
 Pods terete or nearly so,
  Leaves pinnatifid,
   Pods beakless, dehiscent to the tip. *Rorippa.*
   Pods with a flattened, indehiscent beak. *Eruca.*
  Leaves all or mostly entire,
   Herbage stellate pubescent; pods subglobose. *Lesquerella.*
   Herbage glabrous or not stellate,
    Petals white; pods obovoid. *Armoracia.*
    Petals yellow; pods pyriform. *Camelina.*
 Pods compressed at right angles to the partition,
  Cells of pod inflated, each 2-seeded. *Physaria.*
  Cells of pod not inflated,
   Pod winged at apex,
    Cells of pod 1-seeded. *Lepidium.*
    Cells of pod 2–4-seeded. *Thlaspi.*
   Pod wingless,
    Pods cuneate, notched at tip. *Capsella.*
    Pods oval, not notched. *Hutchinsia.*

ALYSSUM

Low branching annual herbs; leaves mostly simple; flowers yellow; filaments often dilated and toothed or appendaged; style slender; pod compressed parallel to the partition; valves convex, nerveless; seeds one or two in each cell. (Name Gr., *a,* non; *lussa,* madness.)

Alyssum alyssoides L. *Small Alyssum.* Plant 5–40 cm. high, stellate-pubescent throughout, usually branched from the base; leaves 6–30 mm. long, silvery, entire, spatulate; flowers in racemes 5–15 cm. long; pedicels 1.5–4 mm. long, ascending; sepals 1.5–2.5 mm. long, lanceolate; petals 2.6–4 mm. long, pale yellow, turning whitish, cuneate; pods 3–4 mm. long, firm, margined, notched at the apex, minutely stellate; seeds 1.2–1.5 mm. long, ovate, brown. A weed, introduced from Europe. Common, dry or stony places, *U. Son., A. T.*

ARABIDOPSIS

Annual or perennial herbs with the aspect of *Arabis,* pubescent with forked hairs, with branched slender erect stems, entire or toothed leaves and small white or pink flowers in terminal racemes; style very short; stigma 2-lobed; siliques narrowly linear, the valves rounded, nerveless or finely nerved, dehiscent; seeds in 1 row in each cell or in 2 rows; cotyledons incumbent. (Named from its resemblance to Arabis; with Gr., *ops,* resemblance.)

Arabidopsis Thaliana (L.) Heynhold. Annual; stem slender, erect, 2.5–26 cm. tall, freely branched, more or less pubescent with short stiff hairs, especially below; basal leaves 2.5–5 cm. long, obtuse, oblanceolate or oblong, narrowed into a petiole, entire or slightly toothed; cauline leaves smaller, sessile, acute or acutish, often entire; pedicels very slender, spreading or ascending, 4–8 mm.

long in fruit; flowers about 3 mm. long; petals about twice the length of the sepals; pods narrowly linear, 8–20 mm. long, acute, often curved upward, glabrous. *Sisymbrium Thalianum* (L.) Gay. A weed introduced from Eurasia or Africa. Rocky places, Latah Co., *A. T. T.*

ARABIS. Rock Cress

Erect annual or perennial herbs; leaves seldom divided, the cauline sessile and usually clasping and auricled at the base; flowers white, yellowish, or purple; pod long-linear, compressed parallel to the partition; seeds flattened, wingless or winged. (Named from the land Arabia.)

Seeds wingless; flowers white,
  Cauline leaves auriculate; plant glaucous, glabrous except near the base. — *A. glabra,* var. *glabra.*
  Cauline leaves not auriculate; plant green, glabrous above, pubescent towards the base,
    Pods 1–2 cm. long; plant hirsute with simple or forked hairs below. — *A. Nuttallii.*
    Pods 2–4 cm. long; plant harshly stellate pubescent below. — *A. crucisetosa.*
Seeds winged or wing-margined,
  Flowers white,
    Herbage finely stellate puberulent. — *A. Holboellii,* var. *retrofracta.*
    Herbage rough hirsute,
      Pods 1.5–2.5 mm. wide; upper stem hirsute. — *A. hirsuta,* var. *Eschscholtziana.*
      Pods about 1 mm. wide; upper stem glabrous. — Var. *glabrata.*
  Flowers red or pink,
    Pod less than 3 mm. wide,
      Cauline leaves not auriculate or cordate; leaves all entire, villous-hirsute. — *A. Cusickii.*
      Cauline leaves auriculate or cordate,
        Basal leaves dentate; whole plant coarsely stellate-pubescent. — *A. sparsiflora,* var. *subvillosa.*
        Basal leaves entire,
          Blades glabrous above. — *A. divaricarpa,* var. *divaricarpa.*
          Blades stellate puberulent. — *A. microphylla,* var. *microphylla.*
    Pods 3–6 mm. wide. — *A. suffrutescens,* var. *suffrutescens.*

Arabis crucisetosa Constance & Rollins. Perennial, the rootstock simple or branched; stems 1–4 dm. tall, usually several; basal leaves 2–6 cm. long, 6–15 mm. wide, spatulate to obovate, obtuse, entire or remotely dentate, the petiole nearly equaling the blade; cauline leaves 1–3 cm. long, sessile, linear-oblong, obtuse, usually entire; petals 5–10 mm. long; pedicels 1–2 cm. long, spreading; pods 2–4 cm. long, 1–1.5 mm. wide, straight or nearly so, glabrous, erect, tipped with a stout capitate style; valves 1-nerved below. Limestone cliff, Lime Pt., Asotin Co., *St. John* 9294.

Arabis Cusickii Wats. Perennial, 1–4 dm. tall; roughly hirsute throughout with simple hairs or somewhat glabrous above; stems usually several; basal leaves 1–4 cm. long, linear or linear-spatulate; cauline 5–30 mm. long, linear or linear-lanceolate, sessile; pedicels ascending; petals 6–9 mm. long; pods 4–8 cm. long, 2 mm. wide, usually curved, ascending or spreading, glabrous; valves 1-nerved below the middle. Deep cayons, *U. Son.,* and scablands, *A. T.*

Arabis divaricarpa A. Nels., var. **divaricarpa.** Biennial, 3–8 dm. tall; stems one to few, usually simple, appressed hirsutulous below, glabrous and glaucous above; basal leaves 2–6 cm. long, oblanceolate or spatulate, acute or obtuse, entire or dentate, loosely stellate pubescent; cauline leaves 2–5 cm. long, narrowly oblong, glabrous, entire or remotely toothed, the base sagittate-auriculate; sepals 4–6 mm. long, oblong; petals 6–8 mm. long, white or pink; pods 2–8 cm. long, 1–1.5 mm. wide, ascending or spreading, glabrous; seeds oblong. Blue Mts., Darlington, in 1913; and Godman Sprs., *Constance et al.* 1,178.

Arabis glabra (L.) Bernh., var. **glabra.** *Tower Mustard.* Biennial, hirsute near the base, glabrous and glaucous above; stems usually simple, 6–15 dm. high; basal leaves petioled, 3–12 cm. long, oblanceolate, dentate or pinnatifid, simple or forked hirsute; cauline 1–5 cm. long, oblong-lanceolate, sessile, sagittate and auricled at base, all but the lower glabrous; petals 4–6 mm. long, yellowish or greenish white; pods 4–10 cm. long, 1–1.2 mm. wide, erect or ascending, linear, tipped with the large 2-lobed sessile stigma; pedicels 4–10 mm. long. *Turritis glabra* L. Meadows and rocky slopes, *A. T.*

Arabis hirsuta (L.) Scop., var. **Eschscholtziana** (Andrz.) Rollins. Biennial or perennial, usually rough-hairy, but shade plants often nearly glabrous; stems erect, 1–8 dm. high, hirsute and stellate; basal leaves 2–10 cm. long, clustered, oblanceolate, entire or somewhat dentate, petioled; cauline 1–6 cm. long, oblong to lanceolate, usually dentate, cordate or auriculate at base; pedicels erect; petals 4–7 mm. long, greenish white; stigma bifid; pods 2–8 cm. long, 1–1.5 mm. wide, erect, tipped with a very short stout style. Open or shady places, *U. Son.* to *A. T. T.*

Var. **glabrata** T. & G. Basal leaves 3–7 cm. long, 1–2.5 cm. wide, obovate to oblanceolate, entire or rarely few-toothed, obtuse, sparsely hirsute to nearly glabrous; cauline leaves obovate to oblong, rarely ovate, entire or rarely few-toothed; pedicels usually divaricately ascending, petals 5–9 mm. long, spatulate, white; pods erect to slightly divaricate. Waitsburg, May, 1897, *R. M. Horner.*

Arabis Holboellii Hornem., var. **retrofracta** (Graham) Rydb. Perennial; stem 1–7 dm. tall, finely stellate; basal leaves 1–3 cm. long, oblanceolate or spatulate, entire or remotely dentate, grayish from the close stellate puberulence, petioled; cauline leaves 1–8 cm. long, lanceolate to linear; pedicels reflexed; petals 7–8 mm. long, magenta to white; pods reflexed, 4–7 cm. long, 1–1.5 mm. wide, stellate to glabrous. Rocky slopes, *A. T., A. T. T.*

Arabis microphylla Nutt. ex T. & G., var. **microphylla.** Loosely matted perennial with numerous sterile basal shoots; plant gray and stellate puberulent below, glabrate above; stems 1–3 dm. tall; basal leaves 5–30 mm. long, oblanceolate, petioled; cauline leaves 8–22 mm. long, linear to lanceolate; pedicels ascending; petals 4–5 mm. long; pods 2–6 cm. long, 1–1.5 mm. wide, slightly curved, ascending, glabrous. Exposed limestone or basalt cliffs, Snake River Canyon down to Bishop, *U. Son.,* and Blue Mts., *Huds.*

Arabis Nuttallii Robins. Perennial with branching rootstocks; stems 10–37 cm. tall, erect or ascending, glabrous above, somewhat hirsute below; basal leaves 1–3.5 cm. long, oblanceolate, entire or sinuate-dentate, petioled, hispid above, glabrous beneath; cauline oblong to elliptical, sessile, smaller; petals 5–9 mm. long, white; pedicels ascending; pods 12–26 mm. long, 1 mm. wide, ascending, beaked with a stout style; valves 1-nerved. Moist meadows, *A. T.*

Arabis sparsiflora Nutt., ex T. & G., var. **subvillosa** (Wats.) Benson. Perennial, 1–11 dm. tall, hirsute, stellate below, glabrous above; stems mostly simple, erect; basal leaves 1–5 cm. long, oblanceolate, dentate, petioled; cauline 1–9 cm. long, lanceolate, subentire, sessile, with a sagittate base; pedicels divergent; petals 6–10 mm. long, purplish; pods 5–11 cm. long, 1.5–2 mm. wide, curved, linear, acute. *A. perelegans* A. Nels. apparently differs in its usually entire basal leaves, its ciliate pedicels and sepals. Basalt ledges, common, *U. Son., A. T.*

Arabis suffrutescens Wats., var. **suffrutescens.** Perennial, 1–4 dm. tall; suffruticose at base; root 2–6 mm. in diameter, woody, brown; stems forking at base; basal leaves 2–5 cm. long, linear to narrowly oblanceolate, rarely stellate on the margin; cauline leaves 1–3 cm. long, oblong to lanceolate, not, or but slightly auriculate; sepals 3–4 mm. long; petals 4–7 mm. long, spatulate, rose to purple; pedicels reflexed; pods 4–6 cm. long, 3–6 mm. wide, glabrous, linear oblong; seeds 2–3 mm. wide, orbicular, broadly winged. Dry ground, Wenaha River Trail, Columbia Co., Blue Mts., *Darlington,* June 19, 1913.

ARMORACIA

Tall glabrous perennials, with fleshy, pungent tap-root and leafy stem; leaves crenate or lobed; flowers white, in dense racemes or panicles; petals clawed; silicle rounded ellipsoid or subglobose; style short; stigma subcapitate. (The classic Lat. name for the horse-radish.)

Armoracia rusticana (Lam.) Gaertn., B. Mey. & Scherb. *Horse-radish.* Plants 2–15 dm. tall, often branched; basal leaves with petioles 15–30 cm. long, the blades 10–30 cm. long, oblong, crenate to pinnatifid; cauline leaves smaller, sessile, narrowly oblong to lanceolate, crenate or dentate; pedicels 4–7 mm. long, slender, ascending; sepals 2.5–3 mm. long, ovate; petals 5–7 mm. long, the limb obovate; pods seldom formed, 4–6 mm. long, obovoid. *A. lapathifolia* (Gilib.) Gilib. A garden plant, of European origin, persisting and spreading by rhizomes from cultivation, Pullman, *G. N. Jones* 2603, *A. T.*

The grated roots furnish the table relish.

ATHYSANUS

Slender annual herbs; flowers minute, in racemes; petals white, linear, or none; stamens nearly equal; style short; ovules 3–11, in some only one maturing; pod flattened, nerveless, 1-celled, 1-seeded, (Name from Gr., *a,* without; *thusanos,* fringe.)

Athysanus pusillus (Hook.) Greene. Annual, 4–30 cm. tall; stems slender, branched from the base, leaves forked hispid; the basal ones 2–15 mm. long, oblanceolate to spatulate, petioled; cauline ones 5–20 mm. long, sessile, oblong-lanceolate, acute, usually coarsely dentate; stems forked hispid; racemes 4–25 cm. long, loose, glabrate above; pedicels 1–4 mm. long, becoming recurved; sepals 0.8–1.3 mm. long, oval, often purple tinged; petals 1.3–1.8 mm. long, spatulate, or in the later flowers wanting; pods 2–3.5 mm. long, apiculate, short stalked, hispid with hooked hairs; seed single, 1.3–1.8 mm. long, oval, flattened, yellow, the raphe prominent. Stony soil, rather common, *U. Son., to A. T. T.*

The pods are variable in pubescence. In addition to the stiff hooked hairs, they are mostly appressed lanate, but often lack this lanate pubescence. Both states are found on plants in the same colony, and even on the same plant, for instance on *St. John,* 5,973 from Wawawai.

BARBAREA. Winter Cress

Somewhat succulent biennials or perennials; stems angled; leaves pinnatifid; flowers yellow, in racemes; pod linear, terete, or somewhat 4-angled, tipped with the short slender style; seeds flat, wingless. (In ancient times named for *St. Barbara* of Asia Minor.)

Barbarea orthoceras Ledeb. Perennial, glabrous; stems 3–8 dm. tall, simple or branched above; basal leaves 3–9 cm. long, simple or with 2–6 lateral leaflets; lower and middle cauline leaves pinnately cleft or divided, with 5–13 leaflets, the terminal segment 1–6 cm. long, orbicular or ovate, entire or nearly so; lateral segments smaller, oblong, entire or toothed; petioles auricled at base; upper cauline leaves obovate, clasping, lyrately pinnatifid; racemes dense; sepals 2–2.5 mm. long, elliptic; petals 2.5–5 mm. long, broadly spatulate; pedicels 3–8 mm. long, ascending; pods 2–3.5 cm. long, 1 mm. wide, erect or ascending, beaked by the slender style 0.5–1 mm. long. *B. americana* Rydb. Meadows and stream banks, *A. T.*

BRASSICA

Erect annual or biennial herbs; lower leaves mostly pinnate or lyrate, with a large terminal lobe; outer sepals more or less gibbous at base; flowers yellow; petals long clawed; pod terete or nearly so, tipped with a slender conical or somewhat flattened beak; seeds globose. (The Lat. name of cabbage.)

Cauline leaves auricled at base and clasping; beak of pod terete. *B. campestris.*
Cauline leaves not auricled or clasping,
  Beak of the pod terete, less than one-fourth the seed bearing part. *B. nigra.*
  Beak of the pod flat, 2-edged, at least half as long as the seed-bearing part. *B. kaber,* var. *pinnatifida.*

Brassica campestris L. *Field Mustard.* Annual, smooth or nearly so, often glaucous, 3–10 dm. high, simple or much branched; lower leaves 10–20 cm. long, petioled, lyrate with a very large terminal lobe; upper leaves smaller, oblong, entire or nearly so, sessile; pedicels spreading; sepals longer than the petal claws, somewhat spreading; petals 6–10 mm. long, the limb obovate, spreading; pods 3–8 cm. long, ascending, narrowed into a slender tipped beak 1–2 cm. long; seeds 1 mm. long, brown. A weed introduced from Europe. Grain fields and waste places, *U. Son., A .T.*

Brassica kaber (DC.) Wheeler, var. pinnatifida (Stokes) Wheeler. *Charlock.* Stout, annual, 3–15 dm. high, usually much branched, hispid with scattered hairs; lower leaves 5–20 cm. long, pinnately parted, consisting of a large ovate-oblong terminal segment and 1 or 2 pairs of much smaller ones, all dentate; uppermost leaves ovate or oblong ovate, sessile; inflorescence of slender racemes; pedicels 3–4 mm. long, ascending; sepals 4–5 mm. long, narrowly oblong; petals 7–10 mm. long, the blade obovate, truncate or emarginate; pods 2–4 cm. long, erect or ascending, the fertile portion torulose, with a beak stout and half as long. *B. arvensis* (L.) Rabenh.; *Sinapis arvensis* L. A weed from Mediterranea, in waste ground, Pullman, *Hardwick,* July 18, 1895, *A. T.*

Brassica nigra (L.) Koch. *Black Mustard.* Annual, 3–40 dm. tall, erect, sparsely hispid; lower leaves green, slender-petioled, the terminal lobe large and coarsely toothed, the few lateral lobes small, the margins dentate or notched; the upper leaves ovate to linear; pedicels 2–4.5 mm. long, erect, glabrous; sepals oblong, longer than the petal claws; petals 5–9 mm. long, the limb obovate, truncate, spreading; pods 15–25 mm. long, erect, 1–2 mm. wide, 4-angled, only the lower fourth seed-bearing, the slender beak 1–3 mm. long; seeds 1–1.5 mm. wide, globose, dark brown. A weed, introduced from Europe. In waste places, *A. T.*

The seeds are the principal source of table mustard.

CAMELINA. False Flax

Erect annual herbs; leaves entire, toothed or pinnatifid; flowers small, yellowish; style slender; pod obovoid or pear-shaped, slightly flattened parallel to the partition, many-seeded. (Name from Gr., *chamae,* dwarf; *linon,* flax.)

Camelina microcarpa Andrz. Stems erect, mostly simple, 30–90 cm. tall, hirsute below with simple and forked hairs, glabrous above; leaves 1–6 cm. long, lanceolate, acute, sagittate at base, half-clasping, erect, entire or nearly so, forked and simple hirsute; racemes many flowered, elongate; pedicels 4–15 mm. long, ascending; sepals 2 mm. long, oval; petals 3–4 mm. long, spatulate, pale yellow; pods 5–8 mm. long, pyriform, strongly margined, coriaceous, beaked by the persistent 2–2.5 mm. style, glabrous; seeds 1.5 mm. long, ovoid, flattened, brown. In fields, introduced from Europe. First collected in 1893, now becoming troublesome, *U. Son., A. T.*

CAPSELLA

Slender annual herbs, with forked pubescence; radical leaves tufted; flowers small, white, in racemes; styles almost none; pod compressed at right angles to the partition, many-seeded; valves boat-shaped; seed wingless. (Name from Lat., *capsa,* box; *ella,* small.)

Capsella Bursa-pastoris (L.) Medic. *Shepherd's Purse.* Annuals or biennials, 1–10 dm. tall, simple or forked hirsute below, glabrous above, sparingly branched; basal leaves 2–13 cm. long, in a rosette, lanceolate or oblanceolate, pinnately-lobed, dentate or entire; cauline leaves few, reduced, lanceolate, auricled and clasping; racemes loose; pedicels 5–20 mm. long, ascending; sepals 1–2 mm. long, ovate; petals 2–3 mm. long, the limb obovate; pods 6–9 mm. long, triangular, cuneate, with a broad shallow notch at the apex. *Bursa bursa-pastoris* (L.) Weber. A common weed, introduced from Europe. Open places, *U. Son., A. T.*

CARDAMINE. Bitter Cress

Mostly glabrous leafy-stemmed perennial herbs, growing in moist places; leaves entire, lobed or divided, all petioled; flowers white or purple; pod elongated, compressed parallel to the partition; seeds wingless. (Name Gr., used by Dioscorides for some cress.)

Leaves all simple,
Stem and leaves glabrous. *C. Lyallii*, var. *Lyallii.*
Stem pilose; leaves sparsely pilose above. Var. *pilosa.*
Leaves or some of them pinnate,
Basal leaves simple; petals 3–4.5 mm. long. *C. Breweri.*
Basal leaves pinnate; petals 1.5–3 mm. long,
Capsules 20–30-seeded; leaflets mostly oblong. *C. pensylvanica.*
Capsules 8–20-seeded; leaflets mostly orbicular. *C. oligosperma*, var. *lucens.*

Cardamine Breweri Wats. Glabrous; stems 1–9 dm. long, sprawling or erect, from running rootstocks; leaves simple, the blades 1–5 cm. long, orbicular, rounded or cordate at base, crenate or crenately lobed, or the leaves pinnately 3–5-parted with more or less oblong divisions and lobes; pedicels ascending; petals white; pods 2–3 cm. long, 1 mm. wide, erect. In springy places near Spokane, *A. T.*

Cardamine Lyallii Wats. var. **Lyallii.** Stems 25–65 cm. tall, erect from running root-stocks; leaves petioled, suborbicular, reniform to cordate, obtuse, sinuate, 2–7 cm. broad; petals 6–10 mm. long, obovate, retuse, white; pods 2–3 cm. long, 1–1.5 mm. wide, ascending, on short pedicels. *C. cordifolia* Gray var. *Lyallii* (Wats.) Nels. & Macbr. Along streams in the Blue Mountains, *Can.*

Var. **pilosa** O. E. Schulz. Differs from var. *Lyallii* only in the pilose stems and upper leaf surfaces. Along streams in woods, Blue Mts., *Piper* 2,455, *Can.*

Cardamine oligosperma Nutt. ex T. & G., var. **lucens** G. S. Torr. Annual, erect, 1–4 dm. high, sparsely hirsute throughout or nearly glabrous; leaflets 5–11, 2–20 mm. long, mostly orbicular, petiolulate, 3–5-lobed or toothed; raceme 3–12-flowered; flowers 2 mm. long; pods erect, 8–20-seeded, 10–25 mm. long, 0.7–1.2 mm. wide, glabrous. *C. lucens* (G. S. Torr.) Suksd. Wet places, infrequent, *U. Son., A. T.*

This variety has been reduced to the species by Detling (1937), but it may well be maintained.

Cardamine pensylvanica Muhl. Annual or biennial, glabrous or nearly so, 2–9 dm. high, simple or branched; leaflets 7–13, sessile, mostly oblong, but those of the lowermost leaves suborbicular; petals 2–3 mm. long, white; pods 12–32 mm. long, less than 1 mm. wide, suberect on spreading pedicels. In moist ground, especially in copses. Ambiguous forms seem to connect with *C. oligosperma. C. flexuosa* With., subsp. *pennsylvanica* (Muhl.) O. E. Schulz. Wet places, *U. Son* to *A. T. T.*

Muhlenberg was a resident of Pennsylvania in the 18th century, and published this valid specific epithet, using an accepted spelling of the geographic name with a single *n*. It should not be "corrected."

CARDARIA

Perennial herbs; cauline leaves dentate, clasping or sessile; flowers small, white, in corymbed racemes; sepals blunt, the outer oblong, the inner oval; petals with narrow claw, and obovate, retuse limb; filaments subtended by a pair of glands; pod cordate to broadly ovate, inflated, indehiscent or nearly so, flattened contrary to the narrow septum; seeds pendulous, wingless. (Named from the heart-shaped fruit.)

Pods cordate at base, papillose; pedicels 6–15 mm. long. *C. Draba.*

Pods subglobose, puberulent; pedicels 2–7 mm. long. *C. pubescens,* var. *elongata.*

Cardaria Draba (L.) Desv. *Hoary Cress.* Perennial, 1–6 dm. tall, from deep seated roots; plant hirsutulous below, glabrous or nearly so above; basal leaves short petioled; cauline leaves 2–6.5 cm. long, oblong or lance-oblong, entire or dentate, sagittate and clasping at base; pedicels 6–12 mm. long, ascending; sepals 1.5 mm. long, oval; petals 3 mm. long, clawed, with an obovate limb; style 0.5 mm. long; pod 3.5–4.5 mm. long, netted veined, glabrous. *Lepidium Draba* L. A weed, introduced from Europe. Grass lands or fields, *A. T.*

Cardaria pubescens (C. A. Mey.) Rollins, var. elongata Rollins. *Siberian Mustard.* Rootstocks strong, long running; plant 1.5–5 dm. tall, grayish green, hispidulous throughout, the hairs simple; cauline leaves 1–6 cm. long, oblong to lanceolate, irregularly dentate, sagittate, the cauline sessile, the basal petioled; racemes congested in flower, in fruit as much as 7 cm. long; pedicels 2–7 mm. long; sepals 1.5 mm. long, white margined, hispidulous on the green back; petals 3–4 mm. long, the blade oval, narrowed to a slender claw; style 1–1.5 mm. long; pods 3–5 mm. long, puberulent; seed 2 mm. long, brown, ovoid. *Hymenophysa pubescens* sensu St. John, not of C. A. Mey. Fields and waste lands, *A. T.*

It forms large patches, spreads rapidly, and is very difficult to eradicate. A Siberian weed, introduced in alfalfa seed in 1917.

CHORISPORA

Annual or perennial, low, branched herbs; leaves entire or pinnatifid; flowers long pedicelled, in racemes; sepals erect, the lateral ones swollen at base; pods slender, cylindric, the cells not dehiscing at maturity; cells numerous, in two rows, at maturity each breaking off with its enclosed single seed. (Name from Gr., *choris,* apart by oneself; *spora,* seed.)

Chorispora tenella (Pall.) DC. Annual, stems 10–35 cm. tall, capitate glandular puberulent; basal leaves 2.5–8 cm. long, numerous in a rosette, lanceolate, sinuately lobed, glandular puberulent, petioled; cauline leaves dentate to entire, short petioled; pedicels 2–4 mm. long, divergent; sepals 3–5 mm. long, oblong, green or purplish; petals 9–13 mm. long, the limb oblanceolate, the claw slender; pods 3–4 cm. long, with 2 rows of polygonal creases bounding the line of dehiscence between the numerous cells; seeds embedded in spongy tissue. Weed, recently introduced from Siberia. Lewiston, *Rodock* 11, April 15, 1929; Spokane, *Bonser,* April 25, 1935; Walla Walla in 1944, *Booth, A. T., U. Son.*

CONRINGIA

Erect glabrous, annual herbs; leaves entire, sessile; flowers yellowish in terminal racemes; sepals and petals narrow; siliques elongate linear, 4-angled, the valves firm, 1–3-nerved; seeds wingless. (Named for *Hermann Conring* of Germany.)

Conringia orientalis (L.) Dumort. *Hare's-ear Mustard.* Stems 1–10 dm. tall, simple or branched; lowest leaves obovate, narrowed to the base, subsessile; cauline leaves 5–13 cm. long, elliptic oblong, glaucous; raceme 10–25-flowered, becoming loose; pedicels ascending; sepals 5–7.5 mm. long, saccate at base, the outer linear, the inner oblong; petals 8.5–13.5 mm. long, pale yellowish; pods 8–13.5 cm. long, 2–2.5 mm. wide, ascending, with a prominent midnerve, the beak 1–2.5 mm. long; seeds 2.5 mm. long, ellipsoid, brown. A weed, introduced from Europe. Along railroad, Pullman, *Wilson,* May 21, 1927, *A. T.*

DRABA

Low annual or perennial herbs; leaves entire or toothed; flowers white or yellow, mainly in racemes; petals entire, notched, or bifid; pod oval to oblong, compressed parallel to the partition; valves nearly flat, nerveless or faintly nerved; seeds few to many. (Ancient Gr. name *drabe, for Cardaria Draba.*)

Perennials, caespitose; leaves linear, rigid. *D. densifolia,* f. *caeruleimontana.*

Annuals, not caespitose; leaves broader, not rigid,
- Petals deeply bifid,
  - Pods broadly elliptic to obovate. *D. verna,* var. *Boerhaavii.*
  - Pods elongate oblong or lance-oblong. Var. *major.*
- Petals entire or merely emarginate,
  - Petals yellow; flowers all alike, not cleistogamous,
    - Pods narrowly elliptic, shorter than the pedicels. *D. nemorosa.*
    - Pods narrowly oblong, as long or longer than the pedicels. *D. stenoloba,* var. *nana.*
  - Petals white; flowers dimorphous,
    - Inflorescence in fruit a long raceme; stems leafy. *D. platycarpa.*
    - Inflorescence in fruit short, compact; leaves mostly basal,
      - Pods hispidulous with mostly simple hairs. *D. reptans,* var. *stellifera,* f. *stellifera.*
      - Pods glabrous. Forma *Hunteri.*

Draba densifolia Nutt. ex T. & G., forma **caeruleimontana** (Payson & St. John) St. John, new comb. Caespitose perennial; leaves all basal or in terminal rosettes on the sterile branches, 8–18 mm. long, 1–1.5 mm. broad, linear or slightly spatulate, rigid, the midrib prominent beneath, sparsely pubescent above and below with mostly stellate hispid hairs, ciliate with remote hispid hairs, cuspidate; scapes 5–16 cm. tall, glabrate above, hispid below with simple and a few forked hairs; raceme 10–20-flowered, loose; pedicels 3–25 mm. long, ascending, glabrous; sepals 3–4 mm. long, oblong-ovate, hispid with simple and forked hairs; petals 4–5.5 mm. long, yellow, the blade suborbicular, emarginate; pods 4–7 mm. long, elliptic- to ovate-lanceolate. *D. caeruleimontana* Payson & St. John, Biol. Soc. Wash., Proc. 43: 119, 1930. Exposed basalt crevices, highest peaks of the Blue Mts., *Huds.*

Described in 1930. The type was from Lewis Peak, *Brode* 3.

Draba nemorosa L. Annual; basal leaves often in a rosette, 2–25 mm. long, oblanceolate to obovate; entire or denticulate, short petioled, forked or stellate hispid; cauline leaves remote, on the lower part of the stem, oblong-lanceolate to elliptic-ovate, sessile, remotely serrate, simply or forked hispid above, forked or stellate hispid below; stems 2–45 cm. tall, glabrate above, simply, forked, or stellate hispid below; raceme loose, elongating in fruit, 4–50-flowered; pedicels 5–25 mm. long, at least the lower much longer than the pods, ascending or divergent; sepals 1.5 mm. long, oblong-ovate, pilose on the back; petals

2.5 mm. long, pale yellow, fading whitish, the limb oblong-spatulate, emarginate; pods 3–10 mm. long, 1.5–2.5 mm. wide, narrowly elliptic-oblong, puberulent with simple ascending hairs. Open slopes, *U. Son., A. T.*

Draba platycarpa T & G. Annual; basal leaves several, 5–20 mm. long, oblanceolate to obovate, few dentate at tip, hispid above at base, generally stellate and forked hispid, the petiole hispid ciliate; cauline leaves 5–26 mm. long, ovate-lanceolate to oblanceolate, remotely dentate, sessile; stems 4–25 cm. tall, pilose with simple, forked, and stellate hairs; racemes loose, 5–50-flowered; flowers dimorphous; terminal raceme with macranthous flowers; pedicels 5–12 mm. long, ascending forked and stellate pilose; sepals 2 mm. long, pilose, the outer lance-elliptic, the inner elliptic-obovate; petals 4 mm. long, white, the limb obcordate; anthers 0.4 mm. long; cleistogamous flowers in short lateral branches; the sepals 1–1.5 mm. long, oblong elliptic; petals unequal, shorter than the sepals; anthers 0.1 mm. long; pods 5–8 mm. long, 2–3 mm. wide, broadly elliptic, puberulent with simple, ascending hairs. *D. viperensis* Payson & St. John. Rocky hillside, Buffalo Rock, Asotin Co., Snake River Canyon, *St. John et al.* 8,251, *U. Son.*

Because of its leafy stem, loose and elongate racemes, straight pods 3–4 mm. wide, this is considered a species, well separated from *D. cuneifolia* Nutt.

Draba reptans (Lam.) Fern., var. **stellifera** (O. E. Schulz) C. L. Hitchc., forma **stellifera.** Slender annual; leaves 2–10 mm. long, at or near the base, often forming a rosette, short petioled or cuneate, obovate, hoary with forked hispid hairs, sparsely hispid ciliate at base; cauline leaves few, on the lower quarter of the stem; inflorescence subumbellate, becoming short racemose, 3–12-flowered; pedicels 1–5 mm. long, ascending, glabrous; flowers dimorphous; petaliferous flowers on the central and on other strong branches; sepals 1.5–2 mm. long, ovate, hispid on the back or glabrous; petals 3.5–4 mm. long, white, the blade obcordate, emarginate; anthers 0.5 mm. long; cleistogamous flowers on weak lateral or on late branches; sepals 1–1.5 mm. long, linear-oblong, hispid at tip or glabrous; petals 1.5 mm. or less in length, spatulate, or wanting; anthers 0.2 mm. long; pods 4–12 mm. long, 1.3–2 mm. wide, linear, often curved, glabrous. *D. carolinana* Walt., subsp. *stellifera* (O. E. Schulz) Payson & St. John. Dry slopes, *U. Son.; A. T.*

It is noteworthy that var. *stellifera* differs from var. *reptans* in that the pods have the simple hairs 0.1 mm. long, intermingled with some forked hairs.

Forma **Hunteri** (Payson & St. John) C. L. Hitchc. It differs from var. *stellifera* only by having the pods glabrous. *D. caroliniana* Walt. var. *Hunteri* Payson & St. John. Sandy soil, Clarkston and Lewiston, in the Snake River Canyon, *U. Son.*

Draba stenoloba Ledeb. var. **nana** (O. E. Schulz) C. L. Hitchc. Winter annual, 5–30 cm. tall; basal leaves numerous, 1–4 cm. long, oblanceolate, usually denticulate, hispid with simple or branched hairs; stems simple or branched, glabrous throughout or only below bearing simple or forked hairs; the 1–8 cauline leaves 3–17 mm. long, lanceolate or oblanceolate; racemes 10–30-flowered; sepals 1–2 mm. long, pilose; petals 2–4 mm. long, spatulate; pods 8–22 mm. long, 1.5–2.3 mm. wide, glabrous. Blue Mts., Oregon Butte, T7N, R41E., *Darlington 349.*

Draba verna L., var. **Boerhaavii** Van Hall. Annual; leaves all basal, in a rosette, 4–20 mm. long, obovate-spatulate, entire or 2–6-dentate, puberulent especially above with mostly branched hairs; petioles simply hispid ciliate; stems 1–10 cm. tall, simply or branched puberulent below; raceme 5–10-flowered, becoming loose; pedicels 5–20 mm. long, ascending, glabrous; sepals 1–1.4 mm. long, obovate; petals 1–2 mm. long, white, cleft to about the middle; pods 3.5–6 mm. long, 3–4 mm. wide, broadly elliptic to obovate. *D. verna* L., var. *aestivalis* Lejeune. Occasional on dry slopes, Snake River Canyon, U. Son.

Fernald in Rhodora 36: 370–371, 1934, accepted the taxon, but rejected the epithet var. *Boerhaavii,* asserting that it was published as a species, though admitting that he had not seen the publication. C. L. Hitchcock accepted var. *aestivalis* on Fernald's authority. Nevertheless, the var. *Boerhaavii* was published as a variety in 1821, and I saw the publication when preparing my revision (1930). Though Fernald was my professor and though I admired him greatly, I maintain that in this instance he was in error.

Var. **major** Stur. *Whitlow Grass.* Differing by having the leaves more coarsely dentate; the petals 2.5–4 mm. long; and the pods 5–8 mm. long, 1.5–2.5 mm. wide, elongate oblong or lance-oblong. *D. verna* L. of Fern., and of C. L. Hitchc. Abundant in open places, *U. Son., A. T.*

ERUCA

Annual or perennial; leaves dentate or pinnatifid; racemes terminal, elongating, bractless; flowers large, whitish, yellow, or purplish; style prominent; stigma undivided; fruit oblong to linear-oblong, turgid, with strongly keeled valves; seeds in two rows, globose, wingless; cotyledons conduplicate. (*Eru-ca,* the Roman name for some plant in the Cruciferae.)

Eruca sativa Mill. *Garden Rocket.* Plant up to 3 dm. tall, somewhat succulent, glabrous; leaves thick, the basal ones 3–20 cm. long, lyrately pinnatifid, the lobes incisely dentate; principal cauline leaves 3–10 cm. long; pedicels thick, only 0.5–5 mm. long; sepals 8–9 mm. long, elliptic; petals 1.5–2 cm. long, whitish or yellowish with violet veins; siliques 1.5–2.5 cm. long, fusiform, 4-angled, erect appressed, with the valves keeled on the back; beak flat, almost as long as the body; seeds 1.5–2 mm. long, ellipsoid, slightly compressed, borne in two rows. Weed, introduced from southern Eurasia or Africa, established in Pullman, 1954, *Abendroth.*

ERYSIMUM

Annual, biennial, or perennial herbs, with 2-branched or stellate pubescence; leaves rather narrow, entire or toothed, not clasping; flowers often large, yellow, brown, or purple; the outer sepals gibbous at base; pod linear, 4-angled or rarely flattened, not stipitate; valves keeled, with a prominent midrib; seeds oblong, marginless. (Name Gr., from *eruein,* to draw blisters.)

Petals less than 1 cm. long; annuals,
  Petals 4–5 mm. long; pods 12–27 mm. long. — *E. cheiranthoides.*
  Petals 6–10 mm. long; pods 4–10 cm. long. — *E. repandum.*
Petals 1.5 cm. or more long; biennials or perennials,
  Pods 4-angled; petals stellate pubescent below at base of limb; leaves broadly linear to lanceolate, usually repand-dentate. — *E. asperum.*
  Pods flattened; petals glabrous; cauline leaves linear, entire or nearly so. — *E. occidentale.*

Erysimum asperum (Nutt.) DC. *Western Wallflower.* Whitish throughout with appressed 2-branched or stellate pubescence; stems simple, stout, 1–12 dm. tall, erect; leaves 7–15 cm. long, linear, lanceolate, or oblanceolate, entire or somewhat dentate, rarely lobed, mostly petioled; sepals 8–12 mm. long, oblong; petals 15–25 mm. long, yellow or orange-yellow; pods 4–12 cm. long, 1–2 mm. wide, spreading or ascending, tipped with a short stout beak and a large 2-lobed stigma. *E. capitatum* (Dougl.) Greene; *Cheirinia aspera* (Nutt). Rydb. Abundant in arid places, *U. Son. to A. T. T.*

Erysimum cheiranthoides L. *Wormseed Mustard.* Erect, 1–10 dm. tall, branched above, minutely rough-pubescent throughout; stems with 2-branched hairs; leaves 2.5–10 cm. long, linear-lanceolate, acute at each end, entire or minutely dentate, sparsely lepidote, and with numerous stalkless stellate hairs, sessile or nearly so; racemes elongate; pedicels stiff, slightly ascending; sepals 2–2.5 mm. long; petals yellow; pod 1–1.2 mm. wide, linear, 4-angled, finely stellate, erect, the beak very short. A weed, introduced from Europe, roadsides and waste places, *A. T.*

Erysimum occidentale (Wats.) Robins. Biennial, 1–6 dm. tall, with 2-branched hairs throughout, simple or branching; leaves 3–10 cm. long; plant often flowering a few cm. from the base; pedicels ascending; sepals 6–11 mm. long, elliptic; petals 15–20 mm. long, yellow; pods 4–10.5 cm. long, 2–3 mm. wide, sharply ascending. Rock Creek, 6 miles below Rock Lake, *Cotton* 955, *U. Son.*

Erysimum repandum L. Plants 1.5–6 dm. tall, lepidote; lower leaves linear-lanceolate, repand-dentate, petioled, rosette-forming; cauline leaves 0.5–9 cm. long, linear or linear-oblanceolate, subsessile, with 2-branched hairs, repand-dentate; racemes terminal, erect; pedicels slightly ascending; sepals 3–5 mm. long, cucullate at tip; petals pale greenish yellow; pods 1–1.5 mm. wide, divergent or slightly ascending. A weed introduced from Europe by 1915. Now abundant in grain fields and waste places, *A. T.*

HESPERIS

Erect biennial or perennial herbs, with forked hairs; leaves simple; flowers purple or white; stigma with 2 erect lobes; siliques elongate, nearly cylindric, the valves keeled; seeds wingless. (Name from Gr., *hesperis,* evening.)

Hesperis matronalis L. *Dame's Violet.* Plants erect, hirsute below, forked hirsutulous above; stem 4–10 dm. tall, simple or branched; lower leaves 7–20 cm. long, petioled, ovate or ovate-lanceolate, acute, dentate, hirsutulous with simple or forked hairs; cauline leaves smaller, ovate-lanceolate, sessile; inflorescence of usually several racemes; pedicels ascending, with paired basal glands; sepals 6–10 mm. long, purplish, the outer saccate at base; petals 15–25 mm. long, purple, pink, or white, the blades broadly ovate, emarginate, rotate; pods 3–4 cm. long, 1.5–2 mm. broad, ascending, contracted between the seeds. An ornamental from Eurasia. Escaped from gardens, Pullman, *St. John* 6,212.

HUTCHINSIA

Low spreading annual with entire or pinnately-lobed leaves; flowers minute, white, in crowded racemes which elongate in fruit; stigma sessile or nearly so; each valve conspicuously 1-nerved; seeds numerous in each cell. (Named for Miss *Hutchins* of Ireland.)

Hutchinsia procumbens (L.) Desv. Branched from the base, glabrous or sparsely puberulent, 5–20 cm. high; stems decumbent or ascending; basal leaves 5–25 mm. long, pinnately lobed or entire, petioled; cauline leaves 5–18 mm. long, spatulate or oblanceolate, sessile, entire, or with a few lobes; pedicels 2–10 mm. long, ascending; sepals 0.5 mm. long, oval; petals 0.5 mm. long, spatulate; pods 3–4 mm. long, smooth. Moist places, *U. Son.*, rare in *A. T.*

IDAHOA

Low acaulescent annual herb; leaves lyrate, few-lobed or entire; flowers minute, white; pods flat, compressed parallel to the partition, suborbicular; seeds broadly winged. (Name from the geographic one, Idaho.)

Idahoa scapigera (Hook.) Nels. & Macbr. *Pepperpod.* Glabrous; scapes 5–10 cm. tall, erect or ascending; leaves basal, the blades 3–15 mm. long, spatulate-oblong, entire or coarsely 3–7-lobed, the petioles 5–20 mm. long, sepals 1.5–1.8 mm. long, oval, purple tinged; petals 2–2.5 mm. long, spatulate; pod 5–11 mm. long, purple spotted, beaked with the 0.5 mm. style; seeds 4–5 mm. broad, the wing 1 mm. or more broad. *Platyspermum scapigerum* Hook. In moist gravelly places in early spring, *U. Son., A. T.*

The peppery pods are eaten by children.

LEPIDIUM. Peppergrass

Erect or diffuse annual, biennial, or perennial herbs; glabrous or with simple hairs; leaves pinnatifid to toothed or entire; flowers small, white, greenish, yellow, or reddish, in racemes or panicles; sepals equal; stamens 2–6; pod orbicular to obovate, much flattened at right angles to the partition, 2-winged at the summit, each valve with one acute nerve forming a keel; seeds flattened. (Name Gr., *lepidion,* a little scale.)

Petals yellow; cauline leaves cordate clasping at base. *L. perfoliatum.*
Petals white or none; cauline leaves not cordate clasping at base,
  Pedicels flat; wing of pod forming acute prominent teeth at tip. *L. dictyotum.*
  Pedicels terete or slightly flattened; wing of pod short and rounded at tip,
    Petals present. *L. virginicum,* var. *medium.*
    Petals wanting,
      Fruits averaging 2.5 mm. in length; pedicels slightly flattened. *L. densiflorum,* var. *densiflorum.*
      Fruits averaging 3 mm. in length; pedicels distinctly flattened,
        Pods glabrous. Var. *macrocarpum.*
        Pods pubescent,
          Pod pubescent only on the margin. Var. *elongatum.*
          Pod pubescent throughout. Var. *pubicarpum.*

Lepidium densiflorum Schrad., var. **densiflorum.** Annual, erect, much branched above, 1–9 dm. tall, minutely puberulent, at least below; lower leaves lanceolate, incisely serrate or lobed, 3–5 cm. long and petioled; upper narrower, sometimes entire; racemes elongate, densely flowered; pedicels 1–4 mm. long, ascending; sepals 0.5–1 mm. long, elliptic; stamens 2–4; pod elliptic-ovate or obovate-rotund, notched, glabrous, narrowly wing margined above, the style minute, shorter than the wings. Open places, Spokane Co., *A. T.*

Var. **elongatum** (Rydb.) Thell. Pods puberulous only on the margins; pedicels somewhat flattened. *L. elongatum* Rydb.; *L. simile* Heller. Grassy or open slopes, *U. Son., A. T.*

Var. **macrocarpum** Mulligan. The pods about 3 mm. in length, obovate-rotund; pedicels somewhat flattened on both sides, but less than twice as broad as thick. *L. densiflorum,* var. *Bougeauanum* sensu C. L. Hitchc., non Thell. Open places, *U. Son., A. T.*

Var. **pubicarpum** (A. Nels.) Thell. Pods averaging 3–3.5 mm. in length, puberulent throughout; pedicels not much flattened. Waitsburg, *Horner* R4B74, *A. T.*

Lepidium dictyotum Gray. Annual, branched from the base, usually spreading; stems 3–18 cm. long, hirsutulous; basal leaves 1–5 cm. long, linear, entire or pinnatifid into few linear lobes; cauline leaves linear, 1–4 cm. long, entire or with one or two teeth, pubescent or glabrous; raceme dense; pedicels 1–4 mm. long, puberulent, flattened; sepals 0.5–0.7 mm. long, ovate, puberulent; petals wanting or rudimentary; pods 2.5–3.3 mm. long, ovate, puberulent, the margin winged at tip, the rounded teeth ¼–⅓ the length of the partition. Arid flats, occasional, *U. Son.*

Lepidium perfoliatum L. *Perfoliate Peppergrass.* Annual or biennial, 1–5 dm. tall, usually branched; lower leaves 2–15 cm. long, petioled, bipinnatifid into linear segments, sparsely hirsutulous; upper leaves 3–30 mm. long, cordate or ovate; pedicels 2–5 mm. long, ascending; sepals 1 mm. long, oval; petals 1.5 mm. long, with a narrow spatulate limb; pods 3–4 mm. long, broadly elliptic to orbicular, minutely wing toothed and notched at tip, the style longer than the teeth. A weed, introduced from Europe by 1918, now abundant. Open places, *U. Son., A. T.*

Lepidium virginicum L., var. **medium** (Greene) C. L. Hitchc. Annual or biennial, 3–6.5 dm. tall, simple or branching, sparsely puberulent, the stems glabrous above; basal leaves 2–8 cm. long, lyrate pinnatifid; cauline leaves 1–6 cm. long, linear or oblanceolate, entire or dentate; pedicels 2–5 mm. long, ascending; sepals 0.6–1 mm. long, oval; petals 1–2 mm. long, spatulate; stamens 2–6; pods 2.7–3.3 mm. long, suborbicular, minutely winged and emarginate at apex, the minute style shorter than the wings, glabrous or puberulent. A form with only two stamens occurs commonly, often growing with the normal 6-stamened plant. This has been named *L. idahoense* Heller. Dry or sandy places, abundant in the Snake River Canyon, *U. Son.*

LESQUERELLA. Bladder Pod

Annual to perennial herbs with branching or stellate hairs; basal leaves entire or pinnatifid; cauline leaves entire; petals yellow, red, purple, or white; pods subspherical, beaked by the persistent style; seeds often flattened or winged. (Named for *Leo Lesquereux* of America.)

Lesquerella Douglasii Wats. Perennial, 1–4.5 dm. tall, closely silvery stellate throughout; basal leaves numerous, 3–10 cm. long, long petioled, the blade obovate to oblanceolate, entire or few-toothed; cauline leaves 1–5 cm. long, linear or nearly so; pedicels 4–16 mm. long, ascending or later recurved; sepals 3–5 mm. long, oblong; petals 6–9 mm. long, broadly spatulate, yellow; pods 3–4 mm. in diameter, not stalked, the cells 2–4-seeded; seeds 1.5 mm. wide, discoid, brown. Rare, on limestone talus, Grand Ronde River, *St. John* 3,482; Lime Pt., Nez Perce Co., *St. John* 4368, *U. Son.*

NASTURTIUM

Aquatic perennials with floating or creeping stems, rooting at the nodes; leaves odd pinnate; flowers in racemes; sepals equal at base, spreading during anthesis; petals white; pods cylindric; styles slender; stigma 2-lobed. (Name from Lat., *nasturcium,* the cress.)

Nasturtium officinale R. Br., *Water Cress.* Plants 1–6 dm. long, glabrous; stems succulent; leaves 2–12 cm. long; lateral leaflets 2–20 mm. long, ovate, slightly sinuate; terminal leaflets slightly larger, suborbicular; pedicels 2–10 mm. long, ascending, later spreading; sepals 2–2.5 mm. long, oval; petals 3–5 mm. long, the limb oval; pods 8–14 mm. long, usually curved and ascending, the valves net-veined. *Radicula nasturtium-aquatica* (L.) Britt. & Rendle; *Sisymbrium Nasturtium-aquaticum* L. Abundant and apparently native, small and cold streams, *U. Son., A. T.*

The young shoots are edible, and are used in salads.

PHOENICAULIS

Low perennial herbs with branching caudex; leaves mostly radical, the cauline, if present, reduced; flowers rose-colored or purple, in racemes on slender scapes; sepals oblong, erect, the outer gibbous at base; petals large, with an obovate blade and a slender claw; seeds orbicular, wingless. (Name Gr., *phaino,* to show; *kaulos,* stem.)

Phoenicaulis cheiranthoides Nutt. ex T. & G. Plant 1–3 dm. tall; root stout vertical; caudex usually branched and covered with the bases of dead leaves; leaves mostly basal, 3–12 cm. long, entire, spatulate to oblanceolate, white with a fine dense stellate pubescence, the petiole about as long as the blade; stems nearly naked; sepals 4–5 mm. long, purple tinged; petals 7–8 mm. long, rose-purple; pedicels 4–25 mm. long; pods 1.5–4 cm. long, linear-lanceolate, flat, glabrous, spreading; seeds 3 mm. long, brown, suborbicular, flat. *P. Menziesii* (Hook.) Steud.; *P. cheiranthoides* Nutt. ex. T. & G., var. *lanuginosa* (Wats.) Rollins; *Parrya Menziesii,* var. *lanuginosa* Wats. On stony slopes in the Blue Mountains, and Rock L., Whitman Co., *A. T.*

PHYSARIA. Double Bladder Pod

Low spreading tufted perennials with stellate pubescence; leaves mostly entire; sepals oblong or elliptical, short; petals longer, spatulate to obovate, entire; style slender; silicle obcordate, compressed at right angles to the partition; valves nerveless. (Name from Gr., *phusa,* bellows.)

Pods obcordate, partition about 5 mm. long; style 6 mm. or more long. — *P. Geyeri*, var. *Geyeri.*
Pods obreniform, partition 8–12 mm. long; style 1–2 mm. long. — *P. oregona.*

Physaria Geyeri (Hook.) Gray, var. **Geyeri.** Whole plant whitish with a dense stellate pubescence; leaves mostly basal, the blades 1–3 cm. long, broadly ovate to orbicular, entire or more or less toothed, long-petioled; cauline leaves spatulate; stems 5–21 cm. long, several, spreading, or ascending at tip; racemes densely flowered, 2–5 cm. long; sepals 5–7 mm. long, lanceolate, the outer saccate at base; petals 8–12 mm. long, spatulate, bright yellow; pods strongly compressed, slightly inflated, broadly notched at apex, narrowed at base, stellate; seeds 2 mm. long, discoid, brown. *P. Geyeri,* var. *typica* Rollins. Common in gravelly or sandy soil about Spokane, where first collected by C. A. Geyer. Apparently spreading southward along the R. R. embankments, to Oakesdale, Garfield, and Pullman, *A. T.*

Physaria oregona Wats. Closely white stellate throughout; leaves mostly basal, the petioles 1–4 cm. long, the blades 6–25 mm. long, lanceolate to orbicular, mostly dentate, or somewhat lobed at base; cauline leaves reduced, linear-oblanceolate; stems 6–17 cm. long, mostly prostrate, or the tips ascending; raceme several flowered, loose; pedicels 5–15 mm. long, ascending; sepals 4–6.5 mm. long, lanceolate, the outer saccate at base; petals 8–9 mm. long, spatulate, yellow; pods broadly notched at apex and base, stellate, much inflated, slightly compressed; seeds 3 mm. long, ovate, flattened, dark-brown. Rocky slopes, higher Blue Mts., and down the Snake River Canyon to Wilma, *U. Son., A. T.*

RORIPPA. Marsh Cress

Aquatic or marsh annual or perennial herbs; leaves usually lyrately or pinnately parted or toothed, auricled at base; flowers small, yellow, in terminal or axillary racemes; sepals spreading during anthesis; style short or slender; pods terete or nearly so; seeds turgid, minute. (From its Saxon name, *Rorippen.*)

Rootstocks present; perennials. — *R. sinuata.*
Rootstocks wanting; annuals or biennials,
  Stems hispid. — *R. islandica,* var. *hispida.*
  Stems glabrous or merely puberulent,
    Pods linear, strongly curved upwards. — *R. curvisiliqua.*
    Pods oblong, straight or rarely curved,
      Leaf segments obtuse. — *R. obtusa.*
      Leaf segments acute. — *R. islandica,* var. *occidentalis.*

Rorippa curvisiliqua (Hook.) Bessey. *Arc Cress.* Glabrous or slightly puberulent, bushy branched, erect or decumbent, 15–40 cm. high; leaves 1–15 cm. long, oblanceolate, sinuate-dentate or pinnately cleft into oblong dentate lobes, the lower petioled, the upper sessile, reduced and less dissected; pedicels 2–6 mm. long, divergent; sepals 1–1.5 mm. long, elliptic; petals 1.5–2.5 mm. long, spatulate; pods 4–16 mm. long, 1–1.5 mm. wide, ascending, styles stout or none. *Radicula curvisiliqua* (Hook.) Greene. Wet places, common and variable, *U. Son., A. T.*

Rorippa islandica (Oeder) Borbas, var. **hispida** (Desv.) Butters & Abbe. Biennial, 3–13 dm. tall, hispid up to the inflorescence; leaves 3–25 cm. long, lyrately pinnatifid, the lower petioled, the upper sessile and less divided; pedicels 3–7 mm. long, divergent; sepals 1.5 mm. long, elliptic; petals 2 mm.

long, spatulate; pods 4–6 mm. long, ovoid, glabrous. *R. palustris* (L.) Bess., var. *hispida* (Desv.) Rydb.; *Radicula hispida* (DC.) Heller. Muddy bank, Rock Creek, *St. John & Warren* 6775, *A. T.*

Var. **occidentalis** (Wats.) Butters & Abbe. Plant 3–10 dm. tall, glabrous or nearly so, erect, branching above; leaves 3–15 cm. long, lanceolate, petioled, pinnately cleft or parted, the segments toothed; sepals 2 mm. long, oval; petals 2.5 mm. long, spatulate; pods oblong, turgid, ascending, 7–10 mm. long, 2–2.5 mm. wide, often curved, equaling the spreading pedicels. *Radicula pacifica* (Howell) Greene. In wet places, Waitsburg, *Horner* 83, *A. T.*

Rorippa obtusa (Nutt. ex T. & G.) Britton. Glabrous or nearly so, branching from the base, spreading; stems 1–3 dm. long; leaves 3–10 cm. long, pinnately parted or divided, or the upper often subentire and oblong; pedicels 2–4 mm. long, ascending or spreading; sepals 0.5–1 mm. long, oval; petals 1 mm. long, spatulate; pods 4–8 mm. long, 1–2 mm. wide, oblong or lanceoloid, straight or curved; style 1 mm. long. *Radicula obtusa* (Nutt. ex T. & G.) Greene. Wet muddy shores, uncommon, *U. Son., A. T.*

Rorippa sinuata (Nutt. ex T. & G.) Hitchc. Stems 1–4 dm. long, prostrate or decumbent, branched, glabrous or slightly scurfy-pubescent; leaves 1–7 cm. long, lanceolate or oblong, pinnatifid into numerous equal linear or oblong lobes, these entire or nearly so; pedicels 2–8 mm. long, slightly ascending; sepals 1.5–2.5 mm. long, oblong; petals 3–5 mm. long; pod oblong, 6–8 mm. long, acute at each end, tipped with the style, 1.5–2 mm. long. *Radicula sinuata* (Nutt. ex T. & G.) Greene. Sandy banks of Snake River at Almota, *Piper* 2653; 2654, *U. Son.*

SISYMBRIUM

Mostly annual or biennial herbs; leaves neither clasping nor auriculate at the base, rarely entire, often finely dissected; flowers small, usually yellow or yellowish; style short or none; stigma 2-cleft; pod linear, short or long, flat or terete; seeds oblong. (Name from Gr. *sisumbrion,* cress, perhaps our *Nasturtium officinale.*)

Leaves pinnatifid or lyrate; pubescence of non-glandular simple hairs, or none,
- Pods tapering from the base to the smaller tip, erect, appressed to the axis. — *S. officinale,* var. *leiocarpum.*
- Pods linear, terete; divergent or on divergent pedicels,
  - Pods 5–10 cm. long; upper leaves linear or the segments linear. — *S. altissimum.*
  - Pods 2–3 cm. long; upper leaves runcinate, the segments broadly triangular. — *S. Loeselii.*

Leaves or the lower ones 2–3-pinnatifid; pubescence stellate or glandular,
- Lower leaves tripinnatifid; partition of pod with 2–3 widely spaced parallel longitudinal nerves. — *S. Sophia.*
- Lower leaves bipinnatifid; partition of pod with 1 or with 2 almost confluent nerves,
  - Fruiting pedicels 1–2 cm. long; inflorescence glabrous or sparsely stellate. — *S. longipedicellatum.*
  - Fruiting pedicels 3.5–8 mm. long; inflorescence stalked glandular. — *S. viscosum.*

Sisymbrium altissimum L. *Jim Hill Mustard.* Annual, erect, much branched, 2–15 dm. tall, glabrous or sparsely hirsute, with simple hairs; basal and lower cauline leaves 3–40 cm. long, petioled, pinnatifid, with linear or lanceolate entire or dentate lobes; upper leaves reduced, sessile, linear or pinnatifid with few linear segments; raceme loose; pedicels 3–9 mm. long, becoming divergent; sepals 4–6 mm. long, oblong, mostly glabrous; petals 6–8 mm. long, pale yellow, later bleaching, the limb obovate; pods 1.2–1.5 mm. wide, slightly 4-angled, divergent; seeds 0.7–1 mm. long, ovate, brown, very numerous. *Norta altissima* (L.) Britton. A noxious weed, introduced from Eurasia. Arid places, *U. Son.* to *A. T. T.*

The mature plant breaks loose and is tumbled along by strong winds, scattering seeds as it bounces.

Sisymbrium Loeselii L. Annual; stem 3–10 dm. tall, usually branched, hirsute, retrorsely so below; basal leaves lyrately pinnatifid, petioled, early withering; cauline leaves 2–4 cm. long, short petioled, runcinate, the terminal lobe oblong-triangular, dentate, the lateral lobes smaller, lanceolate; raceme becoming elongate; pedicels 3–12 mm. long, ascending; sepals 3–4 mm. long, lanceolate; petals 5.5–6 mm. long, the limb obovate; pods ascending, the valves strongly 3-nerved; seeds 0.7–1 mm. long, ellipsoid, brown. *Norta Loeselii* (L.) Rydb. A weed, recently introduced from Europe, as yet uncommon. Tucanon River, *Darlington* 250, *A. T.*

Sisymbrium longipedicellatum Fourn. Annual, 1–6 dm. tall; stem slender, erect or ascending, sparingly branched, usually minutely stellate-pubescent, especially below; leaves 1–7 cm. long, pinnatifid or bipinnatifid, segments of the lower leaves oblong; upper leaves sessile, pinnatifid, the terminal segment usually linear and elongate; racemes loose; pedicels slightly ascending; sepals 1.5–2 mm. long, oval; petals 3–4 mm. long, obovate, clawed; pods 6–17 mm. long, 1–1.5 mm. thick, acute, usually curved, ascending; seeds 1 mm. long, ellipsoid, brown. *Descurainia longipedicellata* (Fourn.) O. E. Schultz; *D. pinnata* (Wats.) Britt., subsp. *filipes* (Gray) Detling; *D. pinnata,* var. *filipes* (Gray) Peck. Open places, abundant, *U. Son., A. T.*

Sisymbrium officinale (L.) Scop., var. leiocarpum DC. *Hedge Mustard.* Annual, hirsutulous, 2–9 dm. tall; upper branches rigid, divergent; leaves 3–6 cm. long, the lower petioled pinnatifid, with 3–4 pairs of ovate, acute, dentate leaflets, the terminal lobe obovate, obtuse, dentate, the upper leaves less dissected and nearly sessile; pedicels 1–2 mm. long, stout, erect; sepals 2–2.5 mm. long, oblong; petals 3–4 mm. long, the limb obovate; pods 1–1.4 cm. long, glabrous or glabrate, tapering from the base to the slender beak; seeds 1–1.3 mm. long, ovoid, brown, shining. A weed, introduced from Eurasia. Rocky hillside and waste places, *A. T.*

Sisymbrium Sophia L. *Flixweed.* Annual, erect, much branched above, 2–10 dm. tall, canescent throughout with short stellate pubescent; lower leaves tripinnatifid, 5–8 cm. long, the segments linear or oblong, small, upper leaves smaller, mostly bipinnatifid; racemes elongated; pedicels 2–8 mm. long, ascending; sepals 1.5–2 mm. long, oblong; petals spatulate, shorter than the sepals; pods 1.5–2 cm. long, curved upwards, torulose; seeds 0.8 mm. long, ellipsoid, brown. *Sophia Sophia* (L.) Britton; *Sophia multifida* Gilib.; *Descurainia Sophia* (L.) Webb. A weed introduced from Eurasia or Africa, now abundant, roadsides or fields, *U. Son., A. T.*

Sisymbrium viscosum (Rydb.) Blank. Annual 2–8 dm. tall; stem forked or stellate pubescent below, stalked glandular above; lower leaves bipinnatifid, 3–15 cm. long, petioled, the segments obovate, obtuse, finely stellate; upper leaves similar, 1–2-pinnatifid, not much reduced; pedicels ascending; sepals 1.5–2 mm. long, oval; petals 2–3 mm. long, obovate, cuneate; pods 8–17 mm. long 0.7–0.9 mm. thick, ascending; seeds 1.2 mm. long, ellipsoid, brown. *Sophia viscosa* Rydb.; *Descurainia Rydbergii* O. E. Schultz; *D. Richardsonii*

(Sweet) O. E. Schultz, subsp. *viscosa* (Rydb.) Detling; *D. Richardsonii*, var. *viscosa* (Rydb.) Peck. Dry or rocky places, Blue Mts., *A. T. T.*

THELYPODIUM

Stout biennial herbs, often succulent; pubescence of simple hairs or none; leaves simple, entire, toothed or pinnatifid, often auricled at the base; flowers usually in rather dense racemes; sepals oblong to linear, rather short; style short; stigma entire; pods stipitate or sessile, slender, terete or slightly flattened parallel to the partition; seeds oblong, wingless. (Name from Gr., *thelus*, female; *podos*, of a foot.)

Sepals whitish; leaves fleshy. *T. laciniatum.*
Sepals all or partly purple; leaves membranous. Var. *streptanthoides.*

Thelypodium laciniatum (Hook.) Endl. Biennial, glabrous and glaucous, rather succulent; stem stout, 3–24 dm. tall, simple or somewhat branched; basal leaves 1–5 dm. long, petioled, deltoid-lanceolate, deeply lobed; cauline leaves irregularly pinnatifid; terminal lobe largest, coarsely dentate or entire; lateral lobes oblong or linear, entire or dentate, sometimes wanting; upper leaves subentire; racemes dense, 1–6 dm. long; pedicels 3–5 mm. long, horizontal; sepals 3–5 mm. long, yellowish-white; petals 7–21 mm. long, 1 mm. wide, linear, white; pod 3–10 cm. long, 1 mm. wide, spreading, or recurved; stipe 2–4 mm. long. Base of cliffs and ledges, Snake River, *U. Son.*, Rock Lake, *Piper* 2792.

Var. **streptanthoides** (Leiberg) Pays. Leaves green, neither glaucous nor succulent, deeply pinnatifid, lobes acute, narrow; sepals purple. *T. streptanthoides* Leiberg. Crevices of basaltic cliffs along Snake River, *U. Son.*

THLASPI

Glabrous annual or perennial herbs; basal leaves entire or toothed; cauline oblong, auricled and clasping; flowers white or purplish; pod more or less compressed at right angles to the partition, the valves often winged at the apex; seeds wingless. (An ancient Gr. name *thlaspi*, for some plant, from *thlaein*, to crush.)

Style 1–3 mm. long, projecting from fruit apex; pod cuneate in outline. *T. glaucum.*
Style 0.1–0.2 mm. long, shorter than deep apical notch of fruit; pod suborbicular in outline,
Pods 10–16 mm. long; leaf auricles deltoid, acute. *T. arvense.*
Pods 4–6 mm. long; leaf auricles ovate, obtuse. *T. perfoliatum.*

Thlaspi arvense L. *Penny Cress.* Annual, 1–8 dm. tall, erect, glabrous, simple or branched; basal leaves oblanceolate, petioled, soon withering; cauline leaves 1–9 cm. long, oblong to oblanceolate, obtuse, sagittate and half-clasping; racemes loose, many flowered; pedicels 3–15 mm. long, ascending; sepals 1.5–2 mm. long, yellowish green, white margined; petals 3–4 mm. long, white, spatulate; pod 10–15 mm. long, broadly winged, deeply notched; seeds 1.5–2 mm. long, ovate, flattened, brown, with rough concentric ridges. A bad weed, introduced from Europe, now becoming troublesome in grain fields, *U. Son., A. T.*

This is reported to produce a bad odor in milk, cream, and butter.

Thlaspi glaucum A. Nels. *Wild Sweet Alyssum.* Perennial, 1–4 dm. tall, glabrous and glaucous, scarcely caespitose; stems usually several from a branched base; basal leaves 5–30 mm. long, obovate or elliptic, petioled, entire or toothed; cauline leaves 5–23 mm. long, oval to elliptic, entire, sessile and somewhat auriculate; raceme many flowered, becoming loose; pedicels 3–11 mm. long, ascending, or later spreading; sepals 1.5–2.5 mm. long, oval; petals 4–6.5 mm. long, the limb oblanceolate, white; pods 4–7 mm. long, winged towards the tip; seeds 1 mm. long, ovate, flattened, brown, smooth. The var. *hesperium* Pays., and var. *pedunculatum* Pays. represent growth stages or conditions, and are not considered worthy of separation. *T. hesperium* (Pays.) G. N. Jones. Moist or rocky places, *A .T. T., Can.*

Thlaspi perfoliatum L. Annual, more or less glaucous, 7–30 cm. tall, without basal sterile shoots; stem erect, simple or branched; leaves of basal rosette petioled, obovate, entire or slightly dentate; cauline leaves sessile, sagittate clasping at base, entire or weakly toothed; racemes finally elongate; sepals 1–1.5 mm. long, elliptic; petals 2–3 mm. long, oblanceolate; style less than 0.5 mm. long; pods 4–6 mm. long, obcordate, cuneate at base, emarginate at apex, the upper side flattened, each margin with a wing ½ to ⅔ width of the elliptic ovate cell; seeds usually 4 in each cell, about 1.5 mm. long, rounded ovoid, compressed, yellowish brown, almost shining. Weed, introduced from Eurasia or Africa. Pullman, *Aase* in 1930, fide G. N. Jones.

T. perfoliatum has the stem terete, the petals 2–3 mm. long, the pod 4–6 mm. long, and the seeds smooth. *T. arvense* has the stem angled, the petals 3–4 mm. long, the pod 10–15 mm. long, and the seeds with rough ridges.

THYSANOCARPUS

Erect and slender, sparingly branched annual herbs; flowers minute, white or rose-colored, in elongated racemes; pod mostly pendulous, on a slender pedicel, 1-celled, disk-shaped or concave, orbicular, winged, 1-seeded; seed flattened but not winged. (Name from Gr., *thusanos,* fringe; *karpos,* fruit.)

Pods puberulent on the face. — *T. curvipes,* f. *curvipes.*
Pods glabrous. — Forma *madocarpus.*

Thysanocarpus curvipes Hook., forma **curvipes.** *Lace Pod.* More or less hirsute below, glabrous above; stem 1–6.5 dm. tall, erect, usually branched above; radical leaves 1–5 cm. long, oblanceolate, obtuse, dentate, narrowed at base, petioled; cauline 1–5 cm. long, linear to oblong, sagittate and clasping at base, acute, usually entire; racemes 5–28 cm. long, loose; pedicels 3–10 mm. long, recurved; pods 4–7 mm. long, notched, apiculate by the persistent style, the entire or crenulate wings as broad as the body, fenestrate; seeds 1.7–2 mm. long, oblong to suborbicular, flat, yellowish, the raphe prominent. Gravelly soils, *U. Son., A. T.*

Forma **madocarpus** (Piper) Macbr. Pods glabrous. Quite as common as the species but growing by itself or mingled with the species. Gravelly soils, *U. Son., A. T.*

CAPPARACEAE. Caper Family

Herbs, shrubs, or trees, with mostly alternate palmate leaves; flowers hypogynous, mostly perfect and regular; sepals usually 4, free or united; petals in the form of a cross; stamens 6 or more but not tetradynamous; ovary sessile or on a gynophore; pod 1-celled with 2 parietal placentae; seeds kidney-shaped. (Named from the genus *Capparis.*)

CLEOME

Annual herbs or low shrubs; with flowers in bracted racemes; calyx 4-parted or of 4 sepals, often persistent; petals entire, with claws; stamens 6, (rarely 4); receptacle produced between the petals and stamens; ovary glandular at base, stipitate; pod linear to oblong, many-seeded. Our species are important honey plants and are eagerly sought by the bees. (Ancient name of unknown derivation, of a European, mustard-like plant.)

Flowers yellow; leaflets mostly 5. *C. lutea.*
Flowers rose-purple; leaflets mostly 3. *C. serrulata.*

Cleome lutea Hook. *Yellow Bee Plant.* Glabrous or nearly so, 2–20 dm. tall; petioles 1–6 cm. long; leaflets 1–6 cm. long, entire, narrowly oblanceolate, short petiolulate; racemes dense; calyx somewhat 2-lipped, the tube 0.5–0.8 mm. long, the unequal, lanceolate, acuminate teeth 1.5–2 mm. long; petals 6–9 mm. long, unequal, 2 obovate, 2 oblanceolate; anthers coiled; pedicel of fruit 8–13 mm. long; stipe 5–12 mm. long; pod 1–5 cm. long, linear-oblong, acuminate; seeds 3 mm. long, ovoid, yellowish, with a median fold, apiculate. *Peritoma luteum* (Hook.) Raf. Arid places, *U. Son.;* also Hangman Creek, *A. T.*

Cleome serrulata Pursh. *Rocky Mountain Bee Plant.* Annual, 2–15 dm. tall; petioles 1–4 cm. long; leaflets 2–6 cm. long, lanceolate, entire or minutely serrulate, remotely pilose; racemes dense; calyx turbinate, the tube 1–1.5 mm. long, the lobes 0.5 mm. long, broadly deltoid; petals 8–11 mm. long, unequal, 2 rhombic-ovate, 2 rhombic-lanceolate and longer clawed; anthers coiled; pedicel in fruit 1–2 cm. long; stipe 5–17 mm. long; pod 2–6 cm. long, linear-oblong, acute; seeds 3 mm. long, ovoid, apiculate, with a median fold, brown, tubercled on the back. *Peritoma serrulatum* (Pursh) DC. Dry places, *U. Son.,* Pomeroy. Spreading along the railways from the sagebrush region.

CRASSULACEAE. Stonecrop Family

Succulent or fleshy plants, mostly herbs; stipules none; flowers regular and symmetrical, in cymes or rarely in racemes or solitary; calyx hypogynous, mostly 4–5-parted or lobed; petals of the same number as the calyx-lobes, distinct or slightly united at the base, rarely wanting; stamens of the same number or twice as many as the petals; carpels of the same number as the sepals, distinct or united below; ovules numerous; follicles 1-celled; seeds minute; endosperm fleshy. (Named from the genus *Crassula.*)

SEDUM. Stonecrop

Fleshy mostly glabrous herbs, erect or decumbent; leaves alternate, entire or dentate, fleshy; flowers perfect, polygamous, or dioecious, in terminal often 1-sided cymes; calyx 4–5-lobed or -parted; petals 4–5, distinct or slightly united; stamens 8–10, perigynous, the alternate ones usually attached to the petals; carpels 4–5, distinct or united at the base; ovules numerous; follicles few–many-seeded. (Name from Lat., *sedere,* to sit.)

Leaves becoming scarious at base when dried; carpels divergent. — *S. Douglasii,* subsp. *Douglasii.*
Leaves not becoming scarious,
- Carpels widely divergent; basal leaves spatulate, petioled. — *S. Leibergii.*
- Carpels erect; basal leaves not so. — *S. lanceolatum.*

Sedum Douglasii Hook., subsp. **Douglasii.** Perennial, 1–3.5 dm. tall, erect, branched at base, from a stout rootstock; leaves 3–32 mm. long, linear-lanceolate, attenuate at the apex, abruptly widened at base, flattened; axils of the cauline leaves commonly bearing short deciduous leafy branches by which the plant is propagated; flowers yellow, sessile, in an open cyme; sepal-lobes 1.5–2 mm. long, triangular to lanceolate, acuminate; petals 6.5–10 mm. long, lanceolate, acuminate, follicles diverging from their united bases, the body 3–4 mm. long, ovoid, asymmetric. Basalt rocks and gravelly soil, common, *A. T., A. T. T.*

Sedum lanceolatum Torr. Perennial, 6–30 cm. tall, from branched rootstocks, glabrous, green; stems erect; leaves 6–25 mm. long, linear to lanceolate, somewhat flattened, acute, sessile; axils not producing leafy propagating shoots; flowers nearly sessile, in a close cyme; sepal-lobes 1.5–2.5 mm. long, deltoid; petals 5–10 mm. long, lanceolate, acuminate, yellow, the midrib green or reddish; follicles erect, only the tips becoming somewhat divergent, the body 4–5 mm. long, elliptic-oblong. *S. stenopetalum* sensu St. John, not of Pursh. On cliffs along the Snake River, and in the Blue Mountains, *U. Son.,* to *A. T. T.*

Sedum Leibergii Britton. Perennial, 6–25 cm. tall, glabrous; stems mostly single, stout, erect; basal leaves 1–2.5 cm. long; cauline 5–15 mm. long, green, lanceolate, somewhat flattened, acute; without propagating branches in the cauline leaf-axils; flowers sessile; sepal-lobes 1.2–2 mm. long, deltoid; petals 5–6 mm. long, lanceolate; folicles divergent, the body lance-ovate, acuminate, 3–4 mm. long. On rock ledges, common in Snake River Canyon, *U. Son,* also in "Scablands," *A. T.*

SAXIFRAGACEAE. Saxifrage Family

Perennial herbs or shrubs; leaves alternate or opposite; stipules usually none; flowers perfect or polygamo-dioecious; calyx usually 5-lobed, free or adherent to the ovary; petals usually 5, sometimes 4–8, perigynous, rarely none; stamens usually definite in number and not more than twice the number of the calyx-lobes, or in some numerous; pistil simple or formed by the partial or complete union of 2–5 carpels; placentae axile or parietal; fruit a capsule, follicle, or berry; seeds usually numerous; endosperm present. (Named from the genus *Saxifraga.*)

Shrubs,
- Leaves alternate; fruit a berry. — *Ribes.*
- Leaves opposite; fruit a capsule. — *Philadelphus.*

Herbs,
  Ovary 2-celled; placentae axile,
    Stamens 5. *Suksdorfia.*
    Stamens 10. *Saxifraga.*
  Ovary 1-celled; placentae parietal,
    Petals pinnatifid or 3-cleft into thread-like divisions. *Mitella.*
    Petals not with thread-like divisions,
      Petals cleft or lobed; carpels 3. *Lithophragma.*
      Petals small and entire or none; carpels 2,
        Stamens 10; carpels unequal. *Tiarella.*
        Stamens 5; carpels equal. *Heuchera.*

HEUCHERA. Alum Root

Perennial herbs with scaly rootstocks; leaves mostly basal, long-petioled, rounded, mostly cordate; cauline, if any, alternate; petioles with dilated margins or adherent stipules; flowers greenish or purple, in clusters which form a long narrow panicle; calyx-tube campanulate or narrower, 5-lobed, coherent with the lower half of the ovary; lobes sometimes unequal; petals 5, small, entire, opposite the sepals, sometimes minute or wanting, or early deciduous; styles 2, slender; ovary with 2 parietal placentae; capsule 1-celled, more or less 2-beaked; seeds numerous. (Named for *Johann Heinrich von Heucher* of Germany.)

Stamens exserted; flowers in loose panicles. *H. micrantha,* var. *micrantha.*
Stamens included; flowers in a spike or spike-like panicle,
  Petioles and scape hirsute. *H. cylindrica,* var. *cylindrica.*
  Petioles glandular puberulent,
    Peduncles 2–3 mm. long; cymules compact. Var. *glabella.*
    Peduncles 5–20 mm. long; cymules looser. Forma *valida.*

Heuchera cylindrica Dougl. ex Hook., var. **cylindrica.** Leaves all basal, 3–8.5 cm. long, oval-orbicular, ciliate, hispidulous on the veins beneath, reniform-cordate, 5–7-lobed, crenate-dentate, the teeth cuspidate, the petioles 4–12 cm. long; stems 2.5–6.5 dm. tall, scapiform, erect; inflorescence 4–12 cm. long; calyx yellowish, 6–10 mm. long, campanulate. *H. columbiana* Rydb. Infrequent, gravelly slopes, Snake River Canyon and Spokane County, *U. Son., A. T.*

Var. **glabella** (T. & G.) Wheelock, forma **glabella.** *Smooth Alum Root.* Leaves all radical, the blades 2–6 cm. long, cordate-orbicular, 5–9-lobed, obscurely crenate, the teeth bristle-tipped; petioles 5–18 cm. long, slender; scapes stout, 30–60 cm. high; inflorescence a spike-like panicle, glandular-puberulent; calyx yellowish, 6–9 mm. long, campanulate. Grassy or gravelly hillsides, common, *A. T., A. T. T.*

The herbage is grazed by sheep.

Var. **glabella** (T. & G.) Wheelock, forma **valida** Butters & Lakela. Cymules 3–12-flowered, looser; lower peduncles sometimes forked. Pullman, *Elmer* 78, and *Piper* 1,497, *A. T.*

Heuchera micrantha Dougl. ex Lindl., var. **micrantha.** Basal leaves with blades 3–9 cm. long, rounded cordate, with 7–9 short rounded lobes, glabrous, minutely ciliate, the broad teeth apiculate; petioles 5–17 cm. long, glabrous or sparingly hairy; stems 3–6 dm. tall, with remote bracts or reduced leaves;

panicle 1–4 dm. long, 5–12 cm. wide, glandular puberulent; calyx 2–4 mm. long, turbinate. *H. Nuttallii* Rydb.; *H. micrantha* Dougl., var. *Nuttallii* (Rydb.) Rosend. Rocky slopes, Tallow Flat, Blue Mts., *Darlington* 240; Arrow, *St. John et al.* 9,533, *U. Son.*, to *A. T. T.*

LITHOPHRAGMA

Perennial herbs; rootstock tuberous or bearing bulblets; stems simple, leafy; leaves mostly basal, the petioles dilated at base; raceme terminal, bearing white or pink flowers; hypanthium adnate to the base or to the lower half of the ovary; calyx-lobes valvate; petals cleft, divided, or entire, inserted in the sinuses of the calyx; stamens 10, included; ovary 1-celled with 3 parietal placentae, 3-valved at apex; style 3, short; seeds many, horizontal. (Name Gr., *lithos,* stone; *phragma,* fence.)

Hypanthium campanulate or hemispheric, rounded at base, adnate only to the base of the ovary; axils of cauline leaves usually producing reddish bulblets. *L. glabra.*
Hypanthium obconic, adnate to at least the lower half of the ovary; axils of leaves without bulblets. *L. parviflora.*

Lithophragma glabra Nutt. ex T. & G. *Baby Face.* Stems 6–35 cm. tall, capitate glandular puberulent; basal leaves several, persistent, the blades 3–5-divided, the divisions 3–15 mm. long, cuneate or obovate, 3-cleft and often again toothed, sparsely puberulent or glabrate; petioles 5–55 mm. long, puberulent to glabrate; cauline leaves smaller, with fewer, narrower lobes, shorter petioled; stipules adnate to the petiole, membranous, pink, rounded, fimbriate; raceme 1–5 cm. long, 2–9-flowered, capitate glandular puberulent, the flowers at times replaced by hirsute bulblets; pedicels 1–15 mm. long; hypanthium and calyx-lobes 3–6 mm. long, capitate glandular puberulent; petals 5–10 mm. long, pink, 3–7-cleft into linear lobes; styles slender, included; capsule exserted; seeds 0.4–0.5 mm. long, ellipsoid, compressed on one side, brown, muriculate ribbed. *L. bulbifera* Rydb. Abundant, rocky, open slopes, *U. Son., A. T.,* open woods *A. T. T.*

Dainty, beloved flower of early spring, blooming two weeks before *L. parviflora.*

Lithophragma parviflora (Hook.) Nutt. *Baby's Breath.* Stems 1–5.5 dm. tall, glandular puberulent or hirsutulous, basal leaves several, the blades 5–25 mm. long, hirsutulous, divided to the base into 3–5 cuneate divisions which are again ternately cleft into oblong or linear lobes; petioles 1–8 cm. long, glandular hirsutulous; cauline leaves 2–3, similar but shorter petioled; stipules adnate to the petiole, green or brown, deltoid, fimbriate; raceme 1–12 cm. long, 3–14-flowered, glandular puberulent; pedicels 2–8 mm. long; hypanthium and calyx-lobes 4–10 mm. long, glandular puberulent; petals 5–8 mm. long, usually white, deeply 3–5-cleft into narrow, oblong lobes; stamens included; styles stout, included; seeds 0.3–0.4 mm. long, ellipsoid, dark brown, minutely reticulate. *Tellima parviflora* Hook. Abundant in spring, open slopes or thickets, *U. Son.* to *A. T. T.*

The second half of the generic name is taken from the Greek word *phragma,* a neuter word. The Montreal International Code, Rec. 75A(2) advises that a generic name thus compounded should be neuter. However, a recommendation is not mandatory, and does not justify changing the feminine gender of the name as chosen by the original authors.

MITELLA. Mitrewort

Low slender perennial herbs, from creeping or ascending rhizomes; leaves simple, mostly radical, long-petioled, ovate or orbicular; cauline rarely few, usually none; flowers small, white, greenish, or purple, in a simple raceme, or cymose at base; calyx short, the broad tube 5-lobed, adherent to the base of the ovary and dilated beyond it; petals 5, slender; stamens 10 or 5, very short; styles 2, short; ovary globose, with 2 parietal almost basal placentae, partly superior; capsule globular or depressed, hardly at all lobed; seeds several to each placenta. (Name from Gr., *mitra,* headband.)

Calyx campanulate; petals 3-fid; anthers oblong. *M. stauropetala.*
Calyx saucer-shaped; petals 7–9-lobed; anthers cordate,
  Stamens opposite the petals; stem scape-like. *M. pentandra.*
  Stamens alternate with the petals; stem leafy,
    Leaves hirsute below. *M. caulescens,* f. *caulescens.*
    Leaves glabrous below. Forma *tonsa.*

Mitella caulescens Nutt. ex T. & G., forma **caulescens.** *Leafy Mitrewort.* Rhizomes ascending or creeping, producing slender leafy runners; stems slender, 1–3.5 dm. high; basal leaf-blades 2–7 cm. long, rounded cordate, 3–5-lobed, crenate-dentate, permanently sparsely hirsute on both sides; petioles 4–13 cm. long, hirsute; cauline leaves 1–3, the petioles 3–18 mm. long, the blades smaller; flowers 9–12 mm. across in anthesis, greenish, in a loose glandular puberulent raceme, 3–14 mm. long; calyx-lobes 1.5–2 mm. long, spreading; petals 3–4 mm. long, pinnately cleft into filiform lobes; seeds 0.9 mm. long, ovoid, black. *Mitellastra caulescens* (Nutt.) Howell. Damp wood, Mt. Carleton [Mt. Spokane] *Kreager* 203, *Can.*

Forma **tonsa** St. John. Differing from *M. caulescens* only in having the leaves glabrous beneath. This condition is not, as Dr. Rosendahl suggests, connected with age.

The type is from Wolf Fork, Touchet R., Blue Mts., *St. John et al.* 6,953. Occasional by shady streams in the Blue Mts., *Can.*

Mitella pendandra Hook. *Bishop's Cap.* Stems scapiform, 1–5.5 dm. high, glandular puberulent; leaves usually all basal, the blades 2.5–9 cm. long, with shallow lobes, broadly cordate, coarsely crenate, sparsely hirsute; the petioles 3–19 cm. long, hirsute; raceme 3–27 cm. long, glandular puberulent, often cymose at base, loose; flowers greenish, 6–9 mm. across in anthesis; calyx saucer-shaped, the lobes 1 mm. long, reflexed; petals 2–2.8 mm. long, pinnately divided into filiform lobes; seeds 0.7 mm. long, ovoid, black. *Pectiantia pentandra* (Hook.) Rydb. Wooded stream banks, Blue Mts., *Can.*

Mitella stauropetala Piper. *Piper's Mitrewort.* Rhizomes slender, creeping; scapes 1–5.6 dm. high, sparsely hirsutulous below, glandular puberulent above; leaves all basal, the blades 2–9 cm. long, orbicular, to cordate-reniform, indistinctly 5-lobed, slightly crenate, sparsely hirsute and ciliate; petioles 5–15 cm. long, hirsute; raceme one-sided, 6–27 cm. long, glandular puberulent; flowers 6–7 mm. broad, violet, later white, nearly sessile; calyx-lobes 2 mm. long, oblong-ovate, crenulate, 3-nerved; petals 3–4 mm. long, 3-parted for half their length into filiform lobes, the lateral lobes divaricate; seeds 1 mm. long, ovoid, brown. *Ozomelis stauropetala* (Piper) Rydb. In coniferous woods in the mountains, *A. T. T., Can.*

The type locality is the Craig Mts.

PHILADELPHUS

Leaves opposite, petioled, entire or toothed, ovate or oblong, without stipules; flowers large, showy, white, solitary or cymose-clustered; calyx-tube top-shaped, 4- or 5-lobed, adherent to the ovary nearly or quite to its summit; petals 4 or 5, large, obovate or roundish; stamens 15–60, on the disk; styles 3–5, more or less united; ovary 3–5-celled, partly inferior; ovules numerous; capsule 3–5-valved, loculicidal; seeds very numerous. (Named after the Egyptian king, *Philadelphus.*)

Disc pilose; anthers hirtellous. *P. confusus.*
Disc glabrous,
 Anthers hirtellous. *P. trichothecus.*
 Anthers glabrous,
  Hypanthia glabrous,
   Petals oblong, the corolla cruciform,
    Leaves ovate, acute. *P. Lewisii,* var. *Lewisii.*
    Leaves oblong, mostly obtuse. Var. *oblongifolius.*
   Petals obovate, the corolla discoid; leaves ovate or oblong-ovate. Var. *Gordonianus.*
  Hypanthia with a few hairs,
   Leaves ovate or oblong, the base rounded Var. *Helleri.*
   Leaves elliptic, the base acute or obtuse. Var. *ellipticus.*

Philadelphus confusus Piper. Erect shrub; second year stems brown, 2–4 mm. in diameter, the bark closed, with a few transverse cracks; current year's stems glabrous, the nodes ciliate; leaves 4–7 cm. long, 2–3 cm. wide, ovate-elliptic, subentire, obtuse at both ends, glabrous except for the principal nerves; inflorescence (3)–5–7-flowered, the lowest ones in axils of normal leaves; hypanthium subglobose, glabrous; sepals 5–6 mm. long, ovate, acute; corolla cruciform, 2.5 cm. across; petals 15 mm. long, 6 mm. wide, oblong, glabrous, the apex rounded; stamens about 25; style short, with the spatulate stigmas 4 mm. long; capsules 6–8 mm. long, broad ellipsoid, the calyx persisting above the middle; seed long caudate; embryo 0.7 mm. long, cylindric. Moscow, Ida., *Murley* 1734.

Philadelphus Lewisii Pursh, var. **Lewisii.** *Syringa.* Shrub up to 3 m. tall; second year's stems mostly brown, the bark closed, with transverse cracks; current year's growth glabrous; the nodes ciliate; buds enclosed; leaves 4–5.5 cm. long, 2–3.5 cm. wide, rounded at base, almost entire or inconspicuously denticulate, ciliate, below the nerves strigose-villous, and the nerve angles stiff hairy; main inflorescences 7–9–11-flowered; rhachis sparsely pilose or glabrescent; pedicels 2–7 mm. long, sparsely pilose, the lowest ones among normal leaves; hypanthium campanulate; sepals 5–6 mm. long, acute or acuminate; petals 13–22 mm. long, the apex rounded and emarginate; stamens 28–35; style glabrous, nearly or quite undivided; capsules 7–10 mm. long, ellipsoid, acute, the calyx persisting above the middle; seeds long caudate; embryo 0.7 mm. long, cylindric. By streams or on springy hillsides, abundant, *U. Son.* to *A. T .T.*

The profuse, snowy flowers have an exquisite fragrance. The *Syringa* is a favorite in cultivation, and this variety has been chosen as the state flower of Idaho. In the new version here given, the treatment by Hu (1955) is followed in general. Though she accepts eight varieties of this species, her var. *a* is var. *Gordonianus,* not the typical variety. Since she has none, though one is required by the 1952 International Code, the var. *Lewisii* was described, and typified by the specimen from Clarck's River, *Capt. M. Lewis.*

Var. **ellipticus** Hu. Branches brown; new growth sparsely pilose, glabrescent; leaves 4–7 cm. long, 2–3.5 cm. broad, acute at both ends, subentire or denticulate, chartaceous, above glabrous or rarely sparsely pilose, below strigose pilose on the nerves; sepals 4–6 mm. long, ovate, acute or acuminate; corolla cruciform; petals 14–16 mm. long, glabrous; stamens about 30; style glabrous. Wawawai, July 4, 1892, *Lake, U. Son.*

Var. **Gordonianus** (Lindl.) Jepson. Leaves of vegetative shoots 5–6 cm. long, 2.5–4 cm. wide, coarsely dentate, with 4–5 teeth to a side, short acuminate, above pilose on the nerves, below sparsely pubescent throughout; leaves of flowering branches 3–5 cm. long, subentire or serrulate, glabrous except for the sparsely pilose nerves; sepals glabrous; petals 1.5–2 cm. long; stamens about 35; capsule 10 mm. long, ellipsoid. Walla Walla Co., June 8, 1883, *Anderson.*

Var. **Helleri** (Rydb.) Hu. Second year's growth with bark brown, closed, with transverse cracks; current growth sparsely pilose, glabrescent; leaves 3–6 cm. long, 1.5–2.8 cm. wide, subentire or serrulate, glabrous or sparsely pilose only on the nerves; inflorescences (5–) 7–9-flowered, the lowest ones in the axils of normal leaves; corolla cruciform, 3–4 cm. across; petals oblong, obtuse; style glabrous. Stream banks, *U. Son., A. T.*

Var. **oblongifolius** Hu. Leaves 4–6.5 cm. long, 1.3–3 cm. wide, entire or rarely crenulate; sepals ovate-lanceolate; corolla 3–4 cm. wide; petals 1.5–2 cm. long, the apex rounded, erose; style glabrous. Wawawai, *Piper* 3838. *U. Son.*

Philadelphus trichothecus Hu. Erect shrub; second year's branches 2–3.5 mm. in diameter, brown, the bark closed; current year's branches strigose-pilose; leaves 3–7 (–8.5) cm. long, 1–3 cm. broad, ovate or oblong-elliptic, the base rounded or obtuse, the apex short acuminate, acute or rarely obtuse, subentire or 2–6-toothed on a side, above pilose only on the nerves, below strigose-pilose on the nerves; inflorescence (3–) 5–7-flowered, leafy bracted; hypanthium sparsely pilose; sepals 8 mm. long, ovate-oblong, caudate; corolla cruciform, 3–4.5 cm. wide; petals 17–20 mm. long, oblong, glabrous; stamens about 30; style glabrous; capsule 7 mm. long, obovoid. Idaho, ("Oregon"), Clearwater, *Rev. Spalding.*

RIBES. Currant, Gooseberry

Low sometimes prickly shrubs; leaves alternate, often fascicled, palmately-veined and lobed; stipules adnate or none; flowers small, solitary or racemose, mostly terminating short and 1–2-leaved axillary shoots; calyx-tube adherent to the globose ovary and more or less extended beyond it, 5-, rarely 4-cleft, commonly colored or petal-like; petals 4–5, small, perigynous; stamens as many as the petals and alternate with them; styles 2, more or less united; ovary 1-celled; ovules few or numerous; berry globose, fleshy, usually many-seeded. (The ancient Arabic name, *ribes.*)

Nodes armed with spines; pedicels not jointed; fruit not disarticulating,
- Stamens longer than the calyx-lobes, exserted,
  - Leaves glabrous above, beneath only the vein axils pilose. — *R. niveum,* f. *niveum.*
  - Leaves appressed puberulent above, pilosulous beneath. — Forma *pilosum.*

Stamens not exserted or longer than the calyx-lobes,
Blades 16 mm. or less in width, 2–4-parted nearly to the base. *R. velutinum,* var. *Gooddingii.*
Blades mostly wider, not parted to near the base,
Hypanthium hirsute; internodes usually very bristly. *R. cognatum.*
Hypanthium glabrous; internodes usually not bristly,
Racemes many flowered; berries bright red. *R. sativum.*
Cymes 1–4-flowered; berries dark red to black,
Stamens twice as long as the petals; peduncles glabrous. *R. inerme,* f. *puberulum.*
Stamens about as long as the petals; peduncles glandular-hairy. *R. irriguum.*
Nodes unarmed; pedicels jointed beneath the ovary; fruit disarticulating,
Stems prickly; calyx saucer-shaped. *R. lacustre.*
Stems not prickly,
Flowers yellow; blades convolute in the bud. *R. aureum,* var. *aureum.*
Flowers not yellow; blades plicate in the bud,
Calyx saucer-shaped; berry glandular dotted. *R. petiolare.*
Calyx campanulate or narrower; berry glandular bristly (or glabrous),
Hypanthium more than twice as long as the calyx-lobes; berries red or orange,
Leaves merely glandular pruinose; bracts usually entire. *R. reniforme.*
Leaves puberulent; bracts dentate or lobed. *R. cereum.*
Hypanthium less than twice as long as the calyx-lobes; berries wine-colored to black,
Branchlets and blades glandular pubescent; hypanthium 5–8 mm. long. *R. viscosissimum.*
Branchlets and blades not glandular; hypanthium 2.5–3.5 mm. long. *R. inerme,* f. *puberulum.*

Ribes aureum Pursh, var. **aureum.** *Golden Currant.* Unarmed, 1–3 m. high, usually glabrous throughout; leaves 2–7 cm. wide, thick, somewhat orbicular, cuneate or rounded or truncate at base, deeply 3-lobed, the lobes coarsely 3–5-toothed or entire, petioles slender, about as long as the blades; flowers in dense ascending or spreading 5–15-flowered racemes; bracts leafy, 5–12 mm. long; hypanthium 5–9 mm. long, slender, cylindrical, calyx-lobes 4–6 mm. long, oval, spreading; in age connivent; petals about 2 mm. long, erose, oblong, frequently dark-red; stamens little exceeding the petals; berries 6–8 mm. in diameter, yellow, also occurring in color forms shading from red, purplish, to almost black. *Chrysobotrya aurea* (Pursh) Rydb. In the warmer valleys, abundant, *U. Son.,* occasional, *A. T.*

Ribes cereum Dougl. *Squaw Currant.* Shrub 1–2 m. high, with numerous short branches, the young puberulent and glandular; leaves 1–4 cm. wide; orbicular to reniform, truncate or subcordate at base, puberulent, glandular dotted, somewhat 3–5-lobed, crenate-dentate; petioles mostly shorter than the leaves, puberulent, remotely glandular; racemes dense, 3–5-flowered, drooping, scarcely exceeding the leaves; bracts 4–7 mm. long, obovate or cuneate, puberulent and glandular; hypanthium white to pink, cylindric, 5–8 mm. long, glandular, the calyx-lobes about 2 mm. long, obtuse; petals 1.5 mm. long, flabellate; stamens attached below, not exceeding the petals; style puberulent near tip; berry 6–8 mm. in diameter, orange, insipid. On warm rocky hillsides, and "Scablands," abundant, *U. Son.,* occasional, *A. T.*

Ribes cognatum Greene. *Black Gooseberry.* Erect shrubs, with long arching branches, 1–3.5 m. high, armed with nodal triple spines 6–17 mm. long, sometimes very prickly as well, the young branches canescent, and often glandular; leaves densely puberulent, more so and more glandular beneath; blades 1.5–7 cm. wide, orbicular to reniform, truncate or cordate at base, 3–5-lobed, incisely dentate; petioles glandular villous, as long as or shorter than the blades; racemes 2–5-flowered, drooping; bracts glandular-pubescent, 1.5–3 mm. long, equaling or shorter than the pedicels; ovary glabrous; hypanthium 3–6 mm. long, cylindrical, whitish, larger than the oblong spreading lobes; petals obovate, truncate or retuse, half as long as the calyx-lobes; style villous below; berry 10 mm. in diameter, purplish-black. *Grossularia cognata* (Greene) Cov. & Britt. Near streams, *U. Son., A. T.*

Ribes inerme Rydb., forma **puberulum** Berger. *White-Stemmed Gooseberry.* Erect shrubs, 1–2 m. high, unarmed or with feeble simple spines and a few prickles; young shoots glabrous, the blades 1–6 cm. wide, somewhat puberulent, orbicular, cordate, 3–5-lobed, coarsely-toothed; petioles shorter than the blades, pilosulous; racemes drooping, shorter than the petioles, 1–4-flowered; ovary glabrous; hypanthium 2.5–3.5 mm. long, narrowly campanulate, green, as long as the obtuse greenish or purplish reflexed lobes; petals 1.5–2 mm. long, pink or white; stamens exserted, nearly as long as the lobes; filaments nearly glabrous; style pilose below; berries 8 mm. in diameter, wine-colored to black, edible. *R. divaricatum* Dougl., var. *inerme* (Rydb.) McMinn. Uncommon, along streams, Pullman, *Sheldon.* April 26, 1897; Wild Rose, *Sprague* 305, *A. T.*

Ribes irriguum Dougl. *Inland Black Gooseberry.* Erect shrub, 1–3 m. tall; young branches finely puberulent, with few or no bristles; nodal spines 1–3, 3–8 mm. long; blades 3–7 cm. wide, suborbicular, truncate or cordate at base, with 3–5 obtuse, incised lobes, glabrous or nearly so above, finely pilosulous and glandular beneath; petioles 1–4 cm. long, pilosulous; peduncles shorter than the leaves 1–3-flowered; bracts glandular ciliate; ovary glabrous; flowers whitish; hypanthium 3–4 mm. long, glabrous, narrowly campanulate; calyx-lobes 4–8 mm. long, oblong, obtuse; petals obovate, half as long as the calyx-lobes; stamens exceeding the petals; style pilose below; berry 7–13 mm. in diameter, reddish purple to black, acid. *Grossularia irrigua* (Dougl.) Cov. & Britt. By streams, *A. T., A. T. T.*

Discovered by Douglas in the Blue Mts. and on the Spokane River.

Ribes lacustre (Pers.) Poir. *Swamp Gooseberry.* Stout, 1–2 m. high, young stems puberulent, armed at the nodes with 3–9 stout spines, united into a semi-circle at base; internodes with numerous bristly prickles; leaves 3–7 cm. wide, orbicular, cordate, deeply 3–5-lobed, incisely dentate, glabrous or nearly so; petioles glandular puberulent and hispid, not exceeding the blades; racemes loose, drooping, 7–20-flowered; bracts ovate, glandular ciliate; flowers greenish or purplish, slender-petioled; hypanthium less than 1 mm. long, calyx-lobes 2–3 mm. long, spreading; petals 1.5 mm. long, fan-shaped, exceeded by the stamens; style glabrous; berries 6–10 mm. in diameter, edible, nearly black, bristly glandular. *Limnobotrya lacustris* (Pers.) Rydb. Moist woods in the mountains, abundant, *A. T. T., Can.*

Ribes niveum Lindl., forma **niveum.** *Snowy Gooseberry.* Shrub 1–3 m. tall; branchlets smooth, reddish-brown; branches gray; nodal spines 1–3, 6–20 mm. long; blades 1–4 cm. wide, suborbicular, cuneate or truncate at base, with 3–5 shallow, few-toothed, obtuse lobes, above glabrous, the margin ciliate, beneath pilose in the lower vein axils; petioles 5–43 mm. long, somewhat pilose or even glandular bristly; racemes 2–4.5 cm. long, drooping, 2–5-flowered; bracts 1–2 mm. long, truncate or rhombic; flowers snowy white, glabrous; ovary glabrous; hypanthium 2 mm. long, narrowly funnelform; calyx-lobes 7–8 mm. long, elliptic-oblong, reflexed; petals 3 mm. long, spatulate; stamens 8–10 mm. long, the anthers 1 mm. long, oval, the filaments pilose; style villous; berries 8–10 mm. in diameter, globose, bluish black with a bloom, acid. *Grossularia nivea* (Lindl.) Spach. Rocky creek banks, Snake River Canyon and tributaries, *U. Son.;* occasional in *A. T.*

A graceful, handsome shrub with attractive, snowy blossoms. Lindley early recommended crossing this with the European gooseberry to improve its flavor.

Forma **pilosum** St. John. Differs from the species by having the blades pilosulous beneath, and appressed puberulent above.

Idaho: n. side of Snake R. Canyon, Lewiston Grade, Nez Perce Co., Oct 12, 1924, *H. St. John* 6,794, (type).

Rocky stream banks or draws, mostly in the Snake River Canyon, *U. Son.,* occasional in *A. T.*

Ribes petiolare Dougl. *Wild Black Currant.* Unarmed, glabrous or nearly so, 1–2 m. high; leaves 4–13 cm. wide, round-cordate, 3–5-lobed, serrate, resinous-dotted beneath, the lobes ovate-deltoid, coarsely doubly dentate, glabrous or sparsely pilose below; the slender petioles puberulent often longer than the blades; racemes erect, 5–12 cm. long; pedicels longer than the bracts; hypanthium 1 mm. long; calyx-lobes whitish, 3–5 mm. long, ascending, puberulent; petals 1.5–2 mm. long, spatulate; style glabrous; ovary resinous-dotted; berry 8–10 mm. in diameter, black. By streams in the Blue Mountains, *Can.*

Ribes reniforme Nutt. Shrub 1.5–2.5 m. tall; young branches glabrous, reddish brown, the older grayish brown; the blades 1–5 cm. wide, reniform to rounded reniform, cordate at base, with 3–5 rounded, crenate lobes, glutinous from the numerous sessile glands, lower surface paler; petioles 6–35 mm. long, slender, with subsessile glands; racemes 2–4 cm. long, 2–7-flowered, drooping, glandular; bracts about 5 mm. long, rhombic-obovate, glandular serrate; flowers creamy white or pink-tinged, stalked glandular, pilosulous towards the tip; hypanthium 6–9 mm. long, cylindric; calyx-lobes 2–3 mm. long, elliptic, nearly erect; petals 1.2 mm. long, the limb transversely oval, tapering below to a cuneate base; stamens 3 mm. long, the oval anthers nearly 1 mm. long; style usually glabrous; berries 6–8 mm. in diameter, red, with a few stalked glands.

Rocky slopes, canyon of the Grande Ronde, and of the Snake, Asotin Co., Craig Mts., and Lewiston, *U. Son., A. T.*

C. L. Hitchcock and Cronquist reduce this to the synonymy of *R. cereum.*

Ribes sativum Syme. *Red Currant.* Erect shrub 1–2 m. tall; leaves commonly 5-lobed, the lateral leaves spreading, slightly pubescent beneath or glabrate, mature blades 3.5–6.5 cm. wide, broadened upward, 3–5-lobed, the lobes mostly short ovate; racemes borne chiefly among the leafy shoots, spreading in anthesis, drooping in fruit, 3–5 (becoming 7) cm. long; the rhachis glabrous, though often glandular; pedicels 2–5 mm. long, mostly glandless; calyx 4–5 mm. long, yellow-green, with its segments oval and abruptly narrowed below the middle; petals 1 mm. long, narrowly cuneate; disks between the stamens very short; the slightly cleft style a high, narrow ring with rounded scalloped margin; berry 4–8 mm. in diameter, red, glabrous. *R. rubrum* of most authors, not of L. European cultivated shrub, escaped and established, Spokane Co., July 11, 1923, *Lackey.*

Ribes velutinum Greene, var. **Gooddingii** (Peck) C. L. Hitchc. *Goodding's Gooseberry.* Shrub 1–2 dm. tall, the arching branches with ferocious spines, forming impenetrable thickets; young branches glabrous or glabrate, reddish brown, the older grayish; nodal spines 1–3, yellow, straight, stiff, sharp, the median ones 7–13 mm. long; blades 6–16 mm. wide, ovate to reniform, 2–4-parted nearly to the base, again cleft or entire, glabrous above, scurfy below; petioles slender, exceeding the blades; peduncles 1–2-flowered, 3–5 mm. long, pilose, jointed; bracts ovate, ciliate; buds pink; flowers white, pilose without, nodding; hypanthium 2.5 mm. long, urceolate; flower parts 5 (or 6–7); calyx-lobes 4–5 mm. long; petals 2–2.5 mm. long, spatulate; stamens nearly as long; style glabrous; berries 3.5–5 mm. in diameter, pilose, cherry red, sweet and pleasant. *R. Gooddingii* Peck.

Base of ledges, low elevations, abundant in upper Snake River Canyon; occasional down to Wilma, *U. Son.*

Ribes viscosissimum Pursh. *Sticky Currant.* Unarmed shrub, 1–2 m. tall; branchlets green to pinkish, finely puberulent and capitate glandular pilose; branches becoming gray or brown, glabrate; blades 3–9 cm. wide, reniform, with 5 shallow, obtuse, doubly crenate lobes, capitate glandular and pilose above, densely so beneath; petioles 1–8 cm. long, dilated at base, densely capitate glandular pilose, and puberulent; racemes 3–11 cm. long, ascending, 3–17-flowered; bracts 4–10 mm. long, oblong-oblanceolate; flowers greenish or pinkish; hypanthium cylindric to narrow campanulate, sparsely pubescent, mostly near the base; calyx-lobes 5–7 mm. long, oblong to elliptic, spreading, glabrous or sparsely glandular; petals 3 mm. long, oval to orbicular; stamens 3.5 mm. long, the anthers 1.5 mm. long, oval, with a terminal gland; style glabrous; berries 8–12 mm. in diameter, ellipsoid, black, puberulent and glandular bristly. Open or moist woods, common, *A. T. T., Can.*

According to David Douglas, if two or three berries are eaten they will cause vomiting.

Ribes ———. Another species, of the affinity of *R. reniforme,* is known only from sterile material. It is a fastigiate shrub, 3 m. tall, that grows along Shumaker Creek, leading into the canyon of the Grand Ronde River. The collection is *St. John & al.* 9,693. It is doubtless an undescribed species.

SAXIFRAGA. Saxifrage

Chiefly perennial herbs; stems short or none; basal leaves in a rosette; the cauline mostly alternate; flowers perfect, solitary, in cymes, or in panicles; calyx adhering to the base of the ovary, or almost free, 5-lobed; petals 5, entire; stamens 10 (or 8), perigynous; carpels 2–4, nearly distinct or more or less united into a 2–4-celled ovary; follicles 2–4 beaked; placentae axile; styles distinct; seeds numerous (Name Lat., *saxum,* rock; *frangere,* to break.)

Leaves sessile, or tapering to petiole-like bases that seldom exceed the blades; blades ovate or narrower,
  Leaves oblanceolate-elliptic, serrate-dentate towards the apex. *S. ferruginea,* var. *foliacea.*

Leaves lanceolate to ovate, entire, denticulate, or if toothed serrate to near the base,
Blades serrate-dentate; gland forming a smooth ring at base of carpel. *S. idahoensis.*
Blades entire to denticulate; gland fleshy, forming a broad, convoluted ring at base of each carpel,
Petals greenish, mostly shorter than the sepals; cymules dense. *S. nidifica.*
Petals white, larger than the sepals; cymules mostly loose. *S. integrifolia.*
Leaves with petioles at least twice their length; blades suborbicular to reniform,
Caudex bulbous; blades coarsely doubly dentate. *S. Mertensiana,* var. *Eastwoodiae.*
Rhizome slender, not bulbous; blades simply dentate. *S. arguta.*

Saxifraga arguta D. Don. Perennial, 2–6 dm. tall; rhizome usually horizontal; leaves basal; blades 2–8 cm. wide, suborbicular to reniform, pale beneath, usually glabrous, coarsely dentate, the teeth ovate to deltoid, glandular apiculate; petioles 4–18 cm. long, the base ciliate with hairs of 1 ranked cells; stipules not evident; scapes above capitate glandular puberulent as is the inflorescence; cymes 7–32 cm. long, paniculate, loose, calyx-lobes 2–3.5 mm. long, elliptic-lanceolate, reflexed, petals 2–3 mm. long, the blade broadly ovate, white with 2 yellow spots at base, short clawed; filaments 3–4 mm. long, white, slender clavate from below the middle; carpels superior, the gland dark rugose, extending up ¼ their length, united to the middle, the beaks slender, erect; stigmas small, truncate; fruiting follicles 6–9 mm. long, often purple, ellipsoid, only the tips of the beaks shortly divergent; seeds 0.7 mm. long, elliptic-ovoid, brown, papillose ribbed. *S. odontophylla* Piper; *S. odontoloma* Piper; *S. punctata* L., subsp. *arguta* (D. Don) Hultén; *Micranthes arguta* (D. Don) Small. Wet or shaded stream banks, Blue Mts., *Can.*

Saxifraga ferruginea Graham, var. **foliacea** A. M. Johnson. Scapes solitary or several, erect or ascending, weak, coarsely glandular hirsute, bearing large leafy bracts to above the middle; leaves large, thin, broadly oblanceolate-elliptic, glabrous or sparingly pubescent, ciliate, rounded at apex, coarsely serrate-dentate, the teeth few; inflorescence prolifically branched, the lateral branches mostly bearing bulblets, the terminal ones with normal flowers.—Description compiled from Johnson.—Plant not seen by the author. Palouse Country and Lake Coeur d'Alene, Aug. 1892, *Sandberg.*

Saxifraga idahoensis Piper. Perennial 15–30 cm. tall; caudex stout, short; leaves basal; blades 2–6 cm. long, ovate, dentate, glabrous above, the margin and lower side somewhat rufous pilose, abruptly tapering into a rufous pilose, petiole-like base about as long as the blade; scapes and inflorescence capitate glandular hirsutulous; cyme 3–12 cm. long, pyramidal, the cymules at first contracted, later loose; calyx-lobes 1.5–2 mm. long, ovate to elliptic, early reflexed; filaments 1.5–4 mm. long, slightly clavate; carpels partly inferior, not united above the receptacle, the gland a broad ring near the base, the free portion of the carpels lance-conic, tapering into the style; stigmas capitate; fruiting follicles 3–5 mm. long, ovoid, the stout beaks divergent. *S. occidentalis* Wats., var. *idahoensis* (Piper) C. L. Hitchc. Rocky slopes, *U. Son.* to *A. T.*

First discovered by the Rev. Henry Spalding at Clearwater [Lapwai, Idaho].

Saxifraga integrifolia Hook. Perennial, 1–5 dm. tall, leaves basal 2–9 cm. long, ovate to elliptic, entire or sinuate, glabrous or nearly so, the petiole-like bases usually shorter than the blades; scapes and inflorescence capitate glandular hirsutulous; bracts small, linear, or occasionally one expanded; inflorescence a pyramidal cyme, loose, elongate, up to 15 cm. (or even 30) in length, or occasionally the cyme compact; hypanthium 1.5–2 mm. long, obconic; calyx-lobes 1–2.2 mm. long, ovate, divergent, tardily reflexed; petals 1.5–3 mm. long, elliptic to oval; carpels conic, divergent, the base imbedded, ringed with a rugose gland; styles short, stout; stigmas capitate; fruiting follicles 3–4 mm. long, depressed conic, the beaks divergent; seeds 0.5–0.7 mm. long, lunate-spindle-shaped, shining brown. Moist ledges and slopes, *U. Son.* to *A. T. T.*

Further studies may prove that these inland plants are better separated as *S. fragosa* Suksd., but until better known the writer prefers to reduce the segregate.

Saxifraga Mertensiana Bong., var. **Eastwoodiae** (Small) Engl. & Irmsch. Perennial 15–54 cm. tall; caudex scaly; leaves all basal or nearly so; blades 1.5–8 cm. wide, orbicular-reniform, glabrous, the dentations ovate; petioles 3–22 cm. long, much exceeding the blades, at base sparsely pilose, the hairs multicellular with the cells in several ranks; stipules membranous, adnate to petiole, enwrapping the stem; cauline leaves, when present, near the base, reduced; scape sparsely pilose at base with hairs of many ranked cells, above and inflorescence capitate glandular puberulent; cyme 6–20 cm. long, paniculate, pyramidal, loose; calyx-lobes 1–2.5 mm. long, lance-elliptic, early reflexed; petals 2–3 mm. long, white, elliptic-ovate, short clawed; filaments 2.5–4.5 mm. long, white, clavate at tip; carpels superior, united to the middle, surrounded by a thick convoluted gland almost to the middle, styles slender, divergent, stigmas oblique; fruiting follicles 6–8 mm. long, forming a broad V-shaped capsule; seeds 1.2–1.6 mm. long, fusiform, olive brown. *Heterisia Eastwoodiae* Small. Blue Mts., *Horner* 299.

Saxifraga nidifica Greene. Erect, 8–45 cm. tall; leaves basal, 2–6.5 cm. long, lanceolate to ovate, gradually narrowed to a petiole-like base, mostly glabrous, the margins ciliolate, entire or denticulate; scapes capitate glandular hirsute, especially above, like the inflorescence; inflorescence paniculate-cymose, 1–7 cm. long, the cymules compact, or rarely the inflorescence longer and looser; hypanthium about as long as the calyx-lobes which are 1.5–3 mm. long, deltoid-ovate, at length reflexed; petals 1–3 mm. long, oblong, elliptic, to spatulate-orbicular, variable; carpels partly inferior, ringed above the receptacle by a conspicuous rugose gland, fruiting follicles 3–4 mm. long, ovoid-fusiform, the short stout beaks wide spreading; seeds 0.5 mm. long, brown, thick fusiform, oblique at each end, striate. *S. columbiana* Piper; *S. integrifolia.* Hook., var. *columbiana* (Piper) C. L. Hitchc.; and var. *claytoniaefolia* (Canby) Rosend; *Micranthes columbiana* (Piper) Small. Rocky slopes and springy places, *U. Son.,* abundant in *A. T., A. T. T.*

A variable species, sometimes with a cluster of bulblets around base of stem. Abrams, (1944) credits *S. debilis* Engelm. to the Blue Mts., Wash., but the writer has not verified this record.

SUKSDORFIA

Slender glandular pubescent perennials with leafy axial flowering shoots from a small, bulblet-bearing rootstock; basal leaves reniform, merely crenate; cauline leaves stipulate; inflorescence a loose, few-flowered panicle; hypanthium at first obconic-campanulate, becoming more deeply campanulate, in fruit urceolate, adnate to the major portion of the ovary; sepals 5, oblong-lanceolate, erect; petals 5, spatulate, entire or 2–3-lobed, clawed, violet or white; stamens 5, alternate; anthers almost sessile; ovary 2-celled, with many-ovuled, axile placentae, almost wholly inferior; styles short, more or less erect; stigmas truncate; capsule dehiscent between the styles. (Named for *Wilhelm Nikolaus Suksdorf* of Bingen, Washington.)

Suksdorfia violacea Gray. Slender perennial; stem 1–3 dm. tall, more or less glandular pubescent; basal leaves reniform with 5–7 rounded lobes, 1–3 cm. wide, more or less puberulent; petioles 2–8 cm. long, glandular puberulent; lower cauline leaves similar, with broad-toothed stipules; the upper ones subsessile, cuneate and ¼-toothed at apex, the broad leaf bases and the foliaceous stipules often coalesced; inflorescence a more or less elongated, panicle; hypanthium obconic-campanulate, becoming broader, densely glandular, 2–3 mm. long, in fruit urceolate and 4–7 mm. long; sepals 2–3 mm. long, often longer than the hypanthium; petals violet, 5–7 mm. long, spatulate, long-clawed and almost erect. Rocky ledge, Spirit L., Kootenai Co., in 1930, *G. N. Jones.*

TIARELLA. False Mitrewort

Slender perennial herbs; leaves palmately lobed, sometimes with small stipules; flowers small, white, in a terminal raceme or panicle; hypanthium campanulate, 5-parted, nearly free from the ovary; petals 5, entire, small, mostly with short claws; stamens 10; filaments long and slender; style 2; ovary 1-celled, 2-horned; placentae parietal, becoming almost basal in fruit; ovules numerous; capsule 1-celled, 2-valved, the valves usually unequal; seeds usually few. (Name Gr., a diminutive of *tiara,* tiara, the head-dress of the Persian king.)

Tiarella unifoliata Hook., forma **unifoliata.** *Nancy-over-the-Ground.* Rhizomes slender, scaly; flowering stems lateral, 1.5–6 dm. tall, below glabrous, above capitate glandular puberulent or hirsutulous; basal leaves several, the blades 2–12 cm. wide, orbicular-cordate, 3–5-lobed, the lobes broadly ovate, doubly ovate-dentate, apiculate, sparingly hirsute; petioles 3–16 cm. long, below glabrous, above capitate glandular hirsutulous; cauline leaves 2 (rarely 1–4), smaller, with much shorter petioles; stipules oblong-lanceolate, brownish, long ciliate; panicle 6–23 cm. long, 35 mm. or less in width; sepals 2–3 mm. long, whitish, lanceolate or elliptic, glandular ciliate; petals 3–4 mm. long, lance-linear, whitish; filaments unequal, exserted; ovaries 2, glandular puberulent; styles filiform, about 2 mm. long; stigmas terminal, minute; fruiting follicles lance-elliptic, flattened, the larger 7–12 mm. long, the smaller 4–9 mm. long; seeds 1.5 mm. long, ellipsoid, somewhat flattened, black, shining. Moist woods, *A. T. T.,* common in *Can.*

ROSACEAE. Rose Family

Herbs, shrubs, or trees; leaves mostly alternate, simple or compound, usually with evident stipules which are often quickly deciduous; flowers regular, mostly perfect, rarely polygamo-dioecious; calyx of 5, rarely 3–8, sepals, united at the base, often surrounded by a row of bractlets; calyx-tube lined by the disk; petals as many as the sepals, rarely wanting, on the edge of the calyx-tube; stamens usually numerous, or few and coherent with the calyx-tube; carpels 1–many, distinct, or few and coherent with the calyx-tube into a 2–several-celled inferior ovary; seeds few or solitary. *Amygdalaceae, Malaceae, Spiraeaceae.* Named from the genus *Rosa,* rose.)

Carpel 1,
  Fruit a drupe; petals present. *Prunus.*
  Fruit an achene,
    Petals present; shrubs. *Purshia.*
    Petals none,
      Shrubs or trees; leaves not lobed. *Cercocarpus.*
      Herbs; leaves lobed,
        Leaves palmately lobed; style nearly basal. *Alchemilla.*
        Leaves pinnate; style terminal. *Sanguisorba.*
Carpels more than 1,
  Carpels united with calyx and receptacle, forming a pome,
    Flowers in racemes; fruits 10-celled. *Amelanchier.*
    Flowers in corymbs or cymes; fruits 1–5-celled,
      Carpel walls papery; plant not thorny. *Pyrus.*
      Carpel walls bony; plant thorny. *Crataegus.*
  Carpels free or partly so, not forming a pome,
    Fruit of 2–several-seeded follicles,
      Pods inflated, dehiscent, 2–4-seeded. *Physocarpus.*
      Pods not inflated,
        Stamens well exserted; pods several-seeded, dehiscent. *Spiraea.*
        Stamens scarcely exserted; pods 1-seeded, not or tardily dehiscent. *Holodiscus.*
    Fruit of numerous 1–2-seeded achenes or drupelets,
      Fruit of drupelets, coherent. *Rubus.*
      Fruit of achenes,
        Receptacle urn-shaped or bell-shaped, enclosing the achenes,
          Receptacle fleshy; achenes numerous; petals present. *Rosa.*
          Receptacle not fleshy; achenes 1–4; petals none,
            Leaves palmately lobed; style nearly basal. *Alchemilla.*
            Leaves pinnate; style terminal. *Sanguisorba.*
        Receptacle broad, flatish or conical,
          Styles elongating, mostly plumose or geniculate. *Geum.*
          Styles deciduous, naked,
            Receptacle conspicuously enlarging, fleshy, edible. *Fragaria.*
            Receptacle not so,
              Styles lateral,
                Stamens 5; pistils 10–15. *Sibbaldia.*
                Stamens about 20; pistils numerous. *Potentilla.*
              Styles terminal,
                Stamens inserted near the base of the calyx-tube. *Potentilla.*
                Stamens inserted well up on the calyx-tube,
                  Filaments dilated, petaloid. *Horkelia.*
                  Filaments filiform. *Ivesia.*

ALCHEMILLA. Lady's Mantle

Low annual or perennial herbs; leaves palmately lobed or compound, with sheathing stipules; flowers perfect, small, greenish, in corymbs; calyx 4–5-lobed, with as many minute bractlets; petals none; stamens 1–4, very small; styles basal or lateral; carpels 1–4, distinct, free from the calyx-tube; achenes 1–4 enclosed in the hypanthium. (Name said to be from Arabic, *alkemelyeh,* the Arabic name, referring to the silky pubescence.)

Alchemilla occidentalis Nutt. ex T. & G. Annual, 3–10 cm. tall, finely appressed hirsute throughout; roots fibrous; stem ascending, simple to freely branched; stipules 2–5 mm. long, connate, lobed into lanceolate divisions; basal leaves with petioles 2–4 mm. long; blades 2–8 mm. long, flabellate, cuneate, deeply parted into ovate or lanceolate divisions; upper leaves 2–4 mm. long, sessile, cuneate, lobed or parted into elliptic or lanceolate lobes; flowers several in the axils; pedicels 1–2 mm. long; hypanthium 0.8–1 mm. long, urceolate, pilose; calyx-lobes 0.5–0.7 mm. long, bristle-tipped, elliptic to lanceolate, longer than the bractlets; stamen 1; the 1–2 achenes 0.9–1 mm. long, ovoid, acute, yellowish, smooth. *Aphanes occidentalis* (Nutt. ex T. & G.) Rydb.; *Alchemilla arvensis occidentalis* (Nutt. ex T. & G.) Piper. Stony soil, *U. Son., A. T.*

AMELANCHIER. Service Berry

Shrubs or small trees, not thorny; leaves alternate, simple, petioled, serrate or entire; flowers white, in racemes; hypanthium campanulate, more or less adherent to the ovary; sepals 5, persistent; petals 5; stamens numerous, on the calyx-tube; styles 2–5, united below or distinct; ovary 5-celled, each cell 2-ovuled; berry-like pome 10-celled, by the growth of a false partition in each cell of the ovary; each cell 1-ovuled. (Savoy name of the medlar.)

Young leaves, buds, and summit of ovary hairy. *A. alnifolia.*
Young leaves and buds glabrous, summit of ovary usually glabrous,
  Blades oval or oblong oval, 2.5–5.5 cm. long. *A. Cusickii.*
  Blades suborbicular, 1.5–3.5 cm. long. *A. basalticola.*

Amelanchier alnifolia Nutt. *Service Berry.* Shrub or small tree, 2–5 m. tall; bark reddish brown, later gray; blades 2–9 cm. long, broadly oval, rounded or subcordate at base, obtuse, coarsely serrate above the middle, tomentose when young, especially beneath, glabrate; petioles 6–25 mm. long, at first villous, glabrate; racemes 2–8 cm. long, 3–14-flowered; calyx-lobes lance-deltoid, more or less villous within; petals 6–10 mm. long, narrowly elliptic, short clawed; fruit 5–12 mm. in diameter, with blue bloom on blackish skin, the pulp sweet, juicy, edible, but variable in quality on different bushes and in different seasons; seeds 4–5 mm. long, asymmetric lanceolate, flattened, brown. *A. florida* sensu St. John, not of Lindl. Abundant, especially in valleys, draws, and springy places, *U. Son.* to *A. T. T.*

The tasty fruits of this and other species were formerly gathered by the Indians and eaten fresh or dried for future use. The herbage is palatable forage for stock.

Amelanchier basalticola Piper. Shrub; bark reddish brown, later gray; blades orbicular or oval-suborbicular, glabrous, serrate from near the base, pale green; petioles 7–20 mm. long, slender, glabrous; racemes 2–4 cm. long, 4–8-flowered; calyx-lobes lance-deltoid, at first pilosulous within, soon reflexed; petals 10–15 mm. long, oblanceolate; fruit said to be 8–9 mm. in diameter and dark purple. On basalt ledges, depths of Snake River Canyon from our S. border, down to Wawawai, *U. Son.*

Described by Prof. C. V. Piper, the type locality being, Bluffs of Snake River, Whitman County, Wash., near Lewiston, Ida., *Hunter.*

C. L. Hitchcock and Cronquist treat this as *A. alnifolia* Nutt., var. *pumila* (Nutt.) A. Nels.

Amelanchier Cusickii Fern. Shrub 2–6 m. tall; bark reddish brown, later gray; blades with the tip obtuse or acute, the base rounded or subcordate, the margin sharply serrate from near the base or from the middle, glabrous; petioles 6–22 mm. long, early glabrate; racemes 2–5 cm. long, 3–8-flowered; calyx-lobes deltoid, pilose within; petals 14–21 mm. long, linear-oblanceolate, obtuse; fruit 6–11 mm. in diameter, globose, edible, with a blue bloom on the blackish skin; seeds 4–5 mm. long, ovate, a little asymmetric, compressed, dark brown. *A. alnifolia* Nutt., var. *Cusickii* (Fern.) C. L. Hitchc. Abundant, stream banks, and moist places, as above outcropping ledges in the deep valleys, *U. Son.* to *A. T. T.*

Discovered first in Union Co., Ore., by the pioneer botanist Wm. C. Cusick, and named in his honor. Blooming about ten days earlier than *A. florida.*

CERCOCARPUS

Shrubs or small trees with alternate simple petioled leaves; flowers perfect, solitary or clustered, axillary or terminal, sessile or nearly so; calyx narrowly tubular, 5-lobed; petals none; stamens 15 or more, inserted at different levels on the calyx-tube; pistil 1, narrow, terete; fruit a coriaceous, linear, terete achene with a long exserted plumose twisted style. (Name from Gr., *kerkis,* shuttle; *karpos,* fruit.)

Blades merely tomentulose beneath, glabrate; petioles 3–5 mm. long. *C. ledifolius,* var. *ledifolius.*
Blades permanently white villous tomentose beneath; petioles 1–2 mm. long. Var. *intercedens.*

Cercocarpus ledifolius Nutt. ex T. & G., var. **ledifolius.** *Mountain Mahogany.* Scraggly shrub, or tree, 3–12 m. high; branchlets canescent, soon glabrous; leaves 1–3 cm. long, 3–10 mm. wide, coriaceous, elliptic, entire, acute, revolute, glabrate above; calyx-tube 3–5 mm. long, campanulate, villous, the lobes 2 mm. long, oblong, obtuse, reflexed; achenes about 1 cm. long, hirsute, the plumose style 5–7 cm. long. Rocky ridges in the Blue Mountains, *A. T.*

The foliage is of slight forage value on the winter range.

Var. **intercedens** Schneid. *White Mountain Mahogany.* Tree, 1–3 m. tall, with a compact, rounded crown; bark smooth, gray; twigs villous, glabrate; leaves 1–5.4 cm. long, 2–9 mm. wide, linear-lanceolate or linear-elliptic, coriaceous, revolute, upper surface at first loosely villous, later glabrate, resinous; flowers 1–3 in axils, sessile; calyx-tube 4–5 mm. long, cylindric, villous, the limb campanulate 2 mm. long, equaling the reflexed lobes; stamens attached ¼ of the distance down the throat; style exserted; achenes 6–8 mm. long, hirsute, the plumose style 5–6 cm. long. *C. hypoleucus* Rydb.; *C. ledifolius* Nutt., var.

hypoleucus (Rydb.) Peck. Extending northward into our limits, gravelly or rocky slopes, Snake River Canyon; Anaconda Cr., T. 7 N., R. 47 E., Asotin Co., *St. John* 4,422; Fisher's Cr., T. 7 N., R. 47 E., Asotin Co., *St. John* 8,175, and *Hardin & English* 930, *U. Son.*

The wood is dark and so hard that a branch can be cut with a knife only with great difficulty.

Recently McVaugh (1942) made this plant a minor, unnamed variant of *C. ledifolius;* Peck (1941) made it a variety *hypoleucus* of the same; and Abrams (1944) made it a synonym of the same species. The writer is unable to agree with any of these treatments.

CRATAEGUS. Thorn. Hawthorn

Shrubs or small trees, mostly thorny; leaves alternate, petioled, entire, serrate, lobed, or pinnatifid; flowers white, rarely rose-colored, in terminal corymbs; calyx-tube cup-shaped or campanulate, adherent to the carpels, the limb 5-lobed; petals 5, roundish, on the calyx-tube; stamens many, or only 5 or 10; styles 1–5, separate; ovary inferior, or its summit free; ovules one to each carpel; pome small, drupe-like, with 1–5 bony carpels, each 1-seeded. (Name from Gr., *kratos,* strength.)

Fruit red; thorns 4–6 cm. long,
- Calyx and fruit glabrous. — *C. columbiana,* var. *columbiana.*
- Calyx and fruit tomentose. — Var. *Piperi.*

Fruit black or chestnut; thorns 2–3 cm. long,
- Stamens 10 (or 8); calyx-lobes linear-lanceolate, glandular dentate,
  - Fruit black. — *C. Douglasii,* var. *Douglasii,* f. *Douglasii.*
  - Fruit chestnut. — Forma *badia.*
- Stamens 20; calyx-lobes short, broadly deltoid, entire. — Var. *Suksdorfii.*

Crataegus columbiana Howell, var. **columbiana** Shrub or tree, 2–5 m. tall; bark light brown, later gray; thorns nearly straight; young branchlets villous; blades 2–6.5 cm. long, thin, oval, or obovate, above pilosulous, later glabrate, the base cuneate, below the middle serrate, above the middle with 5–9 incised shallow lobes which are simply or doubly glandular serrate; petioles 6–35 mm. long, sparsely pilose glandular below the blade; corymbs 4–12-flowered, sparsely pilose; calyx-lobes 4–6 mm. long, lance-linear, glandular serrate, somewhat pilose within; petals 6–8 mm. long, orbicular; stamens 10; fruit 8–12 mm. in diameter, nearly spherical, coral red; nutlets 7 mm. long, lanceoloid, ridged on the back, yellow. Creek banks or springy spots in valleys, infrequent, *A. T.*

Var. **Piperi** (Britton) Eggleston. Differs from the species by having the twigs and lower leaf surface pilose, the calyx and fruit tomentose. Intergrades with the species. *C. piperi* Britton. Springy places, occasional, *A. T.*

Crataegus Douglasii Lindl., var. **Douglasii,** forma **Douglasii.** *Douglas Thorn; Black Hawthorn.* Erect compact shrub, occasionally a tree up to 12 m. tall and 5 dm. in diameter; bark of branchlets smooth, reddish brown, later gray; lenticels elliptic, white; thorns stout, nearly straight; blades 3–8 cm. long, obovate or ovate, acute or obtuse, cuneate, sparsely pilosulous, later glabrate, shiny and dark green above, paler beneath, above the middle sharply serrate, usuallv with shallow lobes also; petioles 5–20 mm. long, winged and ciliate above by the decurrent blade; corymbs pilose or glabrous; calyx-lobes 2–3 mm. long, somewhat pilose within; petals 5–7 mm. broad, orbicular; fruit 9–12 mm. long, oval to subglobose, truncate, dark purplish to black; flesh yellow, sweet; nutlets 5–6 mm. long, yellowish, shaped like the sector of an orange, furrowed on the back, *C. brevispina* (Dougl.) Heller, in ed. 1. Abundant, moist places, *U. Son.* to *A. T. T.*, especially common, forming thickets by streams in *A. T.*

The wood is hard and fine grained, the heartwood brownish rose-red, the sapwood large in proportion. Not used commercially. Named for its collector, the intrepid David Douglas, who collected throughout our area in 1825–27. The foliage is fair forage for stock.

Forma **badia** Sarg. This color form differs from the species only in having chestnut-colored fruit. Occasional, *A. T.*

First collected by Prof. C. V. Piper, the type locality being Union Flat.

Var. **Suksdorfii** Sarg. Differing only in having 20 stamens; and the sepal-lobes short, broadly deltoid, entire. Thicket, head of Hatwai Cr., Whitman Co., *St. John et al.* 9,526, *A.T.*

Crataegus ———. An abundant collection, amply different from our other species. It is in the affinity of *C. macracantha* (Lindl.) Lodd., and *C. succulenta* Schrad., but seems not to be either. Mature fruit is needed. Compact tree 5 m. tall; mature branches all upturned; branchlets glabrous, brown; older stems gray; thorns 6–7 cm. long, stout, slightly curved; blades 3.5–8 cm. long, suborbicular, acute, doubly dentate, beneath pale and pilose on the veins; petioles 3–15 mm. long, pilose above; pedicels and fruit villous; stamens 10; young fruit, globose; calyx-lobes 4–6 mm. long, linear-lanceolate, villous, glandular pectinate-dentate. Near creek, Clearwater River, 2 miles e. of Fir Bluff, Nez Perce Co., *St. John et al.*, 9,529, *U. Son.*

FRAGARIA. Strawberry

Acaulescent perennial herbs, propagating by runners; leaves alternate, basal, tufted, 3-foliolate; flowers polygamo-dioecious, few, in corymbs or racemes, on naked scapes; calyx deeply 5-lobed, with 5 alternate bractlets; petals 5, short-clawed; stamens about 20, in 3 series; style lateral; carpels numerous; receptacle much enlarged and fleshy in fruit, conical scarlet, bearing the small turgid achenes on the surface. (Name from the Lat., *fraga,* strawberries.)

Scapes shorter than the glaucous firm leaves; sepals and bractlets lanceolate. — *F. cuneifolia.*

Scapes longer than the membranous leaves,

- Flowers white, erect, sepals and bractlets almost equal, linear-lanceolate. — *F. bracteata.*
- Flowers rose-colored, on nodding pedicels; sepals lanceolate, bractlets linear. — *F. Helleri.*

Fragaria bracetata Heller. *Wood Strawberry.* Rootstocks stout; runners slender, few; scapes 10–35 cm. high, usually exceeding the leaves, often with a large unifoliate bract; whole plant sparsely silky-villous; petioles 3–24 cm. long, with spreading pilose silky hairs; leaflets 2–9 cm. long, broadly obovate, cuneate at base, coarsely dentate; lateral leaflets oblique; calyx 5–9 mm. long, the lobes ovate-lanceolate, spreading; bractlets oblanceolate; petals 7–11 mm. long, rounded obovate; fruit about 10 mm. long, ovoid or semi-ellipsoid, red, edible, pleasant; achenes 1 mm. long, ovate, narrowed to the tip. *F. crinita* Rydb.; *F. vesca* L., var. *bracteata* (Heller) Davis. In open woods, common, *A. T. T.*

Fragaria cuneifolia. Nutt. ex Howell. *Pale Strawberry*. Rootstocks stout; runners stout; scapes 5–20 cm. high, mostly shorter than the leaves, silky villous; petioles 2–20 cm. long, silky-villous; leaflets 2–8 cm. long, elliptic or obovate, broadly cuneate at base, coarsely toothed above the middle, glabrous or nearly so above, appressed silky-villous below; calyx 6–9 mm. long, villous; petals 7–10 mm. long, nearly orbicular, white; fruit 1–1.5 cm. long, sweet, edible, hemispherical, the achenes 1.5 mm. long, plump, ovoid, sunk in shallow pits. *F. platypetala* Rydb.; *F. Suksdorfii* Rydb. Moist, or grassy places, or open woods, abundant, *A. T., A. T. T.*

Fragaria Helleri Holz. Leaves thin, glabrate above, finely silky beneath, petioles 5–10 cm. long, silky to glabrate; leaflets rounded-obovate, coarsely serrate; runners slender; scape 1.5–2 dm. tall, silky to glabrate, often bracted; flowers 1.5–2 cm. wide; petals suborbicular, almost twice as long as the sepals.

The type collection, *Sandberg et al.* 508, was from Pine Creek, Farmington Landing (Kootenai Co.). No additional collections have been made, nor has the type been available for study. The description here used is compiled. It is quite likely that the plant is only a sporadic color form of *F. bracteata*, but the writer is not in a position to express a definite opinion.

GEUM. Avens

Perennial herbs; leaves pinnate, with stipules, radical clustered; cauline smaller; flowers rather large, solitary or in corymbs; calyx 5-lobed with 5 alternate bractlets; petals 5, obovate; stamens many, on the throat of the calyx; styles terminal, elongated in fruit, usually plumose or jointed, the upper part often deciduous; achenes 2–6, pubescent. (An ancient Lat. plant name.)

Calyx-lobes erect, reddish; leaves 2–3-pinnatifid. *G. triflorum*, var. *ciliatum*.
Calyx-lobes reflexed, green; leaves lyrate,
  Terminal leaflet cuneate; lower style segment glandless. *G. strictum*.
  Terminal leaflet cordate at base; lower style segment minutely capitate glandular. *G. macrophyllum*.

Geum macrophyllum Willd. *Large-leaved Avens*. Stout, erect, 1–10 dm. high, stems and leaves hirsute; basal leaves 10–35 cm. long, petioled, pinnate, the terminal leaflet 5–13 cm. long, broadly ovate or suborbicular, 3–7-lobed, dentate, lateral leaflets 3–8, ovate or elliptic, dentate, 2–3 cm. long, usually with smaller ones interspersed; cauline leaves similar, but nearly sessile, with 1–3 leaflets or segments; stipules 1–2 cm. long, ovate, toothed; flowers several in a loose cyme; bractlets 1–3 mm. long, linear; calyx 4–6 mm. long, the lobes deltoid, lanate within; petals 4–7 mm. long, obovate or obcordate, yellow; fruiting head globose, 1.5 cm. long; styles jointed and twisted near the tip, the deciduous tip short hispidulous at base; achenes 3–4 mm. long, narrowly oblanceolate, more or less hirsute. *G. oregonense* (Scheutz) Rydb.; *G. perincisum* Rydb.; *G. macrophyllum*, var. *perincisum* (Rydb.) Raup. In moist places in open woods, *A. T. T.*

Geum strictum Ait. *Yellow Avens*. Stems 5–15 dm. high, hirsute; basal leaves pinnate, long petioled; cauline 3–5-foliolate; rhachis hirsute; leaflets cuneate-obovate, incisely acutely lobed and toothed, hirsutulous to glabrate, terminal leaflet 6–10 cm. long; bractlets linear, ⅓ to ½ the length of the sepals; calyx 5–8 mm. long, the lobes ovate-lanceolate, acuminate, the inner margins lanate; petals 5–8 mm. long, suborbicular, yellow; receptacle densely hispid; styles glandless, the upper deciduous segment hispid; achenes 3–4 mm. long, unevenly lanceolate, hirsute. Tukanon R., *Lake & Hull* 516; Rock Lake, *Beattie & Lawrence* 2,413. Moist meadows, rare, *A. T. T.*

A collection from Pullman, *Piper,* July 9, 1901, has the heads of *G. strictum,* but the leaves of *G. macrophyllum.* It is perhaps a hybrid.

Geum triflorum Pursh, var. **ciliatum** (Pursh) Fassett. *Tassels.* From stout rootstocks, erect, 1–6.5 dm. high, pilose and silky hirsute throughout; radical leaves tufted, oblanceolate in outline, 1–3 dm. long, pinnate with numerous leaflets, these cuneate, obovate, incisely cleft into acute narrow lobes; cauline leaves 1 or 2, much reduced; stipules 1–3 cm. long, semiorbicular, incised; flowers long-peduncled, 3–7 in a cyme; calyx reddish, the lobes 8–11 mm. long, ovate, acuminate, the linear bractlets exceeding the calyx-lobes; petals oblong, erect, yellowish or pinkish tinged equaling the calyx; styles in fruit 2–6 cm. long, the upper segment 4–6 mm. long, slender, glabrous above, pilosulous below, usually jointed and deciduous, the remainder plumose; achene 3 mm. long, oblanceolate, pilose. *Sieversia ciliata* (Pursh) G. Don; *G. ciliatum* Pursh. Common on grassy hillsides, *A. T., A. T. T.*

First collected by Capt. Meriwether Lewis near Weippe, Ida. The herbage is palatable forage for sheep.

HOLODISCUS

Thornless shrubs; leaves alternate, simply pinnately toothed or lobed, without stipules; flowers numerous, white, in terminal panicles; calyx deeply 5-lobed; petals 5, as long as the calyx-lobes, rounded; stamens 20, on a perigynous entire ring-like adherent disk, scarcely exserted; pistils 5, with 2 ovules, becoming 1-seeded hairy carpels indehiscent or tardily dehiscent. (Name from Gr., *holos,* whole; *diskos,* disk or quoit.)

Holodiscus discolor (Pursh) Maxim., var. **discolor.** *Ocean Spray.* Shrub 1–7 m. high; herbage villous; branchlets brown; older branches smooth, gray to reddish, exfoliating; blades 2–8 cm. long, triangular-ovate, mostly obtuse, truncate or cuneate at base, glabrous above, coarsely doubly dentate or shallowly pinnately lobed, pale beneath; petioles 3–25 mm. long; panicle 1–3 dm. long, villous, broadly pyramidal, much branched, calyx pilosulous, the tube 1.5 mm. wide, saucer-shaped, the lobes 1.2–2 mm. long, ovate-lanceolate; petals 1.5–2.5 mm. long, elliptic, white; mature carpels 3 mm. long, long beaked, the body semi-obovate, hirsute. *Sericotheca discolor* (Pursh) Rydb. Thickets or woods, *A. T. T., Can.*

First collected by Capt. Meriwether Lewis on "the banks of the Kooskoosky" [Clearwater River, Ida.]. The foliage is palatable for stock.

HORKELIA

Perennials with scaly rootstocks or short caudices; leaves pinnate; inflorescence cymose-paniculate; hypanthium campanulate to saucer-shaped; calyx-lobes 5; bractlets 5; petals 5, white to yellowish; stamens 10, on the throat of the hypanthium; filaments dilated, more or less petaloid, persistent, with a distinct midrib; receptacle conic or hemispheric; pistils numerous; styles slender, jointed to the achene, deciduous. (Named for *Prof. John Horkel* of Berlin.)

Horkelia fusca Lindl., var. **capitata** (Lindl.) Peck. Tap-root dark, stout, woody, vertical; caudex short, stout, with a mass of marcescent leaf bases; stems several 30–42 cm. tall, viscid hirsutulous; basal leaves 1–3 dm. long, numerous, stipules viscid hirsutulous, the base 1–2 cm. long, oblong, adnate, the free tips about 1 cm. long, divergent, lance-linear, long caudate; petioles 5–15 cm. long, viscid hirsutulous, as is the rhachis; leaflets 9–13, 8–36 mm. long, 7–30 mm. wide, obovate to suborbicular, the base rounded on the lower half, somewhat truncate on the upper half, coarsely crenate dentate, beneath sparingly glandular, sparingly pilosulous on the veins; cauline leaves 3–4, remote, 6–12 cm. long; the 5–7 leaflets oblanceolate to cuneate, reduced, but more sharply dentate or even laciniate at tip; cymes 1–several, 1.5–3 cm. wide, capitate, equaled or exceeded by the leafy, laciniate sparsely villous subtending bract; calyx rose-purplish, campanulate, viscid hirsutulous; hypanthium 2–2.5 mm. long; calyx-lobes 3–5 mm. long lanceolate, ciliate below; bractlets from ⅔ as long to equaling the calyx-lobes, linear, strongly villous ciliate; petals 6–7 mm. long, 2.5–3 mm. wide, broadly cuneate, retuse, white; outer filaments 1 mm. long, opposite the calyx-lobes, white, thick, broadly deltoid; inner filaments 0.7 mm. long, linear lanceolate, attached at a lower level; anthers cordate; 0.4 mm. long; ovaries semi-obcordate, the margin pale, cartilaginous, the sides darker, membranous. *H. caeruleimontana* St. John; *H. fusca* Lindl., subsp. *capitata* (Lindl.) Keck.
Thickets in Blue Mts., *Can.*

IVESIA

Low perennial herbs; leaves pinnate with numerous small palmately lobed crowded leaflets; flowers yellow, in cymes; calyx bell-shaped to turbinate or saucer-shaped; 5-lobed, with 5 alternate bractlets; stamens 5–20, on the throat of the hypanthium; filaments slender, subulate or filiform; carpels 1–15 on a small villous receptacle; styles filiform, subterminal. (Named for *Lieut. Eli Ives,* American, leader of one of the Pacific R. R. Surveys.)

Ivesia Gordonii (Hook.) T. & G. Root thick, woody; caudex stout, covered with dark marcescent leaf bases; stem 5–20 cm. tall, almost scapose, glandular-puberulent; basal leaves 3–10 cm. long, petiole glandular puberulent; rhachis also pilosulous; leaflets 10–20 pairs, approximate, each 2–5 mm. long, finely pilosulous, divided into 3–5 narrow spatulate, imbricate segments; cauline leaves none or small, pinnatifid; cyme dense; hypanthium 2–4 mm. long, campanulate, glandular puberulent and pilose; calyx-lobes 2.5–4 mm. long, lance-deltoid; petals 3–4 mm. long, spatulate, yellowish; stamens 5, opposite the sepals. *Horkelia Gordonii* Hook. On high ridges in the Blue Mountains, rare, *Huds.*

PHYSOCARPUS. Ninebark

Branching shrubs; leaves alternate, petioled, simple, palmately-lobed; stipules deciduous; flowers in umbel-like corymbs; calyx 5-lobed; petals 5, round, on the calyx-tube; stamens many, distinct, on the calyx-tube; pistils 1–5, more or less united; follicles

1–5, 2-valved; seeds 1–4; endosperm copious. (Name from Gr., *phusa,* bellows; *karpos,* fruit.)

Carpels 3–5, turgid, glabrous at maturity. *P. capitatus.*
Carpels 2, flattened, stellate canescent. *P. malvaceus.*

Physocarpus capitatus (Pursh) Kuntze. *Tall Ninebark.* Shrub 1–4 m. tall; bark brown, exfoliating in long strips; stipules lanceolate; petioles 1–3 cm. long; blades of flowering branches 3–7 cm. long, rounded ovate, 3–5-lobed, doubly serrate, truncate to cordate at base, sparsely stellate beneath; blades of sterile shoots longer, more lobed, corymb many flowered; pedicels 1–2 cm. long, stellate canescent; calyx 4–6 mm. long, stellate canescent; petals 3–4.5 mm. long, ovate to suborbicular, white; carpels 8–10 mm. long, ovate, acuminate; seeds 2 mm. long, narrowly oblique-pyriform, white. Blue Mts., Walla Walla Co., *Piper,* July 1896.

Physocarpus malvaceus (Greene) Kuntze. *Mallow Ninebark.* Erect shrubs, 1–4 m. high; branchlets stellate; branches spreading or recurved, the bark shreddy; blades of flowering branches 2–8 cm. long, broadly ovate or orbicular, bluntly 3–5-lobed, somewhat doubly dentate, rounded or cordate at base, nearly glabrous above, pubescent beneath with stellate hairs; blades of sterile shoots larger; petioles 1–2 cm. long; inflorescence a corymb, stellate canescent throughout; calyx 4–6 mm. long, broadly campanulate, the blunt lobes as long as the tube; petals 3.5–5 mm. long, suborbicular, white; carpels 6–7 mm. long, ovate, beaked, somewhat united at the base; seeds 2 mm. long, pyriform, white. Open or wooded hillsides, a common and beautiful shrub, *U. Son.* to *A. T. T.*

One of the chief forage plants for sheep.

POTENTILLA. Cinquefoil

Annual or perennial herbs or shrubs; leaves alternate, pinnate or palmate, with stipules; flowers perfect, solitary or in cymes; sepals 5, rarely 4, with 5, rarely 4, alternate bractlets; petals 5, seldom 4, rounded, mostly yellow (or red or white); stamens many; style small, terminal or nearly so, not elongating in fruit, neither jointed nor plumose; carpels usually numerous; ovules pendulous, anatropous; receptacles and achenes dry in fruit. (Name from Lat., *potens,* powerful; *illa,* little.)

Petals purplish or maroon; receptacle in fruit enlarged and spongy. *P. palustris.*
Petals yellow or white; receptacle not spongy in fruit,
  Plant stoloniferous; flowers solitary. *P. Anserina,* f. *sericea.*
  Plant not stoloniferous; flowers normally in cymes,
    Styles lateral; ovules ascending; leaves pinnate,
      Stems and buds shaggy viscid villous; principal leaflets suborbicular or rounded at base. *P. Convallaria.*
      Stems loosely viscid villous or pilose; buds pilosulous; principal leaflets cuneate. *P. glandulosa.*
    Styles terminal; ovules pendulous; leaves digitate,
      Cymes leafy,
        Styles fusiform and glandular at base; annuals or biennials,

Stamens 10,
Plant viscid villous; leaflets broadly cuneate obovate. *P. biennis.*
Plant pilose; leaflets cuneate-oblong. *P. millegrana.*
Stamens 15–20,
Floral bractlets oval, obtuse; achenes smooth. *P. leurocarpa.*
Floral bractlets elliptic to lanceolate, usually acute; achenes ridged. *P. norvegica,* var. *hirsuta.*
Styles filiform, not glandular; perennials. *P. argentea.*
Cymes merely bracteate,
Leaflets white on both sides with dense pubescence. *P. permollis.*
Leaflets green, at least on upper surface,
Leaflets deeply cleft into linear, acute lobes,
Leaflets cleft ½ way to midrib. *P. Blaschkeana.*
Leaflets cleft almost to the midrib. *P. flabelliformis.*
Leaflets merely toothed,
Calyx and leaves glandular atomiferous, *P. Nuttallii.*
Calyx and leaves not glandular atomiferous. *P. rectiformis.*

Potentilla Anserina L., forma **sericea** (Hayne) Fern. *Silver-weed.* Villous herb, tufted and spreading by slender runners; leaves all basal, 1–3 dm. long, pinnate; principal leaflets 7–25, white silky on both sides, 10–45 mm. long, obovate to oblanceolate, serrate, with alternating, reduced leaflets interspersed; peduncles 2–7 cm. long, loosely silky pilose; calyx 6–9 mm. long, silky villous, the ovate-lanceolate calyx-lobes and bractlets about ½ as long; petals 6–9 mm. long, ovate, bright yellow; achenes 2–3 mm. long, obliquely ovoid, thickened and deeply furrowed on the back, brown. *Argentina anserina sericea* (Hayne) Piper; *A. argentea* Rydb. In moist spots especially where alkaline, mostly in the "Scablands," *U. Son.*

Potentilla argentea L. *Silvery Cinquefoil.* Stems 1–5 dm. long, ascending or depressed, grayish tomentose or glabrate; blades digitately 3–5-divided; the lobes 8–25 mm. long, oblanceolate, obtuse, cuneate, deeply incised, revolute, above soon glabrate, dark green firm, below white tomentose; petioles tomentose, 2 cm. or less in length; stipules foliaceous, lanceolate, often lobed; cymes leafy; hypanthium tomentose; floral bractlets elliptic-oblong, nearly as long as the calyx-lobes, calyx-lobes 3–4 mm. long, ovate; petals yellow, oval, barely exceeding the calyx-lobes; stamens about 20; achenes 0.7 mm. long, bean-shaped, yellow. A weed, introduced from Europe. On roadside, Spokane, in 1927, *G. N. Jones* 651, *A. T.*

Potentilla biennis Greene. Stems 1 or more, 1–6 dm. tall, little branched, leafy to tip; stipules 3–12 mm. long, lanceolate to ovate, entire or dentate, viscid villous; leaves all 3-foliolate, pilose and capitate glandular, lower petioles 2–12 cm. long, the upper reduced; leaflets 1–4 cm. long; inflorescence somewhat elongate; pedicels 5–20 mm. long; hypanthium 3–4 mm. broad, saucer-shaped, viscid villous; calyx-lobes 2–4 mm. long, deltoid to ovate-oblong, pilose; bractlets elliptic, a little shorter; petals obovate, yellow, shorter than the sepals; achenes 0.6 mm. long, oval, truncate on one side, smooth, white. Pond shores or stream banks, *U. Son.*

Potentilla Blaschkeana Turcz. Perennial; stems 3–10 dm. tall, erect, mostly appressed pilose; lower stipules 1–2 cm. long, ovate-lanceolate, the upper ones longer, toothed; basal petioles 6–25 cm. long, appressed pilose; leaflets about 7, 2–12 cm. long, obovate to oblanceolate, densely white-tomentose beneath; cauline leaves smaller, shorter petioled; hypanthium 5–10 mm. wide, cup-shaped, pilose; calyx-lobes 5–8 mm. long, ovate-lanceolate, acuminate; bractlets oblong-lanceolate, shorter; petals 7–10 mm. long, broadly obovate, with a large shallow notch; achenes 1.1–1.3 mm. long, semicordate, olive-brown, smooth. Prairies and grassy woods, abundant, *A. T., A. T. T.*

Potentilla Convallaria Rydb. Perennial from a thick, woody rootstock; stem 3–10 dm. tall, viscid villous; basal leaves abundant, 1–3 dm. long, 7–11-foliolate; leaflets rhombic-obovate to suborbicular, the larger 2–6 cm. long, viscid villous, the margin coarsely serrate or incised, often 1–2 pairs of reduced leaflets also present; petioles 5–10 cm. long, viscid villous; stipules adnate, the free tips lanceolate; upper leaves much reduced and nearly sessile; stipules foliaceous, more or less lobed; cyme narrow, closely viscid villous; floral bractlets 3–5 mm. long, linear; calyx-lobes 5–8 mm. long, ovate-lanceolate, densely viscid villous; petals 5–8 mm. long, oval to orbicular, yellow; stamens 25; achenes 0.8–1.2 mm. long, lanceoloid, somewhat flattened, smooth, straw-colored. *Drymocallis convallaria* Rydb.; *P. arguta* Pursh, var. *Convallaria* (Rydb.) Th. Wolf. Moist places, *U. Son.* to *A. T. T.*, abundant in "bunch-grass prairies" of *A. T.*

Potentilla flabelliformis Lehm. Perennial, 4–6 dm. tall, appressed white villous, from a woody caudex; lower stipules 13–15 mm. long, lanceolate, the upper ones linear-lanceolate; basal leaves several, the petioles 12–20 cm. long; leaflets 7, palmate, the upper ones 4–5.5 cm. long, oblanceolate, above thinly hairy, but green, below matted white hairy; cauline leaves few, smaller, sessile; hypanthium 7–8 mm. wide, semiglobose; calyx lobes 4–5 mm. long, lanceolate, the bractlets 2.5–3 mm. long, similar; petals 7–10 mm. long, suborbicular; achenes 1.5 mm. long, nearly ovoid. Prairies near Pullman, *A. T.*

Potentilla glandulosa Lindl. Perennial; stems 2–8 dm. tall, viscid or glandular villous, especially upward; basal leaves pinnate, 5–9-foliolate, 1–3 dm. long; leaflets 5–55 mm. long, thin, obovate to rhombic, serrate or doubly serrate, sparsely viscid villous; petioles 1–15 cm. long, viscid villous; lower stipules adnate, viscid villous, the tips lanceolate; upper leaves reduced and shorter petioled; upper stipules free, foliaceous, lance-ovate, often toothed; cyme open; floral bractlets 4–5 mm. long, linear-lanceolate; calyx-lobes 6–10 mm. long, loosely pilose, ovate, acute; petals 5–10 mm. long, broadly obovate to suborbicular, yellow; stamens 20–25; achenes 1–1.3 mm. long, ovoid, asymmetric, flattened, brown, light veined. *Drymocallis glandulosa* (Lindl.) Rydb. Common, grasslands and open woods, *A. T., A. T. T.*

Abundant in the Blue Mts., and sought by the sheep as forage.

Potentilla leurocarpa Rydb. Stem 3–6 dm. tall, often red tinged, softly hirsute; leaves ternate, or some of the basal appearing pinnate by the deep parting of the terminal leaflet; leaflets 1.5–4 cm. long, broadly obovate asymmetric, or rounded, sparingly pilose, coarsely toothed with ovate, obtuse teeth; upper leaves nearly sessile; petioles of lower leaves 3–6 cm. long, hirsute; stipules partly adnate, the tips ovate to lanceolate, acuminate, toothed or entire; inflorescence cymose, rather dense; floral bractlets 3–7 mm. long; calyx-lobes ovate, hirsute, slightly longer; petals 2 mm. long, yellowish, obcordate; stamens 15–20; styles fusiform; achenes 0.7–0.8 mm. long, ovoid, compressed, pale brown. Sandy banks of Snake River, Almota, *Piper 2,734*, *U. Son.*

This collection differs slightly from Dr. Rydberg's description in having some basal leaves falsely pinnate (by parting of the terminal leaflet), and in having somewhat smaller flowers and narrower petals, yet it seems best to retain it within his species which he described from a single collection from Bingen.

Potentilla millegrana Engelm. *Diffuse Cinquefoil.* Plant branched from the base; stems 1–10 dm. long spreading; stipules 3–10 mm. long, ovate to lanceolate, mostly entire; leaves 3-foliolate, light green; lower petioles 1–4 cm.

long, the upper reduced; leaflets 5–35 mm. long, coarsely few dentate; pedicels 5–30 mm. long; hypanthium 3–5 mm. broad, saucer-shaped; calyx-lobes 2–4 mm. long, deltoid-ovate; bractlets nearly as long, lance-elliptic, spreading; petals half as long as the sepals, oblong-ovate, pale yellow; achenes 0.8 mm. long, asymmetric ovate, straw-colored, smooth. *P. rivalis* Nutt., var. *millegrana* (Engelm.) Wats. Dry bed of alkaline pond, near Rock Lake, *St. John & Warren* 6,780, *U. Son.*

Potentilla norvegica L., var. **hirsuta** (Michx.) Lehm. *Rough Cinquefoil.* Plant more or less hirsute throughout; stems one or more, 2–8 dm. tall, often branching above, stout, leafy, erect; stipules 1–3 cm. long, ovate, acute, usually toothed; lower petioles 3–10 cm. long, the upper shorter to subsessile; leaves 3-foliolate, or the lowest sometimes 5-foliolate, green; leaflets 3–10 cm. long, obovate to oblanceolate, serrate; cyme rather dense, leafy; pedicels 4–20 mm. long; hypanthium 6–8 mm. broad, saucer-shaped; calyx-lobes 3–6 mm. long, deltoid to ovate-lanceolate; bractlets 3–5 mm. long; petals pale yellow, obovate, nearly as long as the calyx-lobes; stamens usually 20; achenes 0.8 mm. long, ovate, asymmetric, brownish, the faces sinuous ridged. In meadows, Marshall, Jct., *Piper* 2,256, *A. T.*

Potentilla Nuttallii Lehm. Plant with a short rootstock; stem 6–8 dm high, branched above, sparingly hirsute; basal petioles 1–3 dm. long, appressed hirsute; basal leaves digitate, usually 7-foliolate; leaflets 5–10 cm. long, green, lance-toothed nearly half way to the midrib, prominently veined and sparingly hirsute, glandular granuliferous beneath; cauline leaves smaller and short-petioled; cymes many-flowered; hypanthium 7–10 mm. wide, cup-shaped, like the calyx sparingly hirsute, glandular-granuliferous; calyx-lobes 5–8 mm. long, ovate-lanceolate, long-acuminate; bractlets lanceolate, a little shorter; petals 6–8 mm. long, yellow, obovate, emarginate. *P. gracilis* Dougl. ex Hook., subsp. *Nuttallii* (Lehm.) Keck. Around Lake Waha, *Heller* 3,326; Anatone, *Gessell,* May 15, 1926, *A. T.*

Potentilla palustris (L.) Scop. *Marsh Cinquefoil.* Rootstock cord-like, long creeping; stems 2–6 dm. tall, ascending, glandular pilose above, glabrous below, decumbent at base; leaves pinnate, the lower long petioled, 5–7-foliolate; leaflets oblanceolate, sharply serrate, firm, above dark green, sparsely appressed pilosulous or more commonly glabrous, below glaucous, appressed pilosulous or sericeous; terminal leaflet 2–10 cm. long; petioles 1–15 cm. long, appressed pilosulous; lower stipules membranous, brown, wholly adnate to the petioles; upper stipules foliaceous, more or less pubescent, with free, broadly lanceolate tips; cymes few- to many-flowered (or 1-flowered), loose, glandular pilose; bracts foliaceous; floral bractlets 5–11 mm. long, linear-lanceolate, finally reflexed; calyx-lobes 11–18 mm. long, ovate-lanceolate, acuminate, somewhat pilose, partly maroon; petals 4–5 mm. long, oblanceolate; stamens 20–25, purple, inserted on the large, villous disk; styles lateral; receptacle hemispheric, villous; achenes 1–1.5 mm. long, ovoid, yellowish, shining. *Comarum palustre* L. In swamps, and wet lake or stream banks, from Marshall Jct. and Lake Chatcolet northward, *A. T. T.*

The localities for this plant are all in the glaciated region, north of the terminal morraine.

Potentilla permollis Rydb. Very similar to *P. Nuttallii* but densely hirsute-pubescent throughout with nearly white soft hairs; inflorescence rather dense; teeth of the leaflets rather long, lanceolate. In moist meadows at Endicott, *Elmer,* the type locality.

Potentilla rectiformis Rydb. Whole plant sparsely hirsute; stems 3–10 dm. tall, tufted, erect, not branched below the inflorescence, basal stipules 2–3 cm. long, lanceolate, adnate except the short, divergent tip, the cauline one foliaceous, lacerate dentate; basal petioles 2–3.5 dm. long, appressed hirsutulous; leaves palmately 5–7-foliolate, leaflets 4–8 cm. long, oblanceolate, coarsely and evenly toothed, green on both sides; cymes many-flowered, loose; hypanthium 5–10 mm. broad, saucer-shaped; calyx-lobes 5–8 mm. long, ovate-lanceolate; bractlets linear-lanceolate, nearly as long; petals 4–7 mm. long, broadly obovate, notched; seeds 1.2–1.4 mm. long, ovate, olive-brown, minutely papillose. Low meadows, Pullman, *A. T.*

The type collection, *Elmer* 69, came from Pullman.

PRUNUS. Plum. Cherry

Small trees or shrubs, many with edible fruits; leaves alternate, simple, usually serrulate; flowers perfect, white or rose-colored, solitary or fascicled in the axils or in terminal racemes or corymbs; calyx 5-lobed, free from the ovary; petals 5, on the receptacle-cup; stamens 15–30; pistil 1; style 1; ovary 1-celled, 2-ovuled; fruit a drupe; seeds 1, rarely 2; endosperm none. (The ancient Lat. name, *prunus,* plum tree.)

Flowers racemose, numerous. — *P. virginiana,* var. *melanocarpa.*
Flowers in 3–10-flowered corymbs,
  Terminal bud wanting; branches thorny; drupes blue. — *P. spinosa.*
  Not so,
    Blades subacuminate, the margins salient doubly serrate; calyx tube 5–6 mm. long. — *P. Cerasus.*
    Blades obtuse, the margins appressed serrate; calyx tube 2.5–3 mm. long,
      Leaves glabrous beneath. — *P. emarginata,* var. *emarginata.*
      Leaves pilose beneath. — Var. *mollis.*

Prunus Cerasus L. *Sour Cherry.* Tree, usually with broadly rounded crown, height up to 10 m.; blades 7–10 cm. long, ovate-elliptic to obovate, glabrous beneath, serrate or doubly serrate, the teeth usually wider than high, rounded, bearing a gland near the sinus; flowers several in a sessile umbel, on pedicels 2–3 cm. long; inner scales of overwintering buds erect at anthesis; calyx tube not constricted above, the lobes serrate; sepals glabrous; petals 10–15 mm. long; fruit 10–20 mm. in diameter, globose, red and sour; stone nearly globose. The cultivated cherry, from Eurasia, escaped and established, *U. Son., A. T., A. T. T.*

Prunus emarginata. (Dougl ex. Hook.) A. Eaton, var. **emarginata.** *Wild Cherry.* Shrub or small tree 3–8 m. high; young shoots puberulent, during first winter smooth, reddish; older branches and bark chestnut brown, with prominent roundish lenticels; blades 2–7.5 cm. long, narrow, ovate, elliptical or oblanceolate, usually obtuse, cuneate at base; minutely glandular serrate; glabrous beneath, petioles 3–10 mm. long; corymbs appearing with the leaves; calyx-tube 2.5–3 mm. long, obconic; calyx-lobes 2–2.5 mm. long, oblong, obtuse, reflexed; petals 4–5 mm. long, obovate or broader, white; fruit 8–10 mm. long, ellipsoid, dark red to blackish, bitter; stone with a grooved ridge on one side. In dry open places in the mountains, *A. T., A. T. T.*

The herbage is eaten by sheep.

Var. **mollis** (Dougl. ex Hook.) Brewer. *Hairy Wild Cherry.* Tree up to 20 m. tall; blades 3–12 cm. long; fruit bright red. Thickets, Blue Mts., and Lakes Coeur d'Alene and Chatcolet, *A. T. T.*

Prunus spinosa L. *Sloe.* Shrub or tree to 5 m. tall, producing suckers, very thorny; winter buds ovoid, puberulent; young shoots appressed pilose; petioles

2–10 mm. long, sparsely pilose; blades 2–6 cm. long, oblanceolate or rarely lance-ovate, obtuse, serrate or doubly so, sparsely pilose or glabrate; flowers 1–2 at a node; pedicels 3–8 mm. long, glabrous; calyx lobes 1.5–3 mm. long, deltoid-ovate; petals 5–8 mm. long, elliptic, white; drupe 10–15 mm. long, ellipsoid to subglobose, bluish-black, persistent, the flesh green, sour; stone 7.5–10 mm. long, compressed barrel-shaped, smooth. European fruit tree or ornamental, in cultivation, and established, *A. T.*

Prunus virginiana L., var. **melanocarpa** (A. Nels.) Sarg. *Chokecherry.* Shrub, or rarely a tree up to 10 m. tall; branchlets green, glabrous; older twigs reddish brown to grayish; petioles 1–2 cm. long; blades 3–14 cm. long, firm chartaceous, glabrous, obovate to oval, the apex abruptly acuminate, the base usually rounded, the margin sharply serrulate; racemes on young leafy shoots from the old wood; racemes 3–8 cm. long, usually glabrous; pedicels 2–9 mm. long; petals 3–5 mm. in diameter, suborbicular, white; drupes 6–10 mm. in diameter, globose, black, slightly astringent; stone 5–8 mm. wide, thick lenticular, pale. Common, especially by streams, *U. Son., A. T., A. T. T.*

The foliage is eaten with relish by sheep. Yet the species proper, of eastern U. S., is poisonous, as the leaves when wilted produce prussic acid.

PURSHIA

Shrubs; leaves alternate simple, entire or toothed; flowers axillary, solitary or in a fascicle; hypanthium with a cylindric, persistent tube, widening into a turbinate, deciduous limb; calyx-lobes 5; corolla none; stamens 18–25, attached in 1 row; pistil 1; style elongate, plumose, persistent; achene coriaceous, pubescent; seed basal. (Named for *Frederick Pursh* early botanical explorer in eastern N. Am.)

Purshia tridentata (Pursh) DC. *Antelope Brush, Black Sage.* Shrub, 1–3 m. tall, with arching branches, more or less tomentose throughout; branchlets brownish, but densely white tomentose; branches brown to silvery gray; leaves mostly and flowers wholly from lateral short shoots; leaves 5–30 mm. long, cuneate, 3-lobed at apex, subcoriaceous, above thinly tomentose to green and nearly glabrate, below densely white tomentose, revolute; flowers solitary, short pediceled; hypanthium 3–4.5 mm. long, funnelform, white tomentose, somewhat glandular-puberulent; calyx-lobes 2.5–4 mm. long, elliptic to ovate; petals 6–9 mm. long, yellow, obovate with a slender claw, somewhat pilosulous on the back; achenes pilosulous, fusiform, 12–16 mm. long; seeds 5–7 mm. long, lanceoloid, blackish. *Kunzia tridentata* (Pursh) Spreng. North of Little Spokane, *Bonser* in 1907, *A. T.*

The specimen collected by Spalding and labeled Clear Water, Oregon [Lapwai, Ida.], was without doubt collected on one of his journeys to Walla Walla. It has not been recollected in the Lapwai region. The Indians used a decoction of the ripe fruit as an emetic, and also steeped a dye from the seed coat.

PYRUS

Trees or shrubs, not thorny (in ours); leaves simple or compound; flowers in corymbed cymes; calyx urn-shaped, 5-cleft; petals roundish or obovate, white or pink; stamens numerous; styles 2–5; pome fleshy or berry-like, the 2–5 carpels or cells of a papery or cartilaginous texture, 2-seeded. (The classic Lat. name, *pirus,* pear tree.)

Pyrus scopulina (Greene) Longyear. *Mountain Ash.* Shrub or tree up to 4 m. high; twigs pilosulous, brown, with elongate pale lenticels; older branches gray, smooth; leaves pinnate, 12–25 cm. long; leaflets 11–13, mostly oblong, 2–10 cm. long, acute or acuminate, simply or doubly serrate, glabrous or nearly so, shiny above; cymes dense, compound, 8–15 cm. broad; calyx-tube 1.5–2 mm. long, turbinate, the teeth 1 mm. long, deltoid; petals 3–5 mm. long, the limb ovate-orbicular, white; fruit bright-red, 6–8 mm. in diameter, eaten by birds. *P. dumosa* sensu St. John, non (Greene) Fern.; *Sorbus scopulina* Greene; *S. decora* (Sarg.) Schneid. In mountain woods, *A. T. T.,* abundant in *Can.* It provides fair forage in the fall.

ROSA. Rose

Erect or climbing shrubs, usually with prickly stems; leaves alternate, odd-pinnate, with adherent stipules; leaflets mostly serrate; flowers solitary or in corymbs; hypanthium cup- or urn-shaped, becoming fleshy in fruit; calyx-lobes usually 5, without bractlets; petals 5, large, obcordate; stamens numerous, on the hypanthium; styles distinct or united; carpels numerous, sessile, inserted on the bottom or on the inside walls of the hypanthium; achenes numerous, enclosed in the berry-like fruit. (The ancient Lat. name, *rosa,* rose.)

Styles, calyx-lobes, and upper part of hypanthium deciduous. — *R. gymnocarpa.*
Styles, calyx-lobes, and top of hypanthium persistent,
  Flowers several to many, in corymbs,
    Leaflets at first glandular pruinose beneath,
      Rhachis capitate glandular hispid. — *R. Eglanteria.*
      Rhachis not so. — *R. ultramontana.*
    Leaflets merely pilosulous beneath. — *R. lapwaiensis.*
  Flowers solitary (or few),
    Sepal tips sharply lobed or lacerate,
      Sepal tips mostly narrowly oblanceolate, laciniate; infrastipular prickles much flattened, recurved. — *R. caeruleimontana.*
      Sepal tips mostly broader, pinnately lobed to divided; infrastipular prickles acicular, nearly straight. — *R. Jonesii.*
    Sepal tips entire or slightly serrate,
      Rhachis glabrous except in the upper furrow. — *R. megalantha.*
      Rhachis puberulent,
        Infrastipular prickles of main stems strongly recurved; stipules glabrous. — *R. anatonensis.*
        Infrastipular prickles of main stems nearly straight; stipules puberulent,

Fruit, calyx-lobes, and pedicels not bristly. *R. Spaldingii,* var. *Spaldingii.*
Fruit, calyx-lobes or pedicels bristly,
Fruit and pedicel bristly. Var. *hispida.*
Calyx-lobes bristly or hispid. Var. *Parkeri.*

Rosa anatonesis St. John. Young stems glabrous, internodes unarmed, lower nodes unarmed, upper nodes unarmed or with 1–3 infrastipular prickles, slender, at first twisted; lower stipules oblong, 2.5–3 cm. long, glandular dentate and ciliate; stems 6–10 dm. tall, brown, finally silvery gray and exfoliating; infrastipular prickles 3–6 mm. long, strongly recurved, grayish, from narrowly elliptic bases; flowering branches glabrous, generally unarmed; stipules 5–20 mm. long, obsagittate, glabrous, the margins pilose ciliate, glandular dentate; leaves 2–9 cm. long; petioles and rhachises puberulent and sparsely pilosulous; leaflets 5–7, 4–35 mm. long, 4–21 mm. wide, oval, usually cuneate at base, sharply serrate, upper side green, glabrous, under side paler, pilosulous at least on the veins; flowers solitary, bracted; pedicels 1–3 cm. long, glabrous; hypanthium pyriform, with a neck, glabrous; calyx-lobes 23–34 mm. long, the base lanceolate, lanate on the margins and within, the tips foliaceous linear-oblanceolate, entire, glabrous, persistent in fruit and usually erect; petals 2–3 cm. long, obovate or obcordate, rose-colored; hips 14–20 mm. long, globose or narrowed at base, brown with a bluish bloom; ovules attached on bottom and sides of receptacle, long hispid; fertile seeds few. Woods, Asotin Co., *A. T. T.*

The type is Anatone, *St. John & Palmer* 9,555.

Rosa caeruleimontana St. John. Bush 1–1.5 m. tall; young stems glabrous, with infrastipular stout prickles and often with other weaker ones; older stems brown, smooth; the infrastipular prickles 10–15 mm. long, 12–20 mm. broad at the flattened base, gray, flattened and recurved; flowering branches 7–15 cm. long, glabrous; leaves 4–9 cm. long, 5–9 foliolate; stipules 12–25 mm. long, obsagittate, glabrous on both sides, the margins glandular dentate and pilose; petiole and rhachis pilose mostly in the furrow on the upper side; leaflets 7–30 mm. long, 3–18 mm. wide, sharply serrate or somewhat doubly so, glabrous, dark green above, beneath paler, remotely pilose, especially on the veins; inflorescence 1–3-flowered, leafy bracted; pedicels 1–3 cm. long, glabrous; hypanthium subglobose, glabrous; calyx-lobes 2–3.3 cm. long, capitate glandular prickly, the base linear lanceolate, villous on the margins and within, the median part linear, often pinnately lobed, the tip foliaceous, glabrous, entire or laciniate; petals 18–26 mm. long, obcordate, rose-colored; hips 8–15 mm. long, globose, the calyx-lobes reflexed; seeds attached to bottom and sides of the cavity, 4–6 mm. long, brown, curved, often compressed, hirsute on the back and the tip.

The type is Blue Mts., Asotin Co., woods, *G. N. Jones* 1,892.

Rosa Eglanteria L. *Sweetbriar.* Erect, branching shrubs, 2–3 m. tall, with many strong, hooked prickles and sometimes with bristles; leaflets 5–9, oval to orbicular, on flowering shoots only 7–19 mm. long, but on other shoots up to 37 mm. long, obtuse, doubly glandular serrate, dull green and glabrous above, pale and often pubescent beneath, and above sometimes with sweet smelling glands; pedicels hispid and usually glandular, mostly single; flowers 3.7–5 cm. across, fruits 12–18 mm. long, subglobose to ovoid, orange-red to scarlet, crowned with the long, glandular serrate sepals which are tardily dehiscent. A European species, cultivated here, and escaping and established, *A. T., A. T. T.*

Rosa gymnocarpa Nutt. ex T. & G. *Woodland Rose.* Small shrub 1–3 m. high; branches becoming brown or purplish, armed with numerous slender, straight, broad-based prickles 2–6 mm. long, or partly unarmed; the infrastipular similar or none; stipules 5–15 mm. long, linear to obsagittate, dentate, capitate glandular ciliate; leaflets 5–45 mm. long, 5–9, suborbicular to elliptic, cuneate at base, doubly serrate, green above, paler beneath, nearly glabrous, serratures and petioles glandular; flowers mostly solitary; pedicels 1–3 cm. long, glandular hispid or glabrous; calyx-lobes 5–10 mm. long, ovate, acuminate, appendaged, tomentose on margins and within; hips 5–15 mm. long, ovoid to pyriform, glabrous, scarlet, the tips naked. *R. leucopsis* Greene. Woods, common, *A. T. T., Can.*

Provides good forage, especially in the fall.

Rosa Jonesii St. John. Shrub 6–15 dm. tall; young stems glaucous, glabrous, the nodes with 1–3 subulate, straight infrastipular prickles 5–7 mm. long, from elliptic bases; internodes naked or with acicular prickles or bristles; old stems reddish brown to silvery; flowering branches 5–15 cm. long, glabrous, usually unarmed, occasionally with infrastipular prickles; leaves 3–11.5 cm. long; stipules 1–2.5 cm. long, obsagittate, above glabrous, beneath glabrous, the margins pilose, dark glandular serrate; petiole and rhachis finely puberulent, often stalked glandular, sometimes glandular prickly at base; the 3–7 (usually 5) leaflets 7–55 mm. long, 8–31 mm. wide, oval, coarsely and mostly simply serrate, above dark green and glabrous, beneath pale, sparsely pilosulous on the principal veins, the surface remotely pilosulous or glabrate; pedicels 2–4.5 cm. long, glabrous, the bract stipule-like or occasionally with leaflets; flowers single, all asymmetric; hypanthium glabrous; calyx-lobes 2.2–4.7 cm. long, the base lanceolate pilose on and near the margins, the middle a linear caudate prolongation, pilose, the tip pilosulous below, expanded and foliaceous, from linear-lanceolate to broadly oblanceolate, the narrower ones often entire, the larger ones sharply pinnately lobed, parted, or divided and then the middle section also with linear leaflets; styles slightly exserted; petals 18–27 mm. long, obcordate, deep rose-colored; hips 15–19 mm. long, subglobose, the calyx-lobes connivent, persistent, but the foliaceous tips easily broken; achenes attached both on the bottom and the side walls; achenes 3–5 mm. long, ovoid to ellipsoid, variously compressed, straw-colored, hirsute up one side. Blue Mts., and Moscow Mt., *Can.* The type is summit, Moscow Mt., *St. John & Jones* 9,621.

Rosa lapwaiensis St. John. Young stems not collected; stems about 1 m. tall, brown, glabrous, at first glaucous, infrastipular prickles present at least below, 4–8 mm. long, pale, acicular, subulate towards the tip, slightly flattened below, the narrowly elliptic base 2–6 mm. long; middle and upper internodes unarmed, the lower with a few stiff bristles; lateral and flowering branchlets similar to the main stem, but more often unarmed; the lower with a few stiff bristles; lateral and flowering branchlets similar to the main stem, but more often unarmed; stipules 7–20 mm. long, obsagittate, above glabrous or nearly so, beneath puberulent and more or less short glandular, the margin puberulent, entire or remotely glandular dentate; leaves 3–10 cm. long; petioles and rhachises puberulent or short pilosulous, commonly glandless, occasionally with a few stalked glands; leaflets 5–7, 8–42 mm. long, 3–23 mm. wide, elliptic to oval, often cuneate at base, sharply serrate except in the lower third, upper surface green, remotely and minutely pilosulous, lower surface pale green, pilosulous; flowers few to many in terminal corymbs; pedicels 5–25 mm. long, glabrous at tip, pilosulous below; hypanthium glabrous, globose, often with a short broad neck, to urceolate; calyx-lobes 9–13 mm. long, the base lanceolate, lanate on the margins and within, the back often glandular prickly, the foliaceous tip almost linear, entire, apparently falling from the old fruit; petals 10–20 mm. long, obcordate, rose-colored; hips about 1 cm. long, globose; ovules attached on bottom and sides of the receptacle; seeds 3 mm. long, ovoid, compressed, brown, hispid on one side. Meadows, *U. Son., A. T.*

The type is Nez Perce Co., Lapwai, *St. John et al.* 9,538.

This was reduced by Cole to *R. Woodsii,* var. *ultramontana* (Wats.) Jeps. However, it is easily separated from the pyriform fruited *R. ultramontana,* and *R. Woodsii* of the Great Plains is quite different.

Rosa megalantha G. N. Jones. Shrub 1.5–2.5 m. tall; young branches glabrous, glaucous, with 1–2 straight slender infrastipular spines about 5 mm. long at each node; floral branches glabrous, unarmed; stipules 1–2 cm. long, glandular-dentate; leaflets usually 7, 2.5–5 cm. long, the terminal obovate, the lateral elliptic or oblanceolate, sharply serrate, entire and cuneate at base, glabrous above, paler beneath, glabrous or sparsely appressed pubescent on the veins; flowers solitary or in pairs; pedicels 3–4 cm. long, glabrous; calyx-lobes 2.5–3.5 cm. long, cuneate, the tip linear-lanceolate, villous tomentose on the margins and within; petals 2.5–3.5 cm. long, rose-colored; hips 2 cm. long, globose, glabrous. *R. Spaldingii* Crépin, var. *alta* (Suksd.) G. N. Jones. Open, yellow pine woods, *A. T. T.*

The type collection was made in 1927 at Lincoln Park, Spokane, *G. N. Jones* 614.

Rosa Spaldingii Crépin, var. **Spaldingii.** *Spalding's Rose.* Shrub 1–2 m. tall; young branches with paired, remote, curved, flattened, pale infrastipular prickles 4–8 mm. long, from elliptic bases; also with straight, acicular prickles, few above, many below; flowering branches dark reddish brown, glabrous, with infrastipular prickles, occasionally other prickles, but commonly unarmed; stipules 1–3 cm. long, linear to obsagittate, puberulent, the upper glandular toothed; petioles and rhachis puberulent, somewhat glandular; the 5–9 leaflets 1.5–4 cm. long, oval, coarsely serrate, green and glabrous above, beneath pale, puberulent, at first minutely glandular; flowers solitary or few; pedicels 1.5–4 cm. long, glabrous, often bracted; calyx-lobes 15–28 mm. long, lanceolate, caudate, often with a narrow foliaceous tip, persistent, erect, villous on the margins and within; petals 2–3 cm. long, obcordate, rose-colored; hips 10–25 mm. long, globose to pyriform with a short neck, scarlet. Abundant and showy, *U. Son., A. T.*

Named for its first collector, the Rev. Henry Spalding, early missionary at "Clear Water" [Lapwai] Idaho. It provides good forage when young and tender.

Var. **hispida** (Fern.) G. N. Jones. Differing from the species by having the hypanthium, the calyx-lobes and commonly the tip of the pedicel conspicuously glandular bristly. Forming thickets, *A. T., A. T. T.*

Var. **Parkeri** St. John. Differing from the species in having the calyx-lobes capitate glandular prickly or hispid. Woods or thickets, *U. Son.* to *A. T. T.*

The type came from Latah Co., Grizzly Camp, *Parker* 503.

Rosa ultramontana (Wats.) Heller. Bushes 1–5 m. tall; young stems with the nodes bearing 2–3 yellow, straightish, infrastipular prickles 4–7 mm. long from elliptic bases; internodes with or without weaker, acicular prickles or bristles; flowering branches with infrastipular prickles and a few bristles, or commonly unarmed, glabrate; stipules 1–2.7 cm. long, obsagittate, glandular dentate, on the back puberulent and often pruinose; petiole and rhachis puberulent; the 5–9 leaflets 1–4.5 cm. long, oval to elliptic, serrate, above pale green, beneath paler, puberulent and often pruinose; flowers several, in corymbs; pedicels sparsely pilosulous; bracts foliaceous; calyx-lobes 1–2 cm. long, lanceolate, long caudate with a linear-lanceolate tip, tomentose on the margins and within; petals 1–2.2 cm. long, obcordate, rose-colored; hips 8–15 mm. long, scarlet, pyriform with or without a distinct neck, but when over-ripe wrinkling and becoming ellipsoid. *R. Woodsii* Lindl., var. *ultramontana* (Wats.)

Jepson; *R. pisocarpa* Gray, var. *ultramontana* (Wats.) Peck. Abundant, forming thickets, moist places, *U. Son.* to *A. T. T.*

Rosa ———. Known only in flower. It is closest to *R. megalantha,* but is probably undescribed. Until fruit and more material is available, it is not wise to describe it. The specimens seen, both from the Blue Mts. are Wolf Fork of Touchet River, *St. John, Davison, & Scheibe* 7,023; Stockade Springs, *St. John* 8306.

Rosa ———. A collection that seems closest to *R. hypoleuca* Woot. & Standl., yet is probably distinct. In want of fruit and of additional collections, the writer does not care to describe it as new. Pullman, *St. John* 9,543.

Rosa ———. A 3 m. shrub with large, orange, pyriform fruits, apparently undescribed, and very distinct. Flowers are needed. Bank of Snake River, Buffalo Rock, opposite Capt. John's, *St. John* 9,631.

C. L. Hitchcock and Cronquist (1961) have reduced *Rosa anatonensis, R. caeruleimontana, R. Jonesii, R. megalantha, R. Spaldingii,* var. *hispida,* and var. *Parkeri* to *R. nutkana* Presl; and *R. ultramontana,* and *R. lapwaiensis* to *R. Woodsii* Lindl., a massive lumping with which the writer does not agree.

RUBUS

Perennial herbs or shrubs or vines, often prickly; many species with biennial stems which during the first year vegetate only (*primocane*), then the next season on axillary, lateral branches produce flowers and fruit (*floricane*); leaves alternate, simple, or pinnately 3–7-foliolate, with stipules adherent to the petiole; flowers perfect (rarely dioecious), white or reddish, solitary or in racemes or corymbs; calyx 5-lobed, without bractlets; petals 5 conspicuous; stamens numerous, on the calyx-tube; styles nearly terminal; carpels numerous, on the convex receptacle, ripening into 1-seeded drupelets forming an aggregate fruit. (The ancient Lat. name, *rubus,* blackberry.)

Leaves simple; stems unarmed; styes club-shaped,
  Glands of pedicels short stipitate. — *R. parviflorus,* f. *Nuttallii.*
  Glands of pedicels subsessile, sessile, or none. — Forma *scopulorum.*
Leaves compound; stems prickly or bristly; styles filiform,
  Drupelets persistent on the spongy receptacle and shed with it; petals 8–18 mm. long,
    Leaflets laciniately incised; flowers perfect. — *R. laciniatus.*
    Leaflets doubly dentate, sometimes with a few shallow lobes; flowers functionally unisexual. — *R. macropetalus.*
  Drupelets shedding from receptacle as a thimble-shaped aggregate fruit; petals 6 mm. or less in length,
    Fruit purplish black; inflorescence corymbose; prickles of petioles recurved, strongly flattened at base; leaves when 5-foliolate palmately compound,
      Lower leaf surface, inflorescence, and drupelets tomentose. — *R. leucodermis.*

Lower leaf surface, inflorescence, and drupelets not tomentose. Var. *nigerrimus*.
Fruit red; inflorescence racemose; prickles of petioles straight or nearly so, scarely or not flattened, or wanting; leaves when 5-foliolate pinnately compound.
Plant not glandular hispid,
Young branches, petioles, and inflorescence finely white tomentose. *R. idaeus*.
Young branches, petioles, and inflorescense sparingly pilose, villous, or glabrate. Var. *melanotrachys*.
Plant, and especially the inflorescence, glandular, hispid,
Leaflets white tomentose beneath; calyx - lobes ovate - lanceolate, abruptly acuminate. Var. *strigosus*.
Leaflets green and soon glabrate or nearly so beneath; calyx lobes deltoid-lanceolate, long acuminate. Var. *peramoenus*.

Rubus idaeus L. *Red Raspberry*. Stems 1–1.5 m. tall, erect or arching, biennial; primocanes light colored, glabrate, sparingly armed with bristles or weak prickles; leaves pinnately 5-foliolate; floricanes with the leaves usually 3-foliolate; petiole, rhachis, and midveins finely tomentose and with a few curved prickles; terminal leaflet 3–10 cm. long, broadly ovate, short acuminate, rounded or cordate at base, coarsely doubly serrate, above pilosulous or glabrate, beneath white tomentose, with a slender petiolule; lateral leaflets smaller, subsessile; inflorescences on lateral branches, short racemose, few-flowered, tomentose, and often with sparse bristles; calyx-lobes 5–8 mm. long, ovate-lanceolate, acuminate, tomentose; petals 4–6 mm. long, spatulate, white, erect; fruits thimble-shaped, tomentose, edible. *R. idaeus* L., subsp. *vulgatus* Arrhen. Swamp, Wild Rose Prairie, Spokane Co., *Sprague* 409, *A. T.*

This species is typical, cultivated Red Rasberry, and the source of most of the European cultivated varieties, several of which are common in North America. They persist or spread from cultivation. Our single collection has presumably this history.

Var. **melanotrachys** (Focke) Fern. *Raspberry*. Differs from the species in having: stems glabrous, short bristly; floricanes with leaflets 3–6 cm. long, above green and glabrous, the terminal one broadly ovate or obovate; inflorescence short, almost corymbiform, villous and with numerous, purple, curved, strong bristles or weak prickles; hypanthium purplish, villous, and bristly; calyx-lobes pilose and tomentose; fruits puberulent. *R. melanotrachys* Focke. Moscow Mt., Idaho, *Piper* 2,879; also reported by Prof. Fernald from Spokane, *A. T. T.*

Var. **peramoenus** (Greene) Fern. *Raspberry*. Differs from var. *strigosus* in having: the stems yellowish or brownish, sparingly bristly, in age nearly unarmed; primocanes with leaflets green on both sides, only sparingly grayish puberulent beneath when young; petioles bristly, otherwise glabrous; floricanes with petiole, rhachis, and midrib capitate glandular and sparsely bristly; leaflets thin, green, above glabrous, beneath sparsely tomentose, becoming green and more or less glabrate. *R. peramoenus* Greene. Along wooded stream banks, Spokane, *Piper* 2,268; Salmon River, Blue Mts., *Horner* 295, *A. T. T.*

Var. **strigosus** (Michx.) Maxim. *Red Raspberry*. Differing from the species in having: the stems usually brownish or reddish, sometimes glaucous, but not tomentose; petiole and rhachis often bristly, and sparingly glandular; leaves 3–5-foliolate, the terminal one broadly ovate, sometimes 3-lobed; floricanes having the leaves with the petiole, rhachis, and midribs capitate glandular and somewhat bristly, not tomentose, the terminal leaflet 3–5 cm. long, acute or short acuminate; inflorescence bristly and glandular hispid, not tomentose; calyx glandular, as well as tomentose and bristly; fruit about 1 cm. broad, hemispheric; endocarp 2–3 mm. long, ovoid, straw-colored, reticulately ridged. *R. strigosus* Michx. Common by streams and in springy places, *A. T., A. T. T.*

This excellent wild fruit is the source of dozens of cultivated American varieties, including the Cuthbert.

Rubus laciniatus Willd. *Evergreen Blackberry.* Stems perennial, often 5 m. or more in length, arching, forming impenetrable thickets, green, later purplish, angled, pilose, glabrate, well armed with stout prickles which are 3–8 mm. long, recurved, flattened near the elliptic base; rhachis, petiolules, and midribs with similar or smaller prickles; leaves 5-foliolate, or those of flowering branchlets with 3 or 1 leaflets; leaflets pinnately cleft or parted and laciniately incised, pilose beneath, sparsely so when mature, above dark green and glabrous, the terminal are the larger, 3–10 cm. long, cordate in outline; the lateral ovate in outline; rhachis 1–9 cm. long, pilose; inflorescence paniculate 5–30- or more-flowered, pilose, prickly, and minutely glandular; flowers showy; body of calyx-lobes 5–7 mm. long, ovate-lanceolate, on the outside pilose, sparsely prickly, and densely lanate, on the inside lanate, the tips acuminate, foliaceous, often elongate, entire, toothed, or parted; petals 8–16 mm. long, obovate, often cleft, pilosulous, white or pinkish; fruits globose to ovoid-cylindric, black, sweet, edible; endocarp 3.5 mm. long, ovoid, compressed on inner side, straw-colored, reticulately ridged. Escaping from cultivation, established at Almota, Spokane, and Coeur d'Alene.

This is a prolific yielder of blackberries of good quality. In the humid regions near the Pacific, it takes complete possession of cut-over woodlands. The plant originated in Europe before 1770, it is believed as a mutant from *R. vulgaris* Weihe & Nees.

Rubus leucodermis Dougl. ex T. & G. *Black-cap.* Stems biennial, 1–2 m. tall, erect; primocanes green to yellowish, glaucous, armed, the prickles 2–4 mm. long, straight, much flattened towards the long elliptic base; leaves palmately 3- or 5-foliolate; petiole, rhachis, and midrib with stout recurved prickles; floricanes often brown or purplish; lateral branches with stouter, recurved prickles; leaves 3-foliolate, above sparsely pilosulous, soon glabrate, green, beneath densely white tomentose, terminal leaflet 2–9 cm. long, ovate, broadly so or obovate, acuminate, rounded or cordate at base, deeply doubly serrate, sometimes lobed, lateral leaflets narrower, subsessile; corymbs few-flowered, loosely tomentose, prickly, and minutely atomiferous glandular, not stalked glandular; calyx-lobes 6–12 mm. long, lanceolate, long acuminate, lanate, sometimes with a few prickles; petals 4–5 mm. long, spatulate, white, ascending; fruits edible, pleasant; endocarp 1.5–2 mm. long, bean-shaped, closely reticulately ridged, straw-colored. Moist, especially when shady, places, *U. Son* to *Can.*

Var. **nigerrimus** (Greene) St. John. This plant differs from the species solely in being essentially glabrous throughout. It retains, however, the sparse atomiferous glands of the petioles, inflorescence, and calyx-lobes. The prickles are identical, being straight on the primocanes, but recurved and more flattened on the leaves of the primocanes and the branches, leaves, and inflorescences of the floricanes. *R. hesperius* Piper; *R. nigerrimus* (Greene) Rydb.; *R. transmontanus* Focke; *Melanobatus nigerrimus* Greene. Creek banks and ledges, along Snake River from Truax to Almota, *U. Son.*

First discovered and described by Prof. Piper, from Wawawai and Almota. As this plant differs solely in its lack of pubescence, and as it grows within the range of *R. leucodermis,* it seems more logical to classify it as a variety of that species.

Rubus macropetalus Dougl. ex Hook. *Blackberry.* Stems slightly woody, biennial; primocanes 1–8 m. long, somewhat glaucous, prickly, trailing, usually unbranched, the leaves 3-foliolate, the rhachis and midrib prickly and villous, the leaflet 4–9 cm. long, ovate-deltoid, doubly dentate or lobed, thin, paler beneath; floricanes with slender recurved prickles, villous to glabrate, but bearing numerous short flowering branches with the leaves 3-foliolate, similar or smaller, more obtuse, and less lobed; inflorescence, pedicels and calyx prickly, stipitate glandular, and villous; corymbs 2–14-flowered; flowers unisexual or imperfectly so; sterile flowers with calyx-lobes 8–16 mm. long, lanceolate, caudate; petals 12–18 mm. long, elliptic, white; fertile flowers with calyx-lobes 5–7 mm. long, lanceolate, acute or short acuminate; petals 7–10 mm. long; fruits 10–25 mm. long, ovoid to cylindric, edible, purplish black, glabrous; endocarp 2 mm. long, semicordate, reticulate. *R. ursinus* C. & S., var. *macropetalus* (Dougl.) S. W. Brown. In open woods, especially old "burns," Moscow Mt., and northwards, *A. T. T.*

Rubus parviflorus Nutt., forma **Nuttallii** (T. & G.) Fassett. *Thimble Berry.* Stems 1–2 m. tall, perennial; bark brownish, becoming shreddy; inner bud scales densely villous; blades 1–3 dm. broad, reniform, palmately and acutely 3–5-lobed, irregularly serrate, cordate at base, glabrate or glabrous; stipules lanceolate, glandular; petioles about as long as the blades; corymbs 3–10-flowered; sepals 10–22 mm. long, ovate, long caudate, glandular and woolly; petals 15–30 mm. long, oval, white; fruit red, low convex, juicy, about 2 cm. broad; endocarp 1.5–2 mm. long, oblong, reticulate, the raphe prominent along one edge. *R. parviflorus,* var. *grandiflorus* Farw. Common in open woods; rare away from timber, *A. T. T.*

The handsome fruits are edible but insipid, yet are eagerly gathered. The plant provides fair forage for sheep.

Forma **scopulorum** (Greene) Fassett. *Thimble Berry.* Differs from f. *Nuttallii* by having the leaves glabrous beneath, and the glands of the pedicels subsessile, sessile, or none. *R. parviflorus,* var. *scopulorum* (Greene) Fern. Moist woods or shaded stream banks, *A. T. T., Can.*

There may be several other minor variants to be recognized when all of the local collections are restudied.

SANGUISORBA. Burnet

Annual or perennial herbs; leaves alternate, odd-pinnate; stipules adherent to the petiole; flowers small, perfect (in ours), polygamous, or dioecious, in dense terminal spikes or heads; calyx 4-lobed; petals none; stamens 2–4-many; styles terminal, filiform; carpels 1–3, free from the calyx-tube; achenes usually 1, enclosed in the 4-angled dry closed calyx-tube. (Name from Lat., *sanguis,* blood; *sorbere,* to absorb.)

Leaflets serrate; filaments dilated, long exserted; spikes 2–14 cm. long; calyx-lobes 3 mm. long, elliptic. *S. stipulata.*
Leaflets pinnatisect; filaments filiform, not longer than the calyx-lobes; calyx-lobes 1.5–2 mm. long, broadly ovate. *S. occidentalis.*

Sanguisorba occidentalis Nutt. ex Piper & Beattie. Biennial or perhaps perennial, 2–8 dm. tall, glabrous; tap-root 2–4 mm. in diameter, brown; stems one to several, erect, with ascending branches; stipules of basal leaves entire, lanceolate, those of the cauline leaves foliaceous, palmately cleft into linear segments; leaves odd pinnate, the 5–13 leaflets 5–18 mm. long, oval, pectinately pinnatisect into 9–15 linear divisions; spikes several, terminal, 5–25 mm. long, ellipsoid-cylindric; bracts 2 mm. long, suborbicular, membranous; hypanthium 2–2.5 mm. long, turbinate, 4-sided, the angles with firm wings 0.2 mm. wide, reticulate on the sides; calyx-lobes with the center green, the wide margins white membranous; stamens 2, opposite the inner calyx-lobes; anthers 0.1 mm. long, oval; fruiting hypanthium lance-ovoid, the faces brown. *Poteridium occidentale* (Nutt.) Rydb. Dry or rocky places, *A. T.*

Sanguisorba stipulata Raf. Perennial, glabrous; stems 3–9 dm. high, simple or branched above; leaves as long, with 11–21 leaflets 2–8 cm. long, oval, coarsely serrate, more or less cordate; petiolules 2–30 mm. long; cauline stipules coarsely serrate; spikes 2–14 cm. long; calyx lobed almost to the base, the lobes 2–3 mm. long, oval, green with pale margins; stamens 6–10 mm. long, white, the filaments dilated and flattened above; fruit 3 mm. long, closely enclosed in the 4-winged fertile calyx-tube; achene lanceolate, flattened, yellowish. *S. latifolia* (Hook.) Cov.; *S. canadensis* L., var., *latifolia* Hook. Swampy places, Blue Mts.; Salmon R., *Horner* 415; Table Rock, *St. John et al.* 9,648, *Huds.*

SIBBALDIA

Depressed alpine plants somewhat woody with alternate trifoliolate leaves; flowers in cymes on nearly leafless peduncles; calyx persistent, slightly concave, 5-lobed, with 5 bracts; petals 5, yellow, smaller than the calyx-lobes; stamens 5, on the margin of the villous disk, opposite the calyx-lobes; carpels 5–10, on short pubescent stipes; styles lateral. (Named in honor of *Prof. Robert Sibbald* of Edinburgh.)

Sibbaldia procumbens L. Perennial, tufted, sparsely villous, 5–15 cm. high; leaflets 3, rather thick, 6–20 mm. long, obovate or oblanceolate, cuneate, 3–5-toothed at the truncate apex, petioles 3–50 mm. long; stipules ovate or ovate-lanceolate, membranous, nearly glabrous; cymes compact, 2–9-flowered; peduncles usually shorter than the leaves; calyx-tube 1 mm. long, saucer- to cup-shaped, densely hirsute; calyx-teeth 2–4 mm. long, oblong-lanceolate, the margins glabrous; petals 1.5 mm. long, spatulate; achenes 0.8 mm. long, ovoid, margined at the hilum, pale brown; petals yellow, acute. *Potentilla Sibbaldi* Hall. f. Highest peaks of the Blue Mountains, Walla Walla Co., *Piper,* July 15, 1896; Table Rock, Columbia Co., *Constance et al.* 1,253. *Huds.*

SPIRAEA. Spiraea

Shrubs or perennial herbs; leaves alternate, simple, without stipules; flowers white or pink, perfect, in racemes, cymes, corymbs, or panicles; calyx 4- or 5-lobed; petals 4 or 5, exceeding the calyx and inserted on the calyx-tube; stamens 15–70, distinct, on a disk which is free at the edge and crenate or glandular-toothed; filaments much exserted; pistils commonly 5, (rarely 3–8), superior, alternate with the calyx-lobes; ovules 5–11; follicles not inflated, 1-valved; seeds few–several. (Name probably from Gr., *speira,* anything wrapped around.)

Inflorescence pyramidal or narrower,
  Inflorescence elongate pyramidal; petals rose-colored. *S. Menziesii.*

Inflorescence low pyramidal; petals pinkish or white. *S. × pyramidata.*
Inflorescence broad, dome-shaped,
Petals white; corymb 3–15 cm. wide. *S. lucida.*
Petals rose-colored; corymb 2–6 cm. wide. *S. densiflora.*

Spiraea densiflora Nutt. ex T. & G. Shrub, 2–10 dm. tall, nearly glabrous throughout; branchlets pale, usually pinkish; branches with bark reddish brown to purplish, smooth, exfoliating; petioles 1–5 mm. long; blades 15–45 mm. long, elliptic to oval, paler green beneath, the margin ciliolate at base, at apex crenate or serrate; bracts of inflorescence few, leafy; corymb 2–6 cm. wide, low rounded; pedicels 1–3 mm. long; hypanthium 1 mm. high, broad turbinate; calyx-lobes 1 mm. long, deltoid, erect; petals 1.3–2 mm. long, suborbicular, rose-colored; filaments long exserted; follicles said to be glabrous, the body 3 mm. long, the beak 1 mm. long; and seeds nearly 2 mm. long, linear-lanceolate. Moist bank, 5,800 ft., head N. Fork, Wenaha River, below Table Rock, Columbia Co., Blue Mts., *Constance et al.* 1,248, *Huds.*

Spiraea lucida Dougl. ex Greene. *White Spiraea.* Rootstock woody, erect or creeping; stems 2–10 dm. tall, erect, often with erect branches; bark of twigs glabrous, smooth, reddish or light brown, when older peeling and darkening; blades 1.5–8 cm. long, the lower obovate, the upper oval, thin, glabrous or ciliolate near the base, green above, paler beneath, coarsely serrate or doubly so above the middle; petioles 2–16 mm. long; corymb 3–15 cm. broad, dome-shaped, glabrous; bracts linear, pilose at tip; hypanthium 1.2–1.7 mm. long, hemispheric, glabrous; calyx-lobes 0.5–0.8 mm. long, deltoid to semiorbicular, apiculate, pilose ciliate; petals 1.5–2 mm. long, suborbicular, palmately veined, white; filaments exserted; follicles 2.5–3 mm. long, oblanceoloid, 2-angled, pilose at tip, greenish brown, the stout style forming a beak nearly 1 mm. long; seeds 2–3 mm. long, linear-lanceoloid, attenuate at each end, straw-colored. *S. betulifolia* Pall., var. *lucida* (Dougl. ex Greene) C. L. Hitchc. Abundant in thickets or open woods, upper part of *U. Son.* to *A. T. T.*

The earlier *S. betulifolia* Pallas doubtfully belongs here.

Important forage for sheep on the summer range.

Spiraea Menziesii Hook. *Menzies' Hardhack.* Erect shrub, 1–3 m. high, not much branched; branchlets puberulent, brown; older bark exfoliating; blades 2–9.5 cm. long, elliptic, oval, or oblong, acute or obtuse, rounded or cuneate at base, green above, paler green beneath, coarsely serrate above the middle, rarely entire, beneath glabrous or minutely puberulent on the veins only; petioles 1–8 mm. long; panicles 5–27 cm. long, dense, pilosulous, elongate, pyramidal or the tip almost spike-like; calyx-tube 0.8–1.2 mm. long, the reflexed teeth shorter, ovate; petals 1–1.5 mm. long, suborbicular; carpels 2–3 mm. long, oblong, beaked, nearly glabrous. Wet thickets, Palouse northward and eastward, *A. T. T.*

Provides fair forage in fall for sheep and cattle.

Spiraea × pyramidata Greene (*S. lucida × S. Menziesii*). Apparently a natural hybrid, combining to varying degrees the characters of the two parents. It commonly has leaves like the former; but like the latter has stems puberulent, and calyx-lobes reflexed; and shows a blending of the two in the low broad panicle, and in the color of the petals which are white or pinkish. The single collection from our area has abundant, good pollen, thus indicating that if a hybrid, it is probably a fertile one. It grows where the ranges of the presumed parents overlap, and is less common than either. Low ground, Harrison, Ida., *Sandberg,* July 1893. *A. T. T.*

LEGUMINOSAE

Herbs, shrubs, or trees; leaves alternate, mostly compound, with stipules; flowers irregular (*papilionaceous*) (in ours), perfect or sometimes polygamous, in spikes, heads, racemes, or panicles; calyx 4–5-toothed or -cleft; lobes equal or unequal, sometimes 2-lipped; petals more or less united or separate, perigynous or hypogynous, usually papilionaceous, that is, with the upper or odd petal (*standard*) larger than the others and usually turned backward or spreading; the two lateral ones (*wings*) oblique and exterior to the two lower; the last pair connivent and commonly more or less coherent by their anterior edges, forming the *keel,* which usually encloses the stamens and pistil; stamens 10, rarely 5 or many, monadelphous, diadelphous, or occasionally distinct; pistil 1, 1-celled or several-celled by the intrusion of partitions, free; ovules 1–many; fruit a legume; endosperm usually none. *Caesalpinaceae; Fabaceae; Mimosaceae.* (Name from Lat., *legumen,* pulse or bean plant.)

- Stamens 5, monadelphous; corolla indistinctly papilionaceous; leaves pinnate. — *Petalostemon.*
- Stamens 10; corolla papilionaceous,
  - Stamens distinct. — *Thermopsis.*
  - Stamens monadelphous or diadelphous,
    - Anthers of two forms, round and oblong. — *Lupinus.*
    - Anthers all alike,
      - Leaves palmate or trifoliolate,
        - Foliage glandular-dotted. — *Psoralea.*
        - Foliage not glandular-dotted,
          - Corolla deciduous,
            - Pod curved or spirally coiled. — *Medicago.*
            - Pod straight. — *Melilotus.*
          - Corolla persistent. — *Trifolium.*
      - Leaves pinnate,
        - Foliage dotted with conspicuous glands,
          - Pods spiny. — *Glycyrrhiza.*
          - Pods not spiny. — *Psoralea.*
        - Foliage not dotted with conspicuous glands,
          - Leaves even pinnate, usually with tendrils,
            - Style filiform, hairy around the tip. — *Vicia.*
            - Style flattened, hairy along the inner side. — *Lathyrus.*
          - Leaves odd pinnate; without tendrils,
            - Flowers in umbels or solitary; pods linear. — *Lotus.*
            - Flowers in spikes or racemes, rarely solitary, then the pods not linear,
              - Herbs. — *Astragalus.*
              - Trees. — *Robinia.*

ASTRAGALUS. Milk Vetch

Chiefly perennial herbs (ours all perennials); leaves odd-pinnate, with stipules; flowers in spikes, racemes, or heads; calyx with 5 nearly equal teeth; corolla and its slender-clawed petals usually narrow, spreading; standard equaling or exceeding the wings and blunt keel; stamens diadelphous; anthers all alike; ovary sessile or stipitate; ovules numerous; pod sometimes inflated, 1-celled or 2-celled by a false partition; seeds few or many. (The ancient Gr. name of a leguminous plant.)

Pods membranaceous, thin, much inflated,
  Stipules connate. — *A. Cusickii.*
  Stipules separate,
    Keel gradually up-curved; pods 2-celled. — *A. lentiginosus*, var. *lentiginosus.*
    Keel up-curved at right angles; pod 1-celled. — *A. Whitneyi*, var. *Sonneanus.*
Pod coriaceous or chartaceous, not inflated,
  Herbage densely white-woolly or long hairy,
    Pods 2-celled, straight; corolla 7–10 mm. long. — *A. Spaldingii.*
    Pods 1-celled, curved; corolla 18–26 mm. long,
      Standard ochroleucous; leaflets 7–15. — *A. Purshii*, var. *Purshii.*
      Standard rose to magenta; leaflets 21–25. — *A. inflexus.*
  Herbage and pods either glabrous or short-canescent,
    Pods stipitate, the stipe equaling or exceeding the calyx,
      Pods with both sutures prominent; calyx-tube very gibbous at base. — *A. collinus*, var. *collinus.*
      Pods with the dorsal suture impressed or intruded; calyx-tube little swollen at base,
        Leaflets broadly oval, glabrous; pods oblong. — *A. Beckwithii*, var. *Beckwithii.*
        Leaflets oblong to linear, pubescent at least beneath; pods linear,
          Pods slender, curved, divergent, the stipe much exceeding the calyx; flowers soon reflexed. — *A. Arthuri.*
          Pods stout, straight, ascending, the stipe not exceeding the calyx; flowers erect. — *A. arrectus.*
    Pods sessile or nearly so,
      Pods 2-celled by the intrusion of the sutures,
        Flowers greenish or yellowish; pods oblong, the lower suture slightly sulcate. — *A. Mortoni.*
        Flowers purple or purplish, pod with the lower suture deeply sulcate,
          Pods oblong, appressed black hairy. — *A. striatus.*
          Pods ovoid, long pilose. — *A. agrestis.*
      Pods 1-celled,
        Lower suture not or but slightly intruded,
          Flowers in racemes; leaflets not rigid or prickly-pointed,
            Pod sutures both doubly winged. — *A. riparius.*
            Pod sutures not doubly winged,
              Pods partially 2-celled; leaflets 9–15. — *A. miser*, var. *miser.*
              Pods 1-celled; leaflets 17–37,
                Calyx teeth 1–3 mm. long; corolla whitish, 14–18 mm. long; pod 4–8 mm. wide, ovoid to globose. — *A. reventus.*

Calyx teeth 1 mm. long; corolla yellowish white, 1 cm. long; pod 3–4 mm. wide, lunate-oblong. *A. falcatus.*
Flowers subsessile in the leaf axils; leaflets rigid, prickly pointed. *A. impensus.*
Lower suture strongly intruded. *A. conjunctus.*

Astragalus agrestis Dougl. ex G. Don. Herbage sparsely long strigose; stems slender, procumbent or ascending, 1–3 dm. long; leaflets 13–29, lance-elliptic to oval, 3–20 mm. long, obtuse to retuse; peduncles longer than the leaves; racemes 1.5–3 cm. long, dense; flowers ascending; calyx-tube 4–6 mm. long, narrowly campanulate, white or black pilose; calyx-teeth 3 mm. long, subulate; corolla 15–20 mm. long, purple; pods 7–9 mm. long, the lower suture deeply sulcate; seeds 1.5–2 mm. long, reniform, brown. *A. goniatus* Nutt. Meadows in glacial scab-lands, from Spokane to Revere, *A. T.*

Astragalus arrectus Gray. Erect, 3–6 dm. high, sparsely appressed-pubescent throughout; leaflets 13–21, narrowly oblong to elliptic, somewhat truncate, cuneate at base, glabrous or glabrate above, 1–2 cm. long; peduncles stout, much longer than the leaves; flowers in rather dense erect racemes; calyx black strigose, the tube 3–4 mm. long, campanulate, the acute teeth 1–2 mm. long; corolla whitish, 10–15 mm. long; pods linear, short-beaked, 12–20 mm. long, 3–5 mm. wide, thinly puberulent, slightly roughened, the dorsal suture pressed inwards and except at tip nearly dividing the pod, the stipitate base as long as the calyx; seeds 2 mm. long, bean-shaped, olive-brown. *Tium arrectum* (Gray) Rydb. Grassy hillsides, *U. Son., A. T.*

Described from material collected by C. A. Geyer on the "Kooskooskee" [Clearwater River, Idaho].

Astragalus Arthuri M. E. Jones. Erect or nearly so, pale green, strigose, stems numerous, slender, 3–5.3 dm. tall, leafy only toward the base; leaflets 13–23, oblong, obtuse or retuse, 5–15 mm. long, glabrous above; peduncles 1–3 dm. long, 8–30-flowered; calyx-tube 4–6 mm. long, white or partly black pilosulous, gibbous at base; the teeth 2–3 mm. long, deltoid; corolla 10–13 mm. long, ochroleucous; pods divergent, the stipe 8–10 mm. long, the body 3–4 cm. long, 3 mm. wide, slender, falcate, acuminate, puberulent, triangular in cross section, the dorsal suture deeply intruded; seeds 3–3.5 mm. long, black, bean-shaped, the hilum in a sinus. *Atelophragma Arthuri* (M. E. Jones) Rydb. Grassy slopes, Craig Mountains, and Blue Mountains near the mouth of the Grande Ronde River, *U. Son., A. T.*

The type collection was from Lake Waha, *Heller* 3259.

Astragalus Beckwithii T. & G., var. **Beckwithii.** Glabrous or nearly so; stems ascending or spreading, 2–7.5 dm. long; stipules deltoid, 5 mm. or less in length; leaflets 15–23, oval to nearly orbicular, obtuse or retuse, 4–20 mm. long, pale, thickish; peduncles about as long as the leaves; flowers 5–16, in a short raceme; calyx-tube 4–5 mm. long, pale, glabrous or with a few black hairs, the subulate teeth about equaling the tube; corolla 15–20 mm. long, yellowish; pods 2.5–3 cm. long, glabrous, turgid, coriaceous, the dorsal suture prominent, margined, the ventral curved or even sulcate. *Phaca Beckwithii* (T. & G.) Piper; *Phacomene Beckwithii* (T. & G.) Rydb. Dry slopes, Snake River Canyon and its tributaries, *U. Son.*

Astragalus collinus Dougl., ex G. Don, var. **collinus.** Erect, 30–50 cm. high, greenish, but with an appressed short pubescence throughout; leaflets 9–23, 7–18 mm. long, elliptic-oblong to linear, truncate or retuse, cuneate at base, short-stalked, hirsutulous beneath, smoother above; peduncles mostly terminal, longer than the leaves; flowers reflexed, in racemes, 4–15 cm. long; calyx-tube 7–9 mm. long, pilosulous, the teeth triangular, acute 1–2.5 mm. long; corolla 12–15 mm. long, pale yellowish; pods linear, acuminate, puberulent,

1.5–2.3 cm. long, 3 mm. wide, the stipitate base as long as the calyx; seeds 2.5 mm. long, bean-shaped, olive-brown, smooth. *Homalobus collinus* (Dougl.) Rydb. On warm hillsides, Snake River Canyon and tributaries, *U. Son.*

Astragalus conjunctus Wats. Stems 3–5 dm. tall, glabrous or strigose; leaflets 15–25, linear, 1–2 cm. long, 1–2 mm. wide, strigose beneath, usually glabrate above; peduncles exceeding the leaves; racemes 2–5 cm. long, often crowded; calyx black hispidulous, the tube 5–6 mm. long, campanulate, the teeth 2–3 mm. long, lanceolate or subulate; corolla 15–18 mm. long, white; pod 15–22 mm. long, 4–5 mm. wide, subsessile, oblong, acute, glabrous, reticulate, sulcate on the lower suture; seeds 2.5 mm. long, obliquely reniform, brown. *Tium conjunctum* (Wats.) Rydb.; *A. reventus* Gray, var. *conjuntus* (Wats.) M. E. Jones. Reported by Rydberg from Asotin Co., *Sheldon* 8,236.

Astragalus Cusickii Gray. Perennial 2–6 dm. tall, sparsely strigose, much branched and broom-like; stipules 2–6 mm. long, deltoid; leaves 6–12 cm. long, the 11–17 leaflets 3–25 mm. long, linear; peduncles surpassing the leaves; pedicels 1–2 mm. long; calyx 4–6 mm. long, campanulate, appressed black puberulent, the teeth 1 mm. long, deltoid; corolla 12–15 mm. long, whitish to cream-colored, the standard 2–3 mm. longer than the keel; stipe 2–4 mm. long; pod 2–3.5 cm. long, obovoid, shiny, translucent; seeds 3–3.5 mm. long, reniform, dull olive brown. Slopes of the Blue Mts., Asotin Co., *U. Son.*

Astragalus falcatus Lam. Erect, about 6 dm. tall, from a taproot; herbage green, but strigose with hatchet-like hairs; stipules cauline, free; leaves subsessile; leaflets 25–33, oblong-elliptic, 7–20 mm. long, glabrous above; racemes equalling or surpassing the leaves, dense, many flowered, in fruit becoming 10–15 cm. long; flower and fruit deflexed; calyx tube 4 mm. long, campanulate; banner slightly arcuate; pod 15–25 mm. long, sessile, acute at both ends, laterally compressed, leathery, minutely strigulose with both light and dark hairs, the pod triangular-cordate in cross section, the dorsal suture deeply sulcate, the ventral one very acute. Native to the Caucasus; introduced, and spreading from the Plant Introduction Garden at Pullman.

Astragalus impensus (Sheldon) Woot. & Standl. Stems several, often branched, 3–4 dm. tall; herbage white strigose; stipules 3–10 mm. long, slightly united, scarious at base, partly green, with a strong midrib and a spinulose tip; leaves 1.5–2 cm. long; leaflets 5–7, linear-subulate, 6–15 mm. long, 0.5–2 mm. wide, with a strong spinulose tip; peduncles 5 mm. long or less, mostly 2-flowered; calyx-tube 1.5–2 mm. long, campanulate, strigose, the teeth 1–3 mm. long, subulate; corolla 4–5 mm. long, ochroleucous; pod 6–7 mm. long, lanceoloid, strigose; seeds 3 mm. long, 1.5 mm. wide, brown, usually 2. *Kentrophyta impensa* (Sheldon) Rydb.

Included by Piper & Beattie as *A. viridis* on the basis of *Brandegee* 734. In the Flora of Wash. Piper lists this number as from Walla Walla Region. The writer has borrowed this number from the Brandegee Herb., University of California. He considers it *A. impensus.* Though this specimen has for locality data only "Washington Terr.," the record is included here, trusting Piper's statement" Walla Walla Region," which he interpreted as, "High ridges of the Blue Mountains." No verification of this has been seen, nor are there more recent collections. In any case, the locality might still be outside of our geographic limits. Further collections are much desired. C. L. Hitchcock and Cronquist (1961) reduced this species to *A. kentrophylla* Gray.

Astragalus inflexus Dougl. ex Hook. Densely white-villous throughout; stems prostrate or decumbent, 1–5 dm. long; leaves short-petioled; leaflets 5–16 mm. long, oval to elliptic and acute at each end, short-stalked; peduncles scarcely exceeding the leaves; racemes densely 6–12-flowered, the flowers short-pedicelled; calyx-tube 10–12 mm. long, cylindric, the teeth about half as long, awl-shaped; corolla 20–26 mm. long, rose to magenta; pod 1.5–3 cm. long, lance-ovoid, stout-beaked, curved, villous, often much compressed at the sutures;

seeds 2.5–3 mm. long, obliquely bean-shaped, brown. *Xylophacos inflexus* (Dougl.) Rydb. Dry rocky or sandy slopes, Snake River Canyon and tributaries, *U. Son.*

Astragalus lentiginosus Dougl. ex Hook., var. **lentiginosus.** *Dotted Milk Vetch.* Slightly appressed puberulent but green; stems spreading, 10–35 cm. long; stipules ovate, acuminate; leaflets 11–21, 6–18 mm. long, elliptic to obovate, obtuse or retuse, sparsely puberulent to glabrate; peduncles shorter than the leaves; racemes dense; calyx-tube 2–4 mm. long, appressed strigose, the lobes lance-linear, shorter; corolla 8–10 mm. long, whitish or purple tipped; pods 15–24 mm. long, firm, puberulent, ovate, curved, acuminate, not stipitate; seeds 2–2.5 mm. long, bean-shaped, flat, truncate at one end, olive-brown, minutely pitted. *Cystium lentiginosum* (Dougl.) Rydb. Included on the single, somewhat doubtful, record, Wilson Creek—Whitman Co., *Lake & Hull* 663. Pollen poisonous to honey bees.

Astragalus miser Dougl. ex Hook, var **miser.** Caudex caespitose; stems 5–15 cm. tall, several, slender, ascending, strigose canescent; leaves 4–7 cm. long; leaflets 3–10 mm. long, oblong-linear, obtuse; racemes 2–5 cm. long, few-flowered; calyx 3–5 mm. long, campanulate, strigose with mingled white and black hairs, the teeth half as long as the tube; corolla 8–10 mm. long, yellowish, but the keel purple-tipped; pods 10–20 mm. long, subsessile, narrowly oblong, acute and cuneate, slightly rugulose, strigose, cordate in cross section, the dorsal suture sulcate, the lower one prominent. *A. serotinus Gray.*

The holotype was collected by Douglas, "On low hills of the Spokan River, sixty miles from its confluence with the Columbia."

Astragalus Mortoni Nutt. Erect, 3–11 dm. high, minutely lepidote-pubescent, but green, branching above; leaves 10–15 cm. long; leaflets 11–23, oval or elliptic, apiculate, truncate, or notched at apex, glabrous above, lepidote beneath, 1–4 cm. long; peduncles as long as or longer than the leaves, stout; spikes dense, 3–15 cm. long; flowers reflexed, but the fruit erect; calyx-tube campanulate, gibbous on upper side, 4–6 mm. long, black or white puberulent, the two upper teeth 1.5–2 mm. long, broader than the lower; corolla 12–16 mm. long, greenish; pods 9–23 mm. long, 3–5 mm. thick, narrowly oblong, puberulent, sulcate on the lower suture, sessile, densely crowded, with a slender hooked beak; seeds 2 mm. long, reniform, brown. *A. pachystachys* Rydb. at least as to local specimens; *A. canadensis* L., var. *Mortoni* (Nutt.) Wats.; *A. candensis,* var. *brevidens* (Gandoger) Barneby. Thickets, *U. Son., A. T.,* abundant, open woods, *A. T. T.*

Astragalus Purshii Dougl. ex Hook., var. **Purshii.** Herbage densely white-villous, stems less than 1 dm. long, ascending or spreading, caespitose; leaflets 7–15, sessile, 5–15 cm. long, lance-elliptic, often acute at each end; peduncles shorter than the leaves; racemes short, 5–10-flowered; bracts narrow, exceeding the pedicels; calyx-tube 8–10 mm. long, cylindric, slightly oblique at base, the lobes subulate, subequal, half the length of the tube; corolla 18–25 mm. long, pale yellowish, the tip of the keel purple; pods 1.5–3.5 cm. long, ovoid, densely villous, stout beaked, curved; seeds 3 mm. long, obliquely reniform, brown. *Xylophacos Purshii* (Dougl.) Rydb. Gravelly soil, especially near Spokane, *A. T.*

Collected by Douglas "on the low hills of the Spokan River."

Astragalus reventus Gray. Erect, 3–6 dm. high, strigose; leaves erect or ascending; leaflets 17–37, linear, usually obtuse or truncate, cuneate at base, subsessile, usually glabrate above, 5–20 mm. long; peducles stout, much longer than the leaves; flowers in rather dense erect racemes; calyx long black-pubescent, the tube 4–6 mm. long, campanulate, slightly swollen at base on the upper side, the acute teeth 1–3 mm. long; corolla 14–18 mm. long, whitish; pods 15–24 mm. long, 4–8 mm. wide, transversely roughened, wrinkled, turgid, the dorsal suture not impressed, from narrowly ovoid to globose, stout-beaked;

seeds 2 mm. long, oval, flattened, black. *Cnemidophacus reventus* (Gray) Rydb. Grassy or rocky slopes, Snake River Canyon and low valleys in the Blue Mts., *U. Son.*

Astragalus riparius Barneby. Erect, 1–2 dm. tall, greenish-cinereous strigulose; stipules 3.5–7 mm. long, free, lance-deltoid; leaves 10–21 cm. long; leaflets 21–37, obtuse, subemarginate, or rarely acute, 4–21 mm. long, narrowly oblong or oblanceolate; peduncles 1.4–3 dm. long, erect, stout; raceme 12–21-flowered; calyx black pilosulous, the tube 4.6–5.8 mm. long, campanulate, the teeth 2.2–5.2 mm. long, lance-subulate; corolla cream-colored, drying yellowish; banner ovate-cuneate, 5.5–8.2 mm. broad; keel 10.2–12.8 mm. long; pod 1.7–2.5 cm. long, straight or nearly so, at first fleshy, later leathery or almost woody, brown, rugose, at base rounded or truncate, the apex abruptly narrowed into a triangular beak, both sutures doubly undulate winged; seeds 2.6 mm. long, brown or black. Grassy hillsides or open pine forests, upper slopes of Clearwater and Snake Canyons, *A. T., A. T. T.*

The type is from Wawawai, *Piper 4,133.*

Astragalus Spaldingii Gray. Densely white-villous throughout; stems 1–4 dm. tall, erect or decumbent; leaves 6–10 cm. long, short-petioled; leaflets 15–29, narrowly elliptic to linear-lanceolate, 5–15 mm. long, acute at each end, sessile; peduncles longer than the leaves; flowers in dense spikes, 1–7 cm. long; calyx-tube 3–4 mm. long, campanulate, the linear lobes 2–3 mm. long; corolla whitish, the keel purple-blotched and veined; pod 5–7 mm. long, villous, ovoid, beaked, 1- or 2-seeded, oval in cross-section, the dorsal suture slightly impressed; seeds 2–2.5 mm. long, reniform to oval, flattened, olive-brown. Grasslands, or even alkali flats, *U. Son., A. T.*

The type was collected by Rev. Henry Spalding on the plains of Kooskooskee River [Clearwater R., Lapwai, Idaho].

Astragalus striatus Nutt. ex T. & G. Herbage white strigose or glabrate; stems ascending, 2–5 dm. high; leaflets 13–25, oblong, obtuse, 5–20 cm. long, oblong or elliptic, silky strigose beneath; peduncles exceeding the leaves; racemes 3–6 cm. long, dense, oblong; bracts lanceolate, shorter than the calyx; calyx strigose with black, and white hairs, the tube 4–6 mm. long, the teeth 3–4 mm. long, subulate; corolla 15–18 mm. long, light purple or white; pods 7–10 mm. long, erect, straight, triangular-compressed, with the lower suture deeply sulcate. At Silver Lake, Spokane County, *Henderson* 2,359, also at Medical Lake according to M. E. Jones.

Astragalus Whitneyi Gray, var. **Sonneanus** (Greene) Jeps. *Balloon Plant.* Freely branching from a stout caudex; herbage pale, silky or villous, with a fine whitish pubescence; stems ascending or decumbent, 5–30 cm. long, usually much branched at base; leaflets 11–19, 4–13 mm. long, narrowly elliptic, petiolulate, stipules lanceolate, the lower ones sheathing; peduncles shorter than the leaves, the racemes few-flowered; bracts setaceous, about as long as the pedicels; calyx-tube 3–4 mm. long, pubescent with black and white hairs, the teeth subulate, shorter than the tube; corolla 8–10 mm. long, whitish, the tip purple tinged; pods 2–3 cm. long, pale, mottled with purple splotches, obovoid, thin, inflated, appressed puberulent, the stipe longer than the calyx; seeds 2.5–3.5 mm. long, nearly cordate but one side smaller, flat, pitted, brown. *Phaca Hookeriana* T. & G.; *A. Hookerianus* (T. & G.) Gray, not Dietr.; *A. Sonneanus* Greene. High rocky ridges of the Blue Mountains, *Huds.*

First collected by Douglas in "Interior of Oregon."

GLYCYRRHIZA

Perennial glandular-dotted herbs with long thick sweet roots; leaves odd-pinnate, with minute deciduous stipules; flowers in axillary spikes or heads; calyx with the two upper lobes shorter or partly united; stamens usually diadelphous; anther-sacs confluent at the apex, the alternate ones smaller; pod prickly or glandular, ovate or oblong-linear, compressed, often curved, nearly indehiscent, few seeded. Name from Gr., *glukus,* sweet; *rhiza,* root.)

Glycyrrhiza lepidota (Nutt.) Pursh, var. **glutinosa** (Nutt.) Wats. *Wild Licorice.* Rhizome woody, slender, long creeping at depth of about 3 dm.; stems 5–10 dm. tall, erect, simple or with upper part producing erect branches, with resinous glandular dots, otherwise glabrous, or puberulent but glabrate, or more or less glandular hispid; leaves cauline, 4–20 cm. long, mostly 11–19-foliolate; leaflets 1–6 cm. long, lanceolate, dark green above, puberulent on midrib beneath and margin; petiolules 0.5–2 mm. long, puberulent, stout; stipules linear; peduncles shorter than the leaves, puberulent and commonly glandular hispid; spikes 3–7 cm. long, dense, many flowered; bracts lanceolate, glandular; calyx 6–10 mm. long, cylindric-campanulate, glandular hispid, pale, the linear-lanceolate teeth longer than the tube; corolla greenish yellow or sometimes almost white or bluish; standard 10–14 mm. long, enclosing the wings and keel; pods 1–2 cm. long, lanceolate, almost hidden by the stout, hooked prickles; seeds 2–3 mm. long, bean-shaped, brown. Sandy or gravelly banks, Snake River and principal tributaries, *U. Son.*; also by Latah Creek, Spokane; and south end of Lake Coeur d'Alene, *A. T.*

LATHYRUS. Pea Vine

Mostly smooth, perennial herbaceous vines or erect herbs; leaves pinnate, with tendrils, or tendrils much reduced, or wanting; flowers in racemes or sometimes solitary; peduncles axillary, usually equaling or exceeding the leaves and several-flowered; calyx-teeth nearly equal or the upper ones somewhat shorter than the lower; stamens diadelphous or monadelphous below; styles dorsally flattened near the top, hairy on the inner side; ovules generally numerous; pod flat or terete, 2-valved, 1-celled. (Name Gr., *lathuros,* used by Theophrastus for a leguminous plant.)

Stem with two foliaceous wings; leaflets 2. — *L. latifolius.*
Stems not winged; leaflets 2–12,
  Tendrils simple, mere bristles,
    Leaflets 2 or 4 (–6–8); flowers 8–14 mm. long,
      Leaflets narrowly elliptic to oval. — *L. bijugatus,* var. *bijugatus.*
      Leaflets linear. — Var. *Sandbergii.*
    Leaflets 4–10, ovate to oval. — *L. nevadensis,* subsp. *nevadensis.*
  Tendrils all, or some of them, prehensile, simple or forked,
    Plants glabrous,
      Leaflets elliptic to oval, 1–5 cm. long. — *L. pauciflorus,* var. *pauciflorus.*
      Leaflets linear or lance-linear, 2–9 cm. long. — Var. *tenuior.*
    Plants pubescent,
      Leaflets 3–8-times as long as broad,
        Tendrils prehensile, often 3-forked. — *L. Lanszwertii,* subsp. *Lanszwertii.*

Tendrils mere mucros:—See following heading.
Leaflets broader,
Corollas 17–27 mm. long; tendrils none, or a mucro,
Corollas pink to purple. *L. nevadensis*, subsp. *nevadensis*.
Corollas white or merely pink-lined. Subsp. *Cusickii*.
Corollas less than 17 mm. long; tendrils usually present,
Corollas bluish to reddish-purple. Subsp. *lanceolatus*, var. *lanceolatus*.
Corollas white or but pinkish lined. Var. *Parkeri*.

Lathyrus bijugatus White, var. **bijugatus.** Perenial, rootstocks slender; stems 1–3.5 dm. tall, often branched, somewhat puberulent above; stipules 6–15 mm. long, linear semisagittate, foliaceous; lower leaves with 2 leaflets, the upper with 2 or commonly 4; leaflets 2–5 cm. long, mucronate or acute; peduncles 2–3-flowered, about as long as the rhachis of the subtending leaf; calyx-tube 3–4 mm. long; calyx-teeth triangular, the lower longer, more slender, ciliolate, 2–3.5 mm. long; flowers when dried bluish purple; standard 10–14 mm. long, the limb suborbicular; keel up-turned by 2 rounded bends; narrowed to an obtuse tip; pods 2.5–3.5 cm. long, 4–5 mm. wide, flat, glabrous; seeds 2.5–3 mm. in diameter, discoid, brown. Common in late spring, prairies or thickets, *A. T.*

The type collection was made by J. H. Sandberg in copses in Latah Co.

Var. **Sandbergi** White. Differing in having: narrower stipules; leaflets 3–14.5 cm. long, linear. The flowers are magenta-pink. *L. Sandbergii* (White) Howell. Abundant, prairies or thickets, *A. T.*

The type locality is Latah Co., Ida.

C. L. Hitchcock (1952) reduced this to the synonymy of the species, but the characters here stated distinguish it as a well recognizable variety.

Lathyrus Lanszwertii Kellogg, subsp. **Lanszwertii.** Perennial from rootstocks, 1.5–8 dm. tall, erect or climbing; stems angled; herbage mostly soft pubescent, rarely glabrate or glabrous; stipules ¼–¾ as long as leaflets, mostly linear to linear-lanceolate, 2-lobed or entire; leaflets 4–10 (–12), and 3–10 cm. long, 5–10 mm. broad, mostly elliptic to oblong-elliptic, often coriaceous and heavily veined; tendrils mostly well developed; racemes 2–8-flowered, shorter than or to twice as long as the leaves; flowers 13–16 mm. long; calyx-tube 5–6 mm. long, pubescent; calyx-teeth 1.5–3 mm. long, narrowly lanceolate, ciliate; corolla tinged with pale lavender, pink, or violet, darker lined; pods 4–6 cm. long, 3–6 mm. broad, glabrous; seeds not seen. Blue Mts., *Darlington,* in 1913.

Lathyrus latifolius L. *Everlasting Pea.* Perennial, glabrous; stems 0.5–2 m. tall, ascending or climbing; stipules 10–25 mm. long, semi-sagittate, foliaceous; leaflets coriaceous, 4–9 cm. long, lanceolate to narrowly elliptic, apiculate; tendrils as long as the leaves, divided; peduncles 6–14-flowered, exceeding the leaves, erect, up to 25 cm. in length; calyx-tube 4–5 mm. long; calyx-teeth ciliolate, the upper broadly deltoid, the lower narrower, lanceolate, 3–4 mm. long; plants with either rose-purple or white flowers; standard 2–2.5 cm. long, cordate, reflexed; keel up-turned at right angles at the middle, then reflexed near the obtuse cucullate tip; ovary granular; pod 5–8 mm. long, 6–9 mm. broad, flat, glabrous; seeds 5–6 mm. long, oval, a little flattened, dark mahogany brown, rugose ridged. Spreading from cultivation in Pullman, now well established, *A. T.*

Lathyrus nevadensis Wats., subsp. **nevadensis.** Perennial from slender rootstock; stems 1–4.5 dm. tall, often branched, erect, pilosulous above; stipules 5–10 mm. long, semisagittate or narrowly so; the 4–8 leaflets 2–12 cm. long, ovate to oval, above bright green, glabrous, beneath pale green and at first pilosulous, later mostly glabrate; tendril a minute bristle, or on large plants elongate, even exceeding the leaflets, but unbranched; peduncles 2–5-flowered, well exceeding their subtending leaves, pilosulous or glabrate; calyx-tube 5–6 mm. long, minutely pilosulous, greenish; calyx-teeth lanceolate, subequal, pilosulous or ciliolate; the lower 3–3.5 mm. long; corolla pink, red, blue, or purple; standard 17–27 mm. long, the limb oval, up-turned at an acute angle, white with purple veins down the middle; keel gradually up-turned to nearly a right angle, broad to the abrupt, obtuse tip; pods 2.5–3 cm. long, about 5 mm. wide, flat, glabrous; seeds not seen. *L. nevadensis* Wats., var. *stipulaceus* (White) St. John. Thickets or open woods, *A. T., A. T. T.*

Subsp. **Cusickii** (Wats.) C. L. Hitchc. Stipules 4–8 mm. long, semisagittate, lanceolate; leaflets 2–8, pale green, 3–10 cm. long, beneath sparsely pilose; tendril a minute bristle; peduncles about equaling the subtending leaves, sparsely pilose and rufous glandular narrowly linear to broadly ovate or obovate, up to 12 cm. long and 3 cm. broad, beneath sparsely pilose; tendril a minute bristle; peduncles about equaling the subtending leaves, sparsely pilose and rufous glandular; flower white or white with purple lines, ageing to tan; calyx sparsely pilosulous, the two upper lobes broadly deltoid, connivent, the lower the longer, 2.5–3.5 mm. long; standard 17–22 mm. long; pods 4–5 cm. long, 6 mm. wide; seeds not seen. *L. Cusickii* Wats. Open woods or prairies, *A. T.*

Subsp. **lanceolatus** (Howell) C. L. Hitchc., var. **lanceolatus.** Stems 1.5–2.5 dm. tall, slender, ridged, finely pilose; stipules semisagittate, finely pilose less than half as long as the leaflets; leaflets 8–27 mm. long, elliptic, slightly cuneate at base, apiculate, above bright green and glabrous, beneath pale green; tendril a minute, pilose bristle; peduncles much exceeding, commonly twice the length of, the rhachis of the subtending leaf, sparsely pilose, 3–5-flowered; calyx-tube 3–4.5 mm. long, pilosulous; the two upper calyx-teeth deltoid, incurved; the lower lanceolate, longer, 2–2.5 mm. long, pilosulous; flowers bluish to reddish purple; standard 14–17 mm. long, suborbicular, reflexed; keel up-turned with one rounded bend at right angles, gradually narrowed to an obtuse tip. *L. pedunculatus* St. John. Fir woods, n. w. Ida., *Can.*

Subsp. **lanceolatus,** var. **Parkeri** (St. John) C. L. Hitchc. Glabrous nearly throughout, 2–7 dm. tall; rootstock 1–2 mm. in diameter, pale, creeping; stem erect, striate; leaves numerous, the basal withered at anthesis; stipules 5–8 mm. long, linear-semisagittate, acuminate; petiolules swollen, pilosulous; the 6–10 leaflets 15–45 mm. long, thin chartaceous, oval to ovate, short apiculate, entire, above bright green, beneath pale greenish, the venation reticulate, prominent; tendrils well developed, 3–7 cm. long, simple, or forked, apparently not climbing; peduncles 2–6 cm. long, axillary; racemes 4–7-flowered; calyx with a few pilosulous hairs, the tube 4–5 mm. long on the lower side, the upper teeth 0.5–1 mm. long, deltoid, the lower 1.5–2 mm. long, linear; corolla 13–17 mm. long, cream-colored, white, or with pink lines. *L. Parkeri* St. John.

The type is Grizzly Camp, T. 42N., R. 2W., Latah Co., *Parker* 511. It is named in honor of the late Prof. Charles S. Parker of Howard University, my former student at Pullman. Moist woods, *Can.*

Lathyrus pauciflorus Fern., subsp. **pauciflorus,** var. **pauciflorus.** Perennial, glabrous; rootstock cord-like, branching, erect; stems 2–10 dm. tall, angled; stipules 5–30 mm. long, about half as long as the leaflets, foliaceous, broadly semihastate, the larger ones irregularly toothed or lobed; leaflets at length firm, mucronate, paler green beneath; tendril well developed, forked; peduncles 5–30 cm. long, often exceeding the leaves, 3–8-flowered; calyx-tube 5–6 mm. long; upper calyx-teeth deltoid, connivent, ciliolate, the lower longer, lanceolate, 4–7 mm. long, ciliolate; flowers magenta, showy; standard 18–27 mm. long, reflexed, the limb suborbicular, notched; keel with a rounded upward bend to the broad, obtuse tip; pods 4–6 cm. long, 6–7 mm. wide, flat, narrowed towards the base; seeds not seen. Grassy slopes or open woods, *A. T.*, *A. T. T.*

The type locality is the Snake River Canyon, Almota, *Piper* 1.487.

Var. **tenuior** (Piper) St. John. Differs from the species in having the leaflets 2–9 cm. long, linear or lance-linear. *L. parvifolius* Wats., var. *tenuior* Piper; *L. pauciflorus tenuior* Piper; *L. tenuior* (Piper) Rydb. Grassy slopes, with the species, *U. Son.*, *A. T.*

This plant differs from var. *pauciflorus* only in its longer, narrower leaflets, and intermediate specimens occur. Hence, it is logically classified as a variety. Prof. Piper's final transfer was as a trinomial without indication of category. The type specimen is *Elmer* 52, from the Snake River Bluffs near Almota.

Though C. L. Hitchcock (1952) reduced this variety to the synonymy of the species, it seems a good variety.

LOTUS

Herbs or shrubs; leaves pinnate, 3–many-foliolate; stipules foliaceous, scarious, minute and gland- or spine-like; flowers yellow, or reddish, or white, solitary or in umbels; petals free from the diadelphous stamens; standard ovate or roundish, with a basal lobe approximate to a pocket on the wings; pod linear, compressed or somewhat terete, sessile, several seeded. (Name an ancient Gr. plant name.)

Flowers solitary; roots fibrous, rhizome none; stipules gland-like. *L. Purshianus.*
Flowers in umbels; rhizome or thickened caudex present,
  Stipules scarious; flowers 12–14 mm. long; pod nearly straight. *L. pinnatus.*
  Stipules gland-like; flowers 7–9 mm. long; pod strongly curved. *L. Douglasii.*

Lotus Douglasii Greene. *Trailing Bird-foot.* Main root stout, erect; stems many, caespitose, prostrate, 3–7 dm. long, appressed hirsutulous; leaves 1–2.5 cm. long, pilose; stipules minute, dark; leaflets 3–5, 5–15 mm. long, rhombic obovate to oblanceolate, acute; umbels shorter than the subtending leaves, pilose, 5–10-flowered, subtended by a foliaceous, 1–3-foliolate bract; calyx pilose, the tube 3 mm. long, campanulate, longer than the subulate teeth; corolla bright yellow, with reddish veins towards the base, the claws 3 mm. long; limb of banner suborbicular; wings longer than the gradually up-curved keel; pod indehiscent, pilose, curved into a half or full circle, the beak 5 mm. or less in length, the body 1 cm. or less in length, few seeded; seeds 2 mm. long, bean-shaped, brown. *L. nevadensis* (Wats.) Greene, var. *Douglasii* (Greene) Ottley; *Hosackia decumbens* Benth; *Syrmatium decumbens* (Benth.) Greene. Abundant in pine woods on dry sand plains northward from the terminal morraine near Spokane; Chatcolet; in the channeled scablands south of the morraine, Rock Lake, *Beattie & Lawrence* 2,453; and a single outlying station on thin soil over basalt, Palouse, *St. John & Smith* 8,381, *A. T. T.*

Lotus pinnatus Hook. Perennial; rhizome cord-like; stems 2–4 dm. long, erect or decumbent, appressed puberulent above, glabrate below; leaves 2–8 cm. long, glabrous, 5–9-foliolate; stipules 2–5 mm. long, lance-ovate; leaflets 5–30 mm. long, oblanceolate or obovate, paler green beneath; peduncles 2–13 cm. long, appressed puberulent at tip; leafy bracts usually none, but minute scarious bracts forming a terminal involucre, subtending 3–7 flowers; calyx-tube 4 mm. long, cylindric-campanulate, glabrous; the three lower teeth 2-3 mm. long, subulate, ciliolate, the two upper lanceolate, more united, ciliolate; claws of petals exserted; banner deep yellow, the limb broadly ovate; wings 12–14 mm. long, white; keel shorter, sharply bent or reflexed near the tip; pod 4–8 mm. long, 2 mm. wide, brown; seeds 2–2.5 mm. long, oval, flattened, shining, dark brown with darker spots. *Hosackia bicolor* Dougl. Wet places, rare, Whitman Co. and Latah Co., *A. T.*

Lotus Purshianus (Benth.?) Clem. & Clem. *Spanish clover.* Annual, 1–6 dm. tall, erect or ascending, villous, pilose, puberulent, or nearly glabrous, the larger plants much branched; leaves subsessile or very short petioled, 3-foliolate, or the vigorous branches with some leaves 4–5-foliolate, and the tips of branches occasionally with 1-foliolate leaves; leaflets 5–25 mm. long, ovate, lanceolate, or elliptic, the lateral ones subsessile; peduncles exceeding the leaves, but those of late flowers much reduced; bract leaf-like, 1-foliolate; calyx-tube 1.5–2 mm. long, obconic; calyx-lobes 3–4 mm. long, subulate; corolla 4–9 mm. long, white or cream-colored, veined or tipped with rose; standard suborbicular, only slightly exceeding the wings and keel; wings obliquely elliptic; keel boat-shaped; pods 2–3.5 cm. long, linear, flattened, 3–7-seeded; seeds 2–4 mm. long, oblong, flattened, brown. *Hosackia americana* (Nutt.) Piper; *H. americana pilosa* (Nutt.) Piper; *L. americanus* (Nutt.) Bisch., not Vell.; *L. americanus,* var. *minutiflorus* Ottley. Abundant, usually open soils, hillsides, or in cultivated land, *U. Son.* to *A. T. T.*

This abundant annual is variable in stature, pubescence, and size of peduncle, and flower. Authors have made numerous species and varieties on these. The monographer, Miss Ottley, has shown most of them to be invalid because of the many intermediates. In like manner her var. *minutiflorus* is here reduced, since many plants have flowers of the intermediate length of 5–6 mm., and there seems to be no natural separation on this character.

LUPINUS. Lupine

Annual, biennial, or perennial herbs or shrubs; leaves palmately compound (in ours); petioles short or elongate, dilated at base; stipules present; leaflets 5–17 (or rarely 3–4); flowers racemose, the bracts usually deciduous; calyx 2-lipped, the lips entire, or the upper bifid, often with bractlets between the lips; standard commonly with a ventral, median sulcus, a pair of rounded umbos near the center, the sides commonly reflexed, the back glabrous or more or less pubescent; wings usually glabrous, or sometimes pubescent near the upper distal corner; keel arcuate or nearly straight along the upper edges, ciliate or non-ciliate; stamens 10, monadelphous, the anthers alternately long and short; fruit a flattened legume; ovules 2–12; seeds with a sunken hilum which is often surrounded by a thickened ring. (Name Lat., from *lupus,* wolf; probably because of an erroneous impression that these plants rob the soil.)

The Lupines have been widely advertised as poisonous plants. It is now well known that many species contain poisonous alkaloids in their pods and especially in their seeds. Though the effect of the poison is not cumulative, yet if the grazing animals, especially sheep, eat a large amount of the fruit, fatal poisoning often results. On the other hand they often seek the foliage and graze it with impunity.

Apex of standard normally well reflexed from upper distal corner of wings,
  Standard more or less silky on the back; leaves silky on both sides; keel ciliate on upper edges; pedicels spreading pilosulous. *L. sericeus.*
  Standard glabrous on the back,
    Keel non-ciliate; pods 5–12-seeded; leaflets glabrous or glabrate above. *L. Burkei.*
    Keel more or less ciliate on the upper edges; pods 4–7-seeded,
      Leaflets glabrous or glabrate above; pedicels spreading pilosulous; stems more or less fistulous and villous. *L. Wyethii,* var. *prunophilus.*
      Leaflets permanently pubescent above,
        Petals usually yellow; flowers 15–18 mm. long; largest leaflets 150 x 25 mm. *L. Sabinii.*
        Petals blue; flowers 10–14 mm. long; largest leaflets usually less than 50 x 6 mm.,
          Pedicels spreading pilosulous. *L. minimus.*
          Pedicels appressed pilosulous. *L. ornatus.*
Apex of standard normally not much reflexed from upper distal corner of wings, or pedicels less than 3 mm. long,
  Pedicels less than 3 mm. long, or racemes spicate; keel ciliate; floral bracts more or less persistent,
    Plants 1–4 dm. tall; stems few-leaved; standard glabrous on the back,
      Largest leaflets 14–17 mm. long. *L. lepidus,* var. *medius.*
      Largest leaflets 20–32 mm. long. *Var. aridus.*
    Plants 3.5–10 dm. tall; stems many-leaved; standard usually pubescent on the back,
      Leaflets grayish or greenish, subsericeous to glabrate above; stem hairs 3–4 mm. long, spreading or retrorse. *L. leucophyllus,* var. *retrorsus.*
      Leaflets densely appressed villous on both sides,

Stems densely spreading or appressed pubescent, the longest hairs hardly 1 mm. long. Var. *canescens.*
Stems densely spreading villous, the longest hairs more than 1 mm. in length,
Flowers 11–14 mm. long, longer than tall. *L. leucophyllus,* var. *leucophyllus.*
Flowers 7–10 mm. long, almost as tall as long. Var. *Belliae.*
Pedicels slender, 3–8 mm. long, or keel non-ciliate, or standard not pubescent on the back, or floral bracts deciduous,
Wings more or less pubescent on outer surface near upper distal corner, or standard long clawed and upper calyx-lip very short, much exposed,
Leaflets permanently pubescent above. *L. arbustus,* subsp. *neolaxiflorus.*
Leaflets glabrous or glabrate above. Ssp. *pseudoparviflorus.*
Wings glabrous, or with a few scattered villous hairs near middle or near claws,
Leaflets glabrous or glabrate above; keel more or less ciliate. *L. laxispicatus,* var. *Whithamii.*
Leaflets permanently hairy above,
Petals bright yellow to nearly white; pubescence of stem mostly appressed. *L. sulphureus.*
Petals blue or lilac; stems spreading pubescent,
Standard pubescent on back. Var. *echlerianus.*
Standard glabrous. *L. leucopsis,* var. *mollis.*

Lupinus arbustus Dougl. ex Lindl., subsp. **neolaxiflorus** Dunn. Perennial; stems 3–8 dm. tall, erect or ascending, appressed subsericeous; leaves cauline; petioles 3–15 cm. long; the 7–9 leaflets 3–4 cm. long, oblanceolate, subsericeous above and below; peduncles 3–8 cm. long, racemes 8–18 cm. long, compactly or loosely flowered; bracts early deciduous; flowers 10–12 mm. long, blue, rose, or pale yellow; pedicels 4–8 mm. long; calyx-tube distinctly spurred above, the spur 1 mm. or less in length; upper lip of the calyx green, notched, broad and short, largely or entirely exposed, the lower lip narrower, entire or tridentate; standard long-clawed, more or less hairy near the middle of the back; wings almost always pubescent on the outer surface near the upper distal corner; keel rarely nonciliate; pods 20–25 mm. long, densely yellowish villous, 4–6-seeded; seeds about 5 mm. long, discoid, reddish-brown or pinkish, obscurely spotted. Dry open woods, Blue Mts., and Craig Mts., *A. T. T.*

The foliage is the first choice of grazing sheep, which prefer it to anything else in the area.

Subsp. **pseudoparviflorus** (Rydb.) Dunn. Differs from the above in having the upper surface of the leaflets glabrous or nearly so. The two grow together and look alike, and intermediates occur. *L. pseudoparviflorus* Rydb.; *L. laxiflorus* Dougl., var. *pseudoparviflorus* (Rydb.) C. P. Sm. & St. John. Abundant on grassy slopes or open woods, slopes of Craig Mts. and Blue Mts., *A. T.*, *A. T. T.*

Lupinus Burkei Wats. Perennial; stem 3–8 dm. tall, erect, glabrate except in the inflorescence, usually succulent and fistulous; lower leaves long petioled; petioles 8–20 cm. long; the 5–10 leaflets 4–9 cm. long, oblanceolate, acute, above bright green and glabrous, beneath paler and glabrate; peduncles 5–10 cm. long; racemes 8–30 cm. long, many flowered; bracts villous, rather persistent; flowers 10–13 mm. long, scarcely verticillate, blue, the glabrous standard with a yellow center turning violet; pedicels 2–5 mm. long, spreading pilosulous; calyx loosely pilose, the upper lip entire or notched, the lower entire; keel slender, arcuate; pods 25–35 mm. long, 5–6 mm. wide, villous, 5–8-seeded; seeds 3–4 mm. long, obliquely ovoid, somewhat flattened, dull yellow, marbled with brown. *L. polyphyllus* Lindl. var. *Burkei* (Wats.) C. L. Hitchc. Meadows or moist woods, northward from Cheney and Chatcolet, *A. T.*, *A. T. T.* These localities all lie in the glaciated region north of the terminal morraine. There is a solitary outlying station at Rock Lake, which is in the channeled scablands of the glacial drainage.

Lupinus laxispicatus Rydb., var. **Whithamii** C. P. Sm. Perennial; stems 3–8 dm. tall, erect or ascending, shortly appressed hirsutulous; leaves few, long petioled; petioles slender, the longest 2–3 dm. long; the 7–14 leaflets 5–8 cm. long, oblanceolate to linear, green and glabrous or glabrate above; peduncles 6–10 cm. long; racemes 1–2 dm. long, rather loosely flowered; bracts 5–7 mm. long, lanceolate, deciduous or subpersistent; flowers 8–10 mm. long, blue, spreading, mostly scattered; pedicels 2–4 mm. long, mostly spreading pilosulous; calyx-tube more or less gibbous, the upper lip 3–4 mm. long, notched, the lower 5–6 mm. long, entire, standard more or less pubescent on the back, the orange center turning violet or brown; wings glabrous; keel non-ciliate; pods about 30 mm. long by 7 mm. wide, straw-colored, white villous, 4–5-seeded; seeds not seen.

This was reduced by Dunn (1955) to the synonymy of *L. arbustus*, subsp. *pseudoparviflorus;* and made a subsp. of *L. sulphureus* by Phillips (1955). Spokane Co., moist woods, *A. T. T.*

Lupinus lepidus Dougl. ex Lindl., var. **aridus** (Dougl.) Jepson. Plant 12–22 cm. in height; inflorescence 8–15 cm. long, about half exposed above the foliage; flowers 9–11 mm. long. *L. lepidus,* subsp. *aridus* (Dougl.) Detling. Near Spokane.

Var. **medius** (Detling) St. John. Perennial; stems 1–2 dm. tall, simple or branched, appressed silky and with few or many longer ascending hairs; leaves long petioled, crowded near the base; petioles 5–8 cm. long; the 6–8 leaflets 12–25 mm. long, oblanceolate, silky on both sides; peduncles 2–10 cm. long, ascending or erect; racemes 5–10 cm. long, cylindric, dense; bracts persistent; flowers 6–9 mm. long, violet-blue; pedicels 1–2 mm. long, spreading pilose; calyx villous, the upper lip cleft, the lower bi- or tri-dentate; standard oval or ovate, glabrous, changing from yellow to purple; upper edges of keel nearly straight; pods 10–12 mm. long, silky pilose, 3–4-seeded; seeds about 3 mm. long, whitish or pinkish. *L. lepidus,* subsp. *medius* Detling; *L. aridus,* var. *Torreyi* (Gray) C. P. Sm. Dry hillsides, *A. T.*

Lupinus leucophyllus Dougl. ex Lindl., var. **leucophyllus.** Perennial; stems 6–9 dm. tall, stout, erect, branched, woolly canescent and more or less villous, often densely villous; leaves many, the lower long petioled; petioles

3–15 cm. long; the 7–9 leaflets 3–6 cm. long, villous or woolly on both sides; peduncles 3–8 cm. long; racemes 8–30 cm. long, usually dense and cylindric; bracts shorter or longer than the buds, usually persistent; flowers 12–14 mm. long, white, pinkish, bluish, or purple, often turning brown; standard ovate or oblong, acute or rounded at apex; keel stout; pods 20–25 mm. long, 3–6-seeded; seeds reddish brown and obscurely mottled, or drab and gray spotted. Occasional, dry places, *U. Son., A. T.*

Var. **Belliae** C. P. Sm. Differs in having the stem more or less villous; and the flowers 8–11 mm. long, almost or quite as tall as long, stem 3.5–10 dm. tall. Abundant, dry places, meadows, or open woods, *U. Son.* to *A. T. T.*

This is the commonest of the varieties in our area. It was named in honor of Mrs. May Bell Zundel of Pullman.

Var. **canescens** (Howell) C. P. Sm. Differing in having the stems without spreading hairs; racemes densely or loosely flowered. *L. canescens* Howell. Rock Lake, *Lake & Hull* 432, *A. T.*

Var. **retrorus** (Henders.) C. P. Sm. Differs in having the stems with spreading or retrorse hairs; and the upper surface of the leaflets subsericeous to glabrate, at least greenish or grayish, not whitish. *L. retrorsus* Henders. Open woods, from Moscow northward, *A. T. T.*

The type collection was made by Prof. L. F. Henderson on grassy hills, Lake Coeur d'Alene, opposite Harrison.

Lupinus leucopsis Agardh, var. **mollis** (Heller) C. P. Sm. Stems 3–8 dm. tall, erect, branched, spreading villous and pilose; leaves densely subappressed silky pilose, the lower ones long petioled, the upper with petioles about equaling the leaflets; leaflets 5–8, linear-oblanceolate, 35–60 mm. long; pedicels 4–7 mm. long; calyx 5–6 mm. long, densely subappressed silky pilose, upper lip bidentate, lower lip entire; petals violet blue; keel non-ciliate; pods 20–25 mm. long, 5–6 mm. wide, subappressed silky pilose, 4–6-seeded; seeds 4.5–6 mm. long, obovoid, yellowish, slightly mottled. Medical L., Spokane Co., *Sandberg & Leiberg 49.*

Lupinus minimus Dougl. ex. Hook. Perennial; stems 15–30 cm. long, few, erect to decumbent, unbranched, subappressed silky, with 1–2 reduced leaves; leaves mostly basal; petioles 5–10 cm. long; the 5–9 leaflets 2–3 cm. long; oblanceolate, silky on both sides with somewhat spreading or subappressed villous hairs; peduncles 7–13 cm. long, usually surpassing the foliage and equaling the racemes; bracts deciduous; flowers 8–14 mm. long, deep blue, spreading, loosely subverticillate; pedicels 2–4 mm. long; calyx spreading villous, the upper lip bifid, the lower entire; standard paler at the center, suborbicular, glabrous; keel somewhat curved; pods 20–25 mm. long, spreading villous, 4–5-seeded; seeds 3–4 mm. long, oval, somewhat flattened, flesh-colored with brown spots. *L. piperi* Robins. Gravelly open or wooded slopes, *A. T., A. T. T., Huds.*

Lupinus ornatus Dougl. Differing from *L. minimus* in having the leaflets closely appressed silky pilose; the pedicels appressed pilosulous; the calyx appressed sericeous pilose; and the pods appressed pilose. *L. Hellerae* Heller; *L. minimus* Dougl. ex. Hook., var. *Hellerae* (Heller) C. P. Sm. & St. John. Gravelly or sandy bars of the lower Clearwater River, Ida., *U. Son.*

Lupinus Sabinii Dougl. ex Hook. Perennial; stems 6–12 dm. tall, branched above, appressed subsericeous; leaves cauline, the lower long petioled, scattered; petioles 5–25 cm. long, stout; the 8–12 leaflets 6–15 cm. long, oblanceolate, appressed subsericeous on both sides; peduncles 4–8 cm. long; racemes 15–30 cm. long; bracts early deciduous; flowers 15–18 mm. long, verticillate, bright yellow, or rarely pale purple; pedicels about 10 mm. long, spreading fine puberulent; calyx appressed silky, the upper lip notched, the lower entire or notched; standard suborbicular, glabrous; keel arcuate, densely ciliate on the upper edges; pods about 40 mm. long by 12 mm. wide, 5–7-seeded, dull yellow, densely appressed villous; seeds 6–7 mm. long, oval, flattened, dull reddish brown. *L. sericeus* Pursh, subsp. *sabinii* (Dougl. ex Hook.) Phillips. Very local, open slopes and lower woods Tucanon River, and Blue Mts., Walla Walla Co., *A. T. T.*

The type locality is in the Blue Mts., east of Walla Walla, where this beautiful species was discovered by David Douglas in 1826.

Lupinus sericeus Pursh. Perennial; stems 3–6 dm. tall, erect, simple or branched, appressed sericeous and more or less spreading villous; leaves cauline, the lower long petioled; petioles 3–10 cm. long; the 7–9 leaflets 3–6 cm. long, oblanceolate, acute; peduncles 4–8 cm. long; racemes 12–33 cm. long; bracts longer or shorter than the buds, deciduous; flowers 8–12 mm. long, purple, blue, rose-colored, creamy, or white, verticillate, or scattered below; pedicels 4–6 mm. long; upper calyx-lip bidentate, the lower entire; standard longer than broad; keel arcuate or nearly straight, ciliate on the upper edges; pods 20–25 mm. long, yellow; seeds about 6 mm. long, flesh-colored. *L. flexuosus* Lindl.; *L. ornatus* Piper; *L. ornatus,* subsp. *bracteatus* Robins; *L. subulatus* Rydb. Abundant, in clumps or colonies, dry rocky slopes, grasslands, or open woods. *U. Son.* to *A. T. T.*

Lupinus sulphureus Dougl. ex. Hook. Perennial; stems 4–10 dm. tall, erect, appressed subsericeous; lower leaves long petioled; petioles 4 20 cm. long; the 9–15 leaflets 3–6 cm. long, narrowly oblanceolate, appressed pilose on both sides, yet greenish; peduncles 3–6 cm. long; racemes 1–2 dm. long, rather densely flowered; bracts deciduous; flowers 10–12 mm. long, scattered or subverticillate; pedicels 4–8 mm. long, subappressed pilosulous; calyx more or less gibbous, but not spurred, the upper lip bidentate, the lower entire; standard more or less pubescent on the back; wings glabrous; keel ciliate on upper edges; pods 2–3 cm. long, silky pilose, about 4-seeded; seeds about 5 mm. long, oval, flattened, flesh-colored to brown, unspotted. *L. sericeus* Pursh, subsp. *asotinensis* Phillips, and var. *asotinensis* (L. L. Phillips) C. L. Hitchc. Abundant, open lower slopes to upper wooded slopes, Blue Mts., *A. T. to A. T. T.*

Discovered by Douglas in the Blue Mts., east of Walla Walla, in 1826. The foliage is avoided by grazing sheep.

Var. **echlerianus** C. P. Sm. Differing in having the stems spreading pilosulous; and the petals blue. Openings in dry woods in the Blue Mts., Echler Mt., Columbia Co., July 5, 1927, *Smith & St. John* 4152 (type) and 4164; East Oregon Butte, T. 7 N., R. 41 E., July 23, 1913, *Darlington, A. T. T.*

C. L. Hitchcock and Cronquist (1961) reduce this variety to *L. leucophyllus* Dougl. ex Lindl.

Lupinus Wyethii Wats., var. **prunophilus** (M. E. Jones) C. P. Sm. Perennial; stems 3–6 dm. tall, erect or ascending, stout, more or less fistulous and spreading villous; leaves mostly long petioled; petioles 10–30 cm. long; the 8–12 leaflets 5–10 cm. long, oblanceolate, acute, villous below; peduncles 6–10 cm. long; racemes 10–25 cm. long; bracts often exceeding the buds, often tardily deciduous; flowers 12–16 mm. long, blue, purple, or pink, scattered or subverticillate; pedicels 6–8 mm. long; calyx villous, the upper lip bidentate, the lower entire or tridentate; standard suborbicular; keel arcuate; pods 4–5-seeded, about 30 mm. long, and 8 mm. wide, villous; seeds 5–6 mm. long, oval, flattened, flesh-colored, not or obscurely spotted. *L. prunophilus* M. E. Jones; *L. polyphyllus* Lindl., var. *prunophilus* (M. E. Jones) Phillips; *L. arcticus* Wats., var. *prunophilus* (M. E .Jones) C. P. Sm. Common, rocky slopes, grasslands, or open woods, *U. Son.* to *A. T. T.*

MEDICAGO

Herbs (or rarely shrubs); leaves trifoliolate; leaflets commonly dentate, the veins excurrent from the teeth; flowers small, yellow or violet, in axillary heads or racemes; calyx-teeth short, nearly equal; standard obovate or oblong; wings oblong; keel obtuse; stamens diadelphous; anthers alike; ovary sessile or nearly so, 1–several-ovuled; pod curved or spirally twisted, reticulate or spiny, indehiscent, 1–few-seeded. (Name from the Gr. name of alfalfa, *Medike,* from Medea, its place of origin.)

Flowers blue; pod appressed pilosulous. *M. sativa.*
Flowers yellowish; pod glabrous,
  Corolla 2.5–3 mm. long; calyx appressed pilosulous; pods coiled once, not spiny, *M. lupulina.*
  Corolla 4–5 mm. long; calyx glabrous; pods coiled 2–3-times. *M. hispida.*

Medicago hispida Gaertn. *Bur Clover.* Annual, glabrous or with a few appressed hairs; tap-root 1 mm. in diameter, pale; branches 1–6 dm. long, spreading or procumbent; stipules 4–6 mm. long, laciniate dentate; petioles 1–2 cm. long; leaflets 6–15 mm. long, cuneate-obovate, emarginate, denticulate beyond the middle; peduncles 5–12 mm. long; racemes few-flowered; calyx glabrous, the tube funnelform, 1.5 mm. long, the lobes 1.5–2 mm. long, deltoid-lanceolate; corolla yellowish; pod 2.5–3 mm. wide, strongly reticulate, the dorsal suture thin, with numerous spines 2 mm. long; seeds several, 3–3.5 mm. long, reniform to bean-shaped, brown, smooth. Weed, introduced from Eurasia. In lawn, Pullman, *Swanberg,* Oct. 12, 1922, *A. T.*

Medicago lupulina L. *Black Medic.* Annual, appressed pilosulous throughout; tap-root 2–5 mm. in diameter, blackish; stems 2–9 dm. long, prostrate, angled; stipules 3–12 mm. long, lanceolate to ovate, acuminate, dentate; petioles 4–45 mm. long; leaflets 5–17 mm. long, oval or obovate, cuneate at base, denticulate beyond the middle; peduncles 1–3 cm. long; raceme spike-like, dense, 4–15 mm. long; calyx appressed pilosulous, the tube 0.5 mm. long, campanulate, the lobes 0.5–1 mm. long, lance-subulate; corolla yellow; pods 2 mm. wide, black, heavy reticulate, rounded; seeds 1.7–2 mm. long ovoid-bean-shaped, smooth, yellowish to reddish. Weed, introduced from Eurasia. Waste places, fields, and lawns, *A. T.*

Medicago sativa L. *Alfalfa; Lucerne.* Perennial, 3–10 dm. tall, sparsely appressed pilosulous or in age glabrate; tap-root 3–5 mm. in diameter, pale; stems erect, striate; stipules 5–10 mm. long, lanceolate, laciniate toothed at base; petioles 2–20 mm. long; leaves pinnate; leaflets 1–3 cm. long, oblong- or elliptic-oblanceolate, spinulose denticulate towards the apex; peduncles 1–3 cm. long; racemes 1–5 cm. long, dense; pedicels 1–3 mm. long; calyx-tube 2.5–3 mm. long, campanulate, sparsely pilosulous, the teeth 2.5–3 mm. long, linear-lanceolate, acuminate; corolla 8–10 mm. long; pods 1.5 mm. wide, spiraled 2–3-times; seeds 1.8–2.2 mm. long, bean-shaped, brown. Native of Europe, escaped from cultivation. Roadsides, *A. T.*

MELILOTUS

Annual or biennial herbs; leaves pinnately 3-foliolate; leaflets denticulate; flowers in racemes; calyx campanulate, with short equal teeth; corolla deciduous, free from the stamen-tube; stamens diadelphous; anthers all alike; ovary sessile or stipitate, few-ovuled; pod coriaceous, 1–2-seeded. (Name Gr., *meli,* honey; *lotos,* some leguminous plant.)

Flowers white; standard longer than the wings; pod reticulately ribbed. *M. alba.*
Flowers yellow; standard as long as the wings; pod cross-ribbed. *M. officinalis.*

Melilotus alba Desr. *White Sweet Clover.* Biennial; stems 3–20 dm. tall, branching, glabrous or sparsely appressed puberulent; petioles shorter than the leaflets; leaflets 1–3 cm. long, oblanceolate or oval-oblanceolate, glabrous above, below paler green and appressed puberulent, the margins serrate-dentate except at base; petiolules of lateral leaflets about 1 mm. long; stipules 2–10 mm. long, subulate; racemes 5–20 cm. long, dense in bud, loose in fruit; buds ascending; flowers divergent; fruit pendent; pedicels 1–2 mm. long; calyx-tube 1 mm. long, obconic, green; calyx-teeth longer, lance-subulate, pilose at base; corolla 4–5 mm. long; standard oval, the margins upturned at tip; pods 3–3.5 mm. long, ovoid, glabrous, reticulately ridged, with 1–2 (or 3) seeds; seeds 2 mm. long, ovoid, flattened, truncate at the hilum, greenish yellow. Cultivated and established, *U. Son., A. T.*

This European legume is cultivated as a rotation crop and for forage. At first grazing animals eat it reluctantly, but they can be trained to take it. However, eating of it exclusively is likely to produce poisoning due to its cumarin content.

Melilotus officinalis (L.) Lam. *Yellow Melilot.* Biennial, very similar to *M. alba,* but corolla yellow, 5–7.5 mm. long; calyx-teeth mostly shorter than the tube, usually glabrous; pod 3–4 mm. long, 4–8-seeded, ellipsoid, glabrous, with 5–8 prominent cross ribs, which are reticulately joined. A Eurasian plant, cultivated for forage, rarely persisting or spreading, Tucanon River, in 1913, *Darlington* 185; Spokane in 1927, *Jones, A. T.*

PETALOSTEMON. Prairie Clover

Perennial or rarely annual herbs, glandular-dotted; leaves odd-pinnate; spikes terminal, dense; calyx 10-nerved; corolla indistinctly papilionaceous, white, pink, purple, or yellowish; standard free, long clawed; stamens 5, monadelphous at base; ovary sessile, 2-ovuled; style filiform; stigma minute; pod membranous, somewhat compressed, enclosed in the calyx. (Name Gr., *petalon,* petal; *stemon,* thread, or stamen.)

Petalostemon ornatum Dougl. ex. Hook. Perennial; roots deep, woody, brown; stems 3–5 dm. tall, branched, glabrous, striate; leaves 3–5 cm. long, ascending, 5–7-foliolate; leaflets 10–22 mm. long, obovate or elliptic-oblanceolate, bright green above, beneath paler and prominently dark punctate; spikes 2–5 cm. long, about 15 mm. in diameter, ovoid or short cylindric; bracts ovate-lanceolate, acuminate, longer than the calyces, villous-ciliate; calyx-tube 3 mm. long, campanulate, villous, the teeth 1.5 mm. long, lanceolate, green, very villous; standard with the claw 4 mm. long, the blade 3.5 mm. long, oval, truncate or subcordate at base; other petals with the claw 1.5–3 mm. long, the blades 3–4 mm. long, elliptic; petals rose-colored; stamens about 6 mm. long, exserted; pods with the body 2 mm. long, ovoid, compressed, pilose on the suture, with a slender, pilose beak; seeds 1 mm. long, bean-shaped, brown. *Kuhnistera ornata* (Dougl.) Kuntze. Exposed, rocky slopes, Snake River Canyon, *U. Son.*

Though in the 1961 International Code, Rec. 75A, recommends that this genus be treated as masculine, it was published as a neuter, *Petalostemum,* by Pursh, then emended to *Petalostemon,* also neuter, by Persoon in 1807. Overriding all this, *Petalostemon,* a neuter, is in the official list of nomina conservanda.

PSORALEA

Shrubs or perennial herbs, usually glandular-dotted; leaves mostly 3–5-foliolate, with stipules; flowers blue, red, or white, in spikes or racemes; calyx-lobes 5, the lower longest; corolla papilionaceous; stamens 10, diadelphous or sometimes monadelphous; anthers of two kinds; ovary sessile or short-stalked, 1-ovuled; pod seldom longer than the calyx, thick, often wrinkled, indehiscent, 1-seeded. (Name Gr., *psoraleos,* scurfy.)

Leaves pinnate; leaflets ovate; flowers 10–12 mm. long; calyx accrescent. *P. physodes.*
Leaves digitate; leaflets oblanceolate; flowers 5–6 mm. long; calyx not accrescent. *P. scabra.*

Psoralea physodes Dougl. ex. Hook. Perennial herb with a rootstock; stems 3–9 dm. tall, striate, black hirsutulous, glabrate below; leaves 5–12 cm. long, 3-foliolate; stipules 5–10 mm. long, lance-linear, hirsute; leaflets 2–6 cm. long rhombic-ovate or occasionally suborbicular, acute or mucronate, dark green, hirsutulous on the veins and margin; peduncles 3–12 cm. long, black hirsutulous; racemes 1.5–4 cm. long, dense, black hirsute; calyx campanulate, black hirsute, the tube 4–6 mm. long in flower, in fruit accrescent and 6–8 mm. long, the teeth in flower with the 4 upper deltoid, and the lower larger, lanceolate 3–4 mm. long, in fruit becoming 5–8 mm. long; flowers yellowish white; standard longer than the other perianth parts, enclosing them, and only the margins upturned, the limb elliptic ovate, cordate above the broad claw; limb of wings oblong-semi-sagittate; keel shorter, purple-tipped, the broad tip upturned, eared at base, the claw exceeding the calyx-tube; pod enclosed, 6 mm. long, oval, somewhat compressed, membranous, black hirsutulous; seeds 5 mm. long, bean-shaped, purplish black. *Hoita physodes* (Dougl. ex Hook.) Rydb. Rare, Troy, Ida., *Piper* in 1898; also isolated at Bingham Sprs., Blue Mts. of Ore., otherwise from B. C. to Calif. west of the Cascades.

Psoralea scabra Nutt. ex T. & G. Perennial herb; rootstock long, ropelike; stems 1–6 dm. tall; herbage dark glandular punctate, white strigose; petioles 1–3.5 cm. long; leaves 3-foliolate; leaflets 1.5–4 cm. long, oblanceolate or obovate, mucronate, thick; stipules 3–7 mm. long, lanceolate or subulate; peduncles several, axillary, 1–4 cm. long; racemes 1–2 cm. long, dense; calyx glandular punctate, appressed white strigose, the tube 1.5–2 mm. long, campanulate, the lobes 1 mm. long, ovate, obtuse, densely ciliate; corolla white, tipped with blue; standard with the blade oval, the margins upturned; the wings arcuate-oblong, enclosing the boat-shaped, blue-tipped keel; pods 5–6 mm. long, globose, beaked, glandular punctate, densely white hirsute; seeds 4 mm. in diameter, globose, brown. *Psoralea Purshii* Vail; *P. lanceolata scabra* (Nutt. ex T. & G.) Piper; *P. lanceolata,* var. *purshii* (Vail) Piper; *Psoralidium Purshii* (Vail) Rydb. Dry sandy shores, flats, or dunes by the Snake and Clearwater Rivers, *U. Son.*

ROBINIA

Deciduous trees or shrubs; winter buds naked, hidden by base of petiole, terminal bud wanting; leaves odd pinnate with setaceous or spinulose stipules; stipels present; leaflets opposite, petioluled; flowers white, pink, or pale purple, in pendulous axillary racemes; bracts deciduous; calyx 5-toothed, slightly 2-lipped; standard with a suborbicular, reflexed limb; wings free, curved; keel incurved, united from the middle; upper stamens free or partly free; pod oblong, flat, 2-valved, several seeded, winged or ridged on the upper suture.. (Named in honor of *Jean Robin* and *Vespasian Robin,* herbalists to Henry IV of France, who first cultivated the Locust in Europe.)

Robinia Pseudo-Acacia L. *Black Locust.* Tree 5–27 m. tall; trunk up to 30 cm. in diameter; bark dark brown tinged with red; smooth or later furrowed; branchlets slightly pilosulous, glabrate; the 7–21 leaflets 2–6 cm. long, elliptic to oval, rounded or emarginate and mucronate at apex, when young sericeous, at maturity above green and glabrate, beneath paler and somewhat puberulent; peduncles 2–5 cm. long, appressed pilosulous; racemes 1–2 dm. long, loose, pilosulous; pedicels 5–12 mm. long; calyx-tube 4–5 mm. long, campanulate, the upper lobes broadly ovate-deltoid, mostly united, the lower 2–3 mm. long, lanceolate; flowers 15–22 mm. long, very fragrant; the standard upturned, with a median yellow blotch, emarginate; wings obliquely obovate; limb of keel petals half-moon-shaped; pods stipitate, 5–10 cm. long, 10–30 mm. broad, 4–16-seeded, glabrous, brown; seeds 5–6 mm. long, bean-shaped, greenish with black spots, the hilum near one end. The wood is very hard and strong, close-grained, usually brown, resistant to decay when on the ground; sap wood thin, pale yellow. Native of e. U. S., planted for shade, wind-break, or ornament, spreading and now well established along creeks draining the draws in the Snake River Canyon, *U. Son.*

The bark, and honey from its nectar are poisonous.

THERMOPSIS

Perennial herbs; leaves alternate, palmately 3-foliolate, with large foliaceous stipules; flowers large, yellow, in terminal or axillary racemes; calyx campanulate, the lobes equal and separate or the two upper united; standard nearly orbicular, almost equaling the oblong wings and about equaling the keel; petals of keel obtuse, nearly straight, lightly joined; stamens 10, distinct; ovary sessile or short-stipitate; ovules numerous; pod sessile or short-stipitate in the calyx, flat, linear, straight or curved. (Name Gr., *thermos,* lupine; *opsis,* appearance.)

Thermopsis montana Nutt., ex T. & G., var. **ovata** (Robins.) St. John. *Golden Pea.* Rhizomes stout, long, freely branching; stems 4–10 dm. tall, sparely pilosulous above, glabrate below, leafy, more or less branched above; stipules 1–5 cm. long, lanceolate or commonly ovate, the larger broadly so and cordate at base, glabrous above, appressed pilose beneath; leaflets 3–9 cm. long, oblanceolate to obovate, green above, beneath paler and appressed pilose; petioles 1–5 cm. long; peduncles 2–9 cm. long; racemes 5–20 cm. long, many flowered, pilose, the bracts ovate to lanceolate, green; pedicels 1–10 mm. long; calyx pilose, the tube 5–8 mm. long, the lobes 2–4 mm. long; corolla 17–23 mm. long; keel boat-shaped; pods 4–8 cm. long, 4–6 mm. wide, straight, appressed pilose; seeds 3–4 mm. long, bean-shaped, light brown. *T. montana,* subsp. *ovata* Robins.; *T. ovata* (Robins.) Rydb. Locally abundant, thickets or open woods, from valley of Palouse River southward, *A. T. T.* Occasionally stranded on river banks, as at mouth of Grande Ronde, *St. John & Brown* 4,201, *U. Son.*

T. montana, with narrowly oblanceolate leaflets, of the Rocky Mts., does not occur here.

TRIFOLIUM. Clover

Tufted or diffuse herbs; leaves mostly palmately, sometimes pinnately, 3-foliolate, or in a few palmately 3–9-foliolate stipules foliaceous, united with the petiole; flowers white, yellow, red or purple, in heads or in umbel-like heads, spikes, or racemes; calyx-teeth 5, nearly equal; petals more or less persistent, the claws of all but the standard united more or less below the stamen-tube; stamens 10, diadelphous or the tenth only partly separate; pod small and membranous, 1–6-seeded, included in the calyx. (Name Lat., *tri,* three; *folium,* leaf.)

Leaflets 5 or 7; flowers 2.2–2.5 cm. long. — *T. macrocephalum,* var. *caeruleimontanum.*
Leaflets 3; flowers smaller,
- Heads subtended by a discoid or bowl-shaped involucre,
  - Calyx-lobes 1–3-times 3-forked. — *T. cyathiferum.*
  - Calyx-lobes not 3-forked, mostly simple,
    - Calyx-lobes broadly hyaline, serrate at base; corolla light rose to white; plant pilose. — *T. microcephalum.*
    - Calyx-lobes not so; corolla purplish; plant glabrous,
      - Flowers 12 mm. long; creeping rootstocks present. — *T. fimbriatum.*
      - Flowers 6–7 mm. long; rootstocks wanting, roots fibrous. — *T. variegatum.*
- Heads not subtended by a united involucre,
  - Flowers yellow,
    - Leaflets all sessile; stipules linear. — *T. agrarium.*
    - Terminal leaflet stalked; stipules wedge-obovate. — *T. procumbens.*
  - Flowers not yellow,
    - Peduncles axillary,
      - Flowers pink or purplish; stems not stolon-like at base,
        - Flowers subsessile, in dense heads; calyx-tube pilose-bearded above. — *T. resupinatum.*
        - Flowers slender pedicelled, in loose heads; calyx-tube not bearded above. — *T. hybridum.*
      - Flowers white; stems more or less stolon-like at base,
        - Calyx-teeth 1–2.5 mm. long, glabrous; stems wholly creeping — *T. repens.*
        - Calyx-teeth 3.5–5 mm. long, hirsute; stems creeping at base, then suberect. — *T. latifolium.*
    - Peduncles terminal,
      - Calyx-teeth glabrous, the 4 upper bent down, half encircling the corolla. — *T. Douglasii.*
      - Calyx-teeth long-hairy, nearly straight,
        - Calyx much longer than the corolla; corolla 3–4 mm. long. — *T. arvense.*
        - Calyx shorter than the corolla; corolla 10–16 mm. long,

Heads subtended by two leafy bracts; flowers deep pink to magenta-red; calyx-teeth remotely villous-ciliate. *T. pratense.*
Heads not leafy bracted; flowers white to pink; calyx-teeth densely long villous,
Heads 1–3 cm. long; upper leaflets lanceolate or elliptic-lanceolate. *T. arcuatum*
Heads 2–6 cm. long; upper leaflets linear-lanceolate, acuminate,
Leaflets of basal leaves 2–7 mm. wide, the apices acuminate, apiculate. *T. plumosum,* var. *plumosum.*
Leaflets of basal leaves 8–16 mm. wide, the apices acute, apiculate. Var. *amplifolium.*

Trifolium agrarium L. *Hop Clover.* Annual to perennial, more or less appressed hairy; stems 1–4 dm. long, upright or ascending; stipules 8–15 mm. long, linear-lanceolate, acuminate, strongly nerved; leaflets 12–18 mm. long, obovate or oblong, rounded, truncate, or emarginate at apex, cuneate, finely denticulate; peduncles 1–5 cm. long, axillary; heads 12–20 mm. long, 5–10 mm. in diameter, barrel-shaped or ovoid, dense, many flowered; pedicels shorter than the calyx; calyx-tube 1.2 mm. long, the lower teeth longer than the upper or the tube, pilose to glabrous; corolla yellow, in fruit brown and striate veined; standard 6 mm. long, cuneate, serrate near the middle; wings 5 mm. long; keel 4 mm. long; pod indehiscent; seeds globose. A weed, introduced from Eurasia, Glenwood, Whitman Co., *Kuhnhausen* in 1928, *A. T.*

Trifolium arcuatum Piper. Perennial; tap-root elongate or fusiform; stems often several, 1–4 dm. tall, pilose above, glabrate below; basal leaves long petioled, upper cauline leaves short petioled; basal leaves with leaflets 3–18 mm. long, oval to deltoid-obcordate, obtuse or retuse, glabrous, denticulate; upper cauline leaves with blades 10–45 mm. long, lanceolate or elliptic-lanceolate, serrate with excurrent veinlets, above pilose or glabrate, below pilose; upper stipules 10–45 mm. long, foliaceous, ovate-lanceolate, simple or lobed, denticulate, pilose, adnate for ⅔ their length; peduncles 4–16 cm. long, pilose; heads erect or more or less nodding in age, 10–35 mm. in diameter; flowers numerous, in bud divergent, soon reflexed, subsessile, pinkish or whitish; calyx-tube 2 mm. long, white, membranous, strongly 10-nerved; calyx-teeth subulate, long villous, the lower 4–7 mm. long, the 4 others 3–5 mm. long; standard 10–16 mm. long, oblong, emarginate, a little exceeding the oblong, obliquely acuminate wings which exceed and enclose the boat-shaped keel; pod 4 mm. long 1–2-seeded, obliquely lenticular, villous at tip; seeds 1.5 mm. long, oval truncate, brown. *T. eriocephalum* Nutt., f. *arcuatum* (Piper) McDermott. Woods or subalpine meadows in the Blue Mts., *A. T .T.* to *Huds.*

Trifolium arvense L. *Rabbit-foot Clover.* Annual, pilose throughout with more or less appressed hairs; stems 5–40 cm. tall, simple or bushy branched above; leaves cauline, 1–2.5 cm. long, 3-foliolate; stipules 3–10 mm. long, the adnate base membranous, white, the free tips lance-subulate; leaflets 1–2 cm. long, oblong-oblanceolate or oblanceolate, denticulate at tip; peduncles 1–7 cm. long; heads 1–3 cm. long, 1 cm. broad, globose to cylindric, dense, fuzzy, the lavender hairs drying brown; calyx-tube 1.5–2 mm. long, densely villous, 10-nerved; calyx-teeth 3.5–4 mm. long, subulate, shaggy villous, almost concealing the flowers; standard 3–4 mm. long, white, turning pink, the limb oval, exceeding the oblong wings and semireniform keel; pod ovoid, 1–2-seeded; seeds 1 mm. long, oval, flattened, yellowish. Native of Eurasia, introduced and fugitive with us, Pullman, *Piper* in 1899, *A. T.*

Trifolium cyathiferum Lindl. Annual, glabrous throughout; stems 5–40 cm. long, erect or decumbent; stipules 3–11 mm. long, ovate-lanceolate to ovate, pale, green nerved, more or less laciniately toothed; petioles 1–9 cm. long; leaflets 3–25 mm. long, obovate to oblanceolate, denticulate with excurrent veinlets; peduncles 3–11 cm. long, axillary; involucre 10–22 mm. in diameter, bowl-shaped, pale, strongly veined, the greenish margin unequally, shallowly lobed, laciniately toothed; heads hemispheric, dense, calyx-tube 2–3 mm. long, 20-veined, white, membranous, reticulate; calyx-teeth 4–6 mm. long, 1–3-times 3-forked, the teeth spinulose; corolla white or light pink, drying brown; standard 8–9 mm. long, elliptic, 1 mm. longer than the other corolla parts; wings with the limb lance-oblong, eared; keel with the wing oblong; pods 4 mm. long, 1–2-seeded, oblong, flat; seeds 1.5–2 mm. long, oval, flattened, greenish, brown spotted. Rare, springy, gravelly places, *A. T.*

Trifolium Douglasii House. Perennial, glabrous throughout; tap-root stout, fusiform; stems 3–7 dm. tall, usually several, erect, striate; basal leaves long petioled, the cauline shorter to subsessile; stipules 2–6 cm. long, linear-lanceolate, foliaceous, adnate about ⅔ of length, striate veined, denticulate near the tip; leaflets 2–8 cm. long, linear and acute or lance-elliptic, the upper acute, subcoriaceous, strongly striate veined, serrate by the sharp excurrent veinlets; peduncles 5–15 cm. long, appressed hirsute at tip, usually glabrate; heads 2–4 cm. long, globose or oval, densely flowered; calyx-tube 2.5–3 mm. long, membranous, white, strongly 20-nerved; calyx-teeth lanceolate, hyaline margined, the lower one 5–7 mm. long, straight, the others in pairs, shorter, broader, bent downward around the corolla; flowers magenta, straight; standard 13–16 mm. long, elliptic, contracted at the middle, enfolding the other corolla parts; wings about 2 mm. shorter, the limb oblong, eared, equaling the claw; keel as long as the wings, boat-shaped, eared, the upper margins straight, the petals free at the apex; pod 1–2-seeded; seeds not seen. *T. altissimum* Dougl. In meadows, *A. T.*

Trifolium fimbriatum Lindl. Perennial, glabrous; rhizomes 1–2 mm. in diameter, pale; stems 2–5 dm. long, erect, or decumbent; stipules 7–20 mm. long, lanceolate to semiorbicular, acuminate, deeply lacerate, strongly nerved; petioles 1–6 cm. long; leaflets 1–3 cm. long, narrowly elliptic to oval, acute, the lateral veins excurrent as spinulose teeth; peduncles 2–8 cm. long; involucre about 8 mm. high, cleft about half way, then again incised into lanceolate, spinulose teeth; pedicels 1–2 mm. long; heads dense; calyx-tube 3 mm. long, narrowly campanulate, membranous, strongly ribbed, the teeth 5–7 mm. long, lance-subulate, spine-tipped; corolla 12–15 mm. long, purplish; pod 3 mm. long, with a stipe nearly as long. *T. spinulosum Dougl.; T. Willdenovii* Spreng., var. *fimbriatum* (Lindl.) Ewan. Meadows, *A. T. T.*

Trifolium hybridum L. *Alsike Clover.* Biennial or perennial from a stout tap-root; stems 2–9 dm. long, erect or decumbent, nearly glabrous or sparsely pilose above; stipules 5–23 mm. long, the free tip broadly lanceolate, green veined and margined; petioles 3–10 cm. or more in length, nearly glabrous; leaflets 1–4 cm. long, oval or ovate, cuneate, serrulate especially along the sides, glabrous; peduncles 3–17 cm. long, axillary, glabrous or sparsely pilosulous above; heads 1–2.5 cm. in diameter, globose, with the erect buds white, the expanded flowers horizontal, pink, the withered flowers reflexed, brown; pedicels sparsely appressed pilosulous, the lower shorter, the upper 3–5 mm. long; calyx glabrous or sparsely pilose, the tube 1–1.5 mm. long, white, the teeth 1.5–2.5 mm. long, subulate, green; standard 5–9 mm. long, elliptic; wings 4–7 mm. long, the limb obliquely elliptic, eared at base, equaling the claw; keel scoop-shaped, shorter than the wings; pod ellipsoid, projecting from the calyx, 2–4-seeded; seeds 1.2–1.3 mm. in diameter, oval-cordate, flattened, yellowish green. Forage crop, introduced from Eurasia, occasionally established, *A. T.*

Trifolium latifolium (Hook.) Greene. Perennial; rhizomes slender, creeping; stems 1–4 dm. tall, slender, appressed hispidulous; basal leaves numerous, the petioles 2–15 mm. long, appressed hispidulous, the leaflets 6–32 mm. long, obovate or oval, glabrous above, appressed hispidulous beneath, serrulate; cauline stipules 6–15 mm. long, foliaceous, lanceolate, entire or toothed, united ½ their length; petioles 1–10 cm. long, sparsely hispidulous; leaflets 5–40 mm. long, ovate or lance-ovate, acute or cuspidate, more sharply serrulate, otherwise like the basal ones; peduncles 2–14 cm. long, appressed hispidulous; inflorescence umbel-like, 11–30-flowered, loose; pedicels 1.5–5 mm. long, hirsute, recurving in fruit; calyx-tube 2 mm. long, membranous, white, somewhat hirsute, 10-nerved; calyx-teeth 3.5–5 mm. long, lance-subulate, subequal, hirsute; flowers whitish, withering brown; standard 10–15 mm. long, narrowly elliptic, sharply upturned at the middle; wings oblong-elliptic, eared, 2–3 mm. shorter; keel ⅓ shorter than the standard, scoop-shaped, eared; pod 4 mm. long, oval-lenticular, coriaceous, hirsute, stipitate, 1-seeded; seeds 1.8 mm. long, broadly bean-shaped, brown. *T. Howellii* Wats., var. *latifolium* (Hook.) McDermott; *T. Aitonii* Rydb. Common, evergreen woods, *A. T. T., Can.*

Trifolium macrocephalum (Pursh) Poir., var. **caeruleimontanum** St. John. Perennial; roots large, woody, usually producing short stolons; stems 1–1.5 dm. long, usually decumbent or assurgent, pilose; stipules 7–20 mm. long, the base adnate and membranous, the free tips foliaceous, oblong, the smaller entire, the larger lobed and laciniate serrate, pilose; basal leaves with petioles 1–4 cm. long, glabrous; basal leaves glabrous; cauline leaves with petioles pilose; the 5 or 7 leaflets 1–2 cm. long, obovate, obtuse or apiculate, cuneate, the margin spinulose denticulate by the excurrent veinlets, above glabrous, beneath pilose; peduncles 2–6.5 cm. long, pilose; heads 3.5–5.5 cm. in diameter; globose; pedicels 1–2 mm. long, pilose; calyx-tube 3–4 mm. long, turbinate, white, membranous, pilose, especially on the 10 strong nerves; calyx-teeth 6–9 mm. long, subulate, white pilose with hairs 1–1.5 mm. long; standard 22–25 mm. long, elliptic-ovate, white; wings shorter, with the claw 1 cm. long, obliquely elliptic, eared at base, white; keel with the broad claw 11 mm. long, the limb 8 mm. long, boat-shaped, truncate at base; the outer half magenta-pink or rose; ovary 5 mm. long, glabrous, shaped like a pea-pod; seeds not seen.

Exposed gravelly ridges in the Blue Mts., *Huds.*

The type collection is Wenatchee Ranger Station, Asotin Co., June 19, 1922, *W. T. Shaw.*

Trifolium microcephalum Pursh. Annual, more or less pilose throughout; stems 1–8 dm. long, ascending or procumbent, often branched near the base; stipules 5–15 mm. long, ovate-lanceolate, foliaceous, dentate or lobed; petioles 1–6 cm. long; leaflets 3–25 mm. long, obcordate to oblanceolate, spinose serrate with excurrent veinlets; peduncles 2–8 cm. long, axillary; heads semiglobose to globose, the flowers numerous in 3–4 close whorls; involucre 7–13 mm. in diameter, bowl-shaped, 7–10-lobed, the lobes ovate to lanceolate, spinulose-acuminate, usually entire; calyx-tube 2 mm. long, white, scarious, somewhat pilose towards the mouth, strongly 10-nerved; calyx-teeth 3–4 mm. long, lanceolate, long, spinulose acuminate, towards the base broad hyaline and serrate; flowers light rose to white; standard 3–5 mm. long, elliptic; wings and keel shorter, cleaver-shaped; pod 2 mm. long oval, flattened, 1–2-seeded; seeds 1.5–2 mm. long, ellipsoid, flattened, greenish, with purple spots. Sandy soil or meadows, *A. T.*

Trifolium plumosum Dougl. ex Hook., var. **plumosum.** Perennial, 2–6 dm. tall, densely, almost appressed, pilose throughout; rootstock 1–3 mm. in diameter, yellow to brown, elongate; stems simple; basal leaves with petioles 10–15 cm. long, the stipules lanceolate, glabrate; leaflets 4–6 cm. long, almost linear, 2–5 (–7) mm. wide, the apices slender, acuminate, apiculate, acute, the lateral veins heavy, numerous, arching and almost parallel towards the margin, excurrent; cauline leaves few, smaller, the stipules 15–22 mm. long, foliaceous, adnate most of their length, oval, acute, denticulate; petioles 3–45 mm. long; heads terminal, 2–5 cm. long, 18–30 mm. wide, cylindric, dense; flowers ascending, sessile; calyx densely villous plumose throughout; the tube 4–4.5 mm. long, subcylindric, the lobes 6–8 mm. long, subulate; corolla usually whitish, and pink-tipped; standard 15–18 mm. long, oblanceolate, nearly straight; wings 2–3 mm. shorter; keel 3–4 mm. shorter; pod 4.5–5 mm. long lanceoloid; seeds 1–1.5 mm. long, ovoid. Mountain ridges, Walla Walla Co., *A. T. T.*

The type was collected in the Blue Mts. by Douglas.

Var. **amplifolium** Martin. Leaflets of basal leaves (8)–9–16 mm. wide, the apices broader, acute, apiculate. Nez Perce Co., Ida., *A. T. T.*

Trifolium pratense L. *Red Clover.* Perennial; tap-root stout, dark; stems 15–60 cm. tall, pilose above, glabrate below, the cauline leaves few and internodes elongate; stipules 12–20 mm. long, mostly adnate, white, membranous, green-veined, the free tips ovate-lanceolate, acuminate; petioles 5–30 cm. long, pilose; leaves 3-foliolate; leaflets 1–5 cm. long, ovate or oval, obtuse or emarginate, above often dark green at base and across the middle with a pale chevron-like spot, above glabrous, beneath sparsely pilose, minutely denticulate; heads subtended by 2 sessile leaves; heads 1.5–3 cm. in diameter, globose to ovoid, dense; calyx-tube 3–3.5 mm. long, villous, strongly 10-nerved; calyx-teeth remotely villous ciliate, subulate, 4 of them as long as the tube, but the lower one 5–8 mm. long; corolla pink to magenta-reddish; standard 12–16 mm. long, narrowly elliptic, retuse; wings and keel shorter, each with the claw twice as long as the limb; pod ovoid, 1-seeded; seeds 1 mm. long, ovoid, obliquely truncate above the hilum, dull, light yellow to purple. Native of Eurasia, commonly planted as a leguminous or a rotation crop, persisting in meadows and grasslands, *A. T., A. T. T.*

Trifolium repens L. *White Clover.* Perennial, glabrous or essentially so; branching at base, the branches creeping, often rooting at the nodes; stems 1–4 dm. long; stipules 4–10 mm. long, the free portion ovate-lanceolate, acuminate, membranous, purplish or greenish nerved; petioles 2–20 cm. long; leaflets 8–30 mm. long, obovate or obcordate, usually emarginate, sharply denticulate with excurrent veins; peduncles, exceeding the leaves; heads subglobose, 40–80-flowered, the flowers in bud ascending, in fruit reflexed; pedicels 1–5 mm. long, the upper the longer; calyx-tube 2–3 mm. long, narrowly campanulate, whitish, with 10 green nerves; calyx-lobes 1–2.5 mm. long, lanceolate, acuminate, the midrib strong, green; corolla white (or pinkish); standard 7–9 mm. long, elliptic, ¼ to ⅓ longer than the wings; pod 3–4-seeded, linear, flattened; seeds 1–1.2 mm. long, oval, emarginate, yellow to light brown. A cultivated forage or lawn plant introduced from Eurasia, persisting and spreading, occasional, *U. Son., A. T.*

Trifolium resupinatum L. Annual, glabrous or nearly so; stems 1–5 dm. long, prostrate or assurgent, usually branched; stipules linear-lanceolate, whitish, adnate from ¼ to ¾ their length; lower leaves long petioled; leaflets 5–20 mm. long, obovate, obtuse or acute, serrate, above often white spotted; peduncles mostly exceeding the leaves; heads 7–20 mm. in diameter, hemispheric to spherical, 6–15-flowered; bracts ciliate; calyx pilose on upper margin, the teeth about as long as the tube; corolla pink or purplish, in fruit twisted so that the standard lies below; standard 2–3-times as long as the calyx, narrowly elliptic, the tip emarginate and dentate; wings and keel shorter; fruit ovoid, 1–2-seeded; seeds ovoid, yellowish green, shining. Introduced from Eurasia, Pullman, *Jones* in 1930, *A. T.*

Trifolium variegatum Nutt. ex T. & G. Annual, glabrous throughout; stems 1–6.5 dm. long, often several, decumbent or ascending, slender; stipules 2–10 mm. long, ovate or semiorbicular, laciniately spinulosely lobed; leaflets 2–50 mm. long, obovate to oblong-lanceolate, or even linear, obtuse or retuse, spinulose serrulate; peduncles 1–6 cm. long, axillary; heads subglobose, 6–12 mm. in diameter, few- to many-flowered; involucre 4–10 mm. in diameter, 4–12-lobed, the lobes laciniately 3–7-spinulose-toothed; calyx-tube 2–2.5 mm. long, 15–20-nerved, membranous, subcylindric in flower, broadly campanulate in fruit; calyx-teeth 3–5 mm. long, lance-subulate, simple or one tooth bifid; corolla white-tipped or purple throughout; standard 6–7 mm. long, cuneate, retuse; wings and keel about 1 mm. shorter, cleaver-shaped; pod 2.5 mm. long, obovate, flattened, the sides membranous; seeds 1.5 mm. long, ovoid, flattened, greenish, dark-spotted. Infrequent, moist, even alkaline, places, *U. Son., A. T.*

VICIA. Vetch

Climbing or trailing herbaceous vines; leaves pinnate, tendril-bearing, with half-sagittate or entire stipules; flowers solitary or in loose peduncled axillary racemes; calyx 5-cleft or -toothed, unequal, the two upper teeth often shorter; wings adherent to the short keel; stamens diadelphous or nearly so; style filiform, hairy near the tip only; ovary 2–many-ovuled; pod flat, 2-valved, 2–several-seeded. (The classical Lat. name.)

Racemes 1-sided; calyx saccate at base; plant villous throughout,
  Plant densely villous; all calyx-lobes subulate and longer than the tube. — *V. villosa.*
  Plant sparsely villous; upper calyx-lobes lanceolate, shorter than the tube. — Subsp. *dasycarpa.*
Racemes not 1-sided; calyx not saccate; plant somewhat pilosulous to glabrate,
  Leaflets not truncate or 3–5-denticulate at apex,
    Leaflets elliptic to ovate or oval. — *V. americana,* var. *americana.*
    Leaflets linear. — Var. *angustifolia.*
  Leaflets truncate and 3–5-denticulate at apex. — Var. *truncata.*

Vicia americana Muhl., var. **americana.** *American Vetch.* Perennial, sparsely pilosulous to glabrate; stems 1–16 dm. tall, 4-angled, weak, trailing or climbing by its tendrils; leaves numerous; stipules 5–13 mm. long, foliaceous, mostly sharply palmately lobed; the 8–14 leaflets 12–50 mm. long, mucronate, above bright green, beneath paler green and more or less pilosulous; lower leaves often with narrower, even linear leaflets; tendrils well developed, forking; peduncles 4–9-flowered, usually shorter than the leaves; calyx pilosulous, the tube 3–4 mm. long, somewhat oblique, the upper teeth short deltoid, incurved, the lower longer, 1–2 mm. long, lance-deltoid to lanceolate; corolla 14–22 mm. long, purple-violet drying blue; standard ovate, the tip upturned; keel oblong, straight, with a short upturned, obtuse, scoop-shaped tip; pods 3–5 cm. long, 6–8 mm. wide, flattened, glabrous, with a stipe as long as the calyx; the 4–15 seeds 3–4 mm. wide, globose, somewhat flattened, dull black. Common, grasslands or open woods, *U. Son.,* to *A. T. T.*

An extremely plastic species, showing infinite fluctuations in leaflet shape and size. Certain botanists accept many of these as species. It blends into the following varieties. The plant is much relished by stock.

Var. **angustifolia** Nees. Differs in having linear leaflets and narrower stipules. *V. americana,* var. *linearis* (Nutt.) Wats. Grasslands or open woods, *U. Son.* to *A. T. T.*

Var. **truncata** (Nutt. ex T. & G.) *Brewer.* Leaflets truncate at apex, mucronate and 3–5-denticulate. Occasional, *A. T.*

Vicia villosa Roth. *Hairy Vetch.* Annual or biennial; stem 3–15 dm. tall, angled, branching, weak and climbing by its tendrils; stipules 5–13 mm. long, narrowly semisagittate, entire or toothed; the 10–20 leaflets 15–30 mm. long, narrowly elliptic to linear-lanceolate, mucronate; peduncles shorter or longer than the leaves; racemes 3–30-flowered, 1-sided; calyx-tube 2–3 mm. long, saccate at base very asymmetric; calyx-teeth very unequal, linear, the upper connivent, shorter than the tube, the lower 3–5 mm. long; corolla 12–20 mm. long, violet, narrowly oblong, only the margins of the standard reflexed; keel gradually upturned to the short, broad rounded apex; pods 2–4 cm. long, 5–8 mm. broad, flat, 2–8-seeded, glabrous; seeds 3–4 mm. long, globular, somewhat flattened, brown. Escaped from cultivation to roadsides and fields, *A. T.*

This European plant is grown as a green manure or rotation crop.

Subsp. dasycarpa (Tenore) Cavillier. Stems 3–9 dm. tall; the 14–20 leaflets 1–3 cm. long, 2–5 mm. wide, acute or obtuse; stipules 5–10 mm. long; corolla 12–17 mm. long, purplish violet or white; pod 7–10 mm. wide. Native of Eurasia and Africa, a cultivated legume, escaped, Spokane Co., roadside between Dartford and Mead, 1952, *Gaines et al. 752.*

GERANIACEAE. Geranium Family

Annual or perennial herbs; leaves basal, alternate or opposite, often with stipules; flowers solitary or clustered, perfect, regular, commonly symmetrical, the parts in fives; sepals imbricate, persistent; stamens usually twice as many as the petals, mostly in two sets, those alternate with the petals sometimes sterile; ovary 1, deeply 3–5-lobed and 3–5-celled; ovules 2 in each cavity; fruit a capsule; carpels 1-seeded, separating when mature from the axis. (Named from the genus *Geranium.*)

Anthers 10; leaves palmately parted; styles in fruit nearly glabrous inside. *Geranium.*
Anthers 5; leaves pinnatifid, pinnate, or roundish cordate; styles in fruit bearded on inner side. *Erodium.*

ERODIUM

Herbs, generally with jointed nodes; leaves opposite or alternate, with stipules; flowers regular, in axillary umbels; sepals 5; petals 5, hypogynous; anthers 5, the 5 shorter stamens sterile or wanting; ovary 5-lobed, 5-celled, beaked by the united styles; tails of the carpels becoming twisted. (Name Gr., *erodios,* heron.)

Erodium cicutarium (L.) L'Hér. *Filaree; Alfilaria.* Annual, or perhaps longer lived, hirsute throughout; stems 3–80 cm. long, branched, usually prostrate and mat-forming; basal and lower leaves 7–18 cm. long, petioled, pinnate, the divisions of the leaf pinnatifid almost to the midvein, the lobes oblong-lanceolate, acute; peduncles a little longer than their subtending leaves,

2–12-flowered, rarely glandular; bracts ovate, acuminate, membranous; pedicels 5–25 mm. long, rarely glandular; sepals 3–7 mm. long, lance-elliptic, usually bristle-tipped, with 3–5 green nerves, the hyaline margins often reddish; petals as long or twice as long as the calyx, obovate to spatulate, pink to lilac, sometimes spotted; staminodia linear-lanceolate, membranous, reddish; filaments of perfect stamens broader, longer acuminate, purplish; ovary and style-column hispid; fruit 3–4 cm. long; body of carpel 5–6 mm. long, callous pointed, upwardly hispid, the upwardly hispid tail spiraling when dry; seed 3 mm. long, subcylindric, broadened above, smooth, brown. Native to Eurasia, an introduced weed here. Exceedingly abundant, wet or dry places, *U. Son.* to *A. T. T.*

In 1844 Chas. A. Geyer recorded near Clarkston on the "range of the Nez Percez Indians" . . . "the ground densely covered with . . . *Erodium*." The rough penetrating fruits are harmful to stock, but the strong smelling foliage provides palatable forage in the arid regions throughout the year.

GERANIUM. Geranium, Cranebill

Annual or perennial herbs; leaves palmately-lobed, cleft or divided, with stipules; flowers regular, on axillary 1–2-flowered peduncles; sepals 5; petals 5, hypogynous; stamens 10, rarely 5, generally 5 longer and 5 shorter; ovary 5-lobed, 5-celled, beaked with a compound style; capsule 5-celled, each 1-seeded, and long-tailed, at maturity separating from the long beak of the receptacle, splitting from below upward. (Name from Gr., diminutive of *geranos*, crane.)

Petals 1–3 cm. long; rootstocks woody, enlarged; perennials,
  Stem and petioles glandular villous-hirsute,
    Petals pinkish purple. — *G. viscosissimum*, f. *viscosissimum*.
    Petals white. — Forma *album*.
  Stem and petioles retrorse hairy, not glandular,
    Petioles of upper leaves minutely puberulent. — Var. *nervosum*.
    Petioles of upper leaves long hairy,
      Sepals 11–12 mm. long; petals purple; carpel bodies 5 mm. long, sparingly pubescent. — *G. oreganum*.
      Sepals 6–8 mm. long; petals white; carpel bodies 4 mm. long, hirsutulous. — *G. Richardsonii*.
Petals 2–8 mm. long; roots slender, not enlarged and woody; annuals or occasionally biennials,
  Sepals awnless; seeds smooth or only minutely glandular. — *G. pusillum*.
  Sepals awned or subulate tipped; seeds reticulate,
    Fruiting pedicel much longer than the calyx; beak of style column 2.5–6 mm. long. — *G. Bicknellii*, var. *longipes*.
    Fruiting pedicel shorter or but slightly longer than the calyx; beak of style column 1–2 mm. long,
      Larger mature sepals 5–8 mm. wide, 5-nerved; seeds 2–2.7 mm. in diameter, suborbicular. — *G. sphaerospermum*.
      Larger mature sepals 3–4.5 mm. wide, 3-nerved; seeds 1–1.5 mm. thick, oblong. — *G. carolinianum*.

Geranium Bicknellii Britton, var. **longipes** (Wats.) Fern. Annual or biennial, similar to *G. carolinianum,* but leaves pentagonal, angular in outline; inflorescence loose, the pedicels much longer than the calyx; style column 16–22 mm. long; seeds 3 mm. long, dark reticulate. Infrequent, *A. T. T.*

Geranum carolinianum L. *Carolina Cranebill.* Tap-root slender, the others fibrous; plant 2–7 dm. tall, erect or ascending, hirsutulous throughout; stems retrorsely hirsutulous; stipules 5–11 mm. long, lance-linear; petioles of lower leaves 4–15 cm. long, those of upper leaves mostly longer than their blades; blades 2–6 cm. wide, reniform in outline, deeply 5–9-parted and again pinnately or bipinnately cleft, the ultimate lobes linear-oblong, above sparsely hirsutulous, beneath pale, hirsute on the veins; cyme compact, viscid hirsute; peduncles 2-flowered; sepals ovate, the outer ones with the body 5–7 mm. long, the awn-like tip 1.5–2 mm. long; petals as long as the sepals, pink, oblanceolate, the claws ciliate; style column 12–18 mm. long; body of carpel 3–3.5 mm. long, hirsute and puberulent; seeds 2.2 mm. long, oval in outline, mahogany-colored with pale reticulations. Open places or fields, *U. Son., A. T.*

Geranium oreganum Howell. Similar to *G. viscosissimum,* but the petioles and stems retrorse hirsute, non-glandular. Dayton, *Mrs. J. A. Reehes,* in 1921.

Geranium pusillum Burm. f. *Small-Flowered Cranebill.* Stems 1–5 dm. long, spreading or decumbent, often branched, puberulent; stipules 3–5 mm. long, deltoid to lanceolate, hirsutulous; basal leaves with petioles often 1–2 dm. long, puberulent; blades 1.5–5 cm. wide, reniform to orbicular-reniform, 5–7-cleft, the cuneate divisions with obovate lobes or teeth near the tip, hirsutulous, paler beneath; upper leaves shorter petioled to subsessile, 3–5-cleft, otherwise like the lower ones; peduncles 2-flowered, glandular pilosulous; pedicels up to 15 mm. in length, glandular pilosulous; sepals 2.5–4 mm. long, elliptic to ovate, minutely glandular pilosulous, hirsute on the margins; petals obovate, truncate or emarginate, a little longer than the sepals, violet; the 5 stamens 2.5 mm. long, the filaments lanceolate at base; the 5 staminodia shorter; style column 8–9 mm. long, glandular puberulent; beak less than 1 mm. long; body of carpels 2 mm. long, ellipsoid, compressed, appressed hirsutulous; seeds 1.8 mm. long, ellipsoid, brown. A weed, introduced from Europe. Fields and grass-lands, *A. T.*

Geranium Richardsonii Fisch. & Trautv. Perennial, 7–8 dm. tall; caudex woody, stout, clothed with brown scales or leaf-bases; lower part of stem and petioles sparsely retrorse, white hirsute; first internode 4–5.5 dm. long; first internode densely retrorse hirsute; upper internodes more densely hirsute and capitate glandular hirsute; basal leaves with stipules brown, membranous, the free portion about 15 mm. long, lanceolate, hirsute on the midrib and margins; basal petioles about 3 dm. long; blades of basal leaves 6 cm. or probably more in width, 5-parted, the lobes again pinnately cleft and laciniately toothed, the ultimate lobes ovate, mucronate, above hirsute and with extremely minute pedicellate glands, beneath with similar glands, the veins and veinlets hirsute, the basal leaves mostly withered by anthesis; cauline leaves with the stipules 3–8 mm. long, free, ovate-lanceolate, hirsutulous, hirsute-ciliate; the petioles shorter to subsessile, hirsute and capitate glandular hirsute; blades 5–11 cm. wide, similar to the basal ones; inflorescence densely capitate glandular hirsute and pilose, the peduncles several flowered; pedicels 5–40 mm. long; sepals of equal length, the body 6–8 mm. long, oval, green or with a narrow hyaline margin, the outer hirsute or capitate glandular hirsute on the margins and veins, the inner sparsely so to almost glabrous, sepals with a terminal, dark mucro 1.6–2.8 mm. long; petals 10–15 mm. long, obovate, strongly veined, white; the 5 longer stamens 8–9 mm. long, subulate, thin, ciliate below; stigmas 3–5 mm. long, filiform; fruiting style-column 18–23 mm. long, glandular hirsutulous; fruiting carpel body 4 mm. long, oval in outline, somewhat hirsutulous; seeds 3.7 mm. long, ovoid, brown, elongate reticulate.

G. loloense St. John. Craig Mts., *A .T. T.*

Geranium sphaerospermum Fern. Annual or biennial, 1–4.5 dm. tall; stems minutely retrorse pilose; leaves 2–7 cm. broad, reniform-orbicular, deeply 5-parted, the segments lobed, obtuse, corymb compact; peduncles and pedicels retrorse pilosulous; sepals broadly ovate, mucronate, the nerves and margin pilose-hirsute; petals rose-colored equaling the sepals; carpels long villous; seeds indistinctly reticulate. Clark Springs, Spokane Co., *Kraeger* 132.

Geranium viscosissimum Fisch. & Mey., forma **viscosissimum** Caudex stout, woody, brown; plant 2–6 dm. tall, glandular or viscid villous-hirsute throughout; basal leaves several; the petioles 1–5 dm. long; blades 4–14 cm. wide, pentagonal in outline, 3–5-parted, the divisions sharply and irregularly incised; cauline leaves few, reduced; pedicels up to 3 cm. in length; sepals 8–14 mm. long, elliptic, awn-tipped; petals 17–22 mm. long, obovate, often emarginate, hirsute at base; fruit 3–4 cm. long; body of carpel 4–5 mm. long, ovoid; seeds 4–4.5 mm. long, ovoid, slightly flattened, blackish, with fine reticulate ridges. Abundant, prairies and open woods, *A. T., A. T. T.*

Forma **album** (Suksd.) St. John. Differing only in having the petals white. Occasional, with the species, *A. T.*

The type came from Spangle, *Suksdorf* 8,710.

Var. **nervosum** (Rydb.) C. L. Hitchc. Differs from f. *viscosissimum* in having the petioles of basal leaves and lower stems strigose or retrorsely pubescent with short, whitish, non-glandular hairs. *G. nervosum* Rydb.; *G. strigosius* St. John. Woods, *A. T.*

LINACEAE. Flax Family

Herbs or rarely shrubs; leaves all simple and entire, mostly alternate; stipules none; flowers perfect, regular, symmetrical, in axillary or terminal cymes, or panicles; sepals 5 (rarely 4–6); petals as many as and alternate with the sepals; stamens of the same number and alternate with the petals, united at base, sometimes with alternating staminodia; pistil 1; styles 2–5; ovary 2–5-celled; fruit usually a capsule, often 4–10-celled by false partitions; endosperm fleshy or none. (Named from the genus *Linum.*)

LINUM. Flax

Annual or perennial herbs; bark tough and fibrous; leaves alternate or opposite, sessile, entire; flowers perfect; sepals, petals, stamens, and styles 5, regularly alternate with each other; pistil of 5 united carpels, 5-celled, with 2 seeds in each cell; each cell divided in fruit by a false partition making a 10-celled pod. (The classical Lat. name of the flax.)

Petals yellow; sepals with marginal glands. *L. digynum.*
Petals blue or white; sepals lacking marginal glands,
  Sepals non-ciliate; septa of capsule ciliate,
    Petals blue. *L. Lewisii,* f. *Lewisii.*
    Petals white. Forma *albiflorum.*
  Sepals or at least the inner ciliate; septa non-ciliate. *L. usitatissimum.*

Linum digynum Gray. Annual, glabrous, 5–40 cm. tall; stems simple or branched; leaves mostly opposite, 7–28 mm. long, elliptic, sessile; racemes several, branching; bracts foliaceous, glandular serrate; pedicels 1–2.5 mm. long; outer sepals 2–3 mm. long, ovate, incised or incised-toothed, the teeth gland-tipped ; inner sepals similar but smaller and narrower; petals 3 mm. long, without appendages; filaments lanceolate, glabrous; staminodia none; styles 2, united to near the middle; capsule 2 mm. long, depressed ovoid; seeds 1.5 mm. long, lanceoloid, brown, minutely papillose. *Cathartolinum digynum* (Gray) Small. Uncommon, moist places, *A. T.*

Linum Lewisii Pursh, forma **Lewisii.** *Blue Flax.* Perennial, 2–7 dm. tall, glabrous; crown woody; stems several, often branched at base; leaves 1–4 cm. long linear to lance-linear, pale; bracts similar but smaller; inflorescence a raceme or cymose-racemes; pedicels 5–30 mm. long; sepals 4–7 mm. long, striate veined, hyaline margined, the outer ovate, short acuminate, the inner broader, mucronate; petals 15–20 mm. long, obovate, blue, very delicate; stamens 6–10 mm. long, the filaments filiform but flattened at base; styles distinct; stigmas clavate; capsules 4–6 mm. in diameter, subglobose, acuminate, yellow, dehiscing lengthwise; seeds 3–4.5 mm. long, ovate, flat, cellular reticulate, brown, shiny. Prairies or rocky hills, *A. T.*

Named for its first collector, Capt. M. Lewis. The stems produce excellent fibre, which was used by and traded by the Indians.

Forma **albiflorum** (Cockerell) St. John. *L. perenne* L., f. *albiflorum* Cockerell. Differing in having the petals white. Growing with the blue flowered species, but rare. Colton, *St. John* 9,861.

Linum usitatissimum L. *Common Flax.* Very similar to *L. Lewisii,* but annual; stems 2–8 dm. high; leaves alternate, linear-lanceolate, 2–5 cm. long; flowers in loose cymes; sepals 5–9 mm. long, acute, ciliate, the inner margin scarious; petals 10–15 mm. long, blue; stigmas elongate; capsule nearly indehiscent, as long as the calyx; its septa not ciliate; seeds 4–5 mm. long, elliptic, flat, pale brown, shining. A crop plant, its native home unknown, with us a fugitive escape, Pullman, *Piper* in 1895.

The stem fibres are made into linen. The seed coats become mucilaginous on wetting. The seeds furnish linseed oil, and the residual cake is used as cattle feed.

ZYGOPHYLLACEAE. Caltrop Family

Perennial herbs, shrubs, or trees; leaves alternate or opposite; digitate or even pinnate; leaflets asymmetric, glandular punctate; flowers perfect, regular or nearly so; sepals 5 (rarely 4–6), usually imbricate; petals as many, hypogynous; stamens twice as many; pistil of 2–5 united carpels; ovary 2–5-celled, rarely 10–12-celled, angled or winged; styles united; fruit a capsule or splitting into as many or twice as many nutlets as there are carpels. Named from the genus *Zygophyllum.*)

TRIBULUS

Diffuse herbs; leaves opposite, even pinnate; stipules membranous; flowers solitary on axillary peduncles; sepals 5, soon caducous; petals 5, yellow, orange, or white; ovary 5-celled, sessile, surrounded at base by an urceolate, 10-lobed disk; stigmas 5;

fruit 5-angled, tuberculate or spiny, separating into 5 bony carpels; endosperm none. (The classical *Lat.* name, derived from Gr. *tribolos,* a burr.)

Tribulus terrestris L. *Puncture Vine.* Annual, mat-forming; branches 1–10 dm. long, hispid and puberulent; stipules 2–4 mm. long, lance-subulate, hispid ciliate; leaves 2–5 cm. long; the 10–16 leaflets 6–8 mm. long, obliquely oblong-ovate, acute, pilose on the mid-nerve or glabrate above, hispid ciliate, beneath appressed silky strigose; petiole and rhachis hispid and hirsutulous; pedicels 2–10 mm. long, hispid, hirsutulous; sepals 2.5–3 mm. long, hyaline margined, hispid; petals 4–5 mm. long, oblanceolate, yellow; stamens 3 mm. long; fruit about 8 mm. high; carpels angled, the back puberulent, spiny tuberculate and 2-horned; seeds 2.5–3 mm. long, ovate, brown. Native of Eurasia and Africa, a weed found in gravel on R. R. track, Wawawai, in 1924, *St. John* 6,740, now well established, *U. Son.*

Further south, in the warmer, more arid regions, this weed is one of the most pestiferous known. It thrives by roadsides, and its name comes from the ease with which its spiny carpels puncture auto tires.

SIMAROUBACEAE. Quassia Family

Shrubs or trees with usually bitter bark; leaves alternate (rarely opposite), pinnate or simple; flowers perfect or unisexual, regular, in panicles or spikes; sepals 3–5, more or less connate, imbricate or valvate; petals 3–5 (or none); stamens usually twice as many as the petals; ovary superior, usually surrounded by a disk; carpels 2–5, free at base and united by styles, or connate; carpels 1–several-ovuled; fruit a drupe, berry, or samara; seed with little or no endosperm. (Named from the genus *Simarouba.*)

AILANTHUS

Deciduous trees; buds suborbicular, with 2–4 scales; terminal bud none; leaves alternate, odd-pinnate; leaflets glandular toothed at base; flowers polygamous, in terminal panicles; petals 5–6, valvate; stamens 10, inserted at base of the 10-lobed disk; carpels 5–6 free below or connate; stigmas elongate, spreading; samaras 1–6; endosperm thin. (Named from common name in Molucca, *Ailanto,* tree of heaven.)

Ailanthus altissima (Mill.) Swingle. *Tree of Heaven.* Tree, up to 20 m. tall; bark smooth, pale striped; branchlets puberulent soon glabrate, heavy, often 7–12 mm. in diameter, the bark reddish brown, shining, the lenticels dot-like, white; leaves 45–60 cm. long; the 13–25 leaflets 7–12 cm. long, petiolulate, lance-ovate, usually truncate at base, finely ciliate, each margin with 2–4 coarse teeth, each enclosing a dark, rounded, swollen gland, lower surface pale; panicles 1–2 dm. long; staminate flowers with calyx 1 mm. long, the lobes deltoid; petals 3–4 mm. long, oval, greenish, pilose on lower margins and within near the base; the 10 stamens 3–4 mm. long; disk dark, depressed; pistillate flowers similar but the 5 stamens reduced and the 5-lobed, elongate pistil protruding; samaras 3–4 cm. long, linear-lanceolate, flat, thin, swollen at the middle around the seed, the tip often spiraled. A Chinese tree, cultivated for shade or windbreak, recently established in the Snake River Canyon, *U. Son.*

Its staminate flowers have a strong fetid odor. The fruits are wind-borne and scatter widely. It is very hardy and quick spreading.

EUPHORBIACEAE. Spurge Family

Herbs (in ours), with milky juice, shrubs, or trees; leaves opposite, alternate, or whorled; entire or toothed, sessile or petioled; stipules present or wanting; flowers monoecious or dioecious, often much reduced and subtended by an involucre which resembles a calyx; parts of flowers various, often different in staminate and pistillate flowers; calyx none or minute; petals often wanting; stamens 1–many, free or united; ovary usually 3-celled; fruit a 3-lobed capsule; endosperm copious. (Named from the genus *Euphorbia.*)

EUPHORBIA. Spurge

Perennial or annual herbs (in ours), or shrubs; flowers monoecious, included in a cup-shaped 4–5-lobed involucre resembling a calyx or corolla and usually bearing large thick glands at its sinuses; glands rounded or often petal-like or crescent-shaped; staminate flowers numerous, of a single naked stamen, jointed upon a short pedicel which usually has a minute bract at its base; pistillate flowers solitary in the center of the involucre, pedicelled, and soon exserted; calyx none, or rarely present and minute; styles 3, 2-cleft; stigmas 6; ovary 3-celled, 3-ovuled. (The name *euphorbea* used by Pliny for *E. officinarum,* the name honoring *Euphorbus,* the physician of *King Juba* of Numidia.)

Leaves all opposite; stipules present; glands of involucre with petal-like appendages,
  Seed faces pitted and wrinkled. — *E. serpyllifolia,* var. *serpyllifolia.*
  Seed faces transversely ridged. — *E. glyptosperma.*
Leaves or at least the lower alternate; stipules none; glands of involucre lacking petal-like appendages,
  Leaves serrate; glands elliptic, seeds reticulate. — *E. spathulata.*
  Leaves entire; glands crescent-shaped; seeds smooth,
    Leaves usually broadest above the middle, 2–10 mm. broad. — *E. Esula.*
    Leaves narrowly linear or broadest below the middle, 1–3 mm. broad. — *E. Cyparissias.*

Euphorbia Cyparissias L. *Cypress Spurge.* Perennial, bright green, glabrous throughout; rootstocks woody, horizontal; stems 1–5 dm. tall, clustered, forming dense patches, below brown scaly, above very leafy; stipules none; leaves 1–2.5 cm. long, linear to almost filiform, alternate, but the uppermost whorled; floral bracts 5–8 mm. long, opposite, cordate, green, finally yellow or red; involucres 1.5–3 mm. long, turbinate-campanulate, with 4 unappendaged, crescent-shaped glands; capsule 3 mm. long, glabrous, 3-angled; seeds 2 mm. long, rounded ovoid, gray. *Tithymalus Cyparissias* (L.) Hill. Native of Eurasia, cultivated, especially in graveyards, persisting and spreading, occasional, Pullman, *Piper* in 1897, *A. T.*

Euphorbia Esula L. *Leafy Spurge.* Perennial, 3–15 dm. tall, dull green, glabrous; rootstock running, woody; stems clustered, often branched, stout, very leafy; leaves 12–50 mm. long, alternate, lanceolate or linear-lanceolate; whorl of leaves at base of umbel broadly ovate, acute; lower flowers axillary, the peduncles 2–3.5 cm. long; terminal flowers in an umbel; floral bracts 6–9 mm. long, subcordate, apiculate; involucres 2–3 mm. long, with 4 half-moon-shaped horned glands; stamens 2 mm. long; capsule 2.5–3 mm. long; seeds 1.8–2 mm. long, subspherical, brownish violet, shining. *Tithymalus virgatus* (Waldst. & Kit.) Kl. & Garcke; *E. virgata* Waldst. & Kit.; *E. intercedens* Podp., non Pax; *E. Podperae* Croizat. A weed from Europe, naturalized in cultivated fields, *A. T.*

Euphorbia glyptosperma Engelm. *Ridge-seeded Spurge.* Similar to *E. serpyllifolia,* but often erect; leaves very asymmetric at base, the petiole attachment appearing lateral; white petaloid appendage narrower than the gland; capsules 1–1.5 mm. long; seed faces with strong, transverse ridges. *Chamaesyce glyptosperma* (Engelm.) Small. Canyon walls, and banks of Snake River, *U. Son.*

Euphorbia serpyllifolia Pers., var. **serpyllifolia.** *Thyme-leaved Spurge.* Annual, prostrate or sometimes ascending, glabrous; stems 1–3 dm. long, freely branching; stipules connate, the margin with hair-like fringe; petioles about 1 mm. long, straight, the attachment appearing subbasal; blades 3–12 mm. long, obliquely oval or oblong, unequal at base, serrulate at tip, green or purple-maculate above, whitish beneath; flowers axillary, short pediceled; involucres 0.5–1 mm. long, campanulate, reddish, bearing 4 oval glands, each with a broad white petaloid border or appendage; styles short bifid; capsules 1.5–2 mm. long, smooth; seeds 1–1.3 mm. long, with a heavy white bloom, ovoid, 4-angled, *Chamaesyce serpyllifolia* (Pers.) Small. Occasional, moist pond shores, Revere, and Pullman, Wash.; Moscow Mts., Ida., *A. T.*

Euphorbia spathulata Lam. Annual, 1–6 dm. tall, green and glabrous throughout; leaves 1–3.5 cm. long, spatulate, paler beneath, serrate towards the tip, the middle and upper leaves auricled at base; the upper leaves opposite, reduced; inflorescence umbel-like, the rays 2–3-times forked; involucres 1 mm. long, campanulate, pale, the glands sessile; stamens 1 mm. long; styles bifid; capsules 3–3.5 mm. long, warty; seeds 1.5–2 mm. long, ovoid, pale brown, light reticulate. *E. arkansana* Engelm. & Gray, var. *missouriensis* Norton; *Tithymalus missouriensis* (Norton) Small; *Galarrhoeus missouriensis* (Norton) Rydb. Moist places near or on banks of Snake River, *U. Son.*

CALLITRICHACEAE. Water Starwort Family

Aquatic or rarely terrestrial, usually tufted herbs; leaves opposite, entire, spatulate or linear, without stipules; flowers minute, perfect or monoecious, axillary; perianth none; bracts two, sac-like, or none; stamen 1; pistil 1; styles 2, filiform; ovary 4-celled; ovule 1 in each cell; fruit nutlike, compressed, 4-lobed, 4-celled, more or less winged or keeled; endosperm oily. (Named from the single genus *Callitriche.*)

CALLITRICHE. Water Starwort

Low slender usually tufted herbs; leaves spatulate or linear or both, entire; flowers solitary or 2 or 3 in the axil of a leaf, with or without a pair of membranaceous bracts; staminate flower a single stamen; pistillate flower a single 4-celled ovary, sessile or pedicelled, with 2 distinct sessile stigmas. (Name from Gr., *kalos,* beautiful; *thrix,* hair.)

Callitriche palustris L. Aquatic, glabrous annual, rooting in the mud; roots fibrous; stems if in deep water elongating, 2–10 dm. long, or if stranded on the shore much dwarfed; submersed leaves 1–2 cm. long, narrowly linear, emarginate, membranous, the upper leaves gradually transitional to the floating leaves, these, 5–15 mm. long, the petiolar base as long as the obovate blade, this chartaceous, 3-nerved; the 2 floral scales 0.5–0.8 mm. long, pale, suborbicular; style 1 mm. long; fruit 1–1.3 mm. long, oval, compressed, shallowly notched at apex. *C. verna* L., in part. In ponds and slow streams, *U. Son., A. T.*

LIMNANTHACEAE. Meadow Foam Family

Annual, glabrous herbs, with pungent juice; leaves alternate, pinnately dissected; stipules none; pedicels solitary, bractless, axillary; flowers regular, perfect; hypanthium saucer-shaped; sepals 3–5, persistent; petals 3–5, marcescent; stamens 6–10, more or less perigynous, those opposite the sepals with a gland at base outside; carpels 2–5, distinct but for the basal styles united at tip; fruits separating into tuberculate nutlets; seeds without endosperm. (Named from the genus *Limnanthes.*)

FLOERKEA

Sepals 3, slightly imbricate in bud, spreading in fruit; petals 3, white; stamens 6; filaments subulate; anthers introrse, opening longitudinally; ovaries 2–3; style 2–3-cleft at apex; stigmas capitate. (Named for *Gustav Heinrich Flörke,* of Germany)

Floerkea occidentalis Rydb. *False Mermaid.* Stems 3–20 cm. long, ascending or decumbent; leaves 1–5 cm. long; petioles 1–40 mm. long; the 3–5 leaflets 2–15 mm. long, linear-oblanceolate or linear-lanceolate; peduncles 1–3 cm. long, longer than the petioles, often exceeding the leaves; sepals 2–3 mm. long, ovate-deltoid; petals 1–1.3 mm. long, oblanceolate; stamens 0.4–0.8 mm. long; carpels usually 2, in fruit 2–2.5 mm. in diameter, globose, brown, tubercled. *F. proserpinacoides* of Piper, not of Willd. Meadows or moist woods, common in Blue Mts., and Craig Mts., *Can.*

ANACARDIACEAE. Cashew Family

Trees or shrubs with resinous or milky acrid juice; leaves alternate, usually compound, without stipules, not punctate; flowers regular, small, polygamo-dioecious or perfect; calyx 3–7-lobed;

petals 3–7 or none; disk present; stamens as many or twice as many as the petals, rarely fewer or more; styles 1–3; ovary 1-celled, 1-ovuled, (rarely 2–5-celled); fruit generally a small drupe; endosperm scant or none. (Named from genus *Anacardium.*)

RHUS

Trees or shrubs, some poisonous to the skin; flowers polygamous or dioecious, seldom truly perfect, small, greenish or rarely yellow or rose-color; calyx small, 5-parted; petals 5; stamens 5; styles 3; ovule basal. (The ancient Gr. and Lat. name)

Leaflets 7–21; fruit red, pubescent. *R. glabra.*
Leaflets 3; fruit yellowish to white, glabrous. *R. radicans,* var. *Rydbergii.*

Rhus glabra L. *Smooth Sumac.* Shrub, or small tree, 1–6 m. tall, freely branching, broad-topped; bark smooth, reddish brown with a whitish bloom, becoming gray; winter buds brown woolly, almost surrounded by the cordate leaf-scar; leaves 2–4 dm. long; petiole and rhachis with more or less brown wool, which persists in the upper groove; leaflets 4–10 cm. long, lanceolate or oblong-lanceolate, acuminate, coarsely serrate, above dark green, beneath pale and glaucous; panicle 6–22 cm. long, broad pyramidal, dense, pilose; bracts lance-linear, caducous; calyx-lobes 1.2–2 mm. long, lanceolate to deltoid; pilose at base; petals 3–3.5 mm. long, yellowish, oval, cuneate, pilose up the middle; anthers and filaments each 1.5 mm. long; drupes 3–5 mm. in diameter, spherical, short, acid hirsute; seeds 3–4 mm. long, oval, flattened, brown. *R. glabra,* var. *occidentalis* Torr.; *R. occidentalis* (Torr.) Blank. Forming thickets, near ledges, springy places, or creeks, abundant Snake River Canyon and tributaries, *U. Son.,* occasional, *A. T.*

After the first frost the leaves turn color, painting the hillsides with crimson.

Rhus radicans L., var. **Rydbergii** (Small) Rehd. *Poison Ivy.* Shrub 1–3 m. tall, erect or climbing against bluffs, but with us not really vine-like; bark smooth, brown or gray, lenticels darker, roundish; winter buds brown woolly; leaves 1–3 dm. long; petioles puberulent to glabrate; leaflets 3–15 cm. long, ovate or rhombic, short acuminate, entire, remotely dentate, or lobed, above dark, shining green, beneath pale green, more or less pilose, especially on the veins; calyx-lobes 1.5 mm. long, ovate; petals 3 mm. long, oblanceolate, green; drupes 4–6 mm. in diameter, subglobose, waxy-yellow, white within. *Toxicodendron Rydbergii* (Small) Greene. Abundant, base of ledges and talus slopes, Snake River Canyon, *U. Son.;* locally abundant, *A. T.*

Many people are susceptible to the poison produced by this shrub. It is a non-volatile, liquid resin in the sap. If this sap, liquid or dried, touches a sensitive skin it causes a red painful rash, then watery blisters. A preventative is a wash of 5% ferric chloride in equal parts of glycerine and water. A palliative is 5% permanganate of potash in water.

CELASTRACEAE. STAFFTREE FAMILY

Shrubs, often climbing; leaves simple and undivided, alternate or opposite; stipules none or small and early withering; flowers regular, usually perfect, small; pedicels commonly jointed; calyx 4–5-lobed, the lobes imbricated; petals 4–5, spreading; stamens

4–5, perigynous on a disk, alternate with the petals; ovary 1–5-celled, sessile, free from or confluent with the disk; fruit a somewhat fleshy dehiscent pod; endosperm fleshy. (Named from the genus *Celastrus.*)

Leaves opposite, evergreen; stamens 4; petals 4. *Paxistima.*
Leaves alternate, deciduous; stamens 10, petals 5. *Glossopetalon.*

GLOSSOPETALON

Deciduous shrubs; branches green, glabrous, spinescent; leaves small, alternate, entire, with or without adnate, setaceous stipules; flowers usually 1 in the axils; sepals 4–6, unequal; petals 4–6, oblanceolate, exserted; stamens mostly 5–10, inserted under a crenately 8–10-lobed disk; ovary 1-celled, with 2 basal ovules; follicle oblique, 1–2-seeded; seeds with a thin, white aril or caruncle. (Name from Gr., *glossa,* tongue; *petalon,* petal.)

Glossopetalon stipuliferum St. John. Arching, freely branched, glabrous shrub, 1–3 m. tall; branchlets with decurrent lines from the nodes; leaves 6–17 mm. long, subcoriaceous, entire, oblanceolate, grayish green, the veins rather obscure, the 2–4 lateral veins almost parallel to the margin, leaf subsessile from a dark bulbous, persistent base; stipules 0.5–1 mm. long, dark, lanceolate to subulate; flowers from leafy, axillary, short shoots on the old wood; pedicels 1–5 mm. long; the 5 sepals 1.3–1.8 mm. long, ovate-lanceolate, the 5 petals 6–9 mm. long, 1.1–2.7 mm. broad, white, linear-oblanceolate; stamens 5–8; follicle 4–5 mm. long, ovoid or asymmetrically so, acute, coriaceous, with many heavy, longitudinal ribs; the 1–2 seeds 2.8–3 mm. long, cochleate, brown, cellular roughened, the caruncle prominent. *Forsellesia stipulifera* (St. John) Ensign; *G. nevadense* Gray, var. *stipuliferum* (St. John) C. L. Hitchc. Abundant on slopes of the Snake River Canyon down as far as the region of Colton, Whitman Co., *U. Son.*

The type is Lewiston, Ida., *L. F. Henderson* 4,855.

PAXISTIMA

Low shrubs; leaves smooth, serrulate, coriaceous; stipules minute, caducous; flowers perfect, solitary or in cymes in the axils; calyx-lobes 4, broad; stamens on the edge of the disk; style very short; ovary free; pod small, oblong, 2-celled; seeds 1 or 2, enclosed in a white membranous many-cleft aril. (Name from Gr., *pachus,* thick; *stigma,* stigma.)

Paxistima myrsinites (Pursh) Raf. *Mountain Lover; Oregon Boxwood.* Shrub 3–10 dm. tall, glabrous; bark brown to gray, the lenticels round, whitish; leaves 6–40 mm. long, obovate or oval to oblanceolate or elliptic, subsessile, coriaceous, serrulate; the margins slightly revolute; peduncles and pedicels each 1–3 mm. long; hypanthium saucer-shaped; calyx-lobes 0.7–1 mm. long, ovate; petals 1.2–1.5 mm. long, rhombic-ovate, red; stamens 1 mm. long; capsule 4–6 mm. long, obovoid, beaked; seeds 4.5 mm. long, lanceolate, flattened, brown, shining. *Pachistima myrsinites* (Pursh) Raf.; *Paxistima myrtifolia* (Nutt.) L. C. Wheeler. Abundant, moist woods, *Can.,* occasional, *A. T. T.*

The specific epithet, *myrsinites,* is not invalid under the 1961 rules, even though published with a question mark. L. C. Wheeler (1943) rejected it as invalid under the 1935 rules.

ACERACEAE. Maple Family

Trees or shrubs with watery or sugary sap; leaves simple or pinnately or palmately compound, opposite, without stipules (in ours); flowers small, regular (in ours), polygamous or dioecious; sepals 4–5; petals often none or 4–5; stamens 3–12, inserted on the fleshy disk; ovary 2-celled and lobed (in ours), with 2 ovules in each cell (in ours); endosperm none. (Named from the genus *Acer.*)

ACER. Maple

Flowers polygamo-dioecious, in clusters; calyx colored, usually 5-lobed; petals 5 and equal or none; styles 2; fruit a double samara, 2-winged at apex, separable at maturity, each part 1-seeded. (The classical Lat. name of the maple.)

Leaves palmately lobed or some of them palmately compound; flowers polygamous or perfect. *A. glabrum,* var. *Douglasii.*

Leaves pinnately compound; flowers dioecious. *A. Negundo,* var. *violaceum.*

Acer glabrum Torr., var. **Douglasii** (Hook.) Dippel. *Douglas Maple.* Shrub or small tree, occasionally reaching size of 13 m. tall and 45 cm. in diameter; bark of trunk gray, smooth or slightly furrowed; bark of branchlets at first olive-green, glabrous, turning red or reddish brown, the lenticels few, pale; winter buds ovate, acute, red, the scales lanate within; petioles 1–13 cm. long, glabrous; blades of sprout shoots only 3-parted or 3-divided; other blades 3–12 cm. long, cordate in outline, 3–5-lobed, the lobes coarsely doubly serrate, acuminate, glabrous, above dark green, beneath pale green, sparsely glandular but soon glabrate; cymes 3–10-flowered, glabrous, terminal on lateral leafy branches; pedicels 3–30 mm. long; staminate flowers with calyx-lobes 3 mm. long, oblong-oblanceolate, greenish; petals similar, a little narrower and more cuneate; the 8 stamens 2.5–3 mm. long; ovary rudimentary; perfect flowers with stamens 1–1.3 mm. long; ovary subglobose, beaked, as long as the 2 spreading stigmas; samara with halves spreading at about 30°, each half 2–3.5 cm. long, the back nearly straight, the wing semielliptic, broadest above the middle; seeds 4–5 mm. long, ovoid, brown. Wood whitish, dense, hard; sapwood thick. *A. Douglasii* Hook. By streams, *U. Son., A. T.;* abundant, moist woods, *A. T. T., Can.*

Named in honor of its collector, the Scottish explorer, David Douglas. The herbage is readily eaten by sheep and cattle.

Acer Negundo L., var. violaceum Kirchner. *Box Elder.* Tree up to 15 m. tall, and 1 m. in diameter, much branched near the base, the branches spreading with drooping tips; bark gray or brown, ridged and scaly; bark of twigs green with bluish bloom; lenticels few, elongate, brownish; winter buds ovoid, glaucous, the inner scales pilose; leaves 3–11-, usually 3–7-foliolate; petioles 2–10 cm. long; leaflets 3–10 cm. long, ovate or obovate to lanceolate, entire or irregularly dentate, occasionally lobed, the terminal leaflet sometimes 3-lobed, above at first slightly puberulent, soon glabrate, olive green, beneath puberulent on the veins and hairy tufted in the vein axils, pale green; pedicels slender, drooping, pilose, often 4 cm. long; staminate flowers with calyx-lobes 1 mm. long, pilose, ovate; petals none; filaments slender, exserted; anthers 2.5–3 mm. long, cylindric with a prominent mucronate tip; pistillate flowers with calyx-lobes 0.8–1.5 mm. long, pilose at base and ciliate, obovate; the two stigmas about 5 mm. long; samara with halves shortly connate and narrowed at base, glabrous, the tips diverging at about 45°; the wings with the back nearly straight, the wing semielliptic, broadest near the middle; the half fruit 3–3.5 mm. long; seeds 12 mm. long tapering at each end, red-brown. Wood whitish, light, soft, weak; sapwood hardly separable. *Negundo Nuttallii* (Nieuwl.) Rydb. Native from S. Dak. to Mass., commonly planted, now established, *U. Son., A. T.*

Extensively planted by the homesteader and by the government for windbreak and timber culture.

BALSAMINACEAE. Balsam Family

Usually glaucous succulent herbs with watery juice; leaves alternate, simple, without stipules; flowers irregular with a petal-like imbricated usually spurred calyx; petals 4, united in two pairs; stamens 5, with short filaments and more or less united anthers; ovary 5-celled; seeds without endosperm. (Named from the generic name *Balsamina,* a synonym of *Impatiens.*)

IMPATIENS. Jewelweed

Delicate herbs with translucent stems; leaves coarsely toothed, petioled; flowers axillary or panicled, often of two kinds; large sterile ones and smaller cleistogamous ones which ripen good seed; sepals apparently 3, the posterior one usually spurred; petals 4, united in two pairs; filaments 5, each with a scale-like appendage; appendages united and covering the stigma; pod with evanescent partitions, opening suddenly when touched and projecting the seeds. Named by Dodonaeus from Lat. *impatiens,* impatient.)

Flowers clear or golden yellow. — *I. aurella,* f. *aurella.*

Flowers yellow, but spotted with another color,
  Flowers spotted with crimson. — Forma *coccinea.*
  Flowers spotted with reddish brown. — Forma *badia.*

Impatiens aurella Rydb., forma **aurella.** *Jewelweed.* Annual, glabrous; stems 5–15 dm. tall and as much as 3 cm. in diameter; petioles 1–5 cm. long; blades 2–10 cm. long, ovate or oval, acute, cuneate, coarsely dentate, the teeth mucronate tipped, above bright green, beneath paler; raceme 2–9-flowered; bracts linear; pedicels 5–30 mm. long; lateral sepals 4–7 mm. long, obliquely ovate, acuminate, membranous, posterior sepal about 1 cm. in diameter and in length, conic, clear or golden yellow, tapering gradually into the slender spur of about 8 mm. in length, the dark nectar-bearing tip sharply recurved; lower petals 5 mm. long; pods 10–23 mm. long, fusiform, when ripe exploding and ejecting the seeds at a touch; seeds 4–5 mm. long, ovoid, dark brown. Moist stream banks, Tucanon River; Rock Lake; and abundant from Spokane northward, *A. T.*

Forma **badia** St. John. Flowers with a yellow ground color, but the spur spotted within with reddish brown, and the petals more strongly marked with the same color. Growing with the species. Known only from the type collection from Indian Canyon, Spokane, *St. John* 9,210, *A. T.*

Forma **coccinea** St. John. Flowers with a yellow ground color, but the sac spotted with crimson, and the two lower petals solid crimson. Occasional with the species. The type is from Dartford, *St. John & Warren* 6,749, *A. T.*

RHAMNACEAE

Erect shrubs or small trees; leaves simple; stipules small and early deciduous or none; flowers small and regular, sometimes polygamo-dioecious; disk fleshy; calyx 4- or 5-toothed; petals 4 or 5, on the disk, or sometimes none; stamens 4 or 5, perigynous, opposite the petals; ovary sessile; fruit a drupe or pod, with 1 seed in each cell; endosperm present. (Named from the genus *Rhamnus.*)

Fruit a drupe; petals sessile or none; calyx and disk free from the ovary. *Rhamnus.*
Fruit a capsule; petals clawed; calyx and disk adherent to base of ovary. *Ceanothus.*

CEANOTHUS

Shrubs or small trees, often with spinescent branches, sometimes evergreen; leaves alternate, commonly palmately 3-ribbed, petioled; flowers perfect, in small panicles, cymes, or umbels; calyx-lobes 5, caducous; calyx and disk adherent to the ovary; petals hooded, spreading; stamens 5; ovary 3-lobed; fruit 3-lobed, dry and very oily, splitting into its three carpels when ripe; these elastically dehiscent along inner edge. (An ancient Gr. name, *keanothus,* for some spiny plant.)

Leaves coriaceous, evergreen appearing varnished above, densely hairy beneath; panicles from the leaf axils. *C. velutinus.*
Leaves membranous, deciduous, dull above, sparsely hairy beneath; panicles leafless, from old leaf scars. *C. sanguineus.*

Ceanothus sanguineus Pursh. *Buckbrush; Snowbrush; Chapparal.* Shrub, 1–3 m. tall; young twigs greenish, glabrous, those of second year dark reddish brown, with dark resin blisters; winter buds ovoid, the outer scales dark brown, somewhat resinous, puberulent at tip, the inner pilose; stipules 2–5 mm. long, lance-linear, pilose; petioles 7–25 mm. long, pilose, early glabrate; blades 2–9.5 cm. long, oval or ovate, obtuse or acute, rounded or subcordate at base, glandular crenate or crenate-serrate, above bright green glabrous, beneath puberulent on the veins; panicles 3–15 cm. long, sparingly pilose, leafless, lateral on the old wood; pedicels 3–13 mm. long, glabrous; hypanthium obconic, dark; flowers sweet, almost sickishly so; calyx-lobes 1–1.5 mm. long, broadly ovate, acute, white; petals 2–2.5 mm. long, slender clawed, the limb oval, hooded; stamens 2–2.5 mm. long, styles 3, united part way; capsules 3–4 mm. in diameter;

subglobose, 3-furrowed, the exocarp glutinous, flaking off at maturity; seeds 2–2.3 mm. long, ovoid, hard, shining, olive-green to greenish brown. Occasional in thickets, *A. T.*, abundant, woodlands, *A. T. T.*

First collected by Capt. M. Lewis. The herbage is palatable forage, a favorite with the grazing animals. Sheep seek it out, and fatten upon the oily pods and seeds.

Ceanothus velutinus Dougl. ex. Hook. *Sticky Laurel; Mountain Balm.* Shrub, 1–2 m. tall; young twigs puberulent, green, glutinous; older branches brown, glabrate, more or less glutinous and with resin blisters; stipules 2–3 mm. long, lanceolate, puberulent, dark; petioles 5–17 mm. long, puberulent; blades 2–9 cm. long, oval or elliptic, at base rounded or subcordate, the margin serrate, the teeth tipped with large black glands, above dark green and shining as if varnished, beneath densely white puberulent and somewhat glutinous; panicles 3–18 cm. long, axillary; peduncle puberulent; pedicels 3–12 mm. long, glabrous; flowers very fragrant, sickishly sweet; hypanthium saucer-shaped; calyx-lobes 1.3–1.8 mm. long, white, caducous; petals 2.5–3 mm. long, white, long clawed, the limb scoop-shaped, hooded; stamens 2.5–3 mm. long; capsules 4–5 mm. in diameter, 3-lobed and keeled, the exocarp glutinous, flaking off; seeds 1.8–2.2 mm. long, ovoid or subglobose, shining, brown. Forming dense thickets, in the hills. *A. T. T.*

The whole plant is resinous and strong smelling, and the leaves are tough and leathery. It is unpalatable to grazing animals.

RHAMNUS

Shrubs or small trees; leaves alternate, petioled, pinnately-veined, with small deciduous stipules; flowers greenish, polygamous, perfect, or dioecious, in axillary clusters; calyx 4- or 5-cleft; calyx-tube bell-shaped, lined with the disk, both free from the ovary; petals 4–5, small, oblong, sessile, acute, or none; stamens 4 or 5; ovary 2–4-celled; fruit a berry-like drupe, with 2–4 separate seed-like nutlets. (The ancient Gr. name *rhamnos.*)

Winter buds scaly; leaves with 6–8 pairs of veins; peduncle none; seeds 2-grooved on the back. *R. alnifolia.*

Winter buds naked, hairy; leaves with 8–20, usually 10–20 pairs of veins; peduncles mostly exceeding the petioles; seeds rounded on the back. *R. Purshiana.*

Rhamnus alnifolia L'Hér. *Alder-leaved Buckthorn.* Shrub, 1–2 m. tall; branchlets green, puberulent; branches glabrate, brown, later gray; stipules 2–6 mm. long, subulate or lanceolate, puberulent; petioles 2–11 mm. long; blades 1.5–9.5 cm. long, obovate-elliptic, short acuminate, obtuse, membranous, gland-tipped serrate-crenate, above glabrous, beneath pilose on the veins, scarcely paler green; flowers axillary, solitary or in umbels, mainly dioecious; pedicels 2–7 mm. long, sparsely puberulent or glabrate; hypanthium saucer-shaped; the 5 calyx-lobes 1.5–2 mm. long, ovate-deltoid, green; petals none; stamens 0.7 mm. long; drupes 6–10 mm. in diameter, black, juicy; the 3 nutlets 4–5.5 mm. long, lance-ovate, ridged on the inner angle. Rare, in swamps, Marshall Jct., Wash.; Viola; and Farmington Ldg., Ida., *A. T .T.*

The berries are poisonous.

Rhamnus Purshiana DC. *Cascara; Chittam Bark.* Shrub 2–12 m. tall with us, elsewhere in humid forests a tree up to 14 m. tall and 50 cm. in diameter; branchlets green to reddish, puberulent; branches in second year puberulent or glabrate, reddish brown, the lenticels pale, round or elongate; old bark brown, becoming gray, smooth or scaly; stipules 2–4 mm. long, lanceolate, puberulent, quickly caducous; petioles 5–30 mm. long, puberulent; blades 4–16 cm. long, oval, oblong, or obovate, acute or obtuse, mostly subcordate, serrulate, above bright green, glabrous, or puberulent only on the veins, beneath paler green, generally puberulent or only so on the veins; peduncles 5–40 mm. long, puberulent; pedicels 2–10 mm. long, puberulent; hypanthium campanulate, puberulent; calyx-lobes 2 mm. long, deltoid, puberulent; petals 1 mm. long, orbicular, greenish; stamens 1 mm. long, the filaments subulate; drupes 8–12 mm. in diameter, black, shiny; seeds 5–7.5 mm. long, obovate, compressed, olive-brown, the hilum with a yellow, cartilaginous mouth-like structure. Common, stream banks, *U. Son. to A. T. T.*

The berries are an emetic, but are eagerly eaten by birds, and were used by the Indians. All parts of the plant have cathartic properties, especially the bark. This is stripped, dried, and sold to the drug trade as *Cascara Sagrada.* The type collection was by Capt. M. Lewis on the "Kooskooskee." [That is on the Clearwater River, Ida., at Camp Chopunnish, near Kamiah.]

MALVACEAE. Mallow Family

Herbs, shrubs, or trees; leaves alternate, simple, palmately-veined, often with stellate or branched pubescence; stipules present; flowers regular, showy; peduncles axillary, jointed; calyx valvate; sepals 5, united at base, often subtended by an involucre of numerous bractlets; corolla convolute; petals 5, their bases united with each other and with the stamen-column; stamens numerous; monadelphous, in a column; pistils several, the ovaries united in a ring or forming a several-celled pod; endosperm scant. (Named from the genus *Malva.*)

Bractlets none; stamens in two series. *Sidalcea.*
Bractlets three; stamens monadelphous,
  Stigmas terminal, capitate. *Sphaeralcea.*
  Stigmas on the inner face of the styles. *Malva.*

MALVA

Pubescent or glabrous herbs; leaves dentate, lobed, or dissected; flowers perfect, axillary or terminal, solitary or clustered; involucre 2–3-leaved; calyx 5-cleft; petals 5, obcordate; styles numerous, stigmatic down the inner side; fruit depressed, separating at maturity into as many 1-seeded indehiscent carpels as there are styles. (The ancient Lat. name.)

Malva neglecta Wallr. *Common Mallow; Cheeses.* Annual or biennial, more or less generally hirsute with simple, forked, and stellate hairs; tap-root deep; stems 2–5 dm. long, procumbent or assurgent, branched at base; stipules 2–5 cm. long, ovate-lanceolate; petioles 5–15 cm. long; blades 2–5 cm. wide, orbicular-reniform, cordate, with 5–9 broad, shallow, crenate-dentate lobes; pedicels 12–35 mm long, several clustered in the axils, unequal, reflexed in fruit; bractlets linear; calyx-tube 2–3 mm. long, saucer-shaped, stellate; calyx-teeth 2–3 mm. long, deltoid; corolla-lobes 8–14 mm. long, obovate, emarginate, white or bluish; carpels 1.5 mm. long, suborbicular, flattened at an acute angle, rounded on the back, minutely reticulate, puberulent; seeds 1.2–1.5 mm. wide, discoid with a broad back, brown. *M. rotundifolia* sensu St. John, non L. Native of Eurasia, introduced, a fairly common weed, *U. Son., A. T.*

In the more arid areas it furnishes an excellent forage for grazing animals.

SIDALCEA

Herbs; leaves rounded, mostly palmately lobed or parted; flowers pink to purple, perfect or polygamous, in a narrow terminal raceme or spike; involucre none; stamen-column double, the filaments of the outer series united usually into 5 sets opposite the petals; styles filiform, stigmatic on the inner surface; carpels 5–9, 1-ovuled, separating at maturity from the short axis, indehiscent. (The name Gr. from *Sida,* a genus of this family; *alkea,* mallow.)

Sidalcea oregana (Nutt. ex T. & G.) Gray, subsp. **oregana,** var. **procera** C. L. Hitchc. Perennial from a woody rootstock; stems stellate puberulent erect, 3–15 dm. high, branching above; leaves stellate above; lower leaves orbicular, long petioled, 7–9-cleft, the segments incised; upper leaves more deeply cleft or parted, the segments often linear; stipules linear; racemes strict, 10–30 cm. long; calyx densely stellate, 5–7 mm. long, the lobes deltoid, lanceolate; petals rose, drying magenta, 15–20 mm. long, carpels glabrous, reticulate on the back. A common, very handsome summer flower, meadows and bunch grass hills, *A. T., A. T. T.*

SPHAERALCEA

Herbs or shrubs; flowers in narrow panicle-like racemes; involucre of 3 bractlets which are distinct or united at the base; petals notched or entire; stamens-column anther-bearing at the summit; styles stigmatic at the apex; carpels 2-valved, separating from the axis at maturity; seeds usually 2 or 3 in each cell. (Name Gr., *sphaira,* sphere; *alkea,* mallow.)

Blades 15–70 mm. long; corolla 10–16 mm. long; carpels 3–3.3 mm. long,
- Flowers carmine to salmon-colored. *S. Munroana,* f. *Munroana.*
- Flowers white. Forma *alba.*

Blades 5–17 cm. long; corolla 20–27 mm. long; carpels 7–13 mm. long. *S. rivularis.*

Sphaeralcea Munroana (Dougl. ex Lindl.) Spach, forma **Munroana.** *Salmon Globe Mallow.* Perennial, 2–7 dm. tall, stellate hoary throughout, erect or ascending; root stout, woody, the crown branched; stipules 3–8 mm. long, linear, caducous; lower leaves with petioles 2–6 cm. long, the upper with shorter ones or subsessile; blades ovate to orbicular-cordate, coarsely blunt dentate, and mostly 3–5-lobed; inflorescence thyrsoid, terminal or partly axillary; calyx-tube 2–5 mm. long, campanulate; calyx-teeth 2–6 mm. long, deltoid; corolla-lobes broadly obovate, entire or emarginate; stamen-column stellate; fruiting carpels semicordate, wedge-shaped, hirsute stellate on the rounded backs, upper portion sterile and with the lateral faces smooth, firm, the lower portion fertile and with the lateral faces membranous between heavy reticulate thickenings;

seeds 2 mm. wide, reniform, finely light reticulate and stellate. Locally abundant, Snake River Canyon, *U. Son.*

If *Sphaeralcea* (1827) is not made a nomen conservandum, it will be necessary to make a new combination for this species under *Phymosia* (1825).

Forma **alba** St. John. Differing in having the petals white.

Washington: base of bluffs, Indian, Whitman Co., June 2, 1929, *H. St. John* 9863 (type in Herbarium of Washington State University), *U. Son.*

Sphaeralcea rivularis (Dougl. ex Hook.) Torr. *Maple-leaved Mallow.* Perennial, 1–3 m. tall, stellate hirsute throughout, but green; root woody, stout, stipules 3–8 mm. long, lanceolate; petioles 1–15 cm. long, the upper ones reduced; blades cordate or reniform in outline, sharply 5–7-lobed, the lobes deltoid or ovate-lanceolate, dentate or double dentate; though the leaves on secondary branches produced late in the season may be subsimple and scarcely dentate; thyrse terminal, showy; calyx-tube 3–5 mm. long, campanulate to hemispheric, densely stellate; calyx-teeth 3–7 mm. long, ovate or ovate-lanceolate, densely stellate; corolla lobes broadly obovate, cuneate and ciliate at base, rose-colored; stamen column hirsute; fruiting carpels bean-shaped, compressed, cartilaginous, brown, the outer margins and backs hirsute stellate, the backs hispid also; seeds 2.5–3 mm. long, suborbicular-reniform, papillose, brown, hispidulous stellate. *Phymosia rivularis* (Dougl. ex Hook.) Rydb.; *Iliamna rivularis* (Dougl.) Greene. Open hillsides, or stream banks, *U. Son.* to *A. T. T.*

GUTTIFERAE. Garcinia or St. Johnswort Family

Herbs, shrubs, vines, or trees; leaves opposite or whorled, entire, mostly sessile, often evergreen, or with translucent or dark-colored glandular dots; stipules none; flowers perfect or polygamous, regular; sepals 2–6; petals 2–6, hypogynous; stamens commonly in three or more clusters; styles mostly 2–5, usually distinct or nearly so; capsule 1-celled, with 2–5 parietal placentae, or 3–5-celled; endosperm none. (Name Lat. *gutta,* drop; *ferro,* to bear.)

HYPERICUM. St. Johnswort

Perennial herbs or shrubs; leaves sessile, entire, glandular punctate; flowers perfect, yellow, pink, or purplish, solitary or in cymes; sepals 5, imbricate or valvate; petals 5, convolute; stamens numerous, distinct or in 3–5 bundles; ovary 1-celled, with 3–5 parietal-placentae, rarely 3–5-celled; fruit a septicidal capsule, rarely berry-like. (Ancient Gr. name, *hiperikon,* by Dioscorides applied to *H. crispum.*)

Petals 2.5–4 mm. long, not black dotted; stamens 15–21; leaves not black dotted. *H. anagalloides.*
Petals 8–13 mm. long, black dotted near the margin; stamens 50 or more; leaves usually black dotted,
  Sepals 3–4 mm. long, obtuse; leaves elliptic or broader. *H. Scouleri.*
  Sepals 4–9 mm. long, acute; leaves oblong-elliptic to linear. *H. perforatum.*

Hypericum anagalloides C. & S. *Tinker's Penny.* Annual or perennial, prostrate and mat-forming or ascending, diffuse, glabrous throughout; stems 2–30 cm. long, rooting at the lower nodes; leaves 2–17 mm. long, oval or ovate, remotely punctate; flowers few to several in a leafy cyme; pedicels 0–8 mm. long; sepals 4–9 mm. long, lanceolate to elliptic, cuneate, foliaceous, with glandular ducts near the base; petals oval, yellow; stamens as long as the petals, in 3 clusters; capsule 3–4.5 mm. long, ovoid, 1-celled; seeds 0.6–0.8 mm. long, cylindric, slightly curved, straw-colored to pale brown, 10-ribbed with reticulate cross bars. *H. anagalloides,* var. ***nevadense*** Greene; *H.* ***bryophytum*** Elmer, which is merely a dwarfed alpine phase. Uncommon, grassy stream banks, *A. T., A. T. T.*

Hypericum perforatum L. *Goatweed; Common St. Johnswort.* Perennial, 3–15 dm. tall, glabrous throughout; root woody, thickened, crown often many branched; prostrate, sterile, mat-forming branches produced in the late fall, these with reduced, elliptic or obovate blades; fertile stems erect, woody at base; freely branching above; leaves 10–25 mm. long, pale green beneath, those of the main stem noticeably larger than those of the axillary branches, punctate and more or less black-dotted; cymes leafy; sepals linear-lanceolate, streaked with yellow glandular ducts and black-dotted; petals 10–13 mm. long, elliptic, short clawed, yellow; stamens 7–10 mm. long, 50–60, united at base in 3 groups, the connective black glandular; the 3 styles 5–6 mm. long; capsule 5–10 mm. long, ovoid, 3-beaked, the sides with numerous swollen, yellow, glandular sacs, 3-celled; seeds 0.8–1.3 mm. long, broad, cylindric, apiculate, dark brown, with about 24 rows of cross reticulate bars. Native of Eurasia and Africa, with us an introduced weed. Occasional, *A. T., A. T. T.*

First collected at Colbert, Spokane Co., in 1923, later at Kamiak Butte, and Farmington. It is a colony former, difficult of eradication. It is already a serious pest in the adjacent states and provinces, and is to be feared here. The herbage is poisonous, is rather unpalatable and usually avoided by grazing animals. If much is eaten, it causes blistering and scabbing of the skin, mouth, or eyes, and may cause death. Animals that have white, unpigmented skin, or areas of such, and which are exposed to the sunlight are particularly susceptible.

Hypericum Scouleri Hook. Perennial, 1–6 dm. tall, glabrous throughout; rhizome slender; stems erect, simple or somewhat branched above; leaves 10–33 mm. long, ovate or elliptic to oval, paler green beneath, round or mostly elliptic, punctate, black dotted near the margin; cymes leafy; calyx-lobes ovate, lined with yellow glandular ducts which are often expanded and black spotted at tip; stamens about 75, in 3 clusters, 6–8 mm. long, anthers with connective black glandular; petals 8–11 mm. long, oval or elliptic, yellow, black glandular near the margin or black glandular denticulate; the 3 styles 3–5 mm. long; capsule 4–6 mm. long, 3-celled, somewhat 3-lobed at tip; seeds 0.7–0.9 mm. long, broad cylindric, short apiculate, straw-colored or pale brown, with about 24 longitudinal ribs and numerous reticulate cross bars. *H. formosum* H.B.K., var. *Scouleri* (Hook.) Coult.; *H. formosum* H. B. K., subsp. *Scouleri* (Hook.) C. L. Hitchc. Meadows, grassy hillsides, or moist woods, *A. T., A. T. T.*

ELATINACEAE. Waterwort Family

Low annual marsh herbs or shrubs; leaves opposite or whorled, not punctate, with membranous stipules; flowers minute, axillary, regular, symmetrical; sepals 2–5, free; petals 2–5, hypogynous; stamens as many or twice as many as the petals; ovary 2–5-celled, with a many-ovuled axile placenta; fruit a septicidal capsule. (Named from the genus *Elatine.*)

Sepals obtuse, without a midrib; flowers 2–4-merous. *Elatine.*
Sepals acuminate, the midrib thickened, prominent; flower 5-merous. *Bergia.*

BERGIA

Diffuse or ascending herbs or partly shrubby, often pubescent; leaves opposite, entire or serrate; flowers solitary or clustered; parts of the flower in fives; sepals acute, with a prominent midrib; capsule crustaceous, ovoid, 5-valved; seeds numerous. (Named for *Dr. P. J. Bergius* of Stockholm.)

Bergia texana (Hook.) Seub. Plant prostrate or ascending; roots fibrous; stems 5–30 cm. long, freely branching, reddish, glandular hispidulous; stipules perfoliate at base, the lobe 1–2 mm. long, lanceolate, glandular hispidulous ciliate; petioles 2–5 mm. long, glandular hispidulous; blades 1–3 cm. long, oblanceolate, the margin serrulate towards the apex, ciliate towards the base; pedicels 0.5–1.5 mm. long; sepals 3–4 mm. long, ovate, acuminate, glandular hispidulous, the margins hyaline; petals slightly shorter, oblong, obtuse; stamens included; styles none; capsule 3–4 mm. in diameter, globose; seeds 1.3–1.5 mm. long, pale, shaped like the sector of an orange, papillose on the back. Rare, banks of Snake River, Almota, *Piper,* Sept. 1897, *U. Son.*

ELATINE

Dwarf glabrous plants growing in or near the water, often rooting at the nodes; leaves oposite or whorled; sepals 2–4, obtuse; petals 2–4; stamens as many or twice as many as the petals; styles or sessile stigmas 2–4; pod membranaceous, 2–4-celled, several–many-seeded. (The Gr. name of an obscure herb.)

Elatine californica Gray, var. **Williamsii** (Rydb.) Fassett. Plants 2–5 cm. tall, forming close mats; roots fibrous; stems simple or branched; leaves 2–4 mm. long, firm chartaceous, obovate to oblanceolate, emarginate, sessile or the lower short petioled; stipules lanceolate, minute; pedicels 0.5–1.5 mm. long; the 4 sepals 1–1.5 mm. long, oval; the 4 petals obovate, white; stamens 8; seeds 0.4–0.5 mm. long, brown, longitudinally ribbed and cross reticulate, cylindric but curved and almost cochleate. Rare, edges of ponds, Spokane, *Piper* 2643; Spokane Co., *Suksdorf* 258, *A. T.*

VIOLACEAE. Violet Family

Herbs, shrubs, or trees, with opposite, alternate, or basal leaves, with stipules; flowers perfect, axillary or in cymes, nodding; sepals 5, hypogynous; corolla somewhat irregular, 1-spurred, of 5 petals; stamens 5, hypogynous; anthers connivent over the pistil; ovary 1-celled, with 3–5 parietal several-ovuled placentae; fruit a capsule 3-valved or a berry; endosperm copious. (Named from the genus *Viola.*)

VIOLA. Violet, Pansy

Perennial or annual herbs or tropical shrubs; leaves alternate, with foliaceous stipules; peduncles mostly 1-flowered, axillary; flowers usually of two kinds, the earlier ones perfect and conspicuous, but often sterile, the later (near the ground in stemless species) with small and rudimentary petals, cleistogamous and producing numerous seeds; sepals unequal, more or less auricled; petals unequal, the lower spurred; the two lower stamens spurred. (The ancient Lat. name.)

Corolla mostly violet, blue, or white,
  Spur 5–7 mm. long, equaling the sepals,
    Herbage puberulent. *V. adunca*, f. *adunca.*
    Herbage glabrous or nearly so. Forma *glabra.*
  Spur less than 3 mm. long, shorter than the sepals,
    Plants with erect, leafy stems; leaves mostly hirsutulous on the veins beneath. *V. rugulosa.*
    Plants without erect, leafy stems; leaves glabrous,
      Plants stoloniferous; petals pale violet to nearly white. *V. palustris.*
      Plants not stoloniferous; tufted; petals violet. *V. nephrophylla.*
Corollas mostly yellow or yellowish,
  Stipules pinnate, with a larger terminal lobe; spur 3–6 mm. long, violet. *V. tricolor*, var. *arvensis.*
  Stipules entire or merely toothed; spur shorter and usually not violet,
    Leaves biennial or evergreen,
      Plants with long, prostrate, leafy stolons; leaves evergreen. *V. sempervirens.*
      Plants without prostrate leafy stolons; leaves biennial. *V. orbiculata.*
    Leaves annual,
      Leaves or some of them coarsely dentate. *V. purpurea*, var. *venosa.*
      Leaves entire or crenate or crenate-serrate,
        Upper leaves cordate. *V. glabella.*
        Upper leaves ovate or narrower,
          Lateral petals 15–20 mm. long; sepals 8–15 mm. long, linear. *V. linguaefolia.*
          Lateral petals 7–11 mm. long; sepals 4–7 mm. long, lanceolate. *V. Nuttallii*, var. *vallicola.*

Viola adunca Sm., forma **adunca.** Perennial 4–20 cm. tall, erect or ascending; rootstock slender, mostly brown scaly; stems leafy, often very short at earliest flowering, becoming as much as 13 cm. in length; basal leaves with petioles 1–8 cm. long, the blades or the first ones cordate or cordate-reniform; cauline leaves with stipules 4–15 mm. long, lance-linear, attenuate, foliaceous, mostly spinulose serrulate; petioles 1–10 cm. long; blades 1–3.5 cm. long, cordate to ovate or oblong-ovate, crenate; petaliferous flowers with peduncles 2–15 cm. long, mostly overtopping the leaves; sepals 5–7 mm. long, lanceolate, glabrous; corolla blue to violet; lateral petals 8–15 mm. long, villous bearded near the base within, the other petals glabrous; spur rather variable, straight or curved, obtuse or acute, usually slender, oblong; capsules 5–7 mm. long, ellipsoid, glabrous; seeds 2 mm. long, ovoid, dark brown; cleistogamous flowers on erect peduncles 5–10 mm. long, the sepals 3 mm. long, linear; capsules 3–4 mm. long, globose, glabrous; seeds 2 mm. long, ovoid, dark brown. Common, grasslands, thickets, moist woods, *U. Son.* to *A. T. T.*

The foliage on drying often becomes brown spotted.

Forma **glabra** (Brain.) G. N. Jones. Differing from the species in being glabrous or essentially so. *V. verbascula* Greene. Common, grasslands, thickets, or moist woods, *U. Son* to *A. T. T.*

Viola glabella Nutt. ex T. & G. Perennial, 5–32 cm. tall; rhizomes white, fleshy scaly, 2–5 mm. in diameter; stems erect, with 2–5 nodes, the lowest usually with only lanceolate, membranous bracts, the upper leaf-bearing; basal leaves several; the stipules 8–14 mm. long, broadly deltoid, serrulate, partly adnate to the petiole; petioles 4–17 cm. long, glabrous; the blades 2–7 cm. long, reniform to cordate, obtuse or bluntly short acuminate, glandular crenate-serrate, above puberulent on the veins, otherwise glabrous; cauline leaves with stipules 3–9 mm. long, elliptic to lanceolate, ciliolate, greenish, the petioles 3–30 or even 65 mm. long, glabrous; the blades cordate, acuminate, otherwise similar to the basal ones; petaliferous flowers with peduncles 1–5 cm. long, glabrous; sepals 4–8 mm. long, oblong-lanceolate, glabrous; petals yellow, the 3 lower black-veined towards the base; the lateral ones 10–17 mm. long, bearded towards the base; spur 1–1.5 mm. long, obtuse, scarcely protruding; capsules 5–8 mm. long; glabrous, ellipsoid; seeds 2 mm. long, ovoid, brown; cleistogamous flowers axillary, produced late, on weak peduncles 2–30 mm. long, puberulent; sepals 2 mm. long, lanceolate, ciliolate; capsules 4–5 mm. long, globose-ovoid, glabrous; seeds similar. Abundant, moist woods, *A. T. T., Can.*

Viola linguaefolia Nutt. ex T. & G. Perennial, 1–2 dm. tall, more or less villous throughout; rootstock short erect scaly; roots slender, woody; stems 1–4 cm. tall, with 1–3 nodes, partly subterranean; basal leaves several, the stipules 6–10 mm. long, membranous, largely adnate, the free tips deltoid; petioles 4–7 cm. long, narrowly winged by the decurrent blade; blades 20–37 mm. long, oval, remotely crenate; cauline leaves with stipules 12–15 mm. long, membranous, lance-linear, nearly free; petioles 3–12 cm. long; blades 4–9 cm. long, oval or elliptic, cuneate, irregularly crenate; petaliferous flowers with peduncle 8–19 cm. long, mostly overtopping the foliage; sepals ciliate at least at base; petals yellow, the lower dark veined at base, the lateral ones with a puberulent bearded spot within; spur 1 mm. long, barely protruded, rounded; cleistogamous flowers short peduncled, the sepals 3 mm. long, lanceolate; capsule reported to be globose, usually glabrous; seeds 3 mm. long. *V. Nuttallii* Pursh, var. *linguaefolia* (Nutt.) Jepson; *V. praemorsa* Dougl., var. *linguaefolia* (Nutt.) Peck; *V. Nuttallii* Pursh, var. *major* Hook. Moist, shady places, Blue Mts., and Craig Mts., *A. T. T.*

Viola nephrophylla Greene. Perennial; rootstock short, stout; herbage glabrous or nearly so; stipules 4–7 mm. long, partly adnate to the petiole, lanceolate, remotely laciniate-serrate; petioles 2–25 cm. long; earliest blades orbicular or reniform; later blades 2–6 cm. long, cordate, obtuse or bluntly pointed, the margin crenate-serrate; petaliferous flowers with peduncles 4–20 cm. long, mostly exceeding the leaves; sepals 5–8 mm. long, oblong-lanceolate, hyaline margined; corolla violet, the lower petal dark-lined and villous towards the base; spur 2–3 mm. long, oblong, obtuse; lateral petals 10–17 mm. long, villous at base, the upper ones somewhat so; capsules 7–10 mm. long, short ellipsoid, glabrous, green; seeds 2 mm. long, ovoid, olive-brown; cleistogamous flowers on erect peduncles 2–8 cm. long, sepals 4–5 mm. long. *V. nephrophylla,* var. *cognata* (Greene) C. L. Hitchc. Springy places, rare, *U. Son.,* occasional, *A. T.*

Viola Nuttallii Pursh, var. **vallicola** (A. Nels.) St. John. Perennial, 3–20 cm. tall, puberulent throughout or nearly glabrous; roots fleshy, usually several and fascicled; caudex erect, short, simple or branched; stems in early anthesis very short, largely subterranean, later elongating, with 2–4 nodes, sometimes 15 cm. long; basal leaves with stipules 3–5 mm. long, lanceolate, acuminate, entire or remotely laciniate; petioles 1–8 cm. long, more or less winged by the decurrent blade; blades 12–38 mm. long, subcordate to ovate, obtuse; cauline leaves with stipules 3–10 mm. long, lanceolate, entire, free; petioles 1–8 cm. long; blades 12–50 mm. long, ovate to lance-ovate, usually entire; petaliferous flowers on peduncles 2–10 cm. long, about equaling the leaves; sepals glabrous or rarely papillose puberulous at base; petals yellow, the lower dark- or purple-veined at base; the lateral bearded within near the base; capsules 7–9 mm. long, ellipsoid to ovoid, puberulent; cleistogamous flowers on peduncles 1–3 cm. long, recurved; the sepals 2–3 mm. long, lanceolate, puberulent at base; capsules 3–4 mm. long, globose, puberulent; seeds 2 mm. long, ovoid, pale. *V. vallicola* A. Nels.; *V. subsagittifolia* Suksd.; *V. Russellii* Boivin. Grasslands, *U. Son., A. T.*

This plant is maintained as a species by Nelson and by Rydberg, but Brainerd merely mentions it as a wide leafed form without giving it any status. Its broader, ovate, abruptly cuneate leaves are relatively constant and its range extends further west. To the writer it seems logical to maintain this plant as a variety.

Viola orbiculata Geyer. Perennial; rootstock brown scaly, appearing scaly, older parts white, slender; plant usually appearing stemless; basal leaves several; the stipules 3–5 mm. long, broadly deltoid, brown, nearly free; petioles 1–9 cm. long, glabrous; blades 12–43 mm. long, reniform or ovate-orbicular, more commonly orbicular, deeply cordate at base, the margin glandular-crenate, above puberulent, beneath paler green and glabrous; stems 3–8 cm. long, peduncle-like but with 2 nodes which bear either lanceolate, entire bracts or reduced, short petioled leaves with serrulate stipules; petaliferous flowers 1, rarely 2, terminal, yellow, brown-veined towards the base; sepals 3–4 mm. long, lance-elliptic; lateral petals 6–9 mm. long, bearded near the base; spur 1 mm. long, rounded, scarcely projecting; capsules 5–7 mm. long, ovoid, glabrous, purple spotted; seeds 2 mm. long, ovoid, apparently pale; cleistogamous flowers axillary, the pedicels 1–10 mm. long, the sepals 2–3 mm. long, lanceolate; capsules 3–4 mm. long, globose. Moist woods, *A. T. T., Can.*

Discovered by the early explorer, C. A. Geyer, in 1844 in the Coeur d'Alene Mts., Ida.

Viola palustris L. Perennial, glabrous throughout; rootstock running, slender; stipules 5–9 mm. long, ovate-lanceolate, entire or remotely serrulate, free from the petiole; petioles 2–8 cm. long; leaves 2–4 (or 6), the earliest blades reniform; later blades 2–6 cm. long, cordate or suborbicular-cordate, crenulate; petaliferous flowers on peduncles 3–17 cm. long; sepals 3–5 mm. long, elliptic- or lance-ovate; petals pale violet or lavender, veined with purple, or nearly white; lateral petals 7–13 mm. long, slightly bearded near the base; spur 2–3 mm. long, oblong, obtuse; capsules 5–8 mm. long, ellipsoid, glabrous, green; seeds 1.5 mm. long, dark brown; cleistogamous flowers on erect peduncles 1–3 cm. long; sepals 2–4 mm. long. *V. epipsila* of Fern., probably not of Ledeb. Our northwestern plants blend between the two extremes of petal and style length, and to the writer seem indivisible. Rare, moist stream banks, Moscow Mts., and northward in Idaho, *Can.*

Viola purpurea Kellogg, var. **venosa** (Wats.) Brain. Perennial, 3–20 cm. tall; tap-root slender, woody, the crown producing many slender branches; stems partly subterranean, numerous, tufted; puberulent; basal leaves with stipules 6–13 mm. long, lance-oblong, membranous, adnate to the petiole for most of their length; petioles 2–6 cm. long, puberulent above; blades 11–16 mm. long, orbicular-ovate, nearly entire or coarsely dentate, puberulent at base, elsewhere papillose to glabrous; cauline nodes and leaves several; the stipules, or at least the aerial ones 2–8 mm. long, foliaceous, the free tips lanceolate, entire or lacerate; petioles 1–6 cm. long, puberulent; blades 11–36 mm. long, similar to the basal or some more narrowly ovate; petaliferous flowers with peduncles 2–8 cm. long, puberulent; sepals 3–5 mm. long, lanceolate, more or less puberulent; petals yellow, and especially the upper ones reddish tinged without; the lateral ones 6–10 mm. long, bearded within, and like the lower, purplish veined at base; capsules 5–8 mm. long, globose to ovoid, puberulent; cleistogamous flowers from upper or lower nodes, the peduncles 1–5 mm. long, recurved, puberulent; the sepals 2–3 mm. long, lanceolate, puberulent; capsules 3–4 mm. long, globose, puberulent; seeds 2.5–3 mm. long, globose, brown. *V. atriplicifolia* Greene; *V. purpurea,* var. *atriplicifolia* (Greene) Peck; *V. venosa* (Wats.) Rydb. Upper gravelly ridges, Blue Mts., *Huds.*

Viola rugulosa Greene. Perennial, 2–4 dm. tall; rootstock suberect, somewhat thickened, 2–4 mm. in diameter; producing vigorous subterranean, branching stolons; stems puberulent or glabrous, erect; basal leaves several, the petioles 7–24 cm. long, glabrous or hirsutulous at tip; the blades 2–8.5 cm. long, cordate-reniform, short acuminate, coarsely crenate, ciliolate, above glabrous except on the veins; cauline leaves 3–6, the stipules 7–13 mm. long, greenish, later scarious, the upper linear-lanceolate, the lower asymmetrically ovate-lanceolate; petioles much shorter in the upper leaves; blades from cordate-reniform to ovate-lanceolate, the upper smaller; petaliferous flowers with peduncles 2–8 cm. long, not or but little exceeding the leaves; sepals 6–8 mm. long, linear-lanceolate, ciliolate at base; flowers white, often tinged without with pale violet, the 3 lower dark violet veined towards the base; lateral petals 9–13 mm. long, densely bearded just below the middle, the others glabrous; spur 0.5–1 mm. long, broad, obtuse; capsules 8–10 mm. long, ovoid, 3-angled, puberulent; seeds 2–2.5 mm. long, ovoid, pale brown; cleistogamous flowers produced mostly later than the others, from the upper axils; sepals 2.5–3 mm. long, ovate-lanceolate; petals included; capsules and seeds not seen. *V. canadensis* L. var. *rugulosa* (Greene.) C. L. Hitchc. Moist wooded borders. Waha Cr., Craig Mts., *Henderson* 2,656, *Can.*

Viola sempervirens Greene. Perennial, nearly glabrous; rootstock 1–4 mm. in diameter, short, scaly; stems prostrate, stolon-like, becoming 1–3 dm. long, forming mats; stipules 3–6 mm. long, lanceolate to ovate, acute, brownish glandular streaked; petioles 2–8 cm. long; blades 10–37 mm. long, cordate to orbicular cordate, crenulate, the veins near the margin remotely hirsutulous above, the tissue with glandular streaks that turn brown on drying; petaliferous flowers yellow, with pedicels 3–7 cm. long; sepals 4.5–5 mm. long, asymmetric lanceolate, glandular streaked; lateral petals 8–11 mm. long, spatulate-elliptic, dark-veined and hairy tufted near the base; spur 1 mm. long, rounded, scarcely projecting; capsules 5–6 mm. long, oval; seeds 1.3–1.5 mm. long, ovoid, pale; cleistogamous flowers with pedicels 1–3 cm. long; sepals 2.8–3 mm. long, lanceolate; petals 1.5 mm. long, linear; capsules 5–6 mm. long, globose; seeds 1.8–2 mm. long, ellipsoid, pale. *V. sarmentosa* Dougl., not Bieb. (1808). Six miles n. w. of Hayden Lake, Kootenai Co., *E. Reek,* April 12, 1925, *A. T. T.*

Viola tricolor L., var. arvensis (Murray) Boiss. *Heartsease.* Annual or biennial, 1–2 dm. tall, commonly decumbent, glabrous or sparingly downy; blades of basal leaves, ovate; cauline leaves with stipules coarsely pectinate-pinnate, the 2–4 lateral pairs of divisions linear, the terminal larger, oblanceolate, sparsely crenate; petiole 0.5–2.5 cm. long; blades 1–3 cm. long, ovate to lanceolate, crenate; petaliferous flowers on peduncles 2- or 3-times as long as the leaves; sepals 7–10 mm. long, lanceolate, often violet tinged; corolla shorter or but little longer than the calyx, pale yellowish, the lower one deep yellow at base, the upper sometimes violet tipped, the three lower purple or dark veined at base; spur cylindric, curved, obtuse; capsule about as long as the calyx, ovoid, glabrous; seeds 0.7 mm. long, narrowly obovoid, brown. *V. arvensis* Murray. Native of Europe, an introduced weed here, gravelly slopes, Opportunity, *Nettie M. Cook* in 1928, *A. T.*

V. tricolor is identical in vegetative parts and differs only in having larger and more deeply colored corollas. Its chromosome no. is 2n= 26. It is variable and many natural varieties are recognized. The var. *arvensis* is more vigorous but less variable. Its chromosome no. is 2n= 34. Where the natural ranges of the two meet, the two cross readily and produce fertile offspring. Consequently, the two seem best classified as a species and a variety.

LOASACEAE. Loasa Family

Herbs, shrubs, or trees, mostly with rough-barbed or stinging hairs; leaves alternate or opposite, without stipules; flowers regular, perfect; calyx 4- or 5-lobed, adherent to the ovary; petals 4–5, perigynous, induplicate valvate; stamens usually very numerous, rarely few, free or in bundles opposite the petals, some of the outer occasionally petal-like; ovary inferior, 1–3-celled, with 2 or 3 parietal placentae; ovules one to many, endosperm mostly scanty or none. (Named from the genus *Loasa*.)

MENTZELIA

Annual or biennial erect herbs, more or less rough with rigid barbed hairs, the stems becoming white and shining; leaves alternate, mostly coarsely toothed or pinnatifid; flowers terminal, solitary or in cymose clusters; calyx cylindrical, 5-parted; petals 5 or 10, spreading; stamens perigynous; styles 3, more or less united into one; capsule few–many-seeded. (Named in honor of *Chr. Mentzel* of Germany.)

Stamens 12–30; filaments all filiform; seeds in 1 row, without intervening lamellae; seeds not winged,
  Seeds prismatic, deeply grooved on the angles, minutely muricate; stamens 12–14. *M. dispersa.*
  Seeds irregularly angled, not grooved on each angle, tuberculate; stamens 20–30,
    Leaves or some of them pinnatifid, lanceolate. *M. albicaulis,* var. *albicaulis.*
    Leaves mostly entire and linear. Var. *tenerrima.*
Stamens 100–180; the 5 outer filaments dilated; seeds in 2 rows on each placenta, with intervening lamellae; seeds winged,
  Petals 5–7 cm. long; outer filaments 2–3 mm. wide. *M. laevicaulis,* var. *acuminata.*
  Petals 2.5–4.5 cm. long; outer filaments 1–1.5 mm. wide. *M. Douglasii.*

Mentzelia albicaulis Dougl. ex. Hook., var. **albicaulis**. Annual, 5–40 cm. tall; tap-root slender, with fibrous laterals; stem white, sparsely hirsutulous or glabrate, usually branching; leaves hispid above, beneath hispid with stiff simple hairs and hirsutulous with retrorsely barbed hairs; basal leaves 1.5–10 cm. long, oblanceolate, deeply and remotely pinnatifid, the oblong lobes unequal; middle cauline leaves similar, or less divided, or even linear and entire; upper leaves from linear to ovate-lanceolate, entire or somewhat pinnatifid; flowers sessile in terminal, leafy bracted cymes; calyx-lobes 2.5–4 mm. long, lanceolate, hispid and barbed hirsutulous; petals 2–8 mm. long, obovate, yellow, with several strong longitudinal veins; stamens free, well included; style exceeding the stamens; capsule 10–22 mm. long, 2 mm. in diameter, subcylindric, angled, roughly hairy with pustulate-based hispid hairs and barbed hirsutulous hairs; seeds 1–1.5 mm. long, pale brown. *M. albicaulis,* var. *Veatchiana* (Kellogg) Urb. & Gilg; *Acrolasia albicaulis* (Dougl. ex Hook.) Rydb. Dry talus slope, Palouse Falls, *St. John & Pickett* 6,165, *U. Son.*

The var. *Veatchiana* has been maintained by Macbride, but since there seems to be a complete intergradation in the petal and sepal sizes, and since the extremes are not notable, the variety is here reduced to synonymy.

Var. **tenerrima** (Rydb.) St. John. Stem 5–30 cm. tall, slender, little branched; leaves 10–35 mm. long, linear or narrowly oblong, entire or nearly so; calyx-lobes 1–3 mm. long; petals 2–5 mm. long. *M. tenerrima* Rydb.; *Acrolasia tenerrima* Rydb. Gravelly pine woods, Moscow (Cedar) Mts., *Elmer* 787; Latah Co., *Piper,* 1,626, *A. T. T.*

Though with identical seeds, this plant because of its narrower, mostly entire leaves, and its occurrence in a different habitat at remote localities, seems worthy of varietal rank.

Mentzelia dispersa Wats. *Stick Leaf.* Annual, 1–5 dm. tall; tap-root becoming stout, the laterals fibrous; stem hispid and hirsutulous, with retrorsely barbed hairs, becoming white, freely branching; foliage above hispid with scabrous, pustulate-based hairs, and hirsutulous with minute, retrorsely barbed hairs, beneath hirsute and hirsutulous with retrorsely barbed hairs; basal leaves 3–12 cm. long, narrowly oblanceolate, dentate; cauline leaves 1–15 cm. long, from linear-lanceolate to ovate, and from entire to sharply dentate; flowers sessile, in leafy bracted cymes; calyx-lobes 1–3 mm. long, linear- or elliptic-lanceolate, hispid and hirsutulous with barbed hairs; petals 3–6 mm. long, oval, yellow, with strong longitudinal veins; stamens ⅔ as long, exceeding the style; capsules 15–30 mm. long, 2 mm. in diameter, angled, rough with barbed hispid and hirsutulous hairs; seeds 0.8–1.3 mm. long, gray. *M. integrifolia* (Wats.) Rydb.; *M. dispersa,* var. *latifolia* (Rydb.) Macbr. Common, dry gravelly places, *U. Son.,* glacial sand plains and gravels about Spokane, *A. T.*

Mentzelia Douglasii St. John. Biennial, 3–5 dm. tall; stems divaricately branched, hispid-scabrous, white; leaves hispidulous-scabrous, those of the rosette and the lower cauline ones 6–20 cm. long, oblanceolate or lanceolate, sinuate dentate, tapering into a short petiole; upper cauline leaves gradually decreasing in size, sessile, lanceolate, laciniate dentate; flowers diurnal, terminal or on axillary pedicels shorter than the leaves; calyx-lobes 15–30 mm. long, lance-linear, hispidulous; the 5 petals oblanceolate, yellow; stamens 100–140, in 4–5 series; capsule 2–3.5 cm. long, 7–10 mm. in diameter, subcylindric, tapering at base, angled, hispid-scabrous; seeds 3–4 mm. long, oval-discoid, pale brownish, the margin thin winged. *Bartonia parviflora* Dougl.; *Nuttallia parviflora* (Dougl.) Greene; *M. parviflora* (Dougl.) Macbr., not Heller. Not *M. Brandegei* Wats. Dry slopes near shore of Snake River, Lime Pt., Ida., *St. John* 4,363; and *Ridout & Ransom;* Wawawai, *St. John* 3,386, *U. Son.*

In the monograph by Miss Darlington, *B. parviflora* is reduced to the synonymy of *M. laevicaulis.* Apparently neither of us have seen the Douglas type. I am convinced that my plants have specific characters, and believe them to be the same as those of Douglas.

Mentzelia laevicaulis (Dougl. ex Hook.) T. & G., var. **acuminata** (Rydb.) Nels. & Macbr. *Rough Blazing Star.* Biennial, 3–10 dm. tall; stems stout, much branched, hispid-scabrous, becoming pale; lower leaves and those of the basal rosette 5–20 cm. long, oblanceolate, sinuately dentate, densely hispid-scabrous, tapering into a slender petiole; upper cauline leaves gradually reducing upwards, lanceolate, laciniate-dentate, densely hispid-scabrous; flowers diurnal, terminal or on axillary pedicels shorter than their leaves; calyx-lobes 20–42 mm. long, linear-lanceolate, attenuate, hirsute; the 5 petals 5–7 cm. long, yellow, oblanceolate; stamens 120–180, in 4–5 series; capsule 3–4 cm. long, 8–10 mm. in diameter, angled, hispid-scabrous; seeds 2.5–3.5 mm. wide, discoid, thin winged. *Nuttallia acuminata* Rydb. Abundant, sand plains and river gravels, Spokane River, and Snake River, *U. Son., A. T.;* fugitive at Pullman.

CACTACEAE. Cactus Family

Fleshy and thickened plants; stems flattened, terete, ridged or tubercled, continuous or jointed, leafless or with small leaves, generally spiny; spines from cushions of minute bristles, (*areoles*); flowers solitary, sessile, perfect, regular, showy; sepals and petals intergrading, numerous, in several rows, the bases adherent to the ovary; stamens numerous, on the calyx-tube; style 1; ovary 1-celled, with several parietal placentae; ovules numerous; fruit a 1-celled berry, fleshy or dry; endosperm scanty or copious. (Named from the generic name, *Cactus*.)

OPUNTIA. Prickly Pear

Jointed, much-branched plants; leaves small, terete, subulate, on young branches, early deciduous; areoles axillary, bearing spines, barbed bristles (*glochids*), hairs or glands, and flowers; flowers usually lateral, large; calyx-tube not prolonged beyond the ovary; lobes numerous, spreading; petals numerous, slightly united; stamens very numerous, in several rows; fruit a berry, juicy or dry, often prickly. (Name Gr., *Opuntia,* a town in Greece, where a cactus-like plant is said to have grown.)

Opuntia polyacantha Haw. *Prickly Pear.* Low, spreading plants, forming clumps; stems 2–4 dm. long, the flattened joints 5–15 cm. long, almost circular to obpyriform, light green, the epidermis smooth; areoles 8–15 mm. apart, in spirals; spines pale or brownish, as many as 9 from an areole, the outer ones shorter, the inner ones 2–3 cm. long; glochids 2–3 mm. long, pale, very numerous; ovary spiny and resembling the joints; petals 4–5 cm. long, spatulate-obovate, yellow; anthers 1.5 mm. long; style 2–2.5 cm. long; fruit dry, capsular, 4-sided, obpyramidal; seeds said to be 6 mm. long, white, acute on the margins. Sandy or rocky, arid slopes, *U. Son.*

LYTHRACEAE. Loosestrife Family

Herbs, shrubs, or trees; leaves mostly opposite and entire, with no stipules; flowers axillary or whorled, perfect; calyx enclosing but free from the ovary; petals 4–7, as many as the calyx-teeth, perigynous, or none; stamens 4–14, on the calyx; style 1; stigma capitate or 2-lobed; ovary 2–6-celled; ovules numerous, rarely few; fruit a membranous capsule; endosperm none. (Named from the genus *Lythrum,* loosestrife.)

Leaves auriculate at base; style 1.5–2 mm. long; capsule bursting irregularly. *Ammannia.*
Leaves cuneate at base; style none; capsule septicidally dehiscent. *Rotala.*

AMMANNIA

Low and inconspicuous smooth annual herbs; leaves opposite, sessile, narrow; flowers small, 1–several in the axils; calyx campanulate, globose, or ovoid, usually 4-angled, 4-toothed and with 4 intermediate short tooth-like appendages in the sinuses; petals 4, small, early deciduous, or wanting; stamens 4–8; capsule globular, enclosed in the calyx, 2–4-celled, bursting irregularly. (Named for *Johann Ammann,* of Germany.)

Ammannia coccinea Rottb. Plant 5–30 cm. tall, glabrous; roots fibrous; stems erect or ascending, from simple to freely branched; leaves 2–7 cm. long, sessile, linear to oblong-linear-lanceolate, the base auriculate clasping; flowers 1–5 in the axils, subsessile; calyx 3–4.5 mm. long, campanulate, in fruit becoming globose, the 8 nerves green, somewhat winged, the 4 lobes 0.2–0.4 mm. long, deltoid, the 4 appendages often as large, broadly deltoid; petals 2 mm. long, flabellate, pink; stamens 4–8, barely exserted; stigma capitate; capsule 4–5 mm. long; seeds 0.2–0.3 mm. long, triangular, concave, smooth, yellow. *A. coccinea,* subsp. *purpurea* (Lam.) Koehne. Rare, sandy shore, Snake River, Almota, *Piper,* Sept. 1897, in part; Granite Pt., *St. John* 4,905, *U. Son.*

ROTALA

Low annual mostly glabrous herbs; stems 4-angled; leaves opposite; flowers small axillary, mostly solitary; calyx campanulate or globose, 4-lobed, with appendages in the sinuses; petals 4; stamens 4, short; ovary free from the calyx, globose, 4-celled; capsule globose, enclosed by the membranous calyx, 4-celled, septicidally dehiscent. (Name from Lat., a diminutive of *rota,* wheel.)

Rotala ramosior (L.) Koehne. Plant 8–30 cm. tall, glabrous; roots fibrous; stem from simple to freely branched; leaves 10–42 mm. long, oblanceolate, cuneate at base, subsessile or the larger slender petioled; flowers sessile; calyx 2–3 mm. long, campanulate, in fruit 4–5 mm. long, globose, the lobes 0.3–1 mm. long, deltoid, green, about as long as the broadly deltoid appendages; petals 1–1.5 mm. long, oval, white to pink; stamens included; stigma capitate; capsule 3–5 mm. long, globose; seeds 0.2–0.3 mm. long, rounded, plano-convex, brownish, smooth. Rare, wet shores, Spokane, *Piper* 2,644; Almota, *Piper,* Sept. 1897, in part, *U. Son., A. T.*

HALORAGACEAE. Water Milfoil Family

Herbs or shrubs, mainly aquatic; leaves alternate, opposite, or whorled, the submerged ones divided; stipules none; flowers perfect, monoecious, or dioecious; calyx-tube adnate to the ovary; the lobes 2–4 or 0; petals 2–4 or 0; stamens 1–8; ovary inferior, 1–4-celled; styles 1–4; stigmas papillose or plumose; fruit a nutlet or drupe; the carpels 1-seeded; endosperm fleshy. (Named from the genus *Haloragis.*)

MYRIOPHYLLUM. Water Milfoil

Aquatic herbs with alternate or whorled leaves, the submerged ones pinnatifid into capillary segments, the emersed ones less divided or entire; flowers in a terminal spike, or sessile in the upper axils, perfect or unisexual; calyx of the pistillate flowers entire or 4-toothed, of the staminate 4-lobed; petals 4 or 0; stamens 4 or 8; ovary 2–4-celled; stigmas 4, recurved, plumose; fruit splitting at maturity into 4 bony 1-seeded nutlets. (Name Gr., *murios,* numberless; *phullon,* leaf.)

Leaf segments 7–11 pairs; bracteoles ovate; fruits 2.3–3 mm. long. *M. exalbescens.*
Leaf segments 11–14 pairs; bracteoles linear, dentate or forked; fruits 1.8 mm. long. *M. brasiliense.*

Myriophyllum brasiliense Cambess. *Parrot's Feather.* Glabrous aquatic, all but the tips submerged, or plant decumbent if stranded; stems 1–10 dm. long, often reddish; leaves 4–6 in a whorl, 10–32 mm. long, with filiform segments; flowers unisexual, single in the upper axils; pistillate flowers with 2 linear, dentate bracteoles 0.7 mm. long; pedicel 0.5 mm. long; calyx-tube 1 mm. long, 4-grooved, the 4 linear lobes 0.7–1 mm. long, at length reflexed; petals 0; staminate flowers with 2 linear 2–3-forked bracteoles 3 mm. long; pedicel and calyx-tube 2.5–5 mm. long; calyx-lobes 1.6 mm. long, narrowly deltoid, serrulate; petals 4, elliptic spatulate, 5 mm. long; stamens 8, the anthers 3 mm. long; fruit ovoid, papillose punctate. *M. proserpinacoides* Gill. Native of Brazil, cultivated in water gardens. Established in slough, mouth of Potlatch River, Ida., *St. John et al.* 9743, *U. Son.*

Myriophyllum exalbescens Fern. *Water Milfoil.* Glabrous aquatic, mostly submerged; stems 1–8 or more dm. long, when dry becoming white; leaves 3 or commonly 4 in each whorl, 1.2–3 cm. long, pectinate pinnatifid with capillary, flaccid segments; spikes 1–12 cm. long, terminal; flowers verticillate, the lower staminate, the upper pistillate; bracts rarely equaling the fruit, the lower serrate, the upper entire; bracteoles 0.7–1 mm. long ovate, entire or serrate; pistillate flowers with calyx-tube 0.5–1 mm. long, deeply 4-grooved, the lobes low deltoid; petals 0.5 mm. long, suborbicular; staminate flowers with petals 2–2.5 mm. long, oblong obovate; stamens 8, the anthers 1.2–1.8 mm. long, linear-oblong; fruits subglobose, 4-lobed, the nutlets asymmetric ovoid, brown, smooth or rugulose. *M. spicatum* of Am. authors; *M. spicatum* L., subsp. *exalbescens* (Fern.) Hultén; *M. spicatum,* var. *exalbescens* (Fern.) Jeps. In streams and ponds, *U. Son., A. T., A. T. T.*

ONAGRACEAE. Evening Primrose Family

Herbs, or rarely shrubs or trees, with simple alternate or opposite leaves; stipules none; flowers perfect, symmetrical, the parts in twos or fours; calyx-tube adherent to the ovary; petals on the throat of the calyx or rarely wanting; stamens as many as or twice as many as the petals or calyx-lobes, on the calyx-tube;

style single, slender; stigma-lobes as many as the cells of the ovary; fruit a capsule, berry, or small nut; endosperm none. (Named from the genus *Onagra.*)

Parts of the flowers in fours or more,
  Fruit a capsule, dehiscent,
    Seeds with a silky tuft at one end. — *Epilobium.*
    Seeds not so,
      Calyx-limb divided to the ovary, persistent. — *Ludwigia.*
      Calyx-limb deciduous,
        Anthers attached near the base, erect,
          Calyx-lobes erect; petals clawless. — *Boisduvalia.*
          Calyx-lobes reflexed; petals clawed. — *Clarkia.*
        Anthers versatile,
          Ovary 4-celled; hypanthium prolonged beyond the ovary. — *Oenothera.*
          Ovary 2-celled; calyx-limb divided to the ovary. — *Gayophytum.*
  Fruit indehiscent, nut-like. — *Gaura.*
Parts of the flowers in twos. — *Circaea.*

BOISDUVALIA

Erect leafy annual herbs; leaves alternate, sessile, simple; flowers small, in leafy simple or compound spikes; hypanthium funnelform; calyx-lobes 4, erect; petals 4, 2-lobed, purple or white; stamens 8, those opposite the petals shorter; anthers basifixed, erect; ovary 4-celled, several-ovuled; capsule membranaceous, ovate-oblong to linear, nearly terete, acute, sessile, dehiscent to the base; seeds 3–8, in one row in each cell. (Named for *Jean-Alphonse Boisduval* of France.)

Capsules septifragal; hypanthium funnelform, 1.5–3.5 mm. long,
  Lower leaves lanceolate. — *B. densiflora.*
  Lower leaves lance-linear. — *B. salicina.*
Capsules loculicidal; hypanthium campanulate or ovoid, 0.5–2 mm. long,
  Floral leaves ovate, not reduced; capsule 6–8 mm. long. — *B. glabella,* var. *campestris.*
  Floral leaves lance-linear, reduced; capsules 8–11 mm. long. — *B. stricta.*

Boisduvalia densiflora (Lindl.) Wats. Plants 1.5–6 dm. tall, villous to pilosulous but green; stems simple or branched, very leafy; lower leaves 1.5–5 cm. long, 6–16 mm. broad, denticulate or serrulate; floral leaves shorter, ovate-lanceolate, imbricate; spikes dense, leafy; hypanthium 1.5 mm. long, funnelform, pilose; calyx-lobes 1.5–2 mm. long, lanceolate, unequal; petals 3.5–4 mm. long, oblanceolate, 2-lobed, rose-purple; capsules 8–10 mm. long, 1.2–1.8 mm. in diameter, stout, straight, erect, pilose, septifragal, leaving the septa as 4 wings on the placenta; seeds 1.2–1.7 mm. long, obovoid, dorsiventrally compressed, brown, cellular-pitted. By dried up ponds, *A. T.*

The species is here accepted in the restricted sense, the narrower leaved plants being referred to *B. salicina.*

Boisduvalia glabella (Nutt. ex T. & G.) Walp., var. **campestris** Jepson. Plants 1–2.5 dm. tall, somewhat villous; stems often branched at base and decumbent; lower leaves 8–20 mm. long, 2–5 mm. wide, ovate-lanceolate to lanceolate, serrulate, thick and firm; floral leaves similar, as long and often broader; spikes dense; hypanthium 2 mm. long, ovoid; calyx-lobes 0.5–1 mm. long, deltoid; petals 2 mm. long, violet, 2-lobed; capsules 1–1.5 mm. in diameter, lance-linear in profile, stout, divergent, pilosulous, often curved outward, loculicidal, a portion of the septa adhering to the valves; seeds 1 mm. long, oblanceoloid, irregularly compressed, pale brown. Rare, from dried-up ponds, Pullman, *A. T.*

Boisduvalia salicina Rydb. Plants 1.5–9.5 dm. tall erect, villous to pilosulous, canescent; stem slender, often unbranched; lower leaves 2–6 cm. long, 2–8 mm. wide, denticulate; floral leaves much shorter, lance-ovate to ovate, acuminate, at first imbricate; spikes dense; hypanthium 2–3.5 mm. long, narrowly funnelform; calyx-lobes 2–3 mm. long, deltoid-lanceolate; petals 4–6 mm. long, oblanceolate, 2-lobed, rose-purple; capsules 6–9 mm. long, 1.3–2.2 mm. in diameter, stout, straight, erect, pilose, septifragal, leaving the septa as 4 wings on the placenta; seeds 1.5–1.7 mm. long, obovate, dorsiventrally compressed, dark brown, nearly smooth. In exsiccated spots in valleys, *A. T.*

The specific name cannot be attributed to Nuttall, as his was published in synonymy only.

Because of difference in leaves, hypanthium, calyx, petals, etc., this is still maintained as a species, though Munz (1941) reduced it to a variety.

Boisduvalia stricta (Gray) Greene. Plants 0.5–5 dm. tall, pilose, often canescent, the stem or branches slender, erect, virgate, often flowering almost from the base; lower leaves 1–3 cm. long, 1.5–4 mm. wide, linear or lance-linear, denticulate; floral leaves similar, scarcely broadened or reduced; spikes virgate, rather loose; hypanthium 0.5–1 mm. long, campanulate; calyx-lobes 1–1.5 mm. long, lanceolate; petals 2–4 mm. long, 2-lobed, rose-purple; capsules 1–1.5 mm. in diameter at the base, tapering to the tip, slender, outward curved, loculicidal, a portion of the septa adhering to the valves; seeds 1.2–1.3 mm. long, obovoid, olive brown. *B. parviflora* Heller is based on *Heller* 3411 from Lake Waha. His plants were depauperate *B. stricta.* Wet or exsiccated places, *U. Son., A. T.*

CIRCAEA. Enchanter's Nightshade

Delicate perennial herbs with opposite leaves on slender petioles; flowers white, in terminal or axillary racemes; parts of the flower in twos; calyx-tube prolonged, deciduous; lobes reflexed; fruit indehiscent, small, burr-like, 1–2-celled, with hooked hairs; cells 1-seeded. (Named for *Circe,* the ancient enchantress.)

Circaea alpina L., var. **pacifica** (Aschers. & Magn.) M. E. Jones. Plant 1–4.5 dm. tall; rootstock short, white, tuberous, producing slender rhizomes that form terminal tubers; stems erect, mostly unbranched, puberulent; petioles 5–45 mm. long, puberulent or glabrate; lowest blades orbicular to ovate; upper blades 2–7 cm. long, broadly ovate, acute or acuminate, thin but clammy and fleshy, with linear white cystoliths, above bright green, puberulent on the veins, beneath pale green, glabrous, the margins wavy denticulate; the terminal racemes 2–15 cm. long, pilosulous, the lateral ones shorter; pedicels 1–6 mm. long, glabrate; hypanthium 0.3 mm. long, funnel-shaped; calyx-lobes 1.5–1.7 mm. long, oval to elliptic, concave, white; petals 1–1.5 mm. long, obcordate; stamens as long as the calyx; style slightly shorter; stigma sub-capitate bidentate; fruit 2–2.5 mm. long, clavate, covered with hooked bristles 0.5 mm.

long; seeds 2 mm. long, lanceolate, flattened, brown. *C. pacifica* Aschers. & Magn. Cool moist woods, *A. T. T., Can.*

Several other characters have been alleged, but the fact that the stem below the inflorescence is appressed puberulent, seems to be the only distinction separating the variety from the species.

CLARKIA. Clarkia

Annual, with erect, brittle stems; leaves alternate or opposite, on short slender petioles, the uppermost sessile; flowers in terminal racemes; buds nodding; hypanthium obconic or elongate; calyx-lobes united or distinct; petals 4, with claws at least one-sixth as long as the blades, lobed or entire, pink, purple, or violet; stamens 4, alternate, or 8, or those opposite the petals often sterile; anthers attached at their bases, erect; ovary 4-celled; stigma 4-lobed; capsule linear, attenuate above, coriaceous, erect, somewhat 4-angled, 4-valved to the middle; seeds numerous in 1 row in each cell. (Named in honor of *Capt. Wm. Clark.*)

Petals deeply 3-lobed; fertile stamens 4; capsule 1.5–2 mm. in diameter. *C. pulchella.*
Petals entire; fertile stamens 8; capsule 2–4 mm. in diameter. *C. rhomboidea.*

Clarkia pulchella Pursh. *Deer Horn; Ragged Robin.* Stems 1–5 dm. tall, simple or diffusely branched, appressed cinereous puberulent to strigillose, leafy; petiole up to 1 cm. in length; upper leaves sessile; blades 2–7 cm. long, 2–10 mm. wide, linear-lanceolate to spatulate, entire or remotely denticulate; racemes several-flowered; buds green to lavender, usually with slender tip 1–2 mm. long; hypanthium 2–3 mm. long, glabrous within; calyx-lobes 1–1.5 cm. long, usually united in anthesis, deciduous; petals 1.5–3 cm. long, 1–2.5 cm. wide, lavender to purple with lighter veins, the lobes 6–10 mm. long, the middle one broader, the claw narrow, ⅓ to ½ length of blade, with 2 lateral teeth; stamens 8, the opposite ones reduced, the alternate ones 3–8 mm. long; stigma-lobes 1–3 mm. long, rounded, whitish; capsule 1–2.5 cm. long, linear, 8-ribbed, appressed puberulent; pedicel 3–10 mm. long, seeds about 1 mm. long, cylindric, obliquely truncate, angular, brown, papillose-puberulent. Open or rocky places, abundant, *U. Son., A. T.*

First collected by the Lewis and Clark Expedition "on the Kooskooskee" [or the Clearwater River, opposite Kamiah, Ida.]

Clarkia rhomboidea Dougl. ex Hook. Plant finely appressed puberulent; stem 1–11 dm. tall, simple or few branched; leaves remote, almost opposite, often with smaller axillary ones; petioles 3–30 mm. long; blades 1–7 cm. long, 2–20 mm. wide, lance-ovate to ovate-oblong or elliptic, entire or remotely denticulate, in part glabrate; racemes elongate, 2–11-flowered; buds green, short-beaked; hypanthium 1–3 mm. long, bearing scales and white hairs at the filament base; calyx-lobes 6–12 mm. long, usually distinct in anthesis; petals 5–10 mm. long, 3–6 mm. wide, rose-purple, or lighter at base, limb elongate rhomboidal, 2–4 times the length of the broad claw, which is entire or dentate; stamens 8, the opposite ones reduced, the alternate ones 5–8 mm. long; stigma lobes 0.5 mm. long, white to purple; capsules 1–3 cm. long, 4-ribbed, 4-angled, short beaked; pedicels 1–4 mm. long; seeds 1–1.5 mm. long, obliquely rhomboidal, truncate, brown, puberulent. *Phaeostoma rhomboidea* (Dougl. ex Hook.) A. Nels. Bare hillsides or open woods, Blue Mts., Spokane Co., and Ida., *A. T. T.*

EPILOBIUM. Willow Herb

Annual or perennial herbs; leaves nearly sessile, denticulate or entire, alternate or opposite; flowers axillary or terminal, solitary or in racemes; calyx-tube not or scarcely prolonged beyond the ovary, 4-cleft; petals 4, spreading or somewhat erect; stamens 8, the 4 alternate ones shorter; fruit a dehiscent, elongate capsule; seeds numerous, with a tuft of silky hairs at the end. (Name Gr., *epi,* upon; *lobon,* pod; *ion,* violet.)

Calyx divided to the ovary; stigma deeply 4-cleft. *E. angustifolium.*
Hypanthium prolonged beyond the ovary; stigma entire or notched,
  Bark of base of stem straw-colored, freely flaking off,
    Tube of hypanthium 1–3 (–4) mm. long, funnelform,
      Petals pink or purple, 3.5–7 mm. long,
        Capsules puberulent, at least at first,
          Capsules soon glabrate. *E. paniculatum,* forma *paniculatum.*
          Capsules and pedicels glandular puberulent. Forma *adenocladon.*
        Capsules and pedicels glabrous. Forma *subulatum.*
      Petals white, 2–3 mm. long. Forma *Tracyi.*
    Tube of the hypanthium 4–5 mm. long, cylindric. Forma *laevicaule.*
  Bark of base of stem not straw-colored or flaking off,
    Plant glabrous throughout. *E. glaberrimum,* var. *fastigiatum.*
    Plant somewhat pubescent, at least above,
      Blades pilose or pilosulous,
        Blades 2–12 cm. long, broadest near the base; calyx 4–5 mm. long. *E. exaltatum.*
        Blades 1.2–3 cm. long, broadest near the middle; calyx 2.5–3.5 mm. long. *E. ursinum.*
      Blades glabrous or at most somewhat puberulent,
        Upper internodes with 2 decurrent lines of puberulence,
          Petals purple or lilac, 6–10 mm. long; seeds 1–1.1 mm. long, minutely papillose. *E. Hornemanni.*
          Petals white or pink-tinged, 3–6 mm. long; seeds 1.3–1.5 mm. long, smooth. *E. alpinum,* var. *nutans.*
        Upper internodes with pubescence evenly distributed,
          Calyx 1.5–3 mm. long. *E. minutum.*
          Calyx 3.5–5 mm. long,
            Upper internodes appressed puberulent. *E. boreale.*
            Upper internodes glandular pilosulous,
              Petals 6–9 mm. long; leaves crowded, not conspicuously decreasing in size into the inflorescence. *E. glandulosum,* var. *glandulosum.*

Petals 3–8 mm. long; leaves conspicuously decreasing in size into the inflorescence,
Median leaves narrowly ovate or ovate-lanceolate. Var. *adenocaulon.*
Median leaves narrowly elongate lanceolate. Var. *occidentale.*

Epilobium alpinum L., var. **nutans** Hornem. Perennial; basal stolons short, leafy; roots slender; stems 5–35 cm. tall, erect, nodding at tip, often clustered, with 2 scant lines of puberulence; blades 5–37 mm. long, oval to lance-ovate, obtuse, entire or denticulate, glabrous, mostly opposite and shorter than the internodes; petioles 0–4 mm. long; pedicels 3–30 mm. long, sparsely pilosulous; calyx 2.5–4 mm. long, remotely pilosulous, cleft ⅔ way down into ovate-elliptic, unequal lobes; petals 3–6 mm. long, cuneate, deeply emarginate, white or pink-tinged; capsules 3.5–5 cm. long, at first curved, sparsely pilosulous or glabrate, 4-angled, 4-ribbed; seeds 1.3–1.5 mm. long, oblanceoloid, apiculate, flattened, pale brown, smooth, coma white, persistent. *E. lactiflorum* Haussk. Wet places, Blue Mts., *Huds.*
Munz in 1941 revived *E. lactiflorum* as a species.

Epilobium angustifolium L. *Fireweed.* Stout perennial; rootstock woody, stout, producing creeping rhizomes; stems several, 0.5–3 m. tall, leafy, mostly glabrous; leaves 3–20 cm. long, linear-lanceolate, cuneate, subentire or minutely denticulate, glaucous beneath; racemes terminal, elongate, many flowered, at base leafy-bracted, at first puberulent canescent; later glabrate; pedicels at anthesis 6–14 mm. long, puberulent canescent; ovary canescent; calyx-lobes 8–13 mm. long, oblanceolate, puberulent, purplish; petals 8–16 mm. long, broadly obovate, short clawed at base, rose-purple, the 2 lower slightly smaller; filaments dilated at base, the 4 longer 7–9 mm. long and their anthers 3–3.5 mm. long; style pilose at base, exceeding the stamens; capsules 5–8.5 cm. long, linear, cylindric, canescent puberulent; seeds 1–1.5 mm. long, ellipsoid or oblanceoloid, compressed, pale brown, smooth. *E. angustifolium,* var. *platyphyllum* (Daniels) Fern.; *Chamaenerion angustifolium* (L.) Scop. Occasional, prairies or open woods, *A. T., A. T. T.*

In burnt-over woodlands, one of the first plants to appear, and abundant for several years.

Epilobium boreale Haussk. Perennial; forming basal rosettes; roots numerous, slender; stems 25–65 cm. tall, closely cinereous, appressed puberulent, below glabrate, freely branching, very leafy, even in the inflorescence; blades 2–8 cm. long, ovate to lanceolate, glabrous, denticulate or serrate, rounded at base, opposite below the inflorescence; petioles 2–7 mm. long, at first appressed puberulent; leaves of rosettes oval; pedicels 5–25 mm. long, appressed puberulent; calyx 3.5–4 mm. long, appressed puberulent, cleft ½ way down into lanceolate lobes; petals 4–5 mm. long, obovate, emarginate, rose-colored; capsules 3.5–8.5 cm. long, linear, 4-angled, 4-ribbed, appressed puberulent or glabrate; seeds 1–1.2 mm. long, narrowly ellipsoid, apiculate, dark brown, striate papillose; coma white. Stream banks, *U. Son., A. T.*

Epilobium exaltatum E. R. Drew. Perennial, 6–12 dm. tall; root stout, branching; stem densely white, non-glandular pilose above, and almost to the base; blades 2–12 cm. long, lanceolate, pilosulous and densely so along the veins, the margins sharply callous denticulate, rounded at base, mostly opposite; petioles 1–5 mm. long; pedicels 6–12 mm. long; calyx 4–5 mm. long, pilosulous, cleft ¾ way down into lanceolate lobes; petals 6–9 mm. long, obovate, emarginate, purple; pods 3–5.5 cm. long, linear, pilosulous, 4-angled, 4-ribbed; seeds 1.2–1.4 mm. long, narrowly oblanceoloid, apiculate, brown, striate; coma white, easily detached. *E. cinerascens* Piper; *E. Sandbergii* Rydb. Rich woods, Spokane Co. and Kootenai Co., *A. T. T.*

Prof. Piper's concept of *E. cinerascens* is here altered by the admission of petioled leaves and by the exclusion of *Suksdorf* 2,748 and 2,749, which the writer refers to *E. glandulosum*, var. *glandulosum*. Rydberg's type, *Sandberg et al* 737 is from Mud Lake, Kootenai Co., Idaho.

Epilobium glaberrimum Barbey, var. **fastigiatum** (Nutt. ex. T. & G.) Trel. Perennial, 2–6 dm. tall, glabrous throughout, more or less glaucous; rhizomes 1–2 mm. in diameter, dark; leaves 1.5–4 cm. long, ovate to oblong-ovate, entire or denticulate; pedicels 2–5 mm. long; calyx 3–4 mm. long, cleft half way into lanceolate lobes; petals 4–6 mm. long, white or pale lavender; capsules 4–6 cm. long; seeds 0.9–1 mm. long, oblanceoloid, compressed, brownish, minutely papillose. Moist places, *U. Son.*, *A. T.*

Epilobium glandulosum Lehm., var. **glandulosum.** Differs from var. *adenocaulon* by having the petals 6–9 mm. long; and the leaves crowded, not conspicuously decreasing in size in the crowded inflorescence. Wet places, Pullman, *A. T.*

Var. **adenocaulon** (Haussk.) Fern. Perennial by short stolons which produce fleshy rosettes in the fall; stems 1–10 dm. tall, erect, glandular pilosulous above, mostly branching above; leaves 2–8 cm. long, subsessile, ovate-lanceolate, mostly opposite, callous denticulate, glabrous or somewhat puberulent; pedicels 3–20 mm. long; calyx 4–5 mm. long, pilosulous, cleft ½ to ¾ way down into oblong-lanceolate lobes; petals 3–8 mm. long, obovate, deeply emarginate, pink or rose-purple; stamens included; capsules 3.5–7 cm. long, pilosulous, 4-ribbed and 4-angled; seeds 0.9–1.3 mm. long, ellipsoid, striate, papillose at apex, brown; coma white, easily detached. *E. Palmeri* Rydb.; *E. adenocaulon* Haussk. Wet places or meadows, *U. Son.* to *A. T. T.*

Munz in Fl. Nev., Contr. 32: 19, 1941, has restored *adenocaulon* as a species and placed *occidentale* as a variety under it. Without argument or reasoned deduction, he has rejected Fernald's earlier revision. In these circumstances I prefer to follow Fernald's judgement, and to keep the two as varieties under *E. glandulosum.*

Var. **occidentale** (Trel.) Fern. Differs in having the leaves lanceolate or narrowly so; petals 4–8 mm. long. *E. occidentale* (Trel.) Rydb. Wet places, *U. Son.* to *A. T. T.*

Epilobium Hornemanni Reichenb. Perennial; roots slender; rhizomes slender, creeping; stems 5–40 cm. tall, often clustered, with decurrent lines of crisp puberulent hairs, commonly unbranched; blades 15–45 mm. long, ovate, oval, or lance-elliptic, mostly obtuse and opposite, glabrous, entire or denticulate; petioles 0–7 mm. long; pedicels 3–25 mm. long, viscid pilosulous; calyx 3–6 mm. long, cleft ¾ way down into lanceolate lobes; petals 5–10 mm. long, obcordate, purple or lilac; stamens well included; capsules 3–6 cm. long, linear, 4-ribbed, 4-angled, sparsely pilosulous; seeds 1–1.1 mm. long, oblanceoloid, flattened, pale brown, minutely papillose; coma white. *E. alpinum* of A. H. Moore, not of L. Wet places, *Can., Huds.,* and occasionally stranded by stream banks at lower elevations, as Lewiston, Piper, May 11, 1895.

Epilobium minutum Lindl. ex Hook. Annual, crisp puberulent throughout; roots fibrous; stems 1–3 dm. tall; leaves 7–23 mm. long, narrowly lanceolate or oblanceolate, firm, entire or remotely denticulate; flowers from the upper axils; calyx 1.5–3 mm. long, cleft half way; petals 3–3.5 mm. long, obovate, rose-colored or white; capsules 1–2.5 cm. long, linear, arcuate, torulose; seeds 0.7–1 mm. long, oblanceoloid, compressed, dark brown, smooth; coma easily detached. Infrequent, dry hillsides, *A. T.*

Epilobium paniculatum Nutt. ex T. & G., forma **paniculatum.** Annual; roots fibrous; stems 2–10 dm. tall, paniculately branched above, glabrous below; leaves 1–5 cm. long, mostly alternate, linear or linear-lanceolate, remotely denticulate, glabrous, with axillary clusters of smaller leaves; larger plants with a diffuse many-flowered panicle; pedicels glabrous or nearly so; ovary glandular puberulent, usually soon glabrate; tube of hypanthium 2–3 mm. long, funnelform, the calyx-lobes 2–3 mm. long, lance-deltoid; petals 5–7 mm. long, lilac or rose colored, 2-lobed at apex; stamens included; stigma entire or subentire; capsules 10–25 mm. long, slender clavate, beaked, strongly 4-nerved, 4-angled; seeds 1.7–2 mm. long, broadly oblanceoloid, compressed, muriculate, dark brown; coma easily detached. Dry, open places, *A. T.*

Forma **adenocladon** Haussk. Differing from the species only in having the capsules and pedicels glandular puberulent. *E. adenocladon* (Haussk.) Rydb. Mostly in dry places, *A. T., A. T. T.*

Forma **laevicaule** (Rydb.) St. John. Differing from *E. paniculatum* in having the tube of the hypanthium 4–5 mm. long. *E. laevicaule* Rydb.; *E. paniculatum,* var. *laevicaule* (Rydb.) Munz. Dry ground, Spokane, *Piper* 2361.

Three cotypes were cited from our region: Spokane, *Kreager* 536; 573; Pullman, *Piper* 1631. The duplicates of these in the Herbarium of Washington State University all have the hypanthium tube less than 4 mm. in length, and have the pods glandular puberulent. This clearly indicates the lack of constancy of the characters. Though this plant is another extreme of this polymorphic group, the writer cannot accord it specific rank.

Forma **subulatum** Haussk. Differing from the species only in having glabrous capsules and pedicels. *E. subulatum* (Haussk.) Rydb. Dry open slopes, *A. T., A. T. T.*

Forma **Tracyi** (Rydb.) St. John. Plant glabrous; tube of hypanthium 1–1.5 mm. long, funnelform; expanded portion of calyx 2 mm. long; petals 2–3 mm. long, emarginate, white; capsule glabrous. *E. Tracyi* Rydb. Dry, open places, *U. Son., A. T.*

The writer deems this a minor variant of the polymorphic *E. paniculatum.*

Epilobium ursinum Parish. Perennial with turions; rhizome firm; lateral roots slender; stems 1–3 dm. tall, white pilose, erect, usually simple; leaves 12–30 mm. long, ovate-lanceolate to narrowly oblong-lanceolate, sessile, pilose, densely so when young, mostly opposite; pedicels 1–6 mm. long, pilosulous; calyx 2.5–3.5 mm. long, pilosulous, cleft ⅔ way down into elliptic-lanceolate lobes; petals 4–6 mm. long, oblong-elliptic, narrowly cleft at tip, white or pale lavender; stamens included; capsules 3–4.5 cm. long, linear, 4-angled, 4-ribbed, puberulent; seeds 1.2–1.5 mm. long, ellipsoid, short beaked, brown, roughened; coma white, easily detached. Wet places, Little Potlatch River, Latah Co., Ida., also Squaw Sprs. Trail, Blue Mts., Garfield Co., Wash., *A. T. T.*

GAURA

Annual, biennial, or perennial herbs, somewhat woody at base; leaves alternate, sessile; flowers white or rose-colored, in spikes or racemes; hypanthium much prolonged beyond the ovary; calyx-lobes 4, rarely 3; petals 4, with claws, unequal; stamens mostly 8, bent down; stigma 4-lobed, surrounded by a cup-like border; fruit hard and nut-like, ribbed or angled, indehiscent or nearly so, usually 1-celled and 1–4-seeded. (Name from Gr., *gauros,* superb.)

Gaura parviflora Dougl., ex Hook., var. **parviflora.** *Velvet Weed.* Biennial or rarely perennial; 1–3 m. tall, long white villous and densely glandular pilosulous throughout; tap-root deep; first year plant forming a rosette, the leaves 10–21.5 cm. long, oblanceolate, long cuneate and winged at base; second year the stem erect, leafy, usually branched above, the suffrutescent base pale, smooth due to the exfoliating epidermis; leaves 1–10.5 cm. long, sessile, ovate-lanceolate or lanceolate, entire or denticulate; spikes 5–30 cm. long, at first dense, above almost glabrous; bracts 3–5 mm. long, lance-linear, villous, caducous; flowers ascending, sessile; ovary glabrous; hypanthium 3–3.5 mm. long, subcylindric, pilosulous; calyx-lobes 3–4 mm. long, oblong, acute, glabrous at least towards the tip; petals 2–2.5 mm. long, oblanceolate, crimson; stamens 2.5–3.5 mm. long, the anthers versatile; style 3.5 mm. long; stigma with 4 short lobes; fruits 6–9 mm. long, fusiform, glabrous, 4-sided, 8-ribbed; seeds 2–3 mm. long, ellipsoid, compressed, yellowish. Dry hillsides, *U. Son.*

GAYOPHYTUM

Very slender caulescent branching annuals; leaves alternate, linear, entire; flowers axillary; calyx-tube not prolonged beyond the ovary, 4-parted; petals 4, white or rose-colored, very small, obovate or oval, with a very short claw; stamens 8; anthers broad or rounded, attached by the middle, those opposite the petals on shorter filaments and usually sterile; ovary 2-celled; stigma capitate; fruit a dehiscent, 4-valved capsule; seeds many; naked, in one row in each cell. (Named for *Claude Gay,* the French writer on S. Am.; with *phuton,* Gr., plant.)

Capsule subsessile, not torulose. — *G. humile.*
Capsule pediceled, torulose,
  Seeds strigose canescent. — *G. lasiospermum.*
  Seeds glabrous,
    Petals 2–4 mm. long. — *G. diffusum.*
    Petals 0.5–2 mm. long,
      Capsules longer than the pedicel,
        Capsules erect. — *G. Nuttallii,* var. *Nuttallii.*
        Capsules deflexed. — Var. *intermedium.*
      Capsules shorter than the pedicel. — *G. ramosissimum.*

Gayophytum diffusum T. & G. Plants 1.5–6 dm. tall, profusely branched above, glabrous below, sparsely puberulent above; lower leaves 2–5 cm. long, linear; floral leaves much reduced above; calyx-lobes 2–4 mm. long, oblanceolate, purplish, appressed puberulent; petals obovate, white, turning pink; fruiting pedicels 2–8 mm. long, divergent; capsule 5–12 mm. long, 0.7–1 mm. in diameter, linear or slightly clavate, appressed puberulent canescent; seeds 1–1.2 mm. long, oblanceoloid, brown, smooth. Tchimikane, Spokane Country [Chamokane, Stevens Co., near our boundary], *Geyer* 546, in part, fide Munz. Specimen not seen by the writer.

Gayophytum humile Juss. Plants 5–18 cm. tall, glabrous, or sparsely glandular puberulent above, usually bushy branched from the base; lower leaves 0.7–3 cm. long, linear to oblance-linear, entire, short petioled; floral leaves crowded, slightly shorter; calyx-lobes 0.8–1.2 mm. long, elliptic, rose-purplish; petals 1 mm. long, obovate; stamens and style not equaling the petals; capsule 6–15 mm. long, 0.7–0.8 mm. in diameter, erect, linear flattened; seeds 0.5–0.7 mm. long, ellipsoid, smooth, pale brown, attached obliquely. *G. pumilum* Wats. Gravelly open woods, Blue Mts., Moscow Mt., *A. T. T.*

Although often sparsely glandular puberulent, it does not seen to differ essentially from this species as described. It occurs also in Chili.

Gayophytum lasiospermum Greene. Exactly resembling *G. ramosissimum* except for having the seeds 0.6–0.8 mm. long, narrowly asymmetric oblanceoloid, strigose canescent. Dry hills, *A. T., A. T. T.*

Gayophytum Nuttallii T. & G., var. **Nuttallii.** Plants 1–7 dm. tall, erect, becoming diffusely branched above, glabrous below, the young shoots appressed strigillose; lower leaves 2–4.5 cm. long, 1.5–3 mm. wide, narrowly oblance-linear, glabrate, gradually reduced upwards; floral leaves similar, smaller, sparsely strigillose; calyx-lobes 1.3–2 mm. long, elliptic-lanceolate, glabrous; petals 1–2 mm. long, broadly obovate, turning pink; fruiting pedicel 3–12 mm. long, glabrate, ascending; capsule 5–14 mm. long, 0.5–0.8 mm. in diameter, clavate-linear; tapering at base, appressed canescent strigillose, erect; seeds 1–1.5 mm. long, lanceoloid, compressed, brown, smooth. *G. Nuttallii,* var. *typicum* Munz. Open pine woods, Moscow Mt., Ida., *A. T. T.*

Var. **intermedium** (Rydb.) Munz. Differs in having the puberulence appressed, and the pedicels and fruit deflexed or spreading. Spokane R., Kootenai Co., Ida., *Sandberg, MacDougal & Heller* 652, fide Munz. Specimen not seen by the writer.

Gayophytum ramosissimum T. & G. Plants 1–5 dm. tall, glabrous throughout or somewhat appressed canescent puberulent at the tip, diffusely bushy branched above; lower leaves 1.5–5.5 cm. long, 1–3 mm. broad, linear; floral leaves much reduced; calyx-lobes 1–1.5 mm. long, lanceolate, tinged with reddish brown; petals 1–1.5 mm. long, obovate, white, turning pink; fruiting pedicels 2–7 mm. long, divergent or reflexed; capsules 3–12 mm. long, 0.5 mm. in diameter, linear; seeds 0.8–1.2 mm. long, bluntly oblanceoloid, somewhat compressed, brown, smooth. Dry gravelly slopes, *A. T. T.*

LUDWIGIA

Annual or perennial succulent herbs, often with prostrate, creeping, or floating stems; leaves alternate or opposite, fleshy; flowers perfect, axillary; calyx turbinate, with 4 segments, persistent; petals 4 and small, or none; stamens 4; filaments short; stigma 4-lobed, often nearly sessile; ovary 4-celled, very short; ovules numerous; capsule 4-angled, septicidal; seeds numerous. (Named for *Prof. C. G. Ludwig* of Leipzig, Germany.)

Ludwigia palustris (L.) Ell., var. **pacifica** Fern. & Griscom. *Water Purslane.* Annual or perennial, glabrous throughout or at first somewhat granular; rhizome creeping; stems 1–7 dm. long, often prostrate and rooting at the nodes, or floating, or ascending; leaves 1–4 cm. long, elliptic to obovate, acute, gradually cuneate into the margined petiole, entire; flowers single, sessile, about 4 mm. long, with 2 minute linear bracts at base; calyx-lobes 1.5–2 mm. long, ovate, acute, green; petals none; stamens 1 mm. or less in length; style 0.5 mm. long; stigma subcapitate, 2-lobed; capsule 4 mm. long, obovoid, 4-angled, yellowish, the intervals green; seeds 0.4–0.7 mm. long, whitish, obliquely ellipsoid, the hilum nearly straight. Muddy bank of Touchet River, Waitsburg, *Horner* 585, *A. T.*

OENOTHERA. Evening Primrose

Annual, biennial or perennial herbs; leaves alternate or basal; buds erect or drooping; flowers white, pink, or yellow, blooming by night or by day, axillary or in terminal spikes; hypanthium elongated, terete, filiform or enlarged upward, mostly deciduous; calyx-lobes 4, the tips free or united in bud, finally reflexed; stamens 8, equal, or the opposite ones shorter, with filiform filaments and linear, mostly versatile, anthers; ovules numerous, in one, two, or more rows, horizontal or ascending; capsule 4-celled, 4-angled, loculicidal; seeds numerous, angled or terete, with or without a tubercle. (Name from Gr., *oinos,* wine; *thera,* pursuit.)

Stigma divided into 4 linear lobes,
  Capsules terete or obtusely 4-angled; hypanthium 1.5–5.5 cm. long,
    Petals white, aging rose-colored, stems white. — *O. pallida.*
    Petals yellow, sometimes aging pink; stems not white,
      Petals 10–22 mm. long. — *O. strigosa.*
      Petals 3–4 cm. long. — *O. Hookeri,* subsp. *ornata.*
  Capsules with 8 ribs bearing low tubercles; hypanthium 5–8 cm. long. — *O. caespitosa,* var. *marginata.*
Stigma capitate or discoid,
  Acaulescent; upper part of ovary sterile, tubular, filiform, longer than the fertile part,
    Capsules glabrous; leaves entire or the base coarsely pinnatifid. — *O. heterantha.*
    Capsules appressed puberulent; leaves pinnately parted. — *O. tanacetifolia.*
  Caulescent; ovary fertile throughout,
    Capsules 5–12 mm. long, lance-fusiform; leaves slender petioled. — *O. andina.*
    Capsules 20–35 mm. long, linear; leaves subsessile. — *O. contorta,* var. *contorta.*

Oenothera andina Nutt. ex T. & G. Annual, 2–15 cm. tall, appressed puberulent throughout; stem erect, commonly with numerous spreading branches from near the base, each slender, naked below; leaves 1–3 cm. long, 1–2.5 mm. wide, linear-oblanceolate, crowded at branch tips; spike leafy, condensed; calyx-tube 0.7–1 mm. long, funnelform, narrowly cylindric below; calyx-lobes 1.5–2.5 mm. long, ovate; petals 1.5–2 mm. long, obovate, yellow; opposite stamens ½ the length, alternate ones ¾ the length of the petals; anthers 0.2–0.3 mm. long; pistil nearly equaling the petals; stigma 0.3 mm. broad, thick discoid; capsules 4-angled; seeds 0.6–0.7 mm. long, fusiform, olive-brown, shining, smooth. *O. andina,* var. *typica* Munz; *Sphaerostigma andinum* (Nutt. ex T. & G.) Walp. Alkaline flats, in "scab lands from Spangle southwestward," *U. Son.*

Oenothera caespitosa Nutt., var. **marginata** (Nutt. ex T. & G.) Munz. *Desert Primrose.* Perennial, caespitose or short stemmed, villous-hirsute; taproot woody, stout, dark; stems often becoming 1–18 cm. long; first leaves in a rosette; leaves 5–18 cm. long, 1–5 cm. broad, oblanceolate, variously pinnatifid, the petioles, veins, and margins more densely villous-hirsute; calyx-lobes 2–3 cm. long, the tips scarcely free in bud, later reflexed, pinkish, lanceolate; flowers vespertine, white, turning rose-color; petals 2.5–4 cm. long, broadly obcordate; stamens subequal; anthers 10–14 mm. long, pale; stigma-lobes 5–8 mm. long; pedicels in fruit 1–12 mm. long; capsules 2–4 cm. long, linear-cylindric; seeds about 3 mm. long, obovoid, brown, cellular-roughened, furrowed along the raphe. *Pachylophus marginatus* (Nutt.) Rydb. Arid, rocky slopes, down the Snake River Canyon as far as Wilma, Wn., *U. Son.*

Oenothera contorta Dougl. ex Hook., var. **contorta.** Annual, puberulent to glabrate; stems commonly 5–15 cm., rarely 25 cm. tall, simple or commonly with several subequal branches; leaves 5–25 mm. long, 1–2 mm. wide, lance-linear, denticulate; inflorescence of leafy spikes; hypanthium 1–2 mm. long, narrowly funnelform; calyx-lobes 1.5–3.5 mm. long, lance-ovate; petals 2.5–3.5 mm. long, obovate to obcordate, bright yellow, aging red; opposite stamens ¼, alternate stamens ½ the length of the petals; anthers 0.5–0.7 mm. long; pistil nearly half as long as the petals; stigma 0.5–1 mm. broad, ovoid; capsules 0.6–1 mm. in diameter, terete, torulose, narrowed to a beak, straight or sigmoid; seeds 0.5–1 mm. long, ellipsoid to obovoid, brown, minutely cellular pitted. *Sphaerostigma contortum* (Dougl. ex Hook.) Walp. Ilia, Snake River Canyon, *Lake & Hull,* June 25, 1892, *U. Son.*

Oenothera heterantha Nutt. Perennial, glabrous or nearly so; tap-root woody, stout; leaves 3–30 cm. long, 1–6 cm. wide, radical, numerous; petioles stout, narrowly decurrent winged; blades lanceolate, oblanceolate, or ovate-lanceolate, entire, denticulate, or some or all pinnatifid at base, the margins minutely puberulent; flowers axillary borne on the attenuate, sterile upper part of ovary 3–9 cm. long; calyx-tube 1–2 mm. long, funnelform; calyx-lobes 5–11 mm. long, lanceolate; petals 8–17 mm. long, deltoid-obovate, emarginate, pale yellow; alternate stamens ½ the length of the petals, the opposite ones ¼ the length; anthers of longer stamens 2–3 mm. long; pistil equaling the longer stamens; stigma 1.5–2 mm. wide, discoid; capsules numerous, 12–15 mm. long, lance-ovoid, 4-angled above; seeds 1.5–2 mm. long, ovoid, somewhat curved, brown, deeply cellular pitted. *Taraxia heterantha* (Nutt.) Small; *T. subacaulis* (Pursh) Rydb., perhaps not of Pursh; *T. heterantha taraxicifolia* (Wats.) Small. Meadows, *A. T.*

Oenothera Hookeri T. & G., subsp. **ornata** (Nels.) Munz. Biennial, finely puberulent canescent and often hirsute throughout; plants 3–12 dm. tall; tap-root dark, woody; stem erect, often branched; basal leaves lanceolate, slender petioled; cauline leaves 5–20 cm. long, lanceolate, repand-denticulate; spikes terminal, leafy bracted, many flowered; buds erect; hypanthium 3–5.5 cm. long, linear-cylindric, slightly expanded at tip; calyx-lobes 3–3.5 cm. long, lance-linear, persisting in anthesis, the free tips 1–2 mm. long; petals broadly obovate, cuneate, yellow, aging pink; stamens half or more than half as long as the petals; anthers 9–12 mm. long, linear; pistil nearly equaling the petals, the stigmatic lobes 4 mm. long, linear from a discoid base; capsules 2.5–4 cm. long, narrowly cylindric, 4-ribbed, obtusely 4-angled; seeds 1–2 mm. long, irregularly compressed, reddish brown, cellular pitted. Occasional, Snake River Canyon, Wawawai, Almota, *U. Son.*

Oenothera pallida Lindl. Perennial; rootstock creeping; stems 2–5 dm. tall, glabrous or remotely hirsute above, in age exfoliating; cauline leaves 2–7 cm. long, linear-lanceolate to linear, sessile or short petioled, subentire to sinuate-dentate, glabrous or glabrate; buds nodding; flowers vespertine, fragrant, numerous, borne in the upper axils; ovary and the scarcely distinguishable hypanthium 2–5 cm. long, pedicel-like, glabrous or remotely hirsute, tip of hypanthium slightly funnelform; calyx-lobes 12–23 mm. long, adhering in anthesis, narrowly lanceolate, glabrous or sparsely hirsute, the free tips linear 0.5–3 mm. long; petals 12–25 mm. long, suborbicular, cuneate, sometimes retuse; stamens and pistil equaling the petals; anthers 5–10 mm. long, linear; stigmatic lobes 3–8 mm. long; capsule 1.5–4 mm. long, 1.5–3 mm. in diameter, linear, bluntly 4-angled, 4-ribbed, glabrous or remotely hirsute, usually curved ascending, or sigmoid; seeds in one row, 1.5–2 mm. long, narrowly obovoid, madder-spotted, becoming quite black, very minutely pitted. *Anogra pallida* (Lindl.) Britton. Sandy or gravelly slopes, Snake River Canyon, from Asotin down, *U. Son.*

Oenothera strigosa (Rydb.) Mack. & Bush. Biennial (or annual); tap-root stout, woody; plant 3–12 dm. tall, grayish strigose and hirsute throughout; basal leaves lanceolate, long petioled; cauline leaves 5–15 cm. long, the lower short petioled, the upper sessile, subentire or repand-denticulate; spikes terminal, leafy bracted; buds erect; flowers vespertine, yellow, often aging purplish; hypanthium 15–35 mm. long, gradually widened at the tip; calyx-lobes 9–20 mm. long, lanceolate, the free, linear tips 2 mm. long; petals broadly cuneate-obovate; stamens equaling the petals; anthers 5–9 mm. long, linear; pistil nearly equaling or half as long as the petals, stigmatic lobes 2–4 mm. long; capsules 2–3.5 cm. long, 3–5 mm. in diameter, lance-cylindric, 4-angled, 4-ribbed; seeds 1–1.6 mm. long, irregularly compressed, mahogany-colored, cellular punctate. *Onagra strigosa* Rydb.; *Oenothera Rydbergii* House. Common, especially by streams, *U. Son.* to *A. T. T.*

Oenothera tanacetifolia T. & G. Perennial; tap-root thick; leaves 3–21 cm. long, 5–40 mm. wide, numerous; petioles slender, appressed puberulent; blades lanceolate in outline, sparsely appressed puberulent or glabrate; flowers axillary, borne on the attenuate, filiform sterile, appressed puberulent, upper part of the ovary, 2.5–9 cm. long; calyx-tube funnelform, 2–6 mm. long; calyx-lobes 8–15 mm. long, lanceolate, acuminate, appressed puberulent; petals 1–2 cm. long, obovate, yellow, aging red; alternate stamens ⅔ the length of the petals, the opposite ones ½ the length; anthers 2–3 mm. long, oblong; pistil exceeding the stamens; stigma 2–4 mm. wide, globular or slightly lobed; capsules rare, 17–25 mm. long, 5–6 mm. wide, lanceoloid, quadrangular; seeds about 2 mm. long, oblong, slightly curved, brown, finely pitted in rows. *Taraxia tanacetifolia* (T. & G.) Piper; *T. longiflora* Nutt. ex Small. Meadows in the scab lands, from Marshall Jct. southwestward, *U. Son.*

ARALIACEAE. Ginseng Family

Perennial herbs, shrubs, or trees; leaves alternate or opposite, simple or compound with dilated petioles; flowers umbellate, paniculate, or racemose, perfect or polygamous; calyx adherent to the ovary, the limb entire or toothed; petals usually 5, epigynous; stamens 5, epigynous, alternate with the petals; styles 2 or more; ovary 2 or more celled, crowned with a disk; fruit a few-celled drupe or berry. (Named from the genus *Aralia*.)

Plants unarmed; leaves compound; berry 5-celled. *Aralia.*
Plants ferociously armed; leaves simple; berry 2–3-celled. *Oplopanax.*

ARALIA. Sarsaparilla

Aromatic herbs, shrubs, or trees; leaves alternate or whorled; flowers inconspicuous, regular, the ultimate divisions of the inflorescence usually umbellate; petals 5 or 10, valvate or imbricate; filaments short, distinct; pistil 2–5-celled, ovules solitary in each cell; endosperm fleshy, copious. (Derivation of name unknown.)

Aralia nudicaulis L. *Wild Sarsaparilla.* Perennial herb, nearly acaulescent; stem brown, scaly; roots woody, the main one horizontal, running; leaf 1, ternate, 2–4 dm. tall, long petioled, the primary divisions pinnately 3–5-foliolate; leaflets 4–12 cm. long, oval, ovate, or ovate-lanceolate, the terminal one more cuneate, serrate or doubly serrate, less so at base, pale green beneath; peduncle 1, shorter than the leaf, bearing usually 3, many-flowered umbels; pedicels 5–14 mm. long; hypanthium campanulate; calyx-lobes low deltoid; petals 1.5–2 mm. long, oval, greenish, soon reflexed; stamens 2–3 mm. long, ascending; styles 1 mm. long; berries 5–6 mm. long, subglobose, truncate, 5-lobed, purplish black; seeds 3 mm. long, compressed ovoid, the inner edge nearly straight. Davis Ranch [Mt. Spokane], *Kreager* 206; Newman L., *Suksdorf* 8,802, *A. T. T.*

The underground stem and roots are strongly aromatic. This principle has been used in medicine and for flavoring drinks.

OPLOPANAX. Devil's Club

Stout, very prickly shrubs; leaves simple, long-petioled, suborbicular, palmately-lobed; flowers perfect or polygamous, in numerous umbels which are in racemes or panicles; calyx-margin narrow or obsolete, obscurely crenate-lobed; ovary 2-celled (rarely 3-celled); styles 2 (or 3), filiform; berry laterally flattened. (Name Gr., from *hoplon,* weapon, arms; *Panax,* name of a genus of Araliaceae.)

Oplopanax horridus (Sm.) Miq. *Devil's Club.* Shrub decumbent, loosely branched, 1–4 m. long, armed with abundant, yellowish prickles, 5–13 mm. long; tips of the branches leafy; leaves prickly on petioles, main nerves, and secondary nerves; petioles 1–3 dm. long; blades 1.5–6 dm. broad, nearly orbicular in outline, palmately 3–11-lobed, doubly serrate, the lobes often with coarse shallow lobes, pilosulous on the veins beneath; inflorescence 7–30 cm. long, woolly, terminal; peduncles in axil of a laciniate bract; pedicels 1–10 mm. long; hypanthium campanulate; petals 2.5–3 mm. long, oval, acute, greenish; stamens exserted, anthers 0.8–0.9 mm. long, oval; styles 1.5–2.5 mm. long, united below; berries 5–8 mm. long, oval-suborbicular, scarlet; seeds 4–7 mm. long, oblong, flattened, curved, strongly 5-ribbed, yellowish. *Echinopanax horridus* (Sm.) Dcne. & Planch.; *Fatsia horrida* (Sm.) B. & H. Moist mountain woods, *Can.*

UMBELLIFERAE. Carrot Family

Herbs; stems usually hollow; leaves alternate, compound or simple; petioles often dilated at the base; flowers small, in umbels or rarely heads, the umbels often subtended by primary bracts (the *involucre*), in compound umbels, the secondary rays often subtended by secondary bracts (the bracts of the *involucel*); calyx entirely adherent to the ovary; petals and stamens 5, on the calyx-tube; base of the 2 styles (*stylopodium*) often expanded; ovary 2-celled, inferior, 1-ovuled in each cell; fruit of 2 seed-like dry carpels with contiguous inner surfaces (the *commissure*), each carpel marked lengthwise on the back with 5 primary ribs and often with 4 intermediate secondary ones, in the intervals between which oil-tubes are commonly found; carpels often separating from each other, supported on the summit of a slender axis (the *carpophore*); seeds with copious endosperm.) Mature fruits are almost indispensable for accurate determinations in this family. (Name Lat. *umbella,* sunshade, in botany umbel; *fero,* to bear.)

Key To Plants in Flower

- Flowers yellow,
  - Basal leaves simple, cordate to suborbicular. — *Zizia.*
  - Leaves cleft or compound,
    - Ovary prickly. — *Sanicula.*
    - Ovary not prickly,
      - Calyx-teeth prominent, persistent. — *Pteryxia.*
      - Calyx-teeth minute. — *Lomatium.*
- Flowers white, or pinkish to blue or purplish,
  - Flowers in dense heads; calyx-lobes spine-tipped, persistent. — *Eryngium.*
  - Flowers in umbels; calyx not so,
    - Ovary bristly,
      - Stylopodium obsolete; bristles barbed at tip. — *Daucus.*
      - Stylopodium conical; bristles not barbed,
        - Bristles hooked, spreading. — *Caucalis.*
        - Bristles not hooked, appressed. — *Osmorhiza.*
    - Ovary not bristly,
      - Leaflets 1–3 dm. wide. — *Heracleum.*
      - Leaflets narrower,
        - Stylopodium conical,
          - Leaflets linear. — *Carum.*
          - Leaflets broader,
            - Leaves pinnate,
              - Lower leaves simply pinnate. — *Berula.*
              - Leaves bipinnate or pinnately decompound. — *Conium.*
            - Leaves ternate,
              - Leaflets mostly simple, not parted or cleft. — *Angelica.*
              - Leaflets mostly cleft or parted. — *Ligusticum.*
        - Stylopodium depressed or none,
          - Marsh plants; rootstock with scalariform partitions across the pith,
            - Leaves bipinnate. — *Cicuta.*
            - Leaves pinnate. — *Sium.*
          - Not marsh plants; rootstock not as above,
            - Stems leafy throughout; leaflets broadly lanceolate. — *Osmorhiza.*
            - Stems naked, or if leafy, the leaflets not lanceolate,
              - Plants acaulescent from a corm or tubers; leaves 1–2-ternate, the leaflets elongate linear,
                - Petioles slender,
                  - Leaflets 2–6 mm. wide. — *Orogenia.*
                  - Leaflets 0.5–1 mm. wide. — *Leibergia.*
                - Petioles membranous, expanded. — *Lomatium.*
              - Plants not so. — *Lomatium.*

Key to Plants in Fruit

- Fruit scaly. *Eryngium.*
- Fruit not scaly,
  - Fruit bristly,
    - Bristles covering the whole fruit surface. *Sanicula.*
    - Bristles only on the ribs of the fruit,
      - Stylopodium obsolete, bristles barbed at tip. *Daucus.*
      - Stylopodium conical; bristles not barbed,
        - Secondary ribs the most prominent, with spreading bristles. *Caucalis.*
        - Primary ribs the most prominent, with more or less appressed bristles. *Osmorhiza.*
  - Fruit not bristly,
    - Fruit strongly flattened dorsally,
      - Plants tall caulescent; stylopodium conic,
        - Wings of mericarps contiguous; oil-tubes pendent, shorter than the interval. *Heracleum.*
        - Wings of mericarps distinct; oil-tubes as long as the interval. *Angelica.*
      - Plants mostly acaulescent or short caulescent; stylopodium depressed. *Lomatium.*
    - Fruit not or little flattened dorsally,
      - Oil-tubes none or obscure,
        - Fruit broadly ovate. *Conium.*
        - Fruit almost linear. *Osmorhiza.*
      - Oil-tubes distinct,
        - Oil-tube 1 in each interval,
          - Basal leaves simple, cordate to suborbicular. *Zizia.*
          - Leaves compound,
            - Leaflets lanceolate or broader. *Cicuta.*
            - Leaflets linear,
              - Stylopodium low conic; calyx-teeth prominent. *Carum.*
              - Stylopodium flat; calyx-teeth obsolete. *Leibergia.*
        - Oil-tubes several in each interval,
          - Stylopodium conic,
            - Mericarps subglobose, the ribs obscure. *Berula.*
            - Mericarps oblong or ovoid, the ribs prominent. *Ligusticum.*
          - Stylopodium flat or none,
            - Ribs all winged. *Pteryxia.*
            - Ribs not winged,
              - Plant caulescent; ribs corky. *Sium.*
              - Plant acaulescent; ribs filiform. *Orogenia.*

ANGELICA

Stout perennial herbs; leaves ternate, then pinnate, rarely simply pinnately compound; involucre scanty or none; involucels of small bractlets or none; flowers white, in large terminal compound umbels; calyx-teeth mostly obsolete; stylopodium low conical in American species; fruit flattened dorsally, ovate or oblong, glabrous or pubescent; calyx-tube prominent, crenulate; carpel with strong ribs, the lateral ribs usually broadly winged, distinct from those of the other carpel, forming a double-winged margin to the fruit; oil-tubes 1–several, in the intervals, or indefinite, 2–10 on the commissural side; seed face plane or somewhat concave. (Name Lat. from *angelica,* angelic.)

Ovary puberulent; body of mericarp broader than the wings; styles shorter than the petals. *A. Canbyi.*
Ovary glabrous; body of mericarp narrower than the wings; styles exceeding the petals. *A. Lyallii.*

Angelica Canbyi C. & R. Root dark, woody; plant 4–12 dm. tall, glabrous below the inflorescence; leaves bipinnate; basal leaves 2–4.5 dm. long, with petioles 10–23 cm. long; leaflets 2.5–6.8 cm. long, lanceolate to oval, laciniately toothed and at times lobed; upper leaves much smaller, the petiole reduced to the sheathing base; umbels with 10–20 wide-spreading rays, the marginal ones the longer, 2–5.5 cm. long; pedicels 1–7 mm. long, the marginal the longer, glabrous, at least below; petals 0.6–0.8 mm. long, white, oval to suborbicular, midnerve prominent, puberulent up the back, the tip inflexed; anthers 0.4–0.5 mm. long, oval, much exserted; stylopodium depressed conic; fruit densely puberulent, at length subglabrate; mericarps 4.5–6 mm. long, oval, emarginate at each end, much flattened dorsally, dorsal ribs narrowly winged, especially at either end; lateral wings corky, pale, ⅔ width of the body; oil-tubes 1 in dorsal intervals, 2 in laterals, 6–8 on the flat commissural side. Meadows and woods, Blue Mts., *A. T. T.*

Angelica Lyallii Wats. Root dark, heavy, woody, cylindric; plant 3–18 dm. tall, nearly glabrous to the inflorescence; basal leaves 3.5–7 dm. long, ternate, then 1–2-pinnate, the petiole 1–5 dm. long; leaflets 4–14 cm. long; ovate to lanceolate, sharply dentate, at times incised or lobed; umbel 25–37-rayed, loose; marginal rays the longer, 3–10 cm. long, short connate at base; pedicels numerous, 1–5 mm. long, the marginal the longer; petals 1.3–1.7 mm. long, white, broadly obovate, sharply inflexed and appearing notched at tip, midnerve distinct; anthers 0.6–0.8 mm. long, cordate-oval, well exserted; stylopodium conical, broad margined at base; fruit 4–7 mm. long, oblong-obovate, much flattened dorsally; mericarps with heavy dorsal ribs, winged especially at the ends, covering most of the back, the lateral ribs wing-like, firm, somewhat corky; oil-tubes 1 in each interval, and 1 under the side dorsal ribs; commissural face corky, flat. *A. Piperi* Rydb., fide Piper in mss. Meadows and moist woods, *A. T. T., Can.*

In the N. Am. Fl. 28B(2): 201, 1945, this species is considered a synonym of *A. arguta* Nutt. ex T. & G., but the present writer still considers it distinct.

BERULA

Smooth aquatic or marsh perennials, with simply pinnate leaves; involucre and involucels of conspicuous narrow bracts; umbels compound; flowers white; calyx-teeth minute; fruit somewhat flattened laterally, nearly round, glabrous; carpels nearly globose, pericarp thick and corky, with slender inconspicuous ribs; stylopodium conical; oil-tubes numerous and contiguous, closely surrounding the seed cavity; commissural side flat. (Name Lat., for the water cress.)

Berula pusilla (Nutt. ex T. & G.) Fern. Rootstock fusiform, with fibrous rootlets; plant 1–10 dm. tall; basal leaves 2–4 dm. long, the petioles often exceeding the leafy part, the 7–19 leaflets, mostly opposite, ovate-lanceolate, 20–42 mm. long, serrate, or serrate and laciniately lobed; cauline leaves short petioled, reduced, more deeply laciniately parted, the segments linear or lance-linear; umbels scarcely exceeding their leaves; involucral bracts mostly 5–15 mm. long, linear-lanceolate, occasionally longer and pinnately parted; umbel of 6–12 rays 5–35 mm. long; calyx-lobes obsolete; bracts of the involucels similar to but smaller than the involucre; pedicels 2–8 mm. long; petals with several evident veins, the body 1 mm. long, oval, with an inflexed lanceolate tip nearly as long; anthers nearly 0.2 mm. long, well exserted; styles shorter than the petals, in fruit recurved; fruit 1.5–2 mm. long; mericarp suborbicular in cross-section except for the flat or slightly concave commissural side. Springy places, Spokane, *A. T. T.*

CARUM. Caraway

Smooth erect slender herbs; roots fascicled, tuberous or fusiform; leaves pinnate or ternate and pinnatifid with few linear leaflets; involucels of few to many bracts; flowers white; calyx-teeth persistent; stylopodium conical; fruit compressed laterally, orbicular to oblong, glabrous, carpels with filiform or inconspicuous ribs; oil-tubes large and solitary in the intervals, 2–6 on the flat or slightly concave commissural side. (Name from *karon,* the ancient Gr. name.)

Carum Gairdneri (H. & A.) Gray. *Yampa, Squaw Root.* Perennial, 2–17 dm. tall, often branched above; tubers usually fascicled, 1–7 cm. long; basal leaves 14–35 cm. long, with petioles 10–20 cm. long, pinnate with 3–9 leaflets 5–25 cm. long, linear or lance-linear, soon withering; cauline leaves few, similar, but reduced, shorter petioled, the uppermost simple; involucre 0 or of several linear bracts; umbels flat or round topped with 6–29 rays 5–45 mm. long; involucels of few linear bracts, shorter than the 1–8 mm. pedicels; calyx-teeth ovate-deltoid; petals white, the body suborbicular, 0.8–1 mm. long, with an inflexed lanceolate tip half as long, midnerve evident; anthers 0.3 mm. long, oval, well exserted; styles included; stylopodia low conical; fruits 2–2.5 mm. long, oval-suborbicular, the ribs filiform; mericarps arched, the middle sections becoming separated; commissural side raised near the carpophore, sloping back to the lateral ribs; oil-tubes large, 1 in the intervals, 3 on the commissural side. *Atenia Gairdneri* H. & A.; *Perideridia Gairdneri* (H. & A.) Mathias. Open or gravelly places, *U. Son.* to *A. T. T.*

The sweet, edible tubers were used for food by the Indians. Mathias demonstrates in Brittonia **2**: 243–244, 1936, that the species of *Eulophus* must be placed in *Perideridia,* then she transfers the native species of *Carum* (*Ataenia*) also. She does not list any real generic differences for our species, nor has the writer found any, so he follows Bentham & Hooker, and Engler & Prantl, and Hegi in retaining them in *Carum.*

CAUCALIS

Mostly hispid annuals; leaves pinnately dissected, with very small segments; flowers white, in irregular umbels; calyx-teeth prominent; fruit short, ovate or oblong, flattened laterally; carpel with 5 filiform primary ribs with spreading bristles and 4 prominently winged secondary ones with barbed or hooked prickles; stylopodium thick, conical; oil-tubes solitary in the intervals, that is under the secondary ribs. (Name, *kaukalis,* the Gr. name.)

Caucalis microcarpa H. & A. Plant 6–30 cm. tall, hispidulous, glabrate below; stems simple or divaricately branched above; leaves 1.5–6 cm. long, the petiole slender, sheathing at base; blade pinnately decompound into linear, apiculate segments; involucre prominent, foliaceous, pinnatifid; umbels of 2–5 unequal, divergent rays 5–70 mm. long; bracts of the involucels 1–3 mm. long, entire or few-cleft; pedicels 1–15 mm. long, unequal; calyx-teeth 0.2–0.3 mm. long, narrowly deltoid; petals 0.2–0.3 mm. long, white, elliptic, equaling the stamens; fruit 4–6 mm. long, ellipsoid, the hooked prickles 0.8–1.5 mm. long; the mericarps in cross-section bluntly pentagonal, little flattened, the commisural side grooved. Warm slopes, Snake River Canyon, *U. Son.*

CICUTA. Water Hemlock

Smooth poisonous marsh perennials; leaves pinnately compound; leaflets serrate; involucre of few bracts or none; involucels of several slender bractlets; flowers white; calyx-teeth rather prominent; stylopodium low, sometimes low-conical; fruit flattened laterally, oblong to orbicular, glabrous; carpels with strong flattish corky ribs, the lateral largest, at least in section; oil-tubes solitary in the intervals, two on the commissural side; seed nearly terete or somewhat dorsally flattened with face from plane to slightly concave. (The ancient Lat. name.)

Cicuta Douglasii (DC.) C. & R. Rootstocks fleshy, with scalariform partitions across the pith; plant 6–15 dm. tall, often branched; basal leaves 3–6 cm. long, the petioles long, 2–3-pinnate, the leaflets 3–10 cm. long, lanceolate to ovate-lanceolate, sharply serrate or doubly so; cauline leaves smaller, with shorter petioles dilated at base; umbels numerous, round-topped, with 7–32 rays that are 2–7 cm. long, subequal; bracts of involucre 0 or few, lanceolate, 3–6 mm. long; bracts of the involucels 2–3 mm. long, several, lanceolate, slightly connate; umbelets many flowered; pedicels 2–8 mm. long; calyx-lobes 0.3 mm. long, deltoid; petals white, the body 0.8–1.3 mm. long, suborbicular, with an inflexed lanceolate tip ⅓ as long, midrib evident; anthers 0.2–0.3 mm. long, oval, well exserted; stylopodium depressed; styles well included; fruit 2–3 mm. long, suborbicular, slightly flattened laterally; mericarps oblong-elliptic seen dorsally; ribs corky, broadened, not or little protruding, lateral ribs the largest, oil-tubes filling the intervals, dark brown, evident; commissural side flat, with 2 evident, spindle-shaped oil-tubes. *C. vagans* Greene. Meadows or wet valleys, *A. T., A. T. T.*

CONIUM

Biennial, glabrous, poisonous herbs; stems spotted; leaves pinnately decompound; flowers white, in compound umbels; involucre and involucels of ovate-acuminate bracts; calyx-teeth obsolete; fruit broadly ovate, somewhat flattened laterally; mericarps with strong, wavy ribs; oil-tubes none, but a layer of oil-secreting cells next to the concave commissural side of the seed. (The ancient Gr. name, *koneion.*)

Conium maculatum L. *Poison Hemlock.* Rootstock white, fusiform; plant 4–25 dm. tall; stem usually red spotted; basal leaves 2–5 dm. long, 2–4-pinnately dissected, petioled, the leaflets ovate-lanceolate, dentate or incised; cauline leaves smaller, sessile or nearly so, opposite or 3 in a whorl; umbels with 7–20 rays; bracts of involucels shorter than the pedicels; petals 1.5 mm. long, obovate, with an acute inflexed tip; anthers oval, exserted; styles 1 mm. long; stylopodium low conic; fruit 2.5–3.5 mm. long; mericarps with commissural side convex. Weed, introduced from Eurasia. Wet roadsides, *A. T.*

The plant contains the very poisonous alkaloids conine and coniceine, which cause blindness, paralysis, and death. It was used by the ancient Greeks as a poison for the condemned.

DAUCUS

Annual or biennial herbs, mostly hispid; leaves pinnately decompound into fine segments; umbels compound; flowers white or purplish; involucre and involucels foliaceous, often dissected; calyx-teeth obsolete; petals obovate, those of outer flowers often expanded; stylopodium depressed or none; fruit oblong, somewhat flattened dorsally; ribs 5, slender, bristly; secondary ribs 4, with a row of barbed prickles; oil-tubes 1 under each secondary rib, 2 on the nearly flat commissural side. (The ancient Gr. name.)

Daucus Carota L. *Wild Carrot.* Usually biennial, somewhat hispid; taproot slender, brown; plant 3–15 dm. tall, often branched; basal leaves 1–3 dm. long, slender petioled, 2–4 times pinnatifid into short oblong or linear, acute segments, entire or dentate; peduncles finally elongate; umbels 3–10 cm. wide, the top concave; bracts of the involucre at first equaling the rays, foliaceous, pinnatifid into linear segments; rays numerous, at first hispidulous; involucels often exceeding the pedicels, the bracts lance-linear, usually entire; pedicels 1–4 mm. long, hispidulous; petals 0.5–1 mm. long, the marginal ones enlarged, 1.5–2 mm. long; anthers 0.1 mm. long, exserted; stylopodium discoid; styles erect, included; mericarp body 1.5–2 mm. long; secondary ribs with a row of barbed prickles 0.7–1 mm. long, fused at base. Weed, of fields and roadsides, recently introduced from Eurasia.

This species is the source of the cultivated carrot.

ERYNGIUM

Glabrous perennials; leaves often rigid, coriaceous, entire, spinosely toothed, or divided; flowers white or blue, sessile, in dense bracteate heads; sepals very prominent, rigid and persistent; stylopodium dilated; styles short or long, often rigid; fruit ovoid, flattened laterally, covered with hyaline scales or tubercles; carpel with ribs obsolete; oil-tubes mostly 5, 3 dorsal and 2 commissural; seed face plane. (Gr. name used by Theophrastus.)

Eryngium articulatum Hook. Glabrous perennial, 1–9 dm. tall; stems dichotomous above; lower leaves often aquatic, the petioles elongate, jointed, with or without a lanceolate or ovate blade; upper leaves opposite or alternate mostly sessile, the blades 2–5 cm. long, laciniate, spinulose tipped; heads peduncled, 8–16 mm. high, ovoid-globose, amethyst-blue; bracts foliaceous, lanceolate, laciniate, spinulose tipped; bracteoles coriaceous, cuspidate, with

2 lateral cuspidate teeth, equaling the flowers; calyx-lobes 3–3.5 mm. long, linear-lanceolate, hyaline margined, spine-tipped; petals 1–1.5 mm. long, bluish tinged, deeply bilobed with a prominent midnerve up to the sinus; stylopodium 2-lobed, flat; styles included; stamens exserted; mericarps 2.8–3 mm. long, narrowly lanceoloid, deltoid in cross-section, covered with pale, lanceolate scales. Stream beds or wet shores, *A. T .T.*

First collected by C. A. Geyer, and the species described from his collections from "stony edges of Spokan River, And Skitsoë and Coeur d'Aleine Lakes."

HERACLEUM

Tall stout perennials; leaves large, ternately compound, involucres deciduous; involucels of numerous bractlets; flowers white, in large many-rayed compound umbels; calyx-teeth small or obsolete; petals obcordate, the outer ones often dilated and 2-cleft; stylopodium thick, conical; fruit broadly ovate or oval, very much flattened dorsally, somewhat pubescent; carpel with dorsal and intermediate ribs filiform; the broad lateral wings contiguous to those of the other carpel, strongly nerved toward the outer margin; oil-tubes solitary in the intervals, conspicuous, about half as long as the carpel, 2–4 on the commissural side; seed very much flattened dorsally. (Named Gr., in honor of *Hercules.*)

Heracleum maximum Bartr. *Cow Parsnip.* Plant 1–2.5 m. tall; root stout, elongate, brown; stem hollow, tomentose or villous, especially at the nodes; upper petioles dilated, broad sheathing, lanate; leaflets petioled, 10–25 cm. or more in length, cordate or orbicular in outline, sharply lobed and singly or doubly serrate, glabrate or puberulent but green above, pilosulous and paler beneath; umbels flat-topped, the 8–38 rays subequal, 5–11 cm. long; involucral bracts 5–15 mm. long, lanceolate, long caudate; involucels of narrower bracts; pedicels 4–22 mm. long, pilose; central flowers with petals 1.5–2 mm. long, cuneate obcordate, with evident pinnate nerves; anthers 0.7–0.9 mm. long, oval, exserted; styles stout, shorter than the petals; marginal flowers with petals 4–6 mm. long, the two oblong lobes divergent; mericarps 8–12 mm. long, flat, oval-obcordate, lanate, later subglabrate, the commissural side flat; oil-tubes dark, clavate, not reaching the base. *H. lanatum* Michx. Moist valleys, common, *U. Son.* to *Can.*

The leaves are reported to produce irritation, and yet the Walla Walla Indians ate the stalks like asparagus.

LEIBERGIA

Slender glabrous acaulescent plant from a corm; leaves ternately or pinnately divided into long filiform leaflets; flowers white, in irregular umbels; calyx-teeth obsolete; fruit slightly flattened laterally, linear, beaked; carpels with 5 filiform ribs; stylopodium flat; oil-tubes small, one in each interval, two on the commissural side. (Named for its collector, *John B. Leiberg,* American botanical explorer.)

Leibergia orogenioides C. & R. Corm subglobose, 6–10 mm. in. diameter; plant 1.2–5 dm. tall; basal leaves several, nearly equaling the peduncle, the petiole 6–12 cm. long, dilated and sheathing at base, leaflets 1–7.5 cm. long, entire or with a few teeth or linear lobes; involucre none; umbel loose, of 3–10 unequal rays 2–12 cm. long; involucels of several lanceolate bracts 1–2 mm. long; pedicels 1–2 mm. long; petals with the body 0.7–1 mm. long, broadly elliptic, the lanceolate tip incurved, nearly as long, the midnerve evident; anthers 0.1 mm. long, discoid, well exserted; styles equaling the petals; fruit 6–9 mm. long, the ribs pale; mericarps with lateral ribs thickened, the commissural side channeled. *Lomatium orogenioides* (C. & R.) Mathias. Meadows, Spokane Co., *A. T.*

Very local, found only in n.e. part, but reported locally abundant in early spring.

LIGUSTICUM. Lovage

Smooth perennials from large aromatic roots; leaves usually large, ternately or ternately pinnately compound; flowers white or pinkish in large many rayed compound umbels; involucre usually none; involucels of narrow bracts; calyx teeth small or obsolete; fruit oblong or ovoid, flattened laterally, if at all, glabrous; carpel with prominent and equal ribs; stylopodium conical; oil-tubes 2–6 in the intervals, 6–10 on the commissural side. (Name Gr. *ligustikon,* the ancient name of *Ligusticum levisticum* of Liguria, Italy.)

Pinnules ovate, not laciniately dissected. *L. verticillatum.*
Pinnules lacinately pinnatifid,
  Ultimate pinnule lobes 3–5 mm. wide; fruit 4–5.5 mm. long. *L. Leibergi.*
  Ultimate pinnule lobes 2 mm. wide; fruit 3–4 mm. long. *L. caeruleimontanum.*

Ligusticum caeruleimontanum St. John. Caudex woody, with strong odor of sassafras, stout, dark, the summit clothed with coarse fibers from the old leaf-bases; plant 3.5–6.5 dm. tall, glabrous; stem simple to the inflorescence; basal leaves few, 2–4 dm. long, the petioles 5–26 cm. long, the blades ternate, then tripinnatifid into numerous segments; pinnules ovate-lanceolate in outline, 4–7 cm. long, laciniately pinnatifid into linear-oblong, acute lobes, the ultimate 2 mm. wide; cauline leaves none; bracts of the inflorescence reduced, leafy; umbels several, the terminal the largest, round-topped, with the 8–18 rays subequal, 1.5–4 cm. long; involucre none; involucels 3–6 mm. long, the bracts linear-lanceolate; pedicels 2–6 mm. long; buds purplish tinged; petals purplish-tipped, later white, the body 1 mm. long, suborbicular, with an inflexed acute tip about 0.2 mm. long; anthers 0.2–0.3 mm. long, oval; stylopodium broad conic, with a wide projecting depressed base; styles stout, 0.4 mm. long; fruit 3–4 mm. long, broad oval in outline; mericarps flattened dorsally, 2–3 mm. wide, dorsal and lateral ribs with wings 0.3 mm. wide, corky thickened at base; oil-tubes 4–6 in the intervals, 10–12 in the commissural side which is flat or concave, with a median ridge; seed grooved between the dorsal ribs and channeled on the commissural side.

Gravelly basaltic summits, Blue Mts., *Huds.*

The holotype is: Table Rock, 6,000 ft. alt., Columbia Co., *St. John et al.* 9,654.

The provisions of Article 73, Note 2, of the 1961 International Code of Botanical Nomenclature require the alteration of the epithet *caeruleomontanum* to *caeruleimontanum*.

Ligusticum Leibergi C. & R. Caudex woody, stout, covered with coarse fibers from old leaf-bases; plant 6–10 dm. tall, glabrous; basal leaves with petioles 2–6 dm. long, the blades 15–30 cm. long, ternate, then bipinnatifid, the ultimate leaflets 2–6.5 cm. long, lanceolate, laciniately cleft, the lobes acuminate, lanceolate or oblong; cauline leaves 0 or 1, much reduced, but similar in cutting, the inflorescence with 1 (–2) leafy bracts; inflorescence commonly a long peduncled terminal umbel, surrounded by a whorl of smaller ones; main umbel round-topped with a few lanceolate, early deciduous involucral bracts, the 24–32 rays subequal, 3–5.5 cm. long, slightly scaberulous at each end; outer umbelets with pedicels and lanceolate involucel bracts 3–10 mm. long, the inner with involucels mostly abortive; petals white, the body 1.2–1.4 mm. long, oval, the lanceolate, inflexed tip half as long; anthers 0.2 mm. long, oval, well exserted; stylopodia broad conic; styles included; mericarps with the dorsal and lateral ribs produced into wings 0.2–0.3 mm. wide, the body flattened dorsally, the intervals with 4–6 oil-tubes, the commissural side with 10, and flat, keeled up the middle. Wet meadows, Rock Cr., Spokane Co., and eastward, *A. T. T.*

This species is reported to be poisonous to cattle.

In the N. Am. Fl. 28B(1) : 146, 1944, this species and *L. caeruleimontanum* are reduced to the synonymy of *L. Canbyi* C. & R., but the present writer does not agree with this action.

Though the recent Article 73, Note 3, of the 1961 International Code requires the epithet to end in *–ii,* it is in direct conflict with the long standing Art. 23 under the provisions of which *Leibergi* is valid and not subject to "editing."

Ligusticum verticillatum (Geyer) C. & R. Caudex dark, fibrous, woody; plant 6–9 dm. tall; stems mostly 2-leaved, branching above; basal leaves with sheathing petiole about 5 cm. long, ternate, then bipinnatifid, the leaflets 2.5–7 cm. long, ovate, serrate or doubly dentate or the larger also lobed at base; cauline leaves reduced, subsessile; peduncles 15–23 cm. long; involucral bracts few, linear, early deciduous; umbels in fruit flat-topped, of 16–21 subequal rays, in fruit 4–6 cm. long; involucels 3–4 mm. long, the bracts linear, few; pedicels 1–3, in fruit 5–11 mm. long; petals white, the midrib distinct, the body 1.2–1.4 mm. long, obovate, emarginate, with an inflexed lanceolate tip half as long; anthers 0.4–0.5 mm. long, oval, exserted; styles shorter than the petals; stylopodia broad conic; mericarps 4–6 mm. long, somewhat flattened dorsally, the dorsal and lateral ribs with prominent wings about 0.5 mm. wide, commissural side flat with a median ridge. Open woods, Craig Mts., *Henderson* 2,665, *A. T. T.*

The type locality is "shady grassy borders of pine woods, on high plains of the Nez Perces Indians," *Geyer* 414. This is apparently the Craig Mts., south of Lewiston.

LOMATIUM

Acaulescent or short caulescent or tall, leafy perennials; roots slender fusiform, or tuberous; leaves ternate to dissected, sometimes pinnate; involucre none or of a few bracts; involucels mostly present; flowers yellow, white, or purple, in compound umbels; calyx-teeth mostly obsolete, but sometimes evident; stylopodium depressed; fruit strongly flattened dorsally, linear elliptic, to orbicular; carpel with filiform and approximate dorsal and intermediate ribs, and winged laterals thin or corky, coherent till maturity with those of the other carpel; pericarp thin; oil-tubes 1-several in the intervals (rarely obsolete), 2–10 on the commissural side; seed dorsally flattened, with plane face (rarely slightly concave). (Fide Fernald, named from the Gr., *lomation,* a little border, in reference to the winged fruit.)

Key to Plants in Flower

Flowers white to purple,
  Flowers purple,
    Corm subglobose; pedicels 3 mm. or less long. *L. Gormani,* forma. *purpureum.*
    Root fusiform; pedicels 15–25 mm. long. *L. columbianum.*
  Flowers white, or merely tinged with pink,
    Tap-root cylindric or somewhat swollen in places,
      Bracts of involucels scarious margined. *L. orientale.*
      Bracts of involucels not so,
        Foliage more or less pilose or lanate. *L. macrocarpum.* var. *macrocarpum,*
        Foliage puberulent. Var. *artemisiarum.*
    Roots with one or more subglobose corms,
      Rays puberulent. *L. Gormani.*
      Rays glabrous,
        Corm black, single. *L. Canbyi.*
        Corms brown or yellowish, often moniliform,
          Anthers pink; leaf segments 1–8 cm. long. *L. farinosum.*
          Anthers purple; leaf segments 3–30 mm. long. *L. Geyeri.*
Flowers yellow,
  Stems leafy to the top,
    Foliage glabrous; dorsal ribs strong,
      Leaflets of lower leaves 3–5 mm. wide. *L. ambiguum.*
      Leaflets of lower leaves 0.1–0.3 mm. wide, *L. salmoniflorum.*
    Foliage puberulent; dorsal ribs filiform,
      Plant 2.5–5 dm. tall; blades 2–3-pinnate. *L. Rollinsii.*
      Plant 1–2 m. tall; blades ternate, then 3–4-pinnatifid. *L. dissectum,* var. *multifidum.*
  Stems naked, or plants short caulescent,
    Root producting a corm or rounded enlargements,
      Bracts of involucels linear,
        Leaves 2–3-times ternate, then pinnate; plant 40–60 cm. tall. *L. dissectum.*
        Leaves once ternate, then pinnately decompound; plant 10–45 cm. tall. *L. bicolor.*
      Bracts of involucels broader, scarious margined,
        Plant short caulescent. *L. circumdatum.*
        Plant acaulescent. *L. Cous.*
    Roots terete, not or little enlarged,
      Leaves dissected into very numerous small segments,
        Leaf segments filiform. *L. Grayi.*
        Leaf segments broader,
          Bracts of involucels linear, acuminate. *L. Donnellii.*
          Bracts of involucels linear-lanceolate. *L. serpentinum.*

Leaves 1–2-ternate (rarely more cleft) into large divisions (mostly 2 cm. or more in length),
Leaf segments narrowly linear. *L. triternatum,* f. *triternatum.*
Leaf segments broader,
Root terete, slender. Forma *lancifolium.*
Root stout, often with swellings. *L. ambiguum.*

Key to Plants in Fruit

Stems leafy to the top,
Plants glabrous; lower leaves ternate-pinnate; fruit 8–10 mm. long, oblong. *L. ambiguum.*
Plants puberulent; lower leaves 2–3-pinnate,
Fruits 6–7 mm. long, ovate-oblong; oil tubes 6. *L. Rollinsii.*
Fruits 8–25 mm. long; oil tubes usually none. *L. dissectum.* var. *multifidum.*
Stems naked, or plants short caulescent,
Dorsal ribs winged. *L. anomalum .*
Dorsal ribs filiform or at least not winged,
Lateral wings of fruit thick. *L. salmoniflorum.*
Lateral wings thin,
Roots producing a corm, tubers, or rounded enlargements,
Fruit linear or oblong-linear,
Dorsal ribs prominent,
Leaf segments linear. *L. triternatum,* f. *triternatum.*
Leaf segments broader. Forma *lancifolium.*
Dorsal ribs filiform or nearly so. *L. bicolor.*
Fruit elliptic or broader,
Fruit glaucous. *L. dissectum.*
Fruit not glaucous,
Fruit puberulent, the dorsal ribs prominent,
Intervals with one large oil-tube. *L. Cous.*
Intervals without oil-tubes or with minute abortive ones,
Petals white. *L. Gormani.*
Petals rose-purple. Forma *purpureum.*
Fruit glabrous,
Oil-tubes none. *L. Geyeri.*
Oil-tubes present,
Pedicels 2 or more times the length of the fruit. *L. farinosum.*
Pedicels shorter,
Dorsal ribs filiform,
Pedicels 15–25 mm. long; fruits 16–24 mm. long. *L. columbianum.*
Pedicels 1–12 mm. long; fruits 6–10 mm. long. *L. Canbyi.*
Dorsal ribs prominent. *L. circumdatum.*
Root terete, not or little enlarged,
Oil-tubes 4–6 in the intervals. *L. Donnellii.*
Oil-tubes usually single in the intervals,
Involucel bracts conspicuously scarious margined. *L. orientale.*

Involucel bracts not so,
  Involucels of prominent leaves,
    Foliage more or less pilose or lanate. *L. macrocarpum,* var. *macrocarpum.*
    Foliage puberulent. Var. *artemisiarum.*
  Involucels of small bracts,
    Leaf segments filiform, ill-scented. *L. Grayi.*
    Leaf segments lanceolate or oblanceolate, with pleasant odor of parsley. *L. serpentinum.*

Lomatium ambiguum (Nutt.) C. & R. Root yellowish or brown, slender, vertical, fusiform, or in large plants with one, or several rounded moniliform corms; plant 1–7 dm. tall, glabrous, caulescent, leafy throughout; petioles mostly expanded and sheathing at base; basal and lowest leaves mostly ternate, the leaflets 2.5–4.5 cm. long, 3–5 mm. wide, lanceolate, pale green; middle cauline leaves with the blade 2–8 cm. long, ternate, when 2–3 pinnatifid, the ultimate segments 2–30 mm. long, 1–2 mm. wide, linear-oblong; upper leaves often with more finely cleft, shorter segments; peduncles from several upper axils, shorter than or longer than the leaves; umbels flat-topped, the 6–20 rays 0.5–15 cm. long; involucels wanting; pedicels 2–15 mm. long; calyx-lobes rounded, minute; petals pale yellow, the midnerve evident, the body 0.6–0.8 mm. long, emarginate by the inflexed lanceolate tip ¾ as long; anthers 0.3 mm. long, oval, yellow; styles exserted; fruit 5–12 mm. long, narrowly oblong, narrowed towards the tip; mericarps with the dorsal ribs filiform, the lateral ribs winged, 0.2–0.7 mm. wide; oil-tubes large, 1 in the intervals, 2 on the commissural side. *Cogswellia ambigua* (Nutt.) M. E. Jones. Rocky, often dry slopes, *U. Son.* to *A. T. T.*

Lomatium anomalum M. E. Jones ex C. & R. Tap-root large, about 1 cm. in diameter, sometimes with swellings; plant 3–6 dm. tall, puberulent, caulescent, the nodes at length aerial; basal leaves with petioles 8–20 cm. long, the blades 8–20 cm. long, ternate (though the lateral divisions have basal leaflets, so that the leaf appears 3-pinnate), then 1–2-pinnate, the ultimate segments 1–7.5 cm. long, 2–7 mm. wide, linear; cauline leaves similar, but petioles shorter; peduncles much exceeding the leaves; umbels loose; the 5–15 rays 1–15 cm. long, unequal; involucels 2–7 mm. long, the bracts linear or setaceous; pedicels 2–5 mm. long; calyx-teeth abortive; petals yellow, the midnerve prominent, the body 0.8–0.9 mm. long, emarginate, the inflexed lanceolate tip ⅔ as long; anthers 0.4–0.5 mm. long, oval, yellow, exserted; styles exserted; fruit 12–21 mm. long, oval-oblong, glabrate; mericarps with dorsal ribs forming wings 0.5–1 mm. wide, the lateral ribs with wings 2–4 mm. wide; oil-tubes large, 1 in the intervals, 2 or 4 on the commissural side. *Cogswellia anomala* M. E. Jones; *L. triternatum* (Pursh) C. & R., var. *anomalum* (M. E. Jones) Mathias. Grassy slopes, 900 ft., Rogersburg, Asotin Co., *St. John* 4,194; Lewiston, *Weber* 2,136, *U. Son.*

Lomatium bicolor (Wats.) C. & R. Corm napiform; stem 1–4.5 dm. tall, glabrous or slightly puberulent; petioles wholly dilated; blades ternate, then pinnately decompound, with numerous, filiform divisions; rays few, 1–12 cm. long; involucels 1–3 mm. long, linear; pedicels 0.5–2 mm. long; petals yellow; fruit 10–13 mm. long, oblong-linear, slightly narrowed to the ends, especially the upper; mericarps 2–5 mm. broad, the wing 0.2 mm. wide; oil-tubes 1 in the intervals, 2 on the commissural side. *Cogswellia bicolor* (Wats.) M. E. Jones. Open ground, Blue Mts., *A. T.*

L. leptocarpum (Nutt.) C. & R., adopted by Mathias (1938) is based either on a name published in synonymy and hence invalid, or upon a varietal epithet. The earliest specific epithet is *bicolor.*

Lomatium Canbyi C. & R. Corm black, 1.5–4 cm. in diameter, globose or rounded, forming a thick, black, subterranean, erect stem; plant 7–23 cm. tall, glabrous, short caulescent, but the nodes subterranean; petioles mostly membranous sheathing; blades 2–8 cm. long, ternate then 3-pinnatifid, pale green, thick, the ultimate segments numerous, crowded, 1–6 mm. long, 0.5–1 mm. wide, linear; peduncles exceeding the leaves; umbel flat-topped, slightly asymmetric; the 5–12 rays 3–45 mm. long; involucels 2–4 mm. long, the bracts linear; pedicels 1–12 mm. long; calyx-teeth rounded, evident; petals white, the midrib evident, the body 0.7–0.8 mm. long, oblanceolate or obovate, the inflexed lanceolate tip ¼ as long; anthers 0.2 mm. long, oval, purple, exserted; styles exserted; fruit 6–10 mm. long, oval, glabrous; mericarps with the dorsal ribs filiform, the lateral winged 1 mm. broad; oil-tubes usually 1 in the intervals (the lateral intervals sometimes with 1–2 shorter accessory ones), 2 or 4 on the commissural side. *Cogswellia Canbyi* (C. & R.) M. E. Jones. Rocky slopes, Clearwater and Snake River Canyons, *U. Son.*

Lomatium circumdatum (Wats.) C. & R. Corm napiform to subglobose; plant 2–3.5 dm. tall, short caulescent, glabrous or somewhat puberulent; leaves ternate then 1–2-pinnate, ultimate segments 2–12 mm. long, linear; petiole of cauline leaves broad sheathing; peduncle much exceeding the leaves, it and rays puberulent at tip; the 6–12 rays 1–8.5 cm. long, unequal; involucels 2–4 mm. long, the bracts scarious margined, broadly oblanceolate, often united; pedicels 2–4 mm. long; petals yellow; fruit 6–9 mm. long, oblong-elliptic, glabrous; mericarps with the dorsal ribs raised, submembranous, the lateral ribs winged 1–1.5 mm. wide; oil-tubes large, 1 in the intervals, 4 on the commissural side. *Cogswellia circumdata* (Wats.) M. E. Jones. Uncommon, Lake Waha, and Blue Mts.

Lomatium columbianum Mathias & Constance. Plants 3–5 dm. tall, glabrous, from a stout, thickened root bearing a caudex, wrapped in dilated purplish sheaths; leaves crowded; petioles 10–20 cm. long; blades 5–20 cm. long, ternate, then 2–4-pinnate, the ultimate divisions 3–20 mm. long, 0.5–2 mm. wide, linear to filiform; peduncles exceeding the leaves, stout, fistulose; involucre and involucel of several linear to lanceolate, acute bractlets, shorter than the flowers; rays 6–10, spreading 3–20 cm. long, unequal; umbellets 15–20-flowered; pedicels 15–25 mm. long; flowers purple; fruits 16–24 mm. long, glabrous, oblong-oval, the wings thick and corky, much narrower than the body; oil-tubes 2–3 in the intervals, 4 on the commissural side. *Leptotaenia purpurea* (Wats.) C. & R. Foothills and canyons of the Blue Mts., recorded by E. Lathrop, Walla Walla Coll., Dept. Biol., Publ. 2 (2) : 17, 1952.

Lomatium Cous (Wats.) C. & R. *Cous.* Corm 1–3 cm thick, black coated, from globose to fusiform, simple or branched; plant 5–35 cm. tall, acaulescent, glabrous or slightly puberulent; leaves basal, several, the petioles membranous sheathing nearly throughout; blades 3–10 cm. long, ternate then 2–3-pinnatifid, the ultimate segments 2–20 mm. long, linear-elliptic, rarely elliptic, pale green; summit of scape and rays papillose puberulent; the 6–16 rays at anthesis very short, in fruit very unequal, 5–80 mm. long; involucels 2–5 mm. long, the bracts ovate, acute, cuneate, scarious-margined; pedicels 1–3 mm. long; calyx-lobes ovate, minute; petals yellow, the body 0.6–1 mm. long, oval, cuneate, with a lanceolate inflexed tip ⅓ as long; anthers 0.3–0.4 mm. long, oval, exserted; styles equaling the petals; fruit 6–10 mm. long, lance-elliptic to oval; mericarps with sub-membranous, enlarged dorsal ribs, the intervals papillose-puberulent; oil-tubes large, one in each interval, 4–6 on the commissural side. *Cogswellia Cous* (Wats.) M. E. Jones. Open slopes or woods, *U. Son* to *A. T. T.*

The Indians called this plant *cous* and it furnished them an important article of food.

Daubenmire (1939) reported a collection with albino flowers.

Lomatium dissectum (Nutt. ex T. & G.) Mathias & Const. Acaulescent,

40–60 cm. high; leaves 2- or 3-times ternate, then pinnate; ultimate segments linear and entire or cut into linear lobes, puberulent on the midrib and margins; umbel 8–22 rayed, involucre of several linear bracts; involucels of a few linear bracts; flowers yellow; fruit glaucous, on very short pedicels; sterile flowers on slender pedicels. *Leptotaenia foliosa* (Hook.) C. & R. Lake Waha, Idaho. A poorly known species.

Var. **multifidum** (Nutt. ex T. & G.) Mathias & Const. Root black, stout, woody, napiform or branched; plant 1–2 m. tall, caulescent, much branched, usually several stems in a clump; basal leaves with petioles 1–3 dm. long, the blades 12–40 cm. long, ternate, then 3–4-pinnatifid, the ultimate segments 3–8 mm. long, 0.5–1 mm. wide, linear, scabrous puberulent on the margins and the midribs beneath; cauline leaves 0–3, similar but petioles reduced to broad sheaths; involucre 0, or of few lance-linear bracts; umbels round-topped, the 8–28 rays 2–15 cm. long, the earlier umbels forming perfect flowers and fruit only in the outer margin of the outer umbelets; involucels 3–7 mm. long, of linear, puberulent bracts; pedicels 2–15 mm. long; calyx-lobes minute; petals yellow, the midvein prominent, 2 mm. long, inflexed and bent double at the middle; anthers 0.5–0.6 mm. long oval, well exserted; styles long exserted; fruit 8–12.5 mm. long, flat; mericarps with filiform dorsal ribs; oil-tubes usually none. *Leptotaenia mutifida* Nutt. Loamy or rocky slopes, *U. Son., A. T.*

Described from "Plains of the Oregon, east of Wallawallah, and in the Blue Mountains."

Specimens like *St. John & Pickett* 6,117 from Palouse Falls have shorter, broader fruit with 3 prominent oil-tubes in the intervals. They may be undescribed, but need more field and herbarium study.

Lomatium Donnellii C. & R. Root fusiform; plant 1.5–3 dm. tall, glabrous, short caulescent or acaulescent; leaves ternate, then 2–3-pinnatifid, the ultimate segments short oblong or linear; the 6–12 rays 2.5–10 cm. long, somewhat unequal; the bracts of the involucels linear, acuminate; pedicels 4–16 mm. long; petals yellow; fruit 7–8 mm. long, oval, glabrous; mericarps with prominent dorsal ribs, the lateral ones winged, the wings less than half as broad as the body; oil-tubes small, 4–6 in the intervals, 4–6 on the commissural side. *Cogswellia Donnellii* (C. & R.) M. E. Jones. Not seen, but accepted from report of Coulter & Rose, 750–900 meters, Lewiston, *Heller* in May 1896.

Lomatium farinosum (Geyer) C. & R. Corm brown or yellowish brown, subglobose, single or several and moniliform; plant 1–3 dm. tall, glabrous, short caulescent but the nodes subterranean; basal leaves with membranous sheathing petioles, mostly subterranean, the blades 5–6 cm. long, withered after anthesis; cauline leaves similar, the blades ternate then pinnate or bipinnatifid, the ultimate segments 1–8 cm. long, 0.5–3 mm. wide, glaucous green, linear; peduncles arched ascending; the 5–15 rays unequal, 5–67 mm. long; the bracts of the involucels 1–3 mm. long, lance-linear, more or less united; pedicels in anthesis 1 mm. or more long, in fruit 10–20 mm. long; calyx-teeth ovate, evident; petals white, the midnerve evident, the body 0.8–1 mm. long obovate to suborbicular, the inflexed tip ¼ as long; anthers 0.2 mm. long, oval, pink, exserted; styles exserted; fruit 4–6 mm. long, lance-elliptic, glabrous; mericarps with dorsal ribs filiform, the lateral ribs winged 0.5 mm. wide; oil-tubes slender, 2–4 in the intervals, 4–6 on the commissural side. *Cogswellia farinosa* (Geyer) M. E. Jones. Rocky slopes, *U. Son.,* occasional in *A. T.*

Lomatium Geyeri (Wats.) C. & R. *Biscuit root.* Root producing one or usually several moniliform globose, ovoid or cylindric, brown tubers; plant 1–5.5 dm. tall, glabrous, acaulescent; leaves basal, several, base of the petioles subterranean, membranous sheathing; blades 4–14 cm. long, ternate, then 1–2-pinnate, the ultimate segments 3–30 mm. long, 1–2 mm. wide, divergent, linear; scapes exceeding the leaves; umbels asymmetric; the 7–18 rays 3–67 mm. long;

involucels 1–4 mm. long, the bracts hyaline margined, lanceolate, more or less united; pedicels 0–3 (rarely –8) mm. long; calyx-teeth rounded, minute; petals white, the midnerve evident, the body 0.8–1 mm. long, obovate, with an inflexed lanceolate tip ¼ as long; anthers 0.2 mm. long, oval, purple, exserted; styles exserted; fruit 8–13 mm. long, oval, glabrous; mericarps with dorsal ribs filiform, raised, the lateral ribs or wings 1–1.5 mm. wide; oil-tubes none. *Cogswellia Geyeri* (Wats.) M. E. Jones. Gravelly woods, in Spokane and Kootenai Cos., *A. T. T.*

Apparently used by the Indians as a food root. Named for its collector, C. A. Geyer.

Lomatium Gormani (Howell) C. & R. *Pepper and Salt.* Corm 5–40 mm. in diameter, subglobose, black, with numerous tufts of rootlets; plant 3–22 cm. tall, acaulescent, glabrous below; leaves basal or from one subterranean node, membranous sheathing to the ground surface; blades 1–11 cm. long, ternate, then bipinnate, the ultimate segments 2–15 mm. long, 0.5–1 mm. wide, almost linear, glaucous green; tip of peduncle and umbel more or less puberulent; the umbel asymmetric, with the 3–10 rays 1–40 mm. long; involucels none or a few setaceous bracts; pedicels 3 mm. or less in length; calyx-lobes deltoid, evident; petals 1–1.5 mm. long, white, broadly oblanceolate, the midnerve evident; anthers 0.4–0.5 mm. long, oval to suborbicular, exserted; styles exserted; fruit 5–8 mm. long, oval or elliptic, finely puberulent; mericarps with raised, evident dorsal ribs, the lateral wings 0.7–0.9 mm. broad; oil-tubes obsolete, or 1 little one in each rib, or rarely 2–3 rudimentary ones in the intervals and 2–6 on the commissural side. *Cogswellia Gormani* (Howell) M. E. Jones; *Peucedanum confusum* Piper. Moist rocky slopes, *U. Son., A. T.*

The purplish black anthers against the white petals looks like pepper on salt. It is the earliest spring flower, blooming whenever a few warm days break the winter cold. Collected in flower on Jan. 1st., or even on Dec. 31st.

Forma **purpureum** St. John. Differs by having the petals bright rose-purple. *Cogswellia Gormani* (Howell) M. E. Jones, forma *purpurea* (St. John) St. John. The type is from Pullman, *St. John & B. V. Pickett* 3714.

Lomatium Grayi (C. & R.) C. & R. Root 1–2 cm. in diameter, dark brown, woody, cylindric, the crown clothed with fibrous shredded leaf bases, in age becoming multicipital; plants 1–6 dm. tall, short caulescent, the 1–2 nodes subterranean, enclosed in the leaf bases; leaves 5–40 cm. long, the base of the petiole membranous sheathing, the tip terete, often longer; the blades 4–20 cm. long, 4–6-pinnate, the ultimate segments 2–10 mm. long, 0.1–0.2 mm. wide filiform, very numerous, crowded, scabrous; peduncles soon exceeding the leaves, glabrous; umbel flat-topped, glabrous, the 8–29 rays 0.5–12 cm. long, the fertile ones subequal; involucels 2–4 mm. long, the bracts lance-linear, connate at base; pedicels 2–18 mm. long, glabrous; calyx-teeth rounded, minute; petals yellow, the midnerve thickened, the body 0.8–1 mm. long, obovate, the inflexed tip lanceolate ½ as long; anthers 0.3 mm. long, oval, yellow; styles exserted; fruit 6–16 mm. long, oval, glabrous; mericarps with the dorsal ribs filiform, the lateral ribs winged, 0.7–2.5 mm. wide; oil-tubes 1 (occasionally 2) in the intervals, 2 or 4 on the commissural side. *Cogswellia Grayi* (C. & R.) M. E. Jones. Rocky slopes, *U. Son.* to *A. T.*

Dedicated to the famous Asa Gray of Harvard, though inappropriately so as the plant is vile-smelling, with an odor like that of *Anthemis Cotula.*

Lomatium macrocarpum (H. & A.) C. & R., var. **macrocarpum.** Root brown or yellowish, erect uniform, up to 1 cm. in diameter, or with a deep-seated ellipsoid swelling 1–2.5 cm. in diameter; plant 1–6 dm. tall, more or less pilose or lanate throughout, short caulescent, at maturity the 1–4 nodes aerial; leaves numerous, with broad membranous sheathing petioles; blades 5–15 cm. long, ternate, then 3-pinnatifid, the ultimate segments 1–4 mm. long, 0.5–1 mm.

wide, linear-lanceolate; peduncles exceeding the leaves; umbels flat-topped, the 5–18 rays 5–60 mm. long; involucel 3–12 mm. long, the bracts numerous, conspicuous, linear or lanceolate, more or less united; pedicels 2–10 mm. long; calyx-teeth deltoid, evident; petals white, the midnerve broad, thickened, the body 1–1.3 mm. long, oblanceolate, the inflexed lance-linear tip ⅔ as long; anthers 0.2–0.3 mm. long, oblong-oval, yellowish, exserted; styles exserted; fruit variable, 6–20 mm. long, 3–7 mm. broad, linear-oblong to elliptic, glabrous; mericarps with the dorsal ribs filiform, the wings from very narrow to as broad as the body; oil-tubes 1 (rarely 2–3) in the intervals, 2 or 4 on the commissural side. *Cogswellia macrocarpa* (Nutt.) M. E. Jones. Open or rocky slopes, *U. Son.* to *A. T. T.*

Var. **artemisiarum** Piper. Differing from *C. macrocarpa* only in having the foliage puberulent. *Cogswellia macrocarpa* (Nutt.) M. E. Jones, var. *artemisiarum* (Piper) St. John. Zindel, Asotin Co., *St. John & Brown* 4,159, *U. Son.*

Lomatium orientale C. & R. Tap-root thick, often swollen in places; plant 1–3 dm. tall, short pilose, acaulescent, or short caulescent; leaves 2–3-pinnatifid, the ultimate segments oblong or linear, entire or toothed, short gray pilose, the 4–8 rays 1–3.5 cm. long, unequal; the bracts of the involucel lanceolate; scarious margined; pedicels 5–7 mm. long; petals white or pinkish; fruit 4–6 mm. long, broad oval or suborbicular; mericarps with the dorsal ribs indistinct, the lateral ones winged, narrower than the body; oil-tubes 1 (rarely 2) in the intervals, 4 on the commissural side. *Cogswellia orientalis* (C. & R.) M. E. Jones. Not seen, but accepted on the report by Coulter & Rose, boundary of Idaho and Washington, *Canby* in 1891.

C. L. Hitchcock and Cronquist redetermine this as *L. nevadense* (Wats.) C. & R.

Lomatium Rollinsii Mathias & Const. Plants crisp puberulent, slender, caulescent, alternately branched, 2.5–5 dm. tall; roots elongate, often a tuberous taproot; petioles 5–15 cm. long, narrowly short sheathing at base; blades 5–15 cm. long, oblong in outline, bipinnate or partially tripinnate, the ultimate divisions 2–30 mm. long, 0.5–2 mm. wide, linear, acute or obtuse; involucels minute, filiform; rays unequal; pedicels 6–15 mm. long; umbellets 8–15-flowered; flowers yellow; fruit 3–4 mm. broad, glabrous, the wing half as wide as the body; oil tubes 1 in the dorsal intervals, 2 in the lateral, 4 on the commissural side. Stony hillsides, Nez Perce Co., 7 miles n. of Waha, *Christ* 10877; Webb, *Christ* 10889. *A. T. T.*

Named for the Prof. Reed Rollins, formerly of Washington State University.

Lomatium salmoniflorum (C. & R.) Mathias & Const. Roots woody, black, large; plant 2–4.5 dm. tall, short caulescent; basal leaves 10–33 cm. long, the petioles 5–13 mm. long, broad membranous sheathing at base, the blades ternate, then 3–5-pinnatisect, the ultimate segments 1–3 mm. long, 0.1–0.3 mm. wide, linear, acute; cauline leaves 2–4, much reduced, the membranous sheathing petiole base prominent; peduncles exceeding the leaves; involucre none; umbels round topped, with 7–13 rays, the marginal ones 1–6 cm. long, involucel bracts 2–3 mm. long, filiform; pedicels 2–14 mm. long; calyx-teeth deltoid or ovate, minute; petals salmon-yellow, the body 1–1.2 mm. long, obovate, with an inflexed lanceolate tip nearly as long, the midvein prominent; anthers 0.4–0.5 mm. long, elliptic, exserted; styles exserted; stylopodium depressed; fruit 8.5–12 mm. long, elliptic, flat; mericarps with dorsal ribs filiform; oil-tubes brown, conspicuous, 1 in the intervals, 2 on the commissural side. *Leptotaenia salmoniflora* C. & R. Basalt talus slopes, near bottom of Snake River and Clearwater River Canyons, *U. Son.*

The type collection, *Sandberg* et al. 28, came from "Upper Ferry, Clearwater River, above Lewiston."

Lomatium serpentinum (M. E. Jones) Mathias. Perennial 2–4.5 dm. high; root thick and woody with a multicipital caudex, acaulescent; leaves narrow, 10–30 cm. long; petioles slender 3–12 cm. long dilated and sheathing at base; blades ternate with long slender naked rhachillae then tripinnatifid, the ultimate segments closely massed, lanceolate or oblanceolate, 3–5 mm. long, 1–2 mm. broad, minutely papillose above, shortly callous apiculate the earliest lower leaves smaller and less cleft, their segments elliptic and 4–5 mm. broad; scapes slender glabrous numerous; involucre none; rays 4–12 unequal ascending becoming as much as 6 cm. in length finely puberulent or glabrate; bracts of the involucel 7–11 linear-lanceolate 2–3 mm. long; pedicels 5–15 mm. long ascending or divaricate; flowers bright yellow; fruits broadly elliptic, glabrous 7–10 mm. long, 4–5 mm. broad, the lateral ribs with straw colored wings 1–1.5 mm. broad, the dorsal ribs filiform; oil-tubes large dark-colored and prominent, 2 on the commissural side, 1 in each interval on the dorsal side; whole plant strongly and pleasantly scented with a strong odor of parsley. *Cogswellia fragrans* St. John. Dry gravelly slopes, Grand Canyon of the Snake, down as far as Wild Goose Bar, Asotin Co., *U. Son.*

Lomatium triternatum (Pursh) C. & R., forma **triternatum.** Tap-root 2–10 mm. in diameter, subcylindric, black; plant 2–7.5 dm. tall, acaulescent or short caulescent; leaves 1–3-ternate; petiole 2–20 cm. long, sheathing at base; leaflets 2–10 cm. long, mostly 1–7 mm. wide, linear; peduncle exceeding the leaves, it and the rays puberulent; involucre none; the 5–18 rays 1–5 cm. long; bracts of the involucel 3–4 mm. long, subulate; pedicels 2–7 mm. long; petals yellow; fruit 8–12 mm. long, elliptic, glabrous; lateral ribs with pale wings 1–1.3 mm. wide; oil tubes large, dark, 1 in the intervals, 2 on the commissural side. *Cogswellia triternata* (Pursh) M. E. Jones. Common, prairies or gravelly soil, *U. Son., A. T.*

Forma **lancifolium** (St. John) St. John. Differs in having the leaflets linear-lanceolate, 1 cm. or more in width. *Cogswellia triternata* (Pursh) M. E. Jones, forma *lancifolia* St. John.

Washington: Spokane, May 16 and July 26, 1896, *C. V. Piper* (type); Usk, Pend Oreille Co., May 15, 1923. *R. Sprague, A. T.*

Partly transitional material exists but these broad leaved plants are so extreme that they deserve to be separated as a form.

OROGENIA. Turkey Peas

Small, glabrous, acaulescent perennials; roots tuberous or fusiform; leaves ternate, the segments linear; involucre wanting; umbels compound; the rays unequal; the bracts of the involucels few, linear; flowers white; calyx-teeth minute; stylopodium depressed; fruit oblong, slightly flattened laterally, glabrous; mericarps much flattened dorsally, dorsal and intermediate ribs filiform, the lateral ribs and the middle of the commissure thick, corky, extending towards its mate; oil-tubes small, 3 in the intervals, 2–4 on the commissural side; seed face slightly concave. (Name from Gr., *oros,* mountain; *genos,* race.)

Orogenia linearifolia Wats. Plant 2–15 cm. tall; tubers 6–13 mm. in diameter, globose, the coats pale brownish; sheaths 3–6 cm. long, pale; petioles 3–6 cm. long; leaves 1–2-ternate; leaflets 5–35 mm. long, bright green; rays 1–25 mm. long, very unequal; bracts of the involucels 1–2 mm. long; pedicels 0.3–1 mm. long; petals obovate, white, 1 mm. long, with an acuminate, inflexed tip nearly as long; stamens exserted; anthers 0.2 mm. long, broadly oval, purple; fruit 3–3.5 mm. long, subglobose. Gravelly slopes, near snow banks, Blue Mts., and Moscow Mt., *Huds.*

Also reported from Kamiak Butte, *Warren* 902.

OSMORHIZA. Sweet Cicely

Glabrous to hirsute perennials; roots thick, aromatic; stems leafy; leaves ternately decompound; leaflets broad, ovate to lanceolate, variously toothed; involucre and involucels few-leaved or wanting; flowers white or purple, in few-rayed and few-fruited umbels; calyx-teeth obsolete; stylopodium conical, sometimes depressed; styles mostly short; fruit linear to linear-oblong, attenuate at base, obtuse, acute or beaked at apex, glabrous, or bristly on the ribs; carpels slightly flattened dorsally or not at all, nearly pentagonal in section, with equal ribs and thin pericarps; oil-tubes obsolete in the mature fruit (often numerous in young fruit); seed face from slightly concave to deeply sulcate. (Name Gr., *osme,* scent; *rhiza,* root.)

Fruit glabrous. *O. occidentalis.*
Fruit bristly,
  Fruit obtuse at apex, without a neck. *O. depauperata.*
  Fruit constricted into a short neck. *O. chilensis.*

Osmorhiza chilensis H. & A. Plants 2–7 dm. tall, sparsely hirsute, divaricate branched above; petioles of lower leaves 6–15 mm. long; leaves biternate; leaflets 3–10 cm. long, ovate, acute, often cuneate, coarsely dentate and often incised; umbels long peduncled; the 3–6 rays 3–10 cm. long; pedicels slender, finally 6–25 mm. long; petals 0.5–0.7 mm. long, obovate, white; fruit 15–20 mm. long, including the short beak and the slender, bristly, stalk-like base; mericarps early separating, the ribs strigose hispid, especially at base; stylopodium conic. *O. divaricata* (Britton) Robins. & Fern.; *O. brevipes* (C. & R.) Suksd.; *Washingtonia divaricata* Britton. Moist woods, *Can.*

Osmorhiza depauperata Phil. Plant 1.5–7 dm tall, sparsely hirsute; petioles of basal leaves 5–15 cm. long; leaves biternate; leaflets 2–6 cm. long, ovate to lanceolate, acute or acuminate, coarsely dentate and incised; involucre none; involucels usually none; the 3–5 rays 2–5 cm. long, divergent; pedicels 12–25 mm. long, divergent; petals 0.4–0.6 mm. long, white; fruit 11–16 mm. long, including the hispid bristly, stalk-like base, the ribs sparsely bristly; stylopodium depressed or low conic. *O. obtusa* (C. & R.) Fern.; *Washingtonia obtusa* C. & R. Loamy woods, Mt. Carleton [Mt. Spokane] *Kreager* 267.

Osmorhiza occidentalis (Nutt. ex T. & G.) Torr. Plant 6–12 dm. tall; stems glabrous or early glabrate, simple or with few ascending branches; nodes pilose; rootstock woody, stout, dark, the pith with scalariform partitions; petioles of basal leaves 6–35 cm. long, puberulent or glabrous, expanded, sheathing at base; leaves 2–3-pinnately ternate; leaflets 2–10 cm. long, closely puberulent, sharply serrate dentate or doubly so, occasionally incised at base; umbels 5–12-rayed, the rays 1–10 cm. long, the central ones the shorter, bearing only staminate flowers, the marginal ones having their umbelets with the numerous central flowers all staminate but the marginal ones perfect; involucre 0 or rarely 1–2 bracts; pedicels 2–10 mm. long; involucels 0; staminate flowers with the petals 0.5–0.7 mm. long, broadly obovate, emarginate, white; the stamens equaling the petals; the ovary abortive; the stylopodium discoid; the styles and stigmas 0; perfect flowers with ovary exceeding the corolla; petals 1–1.8 mm. long; stylopodium discoid at base, low conic to the styles; fruit 13–19 mm. long, oblong-linear, but narrowed to a stout beak 2 mm. long, olive-black, shining; mericarps pentagonal in cross-section, the ribs prominent, the commissural face deeply grooved. *Washingtonia occidentalis* (Nutt.) C. & R.; *Glycosma occidentalis* Nutt. ex T. & G. Moist woods, *A. T. T.*

PTERYXIA

Glabrous perennials, clothed at the base by the persistent leaf-sheaths; leaves bright green or somewhat pale, 2–4-pinnatisect into short linear segments; involucre mostly none; involucels of narrow bractlets; flowers yellow or white; calyx-teeth evident; umbels compound; fruit ovate to ovate-oblong, glabrous; carpel usually strongly flattened dorsally, the ribs winged; stylopodium wanting; oil-tubes 3–12 in the intervals, 5–20 on the commissural side; seed-face plane or with a shallow and broad cavity. (Name apparently from Gr. *pteron,* wing; *oxus,* sharp.)

Pteryxia terebinthina (Hook.) C. & R., var. **foeniculacea** (Nutt. ex T. & G.) Mathias. Tap-root stout, brown, woody; plant short caulescent, 1–5 dm. tall; herbage with strong resinous odor; petioles 5–18 cm. long, sheathing; blades 3–16 cm. long, 3–10 cm. broad, 2–4-pinnatisect, the ultimate segments linear, 1–4 mm. long, 0.2–0.5 mm. broad, acute, mucronulate, green; peduncles exceeding the leaves; umbels with 7–12 unequal rays 4–40 mm. long; involucel bracts lanceolate, about equaling the pedicels; pedicels 1–3 mm. long; calyx-teeth lanceolate, nearly equaling the petals; petals pale yellow, the body 0.8–1 mm. long, obovate, the lanceolate inflexed tip ⅔ as long, the midvein swollen glandular; anthers 0.2–0.3 mm. long, elliptic, well exserted; styles twice the length of the petals; fruit 5–10 mm. long, 3–4 mm. broad, oval; mericarps with lateral ribs with firm wings 0.5–0.7 mm. wide, the dorsal wings narrower; oil-tubes 3–12 in the intervals, 6–20 on the commissural side. *P. thapsoides* (Nutt.) C. & R. Rocky slopes, *U. Son., A. T.*

It appears specifically distinct, but the monographer, Mathias, reports intermediate specimens, including *Kreager* 119 from Clarks Springs.

The herbage is grazed by sheep.

SANICULA. Sanicle

Perennial, rather glabrous herbs, with ternate or palmate leaves; flowers perfect and staminate mixed in heads in few-rayed umbels, yellow or green; bracts of the involucres and involucels few; calyx-teeth evident, persistent; fruit globular, the carpels not separating, ribless, the whole surface covered with hooked bristles; oil-tubes 3 to many, commonly 3 dorsal and 2 commissural; commissural side plane or sulcate. (Name Lat., a diminutive from *sanare,* to heal.)

Sanicula nevadensis Wats. var. **septentrionalis** (Greene) Mathias. Plant 1–4 dm. tall. glabrous; root dark, a slender or tuberous thickened tap-root; basal leaves ternate or biternate, the petioles 3–8 cm. long, broad sheathing at base; the blade 2–4 cm. long, the division ovate, coarsely serrate or cleft, firm; upper leaves reduced, less divided, but more sharply serrate or laciniate; an elongate, ascending peduncle from all but the lowest nodes; involucre foliaceous, pinnately laciniate, equaling the flowers; rays 3–6, very unequal, in anthesis very short, in fruit the longer becoming 2.5–8 cm. long; involucels 0.5–2 mm. long, the bracts oblong or lanceolate more or less united; staminate flowers numerous, the pedicels 0.5–1 mm. long; calyx 0.6–0.7 mm. long campanulate, the lance-ovate lobes half as long; petals 1.2–1.5 mm. long, yellow, obovate, 2-lobed, with a prominent midrib up to the sinus, clawed; anthers 0.5 mm. long, ellipsoid, exserted; pistillate flowers 3–5, sessile, the petals 1 mm. long, narrower; styles exserted; body of mericarps 2–3 mm. long, broadly ellipsoid, covered with numerous hooked prickles 1–1.5 mm. long; commissural side flat. *S. septentrionalis* Greene; *S. graveolens* Poepp. ex DC., var. *sepentrionalis* (Greene) St. John. Rocky slopes, Blue Mts., *Can.*

The root is strongly aromatic.

This does not seem to be identical or most closely related to *S. graveolens* Poepp. ex DC. of S. Am. Ours differs in shorter blades, wider leaf segments, staminate petals broader (1.2 mm.), and prickles of the fruit equal, longer, and more slender.

SIUM

Perennial, growing in water or in wet places; leaves pinnate; leaflets serrate or pinnatifid; involucres and involucels of numerous narrow bracts; flowers white; umbels compound, calyx-teeth minute; stylopodium depressed; styles short; fruit flattened laterally, ovate to oblong, glabrous; carpel with prominent corky nearly equal ribs; oil-tubes 1–3 in the intervals, never solitary in all the intervals, 2–7 on the commissural side; seed subangular, with plane face. (Name Gr., *sion,* a certain water plant.)

Sium suave Walt. *Water Parsnip.* Caudex short, larger than the stem, with scalariform partitions; plant 6–20 dm. tall; lower leaves sometimes submersed and finely dissected, the others pinnate, the upper subsessile, the others with petioles 3–4 dm. long, sheathing; leaflets 7–17, lanceolate or linear, 3–14 cm. long, serrate; involucral bracts leafy, lanceolate, often cleft, scarious margined at base; umbels with 12–33 rays, 1–5 cm. long; involucels 3–8 mm. long, lanceolate, scarious margined at base; pedicels 2–9 mm. long; petals 0.7–0.9 mm. long, the incurved lanceolate tip ¼ as long; anthers 0.2 mm. long, exserted; styles almost equaling the petals; stylopodia broad, cushion-like; fruits 2.5–3 mm. long; mericarps with ribs pale, the intervals dark brown. *S. cicutaefolium* Schrank. Wet streams or lake shores, *A. T.*

ZIZIA

Smooth perennials with simple to ternately compound leaves; involucre early deciduous; involucels of small bractlets; flowers yellow, in compound umbels; calyx-teeth persistent; stylopodium none; styles long; fruit somewhat flattened laterally, ovoid to oblong, glabrous; central fruit of each umbellet sessile; carpel with filiform ribs; oil-tubes large and solitary in the broad intervals, 2 on the commissural side and a small one in each rib; seed terete, longitudinally grooved beneath the oil-tubes; commissural side flat. (Named for *I. B. Ziz,* a Rhenish botanist.)

Zizia aptera (Gray) Fern., var. **occidentalis** Fern. *Alexanders.* Root dark, woody, with several tuberous branches; plant 1–9 dm. tall, simple or few branched; nodes few, puberulent; basal leaves except the earliest with petioles 7–20 cm. long, the blades 3.5–12 cm. long, ovate, cordate, or suborbicular, cordate at base, glabrous, membranous, crenate or crenate-serrate; cauline leaves smaller, short petioled to sessile, 3–5-parted, the divisions lanceolate, serrate and at times also incised; peduncles elongated in fruit; umbels in flower round-topped, in fruit concave, 7–19-rayed; involucre of a few lanceolate or elliptic bracts, early deciduous; outer rays in fruit 2–3.5 cm. long, puberulent at apex; umbelets dense; involucels 1–3 mm. long, the bracts lanceolate; pedicels 1–3 mm. long; calyx-teeth ovate, minute; petals yellow, apparently obovate 0.6–0.8 mm. long, but with an acuminate inflexed tip nearly as long, the midnerve prominent; anthers 0.2–0.3 mm. long, ellipsoid, well exserted; styles exserted; central umbelets with staminate flowers only; fruits 2.5–4 mm. long, oval in outline, glabrous; mericarps with the dorsal ribs yellowish, raised; commissural side flat. Thickets by streams, *A. T.*

CORNACEAE. Dogwood Family

Trees or shrubs, rarely herbs; leaves simple, alternate or opposite; flowers regular, perfect, polygamous, or dioecious, in cymes or heads; calyx-tube adherent to the ovary, 4–5-lobed; petals and stamens 4–5, on the margin of the epigynous disk in the perfect flowers; style 1; ovary inferior, 1–4-celled, with one ovule in each cell; fruit a 1–4-seeded drupe, or berry. (Named from the genus *Cornus.*)

CORNUS. Dogwood

Herbs, shrubs, or trees; leaves oposite, sometimes apparently whorled, simple, entire; flowers perfect, in a cyme or head-like cluster; calyx minutely 4-toothed; petals 4, oblong or ovate; stamens alternate, 4, with slender filaments; style 1; drupe ovoid or oblong, with a 2-celled 2-seeded stone. (The ancient Lat. name from *cornu,* horn.)

Herbs with head-like cymes with conspicuous involucre; berries red. — *C. canadensis.*
Shrubs; cymes looser, without an involucre; fruit white to blue,
  Leaves or cymes only appressed pubescent. — *C. stolonifera,* var. *stolonifera.*
  Leaves or cymes with spreading pubescence,
    Either cymes or lower leaf surface with spreading pubescence. — Var. *interior.*
    Both cymes and lower leaf surface shaggy with spreading pubescence. — *C. occidentalis.*

Cornus canadensis L. *Canada Dogwood; Bunchberry.* Rootstock woody, slender, freely branching, elongate; plant herbaceous, 5–30 cm. tall; stems simple or branched at base, above lepidote hairy or glabrate; at maturity middle and lower nodes naked or the middle with opposite bracts or much reduced leaves; summit of stem with several crowded nodes, the commonly

4–6 leaves appearing whorled; leaves lepidote pubescent; petioles 1–3 mm. long; blades 2–8.5 cm. long, oval, ovate, or obovate, acute at each end, beneath paler green and finally glabrate; peduncle 7–45 mm. long; the involucral bracts white, 4, the outer pair the larger, 1–2.5 cm. long, ovate or oval, acute at each end; cyme condensed; flowers subsessile; ovary appressed silvery hirsute; calyx-lobes minute, obtuse deltoid; petals 1–1.5 mm. long, broadly lanceolate, dark purple, with yellowish base; anthers 0.5–0.7 mm. long, oblong; style 1 mm. long, stout; berries 4–6 mm. in diameter, globose; stone 2–3.2 mm. long, ovoid, yellow. *Chamaepericlymenum canadense* (L.) Asch. & Graebn. Moist woods, *Can.*

Cornus occidentalis (T. & G.) Cov. Differs from *C. stolonifera* in having the inflorescence shaggy villous, and the blades loosely pilose beneath. *C. californica* C. A. Mey., var. *pubescens* (Nutt.) Jepson. Woods, Spokane, *A. T. T.*

Cornus stolonifera Michx., var. **stolonifera.** *Red-osier Dogwood.* Shrub 1–3 m. tall, usually stoloniferous; young twigs yellow or green, with appressed 2-branched hairs; older branches with bark red or purplish, glabrous; leaf-scars broad V-shaped; buds puberulent, flattened; plant with appressed 2-branched hairs; leaves opposite; petioles 3–20 mm. long; blades 2–12 cm. long, lance-ovate to oval, short acuminate, whitened beneath; cymes many-flowered, rounded, 2–7 cm. across; ovary canescent with appressed, 2-branched hairs; calyx-lobes 0.2–0.4 mm. long, caducous; petals 2.5–4 mm. long, lanceolate, white; disk prominent; anthers 1.2–1.8 mm. long, oblong; berries 6–8 mm. in diameter, globose, white, lead-colored, or blue; stone 3–6 mm. long, flattened, depressed globose in outline with a short blunt point, furrowed, yellowish. *C. alba* L., ssp. *stolonifera* (Michx.) Wangerin; *Svida instolonea* A. Nels.; *C. sericea* L., forma *stolonifera* (Michx.) Fosb. Abundant, stream banks and woods, *U. Son* to *A. T. T*

The foliage is bitter but is grazed by sheep. The shrub is often cultivated as an ornamental, the bright twigs being attractive in winter.

Var. **interior** (Rydb.) St. John. Differing from the species in having the inflorescence or the lower leaf surface or both somewhat loosely pilose as well as appressed puberulent with 2-branched hairs.; *Suida* (Svida) *interior* Rydb.; *C. interior* (Rydb.) Druce; *C. stolonifera* Michx., forma *interior* (Rydb.) Rickett.

The plants included here form numerous and direct intergrades to *C. occidentalis.*

Rickett has concluded that *C. occidentalis* and *C. stolonifera* hybridize freely (see Brittonia 5: 149–159, 1944). His method of reducing all minor genetic taxa to formae is not concurred with here, any more than the opposite practice now common among western American botanists of making them all subspecies.

ERICACEAE. Heather Family

Perennial herbs or shrubs; leaves simple, commonly alternate, articulated to the stem, without stipules; flowers regular, or nearly so; corolla generally gamopetalous, 4- or 5-lobed or choripetalous; stamens as many or twice as many as the corolla lobes or petals, free from the corolla; anthers often opening by a terminal pore and with 2 awn-like appendages; style single; ovary superior or inferior, with as many or twice as many cells as the corolla lobes, or rarely less; seeds small; endosperm fleshly. (Named from the genus *Erica.*)

Ovary superior,
  Petals distinct; herbs,
    Parasitic leafless plants without chlorophyll. *Monotropa.*
    Herbs with rootstocks, mostly with green color and not parasitic,
      Filaments hairy and dilated near the base. *Chimaphila.*
      Filaments glabrous, subulate. *Pyrola.*
  Petals more or less united,
    Herb, parasitic, without chlorophyll. *Pterospora.*
    Shrub, autophytic, with chlorophyll,
      Fruit a berry-like drupe. *Arctostaphylos.*
      Fruit a capsule,
        Leaves linear, evergreen; buds not scaly. *Phyllodoce.*
        Leaves broader, thin, deciduous; buds scaly. *Menziesia.*
Ovary inferior. *Vaccinium.*

ARCTOSTAPHYLOS

Leaves alternate, usually coriaceous, persistent, entire or with a few irregular teeth; flowers small, nodding, pink or white, in terminal racemes or clusters; calyx free from the ovary; corolla ovoid or urn-shaped, with 5 short teeth; stamens 10; anthers with two reflexed awns on the back; drupe with 4–10 seed-like nutlets. (Name from Gr., *arktos,* bear; *staphule,* grape.)

Arctostaphylos Uva-ursi (L.) Spreng., var. **coactilis** Fern. & Macbr. *Bearberry, Kinnikinnick.* Prostrate, trailing or matted shrub; roots slender, brown; branches 2–8 dm. long, freely branching, leafy well back from tip; old stems with the smooth bark brown or reddish, exfoliating; branchlets greenish to reddish, non viscid, with a persistent canescent-tomentum or somewhat appressed pilosity; winter buds scaly, the outer scales ovate, ciliate; petioles 1–3 mm. long, pilosulous; blades 11–30 mm. long, coriaceous, evergreen, spatulate to obovate, cuneate, obtuse or emarginate at apex, pilosulous on the margin; racemes nodding, shorter than the leaves, 3–9-flowered; bracts heavy, clasping, oblong-lanceolate, obtuse, pilosulous especially on the margins; calyx 2 mm. long, glandular swollen at base, cleft into ovate-suborbicular lobes, the margins hyaline, ciliate; corolla 4–6 mm. long, urceolate, white or pinkish, pilose within, the lobes rounded; filaments 1.5–2 mm. long, with an obovate, pilose dilation at base; anthers 0.5–0.8 mm. long, purple, ovate, with large subterminal pores, the filiform awns often twice as long as the anthers; style straight, included; fruit 5–10 mm. in diameter, globose, the pulp mealy, scarcely palatable; nutlets 3–3.5 mm. long, ovoid, somewhat flattened, 3-keeled, pilosulous. *Uva Ursi Uva-Ursi* (L.) Britton. Open places, or open woods, occasional on basaltic soils, abundant on acid or granitic soils, *A. T. T.*

The leaves contain the glucoside arbutin, gallic acid, and 6% tannin. They are tonic and diuretic. They are nibbled by deer, goats, and sheep. The Indians used them as a substitute or adulterant for tobacco. Grouse and bear are fond of the berries.

CHIMAPHILA. Pipsissewa

Low nearly herbaceous perennials; leaves irregularly opposite, whorled or scattered, coriaceous, short-petioled, serrate; flowers in corymbs or racemes on a leafy stem; sepals 5; petals 5, widely spreading, orbicular; stamens 10; filaments expanded into a disk near the base; anthers 2-horned; style very short, concealed in the umbilicate summit of the ovary; stigma large, peltate, with 5 narrow and conspicuous radiating lobes; valves of the 5-celled capsule naked, dehiscing from the apex. (Named from Gr., *cheima,* winter; *philos,* loving.)

Leaves broadest above the middle; floral bracts subulate, deciduous; anthers pink. *C. umbellata,* var. *occidentalis.*

Leaves broadest at or below the middle; floral bracts rounded, persistent; anthers yellow. *C. Menziesii.*

Chimaphila Menziesii (R. Br. ex D. Don) Spreng. Rhizome extensive; plant 5–22 cm. tall, glabrous; stems simple or few-branched at base, leafy throughout; some of nodes often crowded; leaves alternate; petioles 2–8 mm. long; blades 1–5 cm. long, ovate to lance-elliptic, serrate, above dark green, and the main veins white bordered, beneath pale green, subcoriaceous; peduncles 2–7 cm. long; inflorescence 1–2-flowered; bracts 3–4 mm. long, erose; pedicels 3–22 mm. long; calyx 5–6 mm. long, parted nearly to the base into ovate, erose lobes; petals 5–7 mm. long, oval to suborbicular, white to pink, palmately veined; stamens shorter than the petals; filaments with dilated portion 1.2–1.7 mm. wide, orbicular, villous; anthers 2.5–3 mm. long, oval, narrowed to the large terminal pores; capsules 5–7 mm. in diameter, depressed globose, with 5 rounded lobes; seeds 0.5–0.7 mm. long, the body ellipsoid, brown, long tailed at each end. Deep woods, *Can.*

Chimaphila umbellata (L.) Bart., var. **occidentalis** (Rydb.) Blake. *Pipsissewa.* Rootstock strong, extensive; plant 1–3.5 dm. tall; stems erect, simple or few branched at base, more or less puberulous, the nodes not crowded, the leaves 3–8 in a whorl; petioles 1–7 mm. long; blades 2–9 cm. long, 1.1–2 cm. broad, oblanceolate, acute, coriaceous, above dark shiny green, beneath yellowish green, each margin with 10–18 serrations, the lateral veins obscure; peduncles 3–8 cm. long; raceme 3–7-flowered, puberulous; pedicels 7–28 mm. long, recurved in flower, erect in fruit; calyx-lobes 1–1.5 mm. long, deltoid-oval, erose, ciliolate; petals 4–6 mm. long, oval, pink, ciliolate; stamens shorter than the petals, disc of filaments 1–1.2 mm. in diameter, puberulent near the margin; anthers 2.3–2.8 mm. long, cuneate, the sacs divergent to the nozzle-like pore; stigma 2.3–2.6 mm. in diameter; capsule 5–8 mm. in diameter; depressed globose, 5-lobed; seeds 0.6–0.8 mm. long, the body minute, ellipsoid, brownish, long-tailed at each end. *C. occidentalis* Rydb.; *C. umbellata,* subsp. *occidentalis* (Rydb.) Hultén. Woods, *A. T. T.*

The type was from Pine Cr., near Farmington, Latah Co., Ida., *Sandberg et al.* 519.

MENZIESIA

Shrubs with alternate, hairy leaves; flowers small, nodding, white, greenish or purple, in terminal corymbs or umbels, developed with the leaves; calyx small or minute, flattish, 4-toothed or lobed; corolla cylindrical to urn-shaped, 4-lobed; stamens 8; capsule ovoid, woody, 4-celled, septicidal, 4-valved, many-seeded. (Named for *Dr. Archibald Menzies,* Scotch botanist and surgeon on Capt. Vancouver's voyage.)

Menziesia ferruginea Sm. *Fool's Huckleberry.* Shrub 1–3 m. tall, freely branching; branches often falsely whorled; bark brown, becoming pale and exfoliating in long strips; buds scaly, puberulent; leaves clustered at branch tips; petioles 1–6 mm. long, puberulent or hirsute; blades 2.5–6.5 cm. long, oblanceolate or elliptic, cuneate, apiculate, sparsely glandular hirsute, beneath pale or glaucous, the margin ciliolate, towards the tip serrulate; pedicels 15–35 mm. long, glandular-puberulent or -hirsute or both; calyx 1–1.5 mm. long, with short, rounded lobes, ciliate; corolla 6–11 mm. long, brownish ovoid-campanulate, the lobes ovate or rounded, shallow; stamens 1.5–2 mm. long, brown, linear tapering to the tubular tips; capsule 5–7 mm. long, ovoid, glabrous or sparsely puberulent; seeds 2–3 mm. long, brownish, spindle-shaped, often tailed at one or both ends. *M. ferruginea,* var. *glabella* (Gray) Peck; *M. glabella* Gray. Moist woods, *Can.*

The plant is somewhat fetid. It is unpalatable; it is somewhat poisonous to sheep. *M. glabella* is reduced to synonymy, as the various alleged differences in seeds, sepals, petioles, leaf shape, and size of corolla all intergrade or recombine.

MONOTROPA

Saprophytic herbs, without chlorophyll, from a ball of fibrous roots; the whole plant white to pink, or in age brown to black; stems erect, covered with small scale-like leaves; sepals 2–5 or none, bract-like; petals 4–6, scale-like, saccate at base, tardily deciduous; stamens 8, 10, or 12; filaments subulate; anthers peltate, opening on 2 transverse lines; style columnar; stigma discoid, 4–5-rayed; ovary 4–5 celled; capsule loculicidal, the placentae axile below, parietal above; seeds minute, with a loose testa. (Name from Gr., *monos,* single; *tropos,* a turn, alluding to the reflexed terminal flower.)

Monotropa uniflora L. *Indian-pipe.* Stems simple, often clustered, 1–3 dm. tall; leaves 5–10 mm. long, scale-like; flowers terminal, nodding, but the stem straightening and bearing the fruit erect; sepals present and 14–18 mm. long, oblong or oblong-spatulate, glabrous or ciliate towards the base, and erose towards the apex; petals oblong-spatulate, longer than the sepals, puberulent within; filaments pubescent; capsule 10–15 mm. long, obtusely angled.

In deep humus of cool woods, Moscow Mt., *Can.*

Daubenmire (1939) recorded this species as "not uncommon in our region." But, I cannot confirm or understand his statement that the species was included in the previous book by Piper and Beattie (1914).

PHYLLODOCE

Low alpine heath-like undershrubs; leaves alternate, numerous, obtuse, serrulate; flowers solitary or in umbels at the ends of the branches; calyx 4–6-lobed, free from the ovary; corolla 4–6-toothed; stamens 8–12; anthers pointless, shorter than the filaments; fruit a 5-celled, 5-valved, many-seeded dry capsule. (*Phyllodoce,* a Gr. nymph.)

Phyllodoce empetriformis (Sm.) D. Don. *Pink Heather.* Shrub 15–50 cm. tall, much branched, the branches ascending, broom-like; leaves subsessile,

4–13 mm. long, linear, obtuse, the midrib prominent beneath; umbels terminal, 3–15-flowered, glandular puberulent; pedicels 8–30 mm. long; calyx 1.8–2.5 mm. long, cleft nearly to the base into ovate, ciliolate lobes; corolla 5–8 mm. long, campanulate, commonly magenta-pink, the 5 broadly ovate lobes shorter than the tube, reflexed; stamens included, the filaments pilosulous, the anthers 1.5–1.8 mm. long, oblong; style 5–6 mm. long; stigma hemispheric, with 5 short, erect lobes; capsule 2.5–3.5 mm. long, oval-subglobose, capitate glandular puberulent; seeds 0.4–0.5 mm. long, yellowish brown, ellipsoid, sinuous striate. Moist ground, high peaks, Blue Mts., *Horner* 310; Table Rock, Columbia Co., *Constance et al.* 1,254, *Huds.*

A lovely plant of the high mountains. The color of the flower changes with the aging of the flower, and varies on different plants.

PTEROSPORA

Stout simple purplish-brown viscid-pilose root-parasitic herbs; stem wand-like, with scattered lanceolate scales toward the base, in place of leaves; flowers many, nodding, white, in a long bracted raceme; calyx 5-parted; corolla urn-shaped, 5-toothed; stamens 10; anthers 2-celled, awned on the back; stigma 5-lobed; capsule globose, flattened, 5-lobed, 5-celled; seeds very numerous. (Name from Gr., *pteron,* wing; *spora,* seed.)

Pterospora andromedea Nutt. *Pine Drops.* Roots in a rounded mass up to 7 cm. in diameter; plant 3–13 dm. tall; bracts 1–3.5 cm. long; raceme 1–8 dm. long, many-flowered; floral bracts linear, shorter or as long as the pedicels; pedicels 8–20 mm. long, the tips deflexed; calyx 3–6 mm. long, deeply parted, the lobes lanceolate or ovate-lanceolate, glandular ciliate; corolla 4–6 mm. long; anthers 0.5 mm. long, the awns 1.2–1.5 mm. long; capsule 8–15 mm. in diameter; seeds flat, the body, 2 mm. long, straw-colored, lanceolate, the wings deeply obcordate, each side broader than the seed body. Moist evergreen woods, *A. T. T., Can.*

PYROLA

Low and smooth perennial herbs; leaves roundish, petioled, evergreen, basal or nearly so; flowers nodding, in a simple raceme, on a more or less scaly-bracted scape, or single; corolla of 5 concave petals; stamens 10, the filaments slender; style elongate; stigma 5-lobed or toothed; ovary 4- or 5-celled; ovules very numerous; fruit a flattened globose 5-lobed 5-celled loculicidal capsule. (Name Lat., a diminutive of *Pyrus.*)

Style straight,
  Flower single; petals rotate, 8–12 mm. long. *P. uniflora.*
  Flowers in racemes; flower cup-shaped; petals 4–5 mm. long,
    Style 1–1.5 mm. long; raceme not secund. *P. minor.*
    Style 2–6 mm. long; raceme secund. *P. secunda.*
Style deflexed,
  Green leaves none or very rudimentary; petals pink or greenish white. *P. picta,* forma *aphylla.*

Green leaves present,
  Blades white mottled; petals greenish or reddish. — *P. picta*, f. *picta*.
  Blades green,
    Petals greenish; calyx-lobes broader than long. — *P. virens*.
    Petals pink; calyx-lobes longer than broad,
      Blades firm chartaceous, mostly orbicular. — *P. asarifolia*, var. *incarnata*.
      Blades subcoriaceous, mostly ovate or elliptic. — Var. *bracteata*.

Pyrola asarifolia Michx., var. **bracteata** (Hook.) Jepson. *Wild Beet.* Rootstock cord-like, the apex dark scaly; plant 1.8–5 dm. tall, glabrous; stem leafy and with brown to black scales at base; scape with 1–3 brown, lanceolate bracts 13–20 mm. long; petioles 1.5–9 cm. long; blades 3–8.5 cm. long, denticulate by the excurrent vein tips, mostly ovate or elliptic, acute, cuneate, the lower ones occasionally suborbicular, or cordate at base; raceme 8–23-flowered; bracts 7–17 mm. long, lanceolate, conspicuous, rosy-tinged; pedicels 3–9 mm. long, the tips deflexed; calyx-lobes 3–3.5 mm. long, lanceolate or lance-deltoid; petals 5–8 mm. long, oval, rose-magenta; stamens included; anthers 2.8–3.2 mm. long, pink, oblong-lanceoloid, abruptly narrowed to the pores, at the distal end apiculate; style 6–8 mm. long, nearly uniform; stigma conical, obtuse; capsules 4–6 mm. in diameter, depressed globose, 5-lobed; seeds 0.2–0.3 mm. long, brownish, fusiform, the embryo ellipsoid, with long translucent tail of the seed coat at each end. Moist woods, *A. T. T., Can.*

Var. **incarnata** (DC.) Fern. Resembling var. *bracteata,* but the blades 2–8 cm. long, firm characeous, crenate, obtuse, mostly orbicular, occasionally oval; bracts of the raceme seldom equaling the pedicels; anthers 2–2.5 mm. long. *P. elata* Nutt.; *P. uliginosa* T. & G.

Pyrola minor L. *Lesser Wintergreen.* Perennial, from slender rootstock; plant 7–30 cm. tall, glabrous; petioles 1–3 cm. long; blades 1–4 cm. long, oval or orbicular, firm chartaceous, crenulate to subentire, above dark green, beneath paler; raceme 4–17-flowered, often somewhat secund; bracts lanceolate, exceeding the pedicels; pedicels 2–5 mm. long, deflexed; calyx-lobes 1–1.8 mm. long, ovate-deltoid; petals 4–5 mm. long, suborbicular, white or pinkish, not opening wide; stamens included; anthers 1 mm. long oblong-oval, unevenly truncate at the wide pores; stigma discoid, deeply 5-lobed; capsule 4–6 mm. in diameter, depressed globose, deeply 5-lobed; seeds 0.5–0.6 mm. long, brown, the body orbicular, minute, and with a long tail of the seed coat at each end. *Erxlebenia minor* (L.) Rydb.; *Braxilia minor* (L.) House. Rare, woods, Davis Ranch [Mt. Spokane], *Kreager* 208, *Can.*

Pyrola picta Sm., forma **picta.** *Variegated Wintergreen.* Rootstock slender, creeping; plant 10–37 cm. tall, glabrous; true stem 1–3 cm. long, with numerous lanceolate bracts; scape with 0–3 smaller bracts; petioles 5–60 mm. long; blades 2–8.5 cm. long, lanceolate to ovate, subcoriaceous, entire or denticulate, veins white margined, beneath pale green or purplish; raceme 8–26-flowered; bracts lance-ovate, purplish, shorter than the pedicels; pedicels 2–9 mm. long, recurved; calyx-lobes 1.5–2 mm. long and broad, broadly ovate, acute; petals 6–7 mm. long, oval, greenish, or purplish tinged; anthers 2.8–3.8 mm. long, purplish, oblong lanceoloid, narrowed to distinct tubes with a large pore; style 3–8 mm. long; stigma erect, shallowly lobed; capsule 4–6 mm. in diameter, depressed globose, 5-lobed; seeds 0.5–0.7 mm. long, the embryo ellipsoid, brown, with a pale, translucent, broad tail of the seed coat at each end. Moist woods, *A. T. T., Can.*

Forma **aphylla** (Sm.) Camp. *Leafless Wintergreen.* A leafless parasite; rootstock usually erect and simple; plant 1–5 dm. tall, glabrous; stems often red, erect, usually several from the base, with numerous lanceolate, entire or dentate bracts at base, these occasionally foliaceous, above the scape 1–4 bracted; raceme 5–27-flowered, the bracts firm, lanceolate, pinkish, about equaling the pedicels; pedicels 3–9 mm. long, recurved; calyx-lobes 1.5–2.5 mm. long, ovate, acute, to deltoid; petals 6–8 mm. long, oval or obovate-oblong; anthers 2.5–3.5 mm. long, pink, lanceoloid, ending in distinct tubes with a large pore, the distal end apiculate; style 4–8 mm. long; stigma low conic, truncate, shallowly lobed; capsule 6–8 mm. broad, depressed globose, 5-lobed; seeds 0.3 mm. long, brown, the minute ellipsoid embryo with a long transparent tail of the seed coat at each end. *P. aphylla* Sm. In rich humus in woods, *A. T. T., Can.*

Pyrola secunda L. *One-sided Wintergreen.* Rootstock slender, branching; plant 8–30 cm. tall, glabrous; stem 1–10 cm. long, decumbent, leafy; petioles 2–20 mm. long, alternate; blades 1–5 cm. long, thick chartaceous, ovate or oval, acute, finely serrulate, dark green above, paler beneath; raceme 6–30-flowered; bracts lanceolate, about equaling the pedicels; pedicels 1–7 mm. long, spreading, later reflexed; calyx-lobes 0.5–1 mm. long, ovate, erose ciliolate; petals 4–5 mm. long, ovate or oval, erose, greenish white, with 2 tubercles on the base; disk small, 10-lobed; stamens included; anthers 1.2–1.7 mm. long, tan-colored, papillose, oblong, slightly cuneate, the pores smaller than the apex; stigma hemispheric, grooved; capsule 3–4 mm. in diameter, depressed globose, 5-lobed; seeds 0.2–0.3 mm. long, yellowish brown, the minute ellipsoid embryo central. *Ramischia secunda* (L.) Garcke; *Orthilia secunda* (L.) House. Deep woods, *A. T. T., Can.*

Pyrola uniflora L. *Single Beauty.* Rootstock slender; plant 4–15 cm. tall, glabrous; stem above ground 1–3 cm. long, the crowded nodes with usually 3 leaves, the 1–4 whorls all appearing to be in one whorl; petioles 2–15 mm. long, narrowly winged by the decurrent blade; blades 1–3 cm. long, orbicular or rounded oval, usually rounded at apex, firm, evergreen, the margin crenate or rounded dentate; scape few-bracted; flower deflexed, fragrant; calyx 2–3 mm. long, parted nearly to the base into ovate, ciliolate lobes; petals rhombic-ovate, white, minutely ciliolate; stamens mostly in groups of 2–3; filaments S-shaped, shorter than the petals; anthers oblong, the sacs 2 mm. long, the tubes cylindric, ⅓ as long; style 2–4 mm. long, stout; stigma discoid expanded, with 5 erect lobes; capsule 5–8 mm. long, depressed globose, 5-lobed, dehiscent from apex down; seeds 0.5–0.7 mm. long, brownish, fusiform, pale seed coat tailed. *Moneses uniflora* (L.) Gray. Deep woods, in the mountains, *Can.*

Though *Moneses* is customarily maintained in America, its differences seem too unimportant to be generic. Hence the writer reverts to the usage of the Europeans, as of Engler and Prantl, and restores this species to the genus *Pyrola.*

Pyrola virens Schweigg. *Green Wintergreen.* Rootstock slender, branching; plant 8–32 cm. tall, glabrous; stem above ground very short, occasionally leafless; petioles 8–60 mm. long, alternate; blades 8–35 mm. long, oval, ovate, or orbicular, more or less crenulate, subcoriaceous, above dark green, beneath paler; scape few bracted; raceme loosely 3–12-flowered; bracts lanceolate, shorter than the pedicels; pedicels 3–7 mm. long, deflexed at tip; calyx-lobes 0.5–1.5 mm. long, deltoid to ovate; petals 5–6 mm. long, oval; stamens included; anthers 2.3–3 mm. long, including the 0.5 mm. tubes, oblanceolate, yellow, the distal end mucronate; style 5–9 mm. long, thickened at apex; stigma with 5 ascending lobes; capsule 5–7 mm. in diameter, depressed globose, 5-lobed; seeds 0.4–0.5 mm. long, brownish, the embryo ellipsoid with translucent flat tails of the seed coat at each end. *P. chlorantha* Sw.; *P. chlorantha,* var. *saximontana* Fern. Moist woods, *A. T. T., Can.*

The writer is unable to maintain the var. *saximontana.* He has studied 22 sheets from the Pacific Northwest, and they show every transition in the same collection and on the same plant. Hence only the species is maintained.

VACCINUM. Huckleberry

Branching shrubs with alternate leaves, sometimes coriaceous; flowers solitary or in racemes or clusters; corolla various in shape, epigynous, 4- or 5-cleft or -parted; stamens 8 or 10; anthers sometimes 2-awned on the back; fruit a 4- or 5-celled or sometimes 8–10-celled edible berry; seeds numerous. (The classic Lat. name of *V. Myrtillus.*)

Berry bluish-black with a bloom, 8–12 mm. in diameter; anther body 1–1.3 mm. long; shrub 0.5–3 m. tall. *V. membranaceum.*
Berry of another color, 2–5 mm. in diameter; anther body 0.5–0.9 mm. long; shrub 1–4 dm. tall,
- Branchlets terete, puberulent; leaves broadest above the middle; berries bright blue, with a bloom. *V. cespitosum.*
- Branchlets with several sharp angles, glabrous; leaves broadest at or below the middle; berries wine-colored. *V. scoparium.*

Vaccinium cespitosum Michx. *Dwarf Bilberry.* Bush 1–4 dm. tall, erect, bushy branched; tap-root brown, woody, deep, not thickened; older branches brown, smooth; branchlets green; buds ovoid, compressed, smooth, purplish or brown; leaves 5–36 mm. long, oblanceolate to spatulate-obovate, cuneate, acute or obtuse, serrulate, glabrous; flowers drooping, solitary in the axils; pedicels 1–4 mm. long, recurved, glabrous, glaucous; calyx-limb 0.1–0.4 mm. long, membranous, usually lobed nearly to the base into rounded lobes; corolla 3.5–5 mm. long, ellipsoid to ovoid, pinkish; filaments glabrous, equaling or exceeding the anthers; anthers with the body 0.5–0.8 mm. long, oblong, the tubes and awns terminal opposite each other and divergent, the tubes 0.8–1.3 mm. long, the rigid awns 0.5–1.3 mm. long, scabrous; style equaling the corolla; berry 3–5 mm. in diameter, sweet; seeds 0.5–1 mm. long, brownish, asymmetric ellipsoid, flattened, striate. Moist hillsides, and thickets, *A. T., A. T. T.*

The writer uses the original spelling *V. cespitosum.* Michaux presumably coined his specific name from *cespes,* one of the accepted forms of the Latin word, so a change from his spelling is contrary to rule.

Vaccinium membranaceum Dougl. ex Torr. *Huckleberry.* Shrub nearly glabrous, erect, the branches wide-spreading; branchlets slightly angled, brown, turning gray, puberulous or nearly glabrous; buds lanceolate, smooth; petioles 1–2 mm. long; blades 1–6 cm. long, ovate or lanceolate, acute, thin chartaceous, serrulate, somewhat puberulent on the veins, paler green beneath; flowers axillary, solitary; pedicels 2–9 mm. long, recurved; calyx-limb a truncate, entire collar, 0.5–1 mm. wide; corolla 4–5 mm. long, ovoid to subglobose, pinkish to yellowish white; filaments glabrous, shorter than the anthers; anthers with the body lanceoloid tapering to the tubes 0.5–1 mm. long, the divergent awns 1 mm. long; style exserted; seeds 0.8–1.3 mm. long, yellow or brownish, obliquely ovoid, compressed, striate. *V. macrophyllum* (Hook.) Piper.

The favorite huckleberry of the mountain woods, gathered and eaten fresh or preserved by Indians and whites.

Vaccinium scoparium Leiberg. *Grouseberry.* Shrub 1–3 dm. tall, much branched and bushy, glabrous; tap-root brown, scarcely thickened; older branches terete, brown; branchlets bright green; buds lance-ovoid, brown; leaves subsessile, 5–13 mm. long, ovate or oval, serrulate, bright green; flowers single in the axils; pedicels 1–3 mm. long, recurved; calyx-limb 0.2–0.5 mm. wide, membranous, undulate or slightly lobed; corolla 3–3.5 mm. long, white or pink-tinged, ovoid-urceolate; filaments glabrous, shorter than the anthers; anthers with the body 0.7–0.9 mm. long, oblong, the slender divergent tubes as long, the rigid erect awns nearly as long or half as long; style straight, included; berry 2–5 mm. in diameter; seeds 0.7–1.1 mm. long, asymmetric ovoid, compressed, brown, striate. *V. Myrtillus* L., var. *microphyllum* Hook. Open summit ridges, Blue Mts., *Huds.*

PRIMULACEAE. Primrose Family

Herbs or shrubs; leaves simple, mostly entire, or lobed, alternate, opposite, whorled, or in a basal cluster, without stipules; flowers mostly regular, perfect; corolla gamopetalous, 4–8-, usually 5-lobed or -cleft; stamens as many as and opposite the corolla-lobes, epipetalous; ovary superior or partly inferior, 1-celled, with a free central placenta rising from the base; fruit a capsule; seeds several or many; endosperm abundant, horny. (Named from the genus *Primula.*)

Leaves basal; corolla lobes reflexed. *Dodecatheon.*
Leaves or some of them cauline; corolla lobes erect or spreading,
  Capsule dehiscing lengthwise into valves; flowers not single and subsessile in the axils,
    Normal leaves clustered at apex of stem; corolla lobes 6. *Trientalis.*
    Leaves scattered; corolla lobes usually 5. *Lysimachia.*
  Capsule circumscissile; flowers single and subsessile in the axils. *Centunculus.*

CENTUNCULUS

Small annuals with leafy stems; leaves alternate, entire; flowers small, inconspicuous, calyx 4–5-parted, persistent; corolla rotate, with a short tube, 4–5-cleft, usually withering on the summit of the pod; stamens 4–5; filaments beardless, distinct; stigma capitate; ovules numerous. (Name Lat. *centunculus,* a little patch; also used by Pliny for a diminutive plant.)

Centunculus minimus L. *Chaffweed.* Roots fibrous, small; plant 3–7 cm. tall, glabrous or nearly so; stem erect, simple, or with decumbent branches, leafy throughout; leaves subsessile, 3–6 mm. long, obovate to oval, cuneate; flowers sessile or short pediceled; calyx 2–3 mm. long, the lobes lance-linear; corolla 1–2 mm. long, white or pink, parted to the middle into 4–5 lobes, the tube subglobose, the lobes elliptic-lanceolate, acute; stamens 4, included; capsule 1.5–2 mm. long, globose; seeds 0.3–0.5 mm. long, oval, 3-sided, reddish-brown, papillose. Wet places, near Spokane, *Piper, A. T.*

DODECATHEON. Bird Bill

Perennial smooth or viscid-puberulent stemless herbs; flowers showy, nodding, solitary or in an umbel on a scape, with an involucre; calyx deeply 5-cleft; corolla with a very short tube, 5-parted, purple, pink, or white; filaments distinct and short or united into a tube; style filiform, exserted; capsule ovoid or oblong, many-seeded, in some circumscissile. (Named from Gr., *dodeka,* twelve; *theoi,* gods.)

Filaments united into a tube about ½ length of anthers; capsule dehiscing from apex into valves,
  Plant glabrous. — *D. pulchellum.*
  Plant glandular puberulent at least on scape and pedicels,
    Corolla deep rose to lavender. — *D. Cusickii,* f. *Cusickii.*
    Corolla white. — Forma *album.*
Filaments scarcely united; capsule with a circumscissile lid,
  Plant glabrous. — *D. conjugens,* var. *conjugens.*
  Plant glandular puberulent. — Var. *viscidum.*

Dodecatheon conjugens Greene, var. **conjugens.** Rootstock short, thick, vertical; plant 5–27 cm. tall, glabrous; petioles 5–50 mm. long; blades 1.5–10 cm. long, variable, from oval to lanceolate, oblanceolate, or linear-oblanceolate, thick, pale green; bracts of involucre few, lanceolate, foliaceous, thin margined; umbel 1–4-flowered; pedicels 3–55 mm. long, erect in fruit; calyx 5–8.5 mm. long, parted ½ or ⅔ the way down into lanceolate lobes; corolla tube pale, with a scalloped purple line; corolla lobes 11–22 mm. long, narrowly elliptic-oblong, purple; staminal tube scarcely evident, yellowish; anthers 4–7 mm. long, purple; capsule 7–13 mm. long, cylindric-lanceoloid, with a circumscissile lid, then splitting into usually 8 valves; seeds 0.9–1.2 mm. long, broadly ellipsoid, somewhat flattened, brown. *D. conjugens leptophyllum* (Suksd.) Piper; *D. conjugens* Greene, subsp. *viscidum* (Piper) sensu H. J. Thompson.

Thompson (1953) misapplied the name *viscidum* which must remain attached to the Spangle plant, the type.

Var. **viscidum** (Piper) H. L. Mason. Differs in being glandular puberulent throughout. *D. viscidum* Piper. Meadows or grassy hillsides, Spangle, *A. T.*

The type locality is ten miles west of Spangle, May 24, 1898, *Piper.*

Dodecatheon Cusickii Greene, forma **Cusickii.** Plant heavily glandular puberulent; leaves 5–17 cm. long, 1.5–4.5 cm wide, ovate to oblanceolate, obtuse, entire or rarely dentate; umbel 2–17-flowered; pedicels 1–3 cm. long in flower, later elongating; calyx tube 1.5–2.5 mm. long, the lobes 3–5 mm. long, acute; corolla tube maroon below, yellow above, the lobes 7–12 mm. long, deep rose to lavender; staminal tube 1.5–2.5 mm. long, usually yellow; anthers 3–4.5 mm. long, lanceoloid, the anther sacs yellow or rarely reddish, the connective dark maroon or at base sometimes yellow; capsule 5–8 mm. long, 3–4 mm. wide, ovoid, valvate, the walls usually thick. *D. pauciflorum* (Durand) Greene var. *Cusickii* (Greene) Mason; *D. puberulentum* Heller. Abundant, moist or dry loamy hills, meadows, rocks, or open woods, *U. Son.* to *A. T. T.*

Forma **album** (Suksd.) St. John. Differs from forma. *Cusickii* in having the corolla white. *D. Cusickii* Greene, var. *album* Suksd. Gravelly or stony places, Spokane, Spangle, Rock L., Wn.; Post Falls, Ida. *A. T.*

The type collection is from Spangle, *Suksdorf* 8,601.

Dodecatheon pulchellum (Raf.) Merr. Rootstock short, stout, vertical; plant 13–43 cm. tall, glabrous; petioles 1–8 cm. long, margined; blades 2–14 cm. long, oblanceolate, obtuse, entire or remotely crenulate; bracts of involucre lanceolate, foliaceous; umbel 3–20-flowered; pedicels 1–5 cm. long, erect in fruit; calyx 4–7 mm. long, parted to the middle or below into lanceolate or deltoid lobes; corolla tube yellow, with a scalloped purple ring; corolla lobes 7–18 mm. long, narrowly oblong-elliptic, lilac-purple; staminal tube 1.5–3.5 mm. long, yellow; anthers 4–7 mm. long, purple; style slender, barely exceeding the anthers; capsule 6–12 mm. long, lanceoloid to ovoid, thin, dehiscing from the apex into 5 valves; seeds 0.3–0.7 mm. long, angular, irregularly compressed, brown, papillose, translucent. *D. pauciflorum* sensu St. John, not of (Durand) Greene; *D. radicatum* Greene, subsp. *radicatum.* Low meadows, local, Pullman, *A. T.*

LYSIMACHIA

Erect perennial leafy herb with slender rootstocks; leaves opposite, alternate, or whorled, sessile, punctate; flowers axillary, solitary, in racemes, or spikes; calyx 5–7-divided, corolla deeply 4–9-parted, the tube very short and the segments narrow, the sinus entire or short toothed; stamens 5–7, exserted; filaments slender, united at the base or free; capsule 5–7-valved, few-seeded. (The ancient Gr. name used by Dioscorides for one of the species, *lusimaxeios.*)

Leaves glabrous; flowers in spikes; capsules 1.8–2.5 mm. long. *L. thyrsiflora.*
Leaves or petioles long ciliate; flowers solitary; capsule 4.5–5 mm. long. *L. ciliata.*

Lysimachia ciliata. L. *Fringed Loosestrife.* Rhizome slender, creeping; plant 1–9 dm. tall; stems glabrous, simple or with weak axillary branches; petioles 3–32 mm. long, long ciliate; blades 2–13 cm. long, ovate to lanceolate, acute, the base cuneate, rounded, or subcordate, beneath pale green, the margin ciliolate; peduncles single in the axils, 2–7 cm. long, sparsely brown puberulous; calyx parted nearly to the base, the lobes 3–6.5 mm. long, linear-lanceolate, sharp pointed; corolla yellow, 8–12 mm. long, deeply parted into oval or suborbicular, acute, erose lobes, within atomiferous glandular; filaments 2–2.5 mm. long, subulate, glandular puberulent; anthers 2.3–3 mm. long, linear-oblong; staminodia subulate, delicate, 1 mm. long; style 3.5–4 mm. long; capsule 4.5–5 mm. long, globose-ovoid, acute; seeds 1.5–2 mm. long, oval, 3-sided, brownish papillose. *Steironema ciliatum* (L.) Raf. Wet shores and meadows, *A. T.*

Lysimachia thyrsiflora L. *Tufted Loosestrife.* Rhizome fleshy, usually horizontal, producing branches or tuberous offsets; plant 2–8 dm. tall, glabrous nearly throughout; stem usually simple, bracted below, leafy above; leaves opposite, sessile, 3–13 cm. long, lanceolate or linear-lanceolate, cuneate, dark glandular punctate, pale green beneath; racemes from the middle axils densely many-flowered, 1–2.5 cm. long; peduncle 1–5 cm. long, somewhat pilosulous; floral bracts lanceolate, exceeding the pedicels; pedicels 0.5–2 mm. long; calyx few punctate, parted into 5–6 lance-linear lobes 1.5–2 mm. long; corolla 4–5 mm. long, pale yellow, cleft nearly to the base into 4–9 oblong-linear, punctate lobes, the sinus entire or short toothed; stamens about equaling the corolla, the anthers 0.5–0.8 mm. long, oblong; ovary red glandular warty; style equaling the stamens, capitate; capsule 1.8–2.5 mm. long, globose, black punctate; seeds 1.2–1.5 mm. long, shaped like a sector of a globe, brown, reticulate. *Naumburgia thyrsiflora* (L.) Duby. Wet shores, Rock Lake and northward, *A. T. T.*

TRIENTALIS. Star-Flower

Low glabrous perennial herbs with simple erect stems; leaves few, alternate, minute and scale-like, except a whorl of thin veiny leaves at the summit of the stem; flowers one or few, terminal; sepals narrow, persistent; corolla spreading, flat, without a tube, 5–9-parted; filaments slender; united at the base; anthers oblong, revolute after dehiscing; capsule few-seeded. (Name from Lat., *trientalis,* ⅓ of a foot.)

Trientalis latifolia Hook. *Wild Potato.* Tubers 4–20 mm. long, white, more or less fusiform, warty, and horizontal, forming filiform rhizomes; stem 6–20 cm. tall, sparsely brown puberulent, glabrate below; bracts lanceolate; terminal leaves several, subsessile, 2.5–7 cm. long, oblanceolate to oval or broadly obovate, mostly acute, cuneate, beneath pale green and remotely brown puberulous; flowers 1–6; pedicels 15–50 mm. long, filiform, sparsely brown puberulous or glabrate; calyx 6-cleft almost to the base, the lobes 2–6 mm. long, lance-linear; corolla 4–9 mm. long, rosy or white, saucer-shaped, lobed ½ to ⅔ way down into broadly ovate, acute lobes; filaments 2–3.5 mm. long, from a connate basal ring; anthers 0.8–1 mm. long, linear, coiling at the tip; ovary red glandular puberulous; style equaling the stamens; capsule 3–3.5 mm. long, subglobose, puberulous, 6-valved. Dense woods, Blue Mts., *Can.*

The crisp tubers are bitter.

GENTIANACEAE. Gentian Family

Mostly herbs, but also shrubs and trees; leaves entire or trifoliate; alternate or opposite, sessile or petioled, without stipules; flowers usually perfect, mostly regular, solitary or in bractless cymes or racemes; corolla gamopetalous, 4–12-parted; stamens as many as the corolla lobes and alternate with them, epipetalous; ovary 1-celled, with 2 parietal placentae; fruit a capsule, usually 2-valved; seeds numerous; endosperm copious. (Named from the genus *Gentiana.*)

Leaves simple, opposite or whorled; corolla lobes convolute or imbricate in the bud,
- Corolla lobes 4, with glandular pits at base. *Swertia.*
- Corolla lobes 4–5, lacking glandular pits,
  - Corolla salver-form; anthers twisting spirally on dehiscence. *Centaurium.*
  - Corolla funnelform or campanulate; anthers not twisting spirally on dehiscence. *Gentiana.*

Leaves trifoliolate, basal and alternate; corolla lobes induplicate-valvate in the bud. *Menyanthes.*

CENTAURIUM. Centaury

Low branching annual or biennial herbs; leaves sessile or clasping; flowers yellow, white, or reddish; calyx 4- or 5-parted; corolla

4- or 5-parted, the tube slender, the lobes convolute; anthers exserted, erect, oblong to linear; style filiform, usually deciduous; stigmas capitate or 2-lipped. (Name Lat., *centum,* hundred; *aurum,* gold piece.)

Flowers pink. *C. curvistamineum,* f. *curvistamineum.*

Flowers white. Forma *albiflorum.*

Centaurium curvistamineum (Wittrock) Abrams forma **curvistamineum.** Annual, erect, 5–25 cm. tall, glabrous; roots fibrous; stem simple, or in larger plants freely branched; leaves opposite, 6–20 mm. long, elliptic to lanceolate or oblanceolate; flowers subsessile in the leaf axils or forks of the stem; calyx 7–11 mm. long, subcylindric, with 5 prominent green ribs, excurrent as linear calyx-teeth, the sinus membranous, split by the enlarging fruit; corolla tube 6–9 mm. long, cylindric; corolla lobes 3–4 mm. long, oval to oblong, obtuse or retuse; stamens and style equaling the corolla tube; anthers 0.6–0.8 mm. long, oblong; stigmas 0.5 mm. long, flabelliform; capsule 8–10 mm. long, fusiform; seeds 0.2–0.3 mm. long, semicordate, gray, coarsely cellular reticulate. Moist places, Pullman, *A. T.*

Forma **albiflorum** (Suksd.) St. John, new comb. Differs only in having white flowers. *C. Muehlenbergii* (Griseb.) Wight, var. *albiflorum* Suksd., *Werdenda* **1:** 30, 1927. Wet places, Spangle, Pullman, *A. T.*

The holotype is from Latah Creek, Spangle, *Suksdorf 8,903.* As this plant has no other characters than albinism, and no distinct range, it seems properly classified as a color form.

GENTIANA. Gentian

Herbs; leaves opposite; flowers solitary or in cymes, showy, in late summer or autumn; calyx 4- or 5-cleft; corolla funnelform or bell-shaped, 4- or 5-lobed, regular, without glands, often with intermediate plaited folds which bear appendages at the sinuses; style short or long, simple or 2-parted, persistent or none; stigmas 2. (The ancient Gr. name, *gentiane,* in honor of *King Gentis,* who discovered its tonic properties.)

Corolla 10–14 mm. long, the sinus without plaits, the lobes with fimbriate crown at base within. *G. Amarella.*

Corolla 2–4 cm. long, the sinus with plaits, the lobes without a crown,

Calyx-lobes lanceolate; upper leaves elliptic-lanceolate to ovate. *G. oregana.*

Calyx-lobes linear; upper leaves linear-lanceolate to linear. *G. affinis.*

Gentiana affinis Griseb. in Hook. *Prairie Gentian.* Perennial 1.5–4 dm. tall; root crown dark, with several tuberous secondaries; stems several, usually decumbent, leafy throughout, scaberulous; lower leaves reduced; middle leaves 15–35 mm. long, ovate to lanceolate, glaucous, a little fleshy, the margins scaberulous; flowers few to many, pediceled or racemed from the upper axils; the subtending leaves little reduced; pedicels 1–15 mm. long; calyx-tube 3–6 mm. long, narrowly campanulate, often splitting at the hyaline sinus; calyx-lobes 1–6 mm. long, very unequal; corolla 2–3 cm. long, subcylindric, narrowed to the base, blue, the lobes 3–5 mm. long, somewhat greenish without; plaits of sinus bifid or dentate; stamens about 2 cm. long, filaments subulate at tip, linear-lanceolate expanded near the middle; anthers 1.8–2 mm. long, linear-oblong; stigmas 2 mm. long, strap-shaped; capsule long stipitate, 12–15 mm. long, fusiform; seeds 0.6–0.8 mm. long, flat, oval, the brown body tumid, the pale, thin, cellular reticulate wing nearly as broad as the body. *Dasystephana affinis* (Griseb. in Hook.) Rydb. Spokane Co., *Suksdorf* 938, fide Piper.

Gentiana Amarella L. *Northern Gentian.* Annual or biennial; tap-root slender, with fibrous secondaries; plant 3–60 cm. tall, glabrous; stem erect, often with numerous slender ascending branches; basal leaves spatulate; cauline leaves sessile, 5–45 mm. long, lanceolate to ovate-lanceolate; flowers on axillary peduncles or cymes; calyx 4–9 mm. long, 4–5-parted, the lobes very unequal, lanceolate to linear; corolla cylindric, blue, lobed ⅓ of length into 4–5 lanceolate, acute or obtuse lobes, stamens included in the corolla tube, anthers 0.6–0.8 mm. long, blue, oblong; capsule 12–14 mm. long, lanceoloid, compressed; seeds 0.3–0.5 mm. long, broadly ellipsoid, amber-colored. *G. acuta* Michx.; *Gentianella Amarella* (L.) Börner; *Amarella plebeia* (Cham.) Greene. Open woods of Ida., *A. T. T.*

Gentiana oregana Engelm. ex Gray. Perennial, 1.5–6 dm. tall; root a dark tap-root, with numerous tuberous secondaries often multicipital; stems erect, mostly simple, scaberulous, leafy throughout; leaves sessile, 1–4 cm. long, elliptic-lanceolate to ovate, pale and somewhat glaucous, a little fleshy, paler beneath, the margin scaberulous; flowers few to many, short pediceled or short racemed from the upper axils; upper leaves not conspicuously reduced; pedicels 0–18 mm. calyx-tube 4–7 mm. long, narrowly campanulate, membranous, pale; calyx-lobes 3–10 mm. long, foliaceous; corolla 2.5–4 cm. long, narrowly campanulate, light blue, the lobes 5–10 mm. long, ovate, acute, more or less greenish without; plaits of the sinus acuminate, lobed or dentate; stamens about 2 cm. long, the filaments linear-lanceolate, membranous margined; anthers 2–2.5 mm. long, linear-oblong; stigmas 1.5–2 mm. long, strap-shaped; capsule 2.5–3.5 cm. long, stipitate, oblanceoloid; seeds 1.5–2 mm. long, asymmetric fusiform, caudate at one end, brown, cellular reticulate. *Dasystephana oregana* (Engelm.) Rydb. Prairies or open woods, *A. T., A. T. T., Huds.*

MENYANTHES

Perennial herbs, with thick creeping rootstocks sheathed by the membranous bases of the petioles; leaves all basal, alternate, trifoliolate; calyx 5-parted; corolla funnelform, 5-cleft, bearded within; stamens 5, epipetalous; disk of 5 hypogynous glands; style slender, persistent; stigma 2-lobed; capsule bursting irregularly, many-seeded. (Ancient Gr. name, *men,* month; *anthos,* flower.)

Menyanthes trifoliata L. *Buckbean.* Perennial, 2–3.5 dm. tall, glabrous; rootstock creeping, 1–2 cm. in diameter, green or pale, scaly towards the tip; roots cord-like; petioles 5–25 cm. long, the base broad sheathing, thin; leaflets 3–8 cm. long, obovate to oblanceolate or elliptic; scapes erect, naked; raceme 10–26-flowered, 3–12 cm. long; bracts 2–14 mm. long, foliaceous, ovate or lanceolate, reduced upwards; pedicels 2–22 mm. long; calyx parted almost to the base, the lobes 2–5 mm. long, ovate; corolla white or pinkish, the tube 4–7 mm. long, campanulate, the lobes 4–8 mm. long, lanceolate, the lower two-thirds bearded within with long stout hairs; stamens and styles dimorphic; filaments subulate; anthers 1.3–2 mm. long, purple sagittate; stigmas 1 mm. long, strap-shaped; capsule 7–13 mm. long, ovoid to subglobose, seeds 2.5 mm. long, ellipsoid, brown, shining. Pond shores, in scablands southward as far as Fish Lake and Rock Lake, also at Troy, Ida., *A. T.*

The leaves contains the glucoside menyanthin, and have been used medicinally.

SWERTIA. Columbo

Tall, erect biennial or perennial herbs, with mostly simple stems; leaves opposite or whorled; flowers rather large, numerous, in open cymes arranged in an elongated panicle; calyx deeply 4-parted; corolla rotate, each lobe with 1–2 glandular and fringed pits on its face; anthers 4, oblong, remaining straight with age; stigma 2-lobed. (Named for Emanuel Swert, a Dutch gardener.)

Leaves ternate; corolla lobe with rounded gland. *S. fastigiata.*
Leaves opposite; corolla lobe with linear gland,
  Corolla blue. *S. albicaulis,* f. *albicaulis.*
  Corolla white. Forma *alba.*

Swertia albicaulis (Griseb.) Kuntze, forma **albicaulis.** Perennial 2.5–5 dm. tall, puberulent throughout; tap-root dark, woody, becoming multicipital; stems erect, simple; basal leaves 8–17 cm. long, numerous, linear-oblanceolate to spatulate, 3-ribbed, and with firm whitish margin, tapering at base into the slender petiole; cauline leaves in 1–3 pairs, 4–10 cm. long, mostly sessile; lower bracts foliaceous; panicle 4–25 cm. long, interrupted, densely flowered; bracts linear or lanceolate; pedicel 1–10 mm. long; calyx-lobes 4–8 mm. long, linear-lanceolate, white hyaline margined; corolla tube rotate 0.5–0.7 mm. long; corolla lobes 5–10 mm. long, blue, elliptic, acuminate, the gland linear, extending from the base to the middle of the corolla lobe, thickened, the base covered, the furrow laciniate fringed; scales alternating with the stamens, deeply fimbriate; filaments 4–6 mm. long, subulate; anthers 1.5–2.5 mm. long, linear-oblong; style 2–3 mm. long, stigma subcapitate, scarcely lobed; capsule 6–12 mm. long, elliptic, flattened, beaked; seeds 3.5–4.5 mm. long, elliptic, flat, acute at one end, one edge nearly straight, dark brown, cellular punctate. *Leucocraspedum albicaule* (Dougl.) Rydb.; *Frasera albicaulis* Dougl.; *F. nitida* Benth., var. *albicaulis* (Dougl.) Card. Open slopes, *U. Son.,* abundant in *A. T.*

This is the older species, and must be maintained, even if it and *F. nitida* are considered a species and a variety. The type was collected by D. Douglas, "In the mountain vallies between Spokan and Kettle Falls."

Forma **alba** (St. John) St. John. Differs in having the corolla white. Dry bluff above highway, Pullman to Moscow, Whitman Co., *Parker* 446, the type collection, *A. T.*

Swertia fastigiata Pursh. *American Columbo.* Perennial 4–20 dm. tall, glabrous; tap-root 2–4 (or more) cm. in diameter, fleshy, pale; stem erect, simple, leafy; basal leaves with petioles 5–15 cm. long, the blades 10–30 cm. long, broadly obovate to oblanceolate, chartaceous; cauline leaves mostly in whorls of 3, subsessile 4–21 cm. long, obovate to oblanceolate or elliptic; inflorescence a many-flowered, interrupted, cymose panicle 5–40 cm. long; the peduncles clustered, 2–6 cm. long; bracts 2–10 mm. long, linear, opposite or 3–4 whorled; calyx-lobes 6–9 mm. long, subulate; corolla tube shallow; corolla lobes 8–10 mm. long, pale blue to bluish lavendar, within near the base with a rounded glandular pit about 1.3 mm. long, the membranous, fimbriate border conspicuous; filaments 2 mm. long, subulate; anthers 1 mm. long, oval; stigmas 0.5 mm. long, strap-shaped; capsules 10–15 mm. long, oval to oblanceolate, flattened; seeds 5–6 mm. long, oval to triangular, brown, cellular pitted. *Frasera fastigiata* (Pursh) Heller. Open woods, *A. T. T.*

The original specimen was collected by Capt. M. Lewis, "In moist places on the Squamash Flats," or at Weippe, Ida.

This was the true "Bitter Root", according to the old mountaineers near Mt. Idaho. It is abundant from Camas Prairie to the Salmon R. and the Clearwater R.

APOCYNACEAE. Dogbane Family

Herbs, vines, shrubs, or trees, with acrid milky juice, mostly poisonous; leaves entire, mostly opposite or whorled, without stipules; flowers regular; calyx free from the ovaries; corolla gamopetalous, 4–5-lobed the lobes imbricate; stamens 4–5, alternate with the corolla-lobes, epipetalous; disk or glands usually present; pistils of two carpels, superior, the ovaries separate (in ours), the styles or stigmas united; fruit, entire or a pair of follicles; seeds many, often down-tufted at the apices; endosperm not copious. (Name from the genus *Apocynum.*)

APOCYNUM. Dogbane

Perennial herbs; leaves opposite, entire mucronate-pointed; flowers small, in cymes, on short pedicels; calyx 5-parted, the lobes acute; corolla bell-shaped or cylindric, 5-cleft, bearing 5 triangular appendages below the throat, opposite the lobes; stamens 5, on the very base of the corolla; glands 5; fruit of two long and slender follicles; seed with a tuft of long silky down at the apex; endosperm none. (The ancient Gr. name *apokunon,* from *apo,* from; *kunos,* of a dog.)

Corolla barely exceeding the calyx. *A. cannabinum,* var. *glaberrimum.*
Corolla 2–3-times the length of the calyx,
  Corolla 2–3 mm. long. *A. Suksdorfii.*
  Corolla 4–6 mm. long,
    Corolla subcylindric, about twice as long as the calyx-lobes,
      Plant glabrous. *A. medium,* var. *floribundum.*
      Blades puberulent beneath. Var. *lividum.*
    Corolla campanulate, at least thrice the length of the calyx-lobes,
      Blades glabrous. *A. pumilum,* var. *pumilum.*
      Blades pilosulous beneath. Var. *rhomboideum.*

Apocynum cannabinum L., var. **glaberrimum** A. D.C. *Smooth Indian Hemp.* Plant 3–6 dm. tall, glabrous; stem erect, the upper nodes with ascending axillary branches; petioles 2–5 mm. long; blades 3–10 cm. long, elliptic-oblong, mucronate, acute or obtuse at either end, above green, beneath pale green; cymes terminal; bracts lanceolate, not exceeding the pedicels; pedicels 1–3 mm. long; calyx-lobes 1–2 mm. long, lanceolate; corolla greenish white, the tube 1–1.5 mm. long, ovoid, the lobes 0.8–1 mm. long, deltoid, erect; anthers 1.2–1.4 mm. long, yellowish, lanceolate, the acuminate tip membranous; follicles 10–20 cm. long, pendulous; coma 2.5–3 cm. long, white; seeds 4–5 mm. long, linear, ridged, brown. Wawawai, *Piper* 1,621, *U. Son.*

Not weedy in this area, rather an uncommon plant of stream banks.

Apocynum medium Greene, var. **floribundum** (Greene) Woods. Stems 2–7 dm. tall, above with numerous ascending branches; plant glabrous; petioles 1–5 mm. long; blades 3–10 cm. long, elliptic-lanceolate, above green, below glaucous and pale or reddish, the apex apiculate acute; cymes terminal; bracts lanceolate, shorter than the pedicels; pedicels 1–3 mm. long; calyx-lobes 2–2.5 mm. long, lanceolate; corolla white or pinkish, the tube 2–3.5 mm. long, subcylindric; corolla lobes 1.5–2 mm. long, deltoid, slightly spreading; anthers 2.5 mm. long, tawny, lanceolate; follicles 7–15 cm. long, straight or somewhat falcate, pendulous; coma 2 cm. long, pale tawny; seeds 4 mm. long, cylindric. Guy, Whitman Co., *Elmer* 285, *A. T.*

Var. **lividum** (Greene) Woods. Like var. *floribundum,* but the blades puberulent beneath; the calyx-lobes ciliate-erose. *A. ciliolatum* Piper. Wawawai, *Lake & Hull* 549 (not seen), *U. Son.*

Apocynum pumilum (Gray) Greene, var. **pumilum.** *Small Dogbane.* Rootstock dark, cord-like, running; plant 1–7 dm. tall, glabrous; stem erect, freely dichotomously branched above, the branches ascending or spreading; leaves drooping or spreading, petiolulate or subsessile; blades 2–7 cm. long, ovate to oblong-lanceolate, obtuse or acute, the base commonly rounded or subcordate, above dark green, beneath pale; cymes numerous, terminal and often also in upper axils; pedicels 1–4 mm. long; bracts lanceolate, shorter than the pedicels; calyx-lobes 2–2.5 mm. long, lanceolate; corolla campanulate, white to pink; corolla tube 3.5–5, the lobes 1.5–3 mm. long, oblong, spreading; anthers 2.5 mm. long, lanceolate, brownish; follicles 5–12 cm. long, straight, linear-fusiform, divergent, suberect; coma 1–2 cm. long white; seed 1.5–2.4 mm. long, ovoid-cylindric, brown. Abundant, silty soils, *A. T., A. T. T.*

A persistent and troublesome weed in the wheat fields. Contrary to the opinions of Woodson, the writer considers the flowers as campanulate (not cylindric), and *Piper* 1,620 from Pullman is included here. The writer has seen no *A. androsaemifolium* from this area.

Var. **rhomboideum** (Gray) Greene. Differing only in having the leaves more or less pilosulous beneath. Open hills, *A. T.*, common in open woods, *A. T. T.*

Apocynum Suksdorfii Greene. Plant glabrous, 4–6 dm. tall; stems erect, the upper axils with ascending axillary branches; petioles 1–5 mm. long; blades 3–10 cm. long, elliptic-lanceolate to oblong, above green, beneath pale green, the apex mucronate; cymes terminal; bracts lance-linear, shorter than the pedicels; pedicels 1–4 mm. long; calyx-lobes 1–1.5 mm. long, lanceolate; corolla white, the tube 1.5–2 mm. long, subcylindric; corolla lobes 1–1.5 mm. long, deltoid, erect; anthers 2 mm. long, yellowish, lanceolate with an acuminate membranous apex; follicles 9–10 cm. long, falcate, pendulous; coma 2–2.5 cm. long, white or tawny; seed 3.5–5.5 mm. long, linear, brown. Wawawai, *Piper* Aug. 24, 1894, *U. Son.*

ASCLEPIADACEAE. Milkweed Family

Herbs, vines, or shrubs, mostly with milky juice; leaves entire, simple, opposite, whorled or rarely alternate, without stipules; flowers mostly in cymes or umbels, regular, 5-merous, usually in simple umbels; corolla gamopetalous, 5-lobed; stamens attached to the stigma, all the pollen of each anther-cell in one waxy mass; pistils of 2 carpels with two distinct ovaries, but with a common stigma; fruit a pair of follicles; seeds numerous with a coma of silky down; endosperm scant. (Named from the genus *Asclepias.*)

ASCLEPIAS. Milkweed

Perennial herbs, with copious milky juice; leaves opposite; whorled, rarely alternate; flowers numerous, in umbels, subtended by a whorl of small bracts, the involucre; calyx and corolla deeply 5-parted; stamens on the very base of the corolla, monadelphous, short, crowned behind each anther with a hood-like appendage from the cavity of which rises a horn; anthers adherent to the solid stigma; ovaries with short styles; follicles ovate or lanceolate; seeds numerous, each with a long tuft of down, the coma. (Dedicated to *Aesculapias,* the ancient physician.)

Leaves whorled; corolla lobes 3–4 mm. long. *A. fascicularis.*
Leaves opposite; corolla lobes 10–15 mm. long,
  Plants 6–20 dm. tall, densely white lanate; hoods about as long as corolla lobes, much longer than anther head. *A. speciosa.*
  Plant 1–3 dm. tall, essentially glabrous; hoods about as long as the corolla lobes and as anther head. *A. cryptoceras,* ssp. *Davisii.*

Asclepias cryptoceras Wats., subsp. **Davisii** (Woods.) Woods. Decumbent, herbaceous perennial; stems clustered from base, somewhat flattened, simple or branched from the base, glabrous or nearly so; leaves 4–9 cm. long, the blades very broadly oval or ovate, or suborbicular, the apex broadly rounded, the base rounded to slightly cordate, glaucous, glabrous; petioles 0–5 mm. long; inflorescences terminal and lateral, the lateral ones sessile; terminal peduncles 0–7 cm. long; pedicels 2–4 cm. long, glabrous; calyx lobes 6–7 cm. long, narrowly lanceolate, glabrous; corolla greenish yellow, the lobes 10–15 mm. long; crown pale rose, the hoods 6–9 mm. long, deeply saccate and decurrent, two pointed, the horn inconspicuous and incurved or absent; anther head 3–3.5 mm. long; follicles 4–7 cm. long, broad, fusiform, apiculate, smooth, glabrous; coma 15–25 mm. long, tawny; seeds about 1 cm. long, broadly oval. In the Snake River Canyon, down to the mouth of the Grande Ronde R., *U. Son.*

Asclepias fascicularis Dcne. *Whorled Milkweed.* Plant 3–20 dm. tall, erect, often branching at base; main root horizontal, producing adventitious buds; stems glabrous below, sparsely puberulent above; leaves mostly in whorls of 3–6, subsessile or with petioles up to 6 mm. long; principal blades 5–15 cm. long, linear to linear-lanceolate, flat or folded, glabrous; axillary shoots often present with numerous narrower, smaller leaves; lowest or uppermost leaves sometimes opposite; umbels several, many flowered, subterminal or corymbose; peduncles 15–40 mm. long, sparsely pilosulous; pedicels 5–15 mm. long, pilosulous; bracts 1–4 mm. long, lance-linear, pilosulous; calyx-lobes 1–2.2 mm. long, lanceolate, pilosulous, scarious margined; flowers greenish white to ashy purple; petals 3.5–4.3 mm. long, elliptic, scarious margined, reflexed; crown with hoods oval with concave sides equaling the anther column, the horns projecting; anther tips rounded to ovate, membranous; follicles 4–12 cm. long, spindle-shaped, attenuate, pilosulous; coma 20–25 mm. long, white; seeds 5.5–7 mm. long, ovate, truncate at tip, plano-convex, brown, margined. Gravelly river shores, Spokane R. and Snake R., also at Pullman, *A. T., U. Son.*

Fatally poisonous to sheep, and perhaps to other grazing animals.

Asclepias speciosa Torr. *Purple Top.* Roots cord-like, deep seated, colony forming; plant 6–20 dm. tall, lanate canescent throughout or somewhat glabrate below; stems erect, simple; leaves opposite; petioles 2–12 mm. long; blades 8–21 cm. long, ovate, elliptic, or oblong, the base cuneate to subcordate, thick chartaceous; umbels several, subterminal, many flowered; peduncles 1–11 cm.

long; pedicels 8–40 mm. long; calyx-lobes 3–6 mm. long, lanceolate; corolla rose-purple, the lobes 8–15 mm. long, lanceolate, obtuse, tomentose without; crown with the hoods 9–14 mm. long, the horn about 3 mm. long, inflexed; anther column 3–4 mm. long; subsessile; anther tips membranous, broadly ovate, inflexed, the sides very sinuous; follicles 7–12 cm. long, lanceoloid, erect, densely lanate and with soft spine-like processes; coma 25–30 mm. long, white; seed 6–9 mm. long, ovate, flat, truncate at tip, brown, wing-margined. Hillsides or meadows, *U. Son.*, common in *A. T.*

A somewhat troublesome weed in wheat fields. Experiments have shown it to be non-poisonous.

CONVOLVULACEAE. Morning Glory Family

Mostly twining or trailing herbs, but also shrubs or trees, often with milky juice; leaves, simple, alternate; stipules none; flowers regular, perfect, mostly showy; calyx 5-lobed; corolla gamopetalous, 5-plaited or -lobed, convolute or twisted in the bud; stamens 5, alternate, inserted near the base of the corolla tube; ovary 1–4 celled; ovules 1–2 in each cell, the cells sometimes becoming 4 in the fruit by false partitions; fruit a berry or a globular, 2–6-seed capsule; endosperm mucilaginous. (Named from the genus *Convolvulus.*)

Styles 2, distinct; corolla lobes imbricate; green leaves lacking. *Cuscuta.*
Style 1, entire or cleft only at the apex; corolla plicate and twisted in bud; green leaves present. *Convolvulus.*

CONVOLVULUS. Morning Glory

Herbs or somewhat shrubby plants, twining, erect or prostrate; flowers white, pink, or purple, 1 or 2 in the axils of the leaves; calyx-lobes nearly equal or the outer larger; calyx with or without a pair of enclosing bracts; corolla funnelform to campanulate; stamens included; style undivided, or 2-cleft only at the apex; stigmas 2, filiform to oblong; capsules globose, 2-celled, or imperfectly 4-celled by false partitions between the two seeds or by abortion 1-celled. (Name a diminutive from Lat. *convolvere,* to entwine.)

Calyx enclosed by 2 large bracts; stigmas oval. *C. sepium,* var. *communis.*
Calyx not enclosed by bracts; stigmas filiform. *C. arvensis.*

Convolvulus arvensis L. *Morning Glory.* Perennial; rootstock and roots slender, cord-like, widespreading at several depths, colony forming; plant glabrous or sparsely hirsute; stems 3–12 dm. long, prostrate; petioles 2–25 mm. long; blades 1–4.5 cm. long, elliptic or ovate to oblong, sagittate at base; peduncle 1–6 cm. long, 1–5-flowered, the tip with 2 oblong, ciliate bracts; pedicels 3–20 mm. long; calyx not bracted; calyx-lobes 3.5–5 mm. long, oblong, obtuse, ciliate, the 2 inner smaller; corolla 1.5–2.6 mm. long, broad funnelform, white or the outside pink, with 5 reddish-violet, hirsute stripes; stamens exceeding the style; filaments dilated at base; style 5–7 mm. long, capsule 5–8 mm. long, ovate-orbicular, pointed; seeds 3–4 mm. long, pyriform, one side convex, the other with a longitudinal ridge, dark brown, tubercular roughened. Introduced from Eurasia, abundant, roadsides and fields, *U. Son.* to *A. T. T.*

A noxious weed, one of the most serious pests of the cultivated fields, very difficult to eradicate.

Convolvulus sepium L., var. communis Tryon. *Hedge Bindweed.* Perennial, mostly glabrous; stems 1–3 m. long, twining or climbing; petioles 1–6 cm. long; blades 2–9 cm. long, deltoid to ovate-lanceolate, hastate at base, the lobes entire or somewhat dentate, beneath pale green; peduncles 2–10 cm. long, 1-flowered, bearing at the summit 2 bracts 12–30 mm. long, ovate or cordate-ovate, acute; calyx-lobes cordate-lanceolate or ovate, shorter than the bracts; corolla 4.5–6 cm. long, funnelform, pink with white stripes, the margin entire or nearly so; stamens shorter than the style, the filaments dilated at base; style 2 cm. long; capsule about 1 cm. long, globose, beaked, seeds 5 mm. long, ovoid, dark brown, pustulate, the large pale hilum subbasal. A weed, introduced from e. N. Am. or Eurasia, thickets by Snake River, *U. Son.*

CUSCUTA. Dodder

Leafless annual, indigestible herbs with yellow or red stems, twining and parasitic by haustoria on plants to which they cling; flowers small, clustered; calyx 4- or 5-cleft or of 5 sepals; corolla urn- or bell-shaped, 4- or 5-cleft; stamens inserted in the throat, with a scale-like appendage at the base; ovary 2-celled, 4-ovuled; capsule usually 4-seeded. (Derivation of name uncertain.)

Stigma globose, flattened peltate, or infrequently ligulate,
  Capsules with a definitely thickened stylopodium. — *C. indecora,* var. *neuropetala.*
  Capsules not especially thickened at top,
    Scales below the stamens present. — *C. campestris.*
    Scales below the stamens lacking,
      Anthers oval; corolla lobes spreading, star-like in fruit. — *C. occidentalis.*
      Anthers oblong-linear; corolla not star-like. — *C. californica.*
Stigmas cylindric. — *C. Epilinum.*

Cuscuta californica H. & A. Flowers 3–5 mm. long, on short pedicels, forming loose cymose-paniculate clusters; calyx shorter than or exceeding the corolla tube, the lobes triangular to lanceolate, acute to acuminate, the base overlapping, fleshy; corolla campanulate-cylindric, often saccate between the stamen attachments, the lobes lanceolate, reflexed, longer than the tube; scales represented by ridges forming inverted arches between the stamens; stamens shorter than the lobes; anthers subulate, about equaling the filaments; styles much longer than the globose ovary; stigmas globose; capsule globose, enveloped by the withered corolla; seeds 1 mm. long, oval, somewhat compressed and rostrate. *C. californica,* var. *graciliflora* Engelm. On *Polygonum, Monardella* etc., *A. T., A. T. T.*

Cuscuta campestris Yuncker. Flowers 2–3 mm. long, often glandular, on pedicels mostly shorter than the flowers, in compact globose clusters; calyx about enclosing the corolla tube, the lobes oval, or shorter and orbicular, or sometimes broader than long, mostly overlapping when young, but not protruding to form angles; corolla lobes broadly triangular, acute, about equaling the short campanulate tube; stamens shorter than the lobes; filaments longer than or about equaling the oval anthers; scales ovate, abundantly fringed, exserted, bridged below the middle, ovary globose; styles slender or somewhat subulate;

capsule usually depressed globose, with the withered corolla at its base; seeds 1.5 mm. long, ovoid, flattened usually on one side. *C. pentagona* Engelm., var. *calycina* Engelm. On *Solanum* or many other hosts, Waitsburg, *Horner* 569, *A. T.*

This is reduced to *C. pentagona* Engelm. by C. L. Hitchcock (1959).

Cuscuta Epilinum Weihe. Flowers 3 mm. long, sessile in scattered compact glomerules; calyx as long as the corolla and somewhat loose about it, the lobes broadly ovate, acute; corolla urceolate, early taking the shape of the capsule, the lobes ovate-triangular acute or obtusish, shorter than the tube; scales shorter than the tube, spatulate, truncate, bifid, or entire, sometimes reduced to short wings near the base of the corolla, thin, bridged below the middle; stamens shorter than the corolla lobes; anthers ovate, about as long as the somewhat subulate filaments; ovary depressed globose; styles shorter than or nearly equaling the stigmas; capsule depressed globose, somewhat angled about the seeds, circumscissile, leaving the obcordate dissepiment in the calyx, carrying the withered corolla at the top; seeds 1.2 mm. long, round or ovoid to oval, angular, somewhat scurfy. Introduced from Eurasia or n. Africa. On flax or rarely on other hosts, Deer Park, *Diener* in 1923.

Cuscuta indecora Choisy, var. **neuropetala** (Choisy) Yuncker. Flowers 2 5 mm. long, loose or compacted, whitish, fleshy, broadly campanulate, more or less papillose-hispid; calyx-lobes triangular-ovate, acute or somewhat obtuse, shorter than or equaling the corolla tube; corolla campanulate, the lobes erect to spreading, triangular, acute; scales as long as or longer than the corolla tube, ovate or somewhat spatulate, deeply fringed, bridged at or below the middle; stamens shorter than the lobes; anthers oval, about equaling the filaments; styles becoming divaricate, about as long as the globose, pointed ovary; capsule globose, enveloped by the withered corolla; seeds 1.7 mm. long, globose or broader than long, somewhat scurfy. *C. indecora neuropetala* (Engelm.) Hitchc. Common on *Xanthium* and other herbs, banks of Snake River, *U. Son.*

Cuscuta occidentalis Millsp. Flowers about 3 mm. long, often glandular, mostly sessile in small compact clusters; calyx as long as or longer than the corolla tube, the lobes ovate-lanceolate, acuminate, fleshy and thickened at base; corolla globose, saccate between the stamen attachments, the lobes lanceolate, acuminate, spreading in fruit and so the flower appearing star-shaped; stamens shorter than the lobes; filaments subulate; scales lacking; styles longer than the ovary; stigmas globose; capsule globose, surrounded by the withered corolla; seeds 1.5 mm. long, oval, flattened on two surfaces. *C. californica* Choisy, var. *breviflora* Engelm. Spokane, *Turesson* in 1913.

POLEMONIACEAE. Polemonium Family

Herbs, or rarely shrubs or trees; leaves alternate or opposite, simple or divided, without stipules; flowers regular, perfect, 5-merous, except the pistils; sepals more or less connate; corolla gamopetalous, convolute in the bud; lobes not plaited; stamens epipetalous, alternate with the corolla-lobes, distinct; style 3-lobed; ovary 3-celled, inserted on a disk; style 1, filiform; stigmas 3 (or 2); capsule 3-celled, 3-valved, the valves usually breaking away from the triangular central column; seeds few to many; seed-coats when wetted commonly becoming mucilaginous and developing spiricles; endosperm present. (Named from the genus *Polemonium.*)

Foliage leaves none, plant producing only the cotyledons and a whorl of connate involucral leaves. *Gymnosteris.*
Foliage leaves present,
Calyx wholly herbaceous, more or less accrescent, but not ruptured by the capsule,
Filaments dilated and hairy tufted at base; flowers in terminal or axillary cymes or racemes; corolla usually bluish, exceeding the calyx; biennials or perennials. *Polemonium.*
Filaments not dilated, bearing only a few remote hairs; flowers solitary, oppositifolious; corolla white, included in the calyx; annuals. *Polemoniella.*
Calyx scarious in the sinuses,
Calyx ruptured by the capsule, the sinus not revolute,
Corolla salverform, the tube cylindric; leaves opposite. *Phlox.*
Corolla funnelform to campanulate; leaves alternate or opposite. *Gilia.*
Calyx not ruptured, the sinus more or less distended into a recurved lobe. *Collomia.*

COLLOMIA

Herbs; leaves alternate; flowers in subcapitate, terminal or axillary clusters; calyx turbinate to cup-shaped or obpyramidal, the lobes equal, entire, erect, the sinuses scarious, in age distended into a recurved lobe, not ruptured by the capsule; corolla funnelform, reddish, yellowish, or white, 5-lobed; stamens unequally attached on the tube; seeds 1–2 in each cell, forming mucilaginous spiricles when wetted. (Name from Gr., *kolla,* glue.)

Corolla salmon-colored, the tube 15–20 mm. long; stem glabrous at base. *C. grandiflora.*
Corolla pink, the tube 6–10 mm. long; stem pubescent to the base,
Leaves all entire; nerves of calyx-lobes distinct to base of calyx. *C. linearis.*
At least the basal leaves pinnate; nerves of calyx-lobes uniting in the tube. *C. heterophylla.*

Collomia grandiflora Dougl. ex Lindl. Stem 1–9 dm. tall, leafy, erect, pilose above, simple or branched, late in the season often producing many weak axillary branches with reduced heads and smaller flowers; tap-root strong, pale; leaves 2–7 cm. long, oblanceolate to lanceolate or linear, varying upwards, sessile, puberulent above and on the margins, those of the inflorescence ovate, shorter; flowers numerous, in a dense hemispheric head; calyx viscid villous, obconic, the tube 3–6 mm. long, the lobes 2–4.5 mm. long, 3-nerved, green, deltoid; corolla salmon-colored, funnelform, viscid pilosulous, the lobes 5–8 mm. long, elliptic; anthers 0.5 mm. long, oval, these and the style about equaling the corolla tube; stigmas 1 mm. long, linear; capsule 5–6 mm. long, obconic, 3-angled; seeds 3–3.5 mm. long, elliptic-oblong, furrowed, brown. *C. grandiflora diffusa* (Mulford) Piper. Dry or stony slopes, or open woods, *U. Son.* to *A. T. T.*
Discovered in 1825 by David Douglas.
Collomia heterophylla Hook. Annual, 4–30 cm. tall, viscid villous, simple or branched, often decumbent at base; lower leaves 5–60 mm. long, the petiole

about equaling the pinnatifid to 1–2-pinnate blade, upwards the leaves progressively less dissected; upper leaves ovate often subentire or entire; floral leaves shorter, lance-ovate, subentire or entire; flowers in terminal, hemispheric heads; calyx viscid villous, campanulate, the tube 2.5–3.5 mm. long, each of the 5 angles with 1 thickened broad nerve which divides into 3 in the lobes, the lobes 3–4 mm. long, lanceolate, green, the nerves reuniting at apex; corolla narrowly funnelform, pink or the summit of the tube yellowish, the tube 8–10 mm. long, sparsely pilosulous, the lobes 2.5–3.5 mm. long, elliptic; stamens included in the tube; anthers 0.2–0.4 mm. long, oval; style ⅔ length of corolla tube; stigmas linear; capsule 3–4 mm. long, elliptic-obovoid; seeds 1.3–1.5 mm. long, ovoid to ellipsoid, brown, forming mucilaginous spiricles when wetted. Sandy open thicket, Mica Bay, Lake Coeur d'Alene, *St. John* 9,619, *A. T. T.*

Collomia linearis Nutt. Annual, 8–48 cm. tall; stem white pilose, becoming sparsely so towards the base, at the apex densely viscid pilose, often branched above or simple; leaves 2–7 cm. long, linear-lanceolate, rather uniform in shape, closely puberulent throughout; flowers in terminal heads, those of the lateral branches similar to the main one; floral leaves shorter, more ovate at base, viscid pilosulous; calyx obpyramidal to turbinate, viscid pilose, the tube 3–3.5 mm. long, the lobes 3.5–4.5 mm. long, lanceolate, acuminate, 3-nerved, green-tipped; corolla pink, narrowly funnelform, the tube 6–8 mm. long, puberulous, the lobes 2.5–3.5 mm. long, elliptic; stamens unequal, mostly included; anthers 0.2 mm. long, oval; style included; stigmas linear; capsules 3.5–4 mm. long, obovoid, 3-lobed; seeds 2.2–2.5 mm. long, ellipsoid, brown, forming mucilaginous spiricles when wetted. Moist open places, rocky slopes, or open woods, *U. Son.* to *A. T. T.*

GILIA

Annual or perennial herbs or half-shrubby plants; leaves alternate or opposite; calyx narrow, the lobes acute, the tube scarious below the sinuses; corolla tubular-funnelform, the limb little spreading (in ours); stamens usually equally inserted, often exserted; capsules with 1–many seeds. (Named for *Felipe Gil*, Spanish botanist.)

Shrubby, at least at base,	
Leaf divisions rigid; herbage glandular.	*G. pungens*, var. *Hookeri*.
Leaf divisions flaccid; herbage not glandular.	*G. Nuttallii*.
Herbs,	
Inflorescence capitate,	
Plants glabrous.	*G. capitata*, subsp. *capitata*.
Plants villous,	
Floral bracts not spine-tipped.	*G. congesta*.
Floral bracts spine-tipped,	
Herbage viscid villous; stamens included.	*G. squarrosa*.
Herbage not viscid villous; stamens exserted.	*G. propinqua*.
Inflorescence looser,	
Leaves simple,	
Corolla tube 8–12 mm. long; longer pedicels barely twice as long as the shorter.	*G. gracilis*, var. *gracilis*.

Corolla tube 5–8 mm. long; longer pedicels 4–5-times longer than the shorter. Var. *humilior.*
Leaves compound,
Leaves palmately cleft,
Corolla more than twice as long as the calyx; filaments hairy at base. *G. pharnaceoides.*
Corolla shorter; filaments glabrous at base. *G. septentrionalis.*
Leaves pinnately cleft,
Calyx-lobes longer than the tube; corolla 2.5–3.5 cm. long, usually scarlet. *G. aggregata,* var. *aggregata.*
Calyx-lobes shorter than the tube; corolla 5–8 mm. long, bluish purple. *G. inconspicua,* subsp. *inconspicua.*

Gilia aggregata (Pursh) Spreng. var. **aggregata.** *Fox fire, Honeysuckle.* Biennial, 3–9.5 dm. tall; tap-root yellowish, elongate, up to 8 mm. in diameter; stem erect, usually single and simple to the inflorescence, more or less pilose and glandular puberulent; leaves of rosette sparsely glandular puberulous, hoary pilose with flattened, crinkly hairs, at least below, the petiole 15–25 mm. long, the blade 15–40 mm. long, pinnate, the pinnae linear, spine-tipped, the rhachis margined; cauline leaves alternate, numerous, similar but becoming shorter petioled and less hairy; thyrse 1–5 dm. long, narrow, loose or interrupted, many-flowered, closely glandular puberulent and more or less pilose; pedicels 0–4 mm. long; calyx 4–11 mm. long, campanulate, glandular puberulent and sometimes pilose, the subulate lobes at least twice as long as the tube; corolla tube 15–25 mm. long, gradually narrowed to the base; the lobes 7–11 mm. long, ovate to lanceolate, acute or acuminate; corolla scarlet, varying through pink to white, the lobes and throat maculate with deeper red or orange; stamens exserted; anthers 1–1.7 mm. long, oval; style exceeding the tube; stigmas linear; capsule 4–8.5 mm. long, ellipsoid, apiculate; seeds 2–2.2 mm. long, yellow, ellipsoid, compressed, forming mucilaginous spiricles when wetted. *G. pulchella* Dougl.; *Ipomopsis aggregata* (Pursh) V. Grant. Open places, or open woods, *U. Son.* to *A. T. T., Huds.*

Reported by Pammel to contain saponin and to be poisonous, however, the herbage is grazed by sheep. A species very variable in flower color, the ground color and the spot color occuring in so many shades as to defy classification. The various subdivisions proposed by Brand are inadequate in this area, and are rejected.

Gilia capitata Hook., subsp. **capitata.** Annual, 1–9 dm. tall, glabrate or sparsely puberulent; tap-root pale, up to 6 mm. in diameter; stem erect simple or branched; basal and lower leaves 2–10 cm. long, alternate, pinnate to bipinnate, the lobes variable, mostly linear, 0.5–3 mm. broad; flowers in terminal heads on naked, elongate peduncles; heads dense, usually globose, 7–35 mm. in diameter; calyx 3–4 mm. long, the sinus scarious, the lance-linear, scarious margined lobes equaling the tube; corolla 4–7 mm. long, blue, the tube funnelform, the lobes linear-oblong, equaling the tube; stamens equaling the lobes, the anthers 0.5–0.7 mm. long, oval; style equaling the corolla, trifid; capsule 2–3 mm. long, oval, 3-sided, greenish; seeds 1–1.5 mm. long, ovoid, compressed, reddish brown, rugose, forming mucilaginous spiricles when wetted. Sandy or gravelly places. *U. Son., A. T.*

This species was published in 1826 in the Bot. Mag. In that year Hooker became its editor. In the Kew Herbarium he claimed this species as his own.

Gilia congesta Hook. Perennial, erect or spreading from basal tuft, pubescence arachnoid floccose; petiole broad; blades 1–2-pinnately lobed or laciniate, 1–4 cm. long, those of the inflorescence reduced; flowers capitate in one or more heads; calyx cylindric, densely arachnoid, the lobes flanked by a membranous pseudo-tube; corolla 4–6 mm. long, salverform, white, but the tube 3–4 mm. long, yellow, the lobes 2 mm. long; stamens and pistil exserted; capsule obovoid; seeds 1–2 in a locule. *Ipomopsis congesta* (Hook.) V. Grant. Gravelly hillsides, Spokane Valley, *A. T.*

Gilia gracilis Hook., var. **gracilis.** Annual, 5–25 cm. tall, erect, commonly unbranched below, often bushy branched above; stem viscid pilose, glabrate below; basal leaves ovate; lower leaves 5–35 mm. long, narrowing upwards from spatulate to elliptic to linear, ciliate at base; upper leaves and those of branches alternate, reduced; flowers 2 (–1), supra-axillary; pedicels unequal, the longer barely twice as long as the shorter, viscid pilose; calyx in anthesis 5–8 mm. long, tubular, green, viscid pilose, the lobes lance-linear, exceeding the tube, in fruit 6–10 mm. long, campanulate; corolla tube slightly broadened above, yellowish, the lobes 1.5–2.5 mm. long, rose to lavender, broadly obovate, notched; stamens unequally attached below the throat; anthers 0.5 mm. long, ellipsoid; pistil equaling the calyx; stigmas 0.5 mm. long, linear; capsule 3–4 mm. long, ellipsoid to globose, the cells 1-seeded; seeds 2–2.5 mm. long, brown, thin margined. *Phlox gracilis* (Hook.) Greene; *Microsteris gracilis* (Hook.) Greene. Abundant, open places or open woods, *U. Son., A. T., A. T. T.*

David Douglas discovered it "on light soils, on the banks of the Spoken river, and on high grounds near Flathead river."

This species has been placed in *Phlox* by Mason, and in *Microsteris* by Wherry. The latter placement seems better, if it becomes necessary to split up the genus *Gilia.*

Var. **humilior** (Hook.) St. John. Differs only in being 2–14 cm. tall, commonly diffusely branched from the base, the branches decumbent; longer pedicels 4–5-times longer than the shorter; calyx in anthesis 4–6 mm. long, the lobes shorter than the tube, in fruit 5–7 mm. long; corolla lobes 0.7–1.5 mm. long; capsules 2.5–4 mm. long. *Collomia gracilis, β humilior* Hook.; *G. humilis* (Greene) Piper; *G. gracilis,* subsp. *humilis* (Dougl.) Brand; *G. gracilis,* subsp. *spirillifera* Brand, var. *euphorbioides* Brand; *Microsteris gracilis,* subsp. *humilis* (Greene) V. Grant. Open places, *A. T., A. T. T.*

This plant seems to be a variety as the alleged characters all merge. *Flett* 92 from Hood's Canal and *Sandberg* from Nez Perce Co., June 1892 are intermediate to the species.

Gilia inconspicua (Sm.) Sweet, subsp. **inconspica.** Annual, 1–4.5 dm. tall; tap-root slender; stems simple, erect or branched at base, decumbent, below loosely woolly to glabrate; lower leaves subbasal, clustered, 1–8 cm. long, pinnatifid to tri-pinnatifid, the segments firm, linear to oblong, the apex and teeth lanceolate, rigid pointed, sparsely woolly or glabrate; inflorescence longer than the stem, capitate glandular puberulent, a loose half cyme, the lower nodes with reduced leaves, the upper with bracts; pedicels 1–20 mm. long; calyx campanulate, green to purplish, glandular puberulent, in anthesis 2.5–3.5 mm. long, in fruit 4–7 mm. long, the lobes lanceolate; corolla tube narrowly funnelform, exceeding the calyx, the lobes 2 mm. long, ovate; stamens attached in the throat, not exceeding the lobes; anthers 0.3–0.6 mm. long, oval to cordate; pistil exceeding the tube; stigmas 1 mm. long, linear; capsule 3.5–7 mm. long, elliptic-ovoid; seeds 1.3–1.5 mm. long, ellipsoid, brown, thin margined, irregularly compressed. Rocky slopes, Palouse Falls, *St. John & Pickett* 6,109, *U. Son.*

Gilia Nuttallii Gray. Perennial, 1–3 dm. tall, appressed puberulent; root woody, up to 13 mm. in diameter, crown woody, multicipital, bearing many erect or decumbent branches, simple or branched at base; leaves opposite, sessile, palmately parted into 3–7 segments 5–15 mm. long, linear, somewhat rigid, mucronate, sparsely scabrous puberulent; flowers sessile or nearly so in dense leafy cymes; calyx 5–10 mm. long, pilosulous, the lance-linear lobes about equaling the tube; corolla white, with a yellowish throat; the tube pilose, 5–7 mm. long, including the abruptly funnelform throat; the lobes 6–7 mm. long, oblong-obovate; stamens slightly protruded from the throat; anthers 0.5–0.8 mm. long, oval; pistil included, nearly equaling the tube; stigmas 2 mm. long, filiform; capsule 4–7 mm. long, cylindric; seeds 2.3–2.6 mm. long, greenish, ellipsoid, compressed. *Leptodactylon Nuttallii* (Gray) Rydb.; *Limnanthastrum Nuttallii* (Gray) Ewan; *Linanthus Nuttallii* (Gray) Greene ex Milliken. Open, rocky ridges, Blue Mts., *Huds.*

Gilia pharnaceoides Benth. Annual, 5–35 cm. tall, simple or commonly diffusely branched above, glabrous, or puberulous above; roots slender, fibrous; internodes elongate, often many times exceeding the leaves; leaves 2–22 mm. long, all opposite, 3–7-parted into linear, scabrous-puberulous, acerose-tipped divisions; inflorescence a loose cyme, often half the length of the plant; bracts leafy; pedicels 1–3 cm. long, filiform; calyx 2.5–4 mm. long, campanulate, strigose puberulous, the sinus and margin of lobes hyaline, the lobes lanceolate, nearly equaling the tube; corolla lavender pink, the lobes spreading to almost rotate; corolla tube 1.5–2 mm. long, broad funnelform; the lobes 4–7 mm. long, obovate, truncate, minutely pointed; stamens shorter than the corolla, the filaments attached in the tube, pilose at base; anthers 0.5 mm. long, oval; style equaling the filaments, the stigmas 0.3 mm. long, filiform; capsule 3–5 mm. long, ellipsoid; seeds 0.5–0.6 mm. long, ovoid, yellowish, mucilaginous when wetted. *Linanthus pharnaceoides* (Benth.) Greene; *L. liniflorus* (Benth.) Greene, subsp. *pharnaceoides* (Benth.) Mason; *G. liniflora* Benth., subsp. *pharnaceoides* (Benth.) Brand. On sands or gravels, *U. Son., A. T.*

Gilia propinqua (Suksd.) comb. nov. Annual, 2–33 cm. tall, simple and erect, or branched and often decumbent; stem appressed pilose, at length glabrate below; leaves 5–50 mm. long, alternate, 1–2-pinnatifid, nearly glabrous, the rhachis narrowly winged, the segments linear, heavy nerved, spine-tipped; flowers in heads; bracts 5–10 mm. long, similar to the leaves, the base less dissected and villous at least on the margins, the segments glabrous (or slightly glandular atomiferous), the inner palmately cleft; calyx subcylindric to campanulate, villous at the throat, nearly or quite glabrous below, the tube 4–5 mm. long, the teeth 2–5 mm. long, unequal, spine-like, either all entire or some forked; corolla bluish or whitish, narrow funnelform, the tube 5–9 mm. long, the lobes 0.5–1 mm. long, lanceolate; stamens exserted; anthers 0.2–0.4 mm. long, ellipsoid; style about equaling the corolla; capsule 2.5–3 mm. long, ellipsoid to oblanceoloid, hyaline; seeds 1–1.3 mm. long, ellipsoid, brown, pitted, gelatinous spiriliferous when moistened. *Navarretia propinqua* Suksd., Allgem. Bot. Zeitschr. 12; 26, 1906; *Gilia intertexta* sensu St. John, non (Benth.) Steud.; *N. pilosifaucis* St. John & Weitm. On dried mud, *A. T., A. T. T.*

Gilia pungens (Torr.) Benth., var. **Hookeri** (Dougl. ex Hook.) Gray. Shrub, 1–4 dm. tall, glandular pilosulous above, glabrate below; root woody, brownish, up to 1 cm. in diameter; stem freely branching at base, the stems decumbent or ascending, very leafy; leaves alternate or the lower opposite, 3–10 mm. long, palmately parted into 3–7 acerose, spine-tipped segments, with fascicles of smaller leaves in the axils; flowers solitary in upper axils; calyx 7–11 mm. long, the lobes subequal, heavy ribbed, spine-tipped; corolla rose-colored, white, or yellowish, the tube 8–15 mm. long, the lobes 7–10 mm. long, narrowly obovate; stamens included in the throat; anthers 1 mm. long, oblong-ellipsoid; style and stigmas each about 2 mm. long, linear; capsule about 2.5 mm. long, ellipsoid, yellowish; seeds 0.7 mm. long, brown, ellispoid, somewhat compressed. *Leptodactylon pungens* (Torr.) Jepson., var. *Hookeri* (Dougl. ex Hook.) Jepson; *L. pungens,* subsp. *Hookeri* (Dougl. ex Hook.) Wherry. Dry sand banks, Spokane, *Kreager* 168; Wawawai, *Pickett,* May 13, 1921, *U. Son.*

Gilia septentrionalis (Mason) St. John. Annual, 5–25 cm. tall, puberulous to glabrate; roots fibrous, slender; stem erect, simple, or commonly freely branched above, the branches ascending; internodes elongate, much exceeding the leaves; leaves 2–13 mm. long, opposite, sessile, 3–5-parted, the divisions linear, acerose-tipped, remotely scabrous puberulent; inflorescence a loose cyme, up to ¾ the length of the plant; bracts leafy; pedicels 1–3 cm. long, filiform; calyx 2–4 mm. long, campanulate, the sinus membranous, the lobes lanceolate, strigose at least at tip, half the length of the tube; corolla white, the tube 1.5 mm. long, short funnelform, the lobes 1.5 mm. long, oval, divergent; stamens shorter than the corolla, the filaments pilose at base; anthers 0.1 mm. long, subglobose; style 1 mm. long; stigmas 0.1 mm. long, filiform; capsule 2–3 mm. long, obovoid; seeds 0.5–1.5 mm. long, ellipsoid, greenish, mucilaginous when wetted. *Linanthus septentrionalis* Mason. Open gravels or woods, *A. T. T.*

Gilia squarrosa (Esch.) H. & A. *Skunkweed.* Annual, 5–45 cm. tall, viscid villous and ill-scented; tap-root slender; stem erect, simple or divaricately branched; leaves 5–30 mm. long, 1–2-pinnate, the segments linear, rigid with a heavy nerve, spine-tipped; flowers in dense heads; bracts 1–2 cm. long similar to the leaves or the inner palmately cleft, but with a broad, firm rhachis; calyx-tube 3–4.5 mm. long, campanulate, hyaline except the heavy ribs; calyx-teeth 7–10 mm. long, entire, rigid, spine-like; corolla narrow funnelform, blue, the tube 6–7 mm. long, the lobes 2–3 mm. long, elliptic-oblong; stamens unequal, included in the throat, anthers 0.5 mm. long, ellipsoid; pistil 5 mm. long, stigmas linear; capsule 4–5 mm. long, ellipsoid, the pericarp firm, opaque, dehiscing into 3 valves from the apex; seeds 0.7–0.8 mm. long, ellipsoid, reddish brown, punctate, not cohering at maturity. *Navarretia squarrosa* (Esch.) Hook. Bars of Touchet River, Horner 582, *A. T.*

The local species of this group seem very distinct, and have been commonly maintained as *Navarretia,* but various Californian species break down the generic lines. Hence the writer follows Gray, and Munz in placing them all in *Gilia.*

GYMNOSTERIS

Diminutive annuals, with leaflets, usually simple stems; foliage leaves none, the cotyledons persistent, connate-perfoliate; the 2–4 involucral bracts connate at base; flowers few, in terminal heads; calyx-tube membranous, the teeth lanceolate, somewhat unequal, foliaceous; corolla salverform, pinkish, yellow, or white; anthers sessile in the corolla throat; capsule dehiscent, many-seeded; seeds narrowly margined or winged, mucilaginous when wetted but not forming spiricles. (Name from Gr., *gumnos,* naked; *steris,* robbed.)

Gymnosteris nudicaulis (H. & A.) Greene. Plant 1–12 cm. tall, glabrous; root fibrous, slender; stem almost filiform, simple or with 2 weak axillary branches from the axils of the cotyledons; cotyledons 1–2 mm. long, ovate, fleshy; involucral leaves 5–15 mm. long, oblong-lanceolate, green and fleshy, at base membranous and whitened; heads 3–25-flowered; inner bracts reduced, calyx-tube 2–3 mm. long, campanulate, the lobes 1.5–2 mm. long, aristate, pilose within; corolla "white or yellowish" or lavender-pink, the throat white, and the tube yellowish, the tube 6–12 mm. long, the lobes 3–6 mm. long, cuneate, the summit rounded and more or less sinuate, or truncate and short acuminate; anthers 0.2 mm. long, oval; style about ½ the length of the corolla tube; capsule 4 mm. long, ellipsoid; seeds 0.5–0.7 mm. long, lanceoloid, compressed, brown. *G. pulchella* Greene. Sandy bank of Snake River, Wawawai, *St. John & Merriman* 9,864.

G. pulchella was described as having larger flowers, but as will be noted above, the writer includes this plant within the fluctuations of *G. nudicaulis.* Greene apparently did not verify the original species, for its type, *Tolmie,* Snake Country, has the corolla tube 5–12 mm. long, and the lobes 3–5 mm. long. Exact information on the flower color is still needed.

PHLOX. Phlox

Perennial herbs or half shrubby plants, or rarely annual; leaves opposite, or some of the upper alternate, sessile, entire; flowers in cymes, terminal or in the upper axils, mostly bracted; calyx narrow, 5-ribbed, 5-cleft, the sinuses mostly scarious; corolla salverform with a long tube, a narrow opening and 5 broad or rounded lobes; stamens included, very unequally inserted on the upper part of the corolla tube; ovules 1 or sometimes 2–4; capsule ovoid, 3-celled, with but one seed in each cell; capsule at length distending and rupturing the calyx-tube. (Name Gr. *phlox,* flame, the ancient Gr. name for *Lychnis,* but by Linnaeus transferred to this genus.)

Flowers solitary leaves 2–10 mm. long, rigid, densely crowded. — *P. Douglasii,* var. *rigida.*
Flowers in cymes; leaves longer, not rigid and densely crowded,
  Intercostal portion of calyx replicate,
    Styles 1.5–4 mm. long. — *P. colubrina.*
    Styles 10–14 mm. long,
      Calyx glabrous; leaves glabrous or with sparse basal ciliations. — *P. longifolia.* var. *longifolia.*
      Calyx hirsute; leaves usually so, or puberulent. — Var. *linearifolia.*
  Intercostal portion of calyx plane,
    Style not longer than the ovary; petals notched at tip,
      Corolla pink. — *P. speciosa,* forma *speciosa.*
      Corolla white. — Forma *alba.*
    Style nearly equaling the corolla; petals truncate, erose at tip,
      Herbage tomentose. — *P. mollis.*
      Herbage glandular pilose. — *P. viscida.*

Phlox colubrina Wherry & Const. Woody based perennial; shoots 2–5 dm. tall; herbage glabrous; leaves 4–8 cm. long, 1–2.5 mm. wide, linear; inflorescence 2–12-flowered; calyx 8–12 mm. long, united ½–⅝ way, the lobes subulate, the white interconnecting membranes plicate; flowers pink to white, fragrant; corolla tube 9–15 mm. long, the lobes about 15 X 6 mm., mostly elliptic-oblanceolate, acute or obtuse; styles 1.5–4 mm. long, united ⅜–⅝ way. Abundant on the slopes of the Grand Canyon of the Snake to the south, but occurring downstream as far as the Grande Ronde R., and Lewiston, *U. Son.*

Phlox Douglasii Hook., var. **rigida** (Benth.) Peck. Perennial, caespitose, sparsely glandular pilose throughout, from a woody base; tap-root woody, up to 5 mm. in diameter; stems 6–19 cm. long, bushy branched; leaves numerous, subulate, spine-tipped; flowers subsessile; calyx 7–9 mm. long, glandular hirsute, the teeth subulate, less hairy, spine-tipped; flowers bluish; corolla tube 9–11 mm. long, pilose within below; the lobes 5–7 mm. long, obovate cuneate; anthers 0.7–0.8 mm. long, elliptic-oblong; style shorter than the calyx; capsule 3–4.5 mm. long, ellipsoid; seeds 2.5–4 mm. long, ellipsoid, much flattened or concave, yellow, sinuous reticulate. *P. Piperi* E. Nels.; *P. caespitosa* Nutt., subvar. *Piperi* (E. Nels.) Brand; *P. rigida* Benth.; *P. Douglasii,* subsp. *rigida* (Benth.) Wherry. Gravelly prairies, Spokane and vicinity, *A. T., A. T. T.*

Phlox longifolia Nutt., var. **longifolia.** Perennial, 1–4 dm. tall; tap-root yellowish, elongate, up to 2 mm. in diameter; crown becoming multicipital, stems repeatedly bushy branched, woody, usually glabrous; leaves 1.2–8 cm. long, 0.5–2 mm. wide, linear, acute, glabrous or villous ciliate at base, the midrib heavy; cymes few flowered, loose; pedicels 3–33 mm. long; calyx 7–14 mm. long, the intercostal portions membranous, saccate extruded at base, the lobes subulate, with a strong green rib; corolla deep pink, the tube 12–14 mm. long, slightly broader above, pilose within at base; the lobes 8–16 mm. long, elliptic-oblanceolate, obtuse; anthers 1.5–2 mm. long, elliptic-oblong; style equalling or well exceeding the calyx, shorter than the corolla; capsule 4–5 mm. long, ellipsoid, yellowish, punctate; seeds 2.3 mm. long, ellipsoid, compressed, brown. *P. longifolia,* subsp. *calva* Wherry; and subsp. *humilis* (Dougl.) Wherry. Open or rocky slopes, canyons of Snake and Clearwater Rivers, *U. Son.*

Wherry in 1938 stated that St. John overlooked the word "puberuli" in Nuttall's original description. On the contrary, I read the whole description, and noted also the words, "foliis glabris." In 1936 I gave an account of the type of *P. longifolia* Nutt., four specimens collected by Wyeth in the Rocky Mts. (Gray Herb.). One was white puberulent but three were glabrous or glabrate on calyx, peduncle, and leaf. These three were chosen to represent *P. longifolia* (a lectotype). There are no isotypes in the Academy of Natural Sciences, Philadelphia.

Var. **linearifolia** (Hook.) Brand. Stems glabrous or pilose; leaves 1.2–6.5 cm. long, 0.5–3 mm. wide, glabrous to puberulent or hirsute; calyx hirsute; corolla lobes obtuse or acute. *P. speciosa* Pursh, var. *linearifolia* Hook.; *P. linearifolia* (Hook.) Gray. Dry slopes or prairies, common, *U. Son., A. T.*

The type was collected by Geyer, "in the valley of the Kooskooskie River and the adjoining plains," [or the Clearwater River and the vicinity of Lewiston].

Phlox mollis Wherry. Woody based perennial; shoots 5–15 cm. tall, very tomentose; leaves 15–30 mm. long, 1.5–3 mm. wide; inflorescence of 1–3 flowers, densely tomentose, rarely a few hairs gland-tipped; calyx 12–15 mm. long, united ⅜–½ way, their lobes connected in part by flat membranes; corolla purple to pink, the tube 11–14 mm. long; corolla lobes about 10 X 7 mm., emarginate; styles 7–9 mm. long, free for 1 mm.

Type locality: Zaza, Craig Mts., *Christ 16,557.* Also in Blue Mts. Not seen by the writer, accepted from monograph by Wherry. (1955).

This is tentatively rejected by Cronquist (1959).

Phlox speciosa Pursh, forma **speciosa.** Perennial, 1–7 dm. tall; tap-root brown, up to 7 mm. in diameter; stems at base suffrutescent and usually bushy branched, glabrate below, glandular puberulent above; leaves 11–70 mm. long, 0.5–8 mm. wide, linear to linear-lanceolate, glandular pilosulous or puberulent, finally more or less glabrate, midrib and marginal nerve heavy, pale; cymes loose, few flowered, but plant showy from many flowering stems; pedicels 3–40 mm. long, densely glandular puberulent or pilosulous; calyx 7–11 mm. long, glandular pilosulous or puberulent, the lobes subulate; corolla tube 9–13 mm. long, pilose within at base; the lobes 8–12 mm. long, obcordate, cuneate; anthers 1.3–2.1 mm. long, elliptic-oblong; style shorter than the stigmas, much shorter than the calyx; capsule 4–5 mm. long, ellipsoid, yellow, pitted; seeds 2–2.2 mm. long, ellipsoid, compressed, brown. Meadows and open hillsides, *U. Son.* to *A. T. T.*

Forma **alba** St. John. Flowers white.

Washington: rocky slopes opposite Zindel, T. 7 N., R. 47 E., Asotin Co., May 20, 1922, *H. St. John & Rex H. Brown* 3,879 (type), *U. Son.*

Var. elatior Hook. Though originally recorded from the Spokane region, no material of this, certainly separable from the species has been seen. As a variant, it is rejected by Wherry, and by Cronquist (1959).

Phlox viscida E. Nels. Perennial, 10–18 cm. tall; tap-root dark, woody, up to 1 cm. in diameter; crown multicipital; stems much branched and suffruticose at base, often decumbent at base, very leafy; herbage viscid pilosulous or puberulent throughout; leaves 1–4 cm. long, 1–6 mm. wide, linear or linear-lanceolate, acute, rather rigid; cymes one to few flowered; pedicels 1–25 mm. long; calyx 6–11 mm. long, viscid pilose, the lobes subulate, the pubescence shorter and sparser towards the tips; corolla pinkish, the tube 9–12 mm. long, pilose within near the base; corolla lobes 6–8 mm. long, obovate; anthers 1.2 mm. long, elliptic-oblong; style exceeding the calyx; ovules solitary; capsule 5 mm. long, ellipsoid, yellow, pitted. Dry ridges, Blue Mts., *A. T. T.*

The type locality is Blue Mts., Columbia Co. The type was collected by Prof. C. V. Piper on June 15, 1896.

POLEMONIELLA

Annual, caulescent herbs; leaves odd-pinnate, the lower leaflets usually petiolulate, the upper sessile; flowers pediceled, solitary, oppositifolious; calyx green, herbaceous, accrescent, open campanulate, 5-cleft to or beyond the middle; corolla white, nearly rotate, included in the calyx; filaments not dilated at base, only remotely hairy; capsule 8–10-seeded. (Name Lat., a diminutive of *Polemonium.*)

Polemoniella micrantha (Benth.) Heller. Annual, 3–30 cm. tall, viscid villous, usually loosely branched lower leaves with petioles 5–15 mm. long, the rhachis 10–25 mm. long; the 7–15 leaflets 2–10 mm. long, ovate to elliptic, the terminal the larger; upper leaves gradually smaller and shorter petioled; pedicels 2–15 mm. long, ascending; calyx in anthesis 3–5 mm. long, broad campanulate, viscid villous, in fruit cupuliform, the tube 3–5 mm. long, the lobes 4–7 mm. long, lance-deltoid, the midrib strong, with netted connections to the marginal nerves; corolla broadly campanulate, almost rotate, the lobes almost 2 mm. long, oval; the filaments 1 mm. long, subulate; anther sacs each subglobose, 0.1 mm. long; style and linear stigmas equaling the filaments; capsule 3–4 mm. long, cartilaginous, subglobose; seeds 1.8–2.2 mm. long, shaped like the sector of an orange, dark olive brown, shining. *Polemonium micranthum* Benth. Abundant, open places, also persisting in cultivated ground, *U. Son., A. T.*

The flowering and fruiting characters of this plant combine with the habit to make it appear a good genus, in this family of mostly unsatisfactory, merging genera. Macbride has accepted and transferred to *Polemoniella* the two S. Am. species.

Despite J. F. Davidson's classification (1947), this seems to be a good genus, distinct from *Polemonium*.

POLEMONIUM. Jacob's Ladder

Biennial or perennial, caulescent herbs; leaves alternate; odd-pinnate, the leaflets sessile; calyx herbaceous and green throughout, 5-cleft, campanulate, not ruptured by the fruit; corolla campanulate or rotate-funnelform to tubular-funnelform; filaments dilated and hairy at base, equally inserted, included, more or less declined; ovules usually 3 to many in each cell. (Name from Gr., *polemos*, war.)

Leaflets linear; corolla lobes pilose on the outer face. — *P. pectinatum.*
Leaflets broader; corolla lobes glabrous,
  Inflorescence a narrow thyrse; plant 5–9 dm. tall. — *P. occidentale.*
  Inflorescence a rounded cyme; plant 1–3 dm. tall,
    Corolla 8–11 mm. long. — *P. californicum.*
    Corolla 7–8 mm. long. — *P. delicatum.*

Polemonium californicum Eastw. Perennial, 1.5–3 dm. tall, more or less viscid villous, less so below; rhizome 1–2 mm. in diameter, producing one to several shoots; basal leaves with petioles 1–9 cm. long, the base dilated, villous ciliate; cauline leaves similar but shorter petioled, the rhachis 3–12 cm. long, the 9–23 leaflets 5–30 mm. long, elliptic or lance-elliptic, membranous; cymes compact, several-flowered, each cymule 3–6-flowered; pedicels 2–20 mm. long, ascending, viscid puberulent and sparsely villous; calyx campanulate, not accrescent, the tube 2–3 mm. long, the lobes 2–4 mm. long, lanceolate, viscid puberulent and villous; corolla violet blue with a yellowish eye, broadly campanulate, the tube 2–4 mm. long, the lobes 6–8 mm. long, broadly oval; stamens about equaling the corolla, the filaments fusiform and pilose tufted at base; anthers 0.7–0.8 mm. long, oval; pistil equaling the corolla; stigmas 1 mm. long, linear; capsule 3–4 mm. long, ovoid; seeds 2 mm. long, lanceoloid, brown, shining. *P. columbianum* Rydb.; *P. pulcherrimum* Hook., var. *calycinum* (Eastw.) Brand. Meadows and moist thickets, Blue Mts., *Huds.*

Polemonium delicatum Rydb. Perennial, 1–3 dm. tall, glandular pilosulous, with the odor of skunk; rhizome 1–2 mm. in diameter, producing one or several shoots; basal leaves or those of the sterile shoots with the petiole 2–4 cm. long, dilated and ciliate at base, the rhachis 4–13 mm. long; cauline leaves with shorter petioles, the 9–27 leaflets 5–23 mm. long, mostly lanceolate or lance-elliptic, sometimes broader; cymes many flowered, compact, the cymules 4–8-flowered; pedicels 2–10 mm. long, ascending; calyx campanulate, glandular pilosulous and flattened villous, not accrescent, the tube 2–3 mm. long, the lobes 2–3.5 mm. long, lanceolate; corolla campanulate, blue with a yellow eye, the tube 2–3 mm. long, the lobes 4–6 mm. long, oval, glabrous; the stamens almost equaling the corolla, the filaments attached near summit of tube, the base fusiform, pilose tufted, the anthers 0.6–0.8 mm. long, oval; style exserted; stigmas 1.3 mm. long, linear; capsule 3–4 mm. long, ovoid; seeds 1.3–1.5 mm. long, lanceoloid, compressed, dark brown, shining. *P. pulcherrimum* Hook., subsp. *delicatum* (Rydb.) Brand; *P. pulcherrimum* Hook., var. *delicatum* (Rydb.) Cronq. Moist thickets, Blue Mts.; and Moran Mt., Spokane Co., *Huds.*

Polemonium occidentale Greene. Perennial, viscid villous above, often glabrate below; rhizome horizontal; stem single, erect; foliage mostly cauline; basal leaves long petioled; cauline leaves with petioles from 12 cm. long, reduced upwards, the rhachis 3–14 cm. long, the 13–23 leaflets 1–5 cm. long, lanceolate to ovate, membranous; inflorescence 3–15 cm. long, with mostly short pediceled flowers; calyx campanulate, densely viscid villous, the tube 3–4 mm. long, the lobes 3–4.5 mm. long, lanceolate; corolla blue, campanulate, the tube 3–4 mm. long, the lobes 7–9 mm. long, obovate, glabrous; stamens about equaling the corolla, the filaments hairy tufted, and inflexed at the obcuneate base; style well exserted; stigmas 1.5 mm. long, linear; capsules 3.5–5 mm. long, ovoid; seeds 2–3 mm. long, irregular in shape, but mostly like the sector of an orange, arcuate, dull, dark brown, pale tailed at each end. *P. acutiflorum* Willd. ex L., subsp. *occidentale* (Greene) Hultén; *P. coeruleum,* subsp. *amygdalinum* (Wherry) Munz; *P. coeruleum* L., var. *occidentale* (Greene) St. John; *P. intermedium* (Brand) Rydb. Wet meadows, Cheney, and Marshall Jct.; also Alder Cr., Benewah Co., *A. T., A. T. T.*

Polemonium pectinatum Greene. Perennial, 2.5–6 dm. tall, viscid villous above, glabrous below; root apparently cord-like, slender, erect; stems several; lowest leaves soon withering; middle cauline leaves 5–8 mm. long including the short petiole, the rhachis narrowly winged, the 11–17 leaflets 15–32 mm. long, glabrous; flowers numerous, in a compound cyme; pedicels shorter than the calyx; calyx campanulate, somewhat accrescent, viscid villous, in fruit the tube 2–5 mm. long, the lobes 3–5 mm. long, lanceolate; corolla rose-colored, campanulate, the tube 3–6 mm. long, the lobes 5–10 mm. long, obovate, cuneate; stamens almost equaling the corolla, the filaments fusiform and pilose tufted at base; anthers 1.2–1.5 mm. long, oval; style exserting the 1.5–2 mm. stigmas; capsule 4–5 mm. long, ovoid; seeds 2.5–3.5 mm. long, mostly shaped like slender sectors of an orange, brown, somewhat shiny. Moist places, *A. T.*

A rare species, discovered in 1882 by Hilgard, known definitely only from Oakesdale and Rock Lake; and Hangman Cr., Spokane Co.

HYDROPHYLLACEAE. Waterleaf Family

Herbs or shrubs, mostly caulescent, with watery sap; leaves alternate, rarely opposite or in basal rosettes, without stipules; flowers perfect, regular or nearly so, in more or less scorpioid cymes, which may be congested or capitate, or in racemes, or solitary; calyx 5–7-lobed, often appendaged in the sinuses; corolla 5-lobed, mostly campanulate or funnelform, or even rotate, often appendaged within, the lobes convolute or imbricate in bud; stamens 5, alternate; filaments adnate to base of corolla; styles 2, more or less united; stigmas often capitate; ovary of 2 united carpels, 2- or 1-celled with parietal placentae; ovules few or many; fruit a capsule. (Named from the genus *Hydrophyllum,* the waterleaf.)

Flowers solitary, on naked peduncles; plants acaulescent. *Hesperochiron.*
Flowers more numerous, or axillary; plants caulescent,
  Flowers solitary and axillary; calyx with a reflexed appendage from the sinus. *Nemophila.*
  Flowers in an inflorescence; calyx not appendaged,
    Inflorescence capitate or nearly so; placentae expanded. *Hydrophyllum.*

Inflorescence of scorpioid cymes; placentae thin, intruded in the cell. *Phacelia.*

HESPEROCHIRON

Acaulescent perennial herbs; leaves basal, in a rosette, petioled, entire, spatulate to ovate; flowers single, peduncled; calyx 5–7-parted, the lobes linear-lanceolate, often unequal; corolla campanulate to rotate, deciduous, without appendages; filaments hairy at base; ovary 1-celled with 2 intruding, dilated placentae; ovules 20 or more on each placenta; style 2-cleft; capsule loculicidal; seeds reticulate. (Name from Gr., *hesperos,* western; *Cheiron,* Chiron, a centaur.)

Corolla campanulate; plant hirsute throughout. *H. lasianthus.*
Corolla broadly funnelform or saucer-shaped, rotate; leaves glabrous beneath,
  Blades glabrous on both sides; corolla tube with a pilose fimbriate crown,
    Calyx-tube glabrous. *H. pumilus,* f. *pumilus.*
    Calyx-tube white villous. Forma *nervosus.*
  Blades appressed hispid above; corolla tube densely pilose within, lacking a crown. *H. villosulus.*

Hesperochiron lasianthus (Greene) St. John. Tuber 1–9 mm. in diameter, dark; herbage densely white hirsute; petioles 5–35 mm. long; the 4–11 blades 1–4 cm. long, oval to elliptic or lance-elliptic, thick, the secondary veins obscure; flowers numerous; peduncles 1–5 cm. long, recurved in fruit; calyx white hirsute, 4–9 mm. long, somewhat accrescent, parted nearly to the base into lance-linear lobes; corolla white, veined with lavender, the tube 6–10 mm. long, more or less pilosulous without, villous at base within; the lobes 6–11 mm. long, lance-ovate, spreading, glabrous without, sparingly long villous within; ovary pilose at tip; style shorter than the corolla tube; stigmas 0.5 mm. long, linear; filaments subulate, pilose at base, equaling the corolla tube; anthers 0.7–0.8 mm. long, ellipsoid; capsule 7–9 mm. long, ovoid, hirsute; seeds 1–1.3 mm. long, ovoid, compressed, reddish brown, prominently papillose. *Capnorea lasiantha* Greene; *H. californicus* (Benth.) Wats. var. *Watsonianus* (Greene) Brand, subvar. *lasianthus* (Greene) Brand. Dry open slopes or meadows, *U. Son., A. T.*

Hesperochiron pumilus (Dougl. ex Griseb. in Hook.) Porter, forma **pumilus.** Tuber 1–7 mm. in diameter, brown, cylindric or somewhat moniliform; leaves few, mostly 2–4; petioles 1–5 cm. long, appressed hispid ciliate, membranous dilated at base; blades 1–6 cm. long, mostly lanceolate, but varying from spatulate to obovate, glabrous but for the appressed hispid margin; peduncles 2–9 cm. long, glabrous; calyx 4–8 mm. long, accrescent, parted ⅔ of the way into lanceolate or lance-ovate, hispid ciliate lobes; corolla saucer-shaped, white with red veins, the tube 2–3 mm. long broad funnelform; filaments 5–6 mm. long, pilose at base; anthers 1–1.3 mm. long, ellipsoid, lavender; ovary pilose at tip; style 2 mm. long; stigmas 1 mm. long, linear; capsules not seen. *Capnorea pumila* (Dougl.) Greene. Open pine woods, *A. T. T.*

Forma **nervosus** (Greene) Brand. Differs in having the calyx white villous. *Capnorea nervosa* Greene. Meadows, Moscow, Chatcolet, and Worley, Ida., *A. T.*

The type is from Moscow, *Henderson,* May 13, 1894.

Hesperochiron villosulus (Greene) Suksd. Tuber cylindric or with irregular thickenings up to 5 mm. in diameter, brown; leaves 2–5; petioles 10–25 mm. long, appressed hispid; blades 10–45 mm. long, elliptic to oval, firm, appressed hispid above and on the margin, glabrous beneath; peduncles 1–5 cm. long, appressed hispid, deflexed in fruit; calyx broadly funnelform, 4–8 mm. long, accrescent, generally hispid on surface and margin with somewhat appressed hairs, cleft almost to the base into lanceolate or linear-lanceolate lobes; corolla white, apparently with reddish veins, rotate, the tube 3–4 mm. long, wide funnelform, the lobes 5–8 mm. long, ovate, at base cordate and overlapping; filaments 5–6 mm. long, subulate, villous below; anthers 1–1.3 mm. long, oblong-ellipsoid, lavender; ovary villous; style 1 mm. long; stigmas 1.5 mm. long, linear; capsule 5–8 mm. long, ovoid, villous; seeds 1–1.5 mm. long, ovoid, more or less compressed, reddish brown, strongly papillose. *Capnorea villosula* Greene. Prairies *A. T.*

The type is *Elmer,* 1,001, from Pullman.

HYDROPHYLLUM. Waterleaf

Perennial herbs, mostly caulescent, the sap watery; the blades pinnately parted or divided; flowers in compact, scorpioid racemes or cymes, or solitary; calyx deeply 5-lobed; corolla campanulate or funnelform, 5-lobed, bearing a pair of longitudinal scales meeting over the midrib and forming a nectar-bearing groove; filaments exserted, hairy at the middle; anthers introrse, versatile; ovary of 2 united carpels, 1- or 2-celled, with parietal placentae; styles more or less united; ovules 4, seeds 1–4. (Name from Gr. *hudor,* water; *phullon,* leaf.)

Corolla blue; calyx accrescent; leaflets paler green beneath, the lower margin entire. *H. capitatum,* var. *capitatum.*

Corolla mostly white; calyx not accrescent; leaflets whitened beneath, serrate or incised almost to the base,

- Pubescence harsh; corolla slightly exceeding the calyx. *H. Fendleri,* var. *Fendleri.*
- Pubescence soft; corolla much exceeding the calyx. Var. *albifrons.*

Hydrophyllum capitatum Dougl. ex Benth., var. **capitatum.** *Woollen Breeches; Cat's Breeches.* Plant 1–4.5 dm. tall; roots clustered, tuberous, 1–3 mm. in diameter; basal leaves similar to the upper, but the petioles decreasing upwards; petioles 1–20 cm. long, hirsute; blades 3–13 cm. long, pinnately cleft or at base divided into 5–9 divisions 2–7 cm. long, entire or palmately cleft, the lobes oblong or elliptic-oblong, chartaceous, mucronate, above hirsute, dark green, beneath hirsute; cymes densely flowered, subcapitate, the upper subsessile, the lower on peduncles up to 8 cm. long; pedicels 2–5 mm. long, densely hirsute; calyx in anthesis 3–7 mm. long, campanulate, parted almost to the base into linear-oblong, acute segments, appressed hirsutulous on the back, long hispid ciliate, in fruit 8–11 mm. long; corolla broadly obconic, the tube 3–4 mm. long, the lobes 3–4 mm. long, obovate-oblong, sparsely pilose near the tip without, scales 5–5.5 mm. long; filaments 9–11 mm. long, sparsely villous at the middle; anthers 0.7–1 mm. long, oblong, purplish; ovary hirsute; style 8–11 mm. long sparsely pilosulous below, with two branches 1–1.5 mm. long; stigmas subcapitate; capsule 3–4 mm. long, subglobose; seeds 2.8–3 mm. long, ovoid, more or less compressed, brown, the testa raised in a pale, tiny reticulate network. *H. capitatum,* var. *pumilum* (Geyer) Hook.; and a subvar. *densum* Brand. Open slopes or open woods, *U. Son* to *A. T. T.*

Cusick reports that it is often cooked and eaten as "greens."

Hydrophyllum Fendleri (Gray) Heller, var. **Fendleri.** Perennials, 2.5–9 dm. tall; rhizome scaly, short, with fleshy roots; petioles 3–16 cm. long, retrorse hispid; blades 6–30 cm. long, oblong to oval in outline, pinnatifid, the principal divisions 7–19, ovate to lanceolate, acuminate, 2–12 cm. long, coarsely serrate to incised, strigose, the lower distinct, the upper confluent, below pale and the veins hispid; peduncles 2–13 cm. long, hispid; pedicels 2–6 mm. long; calyx parted almost to the base, the lobes 4–6 mm. long, linear-lanceolate, acute, ciliate and strigose; corolla 6–8 mm. long, white or tinged with violet, the lobes 3–4 mm. long, obtuse; capsules about 4 mm. in diameter; seeds 2.5–3 mm. in diameter, light brown. Tucannon R., *Darington;* and Moscow region.

Hydrophyllum Fendleri (Gray) Heller, var. **albifrons** (Heller) Macbr. Plant 3–6.5 dm. tall; rootstock short, with numerous fleshy rootlets 1–4 mm. in diameter; stems few branched; lower leaves with petioles 15–25 cm. long, these being longer and the leaflets mostly smaller and narrower than those of the upper leaves; upper leaves with petioles 1–20 cm. long, hirsute; the blade pinnate, the 2–3 lateral pairs of leaflets subopposite 3–8 cm. long, obliquely lance-ovate, coarsely serrate and often deeply 2–3-lobed, thin, above bright green and remotely hirsute, beneath appressed pilose; the terminal lobe of 5–7 confluent pinnae; peduncles 1–12 cm. long, densely reflexed hirsutulous; cymes compact, many flowered; pedicels 1–7 mm. long, hirsutulous; calyx 5–7 mm. long, parted almost to the base into linear lobes puberulent on the back and strongly hispid ciliate; corolla broad campanulate, white with a purple blotch near the top of each lobe, the tube 2.5–3 mm. long, the lobes 4–5 mm .long, ovate, softly puberulent up the back, scales 3.5 mm. long; filaments 7–8 mm. long; anthers 1.1–1.5 mm. long, oblong, arcuate; ovary villous; style dimorphic, 2 mm. long with 1 mm. branches, or 10 mm. long with 2 mm. branches; stigmas capitate; capsules 3–4 mm. long, subglobose, hispid; seeds 2.5–3 mm. long ovoid, or the inner sides compressed, brown, with the straw-colored testa raised in a reticulate network. *H. albifrons* Heller. Moist thickets, *U. Son.* to *Can.*

The type specimen, *Heller* 3269 was from Lake Waha.

NEMOPHILA

Diffuse annual herbs; leaves alternate or more commonly opposite, pinnately lobed or pinnatifid; flowers axillary and solitary or racemose; calyx 5-lobed, with a reflexed appendage in each sinus; corolla campanulate to nearly rotate, purple, blue, or white, with 10 minute basal scales between the stamens; stamens included; ovary 1-celled, seeds 1–20. (Name from Gr., *nemos,* grove; *phileo,* to love.)

Leaves alternate; cucullus reduced but persistent. — *N. breviflora*

Leaves all or at least the basal ones opposite; cucullus deciduous,
- Stamens adnate at base of corolla tube; corolla scales broad cuneate, fimbriate; seeds 3–4 mm. long. — *N. Kirtleyi.*
- Stamens adnate ⅓ way up the corolla tube; corolla scales linear and ciliate or reduced to hairy lines or none. — *N. parviflora,* var. *Austinae.*

Nemophila breviflora Gray. Plant 5–30 cm. tall, sparsely hirsute throughout; roots fibrous; stem simple or diffusely branched; cotyledons opposite, petioled, elliptic; leaves with the petioles 3–25 mm. long, the base ciliate, somewhat dilated; blades 6–30 mm. long, membranous, pinnately parted into 5 lobes, the lobes 4–18 mm. long, obliquely oblong-lanceolate, beneath pale green; flowers almost axillary; pedicels short in flower, in fruit becoming reflexed and 6–15 mm. long, retrorse hispid near the middle; calyx campanulate, in flower the tube 1 mm. long, the lobes 3–3.5 mm. long, lanceolate, hispid ciliate, in fruit the tube 1 mm. long, the appendage lanceolate, 2–2.5 mm. long, the calyx lobes 5–6 mm. long; corolla blue or becoming whitish, campanulate, the tube 1–1.5 mm. long, the lobes 1 mm. long, ovate; filaments 1.5 mm. long; anthers 0.1 mm. long, ovoid; style 0.5 mm. long, pilose at base; capsule 3–4 mm. long, subglobose, hispidulous; the 1 seed 2.8–3 mm. long, globose, orange-brown, with a reticulate caruncle. *Viticella breviflora* (Gray) Macbr. Woods, Blue Mts., Moscow Mt., *Can.*

Nemophila Kirtleyi Henders. Plant 5–20 cm. tall, hirsute throughout; roots fibrous; stem purplish below; cotyledons opposite, petioled, oval; lower leaves opposite, the petioles 1–4 cm. long, hirsute and the margins hispid; the blades 12–43 mm. long, hispid ciliate, sinuately 5-cleft, the lobes oblong to ovate; upper leaves alternate, smaller; flowers axillary; pedicels in flower exceeding the petioles, in fruit 3–4.5 cm. long, and recurved, hirsute throughout; calyx broadly campanulate, hirsutulous, the margin hispid, in flower the tube 1–1.5 mm. long, the lobes 2–3.5 mm. long, in fruit the tube 3–4 mm. long, the appendage 2.5–3.5 mm. long, lanceolate, the lobes 7–8 mm. long; corolla whitish, with a blue margin, broadly campanulate, the tube 3 mm. long, bearing flabellate fimbriate scales, the lobes 6–7 mm. long, oval, ciliate; filaments 5 mm. long; anthers 2–2.5 mm. long, oblong, arcuate; style 3 mm. long, hirsute, with 2 branches 1 mm. long; capsule 4–6 mm. long, subglobose, hirsutulous; the 2–4 seeds 2.2 mm. long, ovoid, brown, with a reticulate caruncle. *Viticella Kirtleyi* (Henders.) Macbr. Abundant in talus slopes and thickets, canyons of the Clearwater and the Snake, *U. Son.*

Nemophila parviflora Dougl. ex Benth., var. **Austinae** (Eastw.) Brand. Stems 0.5–3 dm. long, weak, succulent, brittle, sparsely hispid or glabrate; leaves opposite; petioles 4–22 mm. long, margined toward the apex, the upper leaves with longer petioles; blades 8–24 mm. long, 9–24 mm. wide, finely appressed hispid, pinnately 5–7-lobed ⅓–⅔ way to midrib; calyx campanulate, the lobes 1–3 mm. long, lanceolate, the auricles 0.2–0.4 mm. long; corolla 1.5–3 mm. broad white or bluish, the lobes obovate, about as long as the tube, the corolla but slightly longer than the calyx; filaments 0.9 mm. long; anthers 0.2 mm. long; style 0.6–0.8 mm. long; capsule 3 mm. long; seeds 2–2.5 mm. long, yellow to red, smooth but shallowly pitted. Anatone Butte, *F. G. Meyer* 424, *A. T. T.*

PHACELIA

Perennial or annual, caulescent herbs; leaves alternate, or the lower opposite; flowers in scorpioid racemes or cymes; calyx 5-parted, the lobes narrow, somewhat accrescent, without appendages; corolla purple, blue, or white, appendaged within, with 10 vertical plaits in pairs between the filaments; corolla lobes imbricate; ovary 1-celled or nearly 2-celled by the intrusion of the placentae; ovules 4 to many; seeds reticulate, pitted, or corrugated. (Name from Gr. *phakelos,* cluster.)

Plant densely viscid pubescent throughout; stems weak, decumbent. *P. ramosissima,* var. *ramosissima.*

Plant not so,

Stem glabrate below, not hispid; ovules numerous. *P. idahoensis.*

Stem pubescent, hispid at least above; ovules 1–4,

Corolla 7–11 mm. long; leaves sessile, mostly linear and entire, sometimes pinnately parted with 3–5 linear lobes. *P. linearis.*

Corolla 4–6 mm. long; leaves petioled,

Leaves entire; foliage with a closely appressed, sericeous pubescence. *P. leucophylla.*

Leaves usually pinnate, the lateral leaflets much the smaller,

Inflorescence a virgate thyrsus. *P. heterophylla,* var. *heterophylla.*

Inflorescence of loosely divergent cymes. Var. *griseophylla.*

Phacelia heterophylla Pursh var **heterophylla.** Biennial or perennial, 1.5–12 dm. tall, hirsutulous and more or less hispid throughout; root woody, brown, up to 15 mm. in diameter, bearing 1 to several stems; stems erect, leafy; basal leaves often in a dense rosette, more commonly simple, the petiole 5–9 cm. long, the blade 15–75 mm long, oblanceolate, acute, hispid on the margins and veins, closely gray hirsutulous elsewhere, firm chartaceous, less commonly pinnate with 1–3 pairs of smaller lateral leaflets below the terminal one; cauline leaves alternate, reduced and with shorter petioles upwards, simple or pinnate or both; thyrsus narrow, 0.5–9 dm. long; lateral cymes densely flowered, in fruit 2–4 cm. long; pedicels 1–3 mm. long; calyx campanulate, in flower 3–4 mm. long, in fruit 5–7 mm. long, parted almost to the base, the lobes linear-oblong, acute, densely hairy, closely hirsutulous and bristly hispid especially on the margins, the hairs drying yellowish; corolla campanulate, white, the tube 3 mm. long, the scales of the appendage 2.5 mm. long, united for 1 mm.; the lobes 2 mm. long, pilosulous without; filaments 10–14 mm. long, exserted, pilose; anthers 0.5 mm. long, oval; style 7–8 mm. long, pilose, the 2 branches 4–5 mm. long; capsule 3–5 mm. long, ovoid, acuminate, hispid; the 1 seed 1.5–1.8 mm. long, lanceoloid, brown, finely punctate. *P. heterophylla,* var. *typica* Dundas. Dry slopes, in the deep valleys, *U. Son., A. T.*

The type was collected by Capt. M. Lewis, "On dry hills on the banks of the Kooskoosky." [Clearwater River, Ida.]

Var. **griseophylla** (Brand) Macbr. Plant 1–5 dm. tall, the cymes divergent; inflorescence not virgate, mostly round-topped. Open slopes, *A. T., Huds.*

Dundas in 1935 presented a new treatment, reducing var. *griseophylla.* He maintained *frigida* as a variety. Perhaps the dwarfed alpine form may be worthy of a name, but as it is based only on the stature of the plants from alpine habitats, it seems unimportant. The writer considers Macbride's var. *griseophylla,* based on the kind of inflorescence, to be a more natural entity.

Phacelia idahoensis Henders. Perennial, 3.5–10 dm. tall, glabrate at base; rootstock woody, dark, up to 5 mm. in diameter; stem erect, simple, leafy, appressed pilosulous above; petioles alternate, up to 5 cm. in length, reduced upwards; blades 1.5–9 cm. long, pinnately parted into 5–9 oblong or linear, acute segments, strigose, beneath pale green, the lobes entire or again cleft into similar subdivisions; thyrsus 4–30 cm. long, densely flowered, the lateral scorpioid cymes 0.5–2 cm. long; pedicels 0.5–4 mm. long, hirsutulous and hispid; calyx campanulate, 4–5 mm. long, not accrescent, parted almost to the base, the lobes linear, hirsutulous, hispid ciliate, the midrib not prominent; corolla blue, campanulate, the tube 3 mm. long, the appendages 2.5 mm. long, ciliolate, prominent broad scales; the lobes 2–3 mm. long, oblong-ovate, pilosulous without; filaments 5–7 mm. long, the tips incurved; anthers 0.7–0.9 mm. long, broadly ellipsoid; style 3 mm. long, the 2 branches 0.5 mm. long; capsule 5–7 mm. long, ovoid, acute, hirsute; seeds numerous, 1.3–1.7 mm. long, ovate to ellipsoid, short pointed, irregularly compressed, dull brown, covered with rows of coarse pits. Meadows, Craig Mts., *A. T. T.*

Only record is the type collection, *Henderson* 2,770.

Phacelia leucophylla Torr. in Frém. Perennial, 1.5–5 dm. tall, sericeous with a closely appressed, usually grayish white, pubescence, also often hirsute; root brown, woody, up to 7 mm. in diameter; stems 1 to several, often decumbent at base; leaves entire, alternate; the basal leaves forming a rosette, the petioles 1–4 cm. long, the blades 1–10 cm. long, lanceolate to linear-lanceolate, firm chartaceous; cauline leaves gradually smaller and shorter petioled up the stem; inflorescence 3–35 cm. long, in flower compact, in a fruit a loose, divergent cyme; pedicels 0–2 mm. long; calyx campanulate, in flower 2–3 mm. long, in fruit 4–6 mm. long, parted almost to the base, the lobes oblanceolate to linear, hirsutulous and densely hispid ciliate; corolla white, campanulate, the tube 2.5–3.5 mm. long, the appendages 1 mm. long; the lobes 2–3 mm. long, ovate, pilosulous without; filaments 8–10 mm. long, pilose, exserted; anthers 0.5 mm. long, oval; style 5–6 mm. long, pilose, the 2 branches 3–4 mm. long; capsule 2.5–3 mm. long, ovoid, acute, hirsute; the 1 seed 2–2.2 mm. long, lanceoloid, brown, finely pitted. *P. leucophylla,* var. *typica* Dundas; *P. hastata* Dougl. ex Lehm., var. *leucophylla* (Torr.) Cronq. Dry prairies or open woods, *A. T., A. T. T.*

In the type region, east of the Rockies, the flowers are lilac.

Phacelia linearis (Pursh) Holz. Annual, 1–4.5 dm. tall, hirsutulous throughout and more or less hispid; tap-root up to 5 mm. thick, the rootlets fibrous; stems erect, simple or with numerous ascending branches; basal leaves oblanceolate, soon withering; cauline leaves alternate, 15–75 mm. long, sessile, pustulate hirsutulous, hispid ciliate, entire and linear or lance-linear, or pinnately parted into 3–5 linear lobes; racemes often numerous circinate, compact, in fruit 4–11 mm. long; pedicels 0.5–5 mm. long; calyx campanulate, in flower 5–8 mm. long, in fruit 8–12 mm. long, parted almost to the base, the lobes linear, 1-ribbed, hirsutulous, bristly hispid ciliate; corolla rotate-campanulate, violet-blue, the tube 3–6 mm. long, the appendages 4 mm. long; the lobes 4–5 mm. long, suborbicular, sparsely pilosulous without; filaments 8–12 mm. long, sparsely pilose; anthers 0.9–1.2 mm. long, purplish, oblong, arcuate; style 4 mm. long, the 2 branches 2 mm. long; capsule 4–6 mm. long, lanceoloid, hispidulous, somewhat hispid; seeds 1.2–1.5 mm. long, ellipsoid to ovoid or compressed, brown, covered with rows of large pits. Abundant, open or rocky slopes, open woods, *U. Son.* to *A. T. T.*

Phacelia ramosissima Dougl. ex Lehm., var. **ramosissima.** Perennial, 3–9 dm. tall, viscid-hispid, and -hirsutulous, or -pilose throughout; root woody, dark brown, up to 6 mm. in diameter; stems several, decumbent and weak, loosely branched; leaves alternate; petioles 0–5 cm. long; blades 2–15 cm. long, 2–3-pinnatifid, the 5–9 pinnae 2–6 cm. long, ovate or elliptic in outline, with ovate lobes or teeth; spikes scorpioid, densely flowered, in fruit 3–8 cm. long; calyx narrowly campanulate, in flower 4–5 mm. long, in fruit 7–9 mm. long, cleft almost to the base, the lobes viscid hirsutulous, hispid ciliate, with a lanceolate or elliptic tip narrowed to a linear claw nearly as long; corolla whitish or bluish, campanulate, the tube 5–6 mm. long, the appendages 4 mm. long, expanded half way up; the lobes 2–3 mm. long, sparsely pilosulous without; filaments 9–13 mm. long, exserted but incurved; anthers 0.8–1 mm. long, oblong, purplish; style 10 mm. long, 2-branched to the middle, the base pilose; capsule 3–4 mm. long, ellipsoid, hirsutulous; the 2–4 seeds 2.5–3 mm. long, lanceoloid, compressed on the inner sides, brown, finely punctate. *P. ramosissima,* f. *decumbens* (Greene) Brand. Shaded cliffs, Bonnie Lake, *St. John et al.* 3,096; Palouse Falls, *St. John & Pickett* 6,155, *U. Son.*

BORAGINACEAE. Borage Family

Herbs, shrubs, or trees, generally rough hairy; leaves alternate, or sometimes opposite or verticillate, simple; stipules wanting; flowers in scorpioid cymes, racemes, or spikes; calyx 5-lobed; corolla regular, white, or pink in bud changing to blue, 5-lobed, with 5 alternate stamens attached to its tube; ovary superior, usually deeply 4-lobed, with a simple style inserted between the lobes, in fruit splitting into 4 one-sided nutlets; nutlets attached by the base or inner angle or face to the receptacle, which is sometimes elongated (the *gynobase*). Mature fruits are necessary for the successful identification of members of this family. (Named from the genus *Borago,* the borage.)

Style buried in the pericarp of the fruit, falling with it; stigma discoid; corolla plaited in bud. *Heliotropium.*
Style borne between the lobes of the fruit (nutlets) and attached to the gynobase, not falling with the nutlets; stigma capitate; corolla not plaited,
  Corolla irregular, oblique; stamens unequal. *Echium.*
  Corolla regular; stamens equal,
    Attachment of nutlet surrounded by a tumid rim, leaving a pit on the flat or low convex gynobase. *Anchusa.*
    Attachment of nutlet otherwise,
      Stigmas geminate,
        Corolla throat with 5 vertical honey guides (lines of hairs); anthers with a sterile tip; style tip bilobed, prolonged beyond the stigmas. *Buglossoides.*
        Corolla throat with appendages lacking or broad; anther tip rounded or obscurely apiculate; style tip bilobed but shorter than the stigmas. *Lithospermum.*
      Stigmas single, capitate,
        Corolla lobes convolute in bud; nutlets smooth, with a narrow basal attachment. *Myosotis.*
        Corolla lobes imbricate in bud,
          Fruiting calyx markedly accrescent, very veiny, irregularly toothed and lobed, plicate. *Asperugo.*
          Fruiting calyx moderately if at all accrescent, not conspicuously veiny or irregularly lobed, not plicate.
            Cotyledons 2-lobed; corolla orange or yellow, not appendaged. *Amsinckia.*
            Cotyledons not lobed; corollas mostly white or blue, usually appendaged,
              Nutlets with a ventral groove formed by the unfused pericarpial walls. *Cryptantha.*
              Nutlets with the pericarpial walls fused at least above the middle forming a ventral keel,

Back of nutlets not encircled by an upturned rim, not glochidiate prickly,
- Corolla white, the tube not longer than the calyx; style usually shorter than the nutlets. *Plagiobothrys.*
- Corolla blue, the tube usually much longer than the calyx; style usually much longer than the nutlets. *Mertensia.*

Back of nutlets encircled by an upturned, glochidiate prickly rim,
- Nutlets equaling the subulate gynobase, attached for nearly their whole length; style usually surpassing the nutlets. *Lappula.*
- Nutlets twice longer than the stout, pyramidal gynobase, attached above the middle by a broad areola; style shorter than the nutlets. *Hackelia.*

AMSINCKIA. Tar Weed

Annual herbs, rough hispid, from pustulate bases; leaves alternate, oblong, or linear; flowers in naked or sparsely bracted, scorpioid spikes; calyx 5-parted, persistent, two or several of the lobes somewhat united, or the lobes free; corolla yellow to orange, salverform to tubular-funnelform, the limb sometimes plaited at the sinuses; style filiform; stigma capitate or 2-parted; ovules 4; nutlets 1–4, smooth or rough, crustaceous or coriaceous, ovate-triangular, attached by the middle to an oblong-pyramidal gynobase. (Named for *Wilhelm Amsinck,* burgomaster of Hamburg and patron of its Botanic Garden.)

Calyx-lobes 5, distinct; base of the corolla tube 10-nerved; nutlets with a central dorsal ridge, thus 4-sided,
- Corolla glabrous in the throat; stamens attached above the middle of the corolla tube,
  - Corolla 6–7 mm. long; fruiting calyx 6–8 mm. long; nutlets 2.5–3.5 mm. long. *A. retrorsa.*
  - Corolla 4–5 mm. long; fruiting calyx 4–6 mm. long; nutlets 1.8–2 mm. long. *A. micrantha.*
- Corolla with 5 hairy tufts in the throat; stamens attached below the middle of the corolla tube. *A. lycopsoides.*

Calyx-lobes 3, one narrow, and 2 broader ones each formed by the union of two; base of the corolla tube 20-nerved; nutlets low convex on the back, thus 3-sided,
- Nutlets short acuminate, densely low tuberculate, transverse ridges lacking or short and indistinct; corolla 9–10 mm. long. *A. washingtoniensis.*

Nutlets ovoid, with elongate transverse ridges and less prominent papillae; corolla 7–9 mm. long,
Nutlets 2.7–2.9 mm. long, the papillae few and mostly near the margins. *A. densirugosa.*
Nutlets 2.8–3.2 mm. long, the papillae in rows between the prominent transverse ridges. *A. Hendersonii.*

Amsinckia densirugosa Suksd. Plant 3–4 dm. tall; tap-root 2–3 mm. in diameter, brown; stem usually with loose, ascending branches; basal and lower leaves 2–4 cm. long, linear-oblanceolate; upper leaves 2–5 cm. long, lanceolate to linear-lanceolate, sessile; calyx 5–6 mm. long, in fruit 8–13 mm. long, the lobes linear-lanceolate, the 2 connate ones bifid at apex; corolla 7–9 mm. long, glabrous within; stamens attached at one level, above the middle of the corolla tube; nutlets broadly ovoid, covered with prominent transverse ridges. By railroads or cultivated land, Spangle, *Suksdorf* 8718; Waitsburg, *Horner* 146, *A. T.*

Ansinckia Hendersonii Suksd. Plant 1–3 dm. tall; tap-root 1–3 mm. in diameter, brown; stem simple or commonly freely branched; basal and lower leaves 3–8 cm. long, linear-oblanceolate; upper leaves 2–6 cm. long, sessile, lanceolate; calyx 4–5 mm. long, in fruit 9–12 mm. long, the lobes linear-lanceolate, the 2 connate ones short bifid at apex; corolla 7–9 mm. long, pale yellow, glabrous within; stamens attached at one level, above the middle of the corolla tube; nutlets ovoid, with heavy transverse ridges, alternating with rows of low papillae. Rocky slopes, mouth of Grand Ronde, *St. John* 3,530; Indian, Whitman Co., *Spiegelberg* April 21, 1923, *U. Son.*

Amsinckia lycopsoides Lehm. Plant 1–8 dm. tall; tap-root 1–4 mm. in diameter, brown; stem simple or few branched; basal and lower leaves 2–14 cm. long, almost linear; upper leaves 2–6 cm. long, sessile, lanceolate or linear-lanceolate; calyx 4–7 mm. long, in fruit 8–10 mm. long, the lobes linear-lanceolate; corolla 7–11 mm. long, yellow; stamens attached at the same level; nutlets 2.5–3 mm. long, ovoid with many papillae, the larger in rows or ridges. *A. arenaria* Suksd.; *A. simplex* Suksd., *A. hispidissima* Suksd. Dry or cultivated places, *U. Son., A. T.*

This and other species are abundant and troublesome weeds in the grain fields. The hispid hairs are detached during thrashing and float suspended in the warm air. The perspiring workers find them very irritating. The genus is in need of a careful monographic revision.

This, and probably other, species are fatally poisonous, causing cirrhosis of the liver, called "walking disease" of horses, and "hard liver" or "Walla Walla hard liver" in swine.

Amsinckia micrantha Suksd. Plant 2–8 dm. tall; tap-root 1–3 mm. in diameter, brown; stem simple or with ascending branches; basal and lower leaves 3–12 cm. long, narrowly oblanceolate; upper leaves 1–4 cm. long, sessile, lanceolate or lance-linear; calyx 2.5–3 mm. long, the lobes lanceolate; corolla yellow; stamens unequally attached near the throat; nutlets ovoid, closely low tuberculate. Open places, Pullman, Spokane, *A. T.*

Amsinckia retrorsa Suksd. Plant 2–7 dm. tall; tap-root 1–4 mm. in diameter, brown; stem simple or with ascending branches; basal and lower leaves 3–13 cm. long, linear; upper leaves 1–4 cm. long, lance-linear; calyx 4–5 mm. long, the lobes lance-linear; corolla glabrous in the throat; stamens attached at different levels; nutlets broadly ovoid, acute, densely tuberculate, and more or less cross ridged. Open places or cultivated lands, *U. Son., A. T.*

Amsinckia washingtoniensis Suksd. Plant 2–6 dm. tall; tap-root 1–4 mm. in diameter, brown; stem simple or with ascending branches; basal and lower leaves 2–7 cm. long, linear-oblanceolate; upper leaves 2–4 cm. long, sessile, lanceolate to linear-lanceolate; calyx 6–7 mm. long, in fruit 10–13 mm. long, the 3 lobes linear-lanceolate, the two connate ones bifid at apex; corolla pale yellow, glabrous in the throat; stamens attached at the same level, above the middle of the corolla tube; nutlets 2.5–2.8 mm. long, ovoid. Bluffs of Pataha Creek, Dodge, *Courtney,* May 3, 1921, *A. T.*

P. H. Ray and Chisaki (1957) in a cytotaxonomic study have reduced all the species to 13 species, mostly Californian. Their conclusions appear to be preliminary, and are not adopted here.

ANCHUSA. Alkanet

Annual or perennial herbs; leaves opposite; flowers in dense, scorpoid racemes; calyx 5-parted, with small teeth, but little enlarged in fruit; corolla blue, violet, white, or seldom yellow, tubular-funnelform, with 5 scale-like, papillose appendages in the throat; nutlets erect, oblique or incurved, scar basal or oblique, concave, appendaged with a hardened ring. (Name Gr., *agchousa,* the alkanet.)

Nutlets erect; corolla limb 12–20 mm. wide; calyx parted almost to the base. *A. azurea.*
Nutlets horizontal; corolla limb 5–9 mm. wide; calyx lobed to about the middle. *A. officinalis.*

Anchusa azurea Mill. *Alkanet.* Biennial or perennial, densely papillose hispid throughout, 2–10 dm. tall; basal leaves with petioles 5–10 cm. long, the blades 10–20 cm. long, oblanceolate; cauline leaves shorter petioled and the upper sessile, the blades 3–13 cm. long, lanceolate; racemes many flowered, in fruit loose, 10–25 cm. long; pedicels 1–12 mm. long, ascending; calyx cylindric-campanulate, bristly hispid, in anthesis 6–7 mm. long, in fruit 10–13 mm. long, the lobes linear; corolla blue, the tube 8–10 mm. long, the appendages 3 mm. long, ovate, long fimbriate pilose; the lobes oval; anthers 2.5 mm. long, linear; style 1 cm. long; nutlets 6–7 mm. long, yellow, lanceoloid, the back concave, it and the compressed inner faces with mountainous ridges, the scar basal, crater-like. Native of the Mediterranean region, cultivated as an ornamental, escaped rarely. Pullman, escaped in 1916, *A. T.*

Anchusa officinalis L. *Common Bugloss.* Biennial or perennial, 2–10 dm. tall, hispid throughout; root dark, woody, up to 1 cm. in diameter; stem erect, simple, or few branched below the inflorescence; basal leaves with petioles 2–6 cm. long, the blades 3–15 cm. long, lanceolate; upper leaves shorter petioled or sessile 2–10 cm. long, lanceolate or oblanceolate; inflorescence dense, many flowered, in fruit 5–30 cm. long; pedicels up to 2 mm. in length; calyx campanulate, rough hispid, in flower 4–5 mm. long, in fruit 6–8 mm. long, parted to the middle or below into lance-linear lobes; corolla violet to carmine or white, 10–15 mm. long, the appendages ovate, densely papillose puberulent; anthers 2 mm. long, linear; style 7 mm. long; nutlets 2.5–4 mm. long, obliquely ovoid, brown, warty. A European plant, cultivated as an ornamental, escaped near Spokane and Spangle, *A. T.*

ASPERUGO. Madwort

Annual herb; leaves alternate, entire, or the upper sometimes opposite; flowers 1–3 in the upper axils; calyx campanulate, unequally 5-cleft, much enlarged and folded together in fruit, the lobes incised-dentate; corolla blue or whitish, tubular-campanulate, imbricate; stamens included; stigma capitate; the 4 nutlets erect, ovoid, granular-tuberculate, keeled, laterally attached above the middle to the elongate conic receptacle. (Name from Lat., *asper,* rough.)

Asperugo procumbens L. *Madwort.* Plant 1–7 dm. long, often procumbent, rough hispid; roots fibrous; stems weak, often branched, 4–12-angled and retrorse prickly hispid on the angles; lower leaves with petioles 1–4 cm. long, upper leaves sessile; blades 2–6 cm. long, linear-lanceolate to obovate, thin chartaceous, rough hispid; flowers subsessile, but after anthesis the pedicels elongating and reflexing, up to 5 mm. in length; calyx hispid, in flower scarcely 3 mm. long, in fruit 1–1.5 cm. long, with prominent reticulate nerves, the margins hispid-ciliate with ascending curved hairs; corolla 2–3 mm. long; appendages broadly ovate, ciliate; corolla lobes oval; anthers 0.5 mm. long, lance-oblong; style 0.3 mm. long; nutlets 3–3.5 mm. long, compressed, olive to brown, the scar short elliptic. A weed from Eurasia, introduced to cultivated fields, *A. T.*

BUGLOSSOIDES

Annuals or perennials, herbaceous or suffruticose; cymes bracted; calyx 5-parted or 5-lobed; corolla blue to white, funnelform; stamens included; anthers lance-oblong to oblong; nutlets 1–4, erect to strongly divergent, smooth or rough, the attachment basal or oblique, the ventral suture fused; gynobase flat or depressed pyramidal. (Named to show similarity to *Buglossum,* from Gr. *bouglosson,* a plant name used by Dioscorides for a member of the Boraginaceae; and *oides,* from Gr. *eidos,* similar to.)

Buglossoides arvense (L.) Johnston. *Corn Gromwell.* Annual or biennial, 1–7.5 dm. tall, appressed hispidulous strigose throughout; basal leaves long petioled, oblanceolate; cauline leaves short petioled or the upper sessile, 1–5 cm. long, oblanceolate to lanceolate or linear, the hairs pustulate based; flowers sessile or nearly so in spikes, dense in anthesis, in fruit 5–45 cm. long, the fruits remote; bracts leafy, gradually reduced upwards; calyx campanulate, in anthesis 4–5 mm. long, in fruit 7–10 mm. long, strigose hispid, parted to near the base, the lobes linear, acuminate; corolla 5–6 mm. long, narrowly funnelform, the lobes 1.5–2 mm. long, ovate, ascending; anthers 1 mm. long, linear; style 1 mm. long; nutlets brownish, lance-pyramidal, oblique, cartilaginous muriculate below, ridged, and keeled above; the scar basal, nearly oval, narrowly margined. *Lithospermum arvense* L. A Eurasian weed, introduced in our area, Cheney, *Patton* in 1925; near R. R., Spokane, *Milburge* in 1931 (Gray Hb.), *A. T.*

CRYPTANTHA. NIEVITAS

Annual or perennial herbs or shrubs, with stiff pubescence; leaves firm, opposite at base, alternate above; flowers in bractless or bracted spikes or racemes; calyx usually parted to near the base, the lobes erect or connivent, linear or oblong; corolla white, or rarely yellow, the tube short cylindric, with or without scales at the base inside, the throat with intruded appendages, the lobes imbricate, rounded, spreading; style slender; ovules 2–4; the 1–4 nutlets erect, ovoid to triangular, roughened or smooth, margined or marginless, attached laterally through a median ventral and commonly basally forked groove to a usually columnar subulate or pyramidal gynobase. (Named from Gr., *kruptos,* hidden; *anthos,* flower.)

Spikes bracted throughout; nutlets 3–4 mm. long. *C. Sheldonii.*
Spikes naked at least above; nutlets 1.5–3 mm. long,
  Nutlets all smooth,
    Hairs on calyx uncinate or arcuate. *C. flaccida.*
    Hairs on calyx straight,
      Nutlets with excentric groove.
        Blades 2–8 mm. wide, oblanceolate; calyx lobes lanceolate; nutlets 1.8–2.5 mm. long. *C. affinis.*
        Blades 0.6–1.5 mm. wide, linear; calyx lobes linear; nutlets 1–1.2 mm. long. *C. Eastwoodae.*
      Nutlets with the groove central,
        Corolla 4–7 mm. wide; spikes commonly ternate. *C. Hendersoni.*
        Corolla about 1 mm. wide; spikes usually in pairs. *C. Torreyana,* var. *Torreyana.*
  Nutlets or some of them rough,
    Corolla 4–7 mm. wide. *C. Hendersoni.*
    Corolla about 2 mm. wide. *C. simulans.*

Cryptantha affinis (Gray) Greene. Annual; 1–4 dm. tall, hirsute throughout; roots fibrous; stem erect, usually with ascending branches; leaves 1–5 cm. long, oblanceolate, obtuse; spikes solitary or paired, slender, remotely flowered, in fruit 5–15 cm. long, with a few large bracts at base; calyx campanulate, ascending, laterally compressed, hirsute and spreading hispid, the lobes linear-lanceolate; corolla 1–2 mm. long, the limb about 1.5 mm. broad; the 4 nutlets similar, 1.8–2.5 mm. long, ovoid, obliquely compressed, shiny, smooth or granulate, usually mottled green and brown, the back convex, the margins rounded, the groove closed, very eccentric, forked at base; style mostly shorter than the nutlets. Open woods, Kamiak Butte and the Blue Mts., *A. T. T.*

Cryptantha Eastwoodae St. John. Annual, 20–27 cm. tall, white hairy; stems erect, bushy branched, appressed strigilose, sparsely stiff spreading hirsute from pustulate bases; middle and upper leaves 1–2 cm. long, 0.6–1.5 mm. wide, linear, densely spreading hispid from white, large pustulate bases, the blade often more or less hirsute also near the base; spikes several but single, not twinned, the upper branches all spike-bearing; spikes approximate to the vegetative parts, bractless or only the lowest node bracted, 10–55 mm. long, dense; corolla 0.8–1 mm. broad, 2 mm. long, inconspicuous, white; fruiting calyx 3.5–4.5 mm. long, narrowly campanulate, the lobes spreading, linear, spreading pilose and pilose ciliate especially towards the base, the body and especially the much thickened midrib strongly hispid, the hairs yellowish at tip, pustulate at base; nutlets 1–1.2 mm. long, all 4 maturing, ovate, the tip lance attenuate, gray, brown mottled, smooth, shining, minutely granulate, and slightly very minutely tuberculate towards the tip, the back low convex, the edges obtuse, the inner faces almost flat, the groove slightly asymmetric and curved, closed except at the basal fork; gynobase 0.5–0.7 mm. long, narrow, the stigmas attaining about 5/6 the height of the nutlets.

Winona, Whitman Co., *Eastwood & St. John* 13,219 (type), *A. T.*

This was reduced by Abrams (1951) to *C. affinis;* and by Cronquist (1959) to *C. Torreyana* (Gray) Greene.

Cryptantha flaccida (Dougl. ex Lehm.) Greene. Annual, 1.5–4.5 dm. tall, strigose throughout; roots fibrous; stem simple or with ascending branches; leaves 2–6 cm. long, oblance-linear to filiform, firm; spikes quinate to solitary, naked, usually stiffish, 4–16 cm. long; calyx in fruit 2–4 mm. long, oblong-ovoid, asymmetric, appressed, the lobes lance-linear, connivent, but the tips spreading, margins ciliate or strigose, the midrib thickened and armed with pale spreading, encrusted, arcuate or uncinate bristles, the abaxial lobe longest and most hirsute; corolla limb 0.8–2 mm. wide; the 1 nutlet 2–2.2 mm. long, lance-ovoid, acuminate, subterete, smooth or slightly granular, the groove usually closed, at base dilated into a small lanceolate areola. *C. flaccida,* var. *minor* A. Brand. Dry open slopes, *U. Son.*

Cryptantha Hendersoni (A. Nels.) Piper. Annual, 1.5–5 dm. tall, hirsute and hispid throughout; roots fibrous; stems erect, simple or loosely branched; leaves 2–7 cm. long, oblanceolate or linear, acute or obtuse; spikes 2–4 together, naked or sometimes bracted at base, many flowered, in fruit 5–20 cm. long, the fruits ascending, remote; calyx ovoid, densely hispid, in fruit 3–6 mm. long, the lobes linear or lanceolate; corolla tube as long as the calyx, the limb 4–7 mm. broad; nutlets usually 4, ovoid or lance-ovoid, 2–3 mm. long, smooth, granulate, tuberculate, or papillate-muricate, the back low convex, the sides rounded or obtuse, the groove narrow or closed, broadly forked below; gynobase 1.3 mm. long, narrow; style usually 4/5 length of nutlets. *C. grandiflora* Rydb. Dry hills, Palouse, *A. T.,* abundant in *U. Son.*

Named for Prof. L. F. Henderson, of the University of Idaho, later of the University of Oregon. The author is following the judgment of Dr. I. M. Johnston, in grouping plants with smooth, with those tuberculate nutlets. The result is a heterogeneous species.

Cryptantha Sheldonii (Brand) Pays. Perennial, 1–3.5 dm. tall, densely pilosulous and pustulate hispid throughout; root woody, dark, up to 5 mm. in diameter; basal leaves numerous, 2–5 cm. long, spatulate to oblanceolate, mostly obtuse; cauline leaves similar, reduced upwards; inflorescence virgate, densely flowered, mainly uninterrupted, the spikes in fruit 2–9 cm. long; flowers subsessile; calyx campanulate, densely close pilosulous and spreading hispid, in anthesis 3–5 mm. long, in fruit 7–11 mm. long, the lobes linear-lanceolate; corolla white, the tube 4 mm. long, the appendages 0.5 mm. long, creamy, depressed oblong, papillose, the limb 5–10 mm. wide, the lobes broadly oval; anthers 1 mm. long, linear-oblong; style 3 mm. long, subulate; the 4 nutlets commonly maturing, 3–4 mm. long, lanceoloid, subacute, margined, the back convex or somewhat keeled, tuberculate and rugose, the ventral surface flattened, tuberculate, the scar straight, closed. *Oreocarya Sheldonii* Brand. Gravelly hills, near Spokane, *A. T.*

Cryptantha simulans Greene. Annual, 1.5–4.5 dm. tall, strigose throughout; roots fibrous; leaves few, 2–7.5 cm. long, oblanceolate or oblance-linear; spikes solitary or in pairs, often leafy below, usually elongate and sparsely flowered, in fruit 3–10 cm. long; calyx campanulate, slightly asymmetric, ascending, in fruit 3–8 mm. long, the lobes lance-linear, hirsute and spreading hispid; corolla inconspicuous, the limb about 2 mm. wide; the 4 nutlets 2–2.5 mm. long, identical, broadly ovoid, densely granulate or granulate-muriculate, sparsely broad tuberçulate, the black low convex, the margins rounded, groove usually closed, forked at base; gynobase 1–1.5 mm. tall; style 3/4 to 4/5 height of nutlets. *C. ambigua* (Gray) Greene, f. *simulans* (Greene) Brand. Open pine woods, Moscow Mt., Ida., *Henderson* 2811½; Blue Mts., Columbia Co., Wash., *Darlington,* June 1913, *A. T. T.*

Cryptantha Torreyana (Gray) Greene. var. **Torreyana.** Annual, 1–4 dm. tall, hirsute and pustulate hispid throughout; roots fibrous; stem erect, simple or few branched; leaves 2–7 cm. long, oblanceolate or linear, obtuse; spikes mostly in pairs, naked, congested and glomerate, or loosely flowered and projecting, in fruit 3–8 cm. long; calyx lanceoloid, ascending, asymmetric, in fruit 3.5–8 mm. long, the lobes lanceolate or lance-linear; corolla inconspicuous, the limb 1 mm. wide; nutlets usually 4, ovoid, 1.5–2.2 mm. long, smooth, usually mottled, rarely granulate, the back low convex, the sides rounded, the groove closed, forked below; gynobase about 1 mm. tall; style shorter than the nutlets. *C. Torreyana,* var. *genuina* Johnston. Abundant, open slopes, *U. Son.* to *A. T. T.*

ECHIUM

Annual to perennial herbs or shrubs; leaves alternate, rough hairy; racemes scorpioid, leafy; calyx 5-parted, with slender teeth; corollas blue, red, to white, tubular-funnelform, the throat not appendaged; stamens inserted near middle of tube, the longer ones exserted; nutlets erect, rugose, from a flat gynobase. (Name from Gr. *echis,* snake; hence *echion,* several borages with scorpioid inflorescences.)

Echium vulgare L. *Viper's Bugloss.* Biennial herb, 2.5–10 dm. tall, coarsely hispid and appressed puberulent, stem simple or branched; leaves sessile, 2–15 cm. long, narrowly oblong to oblanceolate, the upper ones reduced; cymes axillary, numerous, mostly 1–3 cm. long, but sometimes to 10 cm. long; bracts of cymes 5–10 mm. long; calyx in flower 5–7 mm. long, later 7–13 mm. long; corolla 14–22 mm. long, blue; style pilosulous; nutlets 2.5–3 mm. long, ovoid, triangular, rugose, bony. European weed, introduced in 1948. Roadsides near Spokane, *Gaines* 151.

HACKELIA

Biennial or perennial (rarely annual) herbs; leaves alternate, broad and veiny; flowers in naked or inconspicuously bracted racemes, paniculately arranged; pedicels slender, recurving in fruit; calyx deeply parted; corolla funnelform or campanulate, the lobes rounded, imbricate, the throat with trapeziform, intruded appendages; stamens included, affixed at middle of the tube; filaments slender, short; anthers oblong to elliptic; style slender, scarcely if at all surpassing the nutlets; stigma capitate; the 4 nutlets erect, ovoid, attached ventrally to the pyramidal gynobase by a broad median areola, margin with subulate glochidiate appendages which are frequently confluent at base, the back smooth or with glochidiate appendages. (Named for *Prof. P. Hackel* of Bohemia.)

Marginal prickles of nutlets fused at base; leaves hispid ciliate, cinereous,
  Flowers white; marginal prickles of nutlets united about ½ their length. *H. cinerea.*
  Flowers blue; marginal prickles of nutlets united about ⅓ their length. *H. ciliata.*
Marginal prickles distinct; leaves not ciliate, green, not densely hairy. *H. Jessicae.*

Hackelia ciliata (Dougl. ex Lehm.) Johnston. Perennial, 2–7 dm. tall, cinereous strigose and more or less spreading hispid; root dark, up to 5 mm. in diameter, the crown clothed with old leaf bases; stems 1 or several, erect, loosely branched above; basal leaves 5–14 cm. long, linear, or slightly oblanceolate at tip, appressed pilose and hispid; cauline leaves 2–7 cm. long, sessile, linear; racemes numerous, loose, in fruit becoming 5–22 cm. long; calyx 3–3.5 mm. long, appressed hispid, the lobes lanceolate; corolla blue, broadly funnelform, the tube 2.5 mm. long, the appendages semiorbicular, minutely hairy, the limb rotate, 8–12 mm. broad, the lobes deltoid-obovate; nutlets 5–5.5 mm. long, broadly ovoid, the dorsal side flattened, muriculate and with a few short glochidiate prickles, the inner sides compressed towards a keel, smooth, the scar lanceolate, below the center, the marginal prickles glochidiate, united for nearly half their length. *Lappula ciliata* (Dougl. ex Lehm.) Greene. Gravelly places, Spokane and northward, *A. T.*

Discovered by Douglas at Kettle Falls and Spokane River.

Hackelia cinerea (Piper) Johnston. Perennial, 4–6 dm. tall, appressed cinereous throughout, also hispid and these hairs on the basal leaves pustulate; lower leaves 5–10 cm. long, narrowly linear-lanceolate; cauline ones similar, reduced upwards; inflorescence in fruit loose, 5–20 cm. long; lower pedicels often 10–15 mm. long; calyx lobes linear-lanceolate, acute; corolla white, the tube 2 mm. long, the lobes 3 mm. long, obovate-orbicular; appendages short pilose, its crest semicircular, retuse; nutlets 3–4 mm. long, the marginal prickles unequal, often curving outwards, all glochidiate, the dorsal surface with an indistinct median ridge, muriculate and short glochidiate. Deep Creek Canyon, Spokane Co., *T. Large* 41; and *L. Constance* 1,899, *A. T. T.*

Hackelia Jessicae (McGregor) Brand. Perennial, 5–10 dm. tall, hirsute throughout; root dark, woody, up to 1 cm. in diameter, the crown clothed with persistent leaf bases; stem single, usually simple to the inflorescence; basal leaves with petioles 2–12 cm. long, the blades 4–19 cm. long, oblanceolate; cauline leaves gradually reduced upwards, becoming sessile and lanceolate, the blades 2–10 cm. long; inflorescence loose, the racemes ascending or divergent, in fruit these 5–10 cm. long; calyx 2–3.5 mm. long, mostly appressed hirsutulous, campanulate, parted nearly to the base into lanceolate lobes; corolla blue, the tube 2–2.3 mm. long, appendages broader than long, depressed, short puberulent or glabrate; the limb 7–9 mm. wide, the lobes oblong-obovate; anthers 0.7–0.8 mm. long, elliptic-oblong; style 0.6 mm. long; nutlets 3.5–5 mm. long, broadly ovoid, acuminate, the back nearly flat, muricate, and with a few small glochidiate prickles, the inner angles compressed, more or less muricate or rugose, the scar elliptic, central, the marginal prickles distinct, the bases flattened and thin. *Lappula diffusa* of Piper, not Lehm.; *H. floribunda* of Johnston, not Lehm. Open woods, or moist places, especially common in the Blue Mts., *A. T. T.*

Hackelia hispida (Gray) Johnston. This species was accepted by Piper and Beattie (1914) as *Lappula hispida.* It has not been recollected in the Washington Blue Mts., nor has the *Sheldon,* Asotin Co., specimen been seen and verified.

HELIOTROPIUM. Heliotrope

Annual or perennial herbs, or shrubs; leaves alternate or rarely opposite; flowers perfect, regular, white or violet, in dense 1-sided spikes; calyx-lobes narrow; corolla salverform or funnelform; usually naked in the throat, the sinuses more or less plaited in the bud; stamens included, the anthers acuminate, connivent, nearly sessile; style none; stigma annular or conic; ovary 4-celled or 2-celled with more or less intruding placentae, separating in fruit into 4 one-seeded, closed nutlets. (Name from Gr., *helios,* sun; *trope,* a turning.)

Heliotropium spathulatum Rydb. Perennial, 1–7.5 dm. long, glabrous, fleshy; root 2–4 mm. in diameter, cord-like, dark; branches 1 to several, ascending or prostrate, often branched; leaves cauline, alternate or some subopposite; petioles up to 25 mm. long, reduced upwards and the upper leaves sessile; blades 1–6 cm. long, spatulate to obovate, cuneate, somewhat glaucous; peduncle 1–7 cm. long; spikes 2–5 from near summit of the peduncle, densely flowered, in fruit 2–10 cm. long; calyx campanulate, deeply parted, 2–35 mm. long, the lobes lanceolate, erect or ascending; corolla white or lavender-tinged, funnelform, 4–8 mm. long, the limb almost rotate and 4–8 mm. wide, the lobes ovate; anthers lanceolate 1.5–2.3 mm. long; stigmas nearly 1 mm. broad; nutlets 1.8–2.3 mm. long, triangular, the back concave, the margins with a pale corky thickening, the brown seed exposed on the inside and outside. *H. curassavicum* L., var. *obovatum* DC. Alkali lands, common westward, introduced by R. R., Waitsburg, *Horner* 379, *U. Son.*

LAPPULA. Stickseed

Annual or rarely perennial herbs; leaves alternate, narrow, firm; inflorescence of bracted racemes or spikes; pedicels usually erect; calyx 5-parted into spreading, lanceolate lobes; corolla blue or white, small, the tube short, the lobes rounded, ascending, imbricate, the throat closed by the intruded appendages; stamens included; style short; gynobase subulate-columnar, commonly surpassing the nutlets; stigma subcapitate; the 4 nutlets erect, smooth or verrucose, the scar along the ventral keel, narrow, the back angulate or margined by a single or double row of prickles which are often connate at base, forming a cupulate border. (Name from Lat., a diminutive of *lappa,* burr.)

Fruiting nutlets with marginal prickles subulate and distinct. *L. Redowskii,* var. *Redowskii.*
Fruiting nutlets with marginal prickles dilated and confluent. Var. *desertorum.*

Lappula Redowskii (Hornem.) Greene, var. **Redowskii.** Annual, finely hirsute throughout, 7–60 cm. tall; root dark, elongate, up to 4 mm. in diameter, the rootlets fibrous; stem simple or branched; basal leaves 8–25 mm. long, spatulate, soon withering; cauline leaves numerous, 7–40 mm. long, almost linear, densely pustulate-based hirsute; racemes many flowered, in fruit 5–20 cm. long, the fruits remote; bracts foliaceous, exceeding the flowers or fruit; calyx campanulate, densely hirsute, deeply parted, in flower 2.5–3 mm. long, in fruit 5–7 mm. long, the lobes linear; corolla blue or white, the tube 1.5–2 mm. long; the appendages 0.7 mm. long, deltoid-ovate, membraneous; anthers 0.3 mm. long, ellipsoid; style 0.3 mm. long; nutlets 2–2.5 mm. long, lanceoloid, brown, pale muriculate tubercular throughout, the back with a low rim bounding an area smaller than the nutlet, the marginal prickles distinct, subulate, glochidiate; scar linear, with an ovate base. *L. occidentalis* (Wats.) Rydb. Grassy hills, *A. T.*

Var. **desertorum** (Greene) Johnston. The nutlets having the marginal prickles dilated and confluent at base. *L. columbiana* A. Nels.; *L. cupulata* of Piper, not of Gray. Dry lower slopes, Snake River Canyon, *U. Son.*

LITHOSPERMUM

Annuals or perennials, herbaceous or shrubby; leaves alternate; flowers in leafy spikes in the axils of the upper leaves; calyx 5-parted; corolla greenish, yellowish, or white, salverform or funnelform, the rounded lobes imbricate in the bud, throat with intruded appendages or pubescent or glandular areas; stamens included; filaments short; anthers oblong, usually with apiculate connectives; style filiform; stigmas geminate; nutlets 4, rarely fewer, erect, ovoid or angular, smooth or verrucose, attached by a broad horizontal or oblique basal areola; gynobase flat or broadly pyramidal. (Name from Gr., *lithos,* stone; *sperma,* seed.)

Lithospermum ruderale Dougl. ex Lehm. *Cat's Tooth; Lemonweed.* Perennial, 1–6 dm. tall, rough hairy throughout; caudex 5–45 mm. thick, woody; stems usually numerous, usually simple and virgate to the inflorescence, densely hirsutulous and shaggy hispid; leaves 2–10 cm. long, narrowly lance-linear, sessile, firm chartaceous, closely hirsute; leaves of the inflorescence scarcely reduced; inflorescence dense, many-flowered, in fruit 5–20 cm. long; cymes compact, much exceeded by the leaves; calyx 4–6 mm. long, campanulate, hirsutulous and densely hispid, the lobes linear; corolla hirsute without, campanulate, the tube 4–5 mm. long, glandular area 1 mm. long, ovate; the lobes 3–3.5 mm. long, ovate; anthers 1.2–1.5 mm. long, narrowly oblong; style 4–4.5 mm. long; nutlets ovoid, subacute, the base narrow, inverted saucer-like, the shell white, hard, ivory-like. *L. ruderale,* var. *lanceolatum* (Rydb.) A. Nels.; var. *macrospermum* Macbr. Abundant, open places or open woods, *U. Son.* to *A. T. T.*

MERTENSIA. Bluebells

Glabrous or pubescent perennial herbs; leaves alternate, entire; buds pink; flowers blue or rarely white, mostly bractless, in panicled racemes or in corymbs; calyx deeply 5-cleft or -parted; corolla tubular-funnelform or trumpet-shaped to almost campanulate, the open throat bearing obvious or obsolete transverse folds or crests; corolla lobes imbricate; filaments flattened or nearly filiform; style filiform; stigma entire; nutlets from somewhat fleshy to membranaceous, roughish or wrinkled, attached by a small suprabasal scar to the convex receptacle. (Named for *Franz Carl Mertens* of Germany.)

Aestival plants; leaves with numerous, evident lateral veins,
  Pedicels and leaves pubescent,
    Pubescence of pedicels hispid or hirsute, not dense,
      Blades more or less pubescent on both sides. *M. paniculata*, var. *paniculata.*
      Blades glabrous above. Var. *subcordata.*
    Pubescence of pedicels canescent, dense. Var. *borealis.*
  Pedicels mostly glabrous; leaves glabrous. *M. umbratilis.*
Vernal plants; leaves lacking evident lateral veins,
  Basal leaves numerous, their petioles long persisting and sheathing the stem; leaves pubescent on both sides. *M. oblongifolia*, var. *amoena.*
  Basal leaves none, the base of the stem nearly naked; blades strigose or hirsutulous above. *M. longiflora.*

Mertensia longiflora Greene. *Bluebells.* Root tuberous, dark, often fasciculate forked; plant 1–3 dm. tall; stem glabrous, unbranched; basal leaves none; lower cauline leaves reduced to pale, membranous bracts; next leaves cuneate, short petioled; upper leaves 2–10 cm. long, ovate or elliptic, sessile or half-clasping, strigose or hirsutulous above; cyme terminal, scorpioid, densely many flowered; pedicels 1–4 mm. long, glabrous; calyx cleft nearly to the base, the lobes 2–3 mm. long in anthesis, 3–4 mm. in fruit, lanceolate, scabrous or hispidulous ciliate; corolla tube 8–15 mm. long, 2–3 mm. in diameter; corolla limb funnelform, 4–6 mm. long, cleft nearly half way down into broadly oval lobes; style protruded; nutlets 2–2.5 mm. long, chestnut-brown, muriculate, lanceoloid, the tip incurved, the inner faces concave, the axial angle raised, wing-like, bearing the elongate, pale scar. *M. Horneri* Piper. Abundant, rocky slopes, prairies, or open woods, *U. Son.* to *A. T. T.*

One of the commonest and loveliest of the early spring flowers.

Mertensia oblongifolia (Nutt.) G. Don, var. **amoena** (A. Nels.) L. O. Williams. Caudex vertical, dark, slender, the top clothed with old leaf bases; plant 7–15 cm. tall, densely hirsutulous throughout; stems 1 to several; basal leaves numerous, with petioles 15–55 mm. long, the blades 1–4 cm. long, lanceolate, obtuse, thick; cauline leaves similar, but mostly sessile; cyme terminal, dense, scorpioid; pedicels 2–3 mm. long; calyx-lobes 2–3 mm. long in anthesis, lance-linear, ciliate; corolla tube 6–8 mm. long, 2–2.5 mm. in diameter; corolla limb campanulate 4–5 mm. long with the 1–2 mm. lobes; nutlets unknown. *M. pubescens* Piper. Top of Browns Mt., 5 mi. S.E. of Spokane, *Davison* in 1925, *A. T. T.*

Mertensia paniculata (Ait.) G. Don, var. **paniculata.** Erect, 3–12 dm. tall, branched above; stems hirsute throughout or glabrate below; caudex stout, dark, woody; basal leaves 5–9 cm. long, the oval blades much shorter than the petioles; lower and middle cauline leaves with petioles 1–10 cm. long, the blades 5–18 cm. long, acute, oval to cordate-ovate, appressed strigose above, hirsute beneath; upper leaves 3–9 cm. long, lanceolate or ovate-lanceolate; cyme several to many flowered, becoming loose; pedicels 3–35 mm. long, hirsute; calyx more or less appressed hispidulous or at least ciliate, the lobes deltoid-lanceolate, acute, in anthesis 2–3 mm. long, in fruit 3–5 mm. long;

corolla tube 4.5–6 mm. long, tubular; corolla limb 5–7 mm. long, campanulate, the lobes 1 mm. long; nutlets 3–3.5 mm. long, asymmetric ovoid, buff-colored, the inner sides compressed, the back strongly wrinkled. *M. membranacea* Rydb. Common, moist woods and woodland meadows, *A. T. T.*

Var. **subcordata** (Greene) Macbr. Differs in having the leaves glabrous above. *M. subcordata* Greene. Stream banks and moist woods, *A. T. T.* to *Huds.*

Var. **borealis** (Macbr.) L. O. Williams. Erect, 5–9 dm. tall, caudex black, thickened; stem simple to the inflorescence, sparsely hirsutulous, glabrate below, leafy throughout; basal leaves reduced to scarious bracts; cauline leaves with petioles, the lower 4–6 cm. long, the upper subsessile; blades 3–12 cm. long, acute, oval to lanceolate, membranous, beneath hirsutulous, above papillose puberulent; cyme 5–30 cm. long, many-flowered, loose; pedicels 3–18 mm. long, hirsutulous; calyx-lobes narrowly deltoid, ciliate, otherwise glabrate, in anthesis 1–3 mm. long, in fruit 4–5 mm. long; corolla tube 3–6 mm. long, broad tubular; corolla limb 4–7 mm. long, campanulate, bright blue, the lobes 1–2 mm. long; nutlets 3.5–4 mm. long, broadly ovoid, deeply wrinkled and pitted, gray, scabrous papillose, the scar deltoid, subbasal. *M. pratensis* Heller, var. *borealis* Macbr. Wooded stream banks, *Can.*

Mertensia umbratilis Greenm. Erect, 2–4.5 dm. tall; root slender, running, black, often branched or multicipital; stems simple, glabrous; basal leaves with petiole 4–10 cm. long, the blades 6–12 cm. long, elliptic-oblong to oblong-ovate, narrowed to and decurrent on the petiole; cauline leaves numerous, the petioles 1–2 cm. long, the upper leaves sessile; blades 4–12 cm. long, lanceolate to ovate, acute, papillose, the margins appressed hispidulous; cymes 3–8 cm. long, few-flowered; pedicels 3–12 mm. long; calyx-lobes ciliate, lanceolate, in flower 3–5 mm. long, in fruit 4–8 mm. long; corolla tube 7–10 mm. long, broad tubular; corolla limb bright blue, 6–7 mm. long, the lobes 1.5–2 mm. long; nutlets 3.5 mm. long, whitish, corrugated. *M. arizonica* Greene, var. *umbratilis* (Greenm.) Macbr. Cronquist (1959) published his opinion that this plant is the hybrid *M. oblongifolia* × *paniculata*. Rocky slope, Al Williams ridge, T. 8 N., R. 41 E., Blue Mts., *Darlington* 2.

MYOSOTIS. FORGET-ME-NOT

Annual or perennial caulescent herbs; leaves alternate, entire; flowers perfect, in 1-sided, scorpioid racemes; calyx 5-lobed; corolla salverform, blue, pink, or white, throat with transverse, intruded appendages; filaments filiform; gynobase flat or high convex; the 4 nutlets erect; ellipsoid, smooth and shining, the scar basal. (Name from Gr., *muos,* mouse; *otos,* ear.)

Calyx with short, straight, appressed hairs; corolla limb 5–10 mm. wide. — *M. scorpioides.*
Calyx tube with reflexed, uncinate hairs; corolla limb less than 2 mm. wide,
- Calyx 2-lipped, in fruit 5–7 mm. long. — *M. macrosperma.*
- Calyx equally 5-cleft, in fruit 3–4.5 mm. long. — *M. stricta.*

Myosotis macrosperma Engelm. Annual, hirsute throughout, 7–43 cm. tall; roots fibrous; stem simple or branched; basal leaves few, 1–5 cm. long, spatulate to oblanceolate, tapering at base; cauline leaves 1–5 cm. long, sessile, oblong-oblanceolate, pustulate hirsute; racemes many flowered, leafy bracted only at base, in fruit loose 7–30 cm. long; pedicels 2–6 mm. long; calyx campanulate, in anthesis 1.5–2 mm. long, the tube uncinate hispid, the lobes hispid;

corolla 1.5–2 mm. long, bluish or white, funnelform; anthers 0.1 mm. long, ellipsoid; nutlets 1.5–1.8 mm. long, brownish, shining, ovoid-biconvex, the scar fusiform. *M. virginica* (L.) BSP., var. *macrosperma* (Engelm.) Fern. Moist places, *A. T.*

Myosotis scorpioides L. *Forget-me-not.* Perennial, appressed hispidulous throughout, 1–6 dm. tall; rooting at the nodes, the stem decumbent at base, the tip ascending; leaves 2–8 cm. long, the lower narrowed to the base, the upper sessile, spatulate or linear-lanceolate, soft chartaceous; racemes many flowered, in fruit loose, 5–30 cm. long; pedicels divergent, in fruit 4–5 mm. long; calyx campanulate, appressed hispidulous, in flower 1.5–2 mm. long, in fruit 3–5 mm. long, the deltoid lobes shorter than the tube; corolla in bud pink, in anthesis pale blue, the tube 2–3 mm. long, the appendages transverse, rounded; the lobes oval, retuse; anthers 0.7 mm. long, lanceoloid; style 1.5 mm. long; nutlets 1–2 mm. long, ovoid, compressed with rounded, pale, thin sides. Native of Eurasia, often cultivated, escaped at Walla Walla.

Myosotis stricta Link. ex R. & S. Annual, usually a winter annual, 5–20 cm. tall, hirsute throughout; roots fibrous; stems erect, simple or freely branched; basal leaves 5–15 mm. long, spatulate, short petioled, soon withering; cauline leaves sessile, 5–20 mm. long, narrowly elliptic or lanceolate; inflorescence of racemes, naked above, but axillary flowers produced down almost to the base; pedicels ascending, up to 2 mm. in length, in fruit remote; calyx campanulate, in anthesis 1.5–2 mm. long, lobed beyond the middle into lanceolate, hispid lobes, the tube with reflexed, uncinate hispid hairs; corolla slightly exceeding the calyx, funnelform, blue with a yellow tube, the limb 1.5 mm. wide; nutlets 1 mm. long, brown, shiny, ovoid-biconvex, the scar elliptic. *M. arenaria* Schrad.; *M. micrantha* of Johnston, not of Pallas, which fide Vestergren is not a *Myosotis.* Native of Eurasia, introduced to grasslands and waste places, *A. T.*, *A. T. T.*

PLAGIOBOTHRYS. Popcorn Flower

Annual or perennial herb; leaves linear to oblong, at least the lower opposite, sometimes obscurely so being in a rosette; racemes naked or bracted; calyx parted almost to the base into lanceolate or oblong, erect or connivent lobes; corolla white, the short tube at most barely surpassing the calyx, the lobes spreading, rounded, imbricate, the throat with intruded appendages; style slender, usually short; ovules 4; the usually 4 nutlets erect, ovoid to lanceoloid, smooth or roughened, the areola basal to median, at the lower end of the strong ventral keel or rarely terminating a stipitate prolongation of it, plane or excavate, simple or carunculate; gynobase low convex or pyramidal. (Name from Gr., *plagios,* on the side; *bothrus,* pit.)

Nutlets cruciform, attached at the middle of the ventral side. *P. tenellus.*
Nutlets ovoid or lanceoloid, attached at or near the base,
  Nutlet attachment exactly basal. *P. leptocladus.*
  Nutlet attachment obliquely subbasal,
    Scar in an areola broader than long. *P. cognatus.*
    Scar in an areola longer than broad. *P. hispidulus.*

Plagiobothrys cognatus (Greene) Johnston. Annual, strigose nearly throughout; roots fibrous; stems 5–25 cm. long, usually much branched at base; floriferous to near the base; leaves strigose and pustulate beneath, less so or even glabrous above; lower leaves 2–7 cm. long, linear; upper leaves linear to spatulate-linear; racemes solitary, slender, usually loosely flowered, bracted throughout or only below the middle; pedicels 0.5–1 mm. long; calyx campanulate, parted to near the base, densely hispid, accrescent, in fruit 2–4 mm. long, the lobes lanceolate, ascending; corolla minute, the limb 1–2 mm. wide; style reaching at least the middle of the nutlets; nutlets 1.3–2.2 mm. long, ovoid, somewhat asymmetric, the back keeled near the apex, tuberculate and with irregular ridges, the surface usually granulate, the ventral side with the keel usually folded over near the base, the scar oblique, triangular to ovate, surrounded by ridges usually forming a broad areola; the axial nutlet most firmly attached, and usually with a broader, flatter scar. Muddy places, *U. Son., A. T.*

Plagiobothrys hispidulus (Greene) Johnston. Annual, strigose throughout, 5–40 cm. tall; roots fibrous; stems usually many branched at base, ascending or prostrate; leaves often glabrate above; lower leaves 1–5 cm. long, linear; upper leaves linear to oblanceolate; racemes bracted, extending nearly to the base, loosely flowered, single; pedicels 0.5–1 mm. long, or the lower elongate; calyx campanulate, densely hispid, in fruit 2–3.5 mm. long, parted to near the base, the lobes linear to lanceolate; corolla inconspicuous, the limb 1–2 mm. wide; style ½ to ¾ length of nutlets; nutlets 1.5–2 mm. long, ovoid to lance-ovoid; the back keeled towards the tip, with transverse, often broken ridges, these often papillose or muriculate or minutely hairy, the intervals tuberculate and usually granular, the ventral side keeled to below the middle, the scar narrowly elliptic, lateral, suprabasal, surrounded by a ridge. *Allocarya hispidula* Greene. Muddy banks, Blue Mts.; and Moscow and Craig Mts., Ida., *A. T. T.*

Plagiobothrys leptocladus (Greene) Johnston. Annual, somewhat succulent, strigose throughout, mostly branched at the base and usually prostrate; stems 1–3 dm. long, the lower internodes approximate, floriferous nearly to the base; leaves nearly glabrous above, sparsely strigose and pustulate below; lower leaves rosulate 3–10 cm. long, linear to spatulate-linear; upper leaves oblance-linear; racemes 1-sided, loosely flowered, the pedicels 0–0.5 mm. long; calyx campanulate, strigose, accrescent, in fruit 3–8 mm. long, parted ⅔ way into lanceolate to subulate lobes, strengthened by indurate ribs, connivent but twisted to one side; corolla minute, the limb 1–2 mm. wide; style not ½ length of nutlets; nutlets 1.5–2.5 mm. long, lanceoloid, the back keeled only above the middle, tuberculate, somewhat rugose, smooth or granulate or penicillate hairy, the ventral side angulate, keeled; the scar basal, raised, oval. Rocky shore, Rock Lake, Wash., *Weitman* 275, *U. Son.*

Plagiobothrys tenellus (Nutt. ex Hook.) Gray. *Popcorn Flower.* Annual, 7–25 cm. tall, spreading hirsute throughout; roots fibrous; stems one or commonly several from the base, ascending, loosely branched; basal leaves in a dense rosette, 5–25 mm. long, lanceolate or narrowly elliptic; cauline leaves few, alternate, 4–20 mm. long, lanceolate, pustulate hirsute; spikes dense, few flowered, in fruit 2–10 cm. long; calyx campanulate to subspherical densely hirsute, the hairs at first yellow, later fading, in fruit the calyx 2–4 mm. long, the lobes lanceolate, as long as the tube, finally more or less circumscissile at base; corolla ovoid-funnelform, shorter than the calyx, the limb about 1.5 mm. wide, the appendages elliptic; anthers 0.2 mm. long; nutlets 1.5–2 mm. long, cruciform, brownish, shining, the back rounded, with a low median keel, with numerous fine, muricate transverse ridges, the ventral side with a transverse groove; the caruncle 2-lobed. Dry slopes, *U. Son., A. T.*

VERBENACEAE. Verbena Family

Herbs, shrubs, vines, or trees; leaves opposite or whorled, sometimes alternate, simple or compound, the stipules none; flowers usually irregular, hypogynous; calyx deeply 4–6-cleft; corolla mostly 2-lipped, the 4–5 lobes imbricate in bud; stamens 4 (rarely 2 or 5) attached on the corolla tube, often at different levels and some reduced to staminodia; carpels 2, 4, (or 5); ovary lobed, with as many cells or twice as many by false partitions; stigmas 2; ovules generally 2 for each carpel; fruit dehiscing septicidally into 2 or 4 one-seeded nutlets, or a drupe with succulent exocarp and hard endocarp; endosperm scant or none. (Named from the genus *Verbena,* the vervain.)

VERBENA. Vervain

Herbs or shrubs; leaves usually toothed or dissected; flowers perfect, in dense or loose spikes, or cymes, panicles, or solitary; calyx 5-toothed; corolla funnelform, 5-lobed; stamens 4, included, didynamous; anthers with or without a glandular appendage on the connective; ovary 4-lobed, 4-celled; fruit splitting into 4 dry nutlets. (Derivation of name obscure.)

Bracts exceeding the flowers; nutlets raised reticulate at tip on back of nutlet, striate below; corolla limb 2.5–3 mm. wide. *V. bracteata.*
Bracts shorter than the flowers; nutlets smooth on the back; corolla limb 3–4.5 mm. wide. *V. hastata.*

Verbena bracteata Lag. & Rodr. Apparently perennial; root woody, up to 1 cm. in diameter; stems usually several, 1–6 dm. long, decumbent or ascending, diffusely branched, hirsute; leaves 1–6 cm. long, pinnately incised or 3-lobed, the lateral lobes divaricate, narrow, the terminal one cuneate obovate or sharply incised or cleft, hirsute, narrowed to a short, margined petiole; spikes terminal, conspicuously bracted, in fruit 5–10 cm. long; bracts much longer than the calyx, coarsely hirsute, the lower leaf-like and incised, the upper entire, linear-lanceolate; calyx 3–4 mm. long, in flower fusiform, hirsute, the lobes minute, connivent over the fruit; corolla tube 3–3.5 mm. long; corolla limb blue; anthers 0.2 mm. long ellipsoid; style 0.2 mm. long, the stigmas 0.2 mm. long, obovate; nutlets 1.8–2.2 mm. long, linear-oblong, the inner sides minutely scabrous. Common in dry places, and persisting as a weed in dry farming regions, *U. Son., A. T.*

Verbena hastata L. *Blue Vervain.* Perennial, 4–15 dm. tall; root brown, woody, up to 6 mm. in diameter; stems erect, branched above, rough hispidulous, the hairs more or less appressed; leaves opposite, the lower with petioles 1–3 cm. long, hispidulous, the upper sessile; blades 5–10 cm. long, lanceolate to ovate-lanceolate, acuminate, coarsely or incised serrate, often hastately 3-lobed at base, strigose above, appressed hispidulous below; spikes pedunculate in a panicle, closely many flowered, in fruit 4–23 cm. long; bracts lance-subulate, usually shorter than the calyx; calyx 2.5–3 mm. long, ellipsoid, becoming cylindric, the lobes with short subulate tips connivent; corolla blue, the tube 4–5 mm. long puberulent; the limb deeply parted with oblong-oval lobes; the throat pilosulous; anthers 0.1–0.2 mm. long, oval; style 1 mm. long, the stigmas 0.2 mm. long, obovate; nutlets 1.8–2.1 mm. long, brown, oblong, the back rounded, the inner faces almost smooth. Occasional, meadows or stream banks, *U. Son.*

LABIATAE. Mint Family

Aromatic herbs; shrubs or trees; generally with square stems; and mostly opposite and coarsely pubescent leaves; flowers commonly in cymose clusters in capitate whorls at the nodes or in heads or spikes; bracts and bractlets usually present; flowers usually irregular, hypogynous; calyx 3–5-toothed or -lobed; sometimes 2-lipped; corolla 5–4-lobed, generally strongly 2-lipped; stamens attached on the corolla tube; 4 and didynamous, or 2 with staminodia; disk 2–4-lobed, nectar-bearing; ovary 2-carpellate, 4-lobed or 4-parted by a false septum, each part 1-ovuled; style arising between the lobes of the ovary, bifid at tip; fruit separating into 4 one-seeded nutlets (or rarely drupaceous); endosperm scanty or none. (Name from Lat., *labium,* lip.)

Corolla nearly regular,
  Anther-bearing stamens 2. — *Lycopus.*
  Anther-bearing stamens 4,
    Corolla-lobes 5. — *Trichostema.*
    Corolla-lobes 4. — *Mentha.*
Corolla irregular, distinctly 2-lipped,
  Anther-bearing stamens 2. — *Salvia.*
  Anther-bearing stamens 4,
    Calyx with a protuberance on the upper side. — *Scutellaria.*
    Calyx without a protuberance,
      Calyx-teeth 10, spiny. — *Marrubium.*
      Calyx-teeth 5, not spiny,
        Calyx-teeth unequal, the upper one very large. — *Dracocephalum.*
        Calyx-teeth almost equal,
          Calyx distinctly 2-lipped. — *Prunella.*
          Calyx not 2-lipped,
            Upper corolla-lip flat,
              Plant creeping; flowers axillary. — *Satureja.*
              Plant erect; flowers in whorled heads. — *Monardella.*
            Upper corolla-lip concave,
              Upper pair of stamens longer than the lower,
                Anther-cells parallel. — *Agastache.*
                Anther-cells divergent. — *Nepeta.*
              Upper pair of stamens shorter than the lower,
                Throat of corolla dilated. — *Lamium.*
                Throat of corolla not dilated. — *Stachys.*

AGASTACHE. Giant Hyssop

Tall perennial herbs; leaves petioled, serrate; flowers verticillate-clustered in dense or interrupted, bracted, terminal spikes;

calyx narrowly campanulate, oblique, 5-toothed, the teeth of the upper lip erect, 2-lobed, lower lip spreading, 3-lobed, its middle upper lip erect, 2-lobed, lower lip sreading, 3-lobed, its middle lobe the broader, crenulate; stamens 4, anther-bearing, didynamous, the upper pair longer; anthers 2-celled, their sacs nearly parallel; ovary deeply 4-parted; nutlets ovoid, smooth. (Name from Gr., *agan,* much; *stachus,* ear of corn.)

Agastache urticifolia (Benth.) Kuntze. Plant 5–15 dm. tall, the herbage nearly glabrate; rhizome horizontal, 3–7 mm. in diameter; lower leaves with petioles 1–3 cm. long, the upper sessile; blades 2–10 cm. long, ovate, cordate, obtuse or acute, coarsely dentate; pale green and glandular punctate beneath; spikes 3–11 cm. long, dense; bractlets lanceolate, shorter than the calyx, puberulent; calyx strongly 15-ribbed, puberulent and glandular punctate, the tube 5–6 mm. long, the teeth 4–6 mm. long, lanceolate, acuminate, rose-purple-tinged; corolla light violet purple, puberulent, the tube 10–12 mm. long, the lobes oval; anthers 0.2 mm. long, oval, rosy; style exserted; nutlets 2 mm. long, dull brown, elliptic-oblong, the inner faces flat. Low places or thickets, *A. T., A. T. T.*

DRACOCEPHALUM

Coarse annual or biennial herbs with blue flowers in dense bracteate terminal clusters; calyx tubular, 15-nerved, 5-toothed, the middle one largest and cleft; corolla 2-lipped, the upper lip erect, emarginate, the lower 3-lobed; stamens 4, anther-cells divergent; nutlets ovoid, smooth. (Name from Gr., *drakon* dragon; *kephale,* head.)

Dracocephalum parviflorum Nutt. *Dragon Head.* Annual or biennial; tap-root strong; plant hirsutulous throughout; stem 3–8 dm. tall; basal and lower leaves with petioles exceeding the blades; blades ovate or deltoid-ovate, dentate; middle and upper leaves shorter petioled, the blades 2–6 cm. long, gradually narrower and more sharply serrate or incised, lanceolate to linear; spikes terminal, dense, leafy bracted, and with ovate, pectinate bractlets; calyx scabrous, in the sinuses hirsute, the nerves and reticulate veins raised, prominent, the intervals membranous, the tube 4–6 mm. long, the teeth acerose-tipped, the lower ones lanceolate, the upper ones 5–6 mm. long, ovate, acuminate; corolla 8–10 mm. long, rose purplish, pilose, the upper lobes oval, exceeding the lower; anther cells 0.6 mm. long, oblong; nutlets 2.3–2.5 mm. long, oblong, 3-angled, dark brown. *Moldavica parviflora* (Nutt.) Britton; *Ruyschiana parviflora* (Nutt.) House. Medical Lake, *Henderson;* Blue Mts., Asotin Co., *G. N. Jones* 987.

LAMIUM

Annual or perennial herbs; calyx tubular-bell-shaped, about 5-nerved, with 5 nearly equal awl-pointed teeth; corolla 2-lipped, dilated at the throat, the tube longer than the calyx; the upper lip arched, narrowed at the base; the middle lobe of the spreading lower lip broad, notched at the apex, contracted at the base;

the lateral lobes small, at the margin of the throat; stamens 4, ascending under the upper lip; the lower pair longer; anther sacs divaricate; nutlets smooth or tuberculate. (Name from Gr. *laimos,* throat.)

Upper leaves sessile; calyx-tube hirsute. *L. amplexicaule.*
Upper leaves petioled; calyx-tube glabrous throughout or except at the summit. *L. purpureum.*

Lamium amplexicaule L. *Henbit.* Annual or winter annual; roots fibrous; plant more or less hirsutulous throughout, usually diffusely branched and decumbent at base, 1–4 dm. tall; lower internodes elongate, many times as long as the leaves; leaves 1–3 cm. long, the lower with petioles 1–3 cm. long, the floral ones clasping, deeply crenate dentate; flowers several, sessile in the axils; calyx 5–7 mm. long, pilose, the nerves not prominent, the lanceolate teeth nearly as long as the tube; corolla rose purplish, the tube 8–11 mm. long, the upper lip 2–3 mm. long, pilose; anther sacs 0.2–0.25 mm. long, oblong, widely divergent, hirsute; nutlets 1.8–2 mm. long, oblanceoloid, the inner faces flattened, brown with white spots. Common weed, introduced from Eurasia or n. Africa, cultivated lands. *U. Son., A. T.*

Lamium purpureum L. *Red Deadnettle.* Annual or biennial; roots fibrous; plant more or less hirsutulous throughout, or glabrate below; the young shoots and leaves red-violet; stems 1–3 dm. tall, with weak branches near the base, decumbent at base; lower internodes much elongated, the upper crowded; petioles 5–40 mm. long; blades 5–25 mm. long cordate to ovate-cordate, coarsely crenate, the upper blades smaller, crowded, usually purplish; flowers sessile, in whorls of 6–10 at a node; calyx-tube 3–4 mm. long, campanulate, nearly glabrous, 10-ribbed, the mouth oblique; calyx-teeth 4–6 mm. long, lance-linear; corolla pilosulous, magenta-purple, the tube 10–15 mm. long, the upper lip 3.5–5 mm. long; anther sacs 0.2 mm. long, ovoid, purple, the suture hispid-ciliate; nutlets 1.8–2.2 mm. long, obovoid, the inner faces flattened, gray, white maculate, the large caruncle brown. Weed, introduced from Eurasia or n. Africa. Cultivated fields, Pullman.

LYCOPUS. Water Horehound

Low perennial herbs, glabrous or puberulent, not aromatic; leaves sharply-toothed or pinnatifid; flowers small, mostly white, in dense axillary whorls; calyx bell-shaped, nearly equally 4- or 5-lobed; anther-bearing stamens 2, distant; the upper pair either sterile or wanting. (Name from Gr., *lukos,* wolf*; pous,* foot.)

Corolla nearly twice the length of the calyx; calyx shorter than the nutlets; calyx-teeth obtuse; rhizome tuberous. *L. uniflorus.*
Corolla shorter or barely longer than the calyx; calyx longer than the nutlets; calyx-teeth sharp; rhizomes not (or rarely) tuberous,
Leaves mostly petiolate and incised; calyx 1.5–2 mm. long. *L. americanus.*
Leaves sessile, coarsely serrate; calyx 4–5 mm. long. *L. asper.*

Lycopus americanus Muhl. Rhizome slender, often producing lateral stolons; stems 2–9 dm. tall, simple or branching above, remotely puberulous or glabrous; lower leaves with petioles 5–10 mm. long; blades 3–7 cm. long, lanceolate in outline, deeply incised or laciniate-pinnatifid, glabrous, glandular dotted, the base cuneate; flower heads dense; calyx-teeth as long as the tube, ciliolate, deltoid, acerose tipped; corolla slightly exceeding the calyx, white or lavender-tinged; style included; anthers 0.2 mm. long, horse-shoe-shaped; nutlets 1–1.5 mm. long, cuneate-obovate, the margin corky, the ventral apex glandular. Wet shores or meadows, *U. Son., A. T.*

Lycopus asper Greene. Rhizome slender, or more rarely short and tuberous, without lateral tuber-bearing stolons; stems 2–8 dm. tall, usually simple, hirsute at least on the angles; leaves 3–11 cm. long, lance-oblong or oblong, the margin scabrous, above scabrous hispidulous to glabrate, the tip acute, the base rounded or less commonly cuneate; calyx lobed ⅔ their length into lance-linear, acuminate, ciliolate, herbaceous, teeth; corolla white or lavender-tinged, barely exceeding the calyx; styles exserted; anthers 0.4–0.5 mm. long, horse-shoe-shaped; nutlets 2 mm. long, broadly cuneate, the back with a wide corky margin, almost flat, the pyramidal ventral side coarsely glandular throughout, the apex truncate and low dentate. *L. lucidus,* var. *americanus* Gray; *L. lucidus* sensu Piper, not Turcz. Meadows, *A. T. T.*

L. lucidus Turcz. of central and e. Asia has rhizomes large tuberous, with large tuber-bearing stolons; the internodes only sparingly hirsute; the leaves linear-lanceolate, incised serrate, the calyx-teeth acerose, glabrous; and the nutlets rounded at apex, smaller.

Lycopus uniflorus Michx. *Bugleweed.* Rhizome short tuberous, finally bearing slender stolons, tuber-bearing at tip, or the stem base bearing slender aerial stolons; stem 1–6 dm. tall, puberulous, simple or few branched above; leaves sessile or short petioled, 2–11 cm. long, lanceolate to elliptic-lanceolate, early glabrate, beneath pale greenish, remotely serrate, cuneate at each end; calyx 1.5 mm. long, lobed ½ its length into broadly deltoid, minutely ciliolate lobes; corolla 2.5 mm. long, funnelform, white; style exserted; anthers 0.3 mm. long, cordate-suborbicular; nutlets 1.3 mm. long, obovate, the back nearly flat, with a broad corky rim, the ventral side low pyramidal, glandular throughout, the apex rounded or truncate, dentate. Wet places, mouth of Potlatch R., *St. John et al.* 9,742, *U. Son.*

MARRUBIUM. Horehound

Bitter-aromatic whitish-woolly perennials; leaves rugose, petioled; flowers white or purplish, small, much crowded in axillary whorls; calyx tubular, 5–10-nerved, 5–10-toothed, the teeth spiny-pointed; corolla 2-lipped, the upper lip erect, notched; the lower spreading, 3-cleft, the middle lobe broadest; stamens 4, short, included in the corolla tube, the upper pair longer; style 2-cleft. (Name from Hebrew, *marrob,* bitter juice.)

Marrubium vulgare L. *Common Horehound.* Tap-root strong, brown; stems 1–5 dm. long, often several and decumbent, white lanate; petioles 3–40 mm. long, lanate; blades 1–4 cm. long, rhombic-ovate to suborbicular, firm, deeply rugose, at first densely lanate, later merely pilose above, beneath lanate or pilose, the margin crenate dentate; flowers numerous in dense axillary heads; calyx-tube 3–5 mm. long, subcylindric, contracted towards the apex, villous and stellate; the 10 teeth 2–2.5 mm. long, rigid, acerose, the glabrous tips uncinate; corolla 4–5 mm. long, white, pilose without, the upper lip lanceolate, with minute lanceolate lobes, the lower lip shorter, broadly oblong, the middle lobe flabellate and truncate; nutlets 1.5–2 mm. long, 3-angled, grayish brown, slightly roughened. A weed, naturalized from Eurasia. Open places, *U. Son., A. T.*

MENTHA. Mint

Aromatic fragrant perennial herbs; leaves glandular punctate; flowers very small, in dense clusters, forming false whorls in the axils or in terminal spikes; calyx 10-nerved, bell-shaped or tubular, the 5 teeth equal or nearly so; corolla with a short included tube, almost equally 4-cleft, the upper lobe broadest, entire or notched; anther-bearing stamens 4, equal, erect, distant; anthers with 2 parallel cells; ovary 4-parted; nutlets smooth. (Name Gr., used by Theophrastus, *Minthe,* a nymph changed into the mint.)

Whorls of flowers forming terminal spikes,
  Leaves sessile or subsessile; calyx 1.5–2 mm. long. *M. spicata.*
  Leaves petioled; calyx 2.5–3.5 mm. long. *M. piperita.*
Whorls of flowers in leaf axils,
  Blades ovate or lance-ovate, glabrous. *M. arvensis,* var. *glabrata.*
  Blades usually narrower, cuneate at base,
    Blades glabrous or but sparsely hirsutulous. Var. *canadensis.*
    Blades villous or lanate. Var. *lanata.*

Mentha arvensis L., var. **canadensis** (L.) Briq. Rhizome elongate, producing lateral stolons; stem 1.5–9 dm. tall, hirsute on the angles, often freely branched; petioles 2–20 mm. long, more or less hirsute; blades 2–7.5 cm. long, lanceolate or lance-ovate, more or less hirsutulous, the margin serrate, the apex usually acute, the base cuneate; flowers numerous; pedicels filiform, glabrous, about equaling the calyx 2–3 mm. long, hirsutulous, the deltoid, acuminate lobes ¼ to ⅓ as long as the tube; corolla 3–4.5 mm. long, rose-lavender; stamens exserted; anthers 0.2 mm. long, ovoid, violet; style exserted; nutlets 0.7–0.8 mm. long, olive brown, the inner faces flattened, the scar basal, large. *M. canadensis* L. Wet stream banks, *U. Son., A. T.*

Var. **glabrata** (Benth.) Fern. Differs in having the stems hirsutulous on the angles; the blades ovate or lance-ovate, glabrous except for the puberulent margin and sometimes the veins beneath. *M. Penardi* (Briq.) Rydb.; *M. canadensis borealis* (Michx.) Piper. Wet places, *A. T.*

Var. **lanata** Piper. Differs from var. *canadensis* by having the herbage densely villous or even lanate. *M. canadensis lanata* Piper; *M. arvensis,* forma *lanata* (Piper) S. R. Stewart; *M. lanata* (Piper) Rydb. Moist places, *A. T., A. T. T.*

Mentha piperita L. *Peppermint.* Spreading by rhizomes and stolons; stems 2–10 dm. tall, glabrous; petioles 1–18 mm. long, slender; blades 15–75 mm. long, ovate to elliptic-lanceolate, serrate, acute, glabrous, or sparsely hirsutulous only on the veins beneath; spikes dense or interrupted, densely flowered; pedicels slender, glabrous, nearly equaling the calyx; calyx with the lance-linear, acuminate, rigid, ciliate lobes ½ the length of the tube; corolla 5 mm. long, rose-lavender; stamens included; anthers 0.5 mm. long, oval in outline; style exserted; nutlets usually not matured. Introduced from Europe. Moist places, Almota Cr., *St. John et al.* 9,244, *U. Son.*

It is a cultivated plant, its herbage, rich in aromatic oil, being the source of peppermint flavoring. It also has medicinal value.

Mentha spicata L. *Spearmint.* Spreading by rhizomes and stolons; stems 2–9 dm. tall, glabrous, usually freely branched; leaves sessile or nearly so, 2–9 cm. long, lance-elliptic, acute, serrate, glabrous or merely remotely pilose on the veins beneath; spikes terminal, interrupted, many flowered; pedicels glandular puberulous, mostly shorter than the calyx; calyx with the subulate, hispid teeth as long as the tube; corolla 2.5 mm. long, pale lavender; stamens included; anthers 0.3 mm. long, oblong-oval; style exserted; nutlets seldom if ever matured. *M. viridis* L. Introduced from Europe. Sparingly introduced, in wet places.

The herbage is rich in aromatic oil, which is useful in medicine and for flavoring gum, etc. Foliage is chopped, made into "mint sauce," which is eaten as a flavoring with meat.

MONARDELLA

Erect annuals or woody perennials, with entire, aromatic leaves and small rose-purple or white flowers in terminal glomerules which are subtended by broad thin bracts; calyx 10–15-nerved, tubular, 5-toothed, not 2-lipped; corolla 2-lipped, the upper lip 2-cleft, the lower 3-cleft; stamens 4, the lower pair the longer; anther-cells at length divergent; nutlets oblong-ovoid, smooth. (Name Lat., the genus *Monarda;* plus the diminutive.)

Bracts of inflorescence ovate or rounded, hirsute,
  Flowers rose-purple. — *M. odoratissima,* var. *odoratissima,* f. *odorotissima.*
  Flowers white. — Forma. *alba.*
Bracts of the inflorescence elliptic to ovate, puberulent. — Var. *glauca.*

Monardella odoratissima Benth., var. **odoratissima,** forma **odoratissima.** Perennial forming dense clumps; roots woody, brown; plant somewhat puberulent throughout; stems 1–6 dm. tall, often branched at base; principal leaves 15–30 cm. long, lanceolate, cuneate at base, subsessile, firm, glandular punctate, pale green, often glabrate above, in the axils bearing short shoots with numerous reduced leaves; inflorescence head-like, many flowered; outer bracts equaling the calyx, ovate or rounded, hirsute, purplish tinged; pedicels 1 mm. or less in length; calyx-tube 5–7 mm. long, the heavy nerves puberulent; calyx teeth 1 mm. long, lance-deltoid, hirsute; corolla 13–15 mm. long, rose-purple, funnelform, the lobes 5–6 mm. long, linear; filaments hispidulous at base; stamens long exserted; anther sacs 0.2 mm. long, purple, divergent, the interval filled by the heavy deltoid connective; nutlets 2 mm. long. *Madronella odoratissima* (Benth.) Greene; *Monardella odoratissima* Benth., subsp. ***eudoratissima*** Epling. Arid places, low valleys, also above tree line, Blue Mts., *U. Son., A. T.*

The herbage is rich in an aromatic oil with a penetrating odor.

Forma **alba** St. John. Differs from var. *odoratissima* by having the corollas white. Known only from type locality, 5500 ft., Clayton Spr., Columbia Co., Blue Mts., *St. John et al.* 9,658, *A. T.*

Var. **glauca** (Greene) St. John. Differs from var. *odoratissima* by having the leaves short petioled; the bracts of the inflorescence elliptic to ovate, puberulent. Arid places, Snake River Canyon, *St. John* 3,263.

NEPETA

Perennial herbs; calyx tubular, often incurved, obliquely 5-toothed or slightly 2-lipped; corolla 2-lipped, dilated in the throat; the upper lip erect, rather concave, notched or 2-cleft;

the lower spreading, 3-cleft, the middle lobe largest, notched or entire; stamens 4, ascending under the upper corolla lip, the upper pair longer; anther-cells divergent; nutlets ovoid or ellipsoid, smooth. (Name Lat., from *Nepete,* an Etruscan city.)

Plants creeping; flower clusters loose, axillary; leaves obtuse. *N. hederacea.*
Plants erect; flower clusters in dense terminal spikes; leaves acute. *N. Cataria.*

Nepeta Cataria L. *Catnip.* Perennial, forming clumps; plant closely pilosulous and white throughout or becoming less so and greenish; stems 5–10 dm. tall; petioles 3–40 mm. long, decreasing upwards; blades 2–8 cm. long, cordate to ovate or deltoid, the base cordate, the margin deeply crenate or crenate-dentate; flower clusters in terminal, interrupted, densely flowered spikes, 2–12 cm. long; bractlets lance-linear, shorter than the calyx; calyx 4–7 mm. long, 15-ribbed, the apex oblique and somewhat 2-lipped, the teeth nearly as long as the tube; corolla 7–9 mm. long, white, or purple spotted; anther sacs 0.1 mm. long, ellipsoid, glabrous, widely divergent; nutlets 1.3–1.5 mm. long, oval in outline, the inner faces flattened, brown. Native in Eurasia, cultivated and established near settlements, *U. Son., A. T.*

Nepeta hederacea (L.) Trev. *Ground Ivy.* Perennial; rhizome creeping, 1–2 mm. thick; stems 1–6 dm. long, decumbent with assurgent tips; plant hirsutulous throughout, or glabrate below; lower leaves with petioles 4–7 cm. long, the blades 2–4 cm. in diameter, reniform-suborbicular, deeply crenate-dentate, upright stems with shorter petioles and usually smaller blades; axillary clusters 1–6-flowered, usually 2–3-flowered; pedicels 1–3 mm. long, the opposite bractlets lance-linear; calyx 6–9 mm. long, 15-ribbed, the lanceolate, acuminate teeth ½ the length of the tube, oblique and more or less 2-lipped; corolla 12–20 mm. long, bluish violet; anther sacs 0.2–0.3 mm. long, glabrous, ellipsoid; nutlets 1.8 mm. long, ellipsoid brown. *Glechoma hederacea* L. A weed, introduced from Eurasia. Moist thickets, Colfax, *St. John,* 2,984, *A. T.*

Often separated as the genus *Glechoma,* but I find the calyx no more 2-lipped than in other species of *Nepeta.*

PRUNELLA. Heal-all, Self-heal

Low perennials; flowers in terminal or axillary heads or spikes; calyx tubular-bell-shaped, about 10-nerved, 2-lipped, not gibbous on the upper side, closed in fruit; upper lip broad and flat, truncate, with 3 short teeth, the lower 2-cleft; corolla 2-lipped, slightly contracted at the throat and dilated at the lower side just beneath it; upper lip arched, erect, entire; the lower reflexed-spreading, 3-cleft; stamens 4, ascending under the upper lip, the lower pair longer; nutlets ovoid, smooth. (Derivation of name uncertain.)

Median leaves ovate or ovate-oblong, rounded at base, 2/5 to ⅔ as broad as long. *P. vulgaris,* var. *vulgaris.*
Median leaves lanceolate to oblong, cuneate, 1/5 to ½ as broad as long,
Calyx green or merely purple-tinged. Var. *lanceolata,* f. *lanceolata.*
Calyx purple. Forma *iodocalyx.*

Prunella vulgaris L., var. vulgaris. *Self-heal; Heal-all.* Annual to perennial; caudex short; plant sparsely hispidulous, or becoming glabrate below;

lower leaves on petioles often exceeding the blades; blades 1–5.5 cm. long, ovate, entire or sinuate; median leaves 2–5 cm. long, ovate or ovate-oblong, usually crenate, the petioles 2–25 mm. long; spikes terminal, dense, 15–45 mm. long; bracts reniform, short pointed, hirsute ciliate; calyx 6–8 mm. long, mostly rose-purple, sparsely hirsute, short ciliate, the lower teeth lanceolate; corolla 10–15 mm. long, blue-violet; anthers 0.2 mm. long, ellipsoid; filaments spurred near the tip; seeds 2–2.3 mm. long, somewhat 3-angled, brownish. *Brunella vulgaris* L. A. weed, introduced from Eurasia. In lawns, Spokane, *Keyes* in 1928. *A. T.*

Var. **lanceolata** (Barton) Fern., forma **lanceolata.** Differs in having the bracts and calyces green; the median leaves lanceolate to oblong, cuneate, 1/5 to ½ as broad as long. *P. vulgaris,* subsp. *lanceolata* (Barton) Hultén. Moist places, *A. T., A. T. T.*

Forma **iodocalyx** Fern. Differs from var. *lanceolata* by having the bracts purple-tinged, the calyx purple. Moist thickets, *A. T., A. T. T.*

SALVIA

Shrubs or herbs; flowers numerous, in terminal or axillary heads or spikes; calyx 2-lipped, the upper lip 3-toothed or entire, the lower deeply 2-cleft; corolla 2-lipped, the upper lip straight, concave, falcate or obsolete; the lower spreading, 3-lobed; the middle lobe often cleft or fringed; fertile stamens 2, exserted; anther cells 1 with a prolonged, usually pointed connective; with or without a second but abortive anther cell; upper pair of stamens abortive; ovary deeply 4-parted; style 2-cleft; nutlets smooth, mucilaginous when wetted. (Name from Lat., *salvare,* to save.)

Leaves entire; floral bracts hispidulous ciliate. *S. carnosa.*
Leaves incised or dentate; floral bracts lanate. *S. Aethiopis.*

Salvia Aethiopis L. Biennial, 5–10 dm. tall; herbage densely floccose tomentose; first leaves forming a basal rosette; lower leaves 10–30 cm. long, from deeply incised to dentate, the blade ovate, petioled; cauline leaves the smaller, sessile; inflorescence 20–50 cm. long, an ovoid, open panicle of spikes; floral bracts ovate, acute, clasping the stem and 1–5 flowers; calyx in fruit 12–13 mm. long, campanulate, resin-dotted, lanate below, rigid, the lobes 4–6 mm. long, lanceolate, subulate pointed; corolla 12–18 mm. long, cream-colored, the upper lip strongly arched; nutlets 3 mm. long, olive brown with darker veins, ovoid, but the two inner faces flattened. Weed, introduced from Africa. Dayton, Aug. 21, 1951, *Wolfe.*

Salvia carnosa Dougl. ex Greene, subsp. **carnosa.** *Purple Sage.* Shrub, 3–10 dm. tall, fastigiate, the herbage, aromatic resinous punctate and closely whitish puberulent; branchlets striate; branches fluted, the bark gray, longitudinally fissured; leaves with petioles 5–25 mm. long, the blades 1–4 cm. long, oblanceolate to obovate, obtuse or retuse, entire or rarely crenulate; spikes 2–10 cm. long, the whorls remote; floral leaves and bracts 8–15 mm. long, obovate to oval, suffused with rose purple, white puberulent and hirsute ciliate; calyx glandular punctate, white puberulent, the tube narrowly turbinate; the upper lobes 2.5–3 mm. long, oval; the lower lobes 1 mm. long, ovate, acute; corolla 9–13 mm. long, deep blue; upper lip erect, cleft; lower lip 4–6 mm. long, the lateral lobes incurved, the terminal lobe flabellate, notched, erose; anther cells 1, curved, 1–1.2 mm. long; nutlets 2.5–2.8 mm. long, oval, compressed, brown, punctate. *Ramona incana* (Benth.) Dougl.; *Audibertiella incana* (Benth.) Briq.; not *S. incana* Mart. & Gal. Dry sands and gravels, near Spokane, *A. T.*

SATUREJA

Herbs or shrubs with small leaves; flowers in dense terminal or axillary clusters, or axillary; calyx campanulate, 10–(15-)nerved, campanulate or cylindric, equally 5-toothed, or 2-lipped; corolla 2-lipped, the upper lip flat, erect, entire or emarginate, lower lip with 3 flat lobes; stamens 4, connivent under the upper lip; anthers divergent or rarely parallel; nutlets ovoid, smooth. (The ancient Lat. name.)

Satureja Douglasii (Benth.) Briq. *Tea-vine.* Perennial; roots brown, woody; plant more or less hirsutulous, throughout; stems several, 1–13 dm. long, procumbent; petioles 2–9 mm. long; blades 10–37 mm. long, broadly ovate to suborbicular, remotely crenate, evergreen; pedicels 4–14 mm. long, with a pair of minute linear bracts; calyx cylindric to ellipsoid, 10-nerved, the tube 3–3.5 mm. long, the teeth 0.6–1 mm. long, erect, lanceolate, rigid pointed, nearly equal; corolla 8–11 mm. long, white, puberulent, the lips nearly equal; anther sacs 0.2 mm. long, divergent, lanceoloid; nutlets 1–1.2 mm. long, dark brown. *Micromeria Chamissonis* (Benth.) Greene. Moist mountain woods, *Can.*

The herbage is aromatic and pleasant. In pioneer times it was steeped to make herb tea.

SCUTELLARIA. Skullcap

Annual or perennial herbs, sometimes woody, not aromatic; flowers in axillary or terminal racemes or solitary in the axils of leaves or bracts; calyx bell-shaped in flower, 2-lipped, with a gibbous protuberance on the upper side, splitting to the base at maturity, the upper lip at length usually falling away; corolla with an elongated curved ascending tube, dilated at the throat, 2-lipped, the upper erect, arched or galeate; lower lip 3-lobed, its lateral lobes somewhat connected with the upper lip; stamens 4, ascending under the upper corolla lip, the lower pair longer; style unequally 2-cleft; nutlets papillose, rugose, or winged. (Name from Lat., *scutella,* dish.)

Flowers 5–8 mm. long, in axillary racemes; nutlets 1 mm. long, wingless, yellow; middle leaves with petioles 1–3 cm. long. — *S. lateriflora.*

Flowers 15–30 mm. long, solitary in axils; middle leaves with petioles 0.3 mm. long,

- Flowers 15–20 mm. long; nutlets 1.4–1.6 mm. long, yellow, wingless; middle leaves crenate. — *S. galericulata,* var. *epilobiifolia.*
- Flowers 20–30 mm. long; nutlets 1.2–1.3 mm. long, black, with a narrow appressed wing; middle leaves entire. — *S. angustifolia.*

Scutellaria angustifolia Pursh. Perennial, from freely branching rhizomes, producing white, tuber-bearing stolons; stems often numerous, 1–4 dm. tall, puberulent to subglabrate, commonly few branched at base; lower leaves with petioles up to 1 cm. long, the blades ovate or ovate-lanceolate; middle and upper leaves gradually narrower, 1–4 cm. long, oval to lance-oblong or linear-oblong, firm, puberulent, entire; pedicels 2–5 mm. long, puberulent; calyx 4–6 mm. long, puberulent, the galea 0.5–1 mm. high, truncate on the anterior side; corolla violet-blue, pilosulous, the lip of nearly the same length, the lower pilose within; anthers 1 mm. long, cordate, the sutures hispidulous ciliate; nutlets suborbicular, granular. Abundant, rocky slopes in warmer valleys, *U. Son., A. T.*

The type was collected by Capt. Lewis, "On the river Kooskoosky" [the Clearwater R., Ida.].

Growing with this species are other plants differing by having glandular pubescence. This nameless variant is discussed by Epling in Madrono 5: 64, 1939.

Scutellaria galericulata L., var. **epilobiifolia** (Hamilton) Jordal. *Marsh Skullcap.* Rhizome slender; perennial by slender stolons; stems 1.5–12 dm. tall, often branched, puberulent, especially on the angles, glabrate below; petioles 0–3 mm. long; blades 1–8.5 cm. long, oblong-lanceolate to oblong-ovate, puberulent to glabrate above, closely puberulent below, the base sub-cordate to rounded, the apex acute or obtuse, the upper leaves often entire; pedicels 1–4 mm. long, puberulent; calyx 3–5.5 mm. long, puberulent, the galea a low ridge 0.3–1 mm. high; corolla pilosulous, violet blue, white towards the base, the lower lip slightly the longer; anthers 0.5 mm. long, cordate-oval, the sutures densely villous ciliate; nutlets suborbicular, somewhat compressed, tuberculate. *S. epilobiifolia* Hamilton. From Rock L. and Potlatch, Ida., northward, wet meadows, *A. T.*

Scutellaria lateriflora L. *Mad-dog Skullcap.* Perennial; rhizome slender, producing slender stolons; stems 1–9.5 dm. tall, puberulent on the angles at least above, often branching; petioles 2–30 mm. long; blades 2–9 cm. long, ovate to ovate-lanceolate, almost membranous, glabrous or sparsely puberulent on the veins, the apex acute or acuminate, the margin coarsely dentate or serrate, or the upper entire; racemes secund, mostly shorter than the leaves; bracts foliaceous, lanceolate, as long or longer than the calyx; calyx 3–4 mm. long, puberulent, the galea oblique conic, 0.5–1.5 mm. tall; corolla pilosulous, pale blue to whitish, the lips nearly equal, the upper entire or notched, the lower suborbicular, shallowly 3-lobed; anthers 0.2 mm. long, cordate ciliate; nutlets orbicular, compressed, papillose. Moist thickets, rare.

STACHYS

Annual or perennial, rarely woody, not aromatic; calyx tubular-bell-shaped, 5–10-nerved, equally 5-toothed or the upper teeth united to form an upper lip; corolla purple, red, yellow, or white, not dilated at the throat, 2-lipped, the tube about equaling the calyx; the upper lip concave, often arched, erect or rather spreading, entire or nearly so; the lower usually longer and spreading, 3-lobed, the middle lobe largest and nearly entire; stamens 4, ascending under the upper lip, the lower pair longer. (Name from Gr. *stachus,* a spike; used by Dioscorides for this genus.)

Stachys palustris L., var. **pilosa** (Nutt.) Fern. Perennial herb, hirsute and hirsutulous throughout, 15–90 cm. tall; rhizome white 2–4 mm. in diameter;

stems simple or with weak branches; leaves sessile or a few with petioles 1–5 mm. long; blades 4–11 cm. long, usually ovate-oblong or oblong, rarely elliptic, the base rounded subcordate or truncate, the margin crenate; flowers sessile, usually 6 at each upper node, the subtending leaves much reduced upwards; calyx-tube 4–5 mm. long, hirsute, the lobes 2–4 mm. long, lance-deltoid, acuminate, acerose-tipped; corolla rose purple, maculate, the tube 6–8.5 mm. long, the upper lip 3.5–5.5 mm. long, pilose, the lower lip 6–7.5 mm. long; anther sacs 0.4–0.5 mm. long, ellipsoid, purple, widely divergent; nutlets 2–2.3 mm. long, ovoid, the inner sides much flattened, dark brown. *S. palustris*, subsp. *pilosa* (Nutt.) Epling; *S. scopulorum* Greene; *S. Leibergii* Rydb.; *S. teucriformis* Rydb.; *S. asperrima* Rydb. Meadows and wet shores, *U. Son.*, *A. T.*

TRICHOSTEMA. Blue Curls

Low annual herbs, or somewhat woody perennials; leaves entire; calyx bell-shaped, oblique, deeply 5-cleft, the 3 upper teeth elongated and partly united, the 2 lower very short; corolla small, almost equally 5-parted, the 3 lower lobes more or less united; stamens 4, didynamous, in bud spirally coiled, later long exserted; anther-cells divergent and at length confluent at base; ovary deeply 4-lobed; style 2-cleft; nutlets rugose reticulate, the attachment ventral. (Name from Gr., *thrix*, hair; *stema*, stamen.)

Trichostema oblongum Benth. Annual, 5–40 cm. tall, loosely villous and pilosulous throughout; roots fibrous; stems often diffusely branched; petioles 0–7 mm. long; blades 10–34 mm. long, oblanceolate to elliptic or obovate, subacute or obtuse; cymes axillary, short, few- to many-flowered; pedicels becoming 1–2 mm. long; calyx-tube 1–2 mm. long, the lobes in fruit 3–5 mm. long, lanceolate, acuminate, firm; corollas 3–5 mm. long, violet, deeply lobed; anther cells 0.5 mm. long, oblong, divergent, purple; nutlets 1.5–2 mm. long, obovoid, the apical and dorsal ridges hirsute. Low, moist spots, *A. T.*, *A. T. T.* Herbage rather unpleasantly scented.

SOLANACEAE. Nightshade Family

Herbs or shrubs, vines or trees, commonly rank-scented, many poisonous, with colorless juice; leaves alternate, without stipules; flowers regular, 5-merous, solitary or in cymes, on bractless pedicels; calyx mostly 5-lobed; corolla gamopetalous, mostly 5-lobed, imbricate, valvate, or plaited in the bud; stamens as many as the corolla-lobes and alternate with them; inserted on the tube; anthers 2-celled, apically or longitudinally dehiscent; style 1; ovary entire, superior, 2-celled, or falsely 4-celled, becoming a many-seeded capsule or berry; endosperm fleshy. (Named from the genus *Solanum*.)

Fruit a berry; corolla valvate, the lobes generally induplicate,
  Calyx accrescent and bladder-like in fruit; corolla shallowly lobed. — *Physalis.*
  Calyx not so; corolla deeply 5-lobed. — *Solanum.*
Fruit a capsule; corolla imbricate, not induplicate,
  Capsule circumscissile; corolla irregular. — *Hyoscyamus.*

Capsule dehiscing by valves at the apex; corolla regular,
Calyx prismatic, 5-toothed; capsule prickly, 4-valved. *Datura.*
Calyx tubular-bell-shaped, 5-cleft; capsule not prickly, enclosed in the calyx. *Nicotiana.*

DATURA

Rank narcotic-poisonous annual or perennial herbs, erect, tall, branching; leaves alternate, petioled, ovate; flowers large, showy, solitary, on short peduncles, in the forks of the stems; calyx prismatic, 5-toothed, deciduous; corolla funnelform, with a 5–10-toothed plaited border; fruit a globular prickly 4-valved 2-celled capsule; seeds rather large, flat. (The Arabic name *dhatura,* for *D. Stramonium.*)

Flowers white; stems green. *D. Stramonium.*
Flowers pale-violet; stems purplish. Var. *Tatula.*

Datura Stramonium L. *Jimson weed.* Annual, young shoots sparsely hirsutulous, soon glabrate; roots fibrous; plant 1–2 m. tall, freely branched; petioled 1.5–6 cm. long; blades 5–20 cm. long, ovate in outline, the apex acute or acuminate, the base cuneate, irregularly sinuately lobed, the lobes deltoid, acute, entire or dentate; pedicels 5–18 mm. long; calyx 3–4 cm. long, prismatic, deciduous, the teeth 3–5 mm. long, lanceolate, unequal; corolla 6–10 cm. long, the rounded lobes short acuminate; anthers 5 mm. long, linear-oblong; capsules usually spiny, the spines up to 1 cm. in length, the lower ones sometimes shorter, the body 2–4.5 cm. long, ovoid; seeds 3–4 mm. long, reniform, flattened, brown, cellular-reticulate. Weed, introduced from e. North America. Abundant, dry or moist places, *U. Son., A. T.*

The herbage and especially the seeds contain hyoscyamin, atropin, and scopolamin, all poisonous alkaloids and powerful drugs or medicines. Small quantities if eaten cause sickness or death.

Var. Tatula (L.) Torr. Differs from the species by having the stem purplish; and the corollas violet-purple. *D. Tatula* L.; *D. Stramonium,* var. *chalybea* Koch. Weed, introduced from tropical America. Open places, occasional. *U. Son., A. T.*

HYOSCYAMUS. Henbane

Coarse, viscid pubescent, narcotic, annual to perennial herbs; leaves alternate, mostly lobed or pinnatifid; flowers large, the lower solitary in axils, the upper in more or less 1-sided spikes or racemes; calyx urceolate or narrow campanulate, 5-cleft, striate, accrescent, enclosing the capsule in fruit; corolla funnelform, oblique, 5-lobed, the border plaited, unequal; stamens declined; anthers longitudinally dehiscent; ovary 2-celled; stigma capitate; capsule circumscissile above the middle. (Name used by Dioscorides, from Gr., *hus.,* hog; *kuamos,* bear.)

Hyoscyamus niger L. *Black Henbane.* Biennial, or seldom annual; taproot up to 15 mm. in diameter; plant viscid villous, malodorous, 2–10 dm. tall; leaves 5–20 cm. long, sessile or the upper clasping, lanceolate in outline, acute, irregularly pinnately cleft or lobed; flowers subsessile; calyx in flower 8–11 mm. long, campanulate, the deltoid-lanceolate teeth half as long as the tube, in fruit the calyx 22–25 mm. long, the base globose; corolla 22–30 mm. long, greenish yellow, the center purple and above purple reticulate veined; filaments hirsute; anthers 3 mm long, oblong, purple; capsule 8–15 mm. long, rounded ovoid, pale; seeds 1.2–1.5 mm. long, semireniform, flattened, brown, coarsely pitted. Weed, introduced from Eurasia or Africa. Waste places, Pullman, June 7, 1921, *Ihrig, A. T.*

The herbage and especially the seeds are medicinal and poisonous. It contains hyoscyamin and other poisonous alkaloids. Poisonous to stock, but they usually avoid it. Poisonous to humans, many cases, serious or fatal having been reported.

NICOTIANA. Tobacco

Rank acrid-narcotic herbs; leaves mostly entire; flowers in racemes or panicles, sometimes showy; calyx bell-shaped or oblong, 5-cleft persistent; corolla commonly funnelform or salverform, the plaited border 5-lobed; stigma capitate, somewhat 2-lobed; fruit a smooth 2–4-valved, 2-celled capsule; seeds numerous, small. (Named for *Jean Nicot* of France.)

Nicotiana attenuata Torr. ex Wats. *Coyote Tobacco.* Annual, viscid pilosulous throughout, especially on the stems; tap-root as much as 8 mm. in diameter; basal leaves with petioles 2–5 cm. long, the blades 2–10 cm. long, oval to ovate-lanceolate, entire; cauline leaves with petioles 0–30 mm. long, the upper sessile or short petioled, the blades 3–14 cm. long, lanceolate or the upper lance-linear; panicle 1–3 dm. long, loose, bracteate only below; pedicels 2–10 mm. long; calyx 6–9 mm. long, campanulate, the teeth deltoid half as long as the tube; corolla 2.5–3.5 cm. long, white, vespertine, the limb 8–12 mm. wide, with 5 shallow lobes; anthers 1.5 mm. long, oval; capsule 8–11 mm. long, ovoid, pale, the valves bifid; seeds 0.5–0.7 mm. long, reniform or irregular, brown, deeply pitted. *N. Torreyana* Nels & Macbr. Sandy soil, Snake River Canyon, *U. Son.*

PHYSALIS. Ground Cherry

Annual or perennial herbs; pedicels slender, solitary from the axils (in ours); calyx campanulate, 5-toothed, in fruit accrescent, becoming bladdery inflated, 5-angled or 10-ribbed, membranous, reticulate, enclosing the pulpy edible berry, calyx-teeth mostly connivent; corolla yellow or whitish often with brown or purplish eye, open campanulate or campanulate-rotate, plicate; anthers dehising longitudinally; stigma 2-cleft; seeds pitted. (Name from Gr. *phusalis,* bladder.)

Upper leaves mostly lanceolate, entire; anthers oblong; fruiting calyx indistinctly 10-angled, the base not sunken. *P. lanceolata.*
Upper leaves ovate to cordate, the margin deeply sinuate toothed; anthers lanceoloid; fruiting calyx sharply 5-angled, the base sunken. *P. pruinosa.*

Physalis lanceolata Michx. Perennial; root up to 5 mm. in diameter, pale, cord-like, deep-seated; plant 1–4 dm. tall, sparsely hirsute; petioles 2–20 mm.

long; blades 2–8 cm. long, the lower ovate or oval, the upper lanceolate or oblanceolate, the margin entire or sinuate, the base cuneate, decurrent, the apex obtuse or acute; pedicels 5–35 mm. long, in fruit recurved; calyx usually strigose or villous, in flower 8–11 mm. long, campanulate, the ovate-lanceolate teeth nearly as long as the tube; corolla 9–13 mm. long, dull yellow, with brownish center; anthers 2 mm. long; fruiting calyx 2–4 cm. long, rounded ovoid the teeth erect; berry 10–13 mm. in diameter, yellow or greenish yellow; seeds 2.2–2.5 mm. long, semireniform, flattened, brownish yellow, cellular reticulate. Weed, introduced from the central states. R. R. embankment, Pullman, *Warren* 387, July 2, 1925. *A. T.*

Physalis pruinosa L. *Strawberry Tomato.* Annual; roots fibrous; plant 1–4.5 dm. tall, divaricately branched, densely hirsute or hirsutulous; petioles 5–70 mm. long; blades 2–10 cm. long, the lower the smaller, ovate, entire, the upper obtuse or acute, the base cordate, inaequilateral; pedicels 4–40 mm. long, in fruit recurved; calyx densely hirsute, in flower 4–7 mm. long, the broadly lanceolate lobes about as long as the tube; corolla 7–10 mm. long, yellow, with a purple center; anthers 1.8–2.2 mm. long, lanceoloid; fruiting calyx 2.5–3.5 mm. long, ovoid, the tip acute; berry 10–18 mm. in diameter, yellow or green, or reddish; seeds 1.7–1.9 mm. in diameter discoid or slightly asymmetric, flattened, yellowish, cellular reticulate. Weed, introduced from the central states. Shore of Snake River, Central Ferry, *St. John* 3,092, *U. Son.*

SOLANUM. Nightshade

Herbs or shrubs, many poisonous; calyx and rotate corolla 5-parted or cleft, the latter plaited in the bud; stamens exserted; filaments very short; anthers converging or connate into a cone; styles elongated; ovary 2-celled, rarely more; fruit a berry. (Name from Lat., *solamen,* quieting.)

Vines; flowers purple; berries red. *S. Dulcamara.*
Herbs, not climbing; flowers mostly white or yellow; berries not red,
  Plant prickly; corolla yellowish; calyx prickly, surrounding the berry. *S. rostratum.*
  Plants unarmed; corolla white or purplish tinged; calyx not prickly; berry exposed,
    Leaves pinnatifid; berries green; seeds 2–2.4 mm. long. *S. triflorum.*
    Leaves entire or repand; berries not green; seeds 1–2 mm. long,
      Plant glabrous or sparsely hirsutulous; berries black. *S. nodiflorum.*
      Plant more or less hirsute; berries yellowish. *S. sarrachoides.*

Solanum Dulcamara L. *Bittersweet.* Perennial, half-shrubby, with climbing or spreading branches, sometimes 1 m. or more long, pubescent or glabrate; leaves ovate, acuminate, cordate, simple, 2–6 cm. long, or many of them with 3 lobes or 3-divided at base, the lateral segments smaller; petiole slender, shorter than the blades; cymes loose; petals purple, darker and green bimaculate at base, lanceolate, 6–8 mm. long, reflexed; berries ellipsoid. Introduced from Europe, occasional along streams, *A. T.*

The plant contains poisonous alkaloids, not in great quantities, but it should be avoided.

Solanum nodiflorum Jacq. *Deadly Nightshade.* Annual or perennial; roots fibrous; plant 1–5 dm. tall, freely branched; petioles 5–40 mm. long;

blades 2–7 cm. long, ovate to lanceolate, variable, inaequilateral, acute, entire or irregularly dentate or lobed, at base cuneate and decurrent on the petiole; cymes 3–10-flowered, often umbellate; peduncles 10–23 mm. long; pedicels in fruit deflexed, 7–15 mm. long; calyx 2–4 mm. long, the lobes ovate, about equaling the tube; corolla 5–8 mm. long, white or purplish tinged, the ovate lobes 3 times the length of the tube; anthers 1.5–2.4 mm. long, oblong, dehiscing from the apex; berries 7–10 mm. in diameter, globose; seeds asymmetric ovoid, flattened, pale, cellular reticulate. Weed, naturalized from Europe. Cultivated or waste lands, *U. Son., A. T.*

Dead ripe fruits are edible, but unripe fruits and the herbage contain solanin and other poisonous alkaloids. It is poisonous to humans and stock, producing paralysis, then death.

Solanum rostratum Dunal. *Buffalo Bur.* Annual; roots fibrous, extensive; herbage coarsely hispid stellate, and armed with yellow, subulate prickles 7–15 mm. long; stems 2–5 dm. tall, branching; petioles 2–8 cm. long; blades 5–14 cm. long, irregularly pinnately lobed or 1–2-pinnatifid, ovate or oval in outline, the ultimate divisions mostly ovate or rounded; racemes 3–10-flowered; peduncle 1–4 cm. long; pedicels 3–13 mm. long; calyx densely prickly, in flower the tube 2–3 mm. long, campanulate, the lobes 3–6 mm. long, lanceolate, in fruit the tube globose, enclosing the berry; corolla 2–3 cm. in diameter, stellate without, the margin lobed; stamens and style declined; four stamens similar, the anthers 7 mm. long, linear, dihiscent by apical pores, the lower anther longer and with a deflexed beak; berry 6–8 mm. in diameter; seeds 2–2.5 mm. long, semicordate-orbicular, brown, finely pitted. *Androcera rostrata* (Dunal) Rydb. Weed, introduced from the Great Plains. Dry cultivated lands. *U. Son.*

Solanum sarrachoides Sendt ex Mart. Annual; stems 1–5 dm. long, decumbent or ascending; herbage viscid villous; leaves 2.5–7 cm. long, ovate, gradually or abruptly narrowing to and decurrent on the petiole, acute or obtuse, sinuately toothed; peduncles 3–8-flowered, 5–20 mm. long; pedicels 3–5 mm. long at anthesis, the apex much thickened in fruit; calyx 2–2.5 mm. long and villous at anthesis, in fruit accrescent and partly enclosing the berry; corolla 3–5 mm. in diameter, white, the narrowly deltoid lobes villous without near the apex; anthers 2–2.5 mm. long; berries 6–7 mm. in diameter; seeds 2–2.5 mm. long, yellowish, the testa concentric tessellate. Adventive weed, introduced from Brasil. All plant parts somewhat poisonous.

Solanum triflorum Nutt. *Wild Tomato.* Annual; roots fibrous; plant 1–5 dm. tall, sparsely hispidulous, freely branching, often decumbent; petioles 3–25 mm. long; blades 1–4 cm. long, deeply pinnatifid; the lobes deltoid to lanceolate, the sinuses rounded; peduncles 3–20 mm. long, 1–3-flowered, lateral, not axillary; pedicels in fruit 10–15 mm. long and reflexed; calyx 3–4 mm. long, campanulate, accrescent, in fruit the tube 2–3 mm. long, the lobes 5–7 mm. long, deltoid; corolla 4–5 mm. long, broadly campanulate, the oblong-lanceolate lobes twice as long as the tube, pilosulous without; anthers 2.5 mm. long, linear, dehiscing from the apex; berries 8–11 mm. in diameter, glabrous; seeds obliquely ovate, flattened, pale, cellular reticulate. Occasional, arid places, *A. T.*

Plants and fruits reported to be poisonous.

SCROPHULARIACEAE. Figwort Family

Herbs, sometimes shrubs or trees; leaves alternate, opposite, or whorled, without stipules; flowers perfect, mostly complete and irregular; corolla irregular, more or less 2-lipped, the upper lip 2-lobed, the lower 3-lobed; stamens on the corolla-tube, the fertile

4 and didynamous or only 2, rarely 5 present and all fertile; anthers 2-celled, or confluent, dehiscing lengthwise; style single; stigma entire or 2-lobed, fruit a 2-celled usually many-seeded capsule, with axile placenta; seeds mostly small; endosperm copious. (Named from the genus *Scrophularia.*)

Anther-bearing stamens 5; leaves alternate. *Verbascum.*
Anther-bearing stamens 2 or 4; leaves alternate, opposite, or whorled,
 Fifth sterile stamen present,
  Corolla spurred at the base. *Linaria.*
  Corolla not spurred,
   Sterile stamen elongated, about equaling the others. *Penstemon.*
   Sterile stamen a gland or scale adherent to the upper side of the corolla,
    Peduncles several-flowered. *Scrophularia.*
    Peduncles 1-flowered,
     Middle lobe of lower lip carinate. *Collinsia.*
     Middle lobe of lower lip not carinate. *Tonella.*
 Fifth sterile stamen not present,
  Stamens 2, anther-bearing,
   Calyx 5-parted. *Gratiola.*
   Calyx 4-parted,
    Leaves alternate, mostly basal,
     Corolla campanulate or rotate, not 2-lipped. *Synthyris.*
     Corolla strongly 2-lipped, or none. *Besseya.*
    Leaves all or at least the lower opposite. *Veronica.*
  Stamens 4, either perfect or sterile,
   Anther-bearing stamens 2; sterile stamens 2,
    Sterile filaments 2-forked, exserted. *Lindernia.*
    Sterile filaments simple, included. *Gratiola.*
   Anther-bearing stamens 4,
    Corolla nearly regular. *Limosella.*
    Corolla 2-lipped,
     Stamens not enclosed in the upper lip,
      Calyx sharply prismatic angled. *Mimulus.*
      Calyx campanulate, smooth or slightly sulcate. *Mimetanthe.*
     Stamens enclosed in the upper lip,
      Anther-cells equal,
       Calyx inflated in fruit; leaves opposite. *Rhinanthus.*
       Calyx not inflated; leaves alternate or whorled. *Pedicularis.*
      Anther-cells unequal,
       Galea much longer than the lower lip; calyx usually 2-cleft and the divisions 2-toothed. *Castilleja.*
       Galea little exceeding the lower lip; calyx mostly equally 4-lobed. *Orthocarpus.*

BESSEYA

Perennial herbs from rootstocks; leaves mostly basal, oblong or ovate, crenate; cauline leaves much reduced, alternate, sessile;

flowers perfect, in terminal, leafy bracted spikes; sepals 4 (1–3) -parted; corolla cleft nearly to the base, 2-lipped, the upper lip concave arched, much the longer, lower lip 3-cleft, or corolla none; stamens 2, exserted; anther sacs parallel; capsule flattened, loculicidal. (Named for *Prof. Charles Edwin Bessey* of Nebraska.)

Besseya rubra (Dougl.) Rydb. Plant 3–5 dm. tall, white appressed pilose, glabrate below, caudex short, 3–10 mm. in diameter; basal leaves several, the petioles 4–13 cm. long, the blades 3–10 cm. long, ovate, obtuse, the base subcordate or cuneate, decurrent, the margin coarsely crenate; spike 4–30 cm. long, many flowered, becoming loose below; lower bracts exceeding the flowers; calyx 3–5 mm. long, the divisions lanceolate to ovate, viscid pilose; filaments 2–6 mm. long, red; anthers oval, becoming curved, the sacs 1.1–1.3 mm. long, magenta red; styles 2–4 mm. long; capsule 3–7 mm. long, broadly oval, flattened, emarginate, pilose to glabrous, the lower flowers not fruiting; seeds 1.5–2 mm. wide, saucer-shaped, brownish. *Synthyris rubra* (Dougl.) Benth. Abundant, rocky slopes, prairies, open woods, *U. Son., A. T., A. T. T.*

In this species the corolla is never produced.

CASTILLEJA. Painted Cup, Indian Pink, or Indian Paint Brush

Perennial or annual herbs, sometimes woody at the base; leaves alternate, sessile, entire or cleft into linear lobes; the floral ones usually dilated, colored and more showy than the yellow, red, or whitish spiked flowers; calyx tubular, flattened, 2- or 4-lobed; corolla tube included; upper lip (*galea*) much longer than the lower, narrow, arched and keeled, enclosing the 4 unequal stamens; lower lip short, 3-lobed; anthers unequally 2-celled; capsule loculicidal, many-seeded. (Named for *Domingo Castillejo* of Cadiz, Spain.)

Lower lip of corolla 1/15 the length of the galea; leaves entire. — *C. miniata.*
Lower lip ¼ or more of the length of the galea,
  Lower lip ¾ the length of the galea. — *C. pallescens.*
  Lower lip from ¼–½ the length of the galea,
    Lower lip about ½ the length of the galea,
      Pubescence rough hirsute. — *C. lutescens.*
      Pubescence soft villous. — *C. Cusickii.*
    Lower lip ¼ the length of the galea,
      Leaves, or some of them, cleft,
        Stems densely pilose or hirsute,
          Leaves somewhat puberulent, not rough, the lower ones linear. — *C. angustifolia,* var. *angustifolia.*
          Leaves hispid, rough, the lower ones linear-lanceolate. — Var. *hispida.*
        Stems sparingly pilose; leaves lanceolate to obovate. — Var. *Bradburii.*
      Leaves all entire. — Var. *Whitedi.*

Castilleja angustifolia (Nutt.) G. Don., var. **angustifolia.** Perennial, 1–5 dm. tall, appressed puberulent and hirsute throughout; root 3–10 mm. in diameter, multicipital; stems erect, usually simple; leaves 3–6 cm. long, the lower linear, entire or nearly so, the upper linear-lanceolate or lanceolate, 3–7-cleft, the lobes linear, the lateral lobes divaricate; spikes 3–15 cm. long; bracts not exceeding the flowers, obovate, 3–7-cleft, the tips scarlet or rarely yellowish; calyx 22–25 mm. long, equally cleft, the lateral lobe bifid, the lobes equaling the tube, oblong-lanceolate; corolla 2–3 cm. long; galea 11–13 mm. long; lower lip 2–3 mm. long, trifid; capsule 8–11 mm. long, lance-ellipsoid; seeds 1.2–1.8 mm. long, ovoid, truncate, the brown embryo covered by a transparent testa with pale, deep honey-comb-like reticulations. Open slopes, *U. Son., A. T.*

Pennell (1937) published his interpretation that this species should be given the more recent name *C. chromosoma* A. Nels., and that the name var. *Bradburii* had been misapplied by Fernald. Pennell failed to complete this revision before his death.

Var. **Bradburii** (Nutt.) Fern. Stems sparingly pilose or glabrate below; leaves lanceolate to obovate, nearly all 3–5-cleft; calyx 2–3 cm. long; corolla exceeding the calyx by 5–10 mm. Grassy slopes, *A. T., A. T. T.*

Var. **hispida** Fern. Plant coarse; lower leaves linear lanceolate, attenuate, hispid and hirsutulous; upper leaves oblong or ovate, the 3–5 lobes linear to oblong; calyx 2.5–3 mm. long; corolla 3–4 cm. long. *C. hispida* Benth. Abundant, open slopes, *U. Son., A. T.*

Var. **Whitedi** Piper. Stems 2–3 dm. tall; leaves all entire, 2–3 cm. long, lanceolate; bracts broader than the leaves, the uppermost 3-cleft. Open slopes, Almota, *St. John 3,382, U. Son.*

Castilleja Cusickii Greenm. Perennial, 1.5–6 dm. tall, villous throughout and more or less viscid above; root 3–10 mm. in diameter, multicipital; stems simple, strict; leaves 2–4 cm. long, linear to lanceolate, the lower the narrower, entire, the upper 3–7-cleft above the middle, the lobes linear or lance-linear; spikes 5–27 cm. long; bracts oblong, 3–5-lobed, the tips obtuse yellowish; calyx 2–2.5 cm. long, equally cleft above and below, the lobes equaling the tube, the lateral notched; corolla 2 cm. long, yellowish, the galea 5–6 mm. long; lower lip 3 mm. long, trifid; capsule 8–11 mm. long, ellipsoid; seeds 0.6–1.2 mm. long, ovoid, truncate, the dark embryo surrounded by a greenish, transparent honey-comb-like cellular reticulation. *C. camporum* (Greenm.) Howell; *C. lutea* Heller; *C. pannosa* Eastw. Low meadows, *A. T.*

Castilleja lutescens (Greenm.) Rydb. Perennial, 2–6.5 dm. tall, rough hirsute and puberulent; root 3–10 mm. in diameter, brown, multicipital; leaves 3–7 cm. long, the lower linear or lance-linear, entire, the upper entire or with 3–5 short, linear lobes; spikes 5–15 cm. long; bracts ovate or oblong, trifid, the tips yellowish; calyx 12–15 mm. long, nearly equally cleft, the lance-linear lobes longer than the tube; corolla 20–25 mm. long, yellowish, galea 8 mm. long; lower lip 3 mm. long, trifid, greenish; capsule 11–13 mm. long, ellipsoid, acute; seeds 1.2–1.5 mm. long, ovoid, the brown embryo enclosed by a loose, coarsely cellular reticulate, transparent testa. Dry hillsides or woods, *U. Son., A. T., A. T. T.*

Castilleja miniata Dougl. ex Hook. Perennial, 3–10 dm. tall, glabrous below the inflorescence or nearly so; root 2–10 mm. in diameter, bearing 1 to many stems; leaves 3–6 cm. long, linear to lanceolate, entire; spikes 3–23 cm. long, hirsute; bracts scarlet, entire or the upper 3–5-cleft; calyx 2.5 cm. long, the lance linear lobes longer than the tube; corolla 2.8–3.5 mm. long, yellowish; galea 13–16 mm. long; lower lip 1 mm. long; capsule 13–17 mm. long, ellipsoid; seeds 1.8–2.5 mm. long, ovoid, truncate, the brown embryo covered by the loose, brown irridescent, honey-comb-like, brown testa. Sunny slopes or meadows, *A. T., A. T .T., Huds.*

Castilleja pallescens (Gray) Greenm. Perennial, 1–3.5 dm. tall, finely gray puberulent; root 2–10 mm. in diameter, dark, multicipital; stems in a clump, simple or few branched; leaves 2–6 cm. long, linear, entire or with

3–5 linear lobes; spikes 3–20 cm. long; bracts 3–5-cleft, the lobes linear, the upper whitish or yellowish at tip; calyx 18–20 mm. long, the 2 linear lobes half as long, bifid; corolla 18–24 mm. long, puberulent, yellowish; galea 4–5 mm. long; lower lip 3–4 mm. long; capsule 6–15 mm. long, lanceoloid, oblique; seeds 1–1.4 mm. long, ovoid, the brown embryo covered by a white testa with high, honey-comb ridges. Open places, *A. T.*

COLLINSIA

Annuals; leaves simple, whorled or opposite, sessile or the lowest petioled and the upper whorled; flowers solitary or whorled in the upper axils; calyx deeply 5-cleft; corolla with the tube cylindric or saccate at the base on the upper side, deeply 2-lipped; the upper lip 2-cleft, the lower 3-cleft, the middle lobe keeled and sac-like; anther-bearing stamens 4; staminodium a gland-like structure; capsule 4–many-seeded. (Named for *Zaccheus Collins* of Philadelphia.)

Corolla erect or somewhat deflexed from the calyx, 4–8 mm. long, the lobes obovate or spatulate,
 Corolla blue violet. — *C. parviflora*, forma *parviflora.*
 Corolla not bluish,
  Corolla rose-mallow. — Forma *rosea.*
  Corolla white. — Forma *alba.*
Corolla sharply deflexed, usually at right angles to the calyx, the lobes broadly obovate. — *C. grandiflora.* var. *pusilla.*

Collinsia grandiflora Dougl. ex Lindl., var. **pusilla** Gray. Plant 5–45 cm. tall, puberulent or the leaves sparsely so; differing from *C. parviflora* by the corolla 7–10 mm. long, deflexed almost at right angles to the calyx, the tube more strongly saccate at base above, lips and most of the tube violet, lobes broadly obovate; capsules 4–5 mm. long. Woods, Troy, Gold Hill, and Clearwater River, *A. T. T.*

The Sandberg & Leiberg [31] collection, listed by Newsom, I determine as *C. parviflora.*

Collinsia parviflora Dougl. ex Lindl. forma **parviflora.** *Blue Lips.* Plant 5–50 cm. tall; roots fibrous; stem simple or branched, puberulent or glandular puberulent, or glabrate, cotyledons foliaceous, orbicular, slender petioled; lower leaves with petioles 3–10 mm. long, the blades 5–16 mm. long, often purplish beneath, suborbicular to ovate, entire or crenate, glabrate; cauline leaves short petioled or sessile, 1–5.6 cm. long, opposite or 3–5 in a whorl, oval to linear, crenate or entire, the upper narrower and less toothed; flowers 1 in the lower, 2–5 in the upper axils; pedicels 3–20 mm. long, puberulent or glandular puberulent or glabrate; calyx 3–7 mm. long, campanulate, glandular puberulent or glabrate, the lanceolate lobes about equaling the tube; corolla blue violet, or especially the lips or the lower lip paler, the tube paler or whitish, the tube about as long as the lips, gibbous above near the base, corolla erect or somewhat declined; anthers 0.3–0.4 mm. wide, reniform; capsule 3–4 mm. long, ovoid, acute; seeds 1–2.3 mm. long, ellipsoid, brownish, hilum elongate. *C. tenella* Piper, probably not *Antirrhinum tenellum* Pursh. Very abundant, early spring, open places or thickets, *U. Son., A. T., A. T. T.*

Forma **alba** English. Differs by its white corollas. Prairies, Pullman, *English & Hardin* 985, the type collection, *A.T.*

Forma **rosea** Warren. Differs by its rose-mallow corollas. Moist hillsides, Armstrong, *Warren* 745, the type collection, *A.T.*

Miss Newsom omits color forms in her monograph, but in our species these are few and local. I consider it better to recognize them than to overlook them.

GRATIOLA. Hedge Hyssop

Low mostly perennial branching herbs; leaves opposite, sessile; peduncles axillary, 1-flowered; calyx 5-parted, the narrow divisions usually equal; upper lip of corolla entire or 2-cleft, the lower 3-cleft; anther-bearing stamens 2; sterile filaments 2, simple and included, or none; style dilated or 2-lipped at the apex; capsule 4-valved, many seeded. (Name from Lat., *gratia,* grace.)

Peduncles with 2 terminal bractlets; calyx 3–5 mm. long, connective wider than the anther sacs. *G. neglecta.*
Pedicels bractless; calyx 4–11 mm. long; connective narrower than the anther sacs. *G. ebracteata.*

Gratiola ebracteata Benth. Plant 3–16 cm. tall, sparsely glandular puberulent above; roots fibrous; stems often simple; leaves 7–27 mm. long, lanceolate, entire; pedicels 5–15 mm. long; calyx glandular puberulent, the divisions lanceolate, unequal, herbaceous; corolla of early flowers 6–9 mm. long, tubular, the tube yellow, the lobes purplish tinged, broadly obovate; stamens and style included; anthers 0.3 mm. long, orbicular-reniform; capsule 4–5 mm. long, globose; seeds 0.5–0.7 mm. long, fusiform, brownish, several ribbed and reticulate. Wet places, Cedar [Moscow] Mt., *Sandberg et al.* 430; Pullman, *Hunt & Kimmel* 221, *A. T.*

Gratiola neglecta Torr. Plant 5–35 cm. tall, glandular puberulent above, glabrous below; roots fibrous; stems often freely branching; leaves 1–4 cm. long, lance-linear to lance-ovate, entire or denticulate; peduncles 3–30 mm. long; the 2 terminal bracts leafy, exceeding the calyx; calyx glandular puberulent, the divisions lanceolate, obtuse; corolla of early flowers 8–12 mm. long, glandular puberulent without, the tube cylindric, greenish yellow, pilose within, the lobes obovate, whitish, late flowers smaller; stamens included; anthers 1 mm. long, oval; style included; capsule 4–6 mm. long globose-ovoid; seeds 0.2–0.3 mm. long, cylindric, yellowish, several ribbed and reticulate. Muddy places, *A. T.*

LIMOSELLA

Very small glabrous annual herbs, stoloniferous, rooting and creeping in the mud; leaves entire, fleshy, in dense clusters around the simple 1-flowered naked peduncles; flowers small, calyx campanulate, 5-toothed; corolla rotate-campanulate, 5-cleft, nearly regular; stamens 4, all anther-bearing; style short; stigma capitate; capsule becoming 1-celled, many-seeded. (Name from Gr., *limus,* mud; *sella,* seat.)

Limosella aquatica L. *Mudwort.* Tufted; petioles 2–9 cm. long; blades 7–14 mm. long, narrowly elliptic to oval; peduncles 1–2 cm. long; calyx 2–3 mm. long, membranous at base, the deltoid teeth foliaceous, ½ the length of the tube; corolla 2–3 mm. long, white or purplish tinged; stamens exserted; style 0.2–0.4 mm. long, decurved; capsule 3–4 mm. long, ovoid; seeds 0.4–0.8 mm. long, ellipsoid, curved, brownish, longitudinally angled, and cross ribbed. *L. americana* Glück. Muddy places near Moscow, Ida., *Henderson, A. T.*

LINARIA

Herbs with alternate leaves or the lower opposite; flowers in terminal spikes or racemes; calyx 5-parted; corolla bilabiate, spurred on the lower side, the throat nearly closed; stamens didynamous; capsule thin, opening by pores or slits beneath the summit. (Name from *Linum,* the flax.)

Leaves linear; corolla spur 9–13 mm. long; seeds discoid, smooth, the wing as broad as the seed. — *L. vulgaris.*
Leaves lanceolate or ovate-lanceolate; corolla spur 18–23 mm. long; seeds with ridged angles and reticulate faces, not winged. — *L. dalmatica.*

Linaria dalmatica (L.) Mill. Perennial 3–6 dm. tall, glabrous, pale and glaucous; roots stout, up to 7 mm. in diameter; stems erect, often branched above; leaves 2–7 cm. long, firm somewhat clasping at base, racemes 5–25 cm. long, many flowered, leafy bracted; pedicels 2–6 mm. long; calyx 4–6.5 mm. long, the lobes linear-lanceolate; corolla 15–20 mm. long not including the spur, light yellow, the palate slightly darker, hirsute, anthers 2 mm. wide, reniform; capsule 5–7 mm. long, subglobose; seeds 1–1.5 mm. long, irregular, compressed and several angled, brown. Weed introduced from the Balkan region. First found by Little Spokane R. in 1926, now abundant on roadsides and fields in Spokane Co., and dispersed to Whitman and Latah Cos.

Linaria vulgaris Hill. *Butter-and-Eggs.* Perennial, in tufts, 2–6 dm. tall, pale green or glaucous; roots brown, up to 5 mm. in diameter, spreading, colony-forming; stems erect, usually simple, or branched below, glabrous or sparingly capitate glandular puberulent above; leaves 2–6 cm. long, acute, sessile, numerous; racemes 4–25 cm. long, dense; pedicels 2–10 mm. long; calyx 4–5 mm. long, the lobes ovate-lanceolate; corolla 16–22 mm. long not including the spur, pale yellow, the spur darker, and the palate orange and hirsute; anther cells divergent, ellipsoid, the anthers 1 mm. wide; capsule 7–10 mm. long, ovoid; seeds 1.5–2 mm. in diameter, brown. *Linaria Linaria* (L.) Karst. Weed, introduced from Eurasia. Fields and roadsides, *A. T.*

A troublesome weed, difficult to exterminate. Avoided by grazing animals, and suspected of being poisonous.

LINDERNIA. False Pimpernel

Small and smooth annual herbs; leaves opposite, sessile; peduncles axillary; 1-flowered, the upper becoming racemose; calyx becoming 5-parted, the divisions becoming narrow and nearly equal; upper lip of corolla short, erect, 2-lobed, the lower larger, spreading, 3-cleft; anther-bearing stamens 2 or 4 ; sterile filaments when present unequally 2-forked, exserted; stigma 2-lobed; capsule many-seeded, septicidal. Named for *Franz Balthasar von Lindern* of Strassburg.)

Lindernia dubia (L.) Pennell. Plant, erect or decumbent, often freely branched, the stems 5–35 cm. long; leaves 1–3 cm. long, ovate to oval, sessile, the lower narrowed at base, entire or serrulate; pedicels 5–20 mm. long; calyx 3–5 mm. long, parted almost to the base, the lobes linear, ciliolate; corolla 8–11 mm. long, pale lavender, the lobes rounded; fertile stamens 2, the anthers well included, the sacs 0.5 mm. long, divergent; capsule 4–5 mm. long, ellipsoid; seeds 0.2–0.3 mm. long, cylindric, yellowish lined. *Ilysanthes dubia* (L.) Barnh.; *L. dubia,* ssp. *major* (Pursh) Pennell; *L. anagallidea* (Michx.) Pennell. Wet places, *U. Son., A. T.*

MIMETANTHE

Low, white, viscid villous annual; leaves opposite, sessile; calyx campanulate, 5-cleft, its tube slightly sulcate not prismatic or angled; corolla yellow, obscurely 2-lipped, the lobes plane; stamens 4, of these 2 sometimes sterile; capsule acuminate, loculicidal, dehiscing along the whole upper side and on the lower side towards the apex. (Name from Gr., *mimetes,* imitator; *anthos,* blossom.)

Mimetanthe pilosa (Benth.) Greene. Plant 1–3 dm. tall, usually bushy branched; roots fibrous; leaves 1–6.5 cm. long, lanceolate or oblanceolate, entire; flowers on axillary pedicels 5–40 mm. long; calyx 6–8 mm. long, cleft about ⅓ its length, the teeth deltoid, the 2 upper the longer and less deeply cleft; corolla 9–11 mm. long, subcylindric, the lobes 1.5 mm. long, suborbicular, the lower dark bimaculate; stamens included, the longer with anther sacs 0.3–0.4 mm. long, ovoid, glabrous; capsules 6–8 mm. long, lanceoloid, long acuminate beaked; seeds 0.3–0.4 mm. long, ellipsoid, pointed, yellowish, roughened. *Mimulus pilosus* (Benth.) Wats. Moist places, *U. Son., A. T.*

The herbage has a disagreable odor.

MIMULUS. Monkey Flower

Herbs or shrubs with mostly simple, opposite leaves; flowers axillary; calyx prismatic, or rarely campanulate or tubular, usually 5-angled, 5-toothed, accrescent, the upper tooth usually the largest; corolla irregular, 2-lipped; upper lip erect or reflexed, 2-lobed, the lower spreading, 3-lobed; stamens 4, didynamous, all anther-bearing; stigma 2-lobed; capsule 2-valved, loculicidal, seeds numerous. (Name from Lat., a diminutive of *mimus,* buffoon.)

Flowers rose or purple,
  Corolla 4–5.5 cm. long; calyx 1.5–2.5 cm. long. — *M. Lewisii.*
  Corolla 6–22 mm. long; calyx 5–8 mm. long,
    Corolla 6–10 mm. long. — *M. Breweri.*
    Corolla 17–22 mm. long. — *M. nanus.*
Flowers wholly or partly yellow,
  Calyx teeth unequal, the upper conspicuously larger,
    Pedicels not markedly recurving below base of calyx in fruit; lowest leaf blade 1½–5-times length of petiole. — *M. guttatus,* var. *guttatus.*

Pedicels markedly recurving below base of calyx in fruit; lowest leaf blade shorter or as long as the petiole. Var. *gracilius.*

Calyx teeth subequal,
- Calyx 8–12 mm. long, the teeth 3–5 mm. long. *M. moschatus.*
- Calyx 2–9 mm. long, the teeth 0.5–1.5 mm. long,
  - Corolla 4–5 mm. long. *M. breviflorus.*
  - Corolla 7–20 mm. long,
    - Plant viscid pilosulous; anthers pilose. *M. floribundus.*
    - Plant glandular puberulent; anthers glabrous,
      - Style glandular pubescent; corolla 12–15 mm. long; calyx becoming 6–8 mm. long. *M. washingtoniensis.*
      - Style glabrous,
        - Corolla 14–18 mm. long. *M. ampliatus.*
        - Corolla 7–12 mm. long. *M. patulus.*

Mimulus ampliatus Grant. Annual, 7–15 cm. tall; stems viscid puberulent, erect; petioles mostly longer than the blades; blades 1–2.5 cm. long, broadly ovate, acute, dentate, the base cuneate; pedicels filiform, longer than the leaves; calyx 5–6 mm. long, but in fruit 6–7 mm. long, tubular, the teeth 1 mm. long, broad deltoid, ciliate; corolla 1.2–2 cm. long, yellow, funnelform, the tube exserted, the throat ampliate, the lobes unequal; stamens included, glabrous; style glabrous; capsule narrowly ovoid, shorter than the calyx; seeds longitudinally wrinkled. Lake Waha, Ida., Heller 3330 (holotype); and Lewiston. Wet places, *U. Son.* to *A. T. T.*

Mimulus breviflorus Piper. Annual, the slender stem simple, or more commonly branched from the base, erect, 4–20 cm. high, minutely glandular puberulent throughout or nearly glabrous above; leaves 5–15 mm. long, lanceolate, rarely oblanceolate or ovate, acute or inconspicuously few-toothed, narrowed at base into a short petiole or subsessile, usually shorter than the internodes, gradually reduced above; flowers solitary in the axils, on slender pedicels which about equal the leaves; calyx narrowly campanulate, somewhat constricted above, 2–3 mm. long in flower, 6–8 mm. in fruit, the short acute triangular teeth nearly equal; corolla pale-yellow, tubular, 4–5 mm. long, the lobes short and rounded; stigma scarcely protruding beyond the calyx; capsule 4–5 mm. long, ellipsoid; seeds 0.2 mm. long, ellipsoid, brownish. Moist places, *U. Son.,* to *A. T. T.*

The type was from Pullman, *Piper* 1858.

Mimulus Breweri (Greene) Cov. Annual, glandular puberulent or pilosulous, 2–15 cm. tall, simple or branched; leaves 1–2 cm. long, linear or linear-oblanceolate, entire or denticulate, sessile or the lower short petioled; pedicels 2–13 mm. long; calyx 4–8 mm. long, often reddish, the teeth subequal or the upper longer, 1–2 mm. long, broad deltoid; corolla narrow funnelform, pink to red, glabrous without, the lobes subequal; stamens slightly exserted; style glabrous; capsule 5–8 mm. long, ellipsoid, beaked; seeds 0.2–0.3 mm. long, ovoid, brownish, roughened. Moist banks, Blue Mts., *Huds.*

Mimulus floribundus Lindl. Annual or perennial, 6–50 cm. tall, more or less viscid pilosulous; roots fibrous; stems erect or decumbent, often freely branched; petioles 2–25 mm. long; blades 1–5 cm. long, ovate to subcordate, denticulate; pedicels 10–45 mm. long; calyx 5–8 mm. long, campanulate, often red spotted or tinged, orifice not oblique, the teeth 1–1.5 mm. long, deltoid; corolla 7–14 mm. long, yellow glabrous without, somewhat funnelform, the rounded lobes not wide spreading, the throat often red dotted, pilose within;

stamens included; anther sacs 0.7 mm. long, ellipsoid, divergent; style glabrous; capsule 4–6 mm. long, ellipsoid; seeds 0.2 mm. long, ovoid, brown. Wet places, *U. Son., A. T.*

Mimulus guttatus DC., var. **guttatus.** Annual or perennial, 5–55 cm. long; glabrous, puberulent, or pilosulous; lower leaves slender petioled, the upper sessile; blades 0.5–12.5 cm. long, ovate, palmately nerved, irregularly dentate or denticulate; pedicels 10–60 mm. long; calyx 6–22 mm. long, campanulate, often tinged or dotted with red, orifice not oblique, calyx-teeth ovate-deltoid, the upper longer, in fruit accrescent and inflated, the lower teeth finally infolded; corolla 1–4 cm. long, yellow, often spotted with red, campanulate, glabrous without, the lobes spreading, the throat ridged and pilose within; anthers cordate-suborbicular, 1 mm. long, glabrous; capsule 7–10 mm. long, ellipsoid; seeds 0.2–0.3 mm. long, ellipsoid, roughened, brown. *M. Langsdorfii* Donn; *M. nasutus* Greene. Wet places, *U. Son., A. T. T.*

Var. **gracilius** St. John, nom. nov. It differs in having the stems commonly geniculate, the internodes 1–14 cm. long; blades simple or frequently lyrate, at base rounded or subcordate, leaves and petioles commonly red-tinged; calyx in fruit glabrous or glabrate, usually red-tinged. Wet places, *U. Son., A. T.*

Caulibus plerumque geniculatis, internodis 1–14 cm. longis, laminis simplicibus vel lyratis in basi rotundatis vel subcordatis, calyce glabro vel glabrato. *M. guttatus* DC. var. *gracilis* G. Campbell. Gray did not propose an epithet for this, and Campbell gave no Latin diagnosis. Consequently the plant needs a new name.

Holotype: California, Sonoma, April 23, 1902, *A. A. Heller & H. E. Brown 5,349* (Pomona College Herbarium).

Mimulus Lewisii Pursh. *Crimson Monkey Flower.* Perennial, 3–7.5 dm. tall, viscid pilose; rootstock running; leaves 2.5–8 cm. long, lanceolate to ovate, acute, denticulate, sessile; pedicels 2–8 cm. long; calyx 1.5–2.5 cm. long, tubular or narrow campanulate, sharply angled, often red tinged or dotted, the teeth 4–6 mm. long, subequal, having a deltoid base and subulate tip; corolla rose-red, broad funnelform, glabrous without, pilose within, the lobes rounded or emarginate, the throat ampliate; stamens included; anther sacs 1 mm. long, divergent, ovoid, hirsute; style glabrous, included; capsule 15–20 mm. long, ellipsoid; seeds 0.5 mm. long, ellipsoid, brownish, roughened. Stream banks, Blue Mts., *Huds.*

Named for its first collector, Capt. Meriwether Lewis.

Mimulus moschatus Dougl. ex Lindl. *Musk Flower.* Perennial, 5–100 cm. tall, slimy viscid villous throughout; rhizomes up to 2 mm. in diameter, white, extensively creeping, forking, tuberiferous; stems weak, often decumbent and rooting at base, often branched; lower leaves with petioles 3–15 mm. long; upper leaves shorter petioled or subsessile; blades 1–7 cm. long, ovate or lance-ovate, membranous, pinnately veined, denticulate, the tip acute or obtuse, the base from cuneate to subcordate; pedicels 5–40 mm. long; calyx campanulate the orifice not oblique, the teeth lanceolate, often half the length of the tube; corolla 1.5–3 cm. long, yellow, glabrous without, funnelform, the lobes spreading, rounded notched at the tip, the throat pilose within; stamens included; filaments glabrous; anther sacs 0.8 mm. long, divergent, ovoid, sparsely hirsutulous; capsule 6–9 mm. long, ellipsoid, acute; seeds 0.1–0.2 mm. long, ovoid, roughened, brownish, the inner faces flattened. *M. moschatus,* var. *longiflorus* Gray. Meadows and wet woods, *A. T. T.*

Introduced into cultivation in Europe by Douglas because of its strong odor of musk. Sir A. W. Hill reports that both the wild plants and the cultivated plants have now lost their odor.

Mimulus nanus H. & A. Annual, 2–17 cm. tall, viscid or glandular puberulent; roots fibrous; stems often freely branched; lower leaves 1–2 cm. long, spatulate or oblanceolate, tapering into the short petiole; upper leaves 1–3 cm. long, sessile, lanceolate to ovate, entire; pedicels 2–3 mm. long; calyx 5–8 mm. long, cylindric-campanulate, distended by the capsule, the throat only slightly oblique, the teeth 2 mm. long, deltoid; corolla reddish purple, the tube narrow cylindric, the throat broad funnelform, the lips gaping, somewhat yellowish and purplish spotted, upper lip the longer, lower lip white pilose within; stamens included; anther sacs 0.5 mm. long, somewhat diverging, ovoid; style pilose; stigma funnelform; capsule 7–11 mm. long, lanceoloid, the slender tip curved; seeds 0.3–0.5 mm. long, ovoid, brownish, roughened. *M. clivicola* Greenm. Gravelly banks, Blue Mts., Craig Mts., *A. T. T.*

Minulus patulus Pennell. Annual, glandular pilose, 10–20 cm. tall, erect; petioles 1–2-times length of blades; blades 5–12 mm. long, ovate, undulate, dentate, obtuse or acutish, at base rounded truncate; pedicels 13–20 mm. long; calyx 6–7 mm. long, glandular puberulent, strongly ridged or wing angled, the lobes equal, acute; corolla yellow, the throat narrowly campanulate with two slightly pubescent ridges; style glabrous; capsule 5 mm. long, ellipsoid-ovoid; seeds ellipsoid, smooth. Wawawai, *Elmer* 752 (holotype). *U. Son.*

Mimulus washingtoniensis Gdgr. Annual, 4–18 cm. tall; roots fibrous; stems erect, often bushy branched; petioles 3–18 mm. long; blades 6–15 mm. long, deltoid-ovate to lanceolate, denticulate; pedicels 8–34 mm. long; calyx 4–9 mm. long, almost cylindric in flower, becoming campanulate, the orifice not oblique, the teeth 0.5–1 mm. long, broad deltoid, mucronate; corolla 1–2 cm. long, yellow, glabrous without, funnelform, the lips and lobes spreading, lobes rounded, entire; palate pilose, partly closing the throat; stamens included, the anther sacs 0.3 mm. long, divergent, ovoid; style glabrous or finely pilosulous; capsule 5–6 mm. long, ovoid; seeds 0.1 mm. long, ellipsoid, brown. Moist places, *U. Son., A. T.*

ORTHOCARPUS. Owl's Clover

Low annuals, very similar to *Castilleja;* leaves alternate, sessile, flowers in terminal spikes; bracts usually colored; calyx 4-cleft; corolla tubular; the upper lip (*galea*) scarcely longer and usually much narrower than the inflated 1–3-saccate lower ones; anthers unequally 2-celled, or the smaller anther-cell rarely wanting; capsule many seeded, loculicidal. (Name from Gr., *orthos,* upright; *karpos,* fruit.)

Leaves entire; inflorescence glandular puberulent; galea as wide as long. *O. luteus.*
Leaves or some of them pinnately lobed; inflorescence not glandular; galea lanceolate, attenuate,
  Body of floral bract oval or ovate, tipped with rose purple; lower lip of corolla shallow saccate. *O. tenuifolius.*
  Body of floral bract lance attenuate, green; lower lip of corolla with 3 inflated sacs. *O. hispidus.*

Orthocarpus hispidus Benth. Plant 4–40 cm. tall, simple or few branched, hirsute throughout; roots fibrous; leaves 1–4 cm. long, entire, the lower linear, the upper lance-linear and 3–5-cleft; spike 3–24 cm. long, becoming loose below in fruit, hirsute to pilose, not or slightly glandular; bracts 10–25 mm. long, ovate, 3–7-cleft into linear attenuate lobes; calyx 8–10 mm. long, cleft half way above and below; each lobe cleft into 2 attenuate teeth; corolla 12–20 mm. long, white or yellowish, puberulent; galea 4–5 mm. long, lanceolate, the ventral margins glabrous; lower lip 3–4 mm. long, short trifid, much broader than the galea; anthers 1–1.5 mm. long, lanceoloid, sparsely pilose;

capsule 5–8 mm. long, ovoid; seeds 1–1.5 mm. long, lance-ovoid, brown, with a loose, transparent, coarsely cellular reticulate testa. Meadows, *A. T.*

Orthocarpus luteus Nutt. Plant 8–65 cm. long; roots fibrous; stems soft hirsute; leaves 1.5–4 cm. long, puberulent, the lower linear, attenuate, the upper linear or lance-linear; spike 1–30 cm. long, glandular, soft puberulent; bracts 10–15 mm. long, deeply 3–5-lobed, the lateral lobes lanceolate, divergent, the middle lobe ovate-lanceolate, acute, green; calyx 6–9 mm. long, ellipsoid, the 4 subequal, deltoid teeth 2–3 mm. long; corolla 10–13 mm. long, golden yellow, puberulous, the throat funnelform; galea about 4 mm. long, rounded ovate, minutely hooked at apex; lower lip nearly as long as the galea, rounded, simply saccate, with 3 minute teeth; anthers 1 mm. long, the sacs ellipsoid, villous; capsule 5–7 mm. long, ellipsoid, scabrous puberulent; seeds 1.4–1.8 mm. long, ellipsoid, brown, with a close fitting brown testa with raised longitudinal ridges and lower cross bars. Meadows, Spokane region, *Lake & Hull* 701; *Spalding;* Revere, *Eastwood & St. John* 13,279, *U. Son., A. T.*

Orthocarpus tenuifolius (Pursh) Benth. Plant 1.5–3 dm. tall, hirsutulous throughout; roots fibrous; stems simple or with few erect branches; leaves 1–5 cm. long, linear, coarsely hispid ciliate at base, with 2–4 divergent, linear, lateral lobes; spike 1–9 cm. long, dense, puberulent and hispid ciliate; bracts 1–2 cm. long, with 2–4 attenuate, divergent lateral lobes, the middle lobe oval or ovate, rose purplish at tip; calyx 7–12 mm. long, narrowly ellipsoid, puberulent and hispid, parted nearly ⅔ its length, the teeth linear-lanceolate, half as long as the lobes; corolla 14–20 mm. long, yellow, or the tip purplish; galea 5–6 mm. long, lanceolate, hooked at tip; lower lip 4–5 mm. long, shortly 3 toothed; anthers 2 mm. long, the sacs ellipsoid, ciliolate; capsule 5–9 mm. long, ellipsoid; seeds 1.5–2 mm. long, ellipsoid, brown, with a close fitting opaque testa with honey-comb-like reticulations. Meadows, or prairies, *A. T.*

PEDICULARIS. Lousewort

Perennial herbs; leaves alternate or basal, toothed or pinnatifid, the floral bract-like; flowers rather large, in a bracteate raceme or spike; calyx 2–5-toothed or -cleft, irregular; corolla strongly 2-lipped, the upper lip (*galea*) arched, laterally compressed, sometimes beaked; the lower 3-lobed; stamens 4, enclosed by the upper lip; anthers equally 2-celled; capsule loculicidal, several-seeded. (Name from Lat., *pediculus,* louse.)

Galea prolonged into a proboscis-like beak,
  Leaves simple; lower lip of corolla spreading. — *P. racemosa.*
  Leaves pinnate; lower lip of corolla carinate. — *P. contorta.*
Galea obtuse or merely mucronate beaked,
  Galea obtuse,
    Calyx-lobes with the tip linear, attenuate, glandular puberulent. — *P. bracteosa.*
    Calyx-lobes lanceolate or linear, villous,
      Calyx-lobes linear, longer than the tube. — *P. pachyrhiza.*
      Calyx-lobes lanceolate, shorter than the tube. — *P. latifolia.*
  Galea short beaked. — *P. Canbyi.*

Pedicularis bracteosa Benth. in Hook. *Indian Warrior.* Plant 3–9 dm. tall, nearly glabrous; caudex 5–10 mm. in diameter, the lateral roots tuberous; stems erect, usually single and simple; stem base with brown, oval scales 1–1.5 cm. long; sterile shoots having leaves with petioles 4–9 cm. long; cauline leaves 3–5, sessile or nearly so; blades 5–13 cm. long, broadly ovate in outline, pinnate, the pinnae 2–7 cm. long, narrowly oblong, doubly serrate; spikes 3–25 cm. long, dense, the rhachis pilosulous; bracts oval, caudate, ciliate, shorter than the flowers; calyx 10–12 mm. long, campanulate, capitate glandular puberulent, the lobes lance-linear, attenuate, longer than the tube; corolla 13–16 mm. long, magenta or magenta and yellow, glabrous without, the galea deflexed and hooded at apex, obtuse, the lower lobes ciliate, oval, not equaling the galea; anthers 1.8–2.2 mm. long, glabrous, lanceoloid, pale; capsule 8–12 mm. long, ovoid, curved, brown; seeds 2.5–3.5 mm. long, ovoid, cellular reticulate with large, transverse cells. Woods, *A. T. T.*

Pedicularis Canbyi Gray. Plant 3–7 dm. tall, glabrous or nearly so; caudex woody; lateral roots tuberous; stem base with brown, oval scales 1–1.5 cm. long; cauline leaves 3–4, sessile or nearly so; blades 5–15 cm. long, ovate in outline, pinnate; pinnae 1–4 cm. long, linear-lanceolate, doubly serrate; spikes 3–18 cm. long, dense, glabrous to villous; bracts 1–2 cm. long, lanceolate to ovate, crenulate at tip; calyx 7–9 mm. long, ovoid usually glabrous, lobed about ½ its length above and below, less so laterally, the teeth lanceolate; corolla 16–18 mm. long, greenish yellow, the galea decurved, the apex hooded; lower lip erect, remote from the galea, the short oval lobes entire; capsule 8–10 mm. long, ovoid, the apex curved; seeds 2.5–3.2 mm. long, ellipsoid to ovoid, brown, the loose testa transverse cellular reticulate. *P. siifolia* Rydb. Woods, *A. T. T.*

Pedicularis contorta Benth. ex Hook. Plant 1–4 dm. tall, glabrous; root 2–15 mm. in diameter, woody, multicipital; basal leaves with petioles 2–9 cm. long, the blades 2–8 cm. long, lanceolate in outline, pinnately parted into linear, serrulate divisions; cauline leaves similar but smaller, reduced upwards; raceme 3–18 cm. long; pedicels 2–7 mm. long; calyx glabrous without, puberulent within, campanulate, the tube 4–5 mm. long, cleft on the lower side half its length, on the upper side less so, with 4 lance-linear teeth 2–3 mm. long; corolla yellowish, 10–13 mm. long, the galea proboscis-like, spirally incurved; lower lip 10–12 mm. wide; anthers 2.5–3 mm. long, lance-oblong, glabrous; capsule 8–11 mm. long, oval, compressed; seeds 2–2.5 mm. long, brown, ellipsoid, irregularly compressed. Meadows or woods, Blue Mts., *Can., Huds.*

Pedicularis latifolia Pennell. Plant 4–12 dm. tall, glabrous below the inflorescence; caudex 5–20 mm. in diameter, the lateral roots tuberous; sterile shoots with petioles 2–10 cm. long; stem base with brown, oval scales 1–2 cm. long; cauline leaves 4–8, sessile or nearly so; blades 5–20 cm. long, broadly ovate in outline, pinnate; pinnae 2.5–8 cm. long, linear to lance-linear, doubly serrate or incised; spike 6–40 cm. long, dense; rhachis glabrous or sparsely villous; bracts 11–20 mm. long, lanceolate to ovate, caudate, villous at base; calyx 10–12 mm. long, ovoid, ciliolate, deeply cleft above, the upper tooth much the shorter; corolla 15–20 mm. long, all yellow or partly magenta; galea obtuse, saccate at tip, decurved; the lower lip ascending, not reaching the upper, the lobes broadly oval, erose and remotely ciliolate; anthers 2.8–3 mm. long, oval, cleft at base, glabrous; capsule 8–10 mm. long, ovoid, flattened, the tip curved; seeds 2.8–3 mm. long, ovoid, coarsely transverse cellular reticulate. Woods, *Can.*

Pedicularis pachyrhiza Pennell. Plant 4–10 dm. tall, glabrous to the inflorescence; caudex 5–8 mm. in diameter, lateral roots tuberiferous; leaves of sterile shoots with petioles 3–10 cm. long; basal scales 1–1.5 cm. long, oval, brown; cauline leaves 3–5, the petioles 0–15 mm. long; blades 7–16 mm. long, ovate in outline, pinnate, the pinnae 2–4 cm. long, doubly serrate or shallowly pinnatifid; spike 6–15 cm. long, dense; rhachis villous; bracts 13–25 mm. long, lanceolate, linear-caudate, villous ciliate, the tip crenate; calyx 11–15 mm. long, villous, the upper lobe ½ length of others, from a deep cleft; corolla 15–22 mm. long, greenish yellow; galea obtuse, hooded, slightly decurved;

lower lip erect, nearly touching the galea, the lobes oval, ciliolate only at base; anthers 2–2.5 mm. long, oval, cleft at base, glabrous; capsule 10–11 mm. long, the tip curved; seeds unknown. Woods, Blue Mts., *Can.*

Pedicularis racemosa Dougl. ex Hook. *White Lousewort.* Plant 1–5 dm. tall, glabrous; caudex 1–15 mm. thick, woody, multicipital; stems in clumps, simple or with weak branches; leaves attenuate, the lower with petioles up to 1 cm. in length, the upper sessile; blades 2–8 cm. long, linear-lanceolate or oblong-lanceolate, doubly serrate; leaves of the axillary shoots smaller; racemes 2–9 cm. long, the bracts leafy, not reduced; pedicels 1–3 mm. long; calyx 4–6 mm. long, campanulate, oblique, the lower side deeply cleft, the upper with two lanceolate teeth; corolla white or reddish tinged, 14–18 mm. long; galea proboscis-like, incurved; lower lip reniform, 12–15 mm. wide; anthers 2.5–2.8 mm. long, oblong, glabrous; capsule 11–13 mm. long, lanceoloid, purplish; seeds 2–2.3 mm. long, blackish, ellipsoid, constricted at base. Woods, *A. T. T., Can.*

PENSTEMON. Beard-Tongue

Perennial herbs or half shrubs; leaves opposite or whorled; the upper sessile or partly clasping, the floral reduced to bracts; flowers showy, in a raceme or racemose panicle; calyx 5-parted; corolla red, blue, purple, yellow or white, tubular, more or less inflated or bell-shaped, either decidedly or slightly 2-lipped; the upper lip 2-lobed, the lower 3-cleft; anther-bearing stamens 4; the sterile fifth filament about equaling the others; style long; stigma entire; capsule many-seeded, dehiscent from the apex into two entire or bifid valves. Named from Gr. *pente,* fifth; *stemon,* stamen.)

Anthers long villous throughout, the sacs dehiscent their whole length. — *P. fruticosus.*
Anthers glabrous or merely ciliate on the line of dehiscence,
  Anther sacs dehiscing their whole length, glabrous,
    Corolla 26–35 mm. long,
      Plant glabrous; anther sacs 2–2.6 mm. long,
        Anther sacs glabrous; upper cauline leaves linear-lanceolate, clasping. — *P. speciosus,* subsp. *speciosus.*
        Anther sacs hirtellous on apex and sutures; upper cauline leaves oblong-lanceolate to ovate, caudate-clasping. — *P. Pennellianus.*
      Plant pubescent; anther sacs 1.3–1.5 mm. long. — *P. eriantherus.*
    Corolla 6–20 mm. long,
      Leaves all or some of them toothed,
        Suffrutescent at base; corolla 10–17 mm. long, the tube subcylindric. — *P. deustus,* subsp. *deustus.*
        Herbaceous; corolla 15–20 mm. long, the tube funnelform. — *P. Wilcoxii.*
      Leaves entire,
        Corolla 15–18 mm. long, yellow or blue. — *P. attenuatus,* subsp. *attenuatus.*
        Corolla 6–12 mm. long,
          Inflorescence glandular. — *P. procerus,* var. *procerus.*

Inflorescence not glandular,
Corolla blue-purple; staminodium equalling the corolla tube, dilated at apex. *P. Rydbergii*, var. *varians*.
Corolla yellow; staminodium included, the apex hairy tufted but not dilated. *P. confertus*.
Anther sacs dehiscing only across the apex, the distal part saccate,
Plant viscid pilose throughout; corolla 3–4 cm. long; staminodium glabrous. *P. glandulosus*.
Plant not viscid or glandular below the inflorescence; corolla 12–33 mm. long; staminodium hairy tufted,
Fertile filaments villous; inflorescence glabrous. *P. venustus*.
Fertile filaments glabrous; inflorescence capitate glandular puberulent,
Leaves oposite; corolla 2–3 cm. long; anther sacs 1.3–1.8 mm. long. *P. Richardsonii*.
Middle cauline leaves in whorls of 3; corolla 12–18 mm. long; anther sacs 0.9–1.3 mm. long. *P. triphyllus*, subsp. *triphyllus*.

Penstemon attenuatus Dougl., ex Lindl., subsp. **attenuatus.** Herb, 3–5 dm. tall; root 2–4 mm. in diameter, often horizontal; stem puberulous or glabrous below; basal leaves with petioles 1–3 cm long, blades 2–7.5 cm. long, lanceolate, entire, firm, glabrous; cauline leaves 3–5 pairs, remote, the upper 3–10 cm. long, lanceolate to oblong, sessile, glabrous; inflorescence 5–15 cm. long, interrupted and verticillate, glandular puberulent; pedicels 2–5 mm. long; calyx-lobes 6–8 mm. long, glandular puberulent, lanceolate, acuminate the sides somewhat scarious and sometimes erose; corolla yellow, glandular puberulent without, slightly wider at the throat, lower lip hirsute; fertile anthers glabrous, divergent, the sacs 0.8–1 mm. long, ovoid; staminodium hirsutulous; capsule 6–8 mm. long, ovoid. *P. attenuatus*, var. *varians* A. Nels. Open woods, *A. T. T.*, *Can.*

The type was collected by Douglas in the Craig Mts. This species, with *P. confertus* and varieties are part of a maze of varying forms, from which many species have been described. They have not yet been well classified.

It is avoided by grazing sheep.

Penstemon confertus Dougl. ex Lindl. Herb, 1–5 dm. tall, nearly glabrous; cauline leaves oblong to oblanceolate, 3–6 cm. long; cauline leaves 3–6 pairs, remote, the upper 1–7 cm. long, sessile, ovate to linear-lanceolate, acute, entire; inflorescence 14–40 cm. long, interrupted, with 2–8 whorl-like clusters, minutely puberulous; pedicels 1–3 mm. long; calyx-lobes 4–6 mm. long, lanceolate to ovate, acuminate, scarious margined, erose; corolla yellowish, narrowly cylindric, lower lip hirsute within; fertile anthers glabrous, the sacs 0.3 mm. long, widely divergent, ellipsoid; staminodium hispid tufted, slightly exceeding the tube; capsule 4–5 mm. long, lanceoloid; seeds 0.4–0.6 mm. long, brown to black, irregularly compressed, with pale, thin margins. Rocky slopes, *A. T.*, *A. T. T.*

Penstemon deustus Dougl. ex Lindl., subsp. **deustus.** Plant, 1–6 dm. tall; root 3–10 mm. in diameter, brown, multicipital or forking at apex; stems mostly decumbent at base, minutely glandular puberulous at base, glabrous above; basal leaves with petioles 8–15 mm. long, the blades 1.5–3 cm. long, obovate to spatulate, thick, glaucous, glabrous, salient serrate; cauline leaves numerous, the upper 15–40 mm. long, sessile, elliptic to spatulate, sharply serrate or sinuate dentate; inflorescence 5–20 cm. long, thyrsoid, many flowered, leafy bracted, nearly glabrous; pedicels 2–5 mm. long; calyx-lobes 5–6 mm. long, linear-lanceolate, glabrous or glandular puberulent; corolla white or creamy, sparsely puberulous without, the lower lip puberulous in the throat; fertile anthers reniform, the sacs 0.3 mm. long, ellipsoid, glabrous, confluent; staminodium filiform, glabrous; capsules 4–5.5 mm. long, ovoid, acute; seeds 1.2–1.5 mm. long, rhomboid, brown, irregularly compressed, pitted. Rocky slopes, *U. Son., A. T.*

Herbage avoided by grazing sheep.

Penstemon eriantherus Pursh. Herb, 1–4 dm. tall, viscid villous throughout or puberulent below; roots 3–4 mm. in diameter, brown; basal leaves with petioles 1–3 cm. long, the blades 2–8 cm. long, obovate to lanceolate, thick, mostly entire; cauline leaves 2–3 pairs, the upper 4–7 cm. long, lanceolate to narrowly oblong, acute, entire or dentate; inflorescence 5–12 cm. long, dense, the bracts foliaceous; pedicels 2–4 mm. long; calyx-lobes 9–12 mm. long, lance-subulate, viscid villous; corolla 28–35 mm. long, purplish, funnelform, glandular pilose without, lower lip villous in the throat; fertile anthers horse-shoe-shaped, glabrous, confluent at tip; staminodium exserted, long hispid, the tip clavate; capsule not seen. Gravelly soil, near Spokane and Lime Pt., Nez Perce Co., *U. Son., A. T.*

Penstemon fruticosus (Pursh) Greene. Suffruticose, the branches more or less decumbent at base, or tufted, 1–5 dm. tall, freely branched at base, glabrate below, somewhat glandular puberulent above; leaves 1–4.5 cm. long, short petioled or sessile, elliptic to narrowly so or spatulate, or the smallest sometimes ovate, usually serrate, coriaceous; raceme 5–15 cm. long, capitate glandular puberulent, the lower bracts leafy; pedicels 5–18 mm. long; calyx-lobes 8–13 mm. long, ovate-lanceolate, glandular pilose; corolla 30–42 mm. long, tubular-funnelform, lilac-purple; fertile anthers reniform, the sacs 1–1.3 mm. long; staminodium 1–1.5 mm. wide, villous at apex; capsule 8–12 mm. long, ovoid, acuminate; seeds 0.8–1.2 mm. long, angular, irregular, thin margined brown, roughened. *P. Scouleri* Lindl. Sunny ledges, *Can.*

Herbage avoided by grazing sheep.

Penstemon glandulosus Dougl. ex Lindl. Herb, 4–10 dm. tall; root brown, 3–10 mm. in diameter; stems erect; basal leaves with petioles 3–13 cm. long, the blades lanceolate to oblanceolate, entire or dentate; cauline leaves, 5–9 pairs, the upper, 4–10 cm. long, lance-ovate to cordate, clasping, serrate; inflorescence 8–30 cm. long, interrupted, the bracts foliaceous, cordate, acuminate; pedicels 2–8 mm. long; calyx-lobes 10–15 mm. long lance-linear, viscid pilose; corolla lilac, funnelform, sparsely viscid pilosulous without, nearly glabrous within; fertile anthers reniform, the sacs glabrous except for the hispidulous ciliate suture; staminodium clavate; capsule 9–14 mm. long, ovoid, acute; seeds 2–3.2 mm. long. Moist loamy hillsides, canyons of Clearwater, Snake, and Grand Ronde, from Lewiston and Hatwai south and east, *A. T.*

The type was collected by Douglas in the Craig Mts.

Penstemon Pennellianus Keck. Suffrutescent, 2–5 dm. tall, glabrous; basal leaves 8–25 cm. long including the elongate petiole, lanceolate or narrowly elliptic; cauline blades oblong-lanceolate to ovate, acute, the uppermost cordate-clasping, the floral ones inconspicuous; thyrse 5–25 cm. long, of 3–10 dense, approximate clusters; calyx 5–10 mm. long, the lobes ovate-lanceolate, acuminate or subulate, the margin membranous, entire or moderately erose; corolla 26–32 mm. long, 9–10 mm. wide, bright blue tinged with purple, glabrous, the tube abruptly flaring into the ample throat, limb moderately 2-lipped; anther sacs 2–2.4 mm. long, opening 4/5 way; staminodium filiform or a little dilated at the hirsutulous apex. Blue Mts., *A. T. T., Can.*

Penstemon procerus Dougl. ex R. Graham, var. **procerus.** Slender herb 1–4 (–7) dm. tall, the herbage essentially glabrous, rosette sparse; leaves 2–6 cm. long, including the short slender petiole, thin, the blasal blades lanceolate to oblanceolate; cauline blades broadly oblong to narrowly lanceolate, clasping at base; thyrse strict, of 1–6 dense clusters, the lower ones often well spaced, subsessile or on erect peduncles up to 4 cm. long; calyx lobes 3–6 mm. long, elliptic to obovate, entire, scarious margined, the caudate tip usually equalling or exceeding the basal portion, glabrous or finely white puberulent; corolla 6–10 mm. long, slender, subcylindric but tapering below, bluish purple, bearded on the palate; anther sacs 0.4–0.7 mm. long, opening the full length. *P. procerus,* subsp. *typicus* Keck; *P. micranthus* Nutt. in part; *P. confertus* Dougl., var. *violaceus* Trautv. Meadows or open woods, *A. T., A. T. T.*

Penstemon Richardsonii Dougl. ex Lindl., var. **Richardsonii.** Somewhat suffrutescent at base, 2–8 dm. tall, often bushy branched; root 3–10 mm. in diameter, dark, woody, multicipital; stems numerous, in clumps, often bushy branched, herbage white puberulous or glabrous; cauline leaves numerous, 1–8 cm. long, linear-lanceolate to narrowly ovate, coarsely serrate to pinnately parted, the upper sessile; panicle 5–20 cm. long, loose, pedicels 2–13 mm. long; calyx-lobes 4–7 mm. long, lanceolate to ovate, acute, entire, herbaceous or slightly hyaline at margin, capitate glandular puberulent; corolla crimson or varying from light pink to violet, wide funnelform, glandular puberulent without, glabrous or nearly so within; fertile anthers orbicular-reniform, the sacs glabrous except for the remote hispidulous ciliation on the suture; staminodium exserted, hispid tufted: capsule 5–7 mm. long, ovoid, acute; seeds 1–1.7 mm. long, irregularly flattened thin margined on the angles, brown, cellular pitted. Dry basalt ledges, Spokane and Rock Lake, *A. T.*

Penstemon Rydbergii A. Nels., var. **varians** (A. Nels.) Cronq. Stems 2–7 dm. tall; rosette well developed; herbage bright green and essentially glabrous throughout; leaves thin, the basal ones 3–10 cm. long including the petiole which often equals the blade; blades linear-oblanceolate to elliptic; cauline blades oblong or lance-oblong, the upper clasping; thyrse of 1–6 rather distinct clusters, the lower with short strict peduncles; calyx 3–5 mm. long, the lobes oblong, abruptly narrowed and acuminate, the margin scarious, entire or sometimes erose; corolla 10–13 (–15) mm. long, the limb expanded, the lips equal, the palate prominently yellow bearded; anther sacs 0.5–0.8 mm. long. *P. oreocharis* Greene. Blue Mts., Asotin, Old Mill, *Onstot, A. T. T.*

Penstemon speciosus Dougl. ex Lindl., subsp. **speciosus.** Herb, 2–10 dm. tall; tap-root brown, up to 7 mm. in diameter, multicipital; stems slightly decumbent at base, glaucous; leaves glaucous, the basal with petioles 2–5 cm. long, the blades 3–8 cm. long, spatulate to lanceolate, thick; cauline leaves 3–12 cm. long, narrowly oblong to lanceolate, sessile, many, clasping, entire; panicle thyrsoid, 1–4 dm. long, interrupted, many flowered; pedicels 2–7 mm. long; calyx-lobes 5–9 mm. long, acuminate, the body ovate, broadly scarious margined, usually erose; corolla 3–3.5 cm. long, bright blue, the lobes rounded, subequal, palate glabrous; fertile anthers glabrous, the sacs widely divergent, narrowly ellipsoid, dehiscing from the distal end for nearly the whole length; staminodium glabrous or hispidulous at apex; capsule 8–13 mm. long, ovoid, acuminate; seeds 2–3 mm. long, several angled, irregularly compressed, the angles with broad, thin margins. Gravelly soil, Spokane, Hooper, Whitman Co., and Blue Mts., *A. T.*

Penstemon triphyllus Dougl. ex Lindl., subsp. **triphyllus.** Plants suffrutescent, 2–8 dm. tall; stems numerous, in dense clumps; herbage minutely puberulous; basal leaves 1–2 cm. long, lanceolate, laciniate serrate; cauline leaves numerous, three at a node, or the upper 2–1 at a node, 1–4.5 cm. long, linear to lance-linear, mostly sharply or laciniately toothed, sessile; inflorescence 5–20 cm. long, loosely paniculate, bracts linear, reduced; pedicels 2–12 mm. long; calyx-lobes 4–5 mm. long, ovate-lanceolate, herbaceous, entire, glandular puberulent; corolla slightly enlarged at the throat, bluish-lilac, the throat striped within and glabrous or sparsely hairy, sparsely glandular puberulent without; fertile stamens with filaments glabrous or nearly so, the anthers orbicular-reniform, the sacs oblong-ellipsoid; staminodium exserted hirsute tufted; capsule 4.5–7 mm. long, ovoid, acute; seeds 0.7–1.2 mm. long, brown, elongate, compressed, irregular, papillose. *P. triphyllus,* subsp. *typicus* Keck. Basalt ledges and rocky slopes, canyon of the Snake and its tributaries, *U. Son.*

The type was collected by Douglas in the Blue Mts. on the Walla Walla River.

Penstemon venustus Dougl. ex Lindl. Herb, 3–8 dm. tall; root dark, 2 or more cm. in diameter, multicipital, the stems in dense clumps; stems with 2 lines of appressed puberulence down from each node, or glabrate below; lowest leaves 1–2 cm. long, oblanceolate, serrate; cauline leaves 6–14 pairs, 4–12 cm. long, linear-lanceolate to lance-ovate, firm, glabrous, sessile, sharply serrate or laciniate toothed; inflorescence 1–4 dm. long, many flowered, not appearing interrupted; pedicels 4–25 mm. long; calyx-lobes 3–6 mm. long, lanceolate or ovate, acuminate, the margins hyaline, erose; corolla 25–33 mm. long, funnelform, blue or violet, glabrous without and within, the lobes ciliolate; fertile anthers rounded reniform, glabrous except for the hispidulous ciliate suture, the sacs 1.5–1.8 mm. long; staminodium exserted, villous tufted; capsule 6–9 mm. long, ovoid, acute; seeds 1.2–2.5 mm. long, irregular, compressed and angled brown, cellular papillose. Rocky slopes, from Clearwater River to Craig, and Blue Mts., *U. Son., A. T.*

Penstemon Wilcoxii Rydb. Herb, 3–10 dm. tall; root 2–5 mm. in diameter, the apex often multicipital; stems glabrous or sparingly glandular puberulent below the inflorescence; basal leaves numerous with petioles 3–5 cm. long, the blades 1–8 cm. long, ovate to broadly lanceolate, the tip acute, the base cuneate and decurrent, thick, evergreen, often purplish, mostly entire; cauline leaves 3–5 pairs, sessile the lower oblong or broadly spatulate, the upper 2.5–6 cm. long, lance-ovate or ovate, acute, serrate, clasping; inflorescence 5–60 cm. long, thyrsoid, interrupted, the clusters loose, capitate glandular puberulent; pedicels 1–3 mm. long; calyx-lobes 4–5 mm. long, ovate, acute, capitate glandular puberulent, narrowly scarious margined; corolla blue, but pinkish and yellowish color forms in many shades frequent, glandular puberulent without, lower lip pilose in the throat; fertile anthers with the sacs 1 mm. long, widely divergent, glabrous; staminodium yellow hirsute at apex; capsule 4–5 mm. long, ovoid, acute; seeds 1–1.4 mm. long, fusiform, nearly black, irregularly compressed, with brown thin angles. *P. pinetorum* Piper. Open lower woods, *A. T. T.*

RHINANTHUS. Yellow Rattle

Annual erect, semiparasitic herbs, with opposite leaves; flowers yellowish, in a one-sided spike-like raceme; calyx membranaceous, 4-toothed, inflated and compressed laterally in fruit; corolla 2-lipped, the upper lip arched, ovate, obtuse, with two lateral teeth; the lower lip 3-lobed; stamens 4, all anther-bearing, enclosed in the upper lip; anther-cells equal, pilose; capsule orbicular, flattened, loculicidal; seeds several, orbicular, winged. (Name from Gr., *rhin,* snout; *anthos,* flower.)

Rhinanthus Kyrollae Chabert. Plant 2–7 dm. tall, nearly glabrous; roots fibrous; stems often branched, the upper internodes with puberulent lines, green, not black lined; leaves 1.5–7 cm. long, oblong-lanceolate to lanceolate, hispidulous, firm, coarsely flattened crenate, sessile, the base subcordate clasping; racemes 3–20 cm. long, the bracts foliaceous, ovate-lanceolate, laciniate serrate, the lower exceeding the racemes; calyx reticulate nerved, ovate, in fruit becoming 12–17 mm. long, the margins puberulent, the teeth 2–3 mm. long, deltoid, erect; corolla 8–13 mm. long, pale yellow, puberulent, the teeth of the upper lip whitish, the lobes of the lower lip oblong; anther sacs 1.8 mm. long, oblong, separated at base, pilose; capsule 8–11 mm. long, round, retuse; seeds 4–6 mm. long, ovate, the pale wing 1.5 mm. wide. Gravelly soil, *A. T. T.*

SCROPHULARIA. Figwort

Perennial herbs or shrubs; leaves opposite; flowers small, in loose cymes in a narrow terminal panicle; calyx deeply 5-cleft; corolla 5-lobed, the 4 upper lobes erect, the lower spreading; stamens 4; the fifth sterile one a scale-like rudiment at the summit of the corolla tube; capsule many-seeded. (Name from Lat., *scrophula,* scrophula.)

Scrophularia lanceolata Pursh. Herb, 1–3 m. tall; caudex up to 15 mm. in diameter, brown woody; stem glandular puberulent above, glabrate below; petioles 3–30 mm. long, glandular puberulent; blades 5–13 cm. long, ovate to lanceolate, acuminate, glabrate, the base cuneate or subcordate, the margin coarsely serrate to incised; nodes often with small axillary shoots and reduced leaves; panicle 15–90 cm. long, capitate glandular, puberulent, thyrsoid, the bracts reduced; pedicels 1–18 mm. long; calyx 2–3 mm. long, lobed ⅔ its length, the lobes rounded ovate, glandular puberulent, hyaline margined; corolla 7–11 mm. long, yellowish to brownish or purplish, the tube ovoid, contracted at the throat, the lobes broadly oval; stamens included; staminodium 1.3–1.8 mm. wide, flabellate, usually wider than long; capsule 4–10 mm. long, ovoid-conic; seeds 0.7–0.9 mm. long, cylindric, blackish, reticulate and several ribbed. *S. leporella* Bickn.; *S. occidentalis* (Rydb.) Bickn.; *S. serrata* Rydb. Moist thickets, *A. T., A. T. T.*

SYNTHYRIS

Perennial herbs; leaves alternate, crenate, the radical roundish or cordate or cleft or parted; cauline leaves alternate, bract-like; flowers small, blue or purplish, in a raceme; calyx 4-parted; corolla campanulate or rotate, with 4 more or less unequal lobes; stamens 2, from just below the upper sinuses, exserted; anther sacs parallel or somewhat divergent; stigmas capitate; capsule compressed, loculicidal; seeds few or several, flat or concave. (Name from Gr., *sun,* together; *thuris,* a little door.)

Corollas blue. *S. missurica,* f. *missurica.*
Corollas salmon pink. Forma *rosea.*

Synthyris missurica (Raf.) Pennell, forma **missurica.** Scapes 1–6 dm. tall; caudex 5–10 mm. in diameter; basal leaves glabrous, with petioles 4–19 cm. long; the blades 4–9 cm. wide, orbicular-reniform, doubly dentate or shallow lobed, evergreen; scapes puberulent to glabrous above, glabrous below, naked except for a few ovate bracts near the summit; raceme 2 or more cm. long in flower, 7–30 cm. long in fruit, the lower flowers more remote and sterile; bracts seldom equaling the calyx; pedicels 1–11 mm. long, puberulent or glabrous, calyx 2–4 mm. long, parted almost to the base, the lobes ovate or oblanceolate, foliaceous; corolla 4–7 mm. long, blue, lobed ⅔ their length into obovate, rounded or truncate lobes; stamens exserted; anthers 1.2–1.5 mm. long, oval, later curved; capsule 5–7 mm. wide, suborbicular, flattened, retuse or notched at apex; seeds 1–1.6 mm. wide, discoid, brown. *S. missurica,* subsp. *major* (Hook.) Pennell. Moist deep woods, *A. T. T., Can.*

Forma **rosea** St. John. Differs by having its corollas salmon pink.

Washington: top of north hillside, Kamiak Butte, Whitman Co., April 17, 1921, *H. St. John* 6,069 (type), *A. T. T.*

TONELLA

Annual herbs; leaves opposite, the cauline mainly ternately divided or 3-parted; flowers axillary, perfect, irregular; calyx deeply 5-cleft; corolla obscurely 2-lipped, the 5 more or less unequal lobes somewhat rotately spreading, the tube somewhat gibbous above, stamens 4, the anther sacs confluent; ovules and seeds 1–4 in each cell. (Name unexplained by its author.)

Tonella floribunda Gray. Plant 1–5 dm. tall, glabrous below the inflorescence; roots fibrous; basal leaves 5–10 mm. long, simple, ovate, entire or few toothed slender petioled; cauline leaves with petioles as much as 15 mm. in length or the upper sessile, the blades 3–5-parted or -divided, the leaflets 1–3 mm. long lance-linear or lance-elliptic, remotely hispidulous, glabrate; inflorescence 5–20 cm. long, capitate glandular puberulent, leafy bracted at base; pedicels 5–25 mm. long; calyx 2.5–3.5 mm. long, obconic, becoming campanulate, glandular puberulent, the lance-deltoid teeth herbaceous, as long as the tube; corolla clear blue, with dark violet veins, and a pale tube, 6.5–10 mm. long, the upper lip twice as long as the tube, longer than the lower; anthers reniform, glabrous, the sacs 0.3–0.5 mm. long, dehiscent their whole length; capsule 3.5–5 mm. long, ovoid; seeds 1.3–1.6 mm. long, oblong, finely reticulate, the hilum linear, elongate. Open slopes, abundant, Snake River Canyon and tributaries *U. Son.*

Described from collections from Kooskooskie [Clearwater] River, *Spalding, Geyer,* etc.

VERBASCUM. Mullein

Tall and usually woolly biennial or perennial herbs; leaves alternate, the cauline sessile or decurrent; flowers in large terminal spikes or racemes; calyx 5-parted; corolla purple, red, yellow, or white, 5-lobed, rotate, the lobes slightly unequal; stamens 5, all anther-bearing; style flattened at the apex; capsule globular, 2-valved, many-seeded. (The ancient Lat. name, used by Pliny.)

Middle and lower leaves densely woolly; flowers subsessile;
 stamens unlike; capsule 8–11 mm. long. *V. Thapsus.*
Middle and lower leaves glabrous; pedicels 12–18 mm. long;
 stamens similar; capsule 6–8 mm. long,
 Flowers yellow. *V. Blattaria,* f. *Blattaria.*
 Flowers white. Forma *albiflorum.*

Verbascum Blattaria L., forma Blattaria. *Moth Mullein.* Biennial, 4–20 dm. tall, glabrous below, capitate glandular puberulent above; leaves glabrous, the basal rosulate 4–13 cm. long, short petioled or aparently so from the cuneate, decurrent base, the blade oblanceolate in outline, irregularly doubly dentate or lobed; middle and upper leaves 1–11 cm. long, oblong to lanceolate or ovate, less lobed than the basal ones, sessile, the base often cordate and clasping; racemes 1–7 dm. long, leafy bracted at base; calyx 4–6 mm. long, parted almost to the base into oblong-lanceolate, somewhat unequal lobes; corolla 2–3 cm. in. diameter, lobed about half way into suborbicular lobes, yellow, often brownish without, capitate glandular pilose without; filaments violet, capitate glandular villous; anthers 1.8–2 mm. wide, reniform, orange; capsule globose, apiculate; seeds 0.6–0.8 mm. long, ovoid-cylindric, brown, with several rows of large honey-comb-like pits. Weed, introduced from Europe. Open places or thickets, *U. Son., A. T.*

Forma albiflorum (G. Don) House. Differs by having the flowers white. With the species, rare, *U. Son., A. T.*

Verbascum Thapsus L. *Mullein.* Biennial, 3–22 dm. tall, densely woolly throughout with long, several times branched hairs; basal leaves rosulate, the petiole-like base 2–6 cm. long, the blades 1–5 dm. long, oblanceolate; cauline leaves gradually smaller in size, and becoming sessile, and elliptic or ovate; racemes 5–40 cm. long, spike-like, the bracts inconspicuous; calyx 7–12 mm. long, campanulate, parted almost to the base into lanceolate, slightly unequal lobes; corolla 12–20 mm. in diameter, woolly with branched hairs without; stamens unlike, the three upper smaller, the filament white villous, the 2 lower with the filaments glabrous or with a few remote hairs, the anthers reniform, 1.5 mm. wide; capsule ovoid; seeds 0.4–0.6 mm. long, cylindric, with several lines of large tubercles, brown. Weed, introduced from Eurasia. Open places, *U. Son.,* to *A. T. T.*

VERONICA. Speedwell

Chiefly herbs; leaves opposite or whorled, or the upper alternate; flowers small, in racemes or spikes or sometimes solitary in the axils; calyx 4-parted; corolla rotate, 4-lobed, the lower lobes and sometimes the lateral ones narrower; stamens 2, exserted; style entire; stigma capitate; capsule flattened, usually notched at apex, loculicidal, few–many-seeded. (Derivation of name uncertain.)

Main stem with a terminal inflorescence; upper bracts alternate,
- Leaves all petioled; capsule 4.5–8 mm. wide,
  - Capsule flattened, broadly emarginate; mature pedicels recurving. — *V. persica.*
  - Capsule turgid, the apex rounded; mature pedicels straight, ascending. — *V. hederaefolia.*
- Leaves all or at least the upper ones sessile; capsule 3–5 mm. wide,
  - Leaves linear to narrowly oblanceolate; corolla white. — *V. peregrina,* var. *xalapensis.*
  - Leaves ovate to subcordate; corolla blue,
    - Style as long as the capsule; stems decumbent at base, rooting at the nodes. — *V. serpyllifolia,* var. *humifusa.*
    - Style nearly as long as the notch of the capsule; stems erect or decumbent, not rooting at the nodes. — *V. arvensis.*

Racemes all axillary; leaves all opposite,
Leaves linear to linear-lanceolate; capsule broad reniform, broader than long, flat. *V. scutellata.*
Leaves lanceolate to oval; capsule suborbicular, turgid. *V. americana.*

Veronica americana Schwein. ex Benth. *American Brooklime.* Perennial by decumbent stem base; roots fibrous; plant glabrous; stems 1–13 dm. long, often simple; leaves all with petioles 2–10 mm. long, opposite, the blades 2–7 cm. long, oval, ovate, to lanceolate, crenate; racemes axillary, 4–20 cm. long with the peduncle, the bracts 2–17 mm. long, almost linear; pedicels in fruit 6–17 mm. long; calyx 1.5–3 mm. long, lance-ovate, the lower the larger; corolla 3–4.5 mm. long, light bluish violet, the lower lobe ovate, the other suborbicular; anthers 0.4 mm. long, cordate; capsule 3–4 mm. wide, the apex retuse, rounded, or subacute; style slightly shorter than the capsule; seeds 0.4–0.5 mm. long, oval, flattened, yellowish. Wet places, *U. Son., A. T.*

Pennell (1935) gave the authorities as (Raf.) Schwein. Since Schweinitz gave no basionym or reference, no parenthetical authority is admitted here.

Veronica arvensis L. *Corn Speedwell.* Annual, 5–40 cm. tall, pilose below, viscid pilose above; roots fibrous; stem simple or freely branched; lower leaves with petioles 2–7 mm. long, the blades 5–18 mm. long, broadly ovate or subcordate, crenate; middle leaves similar but sessile; upper leaves in the inflorescence alternate, lanceolate, equaling or exceeding the calyx; flowers subsessile or the lower pedicels shorter than the calyx; calyx accrescent, in fruit 3.5–5 mm. long, the lobes lanceolate, viscid, pilose, the 2 upper shorter; corolla 2–2.2 mm. long, at first blue, often fading to white, the lobes broadly oval; anthers 0.3 mm. long, the sacs divergent at base; capsule 3–4 mm. wide, broadly obcordate, the notch ¼ the length of the capsule; seeds 1–1.1 mm. long, oval, plano-convex, yellowish. Weed introduced from Eurasia. In fields. *U. Son., A. T.*

Veronica hederaefolia L. Annual; stems 5–30 cm. long, ascending or reclining, much branched, pilose; petioles longer or shorter than the blades; blades 8–15 mm. long, reniform-suborbicular, 3–5 (–7)-lobed, the shallow lobes acute; flowers axillary, single; pedicels 8–15 mm. long: calyx lobes ovate, subcordate ciliate; corolla 2–2.5 mm. broad, lilac to blue; capsule 4.5–6 mm. wide, glabrous; seeds 2.5–3 mm. broad. Introduced from Europe, grassy slopes, Almota, *W. C. Muenscher, U. Son.*

Veronica peregrina L., var. **xalapensis** (H.B.K.) St. John & Warren. *Neckweed.* Annual, 5–30 cm. tall, more or less glandular puberulent throughout; roots fibrous; stems simple or bushy branched; leaves 6–30 mm. long, nearly glabrous or glabrate, sessile; upper leaves little reduced; flowers subsessile, in the upper and even in the lower axils; calyx 3–4 mm. long, glandular puberulent, deeply parted, the lobes unequal, lanceolate to elliptic; corolla 2 mm. long, the lobes ovate, acute; anthers 0.3–0.4 mm. long, cordate; capsule 3–5 mm. wide, obcordate; stigma subsessile; seeds 0.6–0.8 mm. long, oval, plano-convex, yellowish. *V. peregrina xalapensis* (H.B.K.) Pennell. Wet places, *U. Son., A. T.*

Veronica persica Poir. Annual, erect or prostrate, more or less hirsute throughout; roots fibrous; stem 1–6.5 dm. long; petioles 1–12 mm. long; blades 8–30 mm. long, ovate to broadly oval, coarsely crenate, the base cuneate, upper leaves alternate; pedicels 1–3 cm. long; calyx 4–6 mm. long, the lobes subequal, lanceolate, ciliate; corolla 5–7 mm. long, the lower lobe ovate, pale blue, the others suborbicular, blue, all blue lined, at base whitish; anthers 1–1.2 mm. long, ovate-sagittate; capsule obreniform, pilosulous, reticulate veined, the wide notch ⅓ the length of the capsule, as long as the style; seeds 1.5–2.5 mm. long, concave, yellowish, many ridged on the back. *V. Tournefortii* Gmel., not Schmidt. Weed, introduced from Asia. Fields, *A. T.*

Veronica scutellata L. *Marsh Speedwell.* Perennial, by the old fallen stems rooting at the nodes, by basal stolons, and by the roots; plant 1–8 dm. tall, glabrous, decumbent and rooting at base; stem simple or with a few weak branches; leaves 1–7.5 cm. long, acute, sessile, entire or denticulate; racemes axillary, with the peduncle 4–8 cm. long, bracts 1–3 mm. long, linear; pedicels in fruit 7–12 mm. long, filiform; calyx 1.5–2.5 mm. long, the lobes oval, equal; corolla 3 mm. long, pale lavender, the lower lobe oval, the others suborbicular; anthers 0.4–0.5 mm. long, cordate; capsule 4–5 mm. wide, the shallow notch about 1/6 the length of the capsule; style longer than the capsule; seeds 1.5–1.8 mm. long, oval, flat, yellowish. Wet places, *A. T.*

Veronica serpyllifolia L., var. **humifusa** (Dickson) Vahl. Perennial by decumbent basal shoots; roots fibrous; plant 1–4 dm. tall; stems puberulent, the tips assurgent; leaves 6–33 mm. long, opposite; ovate to oval, entire or crenulate, glabrous or nearly so, the lower with petioles 1–4 mm. long; racemes 3–25 cm. long, glandular hirsutulous, the bracts alternate, leafy below; pedicels 1–7 mm. long; calyx 2–4 mm. long, glandular hirsutulous, the lobes subequal, oval; corolla 3–4 mm. long, bright blue, the lower lobe ovate, the others larger and suborbicular; filaments 2–4 mm. long; anthers 0.5–0.6 mm. long, cordate; capsule 3–5 mm. wide, orbicular-obcordate, glandular hirsutulous, the apex broad notched ¼ the length of the capsule; seeds 0.6–0.8 mm. long, oval, flattened, yellowish. *V. humifusa* Dickson; *V. serpyllifolia,* var. *decipiens* Boivin. Moist woods, *Can.* Also stranded at Lewiston, *Hunter* 65.

LENTIBULARIACEAE. Bladderwort Family

Small scapose or caulescent herbs, growing in water or wet places; submerged leaves dissected into linear or filiform segments, often bearing bladders; aerial leaves basal and entire or wanting; scapes naked or scaly; flowers perfect, irregular; sepals 2–5, herbaceous; corolla more or less 2-lipped, the tube spurred or saccate; stamens 2, the filaments adnate to upper side of corolla tube, flattened, twisted; pistil solitary, the placenta free, central; style short, thick; stigma with 2 unequal lobes; capsule 2-valved or irregularly dehiscent; seeds numerous; endosperm none. (Named from the genus *Lentibularia.*)

UTRICULARIA. Bladderwort

Aquatic plants; stems mostly submerged; leaves finely dissected, in many species bladder-bearing; bladders urn-shaped, the mouth closed by a lid; flowers single or racemose; calyx 2-lobed, the lobes concave, herbaceous; corolla strongly 2-lipped, the palate of the lower lip prominent, usually 2-lobed; seeds more or less peltate. (Name from Lat., *utriculus,* a little bladder.)

Utricularia vulgaris L. Plant glabrous; stems 0.5–3 m. long, submerged, usually free floating; leaves alternate, 15–35 mm. long, 2–3-pinnately divided into filiform segments; bladders numerous, 3–4 mm. long; scapes 1–3 dm. long, erect, raised above the water; bracts 4–6 mm. long, cordate-oval, clasping at base; racemes 5–15-flowered, 3–15 cm. long; bracts 4–6 mm. long, similar; pedicels 10–18 mm. long; sepals 5–7 mm. long, broadly ovate; corolla yellow, the lower lip 10–15 mm. long, slightly longer than the upper; spur 6–8 mm.

long, tubular, stout, curved; capsule about 4 mm. long, globose; seeds not seen. In water, Spangle, July 21, 1948, *Yocom;* Toolie Lake, Ewan, *Pickett et al.* 1593, *A. T.*

OROBANCHACEAE. Broomrape Family

Root-parasitic herbs, destitute of foliage and green color, yellowish or brownish throughout; leaves reduced to alternate scales; flowers usually perfect, solitary, in loose clusters or in spikes; calyx 4- or 5-toothed or -parted; corolla gamopetalous, tubular, more or less 2-lipped, hypogynous; stamens 4 didynamous, epipetalous; style long; stigma large; capitate or 2-lobed; ovary 1-celled, with 2 or 4 parietal placentae; capsule 1-celled, 2-valved; seeds very numerous, minute. (Named from the genus *Orobanche.*)

OROBANCHE. Broomrape

Parasitic herbs on the roots of various plants, usually yellowish or reddish; calyx deeply 5-cleft; corolla irregular, the tube slightly curved, the upper lip erect or incurved, the lower spreading; stamens included; stigma peltate or 2-lobed; capsule 2-valved. (Name from Gr., *orobos,* vetch; *agchone,* strangler.)

Flowers in a spike. — *O. multiflora,* var. *arenosa.*

Flowers long pediceled,
- Stem usually longer than the pedicels; calyx-teeth not longer than the tube. — *O. fasciculata,* var. *fasciculata.*
- Stem shorter than the pedicels; calyx-teeth longer than the tube.
  - Corolla 22–30 mm. long; anthers pilose. — *O. porphyrantha.*
  - Corolla 14–22 mm. long; anther glabrous. — *O. uniflora,* var. *minuta.*

Orobanche fasciculata Nutt., var. **fasciculata.** Plant 8–18 cm. tall, glandular pilosulous; cauline bracts 6–12 mm. long, ovate to lanceolate; the 4–10 pedicels 2–10 cm. long; calyx campanulate, glandular pilosulous, the tube 3–6 mm. long, the teeth 3–6 mm. long, deltoid, acuminate; corolla 15–25 mm. long, purplish or even yellowish, glandular pilosulous, the lobes 4–5 mm. long, broadly ovate; anthers sparsely villous or glabrous, 1.5–1.8 mm. long, broad oval, the sacs apiculate at base; capsule 8–11 mm. long, ovoid; seeds 0.2–0.3 mm. long, more or less ovoid, dull grayish brown foveate. *Thalesia fasciculata* (Nutt.) Britton; *Anoplanthus fasciculatus* (Nutt.) Walp.; *O. fasciculata* Nutt., var. *typica* Achey. Dry places, *U. Son., A. T.*

Parasitic on roots of *Chrysothamnus, Artemisia,* etc.

Orobanche multiflora Nutt., var. **arenosa** (Suksd.) Munz. Plant 5–22 cm. tall, glandular puberulent throughout; cauline bracts 5–9 mm. long, lance-ovate; spikes 3–12 cm. long, single or more or less fasciculate; calyx 9–16 mm. long, lobed about half its length, the tube campanulate, the lobes lance-linear; corolla 15–23 mm. long, glandular puberulent, yellow or purplish, the upper lip 4–5 mm. long, the lobes rounded; anthers 1.2–1.4 mm. long, sagittate, glabrous; capsule 5–8 mm. long, ovoid; seeds 0.2–0.4 mm. long, more or less ovoid, pale brownish, foveate. Sandy banks of Snake River, *U. Son.*

Parasitic on the roots of *Artemisia, Chrysopsis,* etc.

Orobanche porphyrantha G. Beck. Plant 6–27 cm. tall, glandular pilose; stem usually 2–8 cm. tall and stout (occasionally smaller); cauline bracts 3–12 mm. long, ovate, obtuse or acute; the 1–4 pedicels 3–18 cm. long; calyx campanulate, glandular pilose, the tube 3–5 mm. long, the teeth 4–8 mm. long, subulate or lance-subulate; corolla purple or tinged with violet, decurved, widening above, the throat 5–8 mm. in diameter, the lobes 4–6 mm. long, broadly oval; anthers 1.7–1.9 mm. long, suborbicular, the sacs apiculate at base; capsule 6–8 mm. long, ovoid; seeds 0.3–0.4 mm. long, narrowly ellipsoid, brown, coarsely foveate. *Thalesia purpurea* Heller; *O. uniflora* L., var. *purpurea* (Heller) Achey. Open places, *U. Son., A. T.*

Parasitic on roots of *Lomatium,* etc. The type was from mouth of the Potlatch, Ida., *Heller* 3099.

Orobanche uniflora L., var. **minuta** (Suksd.) G. Beck. Plant 4–15 cm. tall, more or less glandular pilosulous; stem 5–30 mm. long slender; cauline bracts 5–12 mm. long, ovate; the 1–2(–3) pedicels 2–11 cm. long; calyx campanulate, glandular pilosulous, the tube 1.5–3 mm. long, the lobes 3.5–7 mm. long, subulate or lance-subulate; corolla decurved, glandular pilosulous, 3–5 mm. in diameter at the throat, lilac, lavender, or yellowish, the lobes 3–4 mm. long, broadly oval; anthers 1.2–1.5 mm. long, the sacs oval, apiculate at base; capsule 6–8 mm. long, ovoid; seeds 0.3–0.4 mm. long, ovoid, brown, coarsely foveate. *O. Sedi* (Suksd.) Fern.; *O. uniflora* L., var. *Sedi* (Suksd.) Achey; *Thalesia minuta* (Suksd.) Rydb. Thickets or open places, *U. Son., A. T.*

Parasitic on roots of *Lithophragma, Sedum,* etc.

PLANTAGINACEAE. Plantain Family

Annual or perennial herbs; leaves basal or alternate; flowers regular, 4-merous, in spikes; calyx 4-parted, persistent, imbricate; corolla gamopetalous, dry and membranaceous, veinless, with 4 erect or spreading lobes; stamens 4 or rarely 2 on the corolla-tube, alternate with its lobes; ovary 1–2 or falsely 3–4-celled; ovules 1–several in each cavity. (Named from the genus *Plantago.*)

PLANTAGO. Plantain

Leaves nerved or ribbed, alternate or radical; flowers small, in bracted spikes, on naked scapes; calyx of 4 persistent sepals; corolla salverform or rotate, 4-parted; stamens 4, or rarely 2, in all or some flowers with long exserted filaments; ovules 1–several in each cell; capsule 2-celled, 2–several-seeded. (The ancient Lat. name.)

Leaves linear; plant densely white villous throughout; petals obtuse. *P. Purshii,* var. *Purshii.*
Leaves lanceolate or broader; plant sparsely hairy or glabrous, green; petals acute,
Blades ovate; calyx divisions 1.5–2 mm. long, glabrous; capsule 6–16-seeded. *P. major.*

Blades narrowly oblanceolate; calyx divisions 3–3.3 mm. long, pilose on the midrib; capsule 2-seeded. *P. lanceolata.*

Plantago lanceolata L. *English Plantain; Buckthorn Plantain.* Perennial or biennial, 5–75 cm. tall, long villous tufted at base; caudex 2–12 mm. in diameter, brown; leaves basal, green, sparsely pilose; petioles 3–15 cm. long; blades 3–18 cm. long, 3–5-ribbed, minutely denticulate; spikes 1–6 cm. long, at first ovoid, later cylindric; bracts ovate, glabrous, the margins scarious, shorter than the flowers; flowers perfect, proterogynous; calyx divisions ovate, scarious; corolla glabrous, longer than the calyx the lobes 2.3–2.5 mm. long, ovate, rotate; filaments 2–3-times the length of the corolla; anthers 1.5–2 mm. long, lanceoloid, acuminate; capsule 3–4 mm. long, ovoid, circumscissile ⅓ way from the base; seeds 2–2.2 mm. long, dark and pale brown, shining, ellipsoid, flattened and furrowed by the hilum. Weed, introduced from Europe. Meadows, lawns, and cultivated lands, *A. T.*

Declared a noxious weed in Washington.

Plantago major L. *Common Plantain.* Annual or perennial, 1–7 dm. tall, more or less pilose or glabrous; caudex 3–15 mm. in diameter, with many lateral roots; leaves basal, erect or prostrate; petioles 1–30 cm. long; blades 3–25 cm. long, the base cuneate or subcordate, the margin entire or somewhat dentate, palmately 7-nerved; spike 4–45 cm. long, dense; bracts ovate, scarious margined, glabrous, shorter than the calyx; flowers proterogynous; calyx segments oval or ovate, scarious margined; corolla slightly longer than calyx, the lobes about 1 mm. long, lance-ovate, rotate; anthers exserted, 1 mm. long, broadly oval; apiculate at tip, sagittate apiculate at base; capsule 2–4 mm. long, ovoid, circumscissile just below the middle; seeds 0.8–1.2 mm. long, brown, ellipsoid, more or less compressed, roughened. Weed, introduced from Eurasia. Roadside and lawns or cultivated lands, *U. Son., A. T., A. T. T.*

The seeds are used as bird seed.

Plantago Purshii R. & S., var. **Purshii.** Annual, 5–30 cm. tall; tap-root 1–2 mm. in diameter, elongate; acaulescent or nearly so; leaves 3–15 cm. long, 1–4 mm. wide, long acuminate to a callous apex; scapes erect or ascending; spikes 1–15 cm. long, dense; bracts 3–8 mm. in length, foliaceous; calyx divisions 2–3 mm. long, ellipsoid, herbaceous with narrow scarious margins; corolla exceeding the calyx, the lobes 1.8–2.2 mm. long, rotate, broadly ovate, auriculate at base; stamens and styles dimorphic; anthers 0.3 mm. long, lanceolate; capsule 3–4 mm. long, ellipsoid, circumscissile at the middle, 2-seeded; seeds 1.5–2.5 mm. long, ellipsoid, brown, roughened. *P. Purshii* R. & S., var. *typica* Poe. Open places, abundant in *U. Son.; A. T.*

Grazed by stock in arid regions in late summer.

Cronquist (1959) has reduced this to *P. patagonica* Jacq.

RUBIACEAE. Madder Family

Herbs, shrubs, or trees; leaves opposite or whorled, usually entire, with intervening stipules or whorled without stipules; flowers generally perfect; calyx 4- or 5-lobed or -toothed, adnate with the ovary; corolla gamopetalous, epigynous, regular, 4- or 5 (–10)-lobed or -toothed; stamens on the corolla and alternate with its lobes; ovary inferior, 2–5-celled; fruit a capsule, berry, or drupe; endosperm hard, copious. (Named from the genus *Rubia,* the madder.)

Leaves in apparent whorls; corolla rotate. *Galium.*
Leaves opposite; corolla funnelform. *Kelloggia.*

GALIUM. Bedstraw; Cleavers

Annual or perennial with 4-angled stems; leaves whorled, without apparent stipules; flowers small, usually cymose; calyx-teeth obsolete; corolla rotate, 4-parted, rarely 3-parted; stamens as many as the corolla-lobes, short; styles 2, short; stigmas capitate; ovary 2-lobed, 2-celled; ovules solitary; fruit dry or fleshy, globular, separating when ripe into 2 closed 1-seeded carpels. (Name from Gr., *gala,* milk.)

Leaves 2–4 at a node; leaves not cuspidate pointed,
  Flowers dioecious; plant suffrutescent at base,
    Leaves ovate. *G. multiflorum,* var. *multiflorum.*
    Leaves elliptic or lanceolate. Var. *Watsoni.*
  Flowers perfect; plant herbaceous,
    Fruit pubescent; corolla lobes 4,
      Flowers solitary; leaves smooth. *G. bifolium.*
      Flowers in many flowered panicles; leaves hispidulous on midrib and margin,
        Fruit hirsute with straight hairs. *G. boreale,* var. *boreale.*
        Fruit with appressed or incurved puberulence. Var. *intermedium.*
    Fruit glabrous; corolla lobes 3. *G. trifidum,* var. *pacificum.*
Leaves 5–8 at a node; leaves cuspidate pointed,
  Fruit granulate scabrous. *G. asperrimum.*
  Fruit uncinate hispid,
    Leaves elliptic-oblanceolate. *G. triflorum.*
    Leaves linear-spatulate,
      Leaves 2.5–7 cm. long; fruiting carpels 3–4.5 mm. long. *G. Aparine.*
      Leaves 7–20 mm. long; fruiting carpels 1.5–2.5 mm. long. Var. *echinospermum.*

Galium Aparine L. *Cleavers.* Annual, 1–10 dm. tall; roots fibrous; stems weak, often decumbent, sprawling, or climbing, simple or branching, the angles retrorse hispidulous scabrous; leaves 6–8 at a node, linear spatulate, rough hispidulous on margin and veins, sparingly so above; cymes axillary and terminal, leafy bracted, 2–3-flowered; pedicels 3–45 mm. long; corolla white, the lobes 0.5–0.8 mm. long, ovate; fruiting carpels globose, hispid with stout hairs, curved at the tips. Woods and thickets, *U. Son., A. T.*

Var. **echinospermum** (Wallr.) Farw. Annual 1–9 dm. tall; roots fibrous; stems weak, often decumbent, retrorse scabrous hispidulous on the angles; leaves 6–8 at a node, linear to linear-spatulate, retrorse hispidulous on the margin and midrib, sparsely hispidulous above; cymes 2–9-flowered, axillary and terminal, leafy bracted; pedicels 3–26 mm. long; corolla white, the lobes 0.7–1 mm. long, ovate; fruiting carpels broadly ellipsoid, hispid with stout hairs, curved at the tips. *G. Vaillantii* DC.; *G. spurium* L. Thickets, *U. Son., A. T. T.*

Galium asperrimum Gray. Perennial, 3–9 dm. tall; roots filiform, producing rhizomes; stems weak, ascending or decumbent, the angles retrorse hispidulous; leaves 6 in a whorl, 1–3 cm. long, linear-oblanceolate, 1-nerved, the midrib and margin retrorse hispidulous; flowers perfect, few to many in loose, axillary and terminal cymose panicles; bracts foliaceous; pedicels 3–14 mm. long, filiform; corolla white, the 4 lobes 1.5–2.2 mm. long, ovate, acuminate, glandular within along the veins; fruiting carpel 1.2–1.6 mm. long, broadly ellipsoid. *G. asperulum* (Gray) Rydb. Moist thickets, *A. T., A. T. T.*

Galium bifolium Wats. Annual, 4–18 cm. tall, the herbage glabrous; roots fibrous; stem erect or decumbent, weak, smooth, often dichotomously branched; lower and middle nodes with the 4 leaves unequal, the larger pair 8–20 mm. long, elliptic or narrowly so, the other pair smaller; upper nodes with 4 or 2 leaves; pedicels 1-flowered, axillary and terminal, in fruit the axillary 6–20 mm. long, reflexed at apex; corolla white, the lobes 0.3–0.4 mm. long, ovate; fruiting carpels 2.2–2.5 mm. long, broadly ellipsoid, hispid with hooked hairs. Moist open places, or woods, Blue Mts., *Can.*

Galium boreale L., var. **boreale.** *Northern Bedstraw.* Perennial, nearly glabrous; rhizome cord-like, brown, up to 1 mm. in diameter; stems 2–7 dm. tall, erect, simple or branched, the angles smooth or minutely roughened, not hairy; leaves in whorls of 4, subequal, 1.5–6.5 cm. long, 2–7 mm. wide, lance-linear, firm, 3-nerved, obtuse; panicles 4–30 cm. long, terminal; bracts foliaceous, decreasing upwards; pedicels 1–3 mm. long, hirsutulous; flowers perfect, white; corolla lobes 1.5–2 mm. long, ovate; fruiting carpels 1.4–1.6 mm. long, semiorbicular. Meadows and north hillsides, *A. T.*

Var. **intermedium** DC. Differs in having the fruiting carpels with appressed or incurved puberulence. Meadows or moist places, *U. Son., A. T.*

Galium multiflorum Kellogg, var. **multiflorum.** Perennial, 1–4 dm. tall, the herbage glabrous; stems slightly suffrutescent at base, simple or forking, glabrous and smooth; leaves 4 at a node; 6–12 mm. long, roundish ovate, thick, glabrous, 3-nerved; cymes 3–5-flowered, not or little exceeding the leaves; pedicels 1–6 mm. long; staminate flowers greenish, 1.7–2 mm. long, the lobes ovate; pistillate flowers white, the lobes 1.5 mm. long, ovate; fruits concealed by the abundant, white, long hirsute hairs; fruiting carpels 1–2 mm. long, subglobose. Rocky ridges, Blue Mts., *Huds.*

This group has been monographed twice, and is being studied again, so the present treatment is left unchanged for the time being.

Var. **Watsoni** Gray. Differs by having the leaves 8–15 mm. long, elliptic or lanceolate; fruiting carpels 1.5–2.5 mm. long. Stony ridges, Blue Mts., *Horner, 372*; 373, *Huds.*

Galium trifidum L., var. **pacificum** Wieg. Perennial 1–7.5 dm. tall; roots up to 1 mm. in diameter, cord-like, producing rhizomes; stems weak, prostrate or ascending, retrorse scaberulous on the angles, often branched; leaves in whorls of 4, subequal, 5–17 mm. long, linear-spatulate, obtuse, retrorse scaberulous on the midrib and margin 1-ribbed; flowers white, solitary from the axils, or the terminal in 3's, perfect; pedicels 3–12 mm. long, arcuate, filiform, scaberulous; corolla lobes 0.6–0.8 mm. long, ovate; fruiting carpels 0.9–1.2 mm. long, spherical, the pair only slightly joined. Wet places, *A. T.*

Galium triflorum. Michx. *Fragrant Bedstraw.* Perennial, 3–9 dm. tall; roots filiform, reddish; stems simple or few branched, smooth or the angles sparsely retrorse scaberulous; leaves in whorls of 6 (or 5), 1.5–7 cm. long, hispidulous near the margin; peduncles axillary and terminal, 2–3-flowered, 1–6.5 cm. long, bracted at summit; pedicels 3–30 mm. long, glabrous; flowers perfect; corolla white, 4-parted, the lobes 1–1.4 mm. long, ovate, acuminate; fruiting carpels 1.4–2 mm. long, broadly ellipsoid, densely hispid with hooked hairs. Moist thickets or woods, *A. T.*, *A. T. T.*

KELLOGGIA

Small perennial herbs with opposite leaves with stipules; flowers small, in loose cymes or panicles, usually 4-merous; calyx-teeth minute; corolla funnelform; stamens and style more or less exserted; ovary 2-celled; fruit small, dry, coriaceous, with hooked bristles, separating at maturity into 2 closed carpels. (Named for *Dr. Albert Kellogg of San Francisco.*)

Kelloggia galioides Torr. Plant 1–3 dm. tall, the herbage glabrous; stems rooting at base, forking, the root 1–3 mm. in diameter, brown; stem mostly simple; leaves 1.5–5 cm. long, linear-lanceolate entire, pinnately veined; stipules interpetiolar, membranous, lanceolate, entire or bifid; cyme loose, divaricate, few or many flowered; pedicels 1–4 cm. long; corolla pink, puberulent, the tube 2–3 mm. long, the lobes 2–3 mm. long, lanceolate; anthers 1.3–1.6 mm. long, linear, pink; style short bifid; fruiting carpels 3–4 mm. long, broadly ellipsoid, densely uncinate hispid. Woods, *Can.*

Known in our area only at the type locality, Walla-Walla River, *Pickering & Brackenridge.*

CAPRIFOLIACEAE. Honeysuckle Family

Shrubs, trees, or rarely herbs; leaves opposite, without genuine stipules; flowers perfect, generally in cymes; calyx-tube adherent to the ovary, 5-toothed; corolla gamopetalous, epigynous, tubular to rotate, 4- or 5-cleft, sometimes irregular; stamens distinct, inserted on the tube, as many as the corolla-lobes, rarely fewer; ovary 2–5-celled, or not rarely 1-celled; ovules solitary or several; fruit a berry; endosperm fleshy. (Named from *Caprifolium,* the name used by Tournefort for the genus *Lonicera.*)

Corolla rotate or urn-shaped; flowers in compound cymes; style short or none,
- Leaves pinnate; fruit 3–5-seeded. — *Sambucus.*
- Leaves simple; fruit 1-seeded. — *Viburnum.*

Corolla tubular or bell-shaped; flowers not in compound cymes,
- Creeping evergreen vines; stamens 4. — *Linnaea.*
- Shrubs, erect or climbing; stamens 5,
  - Corolla regular, bell-shaped; fruit 2-seeded. — *Symphoricarpos.*
  - Corolla more or less irregular, tubular, commonly 2-lipped, gibbous at base; berry several seeded. — *Lonicera.*

LINNAEA. Twin-flower

Creeping and trailing small evergreen herbs; leaves ovate or orbicular, opposite, petioled; flowers in pairs, on the summit of elongated terminal peduncles; calyx-teeth 5, awl-shaped, deciduous; corolla funnelform, almost equally 5-lobed, pink or purple and whitish, hairy inside; stamens 4, two shorter, all included and inserted near the base of the corolla; ovary and the small dry pod 3-celled, but one seed ripening, the other ovules abortive. (A favorite of *Carolus Linnaeus* of Sweden, and named for him by Gronovius.)

Calyx-segments 1.5–3 mm. long; corolla 8–15 mm. long. *L. borealis*, var. *americana*.
Calyx-segments 3–5 mm. long; corolla 10–16 mm. long. Var. *longiflora*.

Linnaea borealis L., var. **americana** (Forbes) Rehder. *Twin-flower*. Branches trailing, often 1 m. or more long, puberulent or hisutulous; roots fibrous, pale; petioles 2–4 mm. long, hirsute ciliate; blades 7–21 mm. long, evergreen, oval to suborbicular, hirsute ciliate, above sparsely hirsute, impressed rugose nerved, the apex usually few crenate; peduncles 3–9 cm. tall, appressed puberulent and glandular pilose; the 2 apical bracts 2–3 mm. long, linear-lanceolate; pedicels 5–25 mm. long, divergent; flowers pendent; calyx campanulate, glandular puberulent, the segments 1.5–3 mm. long, lanceolate; corolla 8–15 mm. long, pink, funnelform, widening from beyond the calyx, the lobes broadly oval; anthers 1.4–1.7 mm. long, linear; capsule 3 mm. long, lanceoloid, puberulent. *L. americana* Forbes. Cool, evergreen woods, *A. T. T., Can.*

A plant of real beauty, with lovely flowers and an enchanting fragrance.

Var. **longiflora** Torr. Differs from var. *americana* by having the calyx-segments 3–5 mm. long; and the corolla 10–16 mm. long. Cool, evergreen woods, *A. T. T., Can.*

LONICERA. Honeysuckle

Erect or climbing shrubs; leaves opposite, mostly entire; flowers in spikes or pairs, each pair with 2 bracts and 4 bractlets; calyx minutely 5-toothed; corolla tubular or funnelform, often gibbous at the base, more or less irregularly 5-lobed, commonly 2-lipped; stamens 5, epipetalous; stigma capitate; ovary 2- or 3–5-celled, with numerous ovules in each cell; berry several-seeded. The fruits are emetics with bitter, disagreeable taste. (Named for *Prof. Adam Lonitzer,* early German botanist.)

Vine; upper leaves connate. *L. ciliosa*.
Shrub; leaves all separate,
  Berries black; bracts enveloping the flowers; corolla glandular pilosulous without. *L. involucrata*.
  Berries red; bracts not longer than the ovary; corolla glabrous without,
    Blades essentially glabrous beneath. *L. utahensis*, f. *utahensis*.
    Blades hirsute beneath. Var. *ebractulata*.

Lonicera ciliosa (Pursh) Poir. *Orange Honeysuckle*. Vine, climbing on trees as much as 10 m.; branchlets green or purplish glaucous; branches glabrous, yellowish; winter buds ovoid, acute, with lanceolate, decussate, glabrous scales; lower leaves with petioles 1–5 mm. long; blades 4–10 cm. long, oval or ovate, subacute, beneath pale and glaucous, glabrous except the ciliate margin; upper leaves sessile, connate perfoliate; flowers 6–16 in a close terminal cluster; calyx-limb 0.3–0.5 mm. long, shallowly dentate; corolla 2.8–3.5 mm. long, yellow or orange, glabrous or sparsely pilose, at base funnelform, the tube slightly inflated just above the base and slightly enlarged near the throat, the lobes 4–5 mm. long, ovate; anthers 3.5–4 mm. long, linear; berries distinct, 6–8 mm. long, orange-red, subglobose; seeds 4–5.5 mm. long, oval, compressed. Open woods, *A. T. T.*

Lonicera involucrata (Richards.) Banks. ex Spreng. *Black Twinberry.* Shrub 1–3 m. tall; branchlets green, glandular puberulent to glabrous; branches brown or gray, the smooth bark exfoliating; winter buds ovoid, acute, covered by the 2 outer ovate scales; petioles 3–11 mm. long, glandular puberulent; blades 3–15 cm. long, oval to narrowly oblanceolate, obtuse or acute, the base rounded or cuneate, somewhat pilose beneath; peduncles 12–43 mm. long, single, axillary, glandular puberulent; bracts 7–18 mm. long, ovate to flabelliform, glandular pilose, becoming purplish; the pair of flowers yellow, the corolla 12–16 mm. long, yellow or tinged with scarlet; the tube subcylindric, saccate at base, the lobes 2.5–4 mm. long, broadly ovate; stamens not or scarcely exserted; anthers 2.5–3.5 mm. long, linear; berries 8–12 mm. long, globose or ovoid; seeds 2–3 mm. long, oval, compressed, brown, pitted. *Distegia involucrata (Richards.)* Cockerell. Moist thickets, *A. T. T., Can.*

The foliage is grazed by sheep.

Lonicera utahensis Wats., forma **utahensis.** *Red Twinberry.* Shrub, 1–2 m. tall; branchlets glabrous or sparsely glandular puberulent but early glabrate; bark becoming gray or brown, fissured; terminal buds ellipsoid, acute, with several pairs of lance-ovate scales; lateral buds smaller; internodes shorter than the leaves; petioles 2–5 mm. long, sparsely glandular puberulent; blades 15–65 mm. long, ovate, oval, to elliptic-oblong, pale and glaucous beneath, glabrous, rarely with a few hairs beneath and sparsely glandular puberulent ciliate at base; peduncles 1–2.2 cm. long; bracts 2, ovate to lanceolate, shorter than or as long as the ovaries; bractlets smaller, variable; flowers 2, sessile, whitish, fading yellowish; calyx a short membranous collar, entire or lobed; corolla tube 12–17 mm. long, saccate at base; the lobes 5–7 mm. long, ovate; stamens included; anthers 2.5–4 mm. long, lanceolate; style exserted; stigma semi-capitate; berries distinct, 6–9 mm. long, ovoid; seeds 3–4 mm. long, ovoid or compressed, brownish cellular papillose. *Xylosteon utahense* (Wats.) Howell. Woods, *A. T. T., Can.*

Forma **ebractulata** (Rydb.) St. John. Differs in being hirsute on the lower leaf surface and the petioles, and the leaf margin hirsute ciliate near the base. The size of the bracts and bractlets, the size and lobing of the calyx, and particularly the shape and size of the leaves, vary widely. However, there are intermediate conditions on other or on the same individuals, so these do not seem to be of any diagnostic value. Both plants have the same broad range, though locally only one may occur or be predominant. *Lonicera ebractulata* Rydb. Common, woods *A. T. T., Can.*

SAMBUCUS. Elder

Shrubs or trees; leaves opposite, pinnate; leaflets serrate or laciniate; flowers small, white numerous, in terminal compound cymes; calyx-lobes minute or obsolete; corolla open urn-shaped, 5-cleft; lobes broadly spreading; stigmas 3–5; fruits of berry-like drupes with 3–5 nutlets, each with one seed. (The ancient name, used by Pliny, perhaps from Gr., *sambuke,* a musical instrument.)

Cymes flat-topped; berries blue; corolla lobes rotate. *S. glauca.*
Cymes round-topped; berries black; corolla lobes reflexed. *S. melanocarpa.*

Sambucus glauca Nutt. ex T. & G. *Blue Elderberry; Elder Tree.* Shrubs with several stems or tree, up to 7 m. tall and 3 dm. in diameter; bark reddish brown, with squarish scales, deeply fissured; young twigs smooth, brown, somewhat glaucous; young pith white; winter buds small, ovoid, with dark brown decussate scales; leaves glabrous; petioles 1–5 cm. long; the 5–9 leaflets 4–15 cm. long, lanceolate to oblong- or ovate-lanceolate, the tip acuminate, the base asymmetric, the margin serrate, pale green, beneath glaucous; lower lateral petiolules 5–17 mm. long; cymes many flowered, 10–25 cm. wide; flowers white; calyx-teeth 0.2–0.3 mm. long, lanceolate or ovate; corolla lobes 1.5–2 mm. long, oblong-oval; anthers 0.6–0.8 mm. long, ellipsoid; berry 4–5 mm. in diameter, globose, black covered by a sky blue bloom; seeds 2.7–3 mm. long, ovoid, somewhat compressed brown, transversely rugose. Wood light, weak; heartwood yellow tinged with brown; sapwood lighter. *S. coerulea* of recent authors. By streams and springs, *U. Son., A. T., A. T. T.*

The young shoots are favorite forage of sheep on the summer range. *S. cerulea* Raf. is probably this plant but its description is inadequate. Rafinesque never saw it, naming it from a few non-technical words in the Lewis and Clark Travels. No specimen was preserved by the Lewis and Clark Expedition.

Sambucus melanocarpa Gray. *Black-berried Elder.* Shrub, 1–4 m. tall; young twigs green, sparsely hirsutulous, later reddish brown; branches pale brown, smooth; young pith white; winter buds depressed ovoid, acute, with smooth brown, decussate scales; petioles 2–6 cm. long; petiolules of lower pair of leaflets 1–5 mm. long; the 5–9 leaflets 6–15 cm. long, sparsely puberulous or glabrate, lanceolate, acuminate, the base rounded or cuneate, the margin serrate, above dark, beneath pale green; cymes broadly ovoid, many flowered, 4–8 cm. wide, sparsely puberulous or glabrate; flowers creamy white; calyx-teeth 0.3–0.4 mm. long, low, rounded; corolla lobes 2–2.5 mm. long, broadly oval; anthers 0.5–0.8 mm. long, broadly ellipsoid; berry 5–6 mm. in diameter, globose; seeds 2.1–2.8 mm. long, broadly ellipsoid, transversely rugose, the inner faces compressed. *S. racemosa* L., var. *melanocarpa* (Gray) McMinn. Woods, *Can.*

The young shoots are a favorite food of sheep on the summer range.

SYMPHORICARPOS. Snowberry

Low and branching shrubs; leaves deciduous, short-petioled, entire or wavy-margined; flowers 2-bracteolate, in axillary and terminal spikes or clusters, rarely solitary; calyx-teeth 5 or 4, short persistent; corolla regular or nearly so, bell-shaped to salverform, white or pink, 5- or 4-lobed; stamens as many as the corolla-lobes, epipetalous; ovary 4-celled, two of these being 1-ovuled and fertile, the two alternate several-ovuled and sterile; fruit globular and berry-like, containing 2 bony seed-like nutlets. (Name from Gr., *sumphorein,* to bear together; *karpos,* fruit.)

Corolla cylindric-campanulate, 6–9 mm. long. *S. vaccinioides.*
Corolla campanulate, 3–7 mm. long,
  Leaves pilose beneath; corolla 4–7 mm. long; nutlets 4–6 mm. long. *S. rivularis.*
  Leaves pilose on both sides; bracts puberulent; corolla 3–5 mm. long; nutlets 2.5–3 mm. long. *S. hesperius.*

Symphoricarpos hesperius G. N. Jones. Shrub 1–3 m. tall; branchlets gray shreddy, the bark pilosulous to glabrous; petioles 1–2 mm. long, sparsely pilose; blades 1–3 cm. long, oval, tapering to each end, acutish or obtusish, ciliate, entire, above dark green, glabrous or subglabrous, finely reticulate, below pale, reticulate, sparsely short pilose on the veins; bracts ovate, ciliate; calyx with 5 nearly regular teeth 1 mm. long, ciliolate; corolla 3–5 mm. long, campanulate, pink, the lobes about equaling the tube, sparsely pilose at base; stamens and style as long as the corolla; anthers 1 mm. long; berry 5–6 mm. long, subglobose. *S. albus,* var. *mollis* of St. John, not of Keck; *S. mollis* Nutt. ex T. & G., var. *hesperius* (G. N. Jones) Cronq. Blue Mts., Columbia Co., *Piper* 2,412.

Symphoricarpos rivularis Suksd. *Snowberry.* Shrub, 1–2 m. tall, often bushy branched; branchlets brown, smooth; branches gray, the thin bark exfoliating; winter buds ovate, brown, with several pairs of decussate, ovate, loose scales; petioles 2–6 mm. long, pilose or glabrate; blades 15–65 mm. long, oval to suborbicular, glabrous above, beneath paler and more or less pilose, entire or sinuately lobed especially on leaders; bracts ovate, glabrous or merely ciliate; calyx-limb 0.5–1 mm. long, broad campanulate, the teeth ovate, acute; corolla 4–7 mm. long, campanulate, pinkish, lobed ⅓–½ its length into ovate lobes, within hirsute; stamens and style included; anthers 1–2 mm. long, linear; berry 6–15 mm. in diameter, globose, white, pithy; seeds 3–4 mm. long, pale ellipsoid, compressed. *S. albus* (L.) Blake, var. *laevigatus* (Fern.) Blake. Prairies and open woods, *U. Son., A. T., A. T. T.*

The leaves contain a minute amount of the irritant alkaloid loturidine. Nevertheless, it is harmless and palatable, and is extensively grazed.

Symphoricarpos vaccinioides Rydb. Shrub 3–15 dm. tall, bushy branched, puberulent or more or less glabrate; branchlets yellowish brown shining; branches brown or gray, smooth, exfoliating; petioles 1–4 mm. long; blades 15–30 mm. long, elliptic or oval, entire, beneath pale green; calyx-limb 0.9–1.2 mm. long, the lobes broad deltoid; corolla 6–9 mm. long, narrow campanulate, pinkish to white, the lobes 1.5–2 mm. long, suborbicular, within pilosulous only at base of tube; style and stamens included; anthers 1.3–1.5 mm. long, linear; berry 8–10 mm. long, ellipsoid, white; seeds 4–6 mm. long, pale, smooth, ellipsoid, plano-convex. Rocky places, Blue Mts.; Thatuna Hills, Latah Co., *Dillon* 575, *A. T., Can.*

Forage of fair value. The berries are harmless.

VIBURNUM

Shrubs or small trees; leaves simple, commonly toothed, sometimes deeply lobed; flowers white, in flat compound cymes; calyx-lobes 5; corolla saucer-shaped, 5-lobed; stamens 5; stigmas 1–3; fruit a 1-celled, 1-seeded drupe with soft pulp and a flattened crustaceous stone. (The Lat. name, used by Virgil.)

Viburnum pauciflorum Raf. *Squash Berry.* Straggling shrub 1–3 m. tall; branchlets greenish, smooth; branches gray, smooth; winter buds ovate, smooth, brown; leaves opposite on stems, or in 2's on short lateral shoots; petioles 1–3 cm. long, mostly glabrous, glandless; blades 5–11 cm. long, suborbicular to broadly elliptic in outline, commonly with 3 shallow lobes at the apex, the base rounded or subcordate, hirsutulous or glabrate, the margin coarsely dentate; cyme 3–30-flowered, flat-topped or low, 1–3 cm. broad; calyx-teeth 0.2–0.3 mm. long, deltoid; corolla white, 3–5 mm. long, lobed half way into oval lobes; anthers 0.4–0.5 mm. long, ovoid; berries 8–10 mm. long, ovoid or ellipsoid; stone 7–8 mm. long, brown, ovate or oval. *V. eradiatum* (Oakes) House. Occasional, damp woods, *Can.*

VALERIANACEAE. Valerian Family

Herbs; leaves opposite or basal, without stipules; flowers perfect or dioecious, in panicled or clustered cymes; calyx-tube coherent with the ovary; corolla gamopetalous, tubular or funnelform, often irregular, 3–5-cleft; stamens distinct, 1–4, fewer than the corolla lobes, epipetalous; style slender, simple; ovary inferior, with one fertile 1-ovuled cell and two abortive or empty ones; fruit indehiscent; endosperm none. (Named from the genus *Valeriana.*)

Calyx of persistent, plumose bristles; leaves or some of them pinnately divided. *Valeriana.*
Calyx lobes minute or none; leaves entire or merely toothed,
  Corollas spurless, subtended by 2 entire bracts; fertile locule facing the axillary branch or the stronger one when two; ovary 3-celled; stigma 3-lobed. *Valerianella.*
  Corollas spurred, subtended by 6 bracts or a 6-parted bract; sterile locules (wings) facing the subtending bract; ovary 1-celled; stigmas 2-lobed. *Plectritis.*

PLECTRITIS

Annual herbs; leaves entire or with a few serrations or teeth near the base, short petioled, obovate-oblong, the upper sessile, oblong to linear; bracts linear-subulate, the bases of several fused, giving a palmate effect; inflorescence capitate or interrupted spicate; calyx obsolete; corolla 5-lobed or bilabiate, funnelform, with a prominent spur at the base of the throat or rarely merely saccate; stamens 3; stigma lobes 2 (or 3), flat reniform; ovary 1-celled; fruit a winged or wingless achene, the wings being outgrowths from the distal angles of the fertile cell; cotyledons accumbent or incumbent. (Name from Gr., *plektron,* spur.)

Fruit wings with narrow marginal groove; wings spreading at center. *P. macrocera,* var. *macrocera.*
Fruit wings grooveless; wings weakly incurved. Var. *Grayi.*

Plectritis macrocera T. & G., var. **macrocera.** Stems 1–6 dm. tall, glabrous or glandular pubescent in the inflorescence, with tufts of hairs at the nodes; leaves glabrous; diminishing upwards; lower leaves obovate and petiolate; upper leaves obovate to ovate or linear, obtuse to subacute, remotely serrate towards the base; inflorescence capitate or interrupted spicate, 2–3-verticillate; corolla 2–3.5 mm. long, white or pink-tinged, with a short thick spur; fruit 2–3 mm. long, yellowish, pubescent or glabrous, the broad wings about equally connivent at base and apex. Whitman Co., Wawawai, *Piper* 1,505.

Var. **Grayi** (Suksd.) Dyal. Plant 6–45 cm. tall, nearly glabrous; tap-root 0.5–2 mm. in diameter, yellowish; stem erect, simple or with weak branches, the internodes elongate, often much exceeding the leaves; basal leaves with petioles 3–30 mm. long, the blades 5–40 mm. long, oval or obovate; cauline leaves 5–70 mm. long, sessile, elliptic; cymes verticillate at the upper nodes;

bracts 3–5 mm. long, linear, puberulent; corolla 2.5–3 mm. long, funnelform, white to pinkish, the spur 0.5–1 mm. long; fruit 1.7–2 mm. long, orbicular-obcordate in outline, beaked, more or less puberulent, the heavy, incurved wings as broad as the fertile cell. *Aligera Grayi* Suksd. Open places, *U. Son., A. T.*

VALERIANA. Valerian

Tall perennial herbs, with strong-scented thickened roots; leaves simple or pinnate; calyx-lobes of several plumose bristles (*pappus*) which are rolled inward in flower but which unroll as the fruit matures; corolla commonly gibbous near the base, 5-lobed, nearly regular; stamens 3; abortive cells of the ovary small or obscure, obliterated in the achene-like fruit which is therefore 1-celled. (Name mediaeval Lat., perhaps from *valere* to be strong.)

Flowers dioecious; fruit ovate; root thickened, parsnip-like,
  Fruits puberulent. *V. edulis*, f. *edulis.*
  Fruits glabrous. Forma *glabra.*
Flowers perfect; fruit lanceolate; rhizome horizontal, little enlarged,
  Terminal leaflet of cauline leaves sinuate dentate; corolla tube subcylindric, with a reflexed, rounded spur. *V. sitchensis*, subsp. *sitchensis.*
  Terminal leaflet of cauline leaves entire; corolla tube funnelform, slightly gibbous. *V. occidentalis.*

Valeriana edulis Nutt. ex T. & G., forma **edulis.** *Tobacco Root; Ma-sa-wah.* Plant 2–9 dm. tall; tap-root thickened, 1–4 cm. in diameter, brown; basal leaves with petioles 4–13 cm. long; blades 4–20 cm. long, oblanceolate, entire or pinnately parted into 3–5 linear or linear-oblanceolate lobes, somewhat puberulent or glabrous, the margin ciliolate; cauline leaves reduced, usually 1 pair sessile, pinnately 5–11-parted, the divisions linear; panicle 1–3 dm. long, loose, puberulent or glabrate; bracts 2–4 mm. long, lanceolate; staminate flowers sessile, greenish yellow, broad funnelform, the tube 1 mm. long, the rounded lobes 1 mm. long; anthers 0.5–0.8 mm. long, oblong; pistillate flowers greenish yellow, funnelform, the tube 1.3 mm. long, narrow tubular at base, the lobes 0.5–0.7 mm. long, ovate; stigmas bifid, exserted; pappus 5–6 mm. long, the bristles long plumose, membranous margined and united at base; fruit 4–5 mm. long, ovate, puberulent, 1-ribbed on one side, 5-ribbed on the other. *V. ceratophylla* (Hook.) Piper, not of H.B.K. Meadows, *A. T.*

The pungent roots of this ill-smelling plant were used as a food root by the Indians. The baked roots are nutritious. They are brown and smell and taste like chewing tobacco.

Forma **glabra** St John. Differing in having the fruits glabrous. *V. ceratophylla* (Hook.) Piper, not of H.B.K. (1818).

Washington: In low ground, Pullman, July 3, 1893, *C. V. Piper* 1506; only in rich meadows, Pullman, May, 1897, *A. D. E. Elmer* 822 (type) *A. T.*

Valeriana occidentalis Heller. Plant 2–8 dm. tall, nearly glabrous; rhizome 2–5 mm. in diameter, creeping; roots fleshy; basal leaves with petioles 3–10 cm. long, the blades 3–8 cm. long, ovate to lanceolate, thin, entire; cauline leaves in 3 pairs, the lower similar to the basal or somewhat lobed, the upper sessile, reduced, pinnate, 3–9-foliolate; terminal leaflet 2–5 cm. long, linear-lanceolate; nodes hirsutulous; panicle in flower compact, round-topped, 2–3 cm. wide, in fruit 6–18 cm. long, loose, narrow, elongate; bracts 3–4 mm. long, linear; corolla white, the tube 2 mm. long, the lobes 1–1.2 mm. long, oval; stamens and style exserted; anthers 0.8–0.9 mm. long, oblong; pappus bristles 6–7 mm. long, long plumose, membranous margined and connate at base; fruits 5–6 mm. long, flattened, the margins ribbed, one side 1-ribbed, the other 3-ribbed. Prairies, Craig Mts., and north to Benewah Co., *A. T. T.*

The type was from 3500 ft., Craig Mts., *Heller* 2353.

Valeriana sitchensis Bong., subsp. **sitchensis.** *Heliotrope.* Plant 3–10 dm. tall, nearly glabrous; rhizome 2–5 mm. in diameter, creeping; roots elongate somewhat fleshy; basal leaves with petioles 2–10 cm. long, the blades 15–50 mm. long, simple, 3-lobed, 3-foliolate, entire or denticulate; cauline leaves 2–4 pairs, the lower petioles 3–7 cm. long, the upper sessile; blades 3–5-foliolate; terminal leaflet 3.5–8 cm. long, ovate to oblanceolate, the uppermost leaves with narrower and reduced leaflets; nodes hirsutulous; cymes round-topped, in flower 2–5 cm. broad; bracts 3–10 mm. long, lanceolate; corolla white or lavender-tinged, the tube 4–5 mm. long, the lobes 1.5–2 mm. long, ovate; style and stamens long exserted; anthers 1–1.2 mm. long, oblong to oval; pappus bristles numerous, 4–7 mm. long, long plumose, the base lanceolate and connate; fruit 4–6 mm. long, glabrous, one side 1-ribbed, the other with 3 close ribs, the margins ribbed. Meadows, Blue Mts., *Huds.*

The plant has a sour odor.

VALERIANELLA

Low annual dichotomously branched herbs; leaves tender, rather succulent; flowers small, bracted, whitish, cymosely clustered; calyx merely toothed or teeth obsolete; corolla funnelform, manifestly or obscurely 2-lipped; stamens 3, rarely 2; ovary 3-celled, the two lateral ones sterile; fruit 1-celled, 1-seeded. (Name a diminutive of *Valeriana* the valerian.)

Valerianella Locusta (L.) Betcke. *Corn Salad.* Plant 15–30 cm. tall, sparsely hirsute, especially at the nodes; tap-root 0.2–2 mm. in diameter; stem with numerous divergent branches; basal and lower leaves with petioles 5–20 mm. long, the blades 8–20 mm. long, spatulate to elliptic, entire; upper leaves sessile, 15–35 mm. long, oblanceolate to oblong, entire or dentate; cymes 5–20 mm. broad, rounded; bracts leafy, opposite, ciliate; corolla 1.3–2 mm. long, funnelform, white below, the oval lobes bluish; fruit 2 mm. long, sparsely puberulent, oval somewhat compressed, ridged on the sides, the 2 sterile cells corky thickened, connate, the apex apiculate. *V. olitoria* (L.) Poll. River bank, Wawawai Ferry, *St. John* 6,085.

Introduced from Europe, and doubtless escaped from cultivation. The young rosettes are eaten as a salad. They are mild and pleasant to taste.

DIPSACACEAE. Teasel Family

Herbs; leaves opposite or whorled, without stipules; flowers in dense heads, surrounded by an involucre as in the Compositae; calyx-tube adherent to the ovary; corolla epigynous, 2–5-lobed; stamens 2–4, distinct, on the corolla tube and alternate with its lobes; ovary inferior, 1-celled; ovule 1; fruit an achene with persistent calyx-lobes; endosperm fleshy. (Named for the genus *Dipsacus.*)

DIPSACUS. Teasel

Rough-hairy or prickly tall biennial or perennial herbs; leaves large, opposite, the bases somewhat united into a cup; flowers in dense terminal peduncled oblong heads; bracts of the involucre and scales of the receptacle rigid or spiny-pointed; calyx 4-toothed or -lobed; corolla oblique, 2-lipped, 4-lobed; stamens 4; stigma oblique or lateral; achene free or adherent to the involucel. (Name Gr., *dipsen,* to thirst.)

Dipsacus sylvestris Huds. *Wild Teasel.* Biennial, 1–2 m. high, the stems and midribs armed with stout prickles; lower leaves lanceolate, obtuse, crenate, rarely cleft at base, 15–30 cm. long; upper leaves sessile, often cuneate, acuminate, entire; heads ovoid, becoming cylindric, 5–10 cm. long; involucre of linear cuspidate prickly bracts, some of which are longer than the head; bracts of the receptacle ovate, armed with long straight awns; flowers lilac, pilose within, puberulent without, 1 cm. long. First flowers appearing in a ring around the middle of the head, then flowering in two rings, one advancing up and the other down. Introduced from Europe, dominant in moist thickets and river flats, *A. T.*

CUCURBITACEAE. Gourd Family

Herbs or rarely shrubs, climbing; tendrils usually present, spiral; flowers regular, monoecious, dioecious, or rarely perfect; staminate flowers with calyx tubular; corolla polypetalous or gamopetalous; stamens usually 3, free or united, one anther 1-celled, the others 2-celled; pistillate flower with calyx-tube adnate to the ovary; placentae parietal, often meeting in middle, often 3; ovules numerous, styles 1–3; seeds often flattened, without endosperm. (Named from the genus *Cucurbita,* squash.)

ECHINOCYSTIS. Wild Balsam Apple

Climbing herbs; tendrils branched; leaves lobed or angled; flowers monoecious whitish; staminate calyx 5–6-lobed; corolla 5–6 parted; stamens 3, the anthers more or less coherent; pistillate flowers with ovary 2-celled; ovules 2 in each cell; stigma hemispheric or lobed; fruit fleshy, becoming dry, spiny, dehiscent at apex. (Name from Gr., *echinos,* hedgehog; *kustis,* bladder.)

Echinocystis lobata (Michx.) T. & G. *Wild Balsam Apple.* Roots slender, almost fibrous; stems 3–8 m. long, climbing or sprawling, glabrate; nodes villous; cotyledons oval, 3-nerved; leaves alternate, the petioles 1.5–9 cm. long, sparsely pilose or glabrate; blades 4–15 cm. long, more or less scabrous puberulous, orbicular-reniform in outline, 3–7-lobed about half way, the lobes deltoid-lanceolate, acute or acuminate, dentate or denticulate; staminate panicles axillary, somewhat pilosulous; peduncle 1–5 cm. long, rhachis 5–30 cm. long; pedicels 2–8 mm. long; calyx-tube 1–1.5 mm. long, semiorbicular, the teeth 1.5–2 mm. long, linear; anthers 1 mm. long, united; pistillate flowers 1–2, axillary; calyx-tube 0.5 mm. long, saucer-shaped, the lobes 2 mm. long, linear; corolla tube 1 mm. long campanulate, the lobes 5–6 mm. long, lanceolate, acuminate, glandular puberulent; fruit 3–5 mm. long, broad ellipsoid, the slender spines 5–8 mm. long, seeds 15–18 mm. long, flat, narrowly elliptic, brown, mottled. Weed, introduced from eastern N. Am. *Micrampelis lobata* (Michx.) Greene. Stream bank, Albion, *Suksdorf,* Dec. 13, 1925, *A. T.*

CAMPANULACEAE. Bellflower Family

Herbs, rarely shrubs or trees, with milky juice; leaves alternate, simple, without stipules; flowers regular, perfect, generally blue, showy, scattered; calyx-tube adherent to the ovary, the lobes 3–10; corolla 5-lobed, bell-shaped or tubular, valvate; stamens as many as the corolla lobes and alternate with them, usually free from the corolla; anthers distinct; style 1, the upper portion provided with pollen-collecting hairs; stigmas 2 or more; ovary 2–10-celled, with axile placenta; capsule 2–several-celled, many-seeded; endosperm fleshy. (Named from the genus *Campanula.*)

Capsule dehiscing by an apical pore, within the calyx-lobes; inflorescence cymose. *Githopsis.*
Capsule dehiscing laterally, below the calyx-lobes; inflorescence not cymose,
  Flowers long pediceled or peduncled; flowers all alike and not cleistogamous. *Campanula.*
  Flowers sessile, axillary, dimorphic, the earlier ones cleistogamous,
    Capsules opening by irregular fissures; petaliferous flowers with broadly ovate, dentate calyx-lobes and globose-ovoid capsule; open corollas not divided as far as the middle. *Heterocodon.*
    Capsule dehiscing by rounded or oval, uprolling valves; petaliferous flowers with lance-linear, entire lobes and ovoid to cylindric-fusiform capsule; open corollas divided as far as the middle. *Triodanis.*

CAMPANULA. Bellflower

Chiefly perennial herbs; flowers perfect, all alike, showy; leaves alternate or basal; calyx-lobes narrow; corolla campanulate or nearly so, 5-lobed or cleft; stamens 5; filaments dilated at base; capsule turbinate to obpyramidal, 3–5-celled, opening on the side or near the base by 3–5 small uplifted valves leaving round perforations, many-seeded. (Name from Lat., *campana,* bell; *ula,* little one.)

Campanula rotundifolia L. *Scotch Bluebell.* Perennial, 15–100 cm. tall, glabrous or puberulent; tap-root 1–4 mm. in diameter; basal leaves with petioles 1–5 cm. long, the earliest blades 5–35 mm. long, orbicular or ovate, mostly cordate at base, entire or dentate, often withered at anthesis; cauline leaves extremely variable, from lanceolate to linear; flowers nodding, single or several in a loose raceme; pedicels 1–4 cm. long, with linear bracts; calyx-lobes 3–12 mm. long, linear; corolla 12–22 mm. long, blue, the ovate lobes half as long as the tube; anthers 3–4 mm. long, linear, finally spiralling; style included; capsule 4–6 mm. long, obovoid, truncate; seeds 0.8–1 mm. long, ellipsoid, brown, dull. *C. petiolata* A. DC.; *C. intercedens* Witasek. Meadows or rocky slopes, *A. T., A. T. T.*

GITHOPSIS

Low annual herbs, flowers all alike; calyx with a 10-ribbed tube and 5 long and narrow foliaceous lobes; corolla tubular-campanulate, 5-lobed; filaments short, dilated at the base; stigmas 3; ovary 3-celled; capsule strongly ribbed, crowned with the rigid calyx-lobes about as long or longer; seeds very numerous. (Name from *Githago,* corn cockle; *opsis,* Gr., appearance.)

Stem glabrous or only the angles merely sparsely hirsutulous. *G. specularioides, f. specularioides.*

Stems hirsute throughout or nearly throughout. *Forma hirsuta.*

Githopsis specularioides Nutt., forma **specularioides.** Plant 5–20 cm. tall, simple or branched, glabrous or sparsely hirsutulous only on the angles of the stem; roots fibrous; leaves 3–11 mm. long, sessile, narrowly elliptic, coarsely serrate, thick margined; flowers terminal, axillary, or oppositifolious, forming a loose, elongate cyme; lateral peduncles 1–4 cm. long; calyx-lobes 7–15 mm. long, rigid, linear acute, entire or dentate; corolla 8–11 mm. long, blue, the ovate lobes a little shorter than the tube; capsule 8–11 mm. long, clavate, 10-ribbed; seeds 0.3–0.4 mm. long, ellipsoid, brown, shining. *G. specularioides,* subsp. *candida* Ewan. Gravelly open places, Waitsburg, Wawawai, *U. Son., A. T.*

The original description says, "the upper part of the plant smooth." Also one of the Nuttall specimens in the Gray Herbarium matches this description, so it is apparent that Dr. Jepson has misinterpreted the type.

Forma **hirsuta** (Nutt.) St. John. Differs only in being hirsute throughout or nearly throughout. *G. Specularioides* Nutt., *β. hirsuta* Nutt.; *G. specularioides* sensu Jepson, not Nutt. Stony slopes, Waitsburg, *Horner* 158 in part, *A. T.*

As this nearly always grows intermingled with the species, and as it differs only in being more pubescent, it is logical to reduce it to a forma of the species.

HETEROCODON

Delicate little annual; flowers solitary, of two sorts, the lower and earlier ones with merely rudimentary corollas and self-fertilized in the bud; calyx with an obovate or inversely pyramidal tube much shorter than the foliaceous lobes, these broadly ovate, sharply toothed, veiny, 3 or 4 in the earlier; later flowers with 5 calyx-lobes; corolla short-campanulate, 5-lobed; stigmas 3, short; capsule 3-celled, 3-angled; seeds numerous. (Name Gr., *heteros,* different; *kodon,* bell.)

Heterocodon rariflorum Nutt. Plant 5–35 cm. tall, simple to diffusely branched; roots fibrous; stem sparsely hispid on the angles or glabrate; leaves 3–10 mm. long, sessile, broadly oval to suborbicular, the base cordate and clasping, the margin serrate, hispid ciliate; cleistogamous flowers with calyx-lobes 2–2.5 mm. long, hispid ciliate; capsule 2–3 mm. long, obovoid, hispid, several ribbed; seeds 0.5–0.6 mm. long, ellipsoid, brown, polished; later flowers with calyx-lobes 3–5 mm. long; corolla 5–7 mm. long, blue, the lobes ovate; subacute, half the length of the tube; anthers 1–1.2 mm. long, linear; style included; capsule 3–4 mm. long; seeds 0.3–0.4 mm. long, narrowly ellipsoid, brown, polished. *Specularia rariflora* (Nutt.) McVaugh. Gravelly places, *A. T., A. T. T.*

The gender, established in the original publication of this genus, is neuter.

TRIODANIS

Annuals; cauline leaves sessile or clasping; inflorescence spiciform; flowers blue or purple, axillary, of two kinds, the earlier smaller, close-fertilized, with a rudimentary corolla which never opens, with mostly 3–4 calyx lobes; the later with 5 calyx lobes; corolla lobes almost rotate, 5-lobed; filaments abruptly dilated and ciliate at base; ovary 3- (or 2-) celled; capsule ovoid to cylindric-fusiform, opening by valves or irregular ruptures usually over partitions. (Name from Gr., *triodos,* a meeting of three roads.)

Triodanis perfoliata (L.) Nieuwl. *Venus' Looking-glass.* Plant 1–8 dm. tall, somewhat hispidulous throughout; roots fibrous; stem simple or few branched from near the base; leaves 3–20 mm. long, ovate to orbicular-cordate, crenate; cleistogamous flowers with calyx-lobes 2–3 mm. long, oval, dentate, acute; capsules 2.5–3 mm. long, ovoid; seeds 0.2–0.3 mm. long, broadly ellipsiod, brown, smooth; later flowers opening, the calyx-lobes 3.5–7 mm. long; corolla blue-violet, the tube 3–5 mm. long, campanulate, the lobes 5–8 mm. long, ovate; anthers 2.8–3 mm. long, linear; style puberulent towards the apex; capsule 5–7 mm. long; seeds 0.2–0.3 mm. long, broadly ellipsoid, compressed, brown, shining. *Specularia perfoliata* (L.) A. DC. Open places or open woods, *U. Son., A. T., A. T. T*

LOBELIACEAE. Lobelia Family

Herbs, shrubs, or trees; juice often milky; leaves alternate, simple; stipules absent; flowers perfect or rarely unisexual, irregular; calyx adherent to the ovary, 5-lobed; corolla epigynous, 1–2-lipped, of 5 variously united, valvate petals; stamens 5, alternate with the corolla lobes, free or attached to the corolla; filaments free at base; anthers cohering into a tube around the style; ovary more or less inferior, 2–3-celled; style simple, or 2-lobed and with a hairy ring; ovules numerous, axile; fruit a berry or capsule; endosperm copious.

Corolla strongly bilabiate; flowers all alike; pod dehiscing by 1–3 longitudinal slits. *Downingia.*
Corolla scarcely bilabiate; flowers of two kinds, the immersed ones cleistogamous; pod bursting on one side. *Howellia.*

DOWNINGIA

Low annuals, rather succulent and tender; leaves sessile, narrow, entire, the upper reduced to bracts; flowers axillary, sessile;

calyx-tube adherent to the ovary, very long and slender, 3-sided, usually twisted; corolla 2-lipped, with a very short tube; filaments and anthers both united into a tube; capsule long and slender, early becoming 1-celled. (Named for *Andrew Jackson Downing* of New York, an American landscape gardener.)

Corolla 8–14 mm. long; placentae parietal. *D. elegans.*
Corolla 4–7 mm. long; placentae axile. *D. laeta.*

Downingia elegans (Dougl. ex Lindl.) Torr. *Blue Calico Flower.* Plant 5–35 cm. tall, often bushy branched; tap-root white, spongy; leaves 8–26 mm. long, ovate to linear-lanceolate; inflorescence as much as 28 cm. in length, papillose-puberulent, leafy bracted; calyx-lobes 4–10 mm. long, linear, unequal; corolla deep navy blue, with a white, central, chevron-like spot on the lower lip; corolla tube 2–4 mm. long, campanulate; upper lip divided ⅔ its length into lance-linear teeth; lower lip 6–10 mm. long, suborbicular in outline, coarsely tridentate; stamens as long as the corolla, the filament tube glabrous; anther tube 2.5–3 mm. long, arcuate, the lower anthers penicillate and with 2 bristles; capsule 2–5 cm. long, upcurved, tapering; seeds 0.9–1.1 mm. long, fusiform, brown, shining. *Bolelia elegans* (Dougl.) Greene. Exsiccated muddy flats, *A. T., A. T. T.*

A plant with abundant lovely flowers, worthy of frequent cultivation as a border plant. First collected by Douglas "on the plains of the Columbia, near Wallawallah river."

Downingia laeta (Greene) Greene. Slender annuals 5–30 cm. tall, glabrous; leaves 5–25 mm. long, 0.5–2 mm. wide; inflorescence 1–10-flowered; bracts 7–22 mm. long, elliptic to ovate, obtuse or subacute; calyx lobes 3–7 (–9) mm. long, elliptic, the apex rounded or subacute; corolla light blue or purplish, the tube yellow at base, the lower lip with the central area white or yellow and at base with a transverse band of purple or only purple spots, the 3 lobes of the lower lip 1.5–3.5 mm. long, oblong, acute; anther tube 1.3–2 mm. long, capsule 21–43 mm. long, fusiform, terete; seeds not or but slightly twisted. *D. brachyantha* (Rydb.) Nels. & Macbr. Muddy ditch, Plummer, Ida., *Daubenmire* 38,211, *A. T. T.*

HOWELLIA

Aquatic annual herbs; with flaccid stem and narrow leaves; flowers small and in part cleistogamous; hypanthium linear or linear-clavate; calyx lobes 5, narrow; corolla of emersed flowers with its short tube split almost to the base on the upper side, the 5 lobes subequal; anthers 5, unequal, monadelphous; anthers oval, the two smaller with 3 bristles, the three larger naked; ovary 1-celled, with 2 parietal placentae, few seeded. (Named for Thomas Howell, of Portland, Ore.)

Howellia aquatilis Gray. Aquatic, mostly submerged; stems 3–7 dm. long, submerged leaves alternate or verticillate, 1–5 cm. long, less than 1 mm. wide, flaccid, filiform, entire or with a few slender teeth; early flowers submersed, cleistogamous, small, with rudimentary corolla; on emersed stem tips are produced open flowers, the sepals nearly distinct, subequal; corolla 1.5–3 mm. long, whitish or pale lavender; capsule 6–8 mm. long, clavate; seeds 4 mm. long, linear-oblong, smooth. Lake Tesemine (Spirit Lake, Ida.), *Sandberg MacDougal & Heller* 699.

COMPOSITAE. Sunflower Family

Annual, biennial or perennial herbs, shrubs, or trees; leaves without stipules; flowers in a close head on a common receptacle, surrounded by one or more rows of bracts (the *involucre*); heads 1–many, *discoid* when all the flowers bear tubular corollas, *ligulate* when the corollas are all strap-shaped, *radiate* when the outer corollas are strap-shaped and the inner tubular, in which case the outer are *ray flowers* and the inner *disk flowers;* receptacle often with bracts or scales (*chaff*), each subtending a flower; calyx gamosepalous, its tube wholly adherent to the ovary, its limb (*pappus*) none or cup-shaped or developed into teeth, scales, awns or capillary bristles; corollas alike in all the flowers of the head or dissimilar, either *tubular* or strap-shaped (*ligulate*), gamopetalous, epigynous; stamens 5, epipetalous, their anthers usually united into a tube (*syngenesious*); style 2-cleft at the apex or in sterile flowers usually entire; ovary bicarpellary, inferior, 1-celled, 1-ovuled, with basal placenta; fruit an achene sometimes surmounted by the pappus, often compressed at right angles to the subtending chaff (*laterally compressed*) or compressed parallel to the chaff (*obcompressed*). Name Lat., *compositus,* brought together.)

Conspectus of the Tribes of the Compositae

A. Plants without milky sap; heads discoid or radiate,

1. Heads discoid; branches of the style thickened upward or club-shaped, obtuse, minutely and uniformly pubescent; flowers never yellow; leaves opposite or alternate. — Eupatorieae.
   Brickellia, Eupatorium.
2. Heads discoid or radiate; anthers not caudate at base; branches of style in disk flowers flat, smooth up to end of stigmatic lines, above prolonged into a flattened lance-shaped or triangular appendage, hairy on the margin or throughout; receptacle naked; leaves alternate. — Astereae.
   Aster, Chrysopsis, Chrysothamnus, Erigeron, Gutierrezia, Haplopappus, Solidago.
3. Heads discoid; pistillate flowers mostly filiform and truncate; style branches with unappendaged, obtuse or truncate, naked tips; anthers sagittate, the basal lobes attenuate into tails; pappus capillary or none; leaves alternate (but in *Psilocarphus* opposite). — Inuleae.
   Adenocaulon, Anaphalis, Antennaria, Gnaphalium, Psilocarphus.
4. Heads radiate (or rarely discoid); style branches truncate or hairy appendaged; pappus of scales, awns, or cup-like, never capillary; anthers not caudate; involucre of one to several series of bracts, none enfolding the ray achenes; chaff copious; leaves mostly opposite or basal. — Heliantheae.

Balsamorhiza, Bidens, Coreposis, Grindelia, Helianthella, Helianthus, Rudbeckia, Wyethia.

5. Heads radiate; anthers not caudate; involucre of 1 series of bracts, each partly or wholly enclosing a ray achene; chaff commonly a single series between the disk and ray flowers, often united into a cup, or scattered among the disk flowers; disk achenes with pappus of awns, scales, or none; leaves alternate or opposite. MADIEAE.
   Blepharipappus, Hemizonia, Lagophylla, Madia.
6. Heads discoid, unisexual, white or greenish; anthers nearly or quite distinct; pappus none; staminate heads above, in a raceme or spike; receptacle with chaff; pistillate heads few, usually axillary; the corolla none or a mere rudiment; fruit usually a bur; leaves usually alternate. Heads monoecious in *Iva.* AMBROSIEAE.
   Ambrosia, Franseria, Iva, Xanthium.
7. Heads radiate or discoid; pappus a row of several chaffy or bristly-dissected scales; receptacle usually naked; bracts of involucre in one or few series, not or but little imbricated; anthers not caudate; flowers commonly yellow; leaves alternate or opposite. HELENIEAE.
   Chaenactis, Eriophyllum, Gaillardia, Helenium, Rigiopappus.
8. Heads radiate or discoid; anthers not caudate; involucre of dry, scarious, imbricated bracts; pappus a short crown or none; receptacle naked or chaffy; flowers white, yellow, or greenish; leaves strong scented, alternate, usually much divided. ANTHEMIDEAE.
   Achillea, Anthemis, Artemisia, Chrysanthemum, Matricaria, Tanacetum.
9. Heads radiate or discoid; involucre of 1–2 rows of bracts, little or not imbricated, not scarious; receptacle naked; anthers not caudate; pappus capillary; flowers yellow (except in *Petasites*); mostly herbs with alternate or basal leaves. SENECIONEAE.
   Arnica, Crocidium, Petasites, Senecio, Tetradymia.
10. Heads discoid; involucre much imbricated, the bracts usually attenuate into spines; receptacle bristly or hairy; flowers perfect, deeply cleft; anthers caudate at base or long appendaged at tip; style branches short or united, obtuse, unappendaged, smooth, or with a pubescent ring below; pappus mostly bristly or plumose; mostly spiny herbs with alternate leaves. CYNAREAE.
    Arctium, Centaurea, Cirsium, Cnicus, Silybum.

A'. Plants with milky sap; only ray flowers present,

11. Involucre of one or several imbricated series; receptacle flat or nearly so; corollas strap-shaped, 5- or several-toothed at apex; anthers auricled or sagittate at base, the tip not appendaged; style branches filiform, naked, stigmatic along

the inner side or only towards the base; pappus various or none. CICHORIEAE.
Agoseris, Chondrilla, Cichorium, Crepis, Hieracium Hypochaeris, Lactuca, Lapsana, Microseris, Sonchus, Stephanomeria, Taraxacum, Tragopogon.

KEY TO THE GENERA OF THE COMPOSITAE

Corollas tubular in all the flowers of the head or strap-shaped in only the marginal ones; plants without milky juice,
Ray flowers none; corollas all tubular,
Flowers of the heads not all alike, some heads having imperfect flowers,
Perfect and imperfect flowers in the same head,
Marginal flowers neutral and sterile, resembling ray flowers,
Heads not subtended by bristly leaves. *Centaurea.*
Heads sessile, subtended by bristly leaves. *Cnicus.*
Marginal flowers perfect or pistillate and fertile, not resembling ray flowers,
Pappus of capillary bristles,
Involucral bracts in one row,
Herbs; involucral bracts numerous. *Petasites.*
Shrubs; involucre of 4–6 concave bracts. *Tetradymia.*
Involucral bracts in several rows. *Gnaphalium.*
Pappus a short crown or none,
Heads large, 1.5–2 cm. in diameter. *Centaurea.*
Heads small, 5 mm. or less in diameter,
Lower leaves opposite. *Iva.*
Lower leaves alternate,
Leaves entire or nearly so, ovate. *Adenocaulon.*
Leaves incised, or if entire, lanceolate or linear,
Pappus none; heads racemose or corymbed. *Artemisia.*
Pappus a short crown; heads corymbed. *Tanacetum.*
Staminate and pistillate flowers in different heads,
Pappus capillary; fertile involucre not bur-like,
Leaves prickly; heads large. *Cirsium.*
Leaves not prickly; heads small,
Pappus of staminate flowers either club-shaped or barbed at the apex. *Antennaria.*
Pappus of all the flowers alike and neither club-shaped nor barbed. *Anaphalis.*
Pappus none, fertile involucre bur-like,
Bracts of staminate heads separate; bur large, with many prickles. *Xanthium.*
Bracts of staminate heads united; bur small, with 1–4 spines,
Spines in 1 row; pistillate flowers 1 in a head. *Ambrosia.*

Spines in several rows; pistillate flowers 1–4 in each head. *Franseria.*
Flowers of the heads all perfect and alike,
Pappus of separate capillary bristles,
Flowers whitish,
Achenes 10-ribbed. *Brickellia.*
Achenes 5-angled. *Eupatorium.*
Flowers yellowish or brownish,
Involucral bracts in 3–4 rows,
Heads 2–3 mm. in diameter. *Chrysothamnus.*
Heads 20–30 mm. in diameter. *Haplopappus.*
Involucral bracts in 1 row or nearly so,
Involucre campanulate. *Senecio.*
Involucre hemispheric to rotate. *Erigeron.*
Pappus not of separate capillary bristles,
Pappus of numerous bristles united into a ring at the base,
Pappus plumose. *Cirsium.*
Pappus merely rough or denticulate. *Carduus.*
Pappus not of bristles united into a ring,
Pappus of rigid, backwardly-barbed awns. *Bidens.*
Pappus not of backwardly-barbed awns,
Pappus of hyaline or chaffy scales. *Chaenactis.*
Pappus none, or a minute crown, or of short rough bristles,
Pappus of short bristles,
Leaves spineless; filaments separate. *Arctium.*
Leaves spiny margined; filaments united below. *Silybum.*
Pappus none, or a minute crown,
Involucre scarious. *Matricaria.*
Involucre not scarious,
Heads solitary,
Each flower with its own involucre of imbricate scales; flowers blue. *Echinops.*
All flowers in a single involucre; flowers yellow or brown. *Rudbeckia.*
Heads in clusters,
Disk flowers 1–5. *Madia.*
Disk flowers numerous. *Psilocarphus.*
Ray flowers present, at least some of the marginal flowers having strap-shaped corollas,
Pappus none,
Involucre more or less scarious,
Receptacle naked. *Chrysanthemum.*
Receptacle chaffy,
Heads solitary; rays 10–20. *Anthemis.*
Heads corymbed; rays 4–5. *Achillea.*
Involucre not scarious,
Achenes all laterally compressed. *Madia.*
Achenes or at least part of them turgid or obcompressed,

Involucral bracts not at all enclosing the ray achenes. *Balsamorhiza.*
Involucral bracts at least partly enclosing the ray achenes,
Ray achenes turgid, each partly enclosed by the involucral bract. *Hemizonia.*
Ray achenes usually obcompressed, each wholly enclosed by the base of the involucral bract. *Lagophylla.*
Pappus present, at least in the disk flowers,
Pappus of capillary bristles,
Ray flowers not yellow,
Low shrub. *Haplopappus.*
Herbs, woody only at the base,
Bracts of the involucre in many series, their tips spreading. *Aster.*
Bracts of the involucre in 1–5 series, their tips erect,
Flowering shoots without normal leaves, polygamo-dioecious. *Petasites.*
Flowering shoots with normal leaves, the heads all alike,
Bracts in 1–2 series; rays usually narrow and numerous,
Rays in one row. *Senecio.*
Rays in several rows. *Erigeron.*
Bracts in 2–5 series; rays broader, less numerous. *Aster.*
Ray flowers yellow,
Pappus double, the outer row very short. *Chrysopsis.*
Pappus in one row,
Heads in panicles. *Solidago.*
Heads solitary or in corymbs,
Leaves all or mostly opposite. *Arnica.*
Leaves alternate,
Heads large, 1.5 cm. or more broad. *Haplopappus.*
Heads 1 cm. or less broad,
Involucre campanulate. *Senecio.*
Involucre hemispheric or broader,
Bracts of involucre 8–10, broad. *Crocidium.*
Bracts of involucre numerous, narrow. *Erigeron.*
Pappus not of capillary bristles,
Receptacle chaffy,
Pappus of scales or awns,
Scales of the pappus 12–20, thin, fringed. *Blepharipappus.*
Scales of the pappus awn-like, not chaffy,
Awns 2–4, retrorsely barbed. *Bidens.*
Awns 2, subulate. *Coreopsis.*
Pappus crown-like or of short chaffy teeth or awns,
Rays pistillate and fertile. *Wyethia.*
Rays neutral,
Chaff bristle-like. *Gaillardia.*

Chaff scale-like,
Achenes flat and thin. *Helianthella.*
Achenes prismatic. *Helianthus.*
Receptacle not chaffy,
Involucral bracts in several rows,
Involucre hemispheric; pappus of rigid, deciduous awns. *Grindelia.*
Involucre ovoid or narrower; pappus of persistent scales. *Gutierrezia.*
Involucral bracts in 1 row,
Achenes linear, more or less 4-angled,
Involucral bracts united, *Eriophyllum.*
Involucral bracts separate. *Rigiopappus.*
Achenes obpyramidal. *Helenium.*
Corollas strap-shaped in all the flowers of the head; plants with milky juice,
Pappus of scale-like or plumose bristles,
Branches of the plumose pappus interwoven. *Tragopogon.*
Pappus not as above,
Flowers not yellow,
Flowers pink or white; pappus plumose. *Stephanomeria.*
Flowers blue; pappus a crown of short scales. *Cichorium.*
Flowers yellow,
Heads nodding when young; pappus scales 15–20, each with a very plumose awn. *Microseris.*
Heads erect, even when young; pappus scales less than 15, or if more than 15 awnless,
Pappus plumose. *Hypochaeris.*
Pappus not plumose. *Microseris.*
Pappus of capillary bristles, never plumose, or pappus none,
Pappus none; rays yellow, drying white. *Lapsana.*
Pappus present,
Heads solitary; leaves all basal,
Achenes muricate or spinulose at the apex. *Taraxacum.*
Achenes smooth at the apex. *Agoseris.*
Heads several; leaves not all basal,
Achenes flattened,
Achenes beaked, or if short-beaked flowers not yellow. *Lactuca.*
Achenes beakless; flowers yellow. *Sonchus.*
Achenes terete, cylindric, or prismatic,
Achenes not muricate above,
Pappus copious, white and soft. *Crepis.*
Pappus a single row of rough tawny bristles. *Hieracium.*
Achenes muricate above. *Chondrilla.*

ACHILLEA. Yarrow; Milfoil

Usually perennial herbs, rather strong-scented; leaves alternate, serrate or pinnately dissected; heads small, in corymbs, many flowered; involucre campanulate, the bracts imbricate in a few series; receptacle flattish to conical, with thin chaff; flowers yellow, white or sometimes rose-colored, all fertile; ray flowers few or several, mostly short or broad, pistillate, fertile; disk flowers perfect, fertile, yellow, 5-lobed; pappus none; achenes oblong or ovate, obcompressed, surrounded by a narrow and cartilaginous margin. (Name from ancient Gr., called *achilleios* by Hippokrates, in honor of *Achilles.*)

Achillea Millefolium L., var. **lanulosa** (Nutt.) Piper. *Tansy.* Perennial, 15–70 cm. tall, villous throughout; rhizomes 1–3 mm. in diameter, woody, brown; stem erect, simple or branched above, striate; basal and lower leaves with petioles 1–6 cm. long, the blades 3–20 cm. long, oblanceolate in outline, tripinnatifid, the ultimate segments linear, sharp pointed, the rhachis scarcely margined; cymes many flowered, 2–8 cm. broad; peduncles 1–4 mm. long; involucre sparsely villous at base, the bracts greenish or yellowish, the margins pale brownish, more or less hyaline and ciliate, the outer ovate, the inner 4–5 mm. long, elliptic; ray corollas 3–5 mm. long, commonly 5 in number; achenes 2–2.3 mm. long, pale, smooth; chaff 3–4 mm. long, oblanceolate; disk flowers 2 mm. long, funnelform, glandular atomiferous, the lobes ovate; achenes 2 mm. long, thick margined. *A. lanulosa* Nutt. Open or shaded places, *U. Son* to *Huds.*

Reported as somewhat poisonous, yet the herbage is grazed by sheep.

Clausen, Keck, and Hiesey (1939) maintain this as a species, but the whole circumpolar group is still undergoing cytotaxonomic study.

ADENOCAULON

Slender perennial herbs, with alternate petioled leaves, green above, white-woolly beneath; heads few, small, 5–10-flowered, stalked glandular, in a loose panicle; involucral bracts in one row; receptacle flat, naked; flowers white, all tubular, the marginal pistillate, fertile, the central perfect but sterile; achenes elongated at maturity, club-shaped; pappus none. (Name from Gr. *adenos,* gland; *kaulon,* stem.)

Adenocaulon bicolor Hook. *Silver-green.* Plant 2–10 dm. tall, delicate; rootstock 2–8 mm. in diameter, often tuberous, brownish; stem erect, floccose tomentose below, branched, and capitate glandular hirsute above; leaves only at the base and the lower part of the stem, the petioles 3–15 cm. long, margined; blades 2–13 cm. long, deltoid-cordate, the margin coarsely sinuate, above glabrous and bright green, below white tomentose; peduncles filiform; involucre 1.5–2 mm. high, turbinate, the ovate, acute, glabrous bracts connate at base; pistillate corollas 0.8–1 mm. long, white, campanulate; staminate corollas 1.5–2 mm. long, funnelform, white; achenes 7–10 mm. long, capitate glandular hirsute at apex. Moist woods, *Can.*

AGOSERIS. Goat Chicory

Acaulescent annuals or perennials, with milky juice; leaves radical, clustered; heads solitary, on scapes; flowers all ligulate, yellow, rarely orange or purplish; bracts of the campanulate involucre in a few rows; receptacle flat, naked; pappus of copious white capillary bristles, which are not plumose; achenes 10-ribbed, the apex prolonged into a beak. (Name from Gr., *aix,* goat; *seris,* chicory.)

Leaves glaucous; beak of achene ribbed, not more than half the length of the body,
  Involucre glabrous,
    Leaves entire or somewhat dentate. *A. glauca,* var. *glauca.*
    Leaves deeply pinnatifid. Var. *laciniata.*
  Involucre more or less lanate or ciliate. Var. *dasycephala.*
Leaves not glaucous; beak not ribbed, 2–4-times the length of the body of the achene,
  Flowers yellow,
    Outer involucral bracts the broader, ovate, the inner in fruit 30–42 mm. long; corollas 14–18 mm. long; beak of fruit 15–18 mm. long. *A. grandiflora.*
    Outer involucral bracts not broader, all lanceolate, the inner in fruit 12–18 mm. long; corollas 5–7 mm. long; beak of fruit 6–7 mm. long,
      Ribs of fruit straight or but slightly undulate. *A. heterophylla,* var. *heterophylla.*
      Ribs of fruit conspicuously sinuately folded. Var. *Kymapleura.*
  Flowers orange. *A. aurantiaca,* var. *aurantiaca.*

Agoseris aurantiaca (Hook.) Greene, var. **aurantiaca.** Perennial; leaves 10–20 cm. long, oblanceolate or rarely linear, entire or with a few short lobes, glabrate; scapes 1–4 dm. tall, villous at apex; involucre 18–20 mm. high, turbinate-campanulate, imbricate; outer bracts oblong, acute; inner bracts lanceolate; rays orange, turning purple; pappus 10–15 mm. long, white; achene body 3–6 mm. long, linear-fusiform, ribbed, tapering to a slender beak 8–12 mm. long. Salmon River, Blue Mts., *Horner* 346, *Huds.*

Agoseris glauca (Pursh) Raf., var. **glauca.** Perennial; tap-root 3–10 mm. in diameter, brown; leaves 5–35 cm. long, oblanceolate to almost linear, glabrous, entire or somewhat dentate; scape 3–6 dm. tall, glabrous or more or less pilose; involucral bracts glabrous, linear-lanceolate, the outer, few and but little shorter, the inner 12–20 mm. long in flower, in fruit 22–27 mm. long; corollas 15–20 mm. long; pappus 9–15 mm. long, scaberulous; achene 9–17 mm. long, puberulent, fusiform. Grasslands, *A. T.*

Var. **dasycephala** (T. & G.) Jepson. Differs from the species by having the involucre more or less lanate or ciliate. *A. glauca scorzoneraefolia* (Schrad.) Piper; *A. glauca,* var. *villosa* (Rydb.) G. L. Wittrock. Grasslands, *A. T.*

Var. **laciniata** (D. C. Eaton) Smiley. Differs from the species by having the leaves deeply pinnatifid into linear segments. *Stylopappus laciniatus* Nutt. Grasslands, *A. T.*

Agoseris grandiflora (Nutt.) Greene. Perennial, more or less pilose throughout; tap-root 3–15 mm. in diameter, brown; leaves 1–3 dm. long, oblanceolate to spatulate, entire, dentate, or pinnately lobed; scapes 2–7 dm. tall, nearly glabrous; involucre lanate at base, the outer bracts ovate, acute, ⅓ to ½ the length of the inner; inner bracts lance-linear, in anthesis 12–20 mm. long, in fruit 30–42 mm. long; pappus 9–12 mm. long, minutely barbellate; achene body 5–6 mm. long, narrowly fusiform, sharply ribbed, greenish, the beak filiform. *A. obtusifolia* (Suksd.) Rydb. Meadows, *A. T.*

The leaves are exceedingly variable in shape and cutting.

Agoseris heterophylla (Nutt.) Greene, var. **heterophylla.** Annual, pilose nearly throughout; roots fibrous, pale; leaves 4–20 cm. long, linear to oblanceolate, entire to dentate or pinnatifid; scapes 7–40 cm. tall; involucre campanulate, more or less viscid pilose, the outer bracts more pubescent, few, unequal, ascending; inner bracts in flower 7–9 mm. long, hyaline margined; corollas but little longer than the involucre; pappus 5–7 mm. long, minutely barbellate; achene body 3–4 mm. long, fusiform, the prominent ribs straight or slightly undulate, puberulent or glabrous; beak filiform. *A. heterophylla normalis* Piper. Open gravelly slopes, *U. Son., A. T.*

Piper's interpretation is here reversed. Although Nuttall described his achenes as with "the wings undulated," he does not specify the degree. In the Gray Herb. is a type sheet, so marked by Nuttall, with excellent specimens and mature fruit. They have the ribs straight or slightly undulated on the same achene. I consider them inseparable from those with all straight ribs. Then the variety with conspicuously sinuately folded ribs would remain var. *Kymapleura.*

Var. **Kymapleura** Greene. Differs from the species by having its achenes with the wing-like ribs conspicuously sinuately folded. *A heterophylla* sensu Greene, and Piper, not of (Nutt.) Greene. Open gravelly slopes, *U. Son., A. T.*

AMBROSIA. Ragweed

Annual or perennial herbs; leaves pinnately-lobed, at least the lower opposite; heads small, greenish, monoecious; pistillate heads 1-flowered, in terminal bractless racemes or spikes; involucral bracts of staminate heads 5–12, partly united; pistillate flowers enclosed in an achene-like, 1-celled involucre, usually armed with a single series of tubercles or prickles; pappus none; corollas of staminate flowers funnelform, 5-toothed; of the pistillate none; achenes ovoid or obovoid, thick. (A Gr. and Lat. plant name; also the food of the gods.)

Leaves mostly palmately cleft; staminate involucre with upper side hispidulous and 3–4 ribbed; achenes angled from the base. — *A. trifida.*

Leaves mostly 1–3-pinnatifid; staminate involucre generally pubescent, not ribbed; achenes terete at base,

  Leaves pinnatifid; fruit with blunt tubercles. — *A. psilostachya.*

  Leaves 2–3-pinnatifid; fruit with lateral spines,

    Plant glabrous or appressed hirsutulous. — *A. artemisiifolia,* var. *elatior.*

    Plant spreading villous. — Forma *villosa.*

Ambrosia artemisiifolia L., var. **elatior** (L.) Descourtilz. *Roman Wormwood.* Annual 3–9 dm. tall, more or less appressed hirsutulous; lower leaves with petioles 1–2 cm. long; upper leaves sessile; blades 3–8 cm. long, ovate in outline, 2–3-pinnatifid, the ultimate divisions oblanceolate, acute, beneath paler green, the upper leaves sometimes entire; heads in terminal or axillary racemes, the staminate numerous; involucre 3–3.5 mm. wide, cup-shaped, hirsutulous, with 5 shallow lobes; corolla 1.1–1.3 mm. long; receptacle sparsely hirsute; pistillate heads in clusters in axils or at base of the raceme, subtended by ovate-lanceolate bracts; fruit obovoid, the body 2.5–3 mm. long, with 5–7 lateral spines, the slender conical, hirsutulous beak 1–2 mm. long. *A. artemisiaefolia diversifolia* Piper; *A. media* Rydb. A weedy plant, *U. Son., A. T.*

It produces large quantities of pollen, and this is a frequent cause of hay fever.

Forma **villosa** Fern. & Griscom. Differs by having the herbage, especially the stems spreading villous. High timbered regions, Spokan [Spokane], *Spalding, A. T. T.*

Ambrosia psilostachya DC. Perennial, 2–10 dm. tall, rough hirsutulous throughout; rootstock 2–4 mm. in diameter, running, brown; stem often bushy branched; leaves subsessile, 3–8 cm. long, lanceolate in outline, pinnatifid, the divisions linear-lanceolate, acute, entire or few toothed; staminate heads in terminal and axillary racemes 3–12 cm. long; involucre 2.5–3.5 mm. broad, obconic, hispidulous, the margin undulate; corolla 1.7–2 mm. long; pistillate heads 1–3 in upper axils; fruit pilosulous, obovoid, the body 2.5–3 mm. long, rugose, the 4–6 tubercles lateral and tooth-like, the beak 0.8–1 mm. long subcylindric. Weed, introduced from southern U. S. or Mexico. *A. artemisiaefolia,* subsp. *diversifolia* Piper, in part. The other part is *Franseria acanthicarpa.* Roadsides and fields, *U. Son., A. T.*

Ambrosia trifida L. *Great Ragweed.* Annual, 1–7 dm. tall, scabrous hispidulous and somewhat hirsute; tap-root 1–4 mm. in diameter; cauline leaves with petioles 5–15 cm. long, the blades 5–15 cm. long, mostly 3–5-cleft and palmately veined, the lobes oblanceolate or broadly so, acuminate, mucronate crenate; the upper leaves usually simple, lanceolate or ovate; inflorescence of terminal or axillary, spike-like racemes 3–15 cm. long; staminate heads numerous; involucre 2.5–3.5 mm. wide, saucer-shaped, with 6–8 shallow lobes, and 3–4 black ribs on the upper side, and these hispidulous; corolla 1 mm. long; chaff of pilose hairs; pistillate heads clustered at base of the racemes, subtended by trifid, black-lined bracts; fruit turbinate with a conical beak, 2–3 mm. long, and hirsute, the body 6–7 mm. long, with 5–7 prominent ridges, ending in short, conic lateral spines. Weed, introduced from eastern or central N. Am. Walla Walla, *Piper,* Aug. 13, 1897; Palouse, 1907, *Hunter.*

ANAPHALIS. Pearly Everlasting

White-woolly perennial herbs with erect leafy stems and entire leaves; heads numerous, small, discoid, dioecious but usually with a few perfect flowers in the center of the pistillate heads; involucre campanulate, its bracts scarious, in several series; receptacle mostly convex, naked; staminate corolla tubular-funnelform; style undivided; pistillate corolla tubular, the style exserted, 2-cleft; ovary fertile; pappus bristles of staminate flowers little if at all thickened at the apex, that of the fertile flowers not at all united at the base. (The ancient Gr. name of a similar plant.)

Leaves green and glabrous above. *A. margaritacea,* var. *margaritacea.*

Leaves lanate above and below. Var. *subalpina.*

Anaphalis margaritacea (L.) Benth. ex C. B. Clarke, var. **margaritacea.** Plant 2–11 dm. tall; rhizome 2–4 mm. in diameter, strong, forking and colony forming; stems erect, simple, closely white lanate; basal leaves reduced or bracteose; cauline leaves 4–11 cm. long, sessile, linear-lanceolate, 3-ribbed, above bright green and glabrous, beneath densely white lanate, the base auricled; corymb 3–15 cm. broad, white lanate, compound; involucral bracts white, chartaceous, the outer ovate, the inner 5–7 mm. long, ellipsoid; staminate corolla 3.5 mm. long; pappus 4 mm. long, stout capillary; pistillate corollas 5–5.5 mm. long; pappus 5 mm. long, delicate capillary; achenes 0.3 mm. long, ellipsoid, brown. *A. occidentalis* (Greene) Heller; *A. margaritacea,* var. *occidentalis* Greene. Clearings or woods, *A. T. T., Can.*

Var. **subalpina** Gray. Differs in being 1–3 dm. tall; blades permanently thin lanate above; cyme 2–11 cm. broad. Exposed places, Blue Mts., Columbia Co., *Horner* B10, *Can.*

ANTENNARIA. Everlasting

Low white-woolly caespitose perennial herbs; leaves alternate, entire mostly in basal tufts; heads small, solitary or corymbose, completely dioecious; involucral bracts scarious, imbricate in several series; receptacle naked; staminate flowers with corolla filiform, the pappus-bristles thickened or barbellate at the apex; pistillate flowers with corolla tubular, 5-toothed, with the capillary pappus-bristles united at base into a ring. (Name from Lat., *antenna,* sail yard, or antenna.)

Key to Staminate Plants

Note: As the staminate plants of *A. Howellii* are unknown, it is included here on vegetative characters shown by the pistillate plants. These vegetative characters are commonly uniform for the two sexes.

Leaves bright green and glabrous above from the first,
- Basal leaves cuneate-obovate or -oblanceolate. — *A. Howellii.*
- Basal leaves ovate. — *A. racemosa.*

Leaves lanate on both sides,
- Pappus not dilated at tip,
  - Heads several, in a compact, head-like cyme; leaves linear. — *A. stenophylla.*
  - Heads solitary, terminal; leaves spatulate,
    - Leaves linear-spatulate; plant 2–4 cm. tall. — *A. dimorpha.*
    - Leaves broadly spatulate; plant 6–10 cm. tall. — *A. latisquama.*
- Pappus dilated at apex,
  - Lower cauline leaves 4–13 cm. long, linear-oblanceolate,
    - Involucre lanate only at base; pubescence close and silky lanate; principal leaves 1–5 mm. wide. — *A. luzuloides.*
    - Involucre lanate to the middle; pubescence loose, floccose lanate; principal leaves 7–22 mm. wide,
      - Involucral bracts with the body longer than the brownish tip. — *A. pulcherrima.*
      - Involucral bracts with the body shorter than the white tip. — *A. anaphaloides.*
  - Lower cauline leaves 10–27 mm. long, spatulate to linear-oblanceolate,
    - Branches all fertile; erect or ascending; cauline leaves not or little reduced. — *A. Geyeri.*
    - Sterile, prostrate, mat-forming shoots present; upper cauline leaves bracteose,
      - Involucral bracts in about 3 series, the inner 5–7 mm. long. — *A. parvifolia.*
      - Involucral bracts subequal, the inner 4 mm. long. — *A. rosea.*

Key to Pistillate Plants

Heads solitary,
  Heads lanate only at base; plant 2–4 cm. tall. *A. dimorpha.*
  Heads lanate almost to the tip; plant 6–10 cm. tall. *A. latisquama.*
Heads several to many, in cymes, racemes, or panicles,
  Leaves bright green and glabrous above from the first,
    Basal leaves cuneate-obovate or -oblanceolate; involucre 8–10 mm. long. *A. Howellii.*
    Basal leaves ovate; involucre 6–8 mm. long. *A. racemosa.*
  Leaves lanate on both sides,
    Lower cauline leaves 3–13 cm. long, linear to linear-oblanceolate,
      Involucre 3.5–5 mm. long,
        Plant 6–15 cm. tall; leaves linear; involucre 4.5–5 mm. long. *A. stenophylla.*
        Plant 1–5 dm. tall; leaves oblance-linear; involucre 3.5–4.5 mm. long. *A. luzuloides.*
      Involucre 6–10 mm. long,
        Involucre 6–7 mm. long, the bracts subequal. *A. anaphaloides.*
        Involucre 8–10 mm. long, the bracts in 6–7 series. *A. pulcherrima.*
    Lower cauline leaves 10–27 mm. long, spatulate to linear-oblanceolate,
      Branches all fertile, erect or spreading; cauline leaves not or a little reduced; achenes papillose. *A. Geyeri.*
      Sterile, prostrate, mat-forming shoots present; upper cauline leaves bracteose; achenes glabrous,
        Involucre 5.5–7 mm. long. *A. rosea.*
        Involucre 8–10 mm. long. *A. parvifolia.*

Antennaria anaphaloides Rydb. Plant 20–65 cm. tall, loosely white lanate throughout; root 2–3 mm. in diameter, brown; stems erect, single, but often with basal leafy shoots; basal leaves with petiole 1–8 cm. long, the blades 3–10 cm. long, oblanceolate, 3–5-ribbed; lower cauline leaves 4–13 cm. long with the short petiole, oblanceolate, 3-ribbed; upper leaves bracteose; staminate heads in a round topped, finally loose cyme; involucre campanulate, lanate to the middle; outer bracts elliptic, pale brown, the tips brownish, nearly as long as the inner; inner bracts 5–6 mm. long, elliptic-obovate, the body brownish, the tip white, chartaceous, longer than the body, usually erose; pappus 3.5–4 mm. long white, barbed, dilated towards the tip; pistillate heads numerous, in a round topped, finally loose cyme; involucre lanate to the middle, campanulate; outer bracts elliptic-spatulate, nearly as long as the inner, the body brownish, shorter than the white, chartaceous tip; inner bracts 6–7 mm. long, elliptic-oblanceolate the body brownish, shorter than the white, chartaceous, usually erose tip; pappus 4 mm. long, white, capillary; achenes 1.3–1.5 mm. long, fusiform-cylindric, greenish brown, glabrous. Open slopes, *U. Son., A. T.*

Antennaria dimorpha (Nutt.) T. & G. Plants forming dense mats, 2–4 cm. tall, appressed, delicate lanate throughout; rhizomes 1–4 mm. in diameter, brown, short creeping; basal and cauline leaves numerous, 1–2 cm. long, spatulate; staminate heads solitary, terminal, lanate only at base; outer bracts ovate, fuscous; inner bracts 5.5–9 mm. long, elliptic, fuscous throughout or the tip paler; pappus 5 mm. long, white, barbed; pistillate heads terminal, solitary, cylindric-campanulate, glabrous except at base, outer bracts ovate, fuscous, the inner bracts 8–15 mm. long, linear-lanceolate, acuminate, brownish or greenish with white hyaline margins and apex; pappus 10–13 mm. long, white; achenes 3–3.5 mm. long, linear-fusiform, brown, with white puberulence. Dry or stony soils or woods, *U. Son., A. T., A. T. T.*

Antennaria Geyeri Gray. Plant 6–20 cm. tall loosely delicate lanate throughout, erect or decumbent at base; rhizome 2–4 mm. in diameter, woody, brown; stem suffrutescent, usually forked at base, leafy to the tip; basal and cauline leaves 10–25 mm. long, spatulate to linear-oblanceolate; staminate inflorescence a close spike or panicle; heads campanulate, lanate well above the middle; outer bracts ovate, brownish, the inner bracts 6–8 mm. long, narrowly elliptic, brownish, with prominent white (or pinkish) petaloid tips; pappus 4 mm. long, white, barbellate slightly enlarged at tip; pistillate inflorescence a close spike or panicle; heads cylindric-campanulate, lanate to above the middle; outer bracts lanceolate, green on the back, the tips brownish; inner bracts 7–9 mm. long, lance-linear, the prominent tips crimson; pappus 6–8 mm. long, white, capillary; achenes 2 mm. long, narrowly cylindric, brown, papillose. Sand plains or open woods, Spokane Co., *A. T., A. T. T.*

The type no. 542, was collected by C. A. Geyer in "Arid sandy woods near Tschimakaine, Spokan country," that is near the old mission in southern Stevens Co.

Antennaria Howellii Greene. Plant 8–40 cm. tall; rhizome 1–2 mm. in diameter, brown; prostrate, bracted stolons present; the terminal leaves developing after the fruit; stem single, loosely lanate; basal leaves 30–55 mm. long, cuneate-obovate or -oblanceolate, 3-ribbed, apiculate, above bright green, glabrous beneath densely dull lanate; cauline leaves 1–2.5 cm. long, linear bracteose; staminate plants unknown; pistillate heads several, in a round topped, usually close cyme; involucre campanulate, loosely lanate to above the middle, the bracts imbricate in about 3 series; outer bracts lanceolate, the body greenish or pale brownish, longer than the pale thin tips; inner bracts 8–10 mm. long, linear, acuminate, the body greenish or pale brownish longer than the whitish or pale brownish tips; pappus 7–8 mm. long, white, capillary; achenes 1.3–1.5 mm. long, ellipsoid, brown, glandular papillose. *A. neglecta* Greene, var. *Howellii* (Greene) Cronq. Open places or woods, *A. T., A. T. T.*

Antennaria latisquama Piper. Plants tufted, 6–10 cm. tall, white lanate throughout; root 2–4 mm. in diameter, brown; stolons none; stems erect, with brown, marcescent leaf bases below; basal and cauline leaves 2–4 cm. long, spatulate or linear-spatulate; staminate heads single, terminal; involucre campanulate, lanate to the middle, the outer bracts ovate to elliptic, the center dark brown, the margins pale brown, the inner bracts lanceolate to elliptic 7–9 mm. long, dark brown, the margins pale brown; pappus 4–5 mm. long, white, barbed, often branched at apex; pistillate heads solitary, terminal; involucre cylindric, the bracts lanate almost to the apex, the outer ovate lanceolate, dark brown, margined with pink and pale brown; the inner bracts 12–13 mm. long, linear-lanceolate, acuminate, pale brown and pink, the margins brownish; pappus 7–8 mm. long, white, minutely barbellate; immature achenes glabrous. Open slopes, Grande Ronde, *St. John & Brown* 4197, *U. Son.*

Reduced by Cronquist (1955) to *A. dimorpha,* but with this disposition the writer does not agree.

Antennaria luzuloides T. & G. Plant 1–5 dm. tall, closely appressed silky villous throughout; root 2–4 mm. in diameter, brown, creeping, forking; sterile shoots often present, short decumbent; stems usually several, short decumbent at base; basal leaves 3–7 cm. long, oblance-linear, the veins obscure; lower cauline leaves 4–8 cm. long, oblance-linear; upper leaves linear, bracteose; staminate heads numerous in a rounded or pyramidal cyme; involucre campanulate, lanate only at very base; the bracts elliptic, imbricate in several series, the body greenish or brownish, the tips white, chartaceous; inner bracts 4–5.5 mm. long, the tips shorter than the body; pappus 3–3.5 mm. long, white, barbellate the tips spatulate and subentire; pistillate heads numerous in rounded cymes; involucre ovoid or narrowly campanulate, glabrous or lanate only at base; bracts elliptic, imbricate in several series, the body pale yellowish or brownish, much longer than the white, chartaceous tips; inner bracts 3.5–4.5 mm. long; pappus 2 mm. long, white, capillary; achenes 1–1.3 mm. long, narrowly ellipsoid, brown, glabrous. *A. luzuloides*, var. *oblanceolata* (Rydb.) Peck. Open hillsides or woods, *A. T.*, *A. T. T.*

Antennaria parvifolia Nutt. Plant 5–20 cm. tall, lanate throughout; rhizomes 1–2 mm. in diameter, woody, brown; sterile basal shoots numerous, prostrate, but short and leafy throughout, forming a mat, not stolon-like; stems several solitary; basal and rosette leaves 1–2 cm. long, broadly spatulate, apiculate, the nerves hidden by the dense tomentum; cauline leaves 1–2 cm. long, linear to linear-oblong, acute; staminate heads broad turbinate, lanate beyond the middle, the bracts imbricate in about 3 series; outer bracts but little shorter, elliptic, the body green below, brownish or pinkish above as long as the pale, chartaceous tip; inner bracts 5–7 mm. long, oblanceolate, the body greenish, scarcely longer than the white, chartaceous, erose tip; pappus 4.5–5 mm. long, white, barbellate, slightly enlarged at apex; pistillate heads several in a compact, round topped cyme; involucre campanulate, lanate to the middle; bracts oblanceolate, imbricate in 5–6 series; outer bracts with the body greenish or brownish, as long as the chartaceous, erose, whitish or at first pinkish tips; inner bracts 8–10 mm. long, the body greenish or brownish, longer than the chartaceous, erose, white or at first pinkish tips; pappus 7–8 mm. long, white, capillary; achenes 1.3–1.8 mm. long, ellipsoid, brown, glabrous. *A. aprica* Greene. Gravelly places or open woods, Spokane Co., *A. T.*, *A. T. T.*

Antennaria pulcherrima (Hook.) Greene. Plant 2–4 dm. tall, loosely lanate throughout; root 1–3 mm. in diameter, black; stems single, erect, leafy, but the upper leaves much reduced; basal leaves with petioles 1–5 cm. long, the blades 3–9 cm. long, oblanceolate 3-ribbed, tapering into the slender petiole; lower cauline leaves 5–10 cm. long, linear-oblanceolate, mostly sessile; upper leaves bracteose; staminate heads in a compact corymb; involucres campanulate, lanate to the middle; outer bracts ovate, dark brown, the margins pale brown; inner bracts 6–9 mm. long, elliptic, pale brown, the tips whitish, usually entire; pappus 5 mm. long, white, barbed, thickened at apex; pistillate heads in a round topped, compact cyme; involucre campanulate, lanate to the middle, the bracts in 6–7 series; outer bracts ovate, dark brown, the tips pale brown; inner bracts 8–10 mm. long, lanceolate, obtuse, pale brown, or the tips nearly white; pappus 10–12 mm. long, white, capillary; achenes 1.5 mm. long, brown fusiform, glabrous. Prairies, *A. T.*

Antennaria racemosa Hook. Plant 1–6 dm. tall; rhizomes 1–3 mm. in diameter, woody, brown; prostrate leafy stolons well developed; stem floccose lanate below; principal basal leaves with petioles 1–3 cm. long, the blades 25–60 mm. long, ovate, apiculate 3-ribbed, above bright green and glabrous, beneath white lanate; cauline bracts 1–3 cm. long, lanceolate to linear; staminate heads several, in a loose, glandular puberulent raceme or panicle, the lower on peduncles 1–5 cm. long; involucre hemispheric, sparsely lanate to above the middle; bracts nearly equal, elliptic, the body greenish, much longer than the thin, whitish or buff-colored tips, the inner bracts 5–6 mm. long; pappus 3.5 mm. long, white, barbellate, slightly dilated at tip; pistillate heads in a raceme or panicle, finally loose; involucre campanulate, slightly lanate to above the middle; bracts imbricate in about 3 rows, the outer elliptic, greenish with thin white or brownish margins; inner bracts 6–8 mm. long, greenish, the margins thin and whitish; pappus 5–6 mm. long, white, capillary; achenes 1.5–2 mm. long, fusiform, brown, glabrous. Woods, *A. T. T.*, *Can.*

Antennaria rosea Greene. Plant 10–35 cm. tall, closely lanate throughout; rhizomes 1–2 mm. in diameter, woody, brown; sterile basal shoots numerous, short, prostrate, not stolon-like, but forming dense leafy mats; stems erect, simple; basal leaves 15–27 mm. long, spatulate, acute, the nerves obscure; cauline leaves 15–27 mm. long, linear-spatulate to oblong or linear, little reduced upwards; staminate plant rare, the heads in a compact, head-like cyme; involucre turbinate, lanate to the middle; bracts subequal, elliptic, cuneate, the body greenish, nearly as long as the white or pink, chartaceous tip, the inner bracts 4 mm. long; pappus 2.8–3 mm. long white, barbellate, dilated at tip; pistillate heads in a compact, head-like cyme; involucre campanulate, lanate to the middle; bracts elliptic, imbricate in 4–5 series; outer bracts with the body greenish, the short tip white or pinkish; inner bracts 5.5–7 mm. long, the base greenish or pale, slightly longer than the white or pink, chartaceous, serrulate tip; pappus 4.5–5 mm. long, white, capillary; achenes 1.3–1.5 mm. long, terete, linear, brown, glabrous. *A. rosea,* var. *angustifolia* (Rydb.) E. Nels.; subsp. *divaricata* E. Nels. Open places or open woods, *A. T, A. T. T.*

Antennaria stenophylla Gray. Plants tufted, 6–15 cm. tall, white lanate throughout; stems erect, covered with brown, marcescent leaf bases at base, leafy throughout; stolons lacking; basal and cauline leaves similar, 1–8 cm. long, linear or nearly so, white lanate; staminate inflorescence a 4–7-headed, compact cyme; involucre campanulate, glabrous; outer bracts ovate, fuscous; inner bracts 4–6 mm. long, lanceolate, fuscous with a white, scarious tip pappus 3 mm. long, white, barbed; pistillate inflorescence of 3–5 heads in a compact cyme; involucre campanulate, glabrous, brown, the bracts all lanceolate, the inner 4.5–5 mm. long, paler brown at tip; pappus 2.8–3 mm. long, white, barbellate; achenes 1.3 mm. long, cylindric, brown, papillose. Thin soil, in scablands, *A. T.*

ANTHEMIS. Chamomile

Annual or perennial herbs; leaves alternate, mostly tripinnately divided; heads many-flowered; involucre hemispheric, the scales very numerous, scarious margined, imbricated in several series, and appressed; receptacle convex to oblong-conical, chaffy, with slender or thin scales or awns, subtending at least the central flowers; ray flowers numerous, commonly conspicuous, pistillate or sometimes sterile; disk flowers fertile; pappus none, or short chaffy crown; achenes obovoid or oblong, 4- or 5-angled, 8–10-ribbed or many-striate, truncate at the apex. (The ancient Gr. name.)

Herbage not ill-scented; rays fertile; all flowers subtended by lanceolate chaff. *A. arvensis.*
Herbage ill-scented; rays neutral; only the central flowers with bristle-like chaff. *A. Cotula.*

Anthemis arvensis L. *Corn Chamomile.* Annual or biennial, 2–8 dm. tall, sparsely pilose throughout; tap-root 1–4 mm. in diameter, yellowish; stems usually several and bushy branched above, striate; leaves 2–5 cm. long, oblong or ovate in outline, bipinnatifid, the ultimate segments 0.5–1 mm. broad, linear-lanceolate, sharp pointed; heads single and slender peduncled; involucre villous at base, the bracts green up the middle, the outer lanceolate, the inner 4–6 mm. long, elliptic, hyaline margined, lacerate and pilose ciliate; the 15–20 ray flowers white, the rays 6–12 mm. long, linear-lanceolate; chaff 3–4 mm. long; disk flowers 2.5–3 mm. long, funnelform, yellow; achenes 2–3 mm. long, thick wedge-shaped, many ribbed, brown, cellular roughened, sticky when wet. Weed, introduced from Eurasia or Africa. Infrequent, roadsides and barnyards, *A. T.*

Anthemis Cotula L. *Dog Fennel; Mayweed.* Annual, 1–6 dm. tall glabrous or sparsely hirsute throughout; tap-root 1–6 mm. in diameter, brown; stem striate, bushy branched; lower leaves with petioles as much as 2 cm. in length; upper leaves sessile; blades 1–6 cm. long, oval in outline, bipinnatisect, the ultimate divisions filiform or linear, about 0.5 mm. wide, sharp pointed; heads numerous, terminal, on slender peduncles 1–15 cm. long; involucral bracts hirsute, elliptic-lanceolate, the center green, the margin pale, hyaline, the inner 3–4 mm. long; chaff of rigid bristles 3–4 mm. long, subtending only the central flowers; the 10–18 rays white, 5–10 mm. long, narrowly elliptic; disk flowers 2–2.5 mm. long, yellow, glandular atomiferous, funnelform, the lobes deltoid; pappus none; achenes 1–1.5 mm. long, linear-obpyramidal, 10-ribbed, brown, glandular tuberculate. Weed, introduced from Eurasia or Africa. Roadsides, fields and farmyards, *A. T.*

The plant has a strong, fetid odor. The acrid herbage is avoided by grazing animals. It produces blisters on the skin of humans.

ARCTIUM. Burdock

Coarse, branching, rough, mostly biennial herbs; leaves petioled, alternate, broad; heads racemose, corymbose, or paniculate; involucre subglobose, the bracts imbricate in many series, tipped with hooked bristles; receptacle flat, bristly; flowers alike, tubular, purplish or white, 5-cleft; anthers sagittate; pappus of numerous short serrulate scales or bristles; achenes oblong, somewhat compressed and 3-angled, truncate. (The ancient Gr. name *arkteion,* used by Dioscorides, from *arktos,* bear.)

Arctium minus (Hill) Bernh. *Common Burdock.* Biennial 5–20 dm. tall; root thick, fleshy, fusiform, black; stem stout, striate, bushy branched, lanate or glabrate; lower leaves with petioles 1–2 dm. long, lanate, the blades 1–4 dm. long, broadly ovate, cordate at base, above green and glabrate, beneath sparsely white lanate, the margin sinuate dentate; upper leaves smaller, shorter petioled and less cordate; uppermost leaves 2–5 cm. long, ovate, often entire; heads numerous, racemose on the branches; involucre glabrous; the bracts all divergent, lance-linear, the sides serrulate; inner bracts 11–14 mm. long; corolla rose-purple, the tube 7–8 mm. long, funnelform, the lobes 1–1.5 mm. long, narrowly ovate; pappus bristles 1.8–2 mm. long; achenes 4–6 mm. long, oblanceoloid, curved, the surface mottled brown and wave-like. Weed, introduced from Europe or Africa.
Roadsides, *A. T.*

ARNICA

Perennial herbs; stems mostly simple, from creeping rootstocks or a corm-like base; leaves all or some of them opposite, simple, entire, or merely toothed; heads rather large, solitary or few, usually long-peduncled, many-flowered; flowers yellow, all fertile; ray-flowers elongated, pistillate, or sometimes none; involucre broadly campanulate, the bracts in 1–2 rows; receptacle naked, flat; pappus a single series of rather rigid strongly scabrous or barbellate capillary bristles; achenes linear, 5-angled or 5–10-ribbed, somewhat hirsute or nearly glabrous. (Derivation of name obscure.)

Basal and lower leaves oblanceolate,
  Cauline leaves mostly in 5–12 pairs,
    Involucral bracts obtuse or merely acutish, hairy tufted just within the tip. — *A. foliosa.*
    Involucral bracts sharply acute to acuminate, the tip no more hairy than the body. — *A. amplexicaulis,* var. *amplexicaulis.*
  Cauline leaves mostly in 2–4 pairs,
    Heads without rays. — *A. Parryi,* var. *Parryi.*
    Heads with rays,
      Stem base bulbous and densely brown lanate; disk corolla tube pilosulous. — *A. fulgens.*
      Stem base not bulbous or lanate; disk corolla tube glandular puberulent. — *A. sororia.*
Basal and lower leaves cordate to ovate or lanceolate,
  Leaves villous or hirsute; involucre 1–2 cm. high,
    Basal leaves cordate, sharply dentate. — *A. cordifolia.*
    Basal leaves ovate to lanceolate, obscurely dentate or entire. — *A. Hardinae.*
  Leaves sparsely puberulent, — *A. latifolia,* var. *latifolia.*

Arnica amplexicaulis Nutt., var. **amplexicaulis.** Forming large clumps with matted rootstocks; stems 1.5–8 dm. tall, moderately glandular-scabrous or -pilose or subglabrate; innovations occasional; cauline leaves 5–12 pairs, mostly sessile, 5–12 cm. long, narrowly lance-elliptic to lance-ovate, mostly sharply serrate-dentate, the lowest pair petiolate; heads mostly 5–9 broadly campanulate, the base long- and short-stipitate glandular and pilose; involucral bracts 10–12 mm. long, narrowly lanceolate, acute, moderately pubescent towards the base; the 8–14 rays 12–16 mm. long; disk flowers 6.5–7.7 mm. long, the tube pilose; achenes 4.5–5.5 mm. long, sparingly hirsute and frequently glandular; pappus 6.5–7.7 mm. long, tawny. Recorded by Maguire (1943) in the Palouse region.

Arnica cordifolia Hook. Plant 2–5 dm. tall; sparsely villous; rootstock 1–2 mm. in diameter, brown; sterile shoots with petioles 5–16 cm. long; blades 4–10 cm. long, cordate, acute, coarsely dentate; cauline leaves 2–4 pairs, the lower similar but shorter petioled; upper leaves 2–4 cm. long, sessile or nearly so, deltoid-ovate to lanceolate; heads 1 or few; involucre villous, densely so at base, turbinate; bracts oblanceolate; the ray flowers 8–12, the rays 15–30 mm. long; disk flowers 9–11 mm. long; pappus 8–10 mm. long, white; achenes 6–8 mm. long, fusiform, brown, appressed hirsutulous. *A. cordifolia,* var. *macrophylla* (Nutt.) Maguire, from which extreme there is a complete intergradation. Maguire himself (1943) listed *A. macrophylla* Nutt. as a synonym of the species. Abundant, woods, *A. T. T., Can.*

Arnica foliosa Nutt. Stems 1.5–9 dm. tall, often tomentose below, and above short stipitate glandular; cauline leaves in 5–10 pairs, the blades 5–30 cm. long, lanceolate to lance-oblong or oblanceolate, soft tomentose or above rarely glabrate, the lower pairs narrowing to petioles ending in conspicuous sheaths; heads 3–15, hemispheric-campanulate, 15–18 mm. high, the base moderately to densely villous; involucral bracts 2–4 mm. broad, broadly lanceolate, obtuse or acutish; rays 1.2–1.8 cm. long, pale yellow; disk flowers 7.5–8.7 mm. long, villosulous and stipitate glandular; pappus stramineous, barbellate; achenes 4–5.4 mm. long, sparsely hirsute and stipitate glandular. Recorded by Maguire throughout our area.

Following the concept and the definitions of Du Rietz, Maguire in his monograph (1943) used the taxon subspecies for the principal minor taxa with morphologic and geographic distinctions. He made this plant *A. Chamissonis* Less., subsp. *foliosa* (Nutt.) Maguire. Since I prefer to continue the age-old practice of classifying these taxa as varieties, legal under the International Code of Botanical Nomenclature, I would have treated this as one also, but could not certainly determine the status of the earliest described variety. Hence, to avoid ambiguity, this plant is retained as a species under the name given by Nuttall.

Arnica fulgens Pursh. Plant 2–5 dm. tall, viscid pilose throughout; rootstock 3–5 mm. in diameter, brown, usually erect, each year forming an offset close beside the previous one; stem usually leafy in lower half; basal and lowest leaves with petioles 1–5 cm. long, winged; blades 3–9 cm. long, 3–5-ribbed, oblanceolate, remotely serrulate; cauline leaves 2–3 pairs, the upper 1.5–4 cm. long, sessile, linear-lanceolate; heads commonly 1, long peduncled; involucre 11–14 mm. high, hemispheric, densely viscid-hirsute at base and -hirsutulous above; bracts linear-lanceolate, acute; the 10–15 rays 13–18 mm. long; disk flowers 7–8.5 mm. long; pappus 7–8 mm. long, white; achenes 4.5–5.5 mm. long, fusiform, black, hirsutulous. *A. pedunculata* Rydb. Grasslands, *A. T.*

Arnica Hardinae St. John. Plant 1.5–3 dm. tall; rhizome 1–2 mm. in diameter, brown; offsets seldom present, their leaves similar to the basal; stem erect, usually simple, below hirsute, above hirsute and glandular hirsutulous; basal and lowest leaves with petioles 13–40 mm. long, hirsute; blades 2–4 cm. long, lanceolate to ovate, acute, entire, or with 2–3 remote, small teeth on a side, permanently hirsute, but dark green above, paler beneath, firm chartaceous, blade 3-ribbed from the base, but the lateral veins anastomosing with the pinnate venation of the blade; cauline leaves 3–4 pairs; middle leaves similar but with the winged petiole 7–12 mm. long; blades 15–40 mm. long, ovate-lanceolate; upper leaves similar but sessile, 15–35 mm. long; heads 1, or rarely 3 in a cyme; peduncle 4–13 cm. long, hirsute, above hirsute and densely glandular hirsutulous; involucre broadly turbinate, 10–14 mm. high; the 8–15 involucral bracts broadly oblanceolate, acute, hirsute throughout, densely so at base, the tip villous ciliate, the lower margins hyaline, more or less laciniate; the 7–10 rays 20–23 mm. long, 3–6 mm. wide, the tip irregularly several toothed; the nerves puberulent beneath, the tube hirsutulous; the numerous disk flowers 8–9 mm. long, almost cylindric, hirsutulous below, the lobes 1 mm. long, ovate; disk pappus 8–9 mm. long, barbellate, white; disk achenes 6–7 mm. long, linear, brown to black, hirsutulous.

Open woods, Lake Chatcolet, *Can.*

The type specimen was collected there, by *G. Weitman* 226.

The species is named for *Edith Hardin* (Mrs. Carl S. English, Jr.) in recognition of her original botanical work.

Maguire (1943) in his monograph reduced this species to the synonymy of *A. cordifolia,* to which it bears no resemblance.

Arnica latifolia Bong., var. **latifolia.** Plant 2–6 dm. tall; rhizome 2–3 mm. in diameter, brown; stem sparsely villous; offsets bearing leaves with petioles 5–10 cm. long; blades 2–8 cm. long, thin, cordate to ovate, crenate-dentate, remotely puberulent, paler beneath; cauline leaves 3–5 pairs, the lower with petioles as much as 4 cm. in length; blades 3–8 cm. long, dentate or crenate, ovate, the upper sessile; heads 1 or 3–5 in a cyme; involucre broadly turbinate; bracts sparsely villous towards the base, glandular puberulous towards the acuminate tip, oblanceolate, acuminate; the 8–12 rays 15–23 mm. long; disk flowers with the tube pilosulous; pappus 5–7 mm. long, white; achenes 5–6 mm. long, linear, pilosulous. Woods, Blue Mts., *Can.*

Arnica Parryi Gray, var. **Parryi.** Plants 2–6 dm. tall; innovations frequent; stems single, at base lanate-villous and stipitate glandular, above conspicuously long stipitate glandular and somewhat pilose; cauline leaves 2–4 pairs, reduced upwards, the largest with blades 5–20 cm. long, lanceolate, callous denticulate or entire, villous, the upper side short stipitate glandular; petiole narrowly winged; heads 3–9, turbinate campanulate, in bud nodding, the base short stipitate glandular and pilose; involucre 10–14 mm. high; involucral bracts 12–20, lanceolate, acute, sparingly glandular, puberulent and ciliolate towards the tip; rays none; disk flowers 7.5–9 mm. long, pilose; pappus tawny; achenes 4.5–5.5 mm. long, bronze-black, angled, glabrous below, hirsute above. *A. Parryi,* subsp. *genuina* Maguire. Craig Mts. area, recorded by Maguire (1943).

Arnica sororia Greene. Plant 2–6 dm. tall, viscid hirsute throughout; rhizome 1–2 mm. in diameter, brown, elongate, horizontal; stem erect; basal and lower leaves with petioles 1–9 cm. long; blades 5–13 cm. long, 5-ribbed, oblanceolate, obtuse, entire; cauline leaves 2–4 pairs, the upper 2–4 cm. long, sessile, lanceolate; heads 1 or 3–7 in a cyme; involucre 10–15 mm. high, hemispheric, viscid hirsute, densely so at base, at tip viscid hirsutulous; the 15–20 bracts lance-linear; the 12–18 rays 15–18 mm. long; disk flowers 7–8 mm. long; pappus 6–7 mm. long, white; achenes 4–5 mm. long, fusiform, black, hirsutulous. *A. fulgens* sensu Rydb., not of Pursh. Grasslands, *U. Son., A. T.*

ARTEMISIA. Sagebrush. Wormwood

Herbs or undershrubs, usually bitter and odorous; leaves alternate, usually dissected; heads numerous, small, in racemes or panicles, several- to many-flowered; disk flowers yellow or yellowish, all tubular, the outermost series pistillate and with irregular corolla, or all alike, the more numerous perfect flowers either fertile or sterile; ray flowers pistillate and fertile, or wanting; involucral scales dry, imbricated in a few rows, appressed; receptacle flattish to hemispherical, naked, sometimes hairy or chaffy; pappus none; corollas of pistillate flowers slender and small, 2- or 3-toothed, those of the perfect flowers enlarged above, 5-toothed; pappus none; achenes obovoid or oblong, almost always glabrous. (The ancient Gr. name, used by Dioscorides, derived from *Artemis,* the queen of King Mausolus, of Halikarnassos.)

Shrubs,
  Heads single and sessile in the upper leaf axils; leaves entire. *A. rigida.*
  Heads in leafy panicles,
    Leaves deeply divided into 3 linear or narrowly linear-oblanceolate lobes, and often some leaves linear entire. *A. tripartita.*
    Leaves with 3 (or 5) blunt, apical teeth, the blade cuneate to spatulate. *A. tridentata,* subsp. *tridentata.*

Herbs; heads in panicles,
  Plant green and glabrous,
    Lower leaves bipinnatifid; inflorescence subspicate. *A. biennis.*
    Lower leaves trifid; inflorescence paniculate. *A. dracunculoides.*
  Plant more or less lanate,
    Lobes of leaves obtuse; receptacle pilose. *A. Absinthium.*
    Lobes of leaves acute; receptacle glabrous,
      Leaves or some of them bipinnatifid, the pinnules again toothed. *A. Michauxiana.*
      Leaves entire to bipinnatifid; pinnules entire,
        Principal leaves 1–5 cm. wide exclusive of the lobes when present, frequently entire. *A. Douglasiana.*
        Principal leaves 1 cm. or less wide, exclusive of the lobes when present,
          Plant herbaceous to the base,
            Principal leaves entire or merely lobed. *A. ludoviciana,* var. *ludoviciana.*
            Principal leaves more or less parted or divided,
              Involucres 4–5 mm. high, 4–7 mm. wide; leaves white tomentose on both sides. Var. *candicans.*
              Involucres 2.5–3.5 mm. high, 2–3.5 mm. wide. Var. *incompta.*
          Plant suffrutescent, woody at base. *A. Lindleyana.*

Artemisia Absinthium L. *Absinthe.* Perennial herb, or slightly suffrutescent at base, 4–10 dm. tall, subsericeous appressed puberulent throughout; root 5–15 mm. in diameter, woody, brown; stems often several, striate, glabrate below; basal leaves long petioled, the blades 3–5 cm. long, 2–3-pinnatifid, the divisions oblong to oblanceolate, obtuse, often dentate; cauline leaves gradually shorter petioled and less divided; inflorescence 15–40 cm. long, a leafy panicle with ascending branches; heads nodding, short peduncled; heterogamous; involucre 3–4 mm. broad, hemispheric, canescent, the outer bracts linear; the inner bracts oval, obtuse, with broad scarious margins; receptacle densely pilose; ray flowers 9–20, fertile, the corolla 1.5 mm. long, oblique; disk flowers 30–50, fertile, the corolla 1.5–2 mm. long, campanulate; achenes 1.3–1.5 mm. long, subcylindric, glabrous. Weed, introduced from Eurasia. Abundant, roadsides and fields, *A. T.*

It contains oil of absinthe and other poisons. It has been used medicinally, and is still used in the manufacture of absinthe, a powerful intoxicant. When the plant is grazed by horses, it causes paralysis.

Artemisia biennis Willd. Annual or biennial, 3–30 dm. tall, glabrous; tap-root 2–10 mm. in diameter, brown; stem smooth, striate, usually simple; basal leaves 5–15 cm. long, bipinnatifid, the divisions lanceolate, laciniate serrate; cauline leaves 4–8 cm. long, sessile, pinnatifid, the divisions linear-lanceolate, laciniate serrate; inflorescence 1–5 dm. long, 1–2 cm. wide, the heads erect in axillary, clustered spikes; involucre 2–4 mm. broad, the outer bracts elliptic, the inner suborbicular, with broad scarious margins; ray flowers 6–22, fertile, the corolla nearly 1 mm. long, oblique; disk flowers 15–40, fertile, the corolla about 1 mm. long, campanulate; achenes 1 mm. long, ellipsoid, 4–5-nerved, glabrous. A weed, introduced apparently from the Rocky Mountain region. Roadsides and fields, *U. Son., A. T.*

The pollen is a known cause of hay fever.

Artemisia Douglasiana Bess. in Hook. Stems 3–20 dm. tall; lower leaves oblanceolate or obovate, coarsely few lobed; principal leaves 7–15 cm. long, 1.5–10 cm. wide, oblanceolate or oval in outline, sparsely tomentulose and green above, densely white tomentose beneath, acute saliently toothed or cleft, the lobes lanceolate, or blades rarely entire; panicle 2–8 cm. wide; involucre 2.5–3.5 mm. broad, gray tomentose; heads 15–30-flowered. *A. vulgaris* L., subsp. *Douglasiana* (Bess.) St. John; *A. atomifera* Piper; *A. vulgaris,* var. *heterophylla* (Nutt.) Jepson. Banks of Snake River, *U. Son.*

Artemisia dracunculoides Pursh. Perennial, 3–10 dm. tall, glabrous; root 3–10 mm. in diameter, woody, brown; stem smooth, striate, often with many ascending branches; lower leaves 3-cleft, the others simple, entire, 2–6 cm. long, linear, acute; heads numerous, nodding, on short peduncles, in a leafy panicle, heterogamous, 2–3 mm. broad; bracts in about 3 series, glabrous, the outer elliptic, mostly obtuse, the inner oval, broadly scarious margined; ray flowers 10–20, the corollas 1 mm. long, campanulate; disk flowers 10–20, the corollas 1.5 mm. long, campanulate; achenes about 0.5 mm. long. *A. dracunculus glauca* (Pall.) H. & C., minor variation 4 H. & C. Warmer valleys, *U. Son., A. T.*

The author here follows the treatment of Blake, instead of that of Hall & Clements.

Artemisia Lindleyana Bess. in Hook. Plant 2–4 dm. tall; stems several; basal leaves linear-oblanceolate, entire to serrate-dentate at apex or more deeply lobed, often crowded, with fascicles in the axils; upper leaves 2–5 cm. long, mostly less than 1 cm. wide, mostly entire, above tomentulose or glabrate, below tomentose; panicle 1–2 cm. wide; involucre 3 mm. high, 2–3 mm. wide, campanulate, usually lightly tomentulose, involucral bracts 9–13; ray florets 5–9; disk florets 10–30. *A. Leibergii* Rydb. Wawawai, Wash., *U. Son.*

Artemisia ludoviciana Nutt., var. **ludoviciana.** Stems 3–10 dm. tall; lower leaves oblanceolate, lobed or entire; principal leaves linear to oblanceolate in outline, 3–9 cm. long, 0.5–2 cm. wide, gray and loosely floccose or green and nearly glabrous above, beneath white tomentose, mostly cleft into few linear-lanceolate lobes, or dentate, or entire; panicle 1.5–3 cm. broad; involucre 2.5–3 mm. broad, campanulate, more or less tomentose; heads 12–20-flowered. *A. vulgaris* L., subsp. *ludoviciana* (Nutt.) H. & C.; *A. vulgaris,* var. *ludoviciana* (Nutt.) Jepson. Stream banks, *U. Son., A. T.*

Var. **candicans** (Rydb.) St. John. *Sagewort.* Perennial, 5–15 dm. tall, tomentose throughout; root 3–5 mm. in diameter, creeping, brown; stems forming clumps; lower leaves obovate, the lanceolate lobes often again lobed or toothed; principal leaves 4–10 cm. long, 1.5–4 cm. wide, obovate to elliptic, pinnatifid, the segments oblong or lanceolate, often again cleft, or the blade rarely elliptic and nearly entire, densely white tomentose on both sides; panicle cylindric; involucre 4–7 mm. broad, hemispheric, tomentose; heads 20–50-flowered; achenes about 2 mm. long, ellipsoid, glabrous or resinous glandular. *A. candicans* Rydb.; *A. vulgaris* L., subsp. *candicans* (Rydb.) H. & C. Stream banks, Spokane Co., *A. T.*

Var. **incompta** (Nutt.) St. John. Stems 3–9 dm. tall, mostly herbaceous; herbage commonly green, the leaves either glabrate above and white tomentose beneath or densely tomentose or nearly glabrous throughout; lower leaves 2–8 cm. long, mostly obovate to broadly elliptic in outline, rarely fascicled and parted or divided into linear or lanceolate forward lobes, some of which are again toothed or lobed; upper leaves less cut or even entire; panicle narrow and spikelike or even racemose, or broader but not open, not very leafy; heads 3–3.5 mm. high, 2.5–4 mm. wide, the 9–14 involucral bracts sericeous-tomentose or glabrate and shining; ray florets 6–10; disk florets 15–30 (–45). *A. incompta* Nutt.; *A. ludoviciana,* subsp. *incompta* (Nutt.) Keck. Wawawai, *C. V. Piper* 6,466, *U. Son.*

Artemisia Michauxiana Bess. in Hook. Stems 2–8 dm. tall; lower leaves obovate or oblanceolate, cut nearly to the midrib into spreading, toothed or lobed divisions; principal leaves 2–8 cm. long, 1–4 cm. wide, obovate or oval in outline, green and glabrous or glabrate above, white tomentose beneath, dissected nearly to the midrib into linear or lanceolate lobes, these entire, toothed, or lobed; panicle 1–3 cm. broad; involucre 2–4 mm. wide, hemispheric, sparsely tomentose and glabrate, or glabrous, yellowish green; heads 20–50-flowered. *A. discolor* Bess.; *A. vulgaris* L., subsp. *Michauxiana* (Bess.) St. John, var. *discolor* (Bess.) Jepson.

Artemisia rigida (Nutt.) Gray. *Scabland Sagebrush.* Shrub, 1–4 dm. tall; young twigs densely white lanate; older twigs glabrate, yellowish, smooth; old stems with a rough grayish brown bark, exfoliating in strips; principal leaves 1.5–4 cm. long, about 1 mm. wide, silvery lanate canescent, sessile, entire or commonly 3–5-cleft into narrowly linear divisions, axillary shoots with smaller leaves; flowers sessile, usually single in the upper axils; involucre 2.5–3.5 mm. broad, the bracts in 3–4 series, canescent on the backs, the outer elliptic, the inner elliptic to spatulate, obtuse, scarious margined; the 5–15 disk flowers with the corolla 2–2.8 mm. long, funnelform; achenes 1.5 mm. long, prismatic, glabrous. Thin rocky soil, deep canyons and scablands, *U. Son.*

Artemisia tridentata Nutt., subsp. **tridentata.** *Sagebrush.* Shrub 4–50 dm. tall, fastigiately branched, aromatic with a pungent odor; herbage densely white tomentose throughout; branchlets white tomentose; branches with brown to black, fibrous, shreddy bark; leaves 1.5–4.5 cm. long, narrowly cuneate or spatulate, the apex usually 3-toothed; panicles 1–4 dm. long, narrow, dense, leafy; involucre 2–2.5 mm. broad, white lanate, the outer bracts orbicular-ovate, the inner elliptic-spatulate, scarious margined; ray flowers none; disk flowers 4–6; corollas 2–2.5 mm. long, narrowly funnelform; achenes about 1.5 mm. long, subcylindric, resinous dotted. *A. tridentata,* subsp. *typica* H. & C.

The sagebrush is not common or widespread even in the Upper Sonoran Zone of our area. Now (1959) R. F. Daubenmire states to me that it occurs "locally south of the Snake River, but on a wide variety of habitats and in all zones. In the northwest edge of the area it is confined to stony or sandy soils."

Artemisia tripartita Rydb. Shrub 2–8 dm. tall, often root sprouting; bark shreddy; branchlets canescent; leaves of vegetative shoots 0.5–3 cm. long, canescent, deeply divided into 3 linear or narrowly linear-oblanceolate lobes, entire or 3-cleft, or some leaves linear and entire; leaves of flowering shoots deeply 3-cleft; involucre 3–4 mm. high, 2–3 mm. wide; involucral bracts 8–12, canescent, the outer broadly ovate, sometimes with a narrow herbaceous tip, the inner oblong, nearly thrice as long as the outer; florets all disk, 4–7, perfect; corolla 2–2.5 mm. long; achenes resinous granuliferous. *A. tridentata* Nutt., subsp. *trifida* (Nutt.) H. & C.; *A. tridentata,* var. *trifida* McMinn. Scablands, *U. Son.*

ASTER. Aster

Mostly perennial herbs; leaves alternate; heads solitary, corymbed or panicled, many-flowered, radiate; ray flowers several or numerous, in one row, fertile or rarely sterile, white, purple, or blue, never yellow; disk flowers yellow, often turning purple; involucre imbricated; bracts commonly with herbaceous tips; receptacle flat or convex, naked; pappus tawny, simple of copious slender scabrous capillary bristles; anthers tipped with an appendage; styles appendaged; achenes more or less compressed, rarely slender, 4–5-nerved. (Name from Gr., *aster,* a star.)

Pappus in flower nearly as long as the rays, in fruit longer and concealing them. *A. frondosus.*
Pappus shorter than the rays even in fruit,
Involucral bracts glandular puberulent,
Leaves or some of them serrate or dentate; involucral bracts imbricate in 4–8 series,
Involucral bracts with squarrose tips; leaves puberulent, soft. *A. canescens,* var. *viscosus.*
Involucral bracts erect or loosely spreading; leaves scabrous. *A. conspicuus.*
Leaves entire; involucral bracts imbricate in 2–3 series,
Involucre 8–11 mm. high; basal leaves lanceolate; middle leaves oblong-oblanceolate to ovate, auriculate clasping. *A. integrifolius.*
Involucre 5–7 mm. high; basal leaves linear-oblanceolate; linear or oblong-linear, not auricled at base. *A. campestris.*
Involucral bracts not glandular, but often pubescent,
Involucre 3–4 mm. high. *A. ericoides,* forma *Gramsii.*
Involucre 5–15 mm. high,
Outer bracts enlarged, foliaceous, longer than the inner; involucre not regularly imbricate,
Cauline leaves few, little reduced, the upper lanceolate,
Stems monocephalous; outer bracts nearly all foliaceous. *A. foliaceus,* var. *foliaceus.*
Stems with several heads; outer bracts not always foliaceous,
Middle and upper leaves not markedly auriculate; bracts glabrous on the back; stem inconspicuously appressed hairy. Var. *frondeus.*
Middle and upper leaves markedly auriculate; bracts often pubescent on the backs; stem frequently spreading hairy,
Bracts linear, acute or acuminate, the foliaceous ones

linear to narrowly lanceolate.
Var. *Lyallii.*

Bracts broader, the foliaceous ones broadly lanceolate to ovate.
Var. *Cusickii.*

Cauline leaves many, the upper reduced, linear or nearly so,

Outer involucral bracts narrowly linear; plant tufted without rhizomes. *A. cordalenus.*

Outer involucral bracts oblanceolate or narrowly so; plant producing rhizomes,

Blades glabrous above; ray flowers 9–13 mm. long. *A. Eatoni.*

Blades puberulous above; ray flowers 7–9 mm. long. *A. subspicatus.*

Outer involucral bracts the smaller, all regularly imbricate,

Involucral bracts pubescent on the back,

Bracts imbricate in 2 series; middle leaves 15–30 mm. long, linear, rigid, scabrous. *A. stenomeres.*

Bracts imbricate in 3–5 series; middle leaves 3–13 cm. long, lanceolate, chartaceous, softly pilosulous. *A. Jessicae.*

Involucral bracts glabrous on the back,

Herbage glaucous; achenes glabrous. *A. laevis,* var. *Geyeri.*

Herbage not glaucous,

Outer involucral bracts wholly herbaceous, linear. *A. Fremontii.*

Outer involucral bracts cartilaginous at base,

Blades at first appressed pilosulous. *A. columbianus.*

Blades glabrous or merely ciliolate,

Bracts linear-oblong or broader,

Bracts linear-oblong, the apex lanceolate. *A. Pickettianus.*
Bracts oblanceolate, with green tips. *A. adscendens.*
Bracts linear, the tip merely minutely scabrous ciliolate; achenes appressed pilosulous,
Leaves linear or nearly so. *A. occidentalis,* var. *occidentalis.*
Leaves lanceolate or oblanceolate. Var. *intermedius.*

Aster adscendens Lindl. in Hook. Perennial, 1–8 dm. tall, from a rhizome or branching caudex; stem pubescent at least above; lower leaves 2–10 cm. long, oblanceolate, petioled; cauline leaves linear to oblong or lanceolate, about 1 cm. wide and more than 7 times as long, entire, more or less auriculate; heads several; involucre 5–11 mm. high, the bracts well imbricate, the outer ones obtuse; rays 15–40 blue, violet, pink, or white, 5–12 mm. long; pappus white to reddish. *A. chilensis* Nees, subsp. *adscendens* (Lindl. ex DC.) Cronq., var. *euadscendens* Cronq. Wawawai, Wn., Sept., 1903, *Piper & Beattie, U. Son.*

Aster campestris Nutt. Perennial, 1–7 dm. tall; caudex 2–4 mm. in diameter, woody, dark brown, producing slender rhizomes 1–2 mm. in diameter; stems 1 to several, erect or decumbent at base, the base hirsutulous or glabrate, the middle appressed puberulent, the upper part glandular puberulent; lowest leaves linear-oblanceolate, reduced, withered at anthesis; lower leaves 3–7 cm. long, sessile, 3-nerved, oblance-linear, more or less appressed puberulent, the margin hispidulous; middle and upper leaves 1–4 cm. long, similar, linear or oblong-linear, acute; cyme or panicle 3–10 cm. wide, glandular puberulent; involucre broadly campanulate; bracts imbricate in about 3 series, linear, acuminate, the body pale cartilaginous, the tip loose or even somewhat recurved; ray flowers 6–10 mm. long, violet, disk flowers 5–6 mm. long, tubular; pappus 6–6.5 mm. long, brownish; achenes 2 mm. long, linear, brown, appressed pilosulous. Prairies, *A. T.*

Aster canescens Pursh, var. **viscosus** (Nutt.) Gray. Biennial, 2–6 dm. tall, whitish puberulent throughout; stems much branched above; basal and lower leaves 3–8 cm. long, oblanceolate, spinulose dentate, short petioled; upper leaves 1–6 cm. long, sessile, reduced, oblanceolate, more or less dentate, the uppermost bracteose, linear, entire; heads numerous, paniculate or corymbose; involucre 8–10 mm. high, turbinate; bracts imbricate in 5–8 series, linear, acute, the body canescent puberulent, the tips green; ray flowers 8–15 mm. long, violet; disk flowers 6–7 mm. long, tubular; pappus 4–6 mm. long, white; achenes 3–3.7 mm. long, linear-oblanceolate, flattened, yellowish, pilosulous. *Machaeranthera canescens viscosa* (Nutt.) Piper; *M. viscosa* (Nutt.) Greene; *A. leucanthemifolius* Greene. Abundant, dry open places, *U. Son., A. T.*

Aster columbianus Piper. Perennial, 4–7 dm. tall; caudex 2–7 mm. in diameter, brown, often with a slender rootstock; stem erect, appressed pilosulous, below often glabrate and simple; basal and lower cauline leaves withered and gone by anthesis; middle and upper cauline leaves 1–8 cm. long, oblong-linear, firm, acute, acerose-tipped, 1-nerved, appressed pilosulous or glabrate above, the margin appressed pilosulous, sessile by a broad base, the upper leaves much reduced; heads terminal, numerous in a loose cyme; involucre 5–7 mm. high, turbinate; bracts imbricate in 3–4 series, linear-oblanceolate, the body cartilaginous, the tip herbaceous, spreading, acerose-tipped, hirsutulous ciliate; ray flowers 7–8 mm. long, violet; disk flowers 4 mm. long, tubular; pappus 4–5 mm. long, whitish; achenes 1.5 mm. long, linear, brown, appressed pilosulous. *A. ericoides* L., f. *caeruleus* of Blake in part, not of (Benke) Blake. Stream banks, Waitsburg, *Horner* 559 (type), 627; Wawawai, *Piper* 1,602, *U. Son., A. T.*

Cronquist considered this a hybrid of *A. pansus* (Blake) Cronq.

Aster conspicuus Lindl. Perennial, 3–8 dm. tall, spreading by horizontal, forking rhizomes, 2–3 mm. in diameter, brown; stems single or few, simple below, with or without hirsute pubescence, above glandular puberulent; basal and lowest cauline leaves oblanceolate, reduced, almost bracteose, withered at anthesis; lower and middle leaves 6–15 cm. long, sessile, oblanceolate or broadly so, thick and firm, scabrous on both sides, coarsely serrate; upper and bracteal leaves similar, 3–8 cm. long, ovate to lanceolate, acute; cyme 7–25 cm. wide, loose, glandular puberulent; involucre 8–12 mm. high, broadly campanulate; bracts imbricate in 4–6 series, oblong-lanceolate, the body subcartilaginous, pilose ciliate, the tip herbaceous, densely glandular puberulent; ray flowers 15–20 mm. long, blue; disk-flowers 6.5–7.5 mm. long, the tube narrow funnelform, contracting to a tubular base; pappus 7–8 mm. long, brownish; achenes 4–4.5 mm. long, brownish, linear oblanceolate, compressed, several ribbed, appressed pilosulous. Woods, *A. T. T.*

Aster cordalenus Henders. Perennial, in clumps, 2–3 dm. tall; rhizomes 2–4 mm. in diameter, clustered, producing lateral rhizomes; stems ascending, simple and glabrate below, above branched and pilosulous in decurrent lines; basal and lower leaves spatulate, or linear-oblanceolate, entire, the petiole ciliate; middle and upper leaves, 2–12 cm. long, 5–10 mm. wide, but little reduced upwards, linear-oblong-lanceolate, firm chartaceous, the base auriculate clasping, the margin scabrous ciliolate; heads several, in loose cymes or panicles; involucre 5–10 mm. high, semiorbicular; bracts in several subequal series; pilosulous ciliate, the inner linear, acute, mostly cartilaginous with a small herbaceous tip; ray flowers 8–10 mm. long, blue to violet; disk flowers 4 mm. long, tubular; pappus 4–5 mm. long, whitish; achenes 1.5–2 mm. long, oblanceolate, somewhat appressed pilosulous. Gravelly shores, Lake Coeur d'Alene, *A. T. T.*

The type is *Henderson* 2,992, from Fort Sherman.

Cronquist (1943) reduced this to a synonym of the species now called *A. subspicatus* Nees.

Aster Eatoni (Gray) Howell. Perennial, 3–10 dm. tall; rhizome 2–5 mm. in diameter, horizontal, brown, freely producing lateral rhizomes; stem simple and glabrous below, above freely branching and somewhat appressed puberulent; basal leaves with petioles 15–17 cm. long; blades linear-oblanceolate, entire, thin, scabrous on the margin, withered by anthesis; middle and upper leaves 2–12 cm. long, numerous, linear-lanceolate, sessile and auriculate clasping at base, entire, thin chartaceous, glabrous but for the scabrous margin; heads numerous in a panicle 1–4 dm. long; involucre 6–9 mm. high, broadly turbinate; bracts in several series but more or less equal and blending into the bracts of the peduncle, the outer usually the longer, oblanceolate, foliaceous, scabrous ciliate, the inner oblong, acute, the base cartilaginous, the tip foliaceous, scabrous ciliate; ray flowers violet; disk flowers 4.5–5 mm. long, tubular; pappus 4.5–5 mm. long, whitish; achenes 2–2.3 mm. long, oblong-oblanceoloid, compressed, purplish, appressed pilosulous. Stream banks, *U. Son., A. T.*

This species is not clearly separable from *A. Douglasii.*

Aster ericoides L., forma **Gramsii** Benke. Perennial, 3–6 dm. tall, spreading hirsutulous throughout; caudex 4–15 mm. in diameter, short, brown, producing offsets; stems erect, simple below; freely branching above; basal and lower leaves 2–3 cm. long, oblance-linear, withered and gone by anthesis; cauline leaves 10–35 mm. long, linear to oblong-linear, acute, bristle-pointed, 1-nerved; heads numerous, in leafy racemes on the upper branches; involucre turbinate; bracts imbricate in about 3 series, broadly linear, hirsutulous, the body pale, cartilaginous, the tip herbaceous, loosely spreading or somewhat recurved, spine-tipped; ray flowers 4–6 mm. long, white; disk flowers 3–3.3 mm. long, tubular; pappus 2.8–3.3 mm. long, tawny; achenes 1–1.4 mm. long, oblanceoloid, flattened, brown, appressed pilosulous. *A. multiflorus* of Piper, not of Ait.; *A. exiguus* (Fern.) Rydb., not *A. multiflorus,* var. *exiguus* Fern.; *A. ericoides,* var. *prostratus* (Kuntze) Blake. Dry places, *U. Son., A. T.*

Aster foliaceus Lindl., var. **foliaceus.** Perennial, 2–5 dm. tall; rhizome 1–2 mm. in diameter, brownish, freely producing lateral rhizomes; stem erect, simple and glabrous below; basal and lower leaves 1–14 cm. long, spatulate, petioled, withered by anthesis; middle and upper leaves 3–10 cm. long, sessile, auriculate clasping at base, lanceolate or oblong-lanceolate, glabrous except for the scabrous margins; peduncles with decurrent lines of puberulence; heads 1 to several in a corymb; involucre semiorbicular; bracts blending into the bracts of the peduncle, in several series but often subequal, the outer lanceolate, as long as or longer than the inner, the inner linear-oblanceolate, ciliate, the lower margins cartilaginous, the large tip foliaceous; ray flowers 10–14 mm. long, violet; disk flowers 6–6.5 mm. long, yellow; pappus 6 mm. long, white; achenes 2.5 mm. long, oblanceolate, compressed, yellowish, appressed pilosulous. Banks of Spokane River, *Elmer* 866.

Var. **Cusickii** (Gray) Cronq. Perennial, 3–8 dm. tall; rhizome 2–4 mm. in diameter, brown; stem erect, glabrous and simple below, pilose above; basal and lower cauline leaves with petioles 3–5 cm. long, the base dilated, half clasping; blades 4–6 cm. long, oblanceolate, entire, pilosulous, usually withered by anthesis; middle and upper leaves oblanceolate to oblong-lanceolate or ovate, acute, sessile, sometimes the middle ones narrowed above the auriculate clasping base, pilosulous, entire or rarely few toothed, beneath paler green; heads terminal, single or in a few flowered, loose cyme peduncles white pilose; involucre 9–15 mm. high, semiorbicular; bracts in several series but the outermost blending into the bracts of the peduncle, the inner linear, acuminate with the base cartilaginous; ray flowers 15–20 mm. long, pale blue; disk flowers 4.5–5 mm. long, tubular; pappus 5.5–6 mm. long, brownish; achenes 2.3–2.5 mm. long, oblanceoloid, flattened, brownish, glabrous. *A. Cusickii* Gray. By streams, Blue Mts., Garfield Co., *Horner* 363, *Huds.*

Var. **frondeus** Gray. Differs by having the upper leaves 5–15 cm. long, ample; involucre 7–12 mm. high, the outermost bracts often foliaceous and oblong-lanceolate. Moist meadows, Blue Mts., *Huds.*

Var. **Lyallii** (Gray) Cronq. Involucral bracts linear, acuminate or acute, the foliaceous ones tapering from base to apex. It resembles var. *Cusickii,* but frequently has the leaves smaller and not so conspicuously auriculate-clasping. *A. Cusickii* Gray, var. *Lyallii* Gray. Whitman Co., Rock Lake, *Lake & Hull* 813.

Aster Fremontii (T. & G.) Gray. Perennial, 2–9 dm. tall; caudex 2–6 mm. in diameter, woody, brown, usually several branched at apex; stems glabrous below, pilosulous in decurrent lines above; basal and lower leaves often persisting, with petioles 2–9 cm. long, pilose ciliate at base; blades 3–7 cm. long, oblanceolate, glabrous except the scabrous ciliolate margin; middle leaves 3–8 cm. long, oblanceolate to lanceolate, glabrous, the margins scabrous ciliolate, the base half clasping; upper leaves 1–3 cm. long, linear; heads in a loose cyme 3–15 cm. wide, or even single; involucre 5–7 mm. high, suborbicular; bracts narrowly linear, acute, imbricate in 3–5 series, ciliolate, the inner cartilaginous below, herbaceous at tip; ray flowers 10–13 mm. long, violet; disk flowers 4 mm. long, tubular; pappus 4 mm. long, whitish; achenes 1.8–2 mm. long, oblanceolate, compressed, brown, appressed pilosulous. *A. adscendens,* var. *Fremontii* T. & G. Open slopes and thickets, *A. T.*

Cronquist (1943) was of the opinion that this species had no constant differences, and should be reduced to the synonymy of *A. occidentalis,* var. *occidentalis,* but the author disagrees.

Aster frondosus (Nutt.) T. & G. Annual, 3–75 cm. tall, smooth, rather fleshy; tap-root 2–5 mm. wide, brown; stems erect or decumbent, simple or commonly bushy branched, glabrous below, sparsely pilosulous above; basal and lowest leaves 2–4 cm. long, spatulate, the winged petiole hirsute ciliate; cauline leaves 1–4 cm. long, spatulate or oblance-linear, hispid ciliate towards the base; heads numerous in narrow panicles nearly as long as the plant; involucre 5–9 mm. high, broadly campanulate; bracts all foliaceous, blending into the bracts of the peduncle, imbricate in 2–4 series, oblong, acute, or oblance-oblong, the outer ciliolate, the base hyaline margined; ray flowers 5–6 mm. long, pink, barely exceeding the young pappus; disk flowers 4 mm. long, tubular; mature pappus 6–7 mm. long, copious, whitish to brownish; achenes 2–2.3 mm. long, oblanceolate, compressed, yellowish, appressed hirsutulous. *Brachyactis frondosa* (Nutt.) Gray. River or pond shores, especially of alkali ponds, channeled scablands from Medical L. southward; shores of Snake R., *U. Son.*

Aster integrifolius Nutt. Perennial, 1–7.5 dm. tall; rootstock 2–5 mm. in diameter, brown, often forking at top; stems simple and sparsely villous to glabrate below, glandular pilose above; basal and lower leaves with petioles 3–15 cm. long, clasping at base; blades 5–15 cm. long, firm, lanceolate, the base cuneate, decurrent, the midrib and margins pilosulous, usually withered at anthesis; middle and upper leaves 2–12 cm. long, oblong-oblanceolate to ovate, acute, the base auriculate clasping, the upper more or less pilosulous and glandular atomiferous beneath; cymes 4–30 cm. long, racemose, glandular pilosulous; involucre broadly campanulate; bracts densely glandular puberulent, linear-oblanceolate, the body cartilaginous, pale below, often purple above, the tip herbaceous, loosely appressed in about 2 series, or somewhat recurved; ray flowers 13–16 mm. long, violet; disk flowers 6–9 mm. long, tubular; pappus 6–9 mm. long, whitish; achenes 5–6 mm. long, linear-oblanceolate, compressed, brown; appressed pilosulous. Stony hillsides, *Huds.*

Aster Jessicae Piper. Perennial, tufted, 3–15 dm. tall, soft hairy throughout; rhizome 3–5 mm. in diameter, horizontal, brown, freely producing lateral rhizomes; stem erect, simple below, densely white villous; basal and lower leaves with petioles 3–6 cm. long, stout, densely villous, dilated and half clasping at base; blades 3–15 cm. long, ovate or lance-ovate, pinnately veined, obtuse or acute, entire or rarely serrate, pilosulous on both sides, withered by anthesis; cauline leaves contracted to an oblong auriculate base, entire (or rarely serrate), sessile, firm chartaceous, pilosulous on both sides; heads in small clusters on tips of the branches, forming a cyme 15–40 cm. long; involucre 8–10 mm. high, broadly turbinate; bracts blending into the bracts of the peduncle, loosely imbricate, oblong-linear, acute, acerose, tipped, pilose throughout, the lower margins pale, cartilaginous, the middle and upper half thick herbaceous; ray flowers 12–20 mm. long, pale violet; disk flowers 6 mm. long, tubular; pappus 5–6 mm. long, brownish; achenes 3.5–4 mm. long, linear-oblanceoloid, compressed, 4-angled, brown, appressed pilosulous. Low ground, rare, Whitman and Latah Counties, *A. T.*

The type collection is *Piper* 1,604, from Pullman. Prof. Piper named the species for his sister-in-law, Jessie Hungate, later Mrs. C. W. Sampson.

Aster laevis L., var. **Geyeri** Gray. Perennial, 2–11 dm. tall, the herbage glabrous; rootstock 2–4 mm. in diameter, woody, brown, at times producing slender rhizomes; stem erect, simple below; basal leaves 6–10 cm. long, often present at anthesis, oblanceolate, entire or serrulate, tapering into the winged petiole; cauline leaves 1–15 cm. long, sessile, half-clasping by the broad base, oblong-oblanceolate or -lanceolate to ovate, entire or serrulate, the margin scabrous-ciliate; heads numerous, in a cyme or panicle 3–30 cm. long; involucre 6–8 mm. high, broadly campanulate; bracts passing into the bracts of the peduncle, imbricate, in about 4 series, linear, acute, pilosulous ciliate, the body cartilaginous, pale, the tips herbaceous; ray flowers 8–13 mm. long, violet; disk flowers 4 mm. long, tubular; pappus 4–4.5 mm. long, brownish; achenes 2.5–3 mm. long, oblanceoloid, compressed, yellowish. *A. Geyeri* (Gray) Howell; *A. brevibracteatus* Rydb. Open gravels or open woods, especially by streams, *A. T., A. T. T.*

Aster occidentalis (Nutt.) T. & G., var. **occidentalis.** Perennial, 1–6 dm. tall, nearly glabrous; rhizome 1–3 mm. in diameter, brown, producing lateral rhizomes; stem erect, simple below, pilosulous above in decurrent lines; basal and lower leaves with petioles 2–5 cm. long, ciliate at base; blades 2–10 cm. long, oblance-linear to almost linear, 1-nerved, the scabrous ciliolate margin entire or few toothed; cauline leaves 2–15 cm. long, linear or nearly so, scabrous ciliolate, the ciliate base half clasping; cyme many or few headed, loose; involucre 5–7 mm. high semiorbicular; bracts imbricate in about 4 series, acute, pilose ciliate at base, the tip herbaceous, the base pale cartilaginous; ray flowers 7–10 mm. long, lavender; disk flowers 4–4.5 mm. long, tubular; pappus 4–4.5 mm. long, whitish; achenes 2–3 mm. long, oblanceolate, flattened, thick margined, brownish. Spokane Bridge, *Suksdorf* 9,090, *A. T.*

Var. **intermedus** Gray. Differs from the species by having the basal leaves oblanceolate, and the cauline linear-oblanceolate or -lanceolate. Wet places, *U. Son., A. T.*

Aster Pickettianus Suksd. Perennial, 3–6 dm. tall, nearly glabrous; rootstock 2–5 mm. in diameter, short, brown; stem erect, simple below; basal leaves with petioles 1–3 cm. long, hirsute ciliate; blades 3–6 cm. long, oblanceolate, heavy pointed, entire or coarsely crenate; middle leaves 3–8 cm. long, firm chartaceous, sessile, oblanceolate, glabrous, the base auriculate clasping; upper leaves 1–2 cm. long, reduced; heads numerous in a cyme 1 dm. long; peduncles nearly glabrous; involucre 5–7 mm. high, campanulate; bracts imbricate in 4–6 series, mostly pale cartilaginous, only the tip green herbaceous, the outer ones short deltoid, the others oblong-linear, acute, the tip puberulent ciliate; ray flowers 8–14 mm. long, violet; disk flowers 4–4.5 mm. long, tubular; pappus 4.5–5 mm. long, brownish; achenes 2.5–3 mm. long, oblanceolate, compressed, yellowish. Prairies, Spangle, *Suksdorf* 8,984, *A. T.*

Known only from the type collection.

Named for the late Prof. Fermen L. Pickett of Washington State University.

Aster stenomeres Gray. Perennial, 14–31 cm. tall, tufted; caudex 2–5 mm. in diameter, woody, brown, above forking and multicipital; stems erect, simple, pilose; basal leaves reduced to ovate bracts, these and the lowest leaves withered by anthesis; middle cauline leaves subcoriaceous, 1-nerved, acute or bristle-tipped, very scabrous; upper leaves suddenly reduced, leaving a naked peduncle usually 1–5 cm. long; heads terminal, single; involucre 10–17 mm. long, hemispheric, usually purplish; bracts linear, acuminate, dry and scale-like, pilose at base and along the thickened midrib, often purplish, the margins and tip hyaline, puberulent near the tip; ray flowers 15–22 mm. long, violet;

disk flowers 7–8 mm. long, tubular; pappus 7–9 mm. long, whitish, the outer bristles 0.5–1 mm. long, few; achenes 4–5 mm. long, oblanceolate, compressed, thick margined, brown, appressed pilosulous. *Ionactis stenomeres* (Gray) Greene. Grassy summits, Mt. Spokane, Rathdrum Peak, and Mica Peak, *Huds.*

Aster subspicatus Nees. Perennial, 3–15 dm. tall, nearly glabrous; rhizome 2–5 mm. in diameter, horizontal, freely producing lateral rhizomes; stem erect, glabrous below, above somewhat puberulent; basal and lower leaves petioled, oblanceolate, entire, withered by anthesis; middle and upper leaves 2–13 cm. long, sessile, auriculate at base, linear-lanceolate, entire, thin chartaceous; heads numerous, in a panicle 2–4 dm. long; involucre 5–8 mm. high, broadly turbinate; bracts blending into the bracts of the peduncle, imbricate in 3–4 series, oblong-linear, acute, ciliate, the lower margins cartilaginous, the tips herbaceous, a few of the outer bracts foliaceous, enlarged oblanceolate, equaling the inner; ray flowers 7–9 mm. long, violet; disk flowers 4 mm. long, tubular; pappus 4–4.5 mm. long, whitish; achenes 2–2.5 mm. long, oblanceoloid, compressed, somewhat purplish, sparsely appressed puberulent. *A. oregonus* (Nutt.) T. & G. Stream banks, *A. T., A. T. T.*

BALSAMORHIZA. Balsam-root

Perennial herbs; roots fusiform, very large, rich in resin; stems subscapose; leaves mostly basal, or opposite or alternate; heads large, mostly solitary; involucral bracts in 2–4 series; receptacle convex; chaff firm, persistent, folded; ray flowers fertile, yellow or purplish; the achenes 3-angled; disk flowers fertile, yellow, funnelform; pappus none; achenes 4-angled. (Name from Gr., *balsamon,* balsam; *rhizon,* root.)

Leaves all or mostly entire or but shallowly dentate or lobed,
  Blades green on both sides,
    Blades mostly entire, or obtusely dentate or lobed,
      Stems strigose, glandular above; blades entire or rarely dentate, glandular; bracts lanceolate, glandular. — *B. Careyana,* var. *Careyana.*
      Stems pilosulous, and villous at apex; plant not glandular; blades shallowly lobed below the middle; bracts narrowly obcuneate. — B. × *Bonseri.*
    Blades mostly sharply spinulose dentate, rarely lobed; involucral bracts subequal. — *B. serrata.*
  Blades densely white hairy on one or both sides,
    Blades entire, broadly cordate at base. — *B. sagittata.*
    Blades subentire to dentate or shallowly incised, rounded or cuneate at base. — B. × *tomentosa.*
Leaves from deeply incised to bipinnatifid,
  Blades white floccose tomentose on both

sides; involucral bracts tomentose to the tip, the outer the longer. *B. incana.*
Blades green, sparsely hispid; involucral bracts sparsely pilosulous at tip, the outer the shorter,
Rays permanently yellow; achenes glabrous. *B. hirsuta.*
Rays rosy in age; achenes strigose. *B. rosea.*

Balsamorhiza X Bonseri St. John. (*B. rosea* X *sagittata*). Root 1–3 cm. in diameter, brown, woody, the crown simple or divided; stems 2–3.5 dm. tall, appressed pilosulous, near the apex loosely villous also; basal leaves with petioles 4–16 cm. long, appressed pilosulous, the blades 12–19 cm. long, thick chartaceous, greenish on both sides, narrowly deltoid or lance-deltoid, the apex acute, the base broadly cordate, the margin at least below the middle with shallow, more or less ovate lobes, appressed white puberulent above, more densely so beneath; cauline leaves opposite, 1 pair, the petioles 4–6 cm. long, the blades 2.5–5 cm. long, linear-lanceolate; heads solitary; involucre hemispheric, 13–20 mm. in diameter, a little exceeding the disk; involucre bracts 12–18 mm. long, subequal, narrowly obcuneate, loosely white villous; rays 2–3.5 cm. long, yellow or pinkish tinged, pilosulous towards the base, the apex bidentate; the ray achenes 8 mm. long, cuneate, 4-angled, brown, pilosulous below, densely pilose above; disk corollas with the tube 7–8 mm. long, almost tubular, tapering to a constricted base, the lobes subovate, 1 mm. long.

Glacial hill, s. w. of Seven -Mile Bridge, Spokane Co., May 11, 1924, *Bonser* (type). It was named in honor of the collector, the late Thomas A. Bonser, of Spokane. As shown by Ownbey (1943) *B. Bonseri* seems to be the hybrid *B. rosea X sagittata.*

Balsamorhiza Careyana Gray, var. **Careyana.** Root 1–1.5 cm. in diameter, brown, woody; stems bracted, 2–6 dm. tall; strigose hirsute, above also glandular puberulent; basal leaves with petioles 7–30 cm. long, the blades 9–22 cm. long, cordate-ovate to hastate, acute, green, hispidulous, glandular atomiferous; cauline leaves alternate or opposite, bracteose; heads 1–4, paniculate; involucre hemispheric; outer bracts 14–28 mm. long, lanceolate, foliaceous, hirsute and glandular dotted; inner bracts oblong lanceolate, cartilaginous, hirsute at least at tip; rays 2–3 cm. long, elliptic, deciduous; achenes 6–7 mm. long, canescent hirsute, brown, elliptic in outline, curved; chaff 8–12 mm. long, oblanceolate, the acuminate tip hispidulous; disk flowers 7–7.5 mm. long, tubular, narrowed at the base, the lobes ovate; achenes 5–6 mm. long, oblong, the faces hollowed, hispidulous. Arid slopes, Palouse Falls, *St. John & Pickett* 3,238; 6,182; *Brode* 8, *U. Son.*

Described from specimens collected by the Rev. Henry Spalding, labeled Clear Water, Oregon [Lapwai, Ida.]. The specimens were probably collected on a trip westward, perhaps towards Walla Walla.

Balsamorhiza hirsuta Nutt. Root 15–21 mm. in diameter, woody, brown; stems 15–45 cm. tall, subscapose, hirsutulous and glandular puberulent, basal leaves with petioles 2–12 cm. long, hirsute; blades 6–23 cm. long, lanceolate in outline, pinnately divided, hispid and glandular atomiferous, the divisions 1–2-pinnatisect, the ultimate ones linear to linear-lanceolate; heads solitary; involucre hemispheric; involucral bracts hirsute, pilose ciliate, the tips foliaceous, imbricate in 3–4 series, lanceolate, the inner 13–22 mm. long; rays 2.5–4 cm. long, narrowly elliptic, yellow; achenes 6 mm. long, linear spatulate in outline, curved, brown, smooth; chaff 10–12 mm. long, linear-oblong, with an acuminate greenish, pilosulous tip; disk flowers 8–9 mm. long, tubular, the lobes ovate; achenes 5.5–7 mm. long, linear, black, smooth. Dry plains east of Walla-Walla, *Nuttall, A. T.*

The type locality is our only record.

Balsamorhiza incana Nutt. Root 7–15 mm. in diameter, brown; stem 1.5–4 dm. tall, scapose, white lanate; basal leaves with petioles 3–15 cm. long, lanate; blades 8–27 cm. long, lanceolate, pinnately divided, the divisions lanceolate to ovate-lanceolate in outline 5–25 mm. wide, entire, serrate, or incised; heads solitary; involucre turbinate, finally hemispheric, densely white lanate, foliaceous, outer bracts 18–26 mm. long, linear-lanceolate, rays 3–4.5 cm. long, narrowly elliptic, yellow; achenes 7 mm. long, obovate-oblong, brown, smooth; chaff 13–15 mm. long, lance-linear, with an acuminate, green, pilosulous tip; disk flowers 8–10 mm. long, almost tubular, glabrous, the lobes lanceolate; achenes 5.5–7 mm. long, linear-oblong, cuneate at base, brown, smooth. Dry, open slopes, Blue Mts. and Craig Mts., *A. T.*

Balsamorhiza rosea Nels & Macbr. Acaulescent from a short, thick, warty root; leaves 4–10 cm. long including the short petiole, somewhat stiff cinereous, irregularly pinnately divided, the blunt divisions coarsely toothed; scapes about 1 dm. tall, few, with one head, the pubescence longer and looser than on the leaves, at base with a pair of linear, acute bracts 1 cm. long; heads 3–4 cm. broad; bracts lanceolate, acute, rather loose, the outer shorter, densely ciliate with long, glistening, soft hairs, those of the surface subappressed; rays crowded, mostly persistent, rose-colored or purplish, oval, crenately and irregularly few-toothed at apex, pubescent with long, soft hairs, especially on the prominent parallel veins below and around the short tube; chaff scarious margined, and fimbriate near the tip; ray achenes somewhat compressed, carinate on both sides, hirsute near the top, the hairs becoming shorter downwards; disk achenes glabrous or nearly so except at the ciliate summit. Rare and local, Seven Mile, Spokane Co., *F. C. Raney* in 1939.

Balsamorhiza sagittata (Pursh) Nutt. Root 1–5 cm. in diameter, woody, brown; stems 2–6.5 dm. tall, white lanate, densely so at apex; basal leaves with petioles 10–35 cm. long, white lanate; blades 7–27 cm. long, deltoid to ovate, the base broadly cordate, above appressed pilosulous but becoming greenish or green, beneath white tomentulose; heads solitary or occasionally 2–3 in a cyme; involucre hemispheric, densely white lanate; bracts lanceolate to linear-lanceolate, foliaceous, the outer the longer, 13–25 mm. long, the tips spreading; rays 2–4 cm. long, elliptic, yellow; achenes 7–8 mm. long, oblong-oval, curved, brown or grayish, smooth; chaff 13–16 mm. long, oblong, the greenish acuminate tip pilose; disk flowers 7.5–8 mm. long, tubular, the ovate lobes puberulent; achenes 7–8 mm. long, oblong-spatulate in outline, sharply angled, brown, smooth. Abundant, open slopes, *U. Son., A. T.*

The Rev. Henry Spalding recorded that the Indians called this plant *Pash,* and used the tender stalk, the roots, and the seeds for food. Horses graze on it, and the Columbia ground squirrels feed on it.

Balsamorhiza serrata Nels. & Macbr. Root 8–16 mm. in diameter, woody, brown; stems 1–3.5 dm. tall, subscapose, hirsutulous and more or less hirsute; basal leaves with petioles 2–11 cm. long, hirsute, dilated at base; blades 3–17 cm. long, ovate-lanceolate, subcoriaceous, hispidulous, nerves reticulate, the base cuneate decurrent, rarely subcordate, mostly with the margin sharply dentate, rarely somewhat lobed or subentire; heads solitary; involucre hemispheric, the bracts 16–27 mm. long, linear-lanceolate, villous at the base and margins, the back hispidulous, the tip foliaceous; rays 2.5–3.5 cm. long, yellow, narrowly elliptic, pilosulous below; achenes 6 mm. long, oblong-spatulate in outline, curved, brown, smooth; chaff 11–14 mm. long, linear-lanceolate, the acuminate tip green, pilose; disk flowers 8–9 mm. long, broad tubular the ovate lobes puberulent; achenes 6–7 mm. long, linear, brown, smooth. Dry stony land, Blue Mts., *A. T.*

The Rev. Henry Spalding recorded the Indian name *Kayrom,* and stated that the peeled root was baked for food.

Balsamorhiza X **tomentosa** Rydb. emend Weber. (*B. incana* X *sagittata*). Differs from *B. sagittata* in having blades oblong-ovate to oblong-lanceolate, subentire to dentate or shallowly incised. Dry open slopes, Anatone, *St. John & Palmer* 9,549; Zindel, *St. John & Brown* 3,900a, Lake Waha, *Sandberg et al.* 248, *A. T.*

BIDENS. Beggar Ticks

Annual or perennial herbs; leaves opposite; heads small or medium, yellow or sometimes white; ray flowers 3–10, sterile or none, in which case the flowers are all perfect and tubular; receptacle chaffy; pappus 2–4 rigid backwardly-barbed awns; ray and disk achenes obcompressed. (Name from Lat., *bi-*, two; *dens,* tooth.)

Leaves pinnate; rays short and almost hidden or none; achenes flat,
 Outer involucre of 10–16 bracts; inner bracts shorter than the disk. *B. vulgata.*
 Outer involucre of 4–8 bracts; inner bracts equaling the disk. *B. frondosa.*
Leaves simple; rays exserted; achenes 4-angled. *B. cernua.*

Bidens cernua L. Annual herb, 2–24 dm. tall, glabrous or the stems sparsely hirsute; tap-root 1–4 mm. in diameter, pale, the laterals fleshy; stem erect, angled, simple or branched. leaves 3–17 cm. long, sessile, linear-lanceolate, acuminate, remotely coarsely serrate, the base auriculate and more or less connate-perfoliate; heads terminal or in loose cymes, in flower nodding, in fruit erect; outer involucral bracts foliaceous, lanceolate, obtuse, irregular, 1½–3 times the length of the inner; inner bracts 6–10 mm. long, oval, thin, with many brown parallel ribs, the hyaline margin yellow; the 6–10 rays 7–14 mm. long, well exserted, elliptic to oval; chaff lanceolate, dark ribbed; disk flowers 4–5 mm. long, glabrous, the tube cylindric, the throat broadly campanulate; anthers brownish purple; achenes 4–6 mm. long, wedge-shaped, 4-angled, the corky angles retrorsely barbed, the 4 awns 2–3 mm. long. *B. glaucescens* Greene. Wet places, *U. Son., A. T., A. T. T.*

A specimen from Spangle, *Suksdorf* 933, has been determined as the var. *elliptica* Wieg. Though the leaves are broadly lanceolate, yet they are broad at base, so it seems best to retain it in the species.

Bidens frondosa L. Annual, 4–13 dm. tall, nearly glabrous, often purplish, leaves 5–10 cm. long, thin, slender petioled, pinnately 3–5-divided, the segments lanceolate to oblong-lanceolate, sharply serrate, cuneate at base, acuminate at apex; heads numerous long peduncled; outer involucre of ciliate, more or less foliaceous spatulate-linear bracts; inner bracts yellowish, ovate-lanceolate, equaling the disk, scarious margined; rays 1–5, yellow, shorter than or equaling the disk; disk flowers orange; achenes 7–10, flattened, cuneate-oblong, nearly black; the 2 awns barely half as long, retrorsely barbed. Wet places, *U. Son., A. T.*

Bidens vulgata Greene. Annual, 2–30 dm. tall, green and nearly glabrous; tap-root 1–5 mm. in diameter, pale; stem erect, striate, freely branched, smooth or sparsely hirsute; leaves pinnately 3–5-foliolate; petioles 1–6 cm. long; leaflets 2–10 cm. long, linear-lanceolate, serrate, light green, remotely puberulent or glabrate; heads terminal, more or less cymose; involucre hemispheric; outer bracts 10–40 mm. long, foliaceous, oblanceolate, hirsute ciliate; inner bracts 5–8 mm. long, brownish, elliptic or lanceolate, scarious margined; rays 3–3.5 mm. long, yellowish, concealed by the involucre, but usually present;

disk flowers 2.5–3 mm. long, the tube equaling the campanulate throat; achenes 5–11 mm. long, obovate to wedge-shaped, brownish, flat, the margins upwardly barbed, the 2 awns 3–5 mm. long. Near streams, *U. Son., A. T.*

BLEPHARIPAPPUS

Annuals; leaves alternate, linear, entire; heads rather few-flowered; ray flowers 3–6, white, short and broad, pistillate; disk flowers 7–12, white, perfect, some of the central sterile; involucral bracts 6–10, nearly in one row; receptacle convex, chaffy; pappus of 10 or 12 linear hyaline scales, each with a stout awn-like midrib, rarely wanting; ray achenes neither obcompressed nor laterally compressed. (Name from Gr., *blepharis* eyelash; *pappus,* seed down.)

Blepharipappus scaber Hook. Plant 8–30 cm. tall, appressed puberulent below, glandular puberulent and more or less sparsely hirsute above; tap-root 1–2 mm. in diameter, pale; stem simple to freely branched; leaves 5–30 mm. long, numerous, scabrous puberulent; heads few, terminal, or several in loose cymes; involucre 5–7 mm. high, semiorbicular, green, glandular puberulent, sometimes more or less hirsute; bracts lanceolate or oblanceolate, hyaline margined; rays 5–8 mm. long, 3-lobed half their length; disk flowers 2.5–3 mm. long; anthers purple; the pappus 1.8–2 mm. long, brownish; achenes 2 mm. long, wedge-shaped, dark, densely white pilose. *Ptilonella scabra* (Hook.) Nutt. Rocky soil, *U. Son.,* occasional in *A. T.*

BRICKELLIA

Herbs or undershrubs with opposite, alternate, or rarely whorled leaves; heads whitish; involucre campanulate, the scales imbricated, lanceolate or linear, the outer shorter, none herbaceous; receptacle flat, naked; flowers discoid, perfect; corolla slender, 5-toothed; pappus one row of separate capillary barbed or scabrous bristles; achenes prismatic, 10-nerved or -ribbed. (Named in honor of *Dr. John Brickell* of Savannah, Ga.)

Shrubs; involucre with green, squarrose tips; leaves broadly ovate and sessile or subsessile. *B. microphylla,* var. *microphylla.*

Herbs; involucre imbricate, the tips not herbaceous; leaves if sessile not ovate,
- Leaves sessile, elliptic-oblong to linear-lanceolate, entire; outer involucral bracts lanceolate. *B. oblongifolia,* var. *oblongifolia.*
- Leaves petioled, deltoid-ovate or -lanceolate, coarsely serrate or crenate; outer involucral bracts ovate, long caudate. *B. grandiflora.*

Brickellia grandiflora (Hook.) Nutt. Perennial herb, 3–7 dm. tall, appressed puberulent throughout; caudex short, bearing elongate, fusiform tubers 3–10 mm. in diameter, brown; stem striate, often bushy branched above; lower leaves opposite, upper leaves alternate; petioles 1–7 cm. long; blades 3–11 cm. long, glandular punctate, the base cordate; cymes terminal on the branches several to many headed; heads 20–38-flowered; peduncles naked below the heads, outer involucral bracts approximate, puberulent; inner bracts 7–10 mm. long, elliptic-oblong, with 5 heavy nerves, the hyaline margin pilose ciliate; corolla 5–6.5 mm. long, tubular, the lobes 0.2 mm. long, broadly ovate; pappus 5–6 mm. long, white; achenes 4.5–5 mm. long, brown, puberulent, the base callous. *Coleosanthus grandiflorus* (Hook.) Kuntze; *B. grandiflora*, var. *typica* Robins. Banks of Spokane River; and Salmon River, Blue Mts., *Horner* 351, *A. T.*

Brickellia microphylla (Nutt.) Gray., var. **microphylla.** Shrub 4–6 dm. tall, glandular pilose throughout; stems several, in a clump; main stem glabrate below, bushy branched above; leaves 3–20 mm. long, alternate, few toothed or entire, leaves of lateral branches smaller and becoming bracteose; heads about 22–flowered, single and terminal or in compact 2–7-headed cymes; peduncles with bracts blending into the remote lower involucral bracts; involucral bracts greenish, 4-nerved; outer bracts ovate, glandular puberulent; middle bracts oblong-ovate, with a few remote teeth; inner bracts 8–11 mm. long, oblong-linear, acute, glabrous above; corolla 7–8 mm. long, tubular, whitish or purplish, the lobes 0.2 mm. long, ovate; pappus 6.5–7 mm. long, whitish; achenes 4–4.5 mm. long, sharply ribbed, brown, hispidulous. *Coleosanthus microphyllus* (Nutt.) Kuntze. Basalt cliffs, Snake River Canyon, south of Asotin, *St. John et al.* 8,250; 8,258; 9,632, *U. Son.*

Brickellia oblongifolia Nutt., var. **oblongifolia.** Perennial herb, 1–5 dm. tall, glandular puberulent throughout; root 2–15 mm. in diameter, woody, brown, forking; stems numerous, forming a clump, simple to the inflorescence, glabrate below; leaves 1–4 cm. long, alternate, obtuse or acute, 3-nerved; heads 40–50-flowered, terminal or in small terminal cymes; involucral bracts loosely imbricate, viscid puberulent, the outer ones remote, 2–3-nerved; inner bracts 11–16 mm. long, lance-linear, acuminate, 4-nerved, less hairy; corolla 8–10 mm. long, tubular, the lobes 0.2 mm. long, ovate; pappus 7–8 mm. long, whitish; achenes 5–6 mm. long, heavy ribbed, brown, glandular puberulent. *Coleosanthus oblongifolius* (Nutt.) Kuntze; *B. oblongifolia*, var. *typica* Robins. Stream banks, Waitsburg; and Spokane region, *A. T.*

CARDUUS

Annual to perennial, spiny herbs; stem generally winged by decurrent leaf bases; leaves basal and cauline, alternate, serrate to pinnatifid; heads terminal, medium to large, discoid, involucral bracts imbricate in several series, mostly spine-tipped; receptacle flat or convex, densely bristly; flowers all tubular and perfect or the plants sometimes dioecious by abortion; corollas purple, red, white, or yellow, with slender tube and long narrow lobes; filaments pilose; anthers caudate; style with a thickened or hairy ring below the branches; pappus of numerous capillary, minutely barbellate bristles, deciduous in a ring; achenes glabrous, obovate. (The ancient Latin name of the spiny teazel.)

Heads nodding; involucre 3–4 cm. wide, its bracts 2–8 mm. wide. *C. nutans.*
Heads erect; involucre 1.5–2.5 cm. wide, its bracts 1–2 mm. wide. *C. acanthoides.*

Carduus acanthoides L. Biennial, 3–10 dm. tall, slender but tough, both stems and leaves very spiny; stem branched above, many angled and herbaceous winged, the wings narrow but with frequent semicircular expansions; basal leaves 5–20 cm. long, runcinate pinnatifid, sparsely crinkly villous, the margins rigidly spiny, prickles below sparse; cauline leaves much reduced, the upper sessile; heads remote; involucre hemispheric, the bracts imbricate, in many series, green, firm, subulate tipped, somewhat glandular puberulent, the inner 12–15 mm. long; flowers reddish purple; corollas 2-lipped, 14–15 mm. long; seeds 3–3.5 mm. long, olive green, narrowly obovate, punctate. European weed, introduced here. Lake Waha, Nez Perce Co., 1948, *Ownbey & Cronquist 5,633.*

Carduus nutans *L. Musk Thistle.* Usually biennial, 3–10 dm. tall; root fusiform; stem simple or little branched, above woolly and prickly winged; leaves deeply pinnatifid, 10–25 cm. long, 3–10 cm. wide, the lobes spine-tipped, glabrous above and throughout or the midrib and veins villous; heads mostly solitary, long stalked, nodding 3.5–7 cm. long, 1.5–8 cm. wide; outer and middle involucral bracts with the body ovate or lanceolate, 2–8 mm. wide, ending in a long, flat spiny tip which is spreading or reflexed; inner bracts narrower, scarcely spiny, often purple and woolly; achenes 3–4 mm. long; pappus 15–20 mm. long. Introduced from Eurasia; first found in 1939 at Colfax, Wn., by *R. P. Benson.*

CENTAUREA

Herbs; leaves alternate; heads many-flowered; flowers all with tubular and deeply 5-cleft corollas, some of the marginal ones commonly sterile, often much larger and conspicuous, the others perfect and fertile; involucre globose to ovoid or urn-shaped; the bracts imbricate in several series, tipped or margined with spines or scarious appendages; receptacle flat, very bristly; pappus of numerous rigid or sometimes chaffy naked bristles; achenes mostly compressed, the apex crowned. (Name Gr. *Kentaurion,* of the centaur. The name was used by Hippocrates for a healing herb used by the centaur Chiron.)

Involucral bracts entire; achenes attached basally. *C. repens.*
Involucral bracts or some of them with a terminal spine-tipped or lacerate appendage; achenes attached laterally,
  Involucral bracts spine-tipped; corollas yellow. *C. solstitialis.*
  Involucral bracts with the appendage lacerate or pectinate; corollas not yellow,
    Flowers all alike, discoid, not enlarged. *C. nigra.*
    Marginal flowers enlarged, showy,
      Involucral bracts pale, evident, with a narrow whitish, shallowly lacerate apex. *C. Cyanus.*
      Involucral bracts hidden by the brown, flabellate, deeply lacerate appendage. *C. Jacea.*

Centaurea Cyanus L. *Bachelor's Button.* Annual, 2–7 dm. tall, lanate throughout, or in part glabrate; roots fibrous; stems freely branching, many ribbed; leaves 2–10 cm. long, linear or lance-linear, entire or rarely remotely denticulate; heads terminal; involucre campanulate, somewhat lanate, becoming straw-colored; outer bracts ovate, the whole margin lacerate; inner bracts 11–15 mm. long, lanceolate, the tips brown or violet tinged and lacerate; corollas normally blue; the marginal flowers sterile, the corollas 17–22 mm. long, expanded funnelform, asymmetric, showy; central flowers 12–15 mm.

long, symmetrical; pappus awns barbellate, brown, very unequal, the inner longer, 2.5–3 mm. long; achenes 3–3.5 mm. long, ellipsoid, compressed, truncate, pale below, blue-gray at the middle, finely pilosulous, the base pilose tufted, the attachment lateral. Introduced from Eurasia as an ornamental in the gardens. Abundantly escaped to roadsides and fields, *A. T.*

A common cultivated plant with blue or occasionally white flowers. It has spread from cultivation and is abundantly established by roadsides and in fields, and there occurs in a myriad color forms. Among the many different individuals are flowers of every conceivable shade of blue, purple, violet, and pink, to white.

Centaurea Jacea L. *Brown Knapweed.* Perennial, 3–15 dm. tall, more or less lanate, scaberulous, or glabrate; root 2–8 mm. in diameter, brown; stems simple or branched, angled; leaves simple or rarely lobed, the basal with petioles 1–4 cm. long, the blades 5–9 cm. long, lanceolate or oblanceolate; cauline leaves 2–10 cm. long, the upper sessile, lanceolate; heads terminal; involucre 12–20 mm. long, ovoid; outer involucral bracts ovate, the appendage pale, orbicular, pectinate; middle bracts with appendages brown, chartaceous, flabellate and irregularly lacerate; inner bracts lanceolate, the appendages pale, oval, more or less lacerate; flowers rose magenta, the marginal sterile, enlarged, 15–22 mm. long; pappus none or a crown of minute bristles; achenes 2.8–3 mm. long, ellipsoid, compressed, truncate at apex, gray, glabrous, the attachment lateral. Weed, introduced from Eurasia or Africa. Grasslands, Pullman, introduced in 1923.

Centaurea nigra L. *Black Knapweed.* Perennial, 3–8 dm. tall, scaberulous and more or less lanate throughout; tap-root 2–8 mm. in diameter, woody, brown; basal and lower cauline leaves with petioles 1–10 cm. long, the blades 3–10 dm. long, oblanceolate, entire or dentate; middle and upper cauline leaves 3–8 cm. long, sessile, oblanceolate to lanceolate, entire or dentate; heads terminal; involucre 12–15 mm. high, globose; involucral bracts with the body lanceolate, straw-colored, concealed by the brown or black, ovate, long pectinate appendage of the next outer bract; flowers all perfect; corollas 12–20 mm. long, rose-purple; pappus 0.2–0.6 mm. long; achenes 3.5–4 mm. long, oblong, compressed, pale, pilosulous, the attachment lateral. Weed, introduced from Europe. Pullman, in 1895, *Hardwick,* 2,390, *A. T.*

Centaurea repens L. *Russian Knapweed.* Perennial 6–9 dm. tall, somewhat lanate throughout; roots long creeping and colony forming; stems often in dense tufts, freely branched, angled; basal leaves 5–15 cm. long, usually pinnatifid; cauline leaves 2–9 cm. long, sessile, usually entire, linear-lanceolate, lanate but soon glabrate, the margins scabrous; heads terminal; involucre 9–13 mm. long, ovoid, greenish to straw-colored; outer involucral bracts with an orbicular, chartaceous body, and a membranous, acute tip; inner bracts lanceolate, the tip long acuminate, pilose; flowers alike, perfect, rose-colored, the corollas 12–15 mm. long; pappus white, in several series, barbed, unequal, the inner 7–9 mm. long; achenes 3.5–4 mm. long, ellipsoid, compressed, brown, glabrous. *C. Picris* Pallas. Weed introduced from southern Russia or Siberia.

First noted in 1925, now widespread, and already troublesome. It is very difficult to eradicate.

Centaurea solstitialis L. *Barnaby's Thistle.* Biennial or rarely annual, 2–8 dm. tall, lanate throughout; tap-root 2–8 mm. in diameter, brown; stem bushy branched, several winged by the decurrent leaf bases; basal leaves 10–15 cm. long, mostly petioled, lyrate pinnatifid, the lateral lobes oblanceolate, serrate; cauline leaves 1–8 cm. long, linear-oblanceolate to linear, acerose-tipped; heads terminal; involucre 10–12 mm. long, ovoid-globose, yellow, cartilaginous, somewhat lanate, the outer with several spines at tip; inner bracts lanceolate, the appendage suborbicular, hyaline margined, unarmed; middle bracts ovate, the appendage of the central spine 1–2 cm. long, and 4 lateral much shorter spines; flowers fertile, all alike and discoid; corollas 13–15 mm. long; pappus bristles 1–4 mm. long, unequal, rigid, barbellate, lacking on marginal achenes; achenes 2–3.5 mm. long, ellipsoid, compressed, light brown, mottled with dark, the apex truncate, the attachment lateral. Weed, introduced from Eurasia or Africa. In cultivated and waste lands, *U. Son., A. T.*

This has become a bad weed in California.

CHAENACTIS

Annual, biennial, or perennial herbs; leaves alternate, 1–3-pinnately dissected; heads medium or large, peduncled, terminating the branches; ray flowers none; receptacle alveolate, naked or bristly; flowers yellow, white, or flesh-colored, all perfect and tubular but an outer series more or less enlarged simulating rays; involucre campanulate, turbinate, or hemispherical; bracts subequal, in 1–2 rows; pappus of 4–16 awnless and nearly or quite nerveless hyaline or chaffy scales; achenes slender, linear, tapering to the base, more or less 4-angled. (Name from Gr., *chaino,* to gape; *aktis,* ray.)

Stem with several spreading branches; young leaves sparingly lanate. *C. ramosa.*
Stem single, erect, herbaceous; herbage floccose or pilose,
  Herbage pilose. *C. Douglasii,* var. *Douglasii.*
  Herbage densely floccose. Var. *montana.*

Chaenactis Douglasii (Hook.) H. & A., var. **Douglasii.** Biennial or perennial, 1–4 dm. tall; tap-root 2–7 mm. in diameter, brown, elongate; stem erect, often branching, floccose lanate; leaves 5–20 cm. long, bipinnatisect, the lobes short, nearly oblong, 1–2 mm. wide, floccose lanate, finally nearly glabrate; cyme several-headed, loose; the peduncles glandular puberulent; involucre turbinate, in flower 8–10 mm. long, in fruit 12–14 mm. long; involucral bracts oblance-linear, herbaceous, viscid hirsutulous; corolla cream-colored or pinkish, pilosulous, the tube 6–8 mm. long, the lobes 1.5–1.8 mm. long, ovate; pappus scales 4–9 mm. long, elliptic; achenes 6–10 mm. long, densely hirsutulous. Arid or gravelly places, *U. Son., A. T.*

Var. **montana** M. E. Jones. Perennial with branched caudex; leaf segments crowded. *C. douglasii,* var. *achilleaefolia* (H. & A.) A. Nels. Wawawai, *Elmer* 897, *U. Son.*

Chaenactis ramosa Stockwell. Perennial or biennial, with numerous slender, red, basal branches, these branched, leafy, with 1–6 short peduncles; leaves 2–5 cm. long, bipinnate, at first sparingly lanate; heads 12–14 mm. long, slender; involucral bracts 10–12 mm. long, linear to spatulate, glandular puberulent, the margins ciliate below; corollas 6–7 mm. long, sparsely puberulent and glandular; achenes 6–7 mm. long, strigose; pappus of about 16 unequal, oblong, obtuse scales 2–5 mm. long. Blue Mts., Garfield Co., Aug. 7, 1897, *Horner.*

CHONDRILLA

Biennial or perennial herbs, with milky sap; branches virgate; blades usually pinnate except for the upper sessile, entire ones; heads several, and 7–15-flowered; involucre cylindric, of several linear, equal involucral bracts and a row of basal bractlets; achenes terete, angled or with several ribs, smooth below, roughened above by scaly projections, and ending in a beak with an apical pappus-bearing disk; pappus of numerous capillary, white bristles. (A Greek plant name used by Dioscorides for some gum-producing plant.)

Chondrilla juncea L. *Skeleton-weed.* Biennial or perennial, 0.4–1 m. tall; stem bristly hairy below, but above glabrous; basal leaves in a rosette, mostly 5–13 cm. long, 1.5–3 cm. wide, runcinate, glabrous, the cauline ones reduced, 2–10 cm. long, lanceolate or linear; heads remote, on nearly naked branches; heads 1–1.5 cm. long; involucre white tomentose, 9–12 mm. high; ray flowers yellow, in two ranks; achenes about 3 mm. long, tawny, with a beak about as long as the body. Weed, introduced from Eurasia. Greenacres, Spokane Co., Oct. 1938, *J. K. Mahn.*

CHRYSANTHEMUM

Annual or perennial herbs; leaves alternate, dentate or dissected; heads many-flowered; ray flowers usually elongated, numerous, pistillate; involucre hemispherical or flatter, the scales more or less scarious, short-appressed, imbricated in several rows; receptacle flat or convex, naked; ray flowers pistillate, fertile, the rays white, yellow, or rose-colored; disk flowers perfect, fertile, the corollas often flattened or 2-winged below, 4- or 5-toothed; pappus none or a scaly cup; achenes short, nearly terete, several ribbed or angled, truncate at apex. (Name from Gr. *chrusos,* gold; *anthemon,* flower.)

Leaves 2–3-pinnatifid; involucral bracts with firm margins. *C. Parthenium.*
Leaves dentate or pinnatifid; involucral bracts with scarious margins or tips,
  Heads solitary, long peduncled; leaves pinnatifid. *C. Leucanthemum,* var. *pinnatifidum.*
  Heads in corymbs; leaves crenate. *C. Balsamita.*

Chrysanthemum Balsamita L. *Mint Geranium.* Perennial, 6–13 dm. tall, closely appressed puberulent throughout; stem erect, freely branched; basal leaves with petioles 2–3 dm. long; blades 10–20 cm. long, oval, crenate; cauline leaves numerous, mostly sessile, the blades 2–10 cm. long, similar; heads numerous, in corymbs; involucre subglobose; bracts imbricate in 3–4 series, the inner 3–4 mm. long, oblong, appressed puberulent, the tips scarious; ray flowers usually none, or few and white; disk flowers 2 mm. long, tubular, yellow, glandular atomiferous; pappus a short crown; achenes 1.3–1.5 mm. long, cuneate-linear, glandular atomiferous. Escaped from cultivation, established since 1916, *A. T.*

Cultivated in gardens because of its aromatic, fragrant herbage.

Chrysanthemum Leucanthemum L., var. pinnatifidum Lecoq & Lamotte. *Oxeye Daisy.* Perennial 3–10 dm. tall, glabrous or sparsely puberulent; rootstock 1–4 mm. in diameter, brown; stems erect, usually simple; lower leaves with petioles 2–8 cm. long, the blades 1–5 cm. long, obovate, irregularly toothed or lobed; middle cauline leaves oblanceolate to spatulate, pinnatifid or coarsely toothed; upper leaves much reduced, sessile, spatulate, more or less dentate; heads on long naked peduncles; involucre saucer-shaped, glabrous, the bracts imbricate in 3–4 series, lanceolate, obtuse, the inner 6–8 mm. long; the 15–30 ray flowers white, 1–3 cm. long; the pappus usually none; achenes 2–2.2 mm. long, oblanceoloid, black with prominent white ribs; disk flowers with corolla 2.2–2.5 mm. long, funnelform, yellow, the short lobes ovate; pappus none; achenes 2–2.3 mm. long, oblanceoloid, black, with prominent white ribs glabrous. *Leucanthemum Leucanthemum* of Rydb., not of name bringing synonym. Weed introduced from southern Europe. Infrequent, roadsides, *A. T.*

The flowers contain an irritating substance; and the plants are avoided by grazing animals.

Chrysanthemum Parthenium (L.) Bernh. *Feverfew.* Perennial, 3–8 dm. tall, sparsely puberulent or more or less glabrate; rootstock 2–7 mm. in diameter, brown; stem erect, striate, freely branched above; leaves gradually reduced upwards, the basal with petioles 2–12 cm. long, the uppermost sessile; blades 3–13 cm. long, ovate in outline, the 2–5 pairs of pinnae ovate in outline; with rounded, apiculate teeth; heads numerous in corymbs; peduncles 5–20 mm. long; involucre saucer-shaped, yellowish, imbricate in about 2 series, resinous atomiferous and sparsely puberulent, the bracts lanceolate, obtuse, keeled, the inner 3–3.5 mm. long; the 10–20 ray flowers white, 3–5 mm. long; pappus none; achenes 1–1.5 mm. long, oblanceoloid, white ribbed; disk flowers numerous, yellow, 1.8–2 mm. long, tubular, glandular atomiferous; pappus none; achenes 1.3–1.8 mm. long, wedge-shaped, pale ribbed glandular atomiferous. Native of southwestern Asia, cultivated as an ornamental. Occasionally escaped, roadsides and stream banks, Waitsburg, *Horner* R355B293, *A. T.*

CHRYSOPSIS. Golden Aster

Low biennial or perennial herbs; leaves numerous, alternate, sessile; heads solitary or in corymbs with yellow flowers; involucral scales narrow, acute, with scarious margins, in several series; receptacle flat; rays fertile or rarely wanting; disk flowers perfect, fertile; style-branches with appendages; pappus double, of two kinds, the interior of long copious capillary bristles, the exterior of short bristles or chaffy scales; achenes oblong-linear or ovate-oblong, compressed hairy. (Name from Gr., *chrusos,* gold; *opsis,* aspect.)

Stem glandular puberulent and sparingly hispid; leaves hispid. *C. hispida.*
Stem densely long hairy; leaves not hispid,
  Involucral bracts appressed hirsutulous. *C. villosa.*
  Involucral bracts spreading villous. *C. barbata.*

Chrysopsis barbata Rydb. Perennial, 3–4 dm. tall; stem densely hirsute, hirsutulous, and glandular atomiferous; basal and lower leaves oblanceolate, petioled; middle and upper leaves 2–5 cm. long, sessile, lanceolate or broadly oblanceolate, acute, densely white hirsute, hirsutulous, and glandular puberulous; heads several in a cyme; involucre 8–11 mm. high, semiorbicular; bracts imbricate in usually 4 series, linear, acute, scale-like, with a green midrib, resinous atomiferous; ray flowers 10–12 mm. long; disk flowers 6–6.5 mm. long, tubular; pappus 6.5–7.5 mm. long, becoming brownish, the outer series 1–1.5 mm. long; disk achenes 3–4 mm. long, narrowly ellipsoid, compressed, pilosulous. Sandy places by Spokane River, *A. T.*

The type was from Kootenai Co., *Sandberg et al.* 664.

Chrysopsis hispida (Hook.) DC. Perennial, 1–3 dm. tall; tap-root 3–8 mm. in diameter, woody, brown; stems in a clump; lower leaves oblanceolate, short petioled; middle and upper leaves 1–3 cm. long, sessile, oblong-oblanceolate, resinous-atomiferous; heads several in a leafy cyme; involucre 7–9 mm. high; bracts imbricate in 3–4 series, scale-like, the outer lanceolate, the inner oblong-linear, acute, resinous atomiferous, sparingly hirsute to glabrate; ray flowers 8–12 mm. long; disk flowers 5–5.5 mm. long, tubular; pappus 5–5.5 mm. long, whitish to brownish, the outer series 0.5–0.8 mm. long; disk achenes 3.5–4 mm. long, narrowly ellipsoid, flattened, brownish, appressed pilose. *C. villosa,* var. *hispida* (Hook.) Gray ex D. C. Eaton. Clear Water [Lapwai], *Spalding, U. Son.*

Chrysopsis villosa (Pursh) Nutt. Perennial, 1–5 dm. tall; tap-root 2–10 mm. in diameter, brown, woody; stems several, white hirsute and hirsutulous; lower leaves 3–6 cm. long, spatulate; middle and upper leaves 1–4 cm. long, sessile, oblanceolate, both sides resinous atomiferous, subappressed hirsutulous but green, hirsute ciliate towards the base; heads 1 or commonly several in a leafy bracted cyme; involucre 7–9 mm. high; the bracts imbricate in 3–5 series, linear, acute, scale-like, only the midrib green, near the tip resinous atomiferous; ray flowers 9–13 mm. long; disk flowers 5–6 mm. long, tubular; pappus 5.5–6 mm. long, brownish, the outer series about 1 mm. long; disk achenes 3.2–3.5 mm. long, oblanceolate, cuneate, flattened, brown, appressed pilose. Dry slopes, canyons of the Snake and the Clearwater, *U. Son.*

CHRYSOTHAMNUS. Rabbit Brush

Herbs or undershrubs; herbage usually aromatic or resinous punctate; leaves narrow, sessile, entire, alternate; heads mostly small, with 4–20 yellow discoid flowers; heads in corymbs, cymes or panicles; flowers all perfect, with tubular corollas; rays none; involucre cylindric, imbricated in more or less distinct vertical ranks; receptacle flat; pappus simple, of copious capillary, white or reddish bristles; style-branches with appendages; achenes narrow, terete or angular, slightly if at all compressed. (Name from Gr., *chrusos,* gold; *thamnos,* bush.)

Twigs glabrous; plant viscid. — *C. viscidiflorus,* var. *viscidiflorus.*

Twigs with a felt-like tomentum,
- Involucres tomentulose or puberulent. — *C. nauseosus,* var. *albicaulis.*
- Involucres glabrous, though sometimes viscid. — Var. *consimilis.*

Chrysothamnus nauseosus (Pall.) Britton, var. **albicaulis** (Nutt.) Rydb. Shrub 5–10 dm. tall, bushy and round-topped; stems with brown, fibrous bark; twigs white and permanently densely tomentose, leafy to the top; leaves 2.5–4 cm. long, 0.5–1.5 (–3) mm. wide, 1-nerved, permanently floccose, white tomentose; cyme rounded, 2–10 cm. broad; involucre 7–9 mm. long, the bracts in 5 vertical ranks, keeled lanate, the outer lanceolate, the inner oblong-lanceolate, acute; heads 5–6-flowered; corolla 8–10 mm. long, the lobes 1.5–2 mm. long, lanceolate; pappus 8–9 mm. long, whitish, becoming brownish; achenes 5–5.5 mm. long, linear, densely pilose. *C. nauseosus,* subsp. *speciosus* (Nutt.) H. & C. Dry places, *U. Son.*

The various subspecies of this plant are sparingly grazed on the winter range, probably due to the scarcity of other forage. Some cases of poisoning from it have been reported.

Var. **consimilis** (Greene) H. M. Hall. Shrub 6–15 dm. tall; twigs striate, green, with a dense, compact tomentum; herbage with a disagreeable odor; leaves 2.5–5 cm. long, 0.3–0.9 mm. wide, 1-nerved, somewhat resinous and tomentulose to glabrous; cyme 3–8 cm. wide; involucre glabrous, the bracts in fairly distinct ranks, acute, keeled, the outer lanceolate, the inner 7–8 mm. long, oblong-lanceolate; corolla 7–8.5 mm. long, the lobes 1–2 mm. long, linear-lanceolate; pappus 7–8 mm. long, white to brownish; achenes 5–5.5 mm. long, linear, pilose. Dry slopes, Salmon River, Blue Mts., *Horner* 327; Zaza, Craig Mts., *St. John* 9,096, *A. T.*

The cells of the cortex and the rays contain rubber, a possible source of supply in times of national emergency.

Chrysothamnus viscidiflorus (Hook.) Nutt., var. **viscidiflorus.** Shrub 5–12 dm. tall, usually much branched; branchlets striate, with greenish or white, glabrous bark; old stems with brown, fibrous bark; leaves 2–5 cm. long, 2–5 mm. wide, 1–3-nerved, green, glabrous, viscid, the margins entire or scabrous-ciliolate; cyme 2–10 cm. broad; heads about 5-flowered; involucre 5–7 mm. long; bracts in poorly defined ranks, the outer lanceolate, the inner broadly linear, acute; corolla 4.5–7 mm. long, glabrous, the lobes 1–2 mm. long, lanceolate; pappus 6–7 mm. long, brownish; achenes 3–4 mm. long, villous. Arid places, especially in the Scablands, *U. Son.*, *A. T.*

CICHORIUM. Chicory

Erect herbs; leaves basal and alternate, the cauline ones usually reduced; heads large, peduncled or in sessile clusters; flowers all ligulate, matinal, blue, purple, pink, or white; involucral bracts herbaceous in 2 series, the outer of 5 somewhat spreading bracts, the inner of 8–10 erect bracts subtending or partly enclosing the outer achenes; receptacle flat, naked or fimbrillate; anthers sagittate at base; achenes 5-angled or 5-ribbed, truncate; pappus of 2 or 3 series of short, blunt scales. (The ancient Gr. name, *kichora.*)

Cichorium Intybus L. *Chicory.* Perennial, 2–20 dm. tall, sparsely hispid; tap-root brown, often large; stems freely divaricate branched; basal leaves 7–15 cm. long, runcinate pinnatifid; cauline leaves gradually smaller; lanceolate or oblong, lobed or entire, auricled and clasping at base; heads numerous, terminal or 1–4 in sessile, axillary clusters; involucre cylindric to broad campanulate, the flowers almost rotate; outer bracts 8–9 mm. long, capitate glandular hispid ciliate, elliptic, callous thickened at base; inner bracts 10–12 mm. long, linear-lanceolate, often reddish, more or less capitate glandular hispid; rays 18–25 mm. long, blue; pappus scales 0.1 mm. long, lanceolate; achenes 2–3 mm. long, cuneate, yellow or with brown mottling. Weed, introduced from Eurasia. Roadsides and fields, *A. T.*

If grazed by cows, the herbage gives a bitter taste to the milk and butter. It contains the glucoside chicorin. The plant is used as a pot-herb, as a salad, and the roots as an adulterant or substitute for coffee.

CIRSIUM. Thistle

Biennial or perennial herbs; leaves alternate, sessile, mostly pinnatifid and prickly; heads usually large, terminal, many-flowered; flowers all tubular, perfect and all alike, rarely imperfectly dioecious, reddish, yellow, or white; involucre ovoid to spherical; bracts imbricated, in many rows, the tips scarious or prickly; receptacle thickly clothed with soft bristles or hairs; pappus of numerous bristles united into a ring at the base, plumose, deciduous; achenes oblong, flattish, not ribbed. (The Gr. name *kirsion*, used by Dioscorides for a species of thistle.)

Plants dioecious, colony forming by spreading rootstocks; inner involucral bracts 12–15 mm. long. — *C. arvense.*
Plants with perfect flowers, single, not spreading by rootstocks; inner involucral bracts 18–45 mm. long,
- Leaves appressed hispid above. — *C. vulgare.*
- Leaves not hispid,
  - Involucre arachnoid, the outer bracts at least half as long as the inner; leaves membranous. — *C. brevistylum.*
  - Involucre not arachnoid, the outer bracts much less than half as long as the inner; leaves chartaceous,
    - Heads clustered, partly concealed by the upper leaves; involucral bracts not glandular carinate, the inner ones with dilated, lacerate tips. — *C. foliosum.*
    - Heads mostly single on naked peduncles; involucral bracts glandular carinate, the inner ones with acuminate, entire tips,
      - Leaves white lanate above, pinnatifid, the segments deltoid to rhomboidal. — *C. undulatum.*
      - Leaves green and glabrate above, pinnatifid, the segments linear-lanceolate. — *C. brevifolium.*

Cirsium arvense (L.) Scop. *Canada Thistle.* Perennial, 3–15 dm. tall, forming dense patches, spreading by brown horizontal rootstocks 3–6 mm. in diameter; stems erect, glabrous or arachnoid but soon glabrate, branching above, striate; leaves 3–20 cm. long, lanceolate or oblanceolate in outline, deeply pinnatifid, the lobes rhombic or deltoid, contorted, the margin entire or dentate, and armed with rigid spines 2–7 mm. long, green and glabrous on both sides or somewhat lanate beneath; plants dioecious; heads in few to many flowered corymbs; staminate involucres globose, sparsely lanate, the outer bracts ovate, with a terminal spine about 1 mm. long, inner bracts 12–15 mm. long, lance-linear, greenish, with a pale, scarious, lanceolate tip; corolla lilac, the tube 9–11 mm. long, tubular, the lobes 4–5 mm. long, linear; pappus 10 mm. long; pistillate oblong-campanulate but similar; corolla lilac, the tube 9–10 mm. long, filiform, the lobes 3 mm. long; pappus 12–20 mm. long; achenes 3–4 mm. long, brownish, oblanceolate. Weed, introduced from Eurasia or Africa. A noxious weed of roadsides and fields, *A. T., A. T. T.*

Cirsium brevifolium Nutt. Perennial, 3–10 dm. tall; root 3–7 mm. in diameter, brown; stem striate, lanate, usually more or less glabrate; basal leaves with petioles 1–3 cm. long, the blade 15–25 cm. long, oblanceolate in outline, beneath permanently white lanate, pinnately parted, the lobes remote, linear, entire or forked, tipped with spines 1–3 mm. long; cauline leaves similar, but pinnatifid, the lobes linear-lanceolate, the upper leaves sessile, commonly simple; heads mostly terminal on naked peduncles; involucre subglobose, not arachnoid; the bracts imbricate in 11–13 series, sparsely lanate ciliate near the tip, the outer lanceolate, 1-ribbed, tipped with a spine 1–2 mm. long; the middle ones linear-lanceolate, the rib prominent and glandular, the spine 2–3 mm. long; inner bracts 18–27 mm. long, lance-linear, attenuate, glandular atomiferous near the margins, the tip lanceolate, pale, flattened but rigid; corolla yellowish white, the tube 20–22 mm. long, the lobes 7–8 mm. long, linear; pappus 20–22 mm. long, white; achenes 4.5–5 mm. long, ellipsoid, oblique, compressed, brownish, smooth. *C. palousense* Piper. Grasslands, *A. T.*

Cirsium brevistylum Cronq. Biennial, 5–20 dm. tall; tap-root 8–12 mm. in diameter, brown; stem erect, hollow, soft, weak, usually simple, striate, sparsely arachnoid; basal leaves 1–3 dm. long, oblanceolate, tapering into the petiole, above sparsely arachnoid but green, below arachnoid and whitish, pinnatifid about half way to the midrib, the lobes ovate, ciliate with spines 1–3 mm. long; cauline leaves similar but oblanceolate to lanceolate in outline, the marginal spines 1–5 mm. long, the lower leaves auriculate, short decurrent, the upper not decurrent; heads usually several in a terminal glomerule, leafy bracted; involucre subglobose, the bracts loosely imbricate in 5–6 series, the outer linear lanceolate, spiny on the margins and at the apex; middle bracts lance-linear, acuminate, puberulent below the apex, tipped with a spine 2–3 mm. long; inner bracts 2–4 cm. long, linear, long acuminate, puberulent below the apex, the tip flat, thin, brownish, lanceolate erose-ciliolate; corollas rose purple, the tube 12–18 mm. long, the lobes 2–4 mm. long, linear; pappus 15–22 mm. long, brownish; achenes 3–4 mm. long, linear-oblanceoloid, apiculate, compressed, brownish, darker streaked, smooth. Moist woods, *Can.*

The peeled stems are sweet and pleasant. They were eaten raw by the Indians. Horses seek out the plants and eagerly graze the herbage, spines and all.

Cirsium foliosum (Hook.) DC. Biennial, 2–10 dm. tall; tap-root 7–15 mm. in diameter, brown, the top with scalariform partitions; stem striate, crinkly villous, strict, mostly simple; basal leaves with petioles 3–5 cm. long, sparsely crinkly villous, spiny on the margin, the blade oblanceolate in outline, 8–14 cm. long, sparsely crinkly villous above but green, beneath white and finely lanate, pinnatifid about half way to the midrib, the lobes ovate, more or less dentate, the margin and tip with spines 1–3 mm. long; cauline leaves 15–20 cm. long, narrowly lance-oblong, similar to the basal but sometimes less pinnatifid, the spines 2–10 mm. long; inflorescence overtopped by many linear, subentire, foliaceous usually reddish bracts; heads few, in a dense terminal cluster; involucre subglobose, the bracts imbricate in 6–8 series, not arachnoid, the outer lance-ovate, the margin serrulate, the tip with a 2–3 mm. spine; middle bracts lanceolate, spine-tipped, the margins puberulous; inner bracts 22–40 mm. long, lance-linear, attenuate, the apex scarious, brown, dilated, ovate, acuminate, the margin lacerate; corolla pink, the tube 18–24 mm. long, the lobes 4–5 mm. long, linear; pappus 18–25 mm. long, whitish; achenes 4–5 mm. long, ellipsoid compressed, brown, smooth. Grasslands, *A. T.*

Cirsium foliosum × undulatum. Prof. Piper made two collections at Pullman, which recombined in different degrees, characters of the two species. He stated that they were found rarely, and with the supposed parents.

Cirsium undulatum (Nutt.) Spreng. Biennial, 3–20 dm. tall; tap-root 3–7 mm. in diameter, brown; herbage white lanate throughout; stem usually branched, striate; basal leaves with petioles 1–4 cm. long, blades 5–11 cm. long, linear-lanceolate, the margin undulate and with weak spines 1–3 mm. long; cauline leaves similar but 5–34 cm. long, pinnatifid ⅓ to ½ way to the midrib, the lobes deltoid or rhomboid, more or less lobed, the spines 2–7 mm. long, the upper sessile and auriculate at base; heads mostly terminal and remote on naked peduncles; involucre subglobose, the bracts closely imbricate in about 9–11 series, somewhat villous ciliate, the outer ovate, below cartilaginous, above dark and with a glandular keel, tipped with a spine 1–2 mm. long; middle bracts lanceolate, the spine 2–5 mm. long; inner bracts 25–45 mm. long, lance-linear, acuminate, glandular atomiferous near the margin, the tip not dilated; corolla rose-colored, the tube 15–42 mm. long, the lobes 8–10 mm. long, linear; pappus 23–32 mm. long, white; achenes about 7 mm. long, ellipsoid, compressed, brownish, smooth. *C. megacephalum* (Gray) Cockerell; *C. undulatum,* var.

megacephalum (Gray) Fern. Open places, *U. Son., A. T.*

Cirsium vulgare (Savi) Tenore. *Bull Thistle.* Biennial, 5–15 dm. tall; tap-root brown, elongate; stem erect, branched above, striate, lanate, leafy, spiny-winged by the decurrent leaf bases; basal leaves oblanceolate, tapering into the petiole, shallowly lobed; cauline leaves 7–20 cm. long, in outline lanceolate to oblanceolate, beneath lanate and heavy ribbed, above green, pinnatifid, the lobes lanceolate to deltoid, more or less spiny toothed, the tips with rigid spines 5–10 mm. long; heads terminal; involucre subglobose, loosely arachnoid, the bracts imbricate in many series, nearly all rigid spine-tipped, the outer lanceolate, acuminate, the inner 28–40 mm. long, lance-linear, long acuminate, serrulate on the margins below the flat, pale tips; corolla rose-purple, the tube 20–28 mm. long, filiform, the lobes 5–6 mm. long, linear; pappus 20–24 mm. long, whitish; achenes 3–4 mm. long, oblanceoloid, compressed, pale brown, dark lined, smooth. *C. lanceolatum* (L.) Hill. Weed, introduced from Eurasia or Africa. Abundant, roadsides and fields. *A. T., A. T. T.*

CNICUS

Annual herb; leaves alternate, pinnatifid or sinuate, dentate, the lobes or teeth spiny; heads large, discoid, yellow flowered, terminal, sessile and subtended by the upper leaves; involucral bracts imbricate in several series, the outer ovate, the inner lanceolate, tipped by long, pinnately branched spines; receptacle flat, bristly; anther appendages elongate, united to their tips; pappus of 2 series of awns, the outer longer, naked, yellow, the inner fimbriate, white; achenes terete, many striate, laterally attached, 10-toothed at apex. (The ancient Lat. name for one of the thistles.)

Cnicus benedictus L. *Blessed Thistle.* Plant 1–6 dm. tall, hirsute and arachnoid; tap-root 1–4 mm. in diameter, elongate, brown; stem often branched, leafy; leaves 6–30 cm. long, oblong-lanceolate in outline, the lower petioled, coarsely rugose veined; involucre 3–4 cm. long, ovoid to campanulate, cartilaginous; receptacle bristles capillary, as long as the involucre; corolla 3 cm. long, the limb narrow campanulate; outer pappus awns 9–10 mm. long; inner awns 1.5–2 mm. long; achenes 8–10 mm. long, brown, curved. Weed, introduced from Eurasia. Waste places, Waitsburg in 1897; Colfax in 1901, *A. T.*

COREOPSIS. Tickseed

Annual or perennial herbs usually with opposite, or ternate leaves; heads many-flowered, radiate; involucre of 2 rows of about 8 bracts each, the outer spreading and foliaceous, the inner appressed and nearly membranaceous; ray flowers mostly 8, neutral, yellow or purple, rarely wanting; disk flowers perfect, fertile; receptacle flat with deciduous membranaceous chaff; achenes obcompressed, often winged, with 2 barbless subulate awns. (Name from Gr., *koris,* bug; *opsis,* appearance.)

Coreopsis Atkinsoniana Dougl. ex Lindl. Perennial, 3–10 dm. tall, glabrous throughout; tap-root 4–10 mm. in diameter, brown; stem erect, striate, usually branched; leaves opposite or rarely ternate; basal, lower, and middle leaves with petioles 1–9 cm. long; blades 1–2-pinnatifid, the segments linear; upper leaves sessile, reduced; heads hemispheric, slender peduncled in loose cymes; outer bracts 2–3 mm. long, lance-elliptic, greenish; inner bracts 6–10 mm. long, oval, thin, many ribbed, brownish to orange; rays 10–15 mm. long, elliptic or broader towards the tip, yellow throughout or brown at base; chaff 3 mm. long, brown, lance-linear; disk flowers 3 mm. long, glabrous, tubular, brownish purple; anthers purplish black; achenes 2.5–2.7 mm. long, oval, concave, flattened, black, with narrow, pale wings. Wet shores, Lake Coeur d'Alene, and Spokane River, *A. T.*

There is a *Spalding* specimen labeled Clear Water, Oregon [Lapwai, Ida.]. It has not been recollected along this section of the Clearwater or Snake Rivers.

CREPIS. Hawksbeard

Annual, biennial, or perennial plants with milky juice; leaves alternate or basal; heads several-many-flowered; flowers yellow or orange, all ligulate; involucre cylindric to campanulate, usually in 2 series; receptacle flat, naked, or fimbriate, sometimes alveolate; pappus simple, of copious and white capillary bristles; achenes oblong, linear or fusiform, nearly terete or obtusely angled, 10–20-ribbed, generally contracted at base and more tapering at the apex, sometimes even beaked. (Name from Gr. *krepis,* sandal; a plant name used by Theophrastus.)

Stem and leaves glabrous or at least not tomentose; cauline leaves much reduced; involucre turbinate-campanulate. — *C. runcinata,* subsp. *hispidulosa.*

Stem and leaves more or less tomentose; some cauline leaves expanded; involucre narrowly or broadly cylindric,
- Involucre or stem and petioles setose. — *C. barbigera.*
- Involucre or at least stem and petioles not setose,
  - Largest heads 5–10-flowered, with 5–7 inner involucral bracts. — *C. acuminata,* subsp. *acuminata.*
  - Largest heads 9–40-flowered, with 8–13 inner bracts,
    - Achenes greenish, somewhat beaked; leaf lobes linear or narrowly lanceolate, 0.5–2.5 mm. wide, mostly entire. — *C. atribarba.*
    - Achenes yellowish or brownish; leaf lobes broadly lanceolate or deltoid or if narrower usually toothed or lobed,
      - Involucres broadly cylindric, 5–9 mm. broad at anthesis, 9–40-flowered; longest outer bract 5–9 mm. long,
        - Leaves grayish tomentose; peduncles not expanded. — *C. occidentalis,* subsp. *occidentalis.*
        - Leaves green, and glandular pubescent; peduncles dilated at apex. — *C. Bakeri,* subsp. *idahoensis.*
      - Involucres narrowly cylindric, 3–5.5 mm. broad at anthesis, 7–15-flowered; longest outer bract 2–5 mm. long. — *C. intermedia.*

Crepis acuminata Nutt., subsp. **acuminata.** Perennial, 2–6.5 dm. tall; tap root 2–4 mm. in diameter; basal leaves at first tomentose, soon green and glabrate, with petioles 5–10 cm. long; the blades 7–29 cm. long, lance-elliptic, pinnatifid ½ way to midrib into linear-lanceolate, attenuate, entire lobes; 1–3 cauline leaves 6–30 cm. long; inflorescence corymbiform with 20–40 (–100) heads on erect peduncles 2–20 mm. long; involucre narrowly subcylindric; outer bracts 1–3 mm. long, lanceolate, acute; the 5–8 inner bracts 9–11 mm. long, obtuse, greenish, glabrous or tomentulose, ciliate; flowers 5–12, the corollas 10–25 mm. long; pappus 5–10 mm. long, whitish; achenes 5.5–9 mm. long, narrowly fusiform slightly tapering to apiculate apex, 12-ribbed, yellow or brownish. Represented locally by no. 16 apm. *exiloides* of Babc. & Stebb., *U. Son.*

Crepis atribarba Heller. Perennial, 1.5–6 dm. tall; stems 1–2, tomentulose or glabrate; basal leaves 1–3.5 dm. long, lanceolate to linear, pinnate or bipinnate ⅔ way or even almost to midrib into linear or lanceolate segments, grayish tomentulose or glabrate; cyme corymbiform with 3–30 heads; involucre cylindric, tomentose, setose, or glabrous; outer bracts 2–5 mm. long, deltoid, acute, the 6–10 inner bracts 8–15 mm. long, lanceolate, acute or obtuse; flowers 6–35; achenes 3–10 mm. long, greenish, or rarely brownish, fusiform, usually somewhat beaked at apex. Represented locally by subsp. *originalis* Babc. & Stebb., no. 1. apm. *simulans;* no. 2 apm. *breviloba;* no. 3 apm. *longiloba;* no. 8 apm. *paucibarba;* no. 10 apm. *sterilis;* subsp. *typica* Babc. & Stebb., no. 12 apm. *ambigua;* no. 13 apm. *Helleri* of Babc. & Stebb.

Crepis Bakeri Greene, subsp. **idahoensis** Babc. & Stebb. Perennial, 2.5–3 dm. tall; stems 1–3, branched near base, glabrate or glandular hispid; leaves green, sparingly tomentulose, short glandular hispid; basal leaves 15–18 cm. long, 5–5.5 cm. broad, elliptic, pinnatifid ¼ to ½-way into deltoid acuminate, mucronate-dentate lobes, sparingly tomentulose and sparsely glandular hirsutulous; cauline leaves similar; inflorescence of 11–22 heads; peduncles conspicuously expanded at apex; involucres in flower narrowly campanulate, 16–18 mm. high, in fruit 18–21 mm. high; outer bracts 3–5 mm. long, deltoid, acute or acuminate; the 8–13 inner bracts lanceolate or linear, attenuate, conspicuously glandular hispid, at base tomentulose, the midrib becoming thickened at base; florets 11–40 or more, like those of *C. occidentalis;* achenes 7.5–8 mm. long, narrowly fusiform-cylindric, with slight neck below apex; pappus 12–13 mm. long, *A. T.*

Crepis barbigera Leiberg. Perennial, 4–6 dm. tall, sparsely tomentulose tap-root about 4 mm. in diameter, brown; basal leaves with petioles 15–18 cm. long, the blades 15–30 cm. long, green, the pubescence obscure, oblong-lanceolate in outline, runcinately toothed or pinnatifid, usually not more than ½ way to the midrib, the lateral lobes lance-linear, entire, the central portion of blade elliptic; cauline leaves 2–3, reduced, mostly sessile; corymb 10–25-headed, round-topped; peduncles 1–5 cm. long, more or less lanate; outer involucral bracts 2–5 mm. long, lanate and densely setose; inner bracts 9–17 mm. long, lance-linear lanate, the carinate midrib coarsely pale setose; corollas 14–18 mm. long; pappus 5–10 mm. long, barbellate; achenes 6.5–10 mm. long, fusiform, 10–12-ribbed. Rocky or gravelly soil, *A. T.*

Crepis intermedia Gray. Perennial, 2–7 dm. tall, somewhat cinereous lanate; tap-root 2–3 mm. in diameter, dark; stem simple below; basal leaves with petioles 4–10 cm. long, the blades 9–15 cm. long, lanceolate in outline, pinnatifid from ½ to ¾ way to the midrib, the central portion of the blade linear-lanceolate; cauline leaves 7–12 mm. long, sessile, reduced; cyme 4–13 cm. wide, 10–60-headed; peduncles 1–5 cm. long; involucre matted lanate, the outer bracts 2–3 mm. long, lance-linear, the 5–8 inner bracts in flower 9–10 mm. long, in fruit 12–16 mm. long, linear, with or without a few black glandular setae; flowers 5–12; corollas 10–30 mm. long; pappus 6–9 mm.

long, minutely barbellate; achenes 5.5–9 mm. long, minutely puberulous, fusiform, narrowing below the apex. Gravelly slopes, *U. Son., A. T.*

Represented locally by no. 2 apm. *montana,* and no. 10 apm. *angustifolia* of Babc. & Stebb.

Crepis occidentalis Nutt., subsp. **occidentalis.** Perennial, 1–4 dm. tall, tomentose canescent throughout; tap-root 2–5 mm. in diameter, woody, dark brown, often with marcescent leaf bases; stem few branched from near the base; basal leaves sessile or with petioles 3–8 cm. long, the blades 7–13 cm. long, oblanceolate in outline, shallowly or deeply runcinate pinnatifid, acute, the lateral lobes usually more or less contorted, the middle part of blade often oblanceolate and broader than the lobes; cauline leaves 2–3, the upper sessile, 4–9 cm. long, less dissected; corymb 3–18-headed, round-topped; peduncles 5–70 mm. long, often glandular hirsute; involucre broadly cylindric to campanulate, tomentose canescent, with or without black, capitate glandular, hispid hairs, the outer bracts 2–4 mm. long, lanceolate, the inner bracts 9–18, lance-linear, 12–15 mm. long; the 10–30 corollas 14–16 mm. long; pappus 8–10 mm. long, minutely barbellate; achenes 6–10 mm. long, fusiform, narrowed above, with 10 prominent ribs, scaberulous at apex. Rocky slopes, *U. Son. A. T.*

Represented locally by subsp. *occidentalis,* subsp. *conjuncta* of Babc. & Stebb., no. 27 apm. *crassa* Babc. & Stebb.

Crepis runcinata (James) T. & G., subsp. **hispidulosa** (Howell) Babc. & Stebb. Perennial, 4–9 dm. tall, sparsely capitate glandular hirsute throughout; leaves mostly basal, these 6–30 cm. long, thin chartaceous, green, oblanceolate, entire or sinuate denticulate; cauline leaf usually single, 7–12 cm. long, lanceolate; cyme 4–20 cm. broad, round topped, the heads 9–30; involucre green with blackish median stripe, capitate glandular hirsute; outer bracts 2–4 mm. long, lance-linear; inner bracts 9–12 mm. long, lance-linear; corollas 9–11 mm. long; pappus 4–5 mm. long, minutely barbellate; achenes 3.5–5 mm. long, narrowly fusiform, compressed, brown, closely ribbed. Mud flats, Endicott, *Elmer* 1,031; Alkali Lake, Spokane Co., *St. John et al.* 4,880; Rock L., *Weitman* 107, *U. Son.*

CROCIDIUM

Small annual herbs with alternate leaves; heads solitary, terminal, small, radiate, the flowers all fertile; involucre hemispherical, of 8–12 thin herbaceous bracts in one row, slightly united at base; ray flowers about 12, yellow, without pappus or the pappus early deciduous; disk flowers more numerous, yellow; style branches short and broad, with large appendages; pappus of one row of deciduous equal white barbed bristles; achenes ellipsoid, truncate, 5-ribbed. (Name from Gr., *kroke,* loose thread; wool.)

Crocidium multicaule Hook. Plant 5–25 cm. tall, floccose tomentose throughout, or in part glabrate; roots fibrous; stems ascending, simple or branched; basal leaves in a rosette, the petioles 3–25 mm. long, the blades 3–23 mm. long, spatulate to obovate, entire or remotely dentate 6–20 mm. long, mostly on the lower half of the stem; heads solitary; involucre 4–7 mm. high; bracts lanceolate or ovate-lanceolate, thin, several nerved, glabrous except at the apex, greenish or purplish; rays 5–7 mm. long, the pappus as long as the tube, early deciduous; disk flowers 3 mm. long, the tube cylindric, as long as the campanulate throat; pappus 2.5–3 mm. long; achenes 1.5 mm. long, papillose-puberulous between the ribs. Open hillsides. Waitsburg, *Horner* 165, *A. T.*

The only Idaho record is the *Spalding* specimen labeled Clear Water, Oregon. It has not been recollected in the region of Lapwai, Ida., so it is probable that the specimen was really collected on a journey to Walla Walla.

ECHINOPS

Coarse, thistle-like herbs, ours perennial; stems and lower leaf surfaces more or less white woolly; leaves alternate, pinnately toothed or 2–3-pinnatisect, the teeth and lobes prickly; flowers solitary, each with a little involucre of its own consisting of bristle-like outer bracts and linear or lanceolate inner ones which are free or united; flowers aggregated into a dense spherical head, with a small reflexed involucre of linear scales or bristles.

Echinops sphaerocephalus L. *Globe Thistle.* Bushy perennial 1–3 m. tall; stems white woolly and viscid; leaves white arachnoid beneath, the lower ones petioled, 3–4 dm. long, oblanceolate, pinnately dissected; cauline leaves 5–20 cm. long, sessile, lanceolate in outline, very spiny; heads 4–8 cm. in diameter, globose, blue or whitish; involucre of the single flowers 2 cm. long, the outer bracts glandular on the back; corolla tube 5–6 mm. long, cylindric; the lobes 6 mm. long, ligulate, spreading; achenes 7–8 mm. long, cylindric, gray silky. An ornamental, from Eurasia, escaped, Pullman, Sept. 16, 1954, *Rumely* 364.

ERIGERON. Fleabane

Herbs; very similar to *Aster,* but differing in the usually naked-peduncled heads; involucre simpler, of narrow and erect, equal, little if at all imbricated bracts, not coriaceous and without herbaceous tips; rays narrower and usually numerous, often in more than one row; pappus more scanty or fragile, sometimes with a conspicuous short outer row; style appendages very short and roundish or obtuse; achenes mostly 2-nerved. (Name from Gr., *eri,* early; *geron,* old man.)

Rays none, or inconspicuous, at most barely exceeding the disk,
  Stems scapose 5–20 cm. tall; heads solitary; rays none. — *E. Bloomeri,* var. *Bloomeri.*
  Stems leafy, 15–80 cm. tall; heads numerous in corymbs or panicles,
    Involucre 3–4 mm. high; inflorescense a panicle; rays white. — *E. canadensis.*
    Involucre 5–7 mm. high; inflorescence a corymb,
      Rays none; involucral bracts oblong-linear, acute. — *E. inornatus,* var. *inornatus.*
      Rays pinkish; involucral bracts narrowly linear, acuminate. — *E. angulosus,* var. *kamtschaticus.*
Rays conspicuous, much exceeding the disk,

- Rays yellow,
  - Pubescence of stem and leaves appressed. — *E. linearis.*
  - Pubescence of stem and leaves spreading. — *E. chrysopsidis,* subsp. *chrysopsidis.*
- Rays bluish to pinkish or white,
  - Blades 2–4-times ternate into linear lobes. — *E. compositus,* var. *compositus.*
  - Blades simple,
    - Upper cauline leaves ample, ovate to lanceolate, not much reduced,
      - Leaves all or some of them dentate; involucre 3.5–5.5 mm. high,
        - Rays 100–150, pink; pappus of ray and disk flowers alike. — *E. philadelphicus.*
        - Rays 40–70, white or purple-tinged; pappus of ray and disk flowers unlike. — *E. annuus,* var. *annuus.*
      - Leaves entire; involucre 5–8 mm. high,
        - Involucral bracts loosely spreading; rays 30–50; heads commonly single. — *E. peregrinus,* subsp. *callianthemus.*
        - Involucral bracts appressed,
          - Rays 100–150; heads commonly in cymes; stems glabrous except at apex. — *E. speciosus,* var. *macranthus.*
          - Rays 30–100; stem densely hirtellous or canescent. — *E. caespitosus.*
    - Upper cauline leaves reduced, linear to linear-lanceolate,
      - Heads commonly solitary on scapose stems,
        - Herbage spreading hirsute,
          - Achenes hidden by their dense silky hairs; ligules violet. — *E. poliospermus,* var. *poliospermus.*
          - Achenes moderately hirsute, not hidden; ligules white, turning pinkish in age. — *E. disparipilus.*
        - Herbage appressed hirsutulous; rays white

or purplish tinged. *E. Eatoni,* subsp. *villosus.*
Heads commonly in cymes on branching stems,
Basal leaves filiform to linear. *E. filifolius,* var. *filifolius.*
Basal leaves broader,
Involucre 2.8–3.5 mm. high; pappus of ray and disk flowers unlike. *E. strigosus,* var. *septentrionalis.*
Involucre 3.5–6 mm. high; pappus of ray and disk flowers alike,
Basal and lower leaves prominently longitudinally 3-ribbed. *E. corymbosus.*
Basal and lower leaves not 3-ribbed,
Herbage hirsutulous; involucre 3.5–5 mm. high; rays 5–8 mm. long; disk achenes 0.8–1 mm. long. *E. divergens,* var. *divergens.*
Herbage coarsely hirsute; involucre 5–6 mm. high; rays 9–10 mm. long; disk achenes 1.5–1.9 mm. long. *E. pumilus,* var. *intermedius.*

Erigeron angulosus Gaudin, var. **kamtschaticus** (DC.) Hara. Biennial or perennial, 1.5–8 dm. tall, sparingly hirsute; rootstock 2–5 mm. in diameter, brown, often forking above; stems simple below; basal leaves 3–15 cm. long, spatulate, entire; cauline leaves 3–6 cm. long, lanceolate to oblanceolate, the uppermost glabrous or only slightly hirsutulous; corymb loose, glandular puberulent; involucre 5–6 mm. high, hemispheric, the bracts imbricate in about 3 series, glandular puberulent, sparsely hirsutulous on the thickened midrib; rays 5.5–6 mm. long; pappus 5–6 mm. long, tawny, barbellate; achenes abortive; disk flowers 3 mm. long, tubular, whitish, pistillate, then the central ones 5 mm. long, funnelform, perfect; pappus 5.5–6 mm. long, brownish, barbellate, achenes 2 mm. long, brownish, narrowly elliptic, pilosulous, yellow cupulate at base. Open woods, Idaho, *A. T. T.*

Erigeron annuus (L.) Pers., var. annuus. Annual, 3–12 dm. tall, sparsely hirsute below the inflorescence; caudex 3–15 mm. in diameter, brownish; stems erect, simple below; basal and lower cauline leaves with petioles 3–12 cm. long, decurrent winged; blades 5–12 cm. long, obovate or ovate, coarsely dentate; middle and upper leaves 1–6 cm. long, sessile, oblanceolate to lanceolate, more or less dentate; cyme irregular, usually many headed, subappressed hirsutulous; involucre depressed hemispheric, 3.5–4.5 mm. high, the bracts linear, acute, subequal, granular and somewhat hirsute; ray flowers 6–8 mm. long, sterile; pappus 0.1 mm. long; disk flowers 1.8–2 mm. long, yellow, tubular; pappus double 1.5–2 mm. long, whitish, the outer row 0.1 mm. long; achenes 0.8–1 mm. long, ellipsoid, flattened, brown, puberulous. Weed, introduced from eastern or central North America. Waitsburg, *Horner,* July 27, 1896, *A. T.*

Erigeron Bloomeri Gray, var. **Bloomeri.** Perennial, densely tufted; tap-root 2–6 mm. in diameter, brown, usually multicipital; stems appressed pilosulous, scapose; basal leaves 2–11 cm. long, 1–2 mm. wide, linear or slightly spatulate at tip, appressed pilosulous; cauline leaves few, bracteose; involucre turbinate 7–8 mm. long, the bracts linear, acute, subequal, densely villous at base, less so above; disk corollas 4.5–5 mm. long, yellow, narrowly funnelform, the lobes ovate; pappus unequal 1.5–5 mm. long, whitish, barbellate; achenes 4.5–5 mm. long, cuneate, obovoid, compressed, pale, pilosulous above, heavy margined. Stony ridges, Blue Mts., *Huds.*

Erigeron caespitosus Nutt. Perennial 5–30 cm. tall; tap-root stout; caudex usually branched; stems decumbent or curved at base, then erect; leaves hirtellous or canescent; basal ones 5–12 cm. long, oblanceolate or spatulate, more or less 3-nerved, the apex rounded or obtuse, at base tapering into the petiole; cauline leaves from oblong-ovate to linear, reduced; heads one or few; involucre 4–7 mm. high, white canescent and glandular; involucral bracts linear, firm, the back thickened; rays 5–15 mm. long, blue, white, or pink; disk corollas 3.2–4.4 mm. long; inner pappus of 15–25 firm bristles; achenes 2-nerved, hairy. Rock Lake, *Beattie & Lawrence* 2,437, *A. T.*

Erigeron canadensis L. *Horseweed.* Annual, 2–30 dm. tall, sparsely hispid; tap-root 1–7 mm. in diameter, brown; stem simple below; basal and lower leaves 2–8 cm. long, spatulate, often serrate; middle and upper leaves 2–6 cm. long, linear or lance-linear, mostly entire; panicle 5–50 cm. long, narrow; involucre campanulate to semiglobose, the bracts imbricate in 3–4 series, linear, acuminate, glabrous or the outer with a sparse puberulence, the midrib thickened; ray flowers pistillate, not or but little exceeding the involucre; pappus 2 mm. long, white; achenes 1–1.4 mm. long, narrowly oblong, elliptic, finely puberulent; disk flowers 2 mm. long, yellowish, tubular, the 4 teeth obtuse; pappus 2 mm. long, white; achenes similar to the others but broader. *Leptilon canadense* (L.) Britton; *Conyza canadensis* (L.) Cronq. Native, but weedy. Abundant in cultivated lands, *U. Son., A. T.*

Erigeron chrysopsidis Gray, subsp. **chrysopsidis.** Perennial, 3–15 cm. tall; tap-root slender; caudex branched, short; stem scapose above, 4–9 cm. long, narrowly linear-oblanceolate to linear-spatulate, above hirsute, below pustulate hirsute, hispid ciliate towards the base; involucre hemispheric, 5–7.5 mm. high, hispid hirsute and sometimes glutinous; involucral bracts linear, acuminate, subequal; pistillate marginal flowers 20–50, the ligule varying from none to present and 10 mm. long and 2.5 mm. wide, yellow; disk corollas 3–4.5 mm. long; pappus of 15–25 slender inner bristles and a few slender outer setae; achenes appressed pubescent, 2-nerved. Blue Mts., Walla Walla Co., *C. V. Piper* 2,392 in part.

Erigeron compositus Pursh, var. **compositus.** Perennial, 6–22 cm. tall, caespitose, glandular puberulent throughout; tap-root 3–7 mm. in diameter, multicipital, brown; stems scapose, the cauline leaves much reduced, bracteose; basal leaves hirsute, the petioles 1–5 cm. long, the blades 1–3 cm. long, segments elongate; heads terminal, solitary; involucre 6–8 mm. high, hemispheric, the bracts subequal, linear, acute, glandular puberulous, the thickened midrib more or less hirsute; ray flowers 8–12 mm. long, pinkish to white; pappus 2 mm. long, white; achenes 1 mm. long, linear, brown, puberulous; disk flowers 3–3.5 mm. long, yellow, tubular, narrowed to the base; pappus 2–2.5 mm. long, white; achenes 2.5–2.7 mm. long, oblanceoloid, compressed, brown, puberulous. *E. compositus,* var. *typicus* Pays. Gravelly soils, *U. Son., A. T.*

Erigeron corymbosus Nutt. Perennial, tufted, 1–4 dm. tall, spreading hirsutulous throughout; tap-root 2–5 mm. in diameter, woody, brown, multicipital or forked above; stems erect, simple below; basal and lower cauline leaves with petioles 2–7 cm. long, the blades 4–9 cm. long, linear-oblanceolate; middle and upper cauline leaves 1–6 cm. long, sessile, oblance-linear to linear; cyme few to many headed, loose, the heads on slender peduncles; involucre 4–5 mm. high, hemispheric; bracts linear, acuminate, subequal, hirsute along the thickened midrib, densely so at base; ray flowers 8–14 mm. long, lilac to white; pappus 3.5–4 mm. long, white, the outer series 0.3 mm. long; achenes 1.5 mm. long, linear, appressed puberulent; disk flowers 3.8–4 mm. long, yellow, tubular; pappus 3.5–4 mm. long, white, the outer series 0.3 mm. long; achenes 2–2.5 mm. long, linear-oblanceolate, flattened, brownish, appressed puberulent. Prairies and thickets, *A. T., A. T. T.*

Erigeron disparipilus Cronq. Perennial 3–12 cm. tall, from a taproot; caudex branching; stems erect or ascending-spreading, mixed hirsute and puberulent; leaves mostly basal, up to 4 cm. long and 2 mm. wide, linear or linear-oblanceolate, finely hirsute; cauline leaves few and much reduced; head solitary; involucre 5–7 mm. high, more or less spreading hirsute and usually finely glandular or viscid; involucral bracts little imbricated, green, acuminate; rays 30–55, white but in age turning pinkish, 5–10 mm. long, 1.5–2.3 mm. wide; disk corollas 3.5–4.3 mm. long; inner pappus of 15–25 fairly firm, barbellate bristles, the outer of inconspicuous setae. Open rocky places, *A. T.* to *Can.*

The holotype is Asotin Co., 1900 ft., 14 miles w. of Clarkston, *Hitchcock & Samuel* 2,584.

Erigeron divergens T. & G., var. **divergens.** Annual or biennial, 1–4 dm. tall; tap-root 1–4 mm. in diameter, brown, becoming multicipital; stems erect or decumbent usually forked above; basal and lower leaves 2–5 cm. long, spatulate or oblanceolate and tapering into the slender petiole; middle and upper leaves 1–4 cm. long, linear-oblanceolate, cuneate, sessile; heads terminal, remote; involucral bracts subequal, linear, acuminate, conspicuously hirsute, minutely puberulous, the midrib thickened; the ray flowers conspicuous, violet to white; pappus 1.5–2 mm. long, white, with an outer series of much shorter bristles; disk flowers sterile, 1.5–2 mm. long, yellowish, narrowly funnelform; pappus 1.5–2 mm. long, white, with an outer series of much shorter bristles; achenes narrowly ellipsoid, compressed, pilosulous. Sandy soil, *U. Son., A. T.*

Erigeron Eatoni Gray, subsp. **villosus** Cronq. Perennial, 7–20 dm. tall; tap-root 2–4 mm. in diameter, brown, elongate, often forking above; stems simple below, rarely branched above, often decumbent at base; basal and cauline leaves 4–11 cm. long, spatulate-linear to linear, acute, the larger 3-ribbed; other cauline leaves reduced and bracteose; heads terminal, solitary on the scapose stem tips; involucre 5–6 mm. high, hemispheric; bracts linear, acute, subequal, minutely glandular puberulous, the back thickened and more or less hirsute; ray flowers 6–10 mm. long, pappus 2–2.5 mm. long, white; achenes 1.5 mm. long, oblanceoloid, flattened, appressed pilosulous; disk flowers 3.5–4 mm. long, tubular, yellowish; pappus 2.5–4 mm. long, whitish; achenes 2.5–3 mm. long, ellipsoid to oblong, flattened, brown, appressed pilosulous, the margins white, thickened. Gravelly ridges, Blue Mts., *Huds.*

Erigeron filifolius (Hook.) Nutt., var. **filifolius.** Perennial, tufted, 15–30 cm. tall, cinereous throughout with close appressed puberulence; tap-root 2–6 mm. in diameter, brown, forking at apex; stems erect, commonly branched above, the young stems densely white pubescent with almost pilosulous hairs; basal and lower cauline leaves 2–5 cm. long; linear to filiform; middle and upper cauline leaves similar, 1–3 cm. long; heads commonly several in a loose cyme; involucre 5–6.5 mm. high, semiorbicular; bracts imbricate in about 3 series, linear, acuminate, minutely glandular puberulous, more or less hirsute on the thickened midrib; ray flowers 7–9 mm. long, bluish to pink or white; pappus 2.5–3 mm. long, white, in a single series; achenes 1.5 mm. long, oblanceoloid, flattened, brown, pilosulous; disk flowers 4–4.3 mm. long, yellow, tubular; pappus 3–4 mm. long, white, in 1 series; achenes 1.3–1.6 mm. long, oblanceoloid, flattened, brown, somewhat pilosulous. *E. linearis* sensu Piper, Fl. Wash., not as to name bringing synonym, (Hook.) Piper. Gravelly places, *U. Son., A. T.*

Erigeron inornatus (Gray) Gray, var. **inornatus.** Perennial, more or less suffruticose at base, 2–7.5 dm. tall, pale green, nearly glabrous; rootstock 2–3 mm. in diameter, woody, brown, elongate, often forking above; stems often several, usually simple, hispidulous at base; cauline leaves 1–6 cm. long, 2–5.5 mm. wide, linear to oblong-linear or slightly spatulate, 1-nerved, firm, the thick margins hispidulous; corymb 3–13 mm. wide, loose; involucre semiglobose, 5–7 mm. high, the bracts imbricate in 3–4 series, glabrous or somewhat puberulent; disk flowers 4–4.3 mm. long, yellow, subcylindric, the lobes ovate; pappus 1–5 mm. long, unequal, brownish, scaberulous; achenes 2–2.5 mm. long, ellipsoid, cuneate, compressed, yellowish with dark heavy margins, pilosulous above. *E. eradiatus* (Gray) Piper. Dry pine woods, Blue Mts., Garfield Co., *Horner* 349, *A. T. T.*

Erigeron linearis (Hook.) Piper. Perennial, more or less caespitose, 1–2 dm. tall, appressed hirsutulous throughout; tap-root 2–6 mm. in diameter, brown, woody, multicipital or forking above; stems erect, scapose or somewhat leafy; basal and lower cauline leaves 3–8 mm. long, linear to linear-spatulate, numerous; upper leaves reduced; heads terminal, solitary or in a few flowered cyme; involucre 4–6 mm. high, semiorbicular, the bracts subequal or slightly imbricate, oblong-linear, acute glandular puberulous, the thickened back somewhat hirsute; ray flowers 7–9 mm. long; pappus 3–3.5 mm. long, brownish, the slender outer series 0.2 mm. long; achenes 1.8–2 mm. long, oblanceoloid, flattened, brown, pilosulous; disk flowers 3.6–4 mm. long, yellow; pappus 3.7–4 mm. long, brownish, the outer series 0.2 mm. long, slender; achenes 2–2.3 mm. long, oblanceoloid, flattened, yellowish, pilosulous, the heavy margins brown. *E. peucephyllus* Gray; *E. filifolius* sensu Piper, Fl Wash. in large part, not of (Hook.) Nutt. Dry, rocky places, U. Son., *A. T.*

Erigeron peregrinus (Pursh) Greene, subsp. **callianthemus** (Greene) Cronq., var. **callianthemus.** *Mountain Daisy.* Perennial, 1–7 dm. tall; rootstock 2–10 mm. in diameter, brown, often creeping; stems sparsely hirsutulous to glabrate below, densely appressed hirsutulous canescent at apex, simple or simple below; foliage commonly glabrous except for the pilosulous ciliate margins, or occasionally somewhat pilosulous; basal and lower leaves with petioles 1–8 cm. long; blades 1–12 cm. long, oval to oblanceolate, entire; middle and upper leaves 1–10 cm. long, lanceolate to oblanceolate, sessile; heads one to few, terminal, remote on few bracted peduncles; involucre 5–8 mm. long, saucer-shaped, the bracts subequal, linear, acuminate, glandular puberulent, the tips purplish; ray flowers 10–16 mm. long, lavender; pappus 1–3 mm. long, unequal, white; achenes 1.5 mm. long, narrowly oblong, hirsutulous; disk flowers 3–3.3 mm. long, yellowish, almost tubular; pappus 3–3.5 mm. long, white; achenes 2.5–2.8 mm. long, obovoid, compressed, brown, hirsutulous. *E. membranaceus* Greene. Mountain meadows, Blue Mts., *Huds.*

Erigeron philadelphicus L. *Philadelphia Fleabane.* Perennial by offsets, 3–10 dm. tall; hirsute throughout; rootstock 2–5 mm. in diameter, brown; stems erect, simple below; basal and lower leaves 3–18 cm. long, elliptic-oblanceolate, coarsely dentate; middle and upper leaves 2–8 cm. long, elliptic to oblong-oblanceolate or ovate, clasping at base, dentate or the uppermost entire; cyme loose; involucre 4–5.5 mm. high, semiglobose, glandular atomiferous, the bracts subequal, the thickened midrib hirsutulous, the margins membranous; ray flowers 8–11 mm. long, pink; pappus 1.5–4 mm. long, white; achenes 0.5–0.7 mm. long, narrowly ellipsoid, brown, puberulous; disk flowers 2.5–2.8 mm. long, yellow, tubular; pappus 1.5–3 mm. long, white; achenes 0.5–0.7 mm. long, elliptic-obovoid, compressed, brown, puberulous. Meadows or roadsides, infrequent *U. Son., A. T.*

It is possible that this species is introduced in our area, yet it was collected more than a century ago by Spalding.

Erigeron poliospermus Gray, var. **poliospermus.** Perennial, more or less caespitose, 3–16 cm. tall; tap-root 2–5 mm. in diameter, brown; stems simple, ascending, naked and scapose at least above; basal and lower leaves with petioles 5–40 mm. long, margined; blades 10–25 mm. long, spatulate to linear-oblanceolate, acute; upper leaves much reduced; heads terminal, solitary; involucre 6–7 mm. high, hemispheric, the bracts subequal, linear, acuminate, minutely glandular puberulous, densely villous at base, hirsute along the green midrib; ray flowers 9–12 mm. long; pappus 3–3.3 mm. long, white; achenes 2.5–3 mm. long, ellipsoid, compressed, brown, appressed villous; disk flowers 3–3.5 mm. long, yellow, broad tubular; pappus 0.5–4 mm. long, unequal, white; achenes 2.5–3 mm. long, oblanceoloid, compressed, brown, appressed villous. Dry stony places, Spokane, and Bonnie L., and Blue Mts., *A. T.*

Erigeron pumilus Nutt., var. **intermedius** Cronq. Tufted perennial, 10–36 cm. tall; tap-root 2–5 mm. in diameter, brown, multicipital or forking above; stems erect, simple below; basal and lower cauline leaves with petioles 1–4 cm. long, the blades 1–4 cm. long, oblance-linear; middle and upper leaves 1–4 cm. long, sessile, similar; cyme loose, few to many headed; involucre semiorbicular; bracts subequal linear, acuminate, hirsute, often densely so, minutely glandular puberulous, the green midrib thickened; ray flowers violet, pink, or white; pappus 2 mm. long, sparse, white, the outer series 0.2 mm. long, broad based; achenes 1–1.5 mm. long, linear, brown, pilosulous; disk flowers 3.2–3.5 mm. long, yellow, tubular; pappus 2.5–3 mm. long, white, the outer series 0.2 mm. long, broad based; achenes oblanceoloid, flattened, yellowish, puberulous. *E. pumilus,* subsp. *intermedius* Cronq., var. *gracilior* Cronq. Dry or sandy places, *U. Son., A. T.*

Erigeron speciosus (Lindl.) DC., var. **macranthus** (Nutt.) Cronq. Perennial, 4–8 dm. tall; rootstock 2–10 mm. in diameter, woody, brown, forking above; stems glabrous, striate, simple below; basal leaves with petioles 1–7 cm. long, the blades 2–8 cm. long, oblanceolate, entire, hispid ciliate; middle and upper cauline leaves 2–8 cm. long, ovate acute, to narrowly oblanceolate, entire, hispid ciliate; cyme loose, few-headed; heads terminal, on short, bracted peduncles; involucre 5–8 mm. high, depressed subglobose, the bracts imbricate in 2–3 series, minutely glandular puberulous, linear, long acuminate, the middle thickened and greenish, the margins pale; ray flowers 12–16 mm. long, lilac; pappus 2–3 mm. long, whitish, the outer 0.3 mm. long; achenes 1.3–1.5 mm. long, oblong, compressed, brownish, pilosulous; disk flowers 3.8–4 mm. long, yellowish, tubular, narrowed below; pappus 3–4 mm. long, whitish, the outer 0.2 mm. long; achenes 1.5–1.8 mm. long, obovoid, compressed, brown, pilosulous. *E. macranthus* Nutt.; *E. speciosus* of Piper's Fl. Wash. in part, not of DC. Sandy places or thickets, *A. T., A. T. T.*

Erigeron strigosus Muhl. ex Willd., var. **septentrionalis** (Fern. & Wieg.) Fern. *Daisy Fleabane.* Annual or biennial, 3–9 dm. tall, hirsute or appressed hirsutulous throughout; caudex 2–8 mm. in diameter, brownish; stems erect, simple, below; basal and lower leaves 3–11 cm. long, spatulate to obovate, commonly dentate, the petiole exceeding the blade; middle and upper leaves

1.5–8 cm. long, linear-oblanceolate; cyme loose, few to many headed; involucre hemispheric, the bracts subequal, linear, acute, sparsely hirsutulous on the back; ray flowers 5–8 mm. long, white, or rarely purplish; pappus 0.1 mm. long; achenes 0.8 mm. long, brownish, ellipsoid, puberulous; disk flowers 1.6–1.8 mm. long, yellow, tubular; pappus 1.5–1.7 mm. long, sparse, the outer row 0.1 mm. long; achenes 0.4–0.6 mm. long, obovoid, compressed, brown, pilosulous. Meadows or moist places, *A. T., A. T. T.*

ERIOPHYLLUM

Annual or perennial herbs, sometimes shrubby at base; leaves usually alternate and entire or pinnately or ternately parted or lobed; ray flowers present, yellow, fertile; involucre campanulate to hemispheric; bracts erect, commonly united at base, oval or oblong; receptacle convex, usually naked; pappus of nerveless and mostly pointless scales; throat or limb of disk corollas yellow, rather narrow; style-branches truncate or rarely minutely tipped; achenes narrow, mostly 4-angled. (Name from Gr., *erion,* wool; *phullon,* leaf.)

Middle leaves pinnatifid; achenes glabrous. — *E. lanatum,* var. *lanatum.*
Middle leaves usually entire; achenes puberulent. — Var. *integrifolium.*

Eriophyllum lanatum (Pursh) Forbes, var. **lanatum.** *Oregon Sunshine.* Perennial herb, somewhat caespitose, 2–4 dm. tall, densely white lanate throughout; tap-root 2–5 mm. in diameter, woody, brown; stems slightly suffrutescent at base, ascending, often branched; leaves 2–7 cm. long, above greenish, due to the sparser pubescence, oblanceolate in outline, the upper and lower often entire, the middle leaves more or less pinnatifid, the divisions linear or oblong, obtuse, revolute; peduncles 3–10 cm. long; heads solitary; involucre 7–11 mm. high, hemispheric, the 8–15 bracts in 2 series, somewhat connate at base; rays of the same number 10–20 mm. long, elliptic; disk flowers 4–4.5 mm. long, the campanulate throat longer than the glandular puberulent tube; pappus of several scales 0.5–1 mm. long, lanceolate or oblong, lacerate; achenes 4–4.5 mm. long, narrowly oblong, black, glabrous. *E. pedunculatum* Heller. Dry or gravelly soils, *U. Son., A. T.*

The type was collected by Capt. Merriwether Lewis, June 6, 1806, "On the high lands of the Kooskoosky." [Clearwater River, Ida.].

Var. **integrifolium** (Hook.) Smiley. Differs usually in having the leaves entire or merely lobed at apex; pappus 1–1.5 mm. long, subequal; achenes puberulent. *E. multiflorum* (Nutt.) Rydb. Dry soils, *U. Son.*

EUPATORIUM. Thoroughwort

Erect, perennial herbs; leaves alternate, opposite, or whorled, often punctate; inflorescence cymose-paniculate; heads discoid, of white, blue, or purple flowers; involucral bracts imbricate in from 2 to several series; receptacle flat to conic, naked; corolla regular, the tube slender, the limb 5-lobed or 5-toothed; pappus of numerous capillary, usually scabrous bristles; achenes angular or striate. (Names for *Eupator Mithridates,* King of Pontus, who is said to have used this plant as a poison antidote.)

Eupatorium occidentale Hook. Plant 2–5 dm. tall; root woody 3–15 mm. in diameter, brown, stems usually several, forming a clump, appressed puberulent; leaves alternate or a few subopposite or whorled; petioles 3–15 mm. long, appressed puberulent; blades 1–8 cm. long, ovate, sparsely puberulent near the margins, the base cordate to cuneate, the apex acute, the margins entire or dentate; cymes 1–5 cm. broad, compact appressed puberulent; heads 15–25-flowered; involucre campanulate, appressed puberulent, the lance-linear bracts finally divergent; inner bracts 4–4.5 mm. long; corollas 5–6 mm. long, white or purplish tinged, the lobes 0.5 mm. long, ovate; pappus 4–5 mm. long; achenes 2.5–4 mm. long, brown, glandular puberulous. Deep Creek, Spokane Co., *Bonser, A. T.*

FRANSERIA. Sand Bur

Herbs (or shrubs) with mostly alternate leaves; staminate and pistillate heads discoid, separate or sometimes mixed in the inflorescence; staminate heads nodding in terminal spike-like racemes, the involucre bowl-shaped, the corolla funnelform; pistillate involucre 1–4-celled, with a single pistil in each cell, armed with spines in more than one row, bur-like. (Named for *Dr. Antonio Franseri* of Valentia, Spain.)

Franseria acanthicarpa (Hook.) Cov. Annual herb, 2–8 dm. tall, strigose and more or less hirsute throughout; tap-root 1–4 mm. in diameter, elongate, pale; stem striate, bushy branched; cauline leaves with petioles 1–5 cm. long, the blades 2–9 cm. long, ovate in outline, bipinnatifid almost to the midrib, the divisions linear to oblong, acute or obtuse, green; racemes 3–12 cm. long; staminate heads numerous, the involucre 3–5 mm. broad, sparsely hispidulous, with 3 black ribs, lobed half way into ovate, acute lobes; corolla 1.6–2 mm. long; pistillate heads clustered in the upper axils; fruit with the body 3–5 mm. long, ellipsoid, glabrous or sparsely hirsute, reticulate-ridged, the beak 1.5–2.5 mm. long, spine-like; spines 3–5 mm. long, in 3–4 series, lanceolate, flattened. *Gaertneria acanthicarpa* (Hook.) Britton. A weedy plant, shores of Snake River, *U. Son.*

GAILLARDIA

Pubescent, annual to perennial herbs; leaves alternate or basal, entire or incised or even pinnatifid; heads solitary and long-peduncled, large; ray flowers yellow or partly dark-purple, sterile; disk flowers perfect, fertile, usually purplish or brownish; involucral bracts in 2–3 series, the outer larger and foliaceous; receptacle convex or hemispherical, with one or more awns among the flowers resembling chaff; pappus of 6–10 hyaline chaffy scales with a prominent midrib which is prolonged into a naked awn, or in the sterile rays, scales awnless; achenes oblong, top-shaped, each surrounded by a tuft of hairs. (Named for *Gaillard de Charentonneau* of France.)

Gaillardia aristata Pursh. *Blanket Flower.* Perennial, 2–6 dm. tall, villous, hirsutulous, and resinous atomiferous throughout; tap-root 2–8 mm. in diameter, woody, brown; lower leaves 5–20 cm. long, oblanceolate, entire, toothed, or pinnatifid, the blade about as long as the petiole; upper leaves 3–8 cm. long, sessile, lanceolate or oblanceolate, pinnatifid or toothed or more commonly entire; peduncles 1–2 dm. long; involucre saucer-shaped; bracts 10–18 mm. long, subequal, linear-lanceolate, acuminate, villosulous, the lower margins villous ciliate; bristles of the receptacle 2–4 mm. long, awn-like; the 11–17 rays 10–25 mm. long, wedge-shaped, deeply lobed, yellow, with or without purple coloring at base; disk flowers 5–8 mm. long, the throat and purplish lobes villous; pappus scales 6–7 mm. long, straw-colored; achenes 3–4 mm. long, densely hirsute. Common, moist places, *U. Son., A. T.*

GNAPHALIUM. Cudweed

Woolly herbs; leaves alternate, entire; heads small, discoid; pistillate flowers very numerous, in more than one row; perfect flowers fewer in the center; staminate flowers none; all flowers white or yellow; pappus-bristles slender, not thickened above; achenes oblong or ovate. (An ancient Gr. plant name, from *gnaphalon,* wool.)

Involucral bracts lanate almost to the tip; inflorescence with conspicuous leafy bracts. *G. palustre.*
Involucral bracts glabrous or lanate only at base; inflorescence with reduced bracts,
 Leaves above green and glandular puberulent; stems viscid pilose below. *G. Macounii.*
 Leaves above white lanate; stems lanate,
  Upper cauline leaves narrowed to the base; involucral bracts acute; involucres in numerous small glomerules, forming a loose cyme. *G. thermale.*
  Upper cauline leaves not narrowed to the base; involucral bracts obtuse; involucres in few dense heads. *G. chilense.*

Gnaphalium chilense Spreng. Annual or biennial, 1–7 dm. tall, closely white lanate throughout; tap-root 1–4 mm. in diameter, brown; crown usually multicipital; stems erect, usually simple to the inflorescence; basal leaves 1–3 cm. long, spatulate; cauline leaves 2–5 cm. long, narrowly spatulate to linear, the basal auricles short, rounded, decurrent; heads numerous, in compact glomerules, these 1 to several in a panicle 2–5 cm. wide, its bracts leafy, reduced; involucres cup-shaped, lanate at base, glabrous above, the bracts chartaceous, straw-colored or yellowish, the outer ovate, the inner 4–5 mm. long, linear; pappus 3.5–4 mm. long, white, distinct, separately deciduous; achenes 0.5–0.6 mm. long, ellipsoid, brown, smooth. *G. chilense,* var. *confertifolium* Greene; *G. proximum* Greene. Alkali flats or prairies, *U. Son., A. T.*

Gnaphalium Macounii Greene. Annual or biennial, 3–9 dm. tall; tap-root 2–6 mm. in diameter, woody, brown; stem simple or branched, white lanate above; basal leaves 2–9 cm. long, oblanceolate, white lanate; cauline leaves 3–15 cm. long, linear, acute, above green, glandular puberulent, beneath white lanate; heads in glomerules in a round-topped cyme 5–20 cm. wide, its bracts reduced, inconspicuous; involucre ovoid to cup-shaped, glabrous, the bracts white or straw-colored, chartaceous, ovate, acute, the inner 4.5–5 mm. long; pappus 4.5 mm. long, white, the bristles dehiscing separately; achenes 0.5–0.6 mm. long, cylindric, greenish, smooth. *G. viscosum* sensu Cronq.; *G. Ivesii* Nels & Macbr. Open woods, Thatuna Hills, *A. T. T.*

Gnaphalium palustre Nutt. Annual, 5–30 cm. tall, loosely floccose lanate throughout; roots fibrous; stems erect, decumbent, or prostrate, simple or branched; leaves 12–25 mm. long, spatulate to elliptic, sessile; heads in small, terminal clusters; involucre 2.5–3.5 mm. tall, the linear inner bracts woolly except on the white papery tips; pappus 1 mm. long, white, distinct, separately deciduous; achenes 0.3 mm. long, ellipsoid, greenish, smooth. Common, mud flats, *U. Son., A. T.*

Gnaphalium thermale E. Nels. Biennial, 1–6.5 dm. tall, closely white lanate throughout; tap-root 2–8 mm. in diameter, woody, brown; crown multicipital; stems simple; basal leaves 3–5 cm. long, linear-oblanceolate; cauline leaves 2–6 cm. long, linear to linear-oblanceolate, slightly decurrent at base; heads in small glomerules in a cyme 2–15 cm. broad, usually round-topped, its bracts reduced; involucre ovoid to campanulate, lanate at base, glabrous above, the bracts ovate, white chartaceous, imbricate, the inner linear, 4–5.5 mm. long; pappus bristles 4–4.5 mm. long, white, separately deciduous; achenes 0.5–0.6 mm. long, ellipsoid, compressed, brown, smooth. Gravelly soil, *A. T.*

GRINDELIA. Gum Plant

Annual to perennial herbs or shrubs; leaves alternate, the upper sessile or partly clasping; heads yellow, medium or rather large, solitary, terminating leafy branches, or occasionally more or less corymbose, many-flowered, gummy; involucre imbricate, hemispheric or campanulate; rays fertile, numerous, narrow, or rarely none; disk flowers perfect, fertile, 5-toothed; involucral-scales numerous, narrow; receptacle flat or convex; pappus of 2–15 rigid and early-deciduous awns; style-branches tipped with an appendage; achenes compressed or turgid or the outermost somewhat 3–4-angled. (Named for *Prof. David Hieronymus Grindel,* of Russia.)

Middle and upper leaves crenulate-serrate with obtuse teeth,
- Upper and middle leaves 2–4-times longer than broad, ovate to oblong. *G. squarrosa,* var. *squarrosa.*
- Upper and middle leaves 5–8 times longer than broad, linear-oblong to oblanceolate. Var. *serrulata.*

Middle and upper leaves sharply toothed or entire,
- Heads campanulate-hemispheric, broader than high,
  - Stems 25–65 cm. tall; leaves of flowering branches reduced and scattered.
    - Inner involucral bracts 7–10 mm. long, the elongate tips 1.5–3 mm. long. *G. nana,* var. *nana,* f. *nana.*
    - Inner involucral bracts 5–7 mm. long, the revolute tips 0.4–1.5 mm. long. Var. *integerrima,* f. *integerrima.*
  - Stems 3–11.5 dm. tall; leaves of flowering branches not reduced and scattered,
    - Main cauline leaves mostly entire elliptic-obovate to broadly oblong-oblanceolate, mostly 15–23 mm. broad Var. *integrifolia.*

Main cauline leaves serrate to denticulate, oblanceolate to oblong-lanceolate, mostly 7–15 mm. broad.

Var. *nana,* forma *Brownii.*

Heads turbinate to hemispheric, as high as or higher than broad.

Var. *integerrima,* forma *longisquama.*

Grindelia nana Nutt., var. **nana,** forma **nana.** Perennial herb, 8–65 cm. tall, glabrous throughout; root 4–8 mm. in diameter, woody, brown, multicipital; stems erect, branched above, brown to purplish, resinous; leaves firm chartaceous, abundantly resinous punctate and viscid; principal middle and upper leaves 3–9 cm. long, oblanceolate, mostly attenuate to the base; bracteal leaves lanceolate, entire; basal leaves spatulate; heads terminal, several in loose cymes; involucre 7–12 mm. tall, nearly hemispheric, usually broader than high, the bracts imbricate in 5–7 series, resinous throughout from the abundant resin dots, subulate to lance-subulate, the tips squarrose; the 11–18 rays 5–11 mm. long, chrome or lemon-yellow; receptacle foveolate, the ridges scale-like; disk flowers 4.5–5.5 mm. long, narrow funnelform; achenes 3.5–4 mm. long, narrowly oblong, light brown, 1–3-knobed at apex; the 2 awns 3–5.5 mm. long, subulate, curved, entire to remotely serrulate. Grasslands and river banks, *A. T.*

Forma **Brownii** (Heller) Steyerm. Stems 5–8 dm. tall; leaves of floriferous branches scarcely reduced, entire, serrulate, or dentate; principal middle and upper leaves 4–8 cm. long, the upper lanceolate or oblong-lanceolate, clasping at base; involucre 7–17 mm. tall. Sandy bank, Clearwater R., below Lapwai, *Henderson* 4,605, *U. Son.*

Var. **integrifolia** Nutt. Stems 28–44 cm. tall; leaves mostly entire to remotely denticulate; middle and upper leaves 5–8 cm. long, only slightly reduced upwards; lower and middle leaves elliptic-obovate to oblanceolate; involucre 9–12 mm. high, broader than high, the bracts squarrose; rays 10–14 mm. long. *G. nana,* var. *Paysonorum* (St. John) Steyerm. Dry slopes on basalt or limestone, Snake River Canyon, *U. Son.*

Var. **integerrima** (Rydb.) Steyerm., forma **integerrima.** Stems 15–40 cm. tall; leaves salient dentate to entire, the lower and middle ones oblanceolate to oblance-spatulate, narrowed to the slender base; involucre 6–10 mm. tall, usually broader, the bracts lanceolate with squarrose subulate tips; awns 2.5–4.5 mm. long. Sandy places, mostly in Spokane region, *A. T.*

Forma **longisquama** Steyerm. Stems 3.3–6 dm. tall; leaves entire, serrulate, or denticulate, the upper broadly lanceolate or oblong, narrowed to the somewhat clasping base; involucre 10–15 mm. high, hemispheric, the bracts with reflexed or spreading, subulate tips; rays 8–12 mm. long. Known in our area from the type collection, "Stoney land, July 4, Clear Water, Oregon [Lapwai, Ida.] Spalding," *U. Son.*

Grindelia squarrosa (Pursh) Dunal, var. squarrosa. Biennial, 4–10 dm. tall; tap-root 4–10 mm. in diameter, brown; stem often branched, glabrous, whitish or rarely purplish; middle and upper leaves 3–7 cm. long, ovate to oblong, obtuse or acute, the base clasping, dark bluish green, subcoriaceous, glandular punctate crenulate-serrate, the glandular teeth obtuse; lower leaves more obovate- to spatulate-oblong; heads radiate, numerous; involucre hemispheric, 6–11 mm. high, the bracts very glandular, imbricate in 5–6 series, the outer subulate, the inner lance-linear, the tips squarrose; receptacle foveolate; the 24–36 rays 8–10 mm. long, bright or lemon-yellow; disk-flowers 4–6 mm. long, almost tubular; achenes 2.3–3 mm. long, straw-colored to brownish, obovate, 4-angled, compressed; awns 2–3, or even 6, slender 3–5 mm. long, subulate, more or less serrulate. Introduced from the Great Plains. Roadsides, Waitsburg; Pullman, *A. T.*

Var. serrulata (Rydb.) Steyerm. Plant 1.5–6 dm. tall; middle and upper leaves linear-oblong to oblanceolate; rays more golden yellow, 8–14 mm. long. Introduced from the Rocky Mountains. Lake Coeur d'Alene, *Epling & Houck* 10,032.

GUTIERREZIA

Herbs or shrubs, resinous; leaves linear, alternate, entire; heads small, numerous, in cymose panicles, with both ray and disk flowers; involucre campanulate to clavate to linear; involucral bracts coriaceous with green tips, unequal, closely imbricated; receptacle naked, in ours flat; ray flowers 1-11, yellow, short, pistillate, fertile; disk flowers yellow, perfect or the central ones sterile; pappus of 5–9 chaffy or bristleform scales, those of the ray flowers shorter or even none. (Named for *Pedro Gutierrez,* correspondent in the early 19th century of the botanical garden in Madrid.)

Gutierrezia sarothrae (Pursh) Britt. & Rusby. *Brown-weed.* Shrub 2–6 dm. tall, with numerous erect, slender brittle branches; herbage mostly scabrous puberulent; leaves 2–4 cm. long, 1–2 mm. wide, punctate; inflorescence flat- topped; involucre 3–4.5 mm. high, glutinous, subcylindric to narrowly obconic; rays 1–6, the ligules 1.5–2 mm. long; disk flowers 3–7, fertile; pappus scales 9–12, linear-oblong to subulate; achenes silky, shorter than the pappus. Asotin Co., Rogersburg, *L. Constance* 1,813, *U. Son.*

The specific epithet was coined by Pursh as a genitive of the pre-Linnaean generic name *Sarothra.* Since it is a modification of a generic name, the International Code implies, by examples, that it should be decapitalized.

HAPLOPAPPUS

Herbs or low undershrubs; leaves alternate, soft or rigid; heads solitary, terminal, or clustered, many-flowered; involucre imbricated; bracts with or without foliaceous tips, imbricate, or nearly equal; receptacle flat or flattish; ray flowers fertile, or sterile, yellow, or white, or rarely none; disk flowers perfect, usually fertile and numerous, 5-toothed; pappus tawny or reddish, of copious and unequal capillary bristles, somewhat rigid; style-branches with appendages; achenes turbinate and linear, terete, angled or more or less compressed. (Name from Gr., *haplo-,* simple; *pappos,* down.)

Shrubs; leaves linear or linear-spatulate, resinous; involucre 5–8 mm. high. *H. resinosus.*
Herbs; leaves broader, not resinous; involucre 9–20 mm. high,
  Rays none or obscure; involucral bracts rigid,
    Stems puberulent; involucral bracts subequal, lance-

olate. *H. carthamoides,* var. *Cusickii.*
Stems villous; involucral bracts imbricate, oblong. Var. *erythropappus.*
Rays obvious; involucral bracts not rigid,
Plant tomentose; heads solitary; involucre imbricate in about 2 series. *H. lanuginosus,* var. *lanuginosus.*
Plant villous; heads in a raceme; involucre imbricate in 4–5 series. *H. liatriformis.*

Haplopappus carthamoides (Hook.) Gray, var. **Cusickii** Gray. Perennial, 5–40 cm. tall; caudex 5–12 mm. in diameter, woody, black, sometimes forking above; stems several, erect or somewhat decumbent, puberulent, at base finally glabrate; leaves subcoriaceous, pale, entire or merely undulate; the basal and lower cauline with petioles 1–10 cm. long; the blades 3–15 cm. long, oblanceolate, puberulent to glabrate, the margins scabrous; upper leaves similar but sessile, lanceolate, reduced; heads 1–2; involucre 12–20 mm. high, campanulate to hemispheric; bracts loose, subequal, the tips foliaceous and the outer ones foliaceous and blending into the bracts of the peduncle, lanceolate, subulate pointed, puberulent, the inner with scarious margins; ray flowers none; disk flowers 9–13 mm. long, tubular, the lobes puberulent; pappus 10–14 mm. long, brownish; achenes 5.5–7 mm. long, fusiform, flattened, grayish, ribbed. *H. carthamoides, subsp. cusickii* (Gray) H. M. Hall; *Hoorebekia carthamoides cusickii* (Gray) Piper. Rocky slopes above Snake River Canyon, and Blue Mts., *A. T.*

Haplopappus carthamoides (Hook.) Gray, var. **erythropappus** (Rydb.) St. John. Differs from var. *Cusickii,* by having the stems soft villous; heads 1–5; involucral bracts imbricate in about 4 series, oblong, abruptly rounded to the mucronate tip. *Pyrrocoma erythropappa* Rydb.; *Aplopappus carthamoides* (Hook.) Gray, var. *rigidus* (Rydb.) Peck; *H. carthamoides,* subsp. *rigidus* (Rydb.) H. M. Hall. Grasslands, *U. Son., A. T.*

The type was collected by Rev. Henry Spalding at Clear Water, Oregon [Lapwai, Ida.]. Despite the distinctions alleged by Rydberg and Hall, *P. rigida* is indistinguishable. It is an illegitimate name because of the earlier homonym by Philippi (1856), hence the writer selected the equally old *P. erythropappus,* and made the combination. Its characters seem to be of varietal rather than subspecific rank, just as *Cusickii* was classified as a variety by Dr. Gray.

Haplopappus lanuginosus Gray, var. **lanuginosus.** Perennial herb, 7–30 cm. tall, white floccose tomentose throughout; caudex 4–7 mm. in diameter, woody, brown, forking or multicipital at apex; stems striate, erect, almost scapose; basal and lower leaves numerous, crowded, 4–8 cm. long, 2–6 mm. wide, linear-spatulate, narrowed and petiole-like below, obtuse or acute, soft chartaceous, the nerves obscure; cauline leaves similar, crowded on lower part of stem, bracteose; involucre 9–12 mm. high, semiorbicular; bracts linear- to oblong-lanceolate, acute, somewhat lanate, greenish with scarious margins; the 15–20 rays 10–12 mm. long; disk flowers 7–8.5 mm. long, puberulent at base; pappus 7–8 mm. long, white; achenes 4.5–6 mm. long, oblanceoloid, compressed, brown, densely pilosulous. *Hoorebekia lanuginosa* (Gray) Piper. Gravelly slopes, Blue Mts., *A. T.*

Haplopappus liatriformis (Greene) St. John. Perennial herb, 3–7 dm. tall, softly villous throughout; caudex 5–20 mm. in diameter, woody, brown; stems several, erect, often branched; leaves subcoriaceous, entire, the basal and lower with petioles 3–12 cm. long; blades 7–15 cm. long, oblanceolate, subcoriaceous; upper leaves 3–8 cm. long, sessile, linear-lanceolate; raceme 15–25 cm. long; involucre 11–15 mm. high, turbinate, soft villous; bracts loose, the tip foliaceous, the base chartaceous, lance-oblong, acute; the 17–18 rays 5–12 mm. long; disk flowers 9–10 mm. long, almost tubular, sparsely puberulent; pappus 8–10 mm. long, tawny, the bristles deciduous in a group; achenes 5–6 mm. long, fusiform, compressed, brown, appressed pilosulous. *Hoorebeckia racemosa* sensu Piper, not of (Nutt.); *Pyrrocoma liatriformis* Greene. Prairies, *A. T.*

The type was collected in Pullman by Piper in Aug. 1903. This plant is treated by Hall as a subspecies of *H. integrifolius* of Mont., Ida., Wyo. This nearly glabrous plant, with the bracts glabrous on the back, cartilaginous at base, firm herbaceous at tip, subequal or the outer and longer, seems very different from our plant. As the writer has seen no real intermediates, Hall notwithstanding, it is maintained as a species.

Haplopappus resinosus (Nutt.) Gray. Shrub 3–5 dm. tall, round-topped, loosely branched; twigs glabrous, brown, resinous, angled; old bark gray to black, irregularly checked; principal leaves 5–30 mm. long, 0.5–1.5 mm. wide, acute, channeled above, resinous; axils bearing a fascicle of minute secondary leaves; heads 1, or in cymes of 2–3; involucre narrowly turbinate; bracts loosely imbricate in 4–5 series, linear-lanceolate to oblong-linear, acute, the outer somewhat foliaceous, subulate-tipped, the inner chartaceous, ciliolate at tip; the 3–7 rays 3–6.5 mm. long, white (or ochroleucous); the 11–16 disk flowers 5.5–7.5 mm. long, the tube puberulent; pappus 5–7 mm. long, brownish, barbellate. *Ericameria nana* Nutt. Rare, basalt ledges, Snake River Canyon, near Zaza, and mouth of the Tukanon, *U. Son.*

HELENIUM. Sneezeweed

Erect, simple or branching, annual or perennial herbs; leaves all alternate and all but the lower sessile; heads small or large, many-flowered, on naked terminal peduncles; ray flowers yellow, several or numerous, pistillate; disk flowers yellow or turning brownish or purplish, small and very numerous, all fertile; involucral bracts spreading or reflexed at maturity; pappus of 5–12 thin or hyaline chaffy scales with or without midribs; achenes top-shaped, striate-ribbed, hairy on the ribs. (Greek name of some plant, perhaps from *Helenus,* son of *Priam.*)

Helenium autumnale L., var. **montanum** (Nutt.) Fern. Perennial, 3–20 dm. tall, puberulent throughout; caudex 4–10 mm. in diameter, brown, the laterals fleshy; stem erect, glabrate below, branched above; leaves 3–12 cm. long, oblanceolate to lanceolate, more or less serrate, long decurrent on the stems; heads numerous on slender, naked peduncles; involucre 5–6 mm. long, the subulate bracts united at base, puberulent and resin dotted; the 10–18 rays 1–2 cm. long, wedge-shaped, with prominent rounded lobes at tip; disk flowers 3–4 mm. long, puberulent and more or less resinous atomiferous; pappus 1–2 mm. long; achenes 1.5–2 mm. long, brown. *H. montanum* Nutt.; *H. macranthum* Rydb. Sandy banks of Snake River, *U. Son.*

The herbage is acrid, and had been shown to be poisonous to grazing animals.

HELIANTHELLA

Perennial herbs; leaves entire, opposite or the upper alternate; heads large, terminal, chiefly solitary and long-peduncled; flowers yellow; ray flowers neutral; disk flowers perfect; receptacle flat or convex, chaffy; pappus a pair of persistent awns or chaffy teeth, and a crown of intermediate thin chaffy scales or wanting; ray achenes laterally compressed; disk achenes 4-angled or laterally compressed, all very flat, winged when young. (Name a diminutive of *Helianthus, sunflower.*)

Helianthella uniflora (Nutt.) T. & G., var. **Douglasii** (T. & G.) W. A. Weber. Plant 4–12 dm. tall, somewhat hirsute throughout; root 5–12 mm. in diameter, woody, brown, often forking at summit; stems erect, mostly simple, striate; leaves firm chartaceous, scabrous, appressed hirsutulous above; basal and lower leaves with petioles 1–8 cm. long; the blades 12–25 cm. long, linear-oblanceolate, 3-nerved from below the middle; upper leaves similar, 3–13 cm. long, lanceolate; heads terminal, 1–3 depressed semiorbicular; peduncles naked or bracted; involucral bracts 1–2 cm. long, subequal, foliaceous, linear-lanceolate, hirsute ciliate; the 17–26 rays 20–25 mm. long, narrowly elliptic; chaff 10–12 mm. long, linear, carinate, the acute tip hirsutulous; disk flowers 5–6 mm. long, funnelform, the tube 1–2 mm. long, the lobes hirsutulous; anthers brownish; achenes 6–7 mm. long, wedge-shaped, brown, appressed hirsutulous, the 2 awns 3.5–4 mm. long, connected by lacerate scales 1 mm. long. *H. Douglasii* T. & G. Rocky hillsides, *U. Son., A. T., A. T. T.*

David Douglas collected the type "on the subalpine range of the Blue Mountains."

HELIANTHUS. Sunflower

Coarse annuals or perennials; leaves entire or toothed, all or at least the lower ones opposite; head solitary or corymbose, medium or large; involucre hemispheric or depressed; ray flowers yellow, sterile; disk flowers yellow, brownish or dark-purple; receptacle flat or convex; pappus a pair of early-falling chaffy scales or awns; chaff subtending the disk flowers; achenes neither very flat nor winged; ray achenes laterally compressed; disk achenes 4-angled, compressed. (Name from Gr., *helios,* sun; *anthos,* flower.)

Helianthus annuus L. *Common Sunflower.* Annual herb, somewhat hispid and scabrous throughout; tap-root 1–10 mm. in diameter, pale; stem erect, simple or branched, green or purplish mottled; petioles 1–30 cm. long; blades 3–30 cm. long, deltoid-ovate to lanceolate, obtuse or acute, dentate, rounded or subcordate; heads terminal, one to many; disk 1–4 cm. wide; involucre saucer-shaped, the bracts subequal or imbricate in 2–3 rows; bracts 1–2.5 cm. long, foliaceous, ovate, caudate, white hispid; the 8–20 rays 15–35 mm. long, narrowly elliptic; chaff pale, 8–10 mm. long, linear-oblong, the acute tip hirsutulous; disk flowers many, 4–5 mm. long, funnelform, the limb reddish purple; anthers purple; achenes 5–8 mm. long, obovate, gray or mottled, glabrous or somewhat pubescent; the 2 pappus scales 3–4 mm. long, lance-linear. *H. lenticularis* Dougl. Open places or thickets, *U. Son., A. T.*

This is the wild progenitor of the cultivated sunflower, which is important as forage, for silage, for its edible achenes from which oil and oil cake are obtained.

HEMIZONIA

Annual or perennial herbs, usually more or less glandular and viscid and heavy scented; leaves alternate or the lower sometimes opposite; heads not large, many- or sometimes few-flowered; involucral bracts rounded on the back, partly enclosing the turgid more or less oblique ray achenes; receptacle flat or convex, chaff deciduous; flowers yellow or white; ray flowers in 1 series, pistillate, fertile; pappus none; disk flowers tubular, 5-toothed, fertile or sterile; pappus scale-like or wanting. (Name from Gr., *hemi,* half; *zone,* girdle.)

Hemizonia pungens (H. & A.) T. & G., subsp. interior Keck. *Common Spikeweed.* Annual, 1–6 dm. tall, sparsely hirsute and more or less glandular puberulent; tap-root 1–4 mm. in diameter, pale brown; stem freely branching, the branches rigid, ascending, pale, becoming whitish; basal leaves 4–7 cm. long, bipinnatifid; lower cauline leaves 3–6 cm. long, pinnatifid, the segments linear, acerose-tipped; upper leaves 8–20 mm. long, linear, rigid, scabrous, acerose-tipped; heads numerous, terminal, not crowded; involucre globose; outer bracts 5–10 mm. long, squarrose foliaceous, linear, rigid, hispid, acerose-tipped, the inner 3 mm. long, lanceolate, cuspidate, carinate, pilosulous, similar to the chaff; the 25–40 ray flowers 3–5 mm. long, yellow, the ray elliptic; achenes 1.8–2 mm. long, black, semiobovate, ribbed on the inner faces, the back rounded, the tip incurved apiculate; disk flowers 2.5–3 mm. long, yellow, the lobes ovate; pappus none; ovary sterile. Weed, introduced from California. Fields and arid lands, abundant around Walla Walla, *U. Son.*

HIERACIUM. Hawkweed

Hispid and hirsute often glandular or glabrous perennials with milky juice; leaves alternate or basal merely toothed or entire; heads small to medium, corymbose, paniculate, or rarely solitary; flowers yellow, orange, red, or white; bracts of involucre in 1–3 rows; receptacle flat, naked or short fimbrillate; pappus of a single row of rough tawny, persistent bristles; corollas all ligulate; achenes oblong or columnar, terete or 4- or 5-angled, mostly 10-ribbed or striate, the apex truncate. (Name Gr., *hierax,* hawk, used by Dioscorides.)

Involucre with several rows of outer bracts, closely and regularly imbricate, glabrous or glabrate,
  Cauline leaves lance-linear; base of stem sparsely hirtellous, soon glabrate. — *H. umbellatum.*
  Cauline leaves lanceolate to ovate-lanceolate; base of stems pilose,
    Styles yellow. — *H. scabriusculum,* var. *columbianum,* f. *columbianum.*
    Styles brown. — Forma *phaeostylum.*
Involucre with but few, loosely imbricate outer bracts, pubescent,
  Leaves glabrous; inflorescence a loose raceme. — *H. gracile.*
  Leaves pubescent; inflorescence a cyme or panicle,
    Corollas white; stem almost naked above. — *H. albiflorum.*
    Corollas yellow; stem leafy to the inflorescence,

Involucres not pilose, or but sparsely so towards the base. *H. Cusickii.*
Involucres hidden by the long shaggy villosity. *H. albertinum.*

Hieracium albertinum Farr. Perennial, 2–7.5 dm. tall, long white shaggy villous throughout, the hairs drying brownish; caudex 3–8 mm. in diameter, brown; lateral roots cord-like; stem single, simple; basal and lower cauline leaves similar, 10–25 cm. long, linear-oblanceolate, entire, tapering gradually into the winged petiole; upper leaves 3–8 cm. long, lanceolate, sessile; panicle 3–20 cm. long, many headed, puberulous and densely long white villous from black papillose bases; peduncles 3–20 mm. long; involucre campanulate, white, villous; outer bracts few, linear, imbricate; inner bracts 9–12 mm. long, linear, hidden by the pubescence; corollas 10–14 mm. long; pappus 5–6 mm. long; achenes 3 mm. long, cylindric, dark reddish brown. Prairies, *A. T.*
The herbage is grazed by sheep.

Hieracium albiflorum Hook. Perennial, 2–12 dm. tall; caudex 2–10 mm. in diameter, brown; stem long hirsute below, simple, above glabrous; leaves hirsute, chartaceous, pale beneath; basal and lower leaves with petioles 1–6 cm. long, the blades 4–15 cm. long, oblanceolate, sparingly denticulate; middle leaves 3–10 cm. long, sessile; upper leaves few, bracteose; cyme 10–70-headed, round-topped or paniculate, glabrous; peduncles 1–5 cm. long; involucre cylindric or narrowly campanulate, more or less puberulent at base, and sparsely blackish hirsute, loosely imbricate, the outer bracts few; inner bracts 7–9 mm. long, linear; corollas 7–10 mm. long; pappus 5–7 mm. long; achenes 3–3.5 mm. long, brown. *H. Helleri* Gdgr. Woods, *A. T. T., Can.*

Hieracium Cusickii Gdgr. Perennial, 2–9 dm. tall; caudex 3–5 mm. in. diameter, brown, or more slender and creeping; stem villous below, above sparsely so to glabrate, simple; leaves villous, the basal and lower ones with petioles 3–5 cm. long, the blades 4–15 cm. long, oblanceolate or narrowly so, pale beneath, entire; middle leaves 3–8 cm. long, sessile, oblanceolate to lanceolate; upper leaves few, remote, bracteose; cyme paniculate, 10–40-headed, glandular puberulent and more or less villous; peduncles 3–40 mm. long; involucre campanulate, glandular puberulent; outer bracts few, imbricate; inner bracts 8–10 mm. long, linear; corollas 8–12 mm. long, yellow; pappus 4–5 mm. long; achenes 3–3.5 mm. long, reddish black. *H. griseum* Rydb., not Formánek (1896) ; *H. Rydbergii* Zahn. Prairies or thickets, *A. T., A. T. T.*

Hieracium gracile Hook. Perennial, 6–50 cm. tall; caudex 2–8 mm. in diameter, brown; lateral roots fleshy; stems often several, more or less lanate below; the basal leaves with petioles 1–6 cm. long, the blades 2–8 cm. long, spatulate to obovate, apiculate, entire; cauline leaves none or few, bracteose and only subbasal; raceme 1–15-headed, 1–18 cm. long, lanate and black hirsute; peduncles 5–90 mm. long; involucre campanulate, viscid black hirsute and more or less lanate; outer bracts few, lanceolate, imbricate; inner bracts 5–9 mm. long, linear; corollas 7–8 mm. long, yellow; pappus 3–4.5 mm. long; achenes 2–3 mm. long, cylindric, closely ribbed, reddish black. Alpine meadows, Blue Mts., *Huds.*

Hieracium scabriusculum Schwein var. **columbianum** (Rydb.) Lepage, forma **columbianum.** Perennial, 6–12 dm. tall; caudex 5–10 mm. in diameter, woody, short; lateral roots cord-like; stem unbranched below, above glabrous, below densely brown pilose, very leafy; lower leaves 7–13 cm. long, oblanceolate, acute, pale beneath, sparsely hirsute, the margin sparsely salient serrate; middle leaves lanceolate or ovate-lanceolate, smaller; upper leaves bracteose; cyme puberulent, 10–40-headed; peduncles 1–6 cm. long; involucre campanulate, sparsely puberulent, soon glabrate; involucral bracts blackish green, imbricate in about 3 series, the inner 7–10 mm. long; corollas 9–14 mm. long, yellow; pappus 4–7 mm. long; achenes 3–3.5 mm. long, reddish black. *H. Columbianum* Rydb.; *H. canadense columbianum* (Rydb.) Piper; *H. canadense* var. *columbianum* (Rydb.) St. John. Thickets, northeast part, *A. T. T.*

Forma **phaeostylum** Lepage. Differs by having the styles brown. The holotype is Dartford, Sept. 12, 1902, *F. O. Kreager*.

Hieracium umbellatum L. Perennial, 3–10 dm. tall; caudex 3–15 mm. in diameter, woody, short, the lateral roots numerous, fleshy; stems often several, densely leafy throughout, simple below; leaves 2–9 cm. long, mostly lance-linear, pale, scabrous, coarsely few serrate along the sides; cyme 3–30-headed, glandular puberulent; peduncles 1–5 cm. long; involucre semiorbicular; involucral bracts glabrous, blackish green, linear, regularly imbricate in about 4 series, the inner 9–11 mm. long; rays 12–15 mm. long, yellow; pappus 6–7 mm. long; achenes 3–4 mm. long, reddish brown to black. Low places, *A. T.*

HYPOCHOERIS

Herbaceous, and usually perennial, plants; leaves radical in a rosette, entire or commonly pinnatisect; scapes simple or branching, naked or with a few reduced bracts; involucre oblong-cylindric or campanulate, the bracts herbaceous, imbricate in two or more ranks; receptacle flat, bearing chaff; flowers yellow, all ligulate, dentate at apex; anthers sagittate; style branches filiform, obtuse; achenes oblong to linear, contracted into an apical beak; pappus plumose, borne in one rank or sometimes with a shorter second rank or sometimes none on the marginal achenes. (Name from Gr., *hupo,* beneath; *choiros,* pig.)

Hypochoeris radicata L. *Gosmore.* Perennial herb; stems 2–7 dm. tall, scapose, one to several, more or less branched, glabrous; leaves 4–20 cm. long, basal, broadly oblanceolate in outline, pinnatifid, with divergent lobes, to merely sinuate-dentate, rough hirsute; involucre 15–20 mm. high, campanulate; ligules showy, much exceeding the involucre; bracts, glabrous or strigose; achenes brown, the fusiform body about 3.5 mm. long, rough, all with very slender beaks longer than the body; pappus about 1 cm. long. Weed, introduced from Europe, Latah Co., 1942, *Daubenmire*.

IVA

Herbs; leaves simple, at least some of the lower opposite; heads small, nodding, in the axils of the leaves or in terminal spikes or panicles; flowers monoecious, in the same head; a few marginal ones pistillate, the others staminate and more numerous; bracts of the involucre few, commonly united into a cup; receptacle chaffy, with scales subtending the sterile flowers; pappus none; achenes obovate, thick. (Derivation of name unknown.)

Heads axillary, peduncled; leaves entire; pistillate corollas present. *I. axillaris.*

Heads in spikes or panicles, sessile; leaves double dentate; pistillate corollas none or rudimentary. *I. xanthifolia.*

Iva axillaris Pursh. *Poverty Weed.* Perennial, 2–6 dm. tall; roots 3–4 mm. in diameter, creeping, colony-forming; stems suffrutescent at base, often bushy branched, hirsutulous, below glabrate; leaves opposite or alternate below, alternate above, 1–4 cm. long, sessile or subsessile, narrowly elliptic to oblanceolate or obovate, thick, pale, glandular-atomiferous, 1–3-nerved;

heads nodding, peduncle 2–4 mm. long, hirsutulous; involucre 3–4 mm. long, hemispheric, more or less hirsutulous, the 4–5 bracts more or less united, the sinuses extending down from 1/6 to 3/4 the length of the involucre; staminate corollas 3–4 mm. long, funnelform; pistillate corolla 0.7 mm. long, cylindric; achenes 2–3.2 mm. long, glandular atomiferous, obpyramidal, obtuse, somewhat compressed. Arid places, especially if alkaline, *U. Son., A. T.*

Iva xanthifolia Nutt. *Careless Weed.* Annual, 5–20 dm. tall; tap-root 3–10 mm. in diameter, strong, brown; stem glabrous below; leaves opposite or only the upper alternate; petioles 2–25 cm. long, narrowly margined, hirsutulous, glabrate towards the base; blades 5–30 cm. long, lance-ovate to subcordate, palmately 3-ribbed, sometimes shallowly lobed, above short hispidulous, beneath pale or cinereous appressed puberulent; inflorescence hirsutulous, the heads numerous, in axillary or terminal spikes or panicles; involucre 2–4 mm. long, turbinate, hirsutulous, the 5 bracts obovate, acute, erose; staminate corollas 1–1.5 mm. long, narrowly obconic; achenes 2–3 mm. long, obovoid, compressed, brown to black, cellular reticulate. *Cyclachaena xanthifolia* (Nutt.) Fresen. Weed, introduced from the Great Plains. By railways and roads, *U. Son., A. T.*

The pollen is a cause of hay fever.

LACTUCA. Lettuce

Leafy-stemmed annual or biennial herbs, with milky juice; leaves alternate; flowers all ligulate, yellow or blue or whitish, in paniculate few- to many-flowered heads; involucre cylindric, or in fruit conical; bracts of the involucre in 2-few rows of unequal length, the outer shorter; receptacle flat, naked; pappus of copious very short and fine capillary bristles falling separately; achenes obcompressed, flat or flattish, narrowed at the summit or beaked. (The ancient Lat. name, used by Pliny, from *lac,* milk.)

Pappus brownish; beak less than 1/10 as long as the body of the achene; leaves pinnatifid, not prickly. *L. biennis.*
Pappus white; beak at least 1/2 as long as the body of the achene,
  Leaves not prickly, the upper entire; corollas 15–26 mm. long; the stout beak half as long as the body of the achene. *L. pulchella.*
  Leaves prickly margined; corollas 8–9 mm. long; the filiform beak about 2/3 the length of the body of the achene.
    Leaves pinnatifid. *L. Scariola,* var. *Scariola.*
    Leaves entire. Var. *integrifolia.*

Lactuca biennis (Moench) Fern. Biennial, nearly glabrous, 1–3.5 m. tall; tap-root 5–12 mm. in diameter; leaves 5–30 cm. long, pinnatifid or runcinate, the lower petioled, the upper sessile and clasping, smaller, glabrous or the veins pilose beneath, the lobes oblong to deltoid, sinuate dentate, panicle 1–6 dm. long, narrow, many flowered; peduncles 2–20 mm. long, with several lanceolate bracts; involucral bracts linear-lanceolate, regularly imbricate and not clearly separable into series, 7–12 mm. long, often purplish; heads 17–30-flowered; rays 9–10 mm. long, white or blue; pappus 5–7 mm. long, minutely barbellate; achenes 4.5–5.5 mm. long, elliptic, flattened, brown with dark mottlings, 5-ribbed on each side, punctate, narrowed to a stout beak, less than 0.5 mm. long. Moist woods, *A. T. T.*

Lactuca pulchella (Pursh) DC. *Blue-flowered Lettuce.* Perennial, glabrous, 3–10 dm. tall; roots 2–4 mm. in diameter, white, long spreading, colony-forming by upright shoots; stems usually simple to the inflorescence; lower leaves 9–15 cm. long, often short petioled, entire or pinnatifid, linear-lanceolate to lanceolate, acute, pale, more or less glaucous beneath; upper leaves 4–10 cm. long, entire, linear-lanceolate; panicle 5–30 cm. long, loose, the bracts lanceolate; peduncles 3–35 mm. long, with several lanceolate bracts; head 18–25-flowered; involucral bracts often rose-colored, regularly imbricate, not clearly separable into series but the outer ovate-lanceolate, the inner 15–17 mm. long, lanceolate, not accrescent; pappus 8–9 mm. long, barbellate; achene body 4–5 mm. long, brownish, with minute transverse corrugations, fusiform, flattened, 3-ribbed on the faces, the stout beak 2–2.5 mm. long, terete. *L. tatarica* (L.) C. A. Mey., subsp. *pulchella* (Pursh) Stebbins. Low ground, *U. Son., A. T.*

A native plant, but persisting as a weed in cultivated land, especially in the somewhat alkaline flats. Avoided by grazing animals, probably bitter, and perhaps somewhat poisonous.

Lactuca Scariola L., var. Scariola. *Prickly Lettuce.* Winter annual or biennial, 3–15 dm. tall; tap-root 3–8 mm. in diameter, pale; stem erect, glabrous, more or less glaucous, usually simple to the inflorescence; leaves 5–20 cm. long, somewhat glaucous, prickly on the midrib, pinnatifid, with few divergent lobes, the margin spinose doubly dentate, the blades commonly turning one edge up, and said to point north or south, the base sagittate and clasping; panicle 2–4 dm. long, pyramidal, glabrous; peduncles 2–10 mm. long, lance-cordate bracted; heads 6–12-flowered; involucral bracts lanceolate, regularly imbricate, the inner in flower, 7–8 mm. long, in fruit 12–15 mm. long; corollas yellow, drying bluish; pappus 4–4.5 mm. long, minutely barbellate; achene body 3–3.8 mm. long, broadly fusiform, flattened, greenish brown, 5–7-ribbed on the sides, somewhat scabrous or hispidulous above; the beak 4–5 mm. long, filiform. Weed, introduced from Europe. Waste grounds, Spokane; Palouse Falls, *A. T.*

Var. integrifolia Bogenhard, Fl. Jena 269, 1850. Differs from the species in having the leaves all entire, and sometimes the midrib unarmed. *L. Scariola,* var. *integrata* Gren. & Godr. Weed, introduced from Europe. Excessively abundant, waste lands and cultivated lands, *U. Son., A. T.*

LAGOPHYLLA

Annuals; stem slender, much-branched; leaves alternate or the lower opposite, mostly entire; heads small, subtended by foliaceous bracts, few-flowered; ray flowers about 5, yellow, pistillate, fertile; disk flowers as many, perfect but sterile; involucre of as many scales as and enclosing and dehiscent with the ray achenes, which are obovate, obcompressed, smooth; receptacle small and flat; pappus none; disk flowers about 5, sterile, surrounded by a single row of chaffy bracts; disk achenes slender and abortive; pappus none. (Name from Gr., *lagos,* hare; *phullon,* leaf.)

Lagophylla ramosissima Nutt. Annual, 2–11 dm. tall; tap-root 1–4 mm. in diameter, brown; stems one or more, erect, diffusely branched, at first pilose, later glabrate and pale brownish, shining; leaves densely silky pilose; basal leaves 2.5–6 cm. long, spatulate to oblanceolate tapering into a slender petiole; upper leaves 5–40 mm. long, elliptic to oblanceolate, rarely serrulate, sessile; heads clustered on the branches; involucre ellipsoid, the bracts 4–5 mm. long, lanceolate, foliaceous, silky pilose; rays 4–6 mm. long, white with heavy purple, pilose nerves, flabellate, 3-lobed, the short tube pilose; achenes 3–3.2 mm. long, oblanceolate, 3-angled, brown, smooth and shining; chaff 5–5.5 mm. long, lanceolate, pilose, the acuminate tip green; disk flowers 3 mm. long, yellowish, funnelform, the lobes ovate. *L. ramosissima,* subsp. *typica* Keck. Dry soils, *U. Son., A. T.*

The type specimen was collected by Nuttall, "In the prairies near Walla-Walla."

LAPSANA

Annual herbs, with milky sap; leaves alternate; heads few, slender stalked in a panicle; flowers all ligulate, yellow; involucre subcylindric, its bracts equal, in 1 series except for the minute basal ones; receptacle naked; achenes arcuate fusiform, about 20-nerved; pappus none.

Lapsana communis L. *Nipplewort.* Stem usually simple at base, branched above, 15–150 cm. tall, glandular hirsute; petioles up to 4 cm. long, the lower ones lyrate, the upper leaves sessile; blades 1.5–6 cm. long, mostly dentate, ovate to lanceolate, membranous, hirsute; inflorescence 6–50 cm. long; involucre 6–8 mm. high; principal bracts narrowly oblanceolate, bright green, strongly keeled at base; flowers 8–15, the rays 4–6 mm. long, oblong, 5-toothed; achenes 3.5–5 mm. long, stramineous. An European weed, established in 1946 at Pullman, and Moscow.

MADIA. Tarweed

Annual or perennial herbs; leaves linear or lanceolate, entire or slightly toothed, viscid, at least the upper alternate; heads peduncled, axillary and terminal; receptacle flat or convex, chaffy only at the margin; flowers yellow; ray flowers 1–20 and pistillate, or rarely wanting; disk flowers 1–5, perfect; pappus none or of several small scales in the sterile flowers; ray achenes laterally compressed, enclosed in the deciduous infolded involucral scales; disk achenes laterally compressed. (Name from *Madi,* the Chilean common name.)

Involucre 2.5–4 mm. tall; disk flower single. *M. exigua.*
Involucre 5.5–11 mm. tall; disk flowers several,
  Heads in glomerules; involucre longer than broad, laterally compressed. *M. glomerata.*
  Heads not in glomerules; involucre as broad or broader than long, not laterally compressed,
    Chaff united into a cup around the disk flowers; ray achenes with the back broad and rounded. *M. citriodora.*
    Chaff separate; ray achenes flattened. *M. gracilis.*

Madia citriodora (Gray) Greene. Annual, 1–5 dm. tall, pilose below, pilose and glandular puberulent above, lemon-scented; tap-root 1–4 mm. in diameter, brown; stem simple below, finally diffusely branched above; leaves 15–80 mm. long, sessile, 1-nerved, linear to linear-oblong, acute, the lower opposite, the middle and upper alternate; heads terminal on the divergent, leafy bracted branches; involucre 5.5–7.5 mm. tall, subglobose, taller than wide, dark glandular puberulent and densely pilose, the bracts oblanceolate, long acuminate; the 8–9 ray flowers 6–8 mm. long, the tube pilose, the ray broadly oblong, deeply lobed; pappus none; achenes 3.2–3.5 mm. long, semiobovoid, brown, smooth, 2 inner faces flat; disk corollas 2.5 mm. long, yellow, tubular, the tube pilosulous; pappus none; achenes 4 mm. long, semiobovoid, blackish, cellular reticulate. Open, lower slopes of Blue Mts.; and Little Almota Cr., *U. Son., A. T.*

Madia exigua (Sm.) Gray. Annual, 4–50 cm. tall, hirsute and glandular puberulent throughout; tap-root 0.5–5 mm. in diameter, brown; stem becoming bushy branched; leaves 10–55 mm. long, sessile, linear, acute, 1-nerved; peduncles 10–55 mm. long, filiform, naked; involucre turbinate in outline, hirsute and glandular puberulent, green, the 4–8 bracts lanate, strongly connate, acuminate; ray flowers 1.2–1.5 mm. long, yellowish or purplish, the ray flabellate; pappus none; achenes 2.5–3 mm. long, black, granular, obovate-lunate; disk flower 1.2–1.4 mm. long, funnelform, yellowish, the lobes ovate; pappus none; achenes 1.5–2 mm. long, oblanceolate, compressed, black, cellular reticulate. *Harpaecarpus exiguus* (Sm.) Gray. Open places, *A. T.*

Madia glomerata Hook. Annual, 3–10 dm. long, hirsute below, hirsute and glandular puberulent above; tap-root 0.5–5 mm. long, brown; stem usually strict and simple to the inflorescence; leaves 15–85 mm. long, sessile, linear, acute, 1-nerved, the lower opposite, the upper alternate; heads glomerate in a panicle, usually narrow and compact, but occasionally diffuse and loose; involucre 6–11 mm. long, elliptic, compressed, the bracts semioblanceolate, acuminate, with large-capitate glandular puberulence; the 2–5 rays with the corollas 3–4 mm. long, yellow, the tube puberulent; pappus none; achenes 4.3–5 mm. long, oblanceolate-lunate, brown to blackish, angled and cellular reticulate; disk flowers 2.5–3 mm. long, tubular, yellow, puberulent, the lobes ovate; pappus none; achenes 4–4.3 mm. long, linear-oblanceolate, angled, cellular reticulate, brown to blackish. *M. glomerata,* var. *ramosa* (Piper) Jepson. Abundant, dry places, *U. Son., A. T., A. T. T.*

The seeds were used for food by the Klamath Indians.

Madia gracilis (Sm.) Keck. Annual, 1–9 dm. tall, hirsute below, hirsute and glandular puberulent above; tap-root 1–4 mm. in diameter, brown; stem erect, simple or diffusely branched; leaves 1–11 cm. long, linear or linear-oblong, 1-nerved, the lower opposite, the middle and upper alternate; inflorescence paniculate or loosely racemose; involucre 7–9 mm. tall, ovoid-subglobose, hirsute and capitate glandular puberulent, the bracts obovate-lunate, long acuminate; the 5–8 ray flowers with the rays 3–4 mm. long, yellow, the tube pilose, the ray broadly oblong; pappus none; achenes 3.5–4 mm. long, lunate-oblanceolate, blackish, cellular reticulate; disk flowers yellow, the corolla 2–2.5 mm. long, pilose, the lobes ovate; pappus none; achenes 3.3–3.7 mm. long, obliquely oblanceolate, flattened, blackish, cellular, reticulate. *M. racemosa* (Nutt.) T. & G.; *M. sativa,* subsp. *dissitiflora* (Nutt.) Keck; *M. sativa* Molina, var. *dissitiflora* (Nutt.) Gray. Common, open places, *U. Son., A. T.*

MATRICARIA

Annual or perennial herbs; leaves alternate, pinnately dissected; heads peduncled, discoid and radiate, or the rays wanting; involucre hemispheric, the bracts imbricate in a few series, appressed; receptacle conic to hemispheric, naked; rays white, pistillate, fertile; disk flowers yellow, perfect, fertile, 4–5-toothed; pappus none, or a mere crown; achenes 3–10-ribbed or angled. (Name from Lat., *mater,* mother.)

Matricaria Matricarioides (Less.) Porter. *Pineapple-weed.* Annual, 5–40 cm. tall, glabrous, strongly aromatic with an odor like pineapple; tap-root 1–7 mm. in diameter, brown; stem erect, mostly bushy branched, striate; leaves 2–7 cm. long, 2–3 pinnatisect, the segments 0.5–1 mm. broad, linear, acute; heads numerous; involucral bracts oval, 1-nerved, with broad scarious margins, the inner 2.5–4 mm. long; disk 5–9 mm. high, conical; rays none; disk corollas 1.3–1.5 mm. long, greenish yellow, the short lobes ovate; pappus crown nearly obsolete; achenes 1–1.5 mm. long, 5-ribbed, oblanceoloid, oblique, brownish. *M. suaveolens* (Pursh) Buch.; *Chamomilla suaveolens* (Pursh) Rydb.; *M. discoidea* DC. A native, but weedy plant, roadsides and fields, *U. Son., A. T.*

MICROSERIS

Perennials with linear-attenuate wavy radical leaves with white-tomentulose margins; heads solitary on scape-like peduncles; involucre campanulate; bracts in 2–3 rows, narrowly lanceolate, nearly equal; flowers ligulate, yellow; receptacle flat, naked; achenes fusiform, contracted or beaked at the summit, 10-ribbed; pappus parts 5-many, with scale-like base, bristle tip, or bristles, naked, or with plumose tips. (Name from Gr., *mikros,* small; *series,* endive.)

Heads erect; pappus scales less than 15, or if more than 15, awnless,
  Pappus scales 20–24, awnless. — *M. troximoides.*
  Pappus scales 5–10, with an awn from the notched apex. — *M. linearifolia.*
Heads nodding when young; pappus scales 15–20, each with a very plumose awn. — *M. nutans.*

Microseris linearifolia (DC.) Schultz-Bip. Plant 1–6 dm. tall; tap-root slender, roots fibrous; leaves 5–20 cm. long, linear and entire, or laciniate pinnate with few linear lobes, more or less pilose; peduncles glabrous or merely capitate glandular at apex; involucre glabrous; outer bracts few, lanceolate; inner bracts 12–15 mm. long, becoming in fruit 20–25 mm. long, linear-lanceolate, thin, scarious margined; corollas 5–7 mm. long; pappus scales 6–10 mm. long, linear-lanceolate, white, the margins erose; the awn 3–5 mm. long; achenes 10–14 mm. long, brown to blackish, serrulate. *Uropappus linearifolius* (DC.) Nutt. Dry, lower slopes, Snake River Canyon, *U. Son.*

Microseris nutans (Geyer) Schultz-Bip. Plant 5–60 cm. tall, glabrous or nearly so; tubers 1 or more, 3–7 mm. in diameter, pale; basal leaves several, usually persistent, exceedingly variable, 4–20 cm. long, from linear to ovate, sessile to long petioled, entire to laciniate pinnatifid; cauline leaves few, similar to the basal and equally variable, reduced upwards; peduncles 3–30 cm. long, the tip nodding in bud; heads 8–20-flowered; involucre usually granular and more or less puberulent at apex, the outer bracts linear-lanceolate, much the shorter; the 7–11 inner bracts 10–22 mm. long, linear-lanceolate, imbricate, thin; corollas 10–16 mm. long; pappus 5–8 mm. long, the scale-like base 0.8–2 mm. long, lanceolate, truncate or abruptly contracted into the plumose bristle; achenes 4–8 mm. long, cylindric, puberulent at apex. *Ptilocalais nutans* (Geyer) Greene; *P. major* (Gray) Greene; *Scorzonella nutans* Geyer, var. *laciniata* (Gray) Jepson; *Microseris nutans* (Geyer) Schultz-Bip., var. *major* (Gray) Nels. & Macbr. Open slopes or open woods, *A. T., A. T. T.*

Geyer reported the root to be "succulent and almost transparent, full of a bitterish, milky juice, eaten raw by the Indians."

Microseris troximoides Gray. Plant 1–3 dm. tall; root 4–15 mm. in diameter, black, woody; leaves 1–2 dm. long, linear, the margins often undulate; scapes glabrous or somewhat pilosulous, erect; heads 20–30-flowered; involucre 15–24 mm. long, glabrous or pilosulous at base, the bracts imbricate, scarious margined with a dark purplish midrib; receptacle alveolate, glabrous; corollas 18–25 mm. long; pappus of 10–30 scales 9–15 mm. long, linear lanceolate, attenuate into a bristle-like tip; achenes 9–11 mm. long, pale brownish, puberulent towards the tip, the inner the narrower, the base with a bulbous callosity. *Nothocalais troximoides* (Gray) Greene; *Scorzonella troximoides* (Gray) Jepson. Open slopes, *U. Son., A. T.*

PETASITES. Sweet Coltsfoot

Perennial herbs with creeping rootstocks; leaves large, radical, appearing later than the flowers, the cauline reduced to bracts; heads numerous in a raceme or corymb on the end of a scape-like stem, subdioecious; flowers white or purplish, those of fertile plants all or mostly pistillate, fertile; corolla irregularly 2–5-toothed and cylindric, or ligulate; flowers of substerile plants mostly perfect but sterile, with a few marginal pistillate ones; corolla of sterile flowers tubular, 5-toothed; involucral bracts in one row; achenes narrow, 5–10-ribbed, pappus of soft white bristles. (The Gr. name, used by Dioscorides, derived from *petasos,* a broad-brimmed hat.)

Petasites sagittatus (Banks) Gray. Perennial, lanate throughout; rhizome 3–10 mm. in diameter, horizontal; basal leaves few, from distinct sterile shoots; petioles 1–4 dm. long; blades 10–25 cm. long, deltoid sagittate, coarsely dentate, above sparsely lanate and more or less glabrate, beneath densely white lanate; scape 2–9 dm. tall, with numerous lanceolate, sheathing bracts 3–7 cm. long; raceme 3–15 cm. long; peduncles 15–50 mm. long; hermaphrodite heads with involucre 8–10 mm. high, the bracts oblong-lanceolate, acute, ray flowers in 1 series, sterile; disk flowers numerous, sterile, the tube 5 mm. long, the throat 3 mm. long, funnelform; pistillate heads with involucre 7–8 mm. long, the flowers nearly all pistillate and fertile, the outer with a ligule 3–4 mm. long, the inner with ligules smaller or none; pappus in anthesis 6–8 mm. long, in fruit 15–25 mm. long; achenes 2–4 mm. long, fusiform, brown, glabrous. Swamps, Marshall Jct. *Piper,* July 2, 1896, *A. T. T.*

PSILOCARPHUS

Low woolly annuals; leaves entire, mostly opposite; heads small, discoid, many-flowered, in terminal capitate clusters and in the forks of the branches, surrounded by the upper leaves, but with no other involucre; fertile flowers numerous, in several series on the globular chaffy receptacle; chaff of pistillate flowers semi-ovate, cucullate, with a hyaline appendage; perfect flowers without chaff; marginal flowers pistillate, fertile, the corolla filiform; perfect and sterile flowers few, the corolla tubular, 5-toothed; pappus none; achene loose in the bladder-like bract, oblong or narrower, slightly compressed. (Name from Gr., *psilos,* slender; *karphos,* chaff.)

Well developed chaff 2.5–4 mm. long; leaves linear to linear-oblong; achenes cylindric-fusiform. *P. elatior.*
Well developed chaff 1.2–2.7 mm. long,
 Leaves linear to linear-oblanceolate, mostly 6–12 times as long as wide; achenes fusiform. *P. oregonus.*
 Leaves spatulate, oblanceolate, or oblong, mostly 1.5–6 times as long as wide; achenes broadly oblanceolate to narrowly obovate. *P. tenellus,* var. *tenellus.*

Psilocarphus elatior Gray. Plant 3–15 cm. tall, loosely but abundantly white lanate throughout; roots fibrous; stems simple and erect or branched and more or less decumbent; leaves 5–18 mm. long, 0.6–3.5 mm. wide, sessile, linear to linear-oblong, acute; heads 5–8 mm. broad, globose, white throughout with a loose tomentum; sterile flowers with corolla tubular, the short ovate lobes purplish; chaff 2–2.5 mm. long to the bend; achenes 1.2–1.4 mm. long, cylindric-fusiform brown, shining. Moist or exsiccated spots, *A. T., A. T. T.*

Psilocarphus oregonus Nutt. Plant 2–8 cm. tall or long, white, abundantly but loosely lanate throughout; roots fibrous; stems simple and erect, or commonly diffusely branched and more or less decumbent; leaves 10–17 mm. long, 0.8–2 mm. wide, sessile, linear to linear-spatulate, acute; heads 3–5 mm. wide, globose, white with a close felt-like tomentum; sterile flowers with corolla tubular-funnelform, the short ovate lobes purplish; chaff 1.8–2 mm. long to the bend; achenes 0.6–0.9 mm. long, fusiform, brown, shining. Exsiccated mud, Spokane Co., *Suksdorf* 931; 932, *A. T.*

Psilocarphus tenellus Nutt., var. **tenellus.** Slender annual, erect, later much branched and prostrate, tomentum thin, loose and partly deciduous; leaves 4–15 mm. long, 1–6 mm. wide; heads with 25–46 pistillate flowers; achenes 0.6–1.2 mm. long, moderately compressed, turgid. West slope of Moscow Mts., *Gaines* 64.

RIGIOPAPPUS

Annuals; leaves alternate, linear, entire; heads rather many-flowered; ray flowers 5–12, pistillate; all the flowers fertile; disk flowers perfect, 3–4-toothed; involucre of one or two rows of rather rigid herbaceous erect subulate-linear bracts, folded and partly enclosing the ray flowers; receptacle flat, naked; pappus of 3–5 rigid opaque subulate awn-shaped scales; achenes linear, slender, compressed, those of the disk more or less 4-angled. (Name from Gr., *rigios,* stiffened; *pappos,* pappus.)

Rigiopappus leptocladus Gray. Herb, 4–40 cm. tall, puberulent throughout; tap-root 0.5–2 mm. in diameter, pale, the laterals fibrous; stem erect, becoming pale or reddish and more or less glabrate, the lateral branches filiform, ascending, naked below; leaves 1–3 cm. long; involucre narrowly turbinate, puberulent, of the main terminal head 6–8 mm. high, of the lateral heads 5–6 mm. high; rays 3–4 mm. long, yellowish or purplish, not exserted; disk flowers 3–3.5 mm. long, tubular, glabrous, yellowish or purplish tipped; pappus 4.5–5 mm. long, straw-colored; achenes 5–6 mm. long, greenish to purplish or black, transversely ridged, hirsutulous. Dry, open soils, *U. Son., A. T.*

RUDBECKIA

Biennial or perennial herbs; leaves alternate; heads many-flowered, mostly with sterile ray flowers but rayless in ours; disk flowers 5-lobed, perfect; receptacle elongated, becoming columnar; chaff of concave scales, subtending or enveloping the disk flowers; pappus a chaff-like cup or 4 chaffy teeth more or less united into a cup; achenes quadrangular and mostly laterally compressed. (Named by Linnaeus for the two professors at Uppsala, Sweden, *Olaf Rudbeck,* father and son.)

Rudbeckia occidentalis Nutt. *Nigger-head.* Perennial, 5–20 dm. tall, smooth and nearly glabrous; root 1–2.5 cm. in diameter, woody, brown; stem striate, erect, simple or few branched; basal leaves with petioles 15–35 cm. long, the blades 15–20 cm. long, lanceolate to ovate, more or less dentate, beneath sparsely hirsute; cauline leaves with shorter petioles and the upper sessile; the blades 7–20 cm. long, ovate to lanceolate, acuminate, dentate or entire, above smooth and glabrous, beneath more or less hispid; peduncles 1–3 dm. long; involucral bracts 1–3 cm. long, spreading, foliaceous, linear-lanceolate; somewhat hirsutulous; disk 1–5 cm. high, ovoid-conical, madder-brown; chaff 5–6 mm. long, puberulent on the back, the margins with 1 dark nerve, the tip deltoid; corollas 4–4.3 mm. long, the short lobes ovate; pappus crown 0.5–1 mm. long; achenes 3–4.5 mm. long, black, smooth, narrowly oblong. Thickets or open woods, *A. T. T.*

SENECIO. Ragwort; Groundsel

Herbs or shrubby plants; leaves basal or alternate; heads usually solitary or in corymbs, many-flowered; involucre cylindric or campanulate, the bracts herbaceous, mostly narrow, equal, in one row, or with a few short outer bracts; receptacle flat or merely convex, naked; flowers yellow, all fertile; ray flowers pistillate or occasionally none; disk flowers perfect, tubular, 5-toothed; pappus of very numerous and mostly white, fine and soft capillary and merely scabrous bristles; achenes terete or somewhat angled, usually 5–10-ribbed. (Name used by Pliny, from Lat., *senex,* old man.)

Leaves all pinnatifid; rays usually wanting. *S. vulgaris.*
Leaves all or some of them entire; rays usually present,
  Stems very leafy to the inflorescence, the upper leaves but slightly the smaller,
    Blades deltoid; pappus 4.5–5 mm. long. *S. triangularis,* var. *triangularis.*
    Blades lanceolate; pappus 6.5–7 mm. long,
      Lower blades sharply serrate. *S. serra,* var. *serra.*
      Lower blades entire or nearly so. Var. *integriusculus.*
  Stems leafy at base, the upper leaves much reduced, bracteose,
    Rays white. *S. integerrimus,* var. *ochroleucus.*

Rays yellow,
Plant glabrous and somewhat glaucous. *S. hydrophilus.*
Plant more or less pubescent,
Plant closely white tomentose; upper leaves usually pinnatifid. *S. canus.*
Plant loosely floccose lanate; upper leaves entire or dentate. *S. integerrimus,* var. *exaltatus.*

Senecio canus Hook. Perennial herb, 1–3 dm. tall, loosely caespitose, in age sometimes partly glabrate; rootstock 2–5 mm. in diameter, brown; stems erect, simple or branched; basal and lower leaves with petioles 3–8 cm. long, the blades 15–45 mm. long, elliptic-lanceolate to oblanceolate, usually entire; upper leaves reduced, sessile; heads 5–20 in a cyme; involucre 6–10 mm. high, campanulate, sparsely tomentose; outer bracts few, greenish; inner bracts about 13, lance-linear, the midrib glandular thickened; the 8–12 ray flowers 7–12 mm. long; the 30–40 disk flowers 5–6 mm. long; pappus 5–6 mm. long; achenes 2–2.3 mm. long, oblanceoloid, brownish, ribbed. *S. Howellii* Greene. Gravelly soils, especially near Spokane, *A. T.*

Senecio hydrophilus Nutt. Perennial herb, 4–20 dm. tall; root 5–15 mm. in diameter, short, with many fleshy laterals; stem erect, striate, hollow, simple; leaves fleshy, entire or denticulate; basal and lower leaves with petioles 3–20 cm. long; blades 3–20 cm. long, oblanceolate; middle and upper leaves sessile sharply reduced, linear-lanceolate; heads numerous in cymes 3–10 cm. wide; involucre 6–8 mm. high, cylindric-campanulate, the outer bracts 1–3 mm. long, subulate; inner bracts greenish or yellowish, oblong-lanceolate, the tip blackened and pilosulous, the margins hyaline; rays 0 or 3–5, and 5–6 mm. long; disk flowers 5.5–6.5 mm. long, glabrous, the tube shorter than the slender throat; pappus 5–6 mm. long; achenes 4–5 mm. long, linear, tapering to the tip, brown, glabrous. *S. hydrophiloides* Rydb. Moist places, especially by Spokane River, *A. T.*

Senecio integerrimus Nutt., var. **exaltatus** (Nutt.) Cronq. Perennial herb, 2–6 dm. tall, somewhat floccose lanate, at least when young; root 4–7 mm. in diameter, short, with numerous fleshy laterals; stem erect, striate, simple; basal and lower leaves with petioles 3–9 cm. long, the blades 3–10 cm. long, oblanceolate, entire or denticulate; middle leaves 5–9 cm. long, sessile, lanceolate, somewhat auriculate; upper leaves much reduced, bracteose, linear-lanceolate; heads numerous in a loose or condensed cyme, 3–10 cm. wide; involucre 6–11 mm. high, campanulate, the few outer bracts 2–4 mm. long, subulate; inner bracts linear, more or less lanate, the acute tip black and pilosulous; the 7–9 rays 5–8 mm. long; disk flowers 7–8 mm. long, glabrous, the tube equaling the slender throat; pappus 6–7 mm. long; achenes 4.5–5 mm. long, linear, glabrous, yellowish. *S. atriapiculatus* Rydb.; *S. condensatus* Greene; *S. columbianus* Greene. Common, rocky slopes, grasslands, open woods, *U. Son., A. T., A. T. T.*

S. condensatus from the Blue Mts. has a condensed cyme, but it occurs with the loose form (shown by *St. John & Palmer* 9608), so it does not merit recognition.

Var. **ochroleucus** (Gray) Cronq. Perennial, 2–6.5 dm. tall, white arachnoid throughout; caudex 3–5 mm. in diameter, the lateral roots numerous, somewhat tuberous; stem simple, erect, striate; basal and lower leaves with petioles 3–8 cm. long, the blades 3–8 cm. long, lanceolate to ovate, denticulate; upper leaves reduced to linear-lanceolate, sessile bracts; heads 15–35 in a loose, simple or frequently compound umbellate cyme; peduncles 2–8 cm. long; involucre 6–7 mm. high, campanulate; the few outer bracts 2–4 mm. long, subulate, black-tipped; inner bracts oblong-linear, acuminate, sparsely pilosulous or puberulent, and glandular, so at least on the black tips; the 6–8 rays 8–13 mm. long, narrowly elliptic, white; disk flowers 7–8 mm. long, almost tubular, yellow, glabrous; pappus 4.5–6 mm. long; achenes 3–3.5 mm. long, linear, brown, sparsely puberulent at apex. *S. Leibergii* Greene; *S. exaltatus ochraceus* (Gray) Piper. Gravelly clearing, Eloika L., Spokane Co., *St. John et al.* 7,046; Mt. Spokane, June 22, 1935, *Clarke;* T8N., R41E., Blue Mts., *A. T.*, *A. T. T.*

Senecio serra Hook. var. **serra.** Perennial herb, 6–15 dm. tall, glabrous or nearly so; rootstock 10–15 mm. thick, brown, woody; stem erect, striate, simple below; lower cauline leaves with petioles 1–3 cm. long, the upper sessile; blades 3–15 cm. long, lanceolate to linear-lanceolate, sharply serrate, the uppermost sometimes entire; heads many, in rounded cymes 5–20 cm. wide; involucre 5–7 mm. high, subcylindric, the few outer bracts subulate, sometimes nearly as long as the inner; inner bracts yellowish, linear-oblong, thin, the acute tip pilosulous and dark, the midrib glandular thickened; the 5–10 rays 6–10 mm. long; disk flowers 7–8 mm. long, glabrous, the tube as long as the narrow throat; achenes 3–3.5 mm. long, linear, angled, yellowish. Low places, common, *U. Son., A. T.;* also Blue Mts.

Var. **integriusculus** Gray. Leaves all entire or nearly so. *S. serra lanceolatus* (T. & G.) Piper; Waitsburg, *Horner* 572, *A. T.*

Senecio triangularis Hook., var. **triangularis.** Perennial herb, 5–15 dm. tall, glabrous or nearly so; rootstocks 5–15 mm. in diameter, clustered; stems in a clump, erect, striate; lower leaves with petioles 1–6 cm. long; upper leaves with petioles shorter or sessile; blades 3–13 cm. long, thin chartaceous, sharply dentate; heads 5–40 in a loose cyme; involucre 6–9 mm. high, campanulate, glabrous; outer bracts few, linear; inner bracts thin, linear-oblong, the narrowed tip pilosulous, the midrib glandular thickened; the 6–12 rays 8–10 mm. long; disk flowers 6–7 mm. long, glabrous, the throat longer than the tube; achenes 3.5–4 mm. long, linear, prismatic, brown, glabrous. Meadows, *Can., Huds.*

Senecio vulgaris L. *Common Groundsel.* Annual or biennial herb, 8–50 cm. tall, with a few woolly hairs or glabrate; tap-root 1–4 mm. in diameter, pale, the laterals fibrous; stem erect, hollow, simple or branched; leaves 5–15 cm. long, oblanceolate in outline, the segments oblong and irregularly dentate; heads in corymbs; involucre 6–8 mm. long, cylindric-campanulate, the outer calyculate bracts 1–2 mm. long, black tipped, the inner greenish, often black tipped; disk flowers 5–6 mm. long, glabrous, the throat slightly enlarged; pappus 5–6 mm. long; achenes 2.3–3 mm. long, fusiform, brownish, appressed puberulous. Weed, introduced from Eurasia. Roadsides, *A. T.*

SILYBUM

Annual or biennial herb; leaves alternate, clasping, pinnatifid; heads discoid, terminal; involucre, subglobose, the bracts rigid, imbricate in many series, the lower ones fimbriate-spinulose at the broad triangular summit, but armed with long, spreading or recurved spines, the inner bracts lanceolate, acuminate; receptacle flat, bristly; corolla purple, the tube slender, the limb narrowly campanulate, deeply 5-cleft; anthers sagittate at base; style nearly entire; achenes obovate-oblong, with a hairy crown at apex; pappus in several series, flattish, barbellate or scabrous. (Ancient Gr. name for a land of thistle.)

Silybum marianum (L.) Gaertn. *Milk Thistle; Lady's Thistle.* Plant 6–15 dm. tall, glabrous or sparsely lanate; root fusiform; leaves up to 3 dm. in length, shining green, with white blotches along the nerves, spiny on the margins; heads 4–5 cm. high; involucre glabrous; achenes 6–7 mm. long, brown spotted; pappus white. *Mariana Mariana* (L.) Hill. Weed introduced from the Mediterranean region. Established, waste ground, Pullman, as early as 1926, *A. T.*

SOLIDAGO. Goldenrod

Perennial herbs; leaves alternate; heads small, mostly in panicles or panicled racemose cluters, radiate, the ray flowers fertile, yellow; disk flowers perfect, fertile, yellow; involucre imbricated, the bracts usually without herbaceous tips; style-appendages lanceolate or triangular-subulate; pappus simple, of a single series of mostly equal and slender scabrous capillary bristles; achenes terete or angular, 5–10-ribbed. (Name from Lat., *solidare,* to make whole.)

Ray flowers more numerous than the disk flowers; heads in a corymb, often flat-topped. *S. occidentalis.*
Ray flowers not more numerous than the disk flowers; heads in a panicle,
- Basal leaves petioled, much larger than the cauline. *S. missouriensis.*
- Basal leaves similar to, and but little larger than the cauline,
  - Inflorescence narrowly pyramidal, the branches ascending, little if at all secund. *S. lepida,* var. *elongata.*
  - Inflorescence broadly pyramidal, the branches spreading or recurved, secund,
    - Leaves scabrous. *S. gigantea.*
    - Leaves not scabrous. Var. *leiophylla.*

Solidago gigantea Ait. Differs in having the leaves scabrous, and the veins puberulent beneath. *S. salebrosa* (Piper) Rydb. Stream banks or grasslands, *U. Son., A. T.*

Var. **leiophylla** Fern. Plant 5–25 dm. tall, nearly glabrous; rhizome 3–6 mm. in diameter, brown, producing lateral rhizomes; stem erect, glabrous and simple below, above puberulent; leaves 5–15 cm. long, sessile, lanceolate to oblanceolate, serrate, 3-nerved at least above the middle, the margin scabrous ciliolate, the upper ones scarcely reduced; panicle 5–30 cm. long, rhomboidal, involucre 3.5–5 mm. high, campanulate; bracts imbricate in 3–4 series, lanceolate to linear-oblong, obtuse, thin, ciliolate; the 7–15 ray flowers 4–5 mm. long; the 8–12 disk flowers 3 mm. long; pappus 2.5 mm. long, white; achenes 1.2–1.4 mm. long, linear, brown, sparsely appressed pilosulous. *S. serotina* Ait. Thickets by streams, *U. Son., A. T.*

Solidago lepida DC., var. **elongata** (Nutt.) Fern. Plant 3–10 dm. tall; rhizome 3–5 mm. in diameter, brown; stem erect, simple and glabrous below, puberulent above; basal leaves like the cauline, withered at anthesis; cauline leaves 2–10 cm. long, gradually reduced upwards, sessile, 3-nerved from below the middle, linear-oblanceolate, acute, serrate, the margins scabrous puberulent, the veins puberulent or glabrous; panicle 1–2 dm. long, involucre 3–5 mm. high, campanulate; bracts imbricate in 3–4 series, linear, acute, ciliolate; the 10–16 ray flowers 1.5–3 mm. long; the 12–15 disk flowers 3–3.4 mm. long;

pappus 2.8–3 mm. long, white; achenes 1 mm. long, linear, brown, appressed pilosulous. *S. elongata* Nutt.; *S. canadensis* L., var. *salebrosa* sensu Cronq. in part. Wet places, *A. T.*

Solidago missouriensis Nutt. Plant 2–10 dm. tall, glabrous; rhizome 2–7 mm. in diameter, brownish, often forking; stems simple below; basal and lower leaves with petioles 2–6 cm. long; blades 3–13 cm. long, oblanceolate or broadly so, 3–5-ribbed, thick chartaceous, serrate, scabrous ciliolate; middle leaves 4–12 cm. long, obscurely 3-nerved, oblanceolate more or less serrate, scabrous ciliolate; upper leaves much reduced, almost linear; panicle 3–20 cm. long, ovoid, the branches racemose, dense, divergent; involucre 3.5–5 mm. high, campanulate; bracts imbricate in 3–4 series, ovate to ovate-lanceolate, the obtuse tip ciliolate; the 6–13 ray flowers 4–5 mm. long; the 12 or so disk flowers 3 mm. long; pappus 3 mm. long, white; achenes 1.5 mm. long, linear, appressed pilosulous. *S. glaberrima* Martens. Grasslands or open woods, *A. T., A. T. T.*

Solidago occidentalis (Nutt.) T. & G. Plant 3–20 dm. tall, nearly glabrous; rhizome 2–8 mm. in diameter, pale, freely producing lateral rhizomes; stem erect, simple below; principal leaves 4–13 cm. long, 3-nerved, linear, acute, the margins scabrous ciliolate; leaves of axillary branches numerous, much smaller; corymb flat or round-topped, 4–20 cm. wide; involucre 4–5 mm. high, campanulate; bracts imbricate in 3–4 series, lance-linear, acute, ciliolate, the tip more or less glutinous; the 16–20 ray flowers 5–6 mm. long; the 8–14 disk flowers 4–4.5 mm. long; pappus 4 mm. long, whitish; achenes 1.5 mm. long, linear, appressed pilosulous. *Euthamia occidentalis* Nutt. Stream banks and meadows, *U. Son., A. T.*

SONCHUS. Sow Thistle

Leafy-stemmed annuals or perennials, mostly glabrous, generally coarse herbs, with milky juice; leaves alternate, prickly margined; flowers all ligulate, yellow, in corymbose or paniculate heads; involucre ovoid or campanulate, enlarged at base in age; bracts of the involucre imbricated, the outer shorter; receptacle flat, naked; pappus of copious very fine and short capillary bristles, mostly falling in one group; achenes flat or flattish, truncate, not beaked. (The ancient Gr. name, *sogchos,* used by Theophrastus.)

Sonchus asper (L.) Hill. *Spiny Sow Thistle.* Annual, 3–17 dm. tall, nearly glabrous; tap-root 3–8 mm. in diameter; stem glabrous, usually unbranched to the inflorescence; lower leaves 1–2 dm. long, oblanceolate, glaucous beneath, glabrous, simple or pinnatifid, the margin spinose dentate; upper leaves 5–15 cm. long, lanceolate to ovate-lanceolate, the clasping base with rounded auricles; cyme 5–20 cm. long, loose, capitate glandular hirsute; peduncles 1–5 cm. long, usually bractless; involucral bracts glabrous, lance-linear, the inner 9–12 mm. long; heads with about 120 flowers; corollas 11–17 mm. long; pappus 7–8 mm. long, white, minutely barbellate; achenes 2.4–3 mm. long, oblanceolate, flattened, brown, ribbed on the margins, 3-ribbed on the sides. Weed, introduced from Eurasia or Africa. Roadsides and cultivated lands, *A. T., A. T. T.*

STEPHANOMERIA. Flowering Straw

Leafy-stemmed and branching herbs, with milky juice; upper leaves often reduced to bracts; heads small; flowers pink or white;

involucre cylindric, rarely campanulate, 3–20-flowered; involucral bracts few, in 1 series, or with a few small outer calyculate ones; receptacle naked; pappus of plumose or partly plumose bristles; corollas all ligulate; achenes short, truncate at both ends, about 5-ribbed or angled. (Name from Gr. *stephane,* wreath; *meros,* part.)

Pappus fuscous; inflorescence paniculate, the numerous heads almost virgate on the branches; achenes transversely rugose. *S. paniculata.*
Pappus white; heads usually remote and terminal on the branches; achenes rugulose between the angles. *S. tenuifolia.*

Stephanomeria paniculata Nutt. Annual, 3–11 dm. tall, glabrous; taproot 2–6 mm. in diameter, elongate; stem erect, usually simple below; lower leaves 2–9 cm. long, oblanceolate to spatulate, runcinate to dentate; upper leaves linear, much reduced and bracteose; inflorescence 3–6 dm. long; peduncles 2–15 mm. long, with several minute bracts; heads 3–8-flowered; involucre cylindric, the outer bracts foliaceous, elliptic, ciliolate, the inner bracts usually 5, lance-linear, 7–9 mm. long, imbricate, hyaline margined; ray corollas 9–11 mm. long, pink; pappus bristles 5–6 mm. long, plumose to the base, the lateral branches 0.3 mm. long, shorter below; achenes 4–5 mm. long, yellowish, pentagonal, narrowly wedge-shaped. *Ptiloria paniculata* (Nutt.) Greene. Warm slopes, Snake River Canyon, and lower tributaries, *U. Son.*

Stephanomeria tenuifolia (Torr.) H. M. Hall. Perennial, 1–7 dm. tall, glabrous or nearly so; root 3–15 mm. in diameter, dark, woody; stems often several, diffusely branched, often from near the base; lower leaves 2–5 cm. long, oblanceolate, more or less pinnatifid; upper leaves 1–8 cm. long, linear, entire, mostly reduced and bracteose; heads 4–5-flowered; involucre cylindric, the outer bracts foliaceous, lanceolate, more or less ciliolate, the inner bracts usually 5, linear-lanceolate, obtuse, 6–8 mm. long, imbricate, the margins hyaline, the back granular puberulous, heavy ribbed, thickened at base; corollas 7–9 mm. long, pink; pappus bristles 5 mm. long, plumose to the base, the median branches 0.3 mm. long, shorter below; achenes 4 mm. long, linear, 5-angled, yellowish. *Ptiloria tenuifolia* (Torr.) Raf. Rocky or dry places, *U. Son., A. T.*

TANACETUM

Aromatic, mostly perennial herbs; leaves alternate, 1–3-pinnatifid; heads corymbose, discoid or with short rays not exceeding the disk; involucre hemispheric or depressed, the bracts appressed, imbricate in several series; receptacle naked, flat or convex; marginal flowers pistillate, fertile, the corollas radiate or 2–5-toothed; disk flowers perfect, fertile, the corollas 5-lobed; achenes 3–5-angled or -ribbed; pappus none or a short crown. (Name from its old French name, *tanasie.*)

Tanacetum vulgare L. *Tansy.* Perennial, 4–16 dm. tall, green and nearly glabrous; root 3–30 mm. in diameter, creeping, cord-like or tuberous thickened, woody, brown; stem striate, angled, glabrate; leaves 1–3 dm. long, oval in outline, bipinnatifid, glandular punctate, more or less puberulent, the primary divisions 2–5 cm. long, linear-lanceolate, acute, decurrent on the rhachis, the secondary divisions lanceolate, laciniate serrate; heads numerous in compound corymbs, 200–400-flowered; involucre 4 mm. high, 6–8 mm. broad; bracts sparsely puberulent or glabrate, the outer lanceolate, the inner elliptic; ray flowers in 1 series; disk corollas 2 mm. long; pappus a 3–5-toothed crown; achenes 1–1.5 mm. long, 3–5-angled. *Chrysanthemum Tanacetum* Karsch. A weed, introduced from Europe. Roadsides, *A. T.*

Very poisonous to animals, producing madness; less poisonous to people. The distinctness of the genus is open to question. Many botanists reduce it to *Chrysanthemum.*

TARAXACUM

Acaulescent biennials or perennials, with milky juice; leaves radical, pinnatifid; heads large, on hollow scapes; flowers yellow; receptacle flat, naked; involucre of 2 rows of bracts, the outer short; pappus of copious and white-capillary bristles which are not plumose; corollas all ligulate; achenes fusiform, angled, about 10-ribbed, attenuate at base, with a long filiform beak at the apex. (Name perhaps from Gr., *tarassein,* to disquiet; or from Arabic *Tharakhchakon,* the name of a related plant.)

Terminal lobe of leaf the largest; involucral bracts not appendaged; achenes gray to olive-brown. *T. officinale.*
Terminal lobe of leaf not the larger; involucral bracts usually corniculate appendaged; achenes red. *T. laevigatum.*

Taraxacum laevigatum (Willd.) DC. *Red-seeded Dandelion.* Plant 5–30 cm. tall; root 3–7 mm. in diameter, brown; leaves 5–20 cm. long, glabrous or sparsely pilosulous, oblanceolate, deeply incised pinnatifid almost to the midrib, the lateral lobes narrow, linear to narrowly deltoid, often with intermediate smaller teeth, the base petiole-like; scape more or less lanate at apex; heads campanulate, 70–90-flowered; involucre campanulate, glabrous, more or less glaucous; outer bracts lanceolate, ascending or spreading; inner bracts 12–20 mm. long, linear-lanceolate, blackish green; corollas 8–13 mm. long; pappus 4–7 mm. long, white; body of achene 3–4 mm. long, linear-oblanceoloid, muricate near the apex, the beak 5–7 mm. long. *T. erythrospermum* Andrz.; *Leontodon erythrospermum* (Andrz.) Eichw. Weed, introduced from Eurasia or Africa. Lawns, Pullman; and Alder Cr., Benewah Co., *A. T.*

Taraxacum officinale Weber in Wiggers. *Dandelion.* Plant 1–10 dm. tall; root 3–15 mm. in diameter, brown, often multicipital; leaves 5–35 cm. long, oblong or spatulate in outline, sinuate pinnatifid, the lateral lobes toothed and with frequent intermediate smaller lobes, the base tapering slender, winged, petiole-like, glabrous or somewhat pilose; scapes somewhat pilose near the apex; heads broad, with 150–200 flowers; involucral bracts usually all unappendaged, the outer conspicuously reflexed, the inner 12–20 mm. long, lance-linear; corollas 12–15 mm. long; pappus 5–6 mm. long, white; body of achene 3–4 mm. long, oblanceoloid, muricate at apex, the beak 8–10 mm. long. *T. Taraxacum* (L.) Karst., in ed. 1; *T. vulgare* (Lam.) Schrank; *Leontodon Taraxacum* L.; in part *T. palustre* (Lyons) Lam., var. *vulgare* (Lam.) Fern., not *Leontodon palustre* Lyons. Weed, introduced from Eurasia. Roadsides, cultivated lands, and lawns, *U. Son., A. T., A. T. T.*

The first leaves are used as spring greens. The root contains a bitter substance taraxacin, and is used in medicine.

TETRADYMIA

Low rigid shrubs, more or less white tomentose; leaves alternate, entire, with smaller fascicled axillary ones, or the primary ones becoming spines; heads discoid; involucre cylindric, the 4–6 bracts concave, overlapping; receptacle flat, naked; disk flowers yellow, 4–9; pappus of numerous capillary, scaberulous bristles; achenes 5-nerved, short. (Name from Gr., *tetradumos,* four together.)

Tetradymia canescens DC., var. **canescens.** Shrub 1–10 dm. tall; branchlets white floccose lanate; old stems roughened by the axillary short branches, the bark brown, thick, longitudinally fissured; primary leaves 13–40 mm. long, linear or somewhat spatulate, acute, densely white lanate; secondary leaves smaller; heads several, in close corymbs; involucre 7–11 mm. high, white tomentose; bracts narrowly oblong-elliptic, keeled, the margins hyaline; the 4 disk flowers 10–12 mm. long, the tube slender, the throat campanulate, the linear lanceolate lobes 3–4 mm. long; pappus 10–11 mm. long, white; achenes 3.5–4 mm. long oblong, densely villous. *T. canescens,* var. *typica* Pays. Arid places, in "channeled scablands" from Spokane southward, *U. Son.*

The Spalding specimen, labeled Clear Water, Oregon, probably came from elsewhere, perhaps Spokane or Walla Walla, as the shrub has not been recollected near Lapwai.

TRAGOPOGON

Stout leafy-stemmed and usually branching biennials or perennials, with milky juice; leaves entire, grass-like, clasping; flowers yellow or purple, matinal, in large solitary heads; involucre simple, of several equal bracts in 1 series, united at base; receptacle naked, alveolate; pappus of numerous long-plumose bristles; corollas all ligulate; achenes narrowly fusiform, muricate, 5–10-ribbed, long-beaked. (The Gr. name used by Theophrastus, from *tragos,* goat; *pogon,* beard.)

Rays violet-purple, 20–24 mm. long. *T. porrifolius.*
Rays yellow, 10–18 mm. long,
  Peduncle in fruit 3–6 mm. in diameter at apex; bracts in fruit 25–40 mm. long; achene body 9–13 mm. long. *T. pratensis.*
  Peduncle in fruit 8–10 mm. in diameter at apex; bracts in fruit 5–6 cm. long; achene body 13–20 mm. long. *T. dubius,* subsp. *major.*

Tragopogon dubius Scop., subsp. major (Jacq.) Schinz & Keller. Biennial, 2–10 dm. tall, glabrous or at first somewhat floccose lanate; tap-root 5–10 mm. in diameter, brown; stem simple or branched; leaves 7–30 cm. long, glaucous green, linear, long acuminate, keeled, the base lanceolate, clasping; peduncles 5–20 cm. long, in fruit inflated and fistulose at apex; heads cylindric to ovate, in anthesis almost rotate; involucral bracts 8–12, in 2 rows, in anthesis 25–35 mm. long, rays 13–15 mm. long; pappus 25–32 mm. long, brownish; achenes 25–35 mm. long, angled, brownish, the body linear, the inner nearly smooth, the outer sharply muricate. Weed introduced from Europe. Roadsides and fields, *U. Son., A. T.*

Tragopogon porrifolius L. *Salsify, Oyster Plant.* Annual or biennial, 6–12 dm. tall, glabrous; tap-root 5–20 mm. in diameter, brown; stem simple or few branched; leaves 5–35 cm. long, glaucous green, lance-linear, gradually acuminate, keeled, clasping at base; peduncles 10–25 cm. long, becoming conspicuously enlarged and fistulose at apex; heads subcylindric to campanulate, in anthesis almost rotate; involucral bracts usually 8, in 2 rows, 3–4 cm. long in anthesis, 4–6 cm. long in fruit, linear-lanceolate, glabrous; pappus 17–22 mm. long, brownish; achenes 2.5–3 cm. long, straw-colored to brown, the body 12–15 mm. long, linear-lanceoloid, tapering into the beak, the inner ones almost smooth, the outer coarsely muriculate and curved. Weed, native of the Mediterranean region. Abundant, roadsides and cultivated lands, *A. T.*

This has probably escaped from cultivation. The roots furnish the well-known vegetable.

Tragopogon pratensis L. *Meadow Salsify.* Biennial or perennial, 3–7 dm. tall, glabrous; tap-root 3–15 mm. in diameter, brown; stem erect, simple or few branched; leaves 5–25 cm. long, linear, glaucous green, keeled, the base lanceolate and clasping, the upper reduced; peduncles 5–15 cm. long, only slightly enlarged at summit; heads narrowly ovate, in anthesis almost rotate; involucral bracts usually 8, in 2 series, in anthesis 15–30 mm. long, somewhat lanate at base, later glabrate, lanceolate, long acuminate; rays 10–18 mm. long; pappus 16–20 mm. long, brownish; achenes 16–25 mm. long, straw-colored to brownish, narrowly lance-linear in outline, tapering into the beak, the inner straight, nearly smooth, the outer curved, muricate. Weed, introduced from Eurasia. By roads and railroads, *A. T.*

WYETHIA

Perennial herbs; stems simple, rarely branching; leaves alternate, mostly entire and ample; heads many-flowered, solitary or few, medium or large; flowers yellow; ray flowers elongated, pistillate or fertile; pappus a chaffy crown or cup; ray achenes neither obcompressed nor laterally compressed; disk flowers perfect, fertile; chaff lanceolate, partly enclosing the achenes; disk achenes 4-angled. Named for *Capt. Nathaniel J. Wyeth,* early explorer and trader in the Pacific Northwest.)

Wyethia amplexicaulis Nutt. *Pe-ik, Pik.* Plant 3–6.5 dm. tall, glabrous throughout; root 10–15 mm. in diameter, woody black; basal and lower leaves with petioles 1–10 cm. long, the blades 15–40 cm. long, oblanceolate to elliptic, entire or denticulate, with numerous glands producing resin, often the whole surface shining with it; upper leaves similar, 6–15 cm. long, oval to lanceolate; heads 1–5, approximate, semiorbicular; outer bracts 2–3 cm. long, foliaceous, lance-oblong, glabrous; inner bracts 17–20 mm. long, lanceolate, firm, scale-like; the 11–17 rays 2–3.5 cm. long, narrowly elliptic; pappus 1–3 mm. long of irregular fimbriate scales; achenes about 9 mm. long, 4-angled, wedge-shaped; disk flowers 9–10 mm. long, tubular; anthers blackish; achenes 8–10 mm. long, wedge-shaped, the pappus crown 1–2 mm. long, cut into small scales, rarely bearing 1–2 awns. Low meadows, *U. Son., A. T.*

The Nez Perce name "*Pik*" was recorded by the Rev. Henry Spalding. The plant is one of the so-called "compass-plants." It is not poisonous.

XANTHIUM. Cocklebur

Annual herbs; leaves alternate, petioled; heads monoecious, in axillary or terminal clusters or short interrupted spikes; the pistillate heads 2-flowered and below the several-flowered staminate ones; involucre of the staminate heads of several distinct narrow scales; involucre of the pistillate heads bur-like, ovoid or oblong, closed, indurated, 2-celled, 2-flowered, 2-beaked at apex, armed all over with strongly hook-tipped bristles; pappus none; corolla none; achenes compressed, beaked. (Name from Gr. *xanthion,* a plant (perhaps *X. strumarium*) yielding a yellow hair dye; *xanthos,* yellow.

Seeds and seedlings in the cotyledon stage of some and perhaps all the species are very poisonous. They will kill pigs, sheep, cattle, and chickens.

Nodes armed with 3-forked spines; blades densely white hairy beneath; fruit body 9–12 mm. long. — *X. spinosum.*
Nodes unarmed; blades green, and but sparsely hispidulous beneath; fruit body 10–25 mm. long,
  Fruit body 20–25 mm. long, oblong-ovoid. — *X. oviforme.*
  Fruit body 10–20 mm. long, ellipsoid,
    Fruit body and prickles hispid. — *X. italicum.*
    Fruit body and prickles glabrous or puberulent. — *X. pensylvanicum.*

Xanthium italicum Mor. Plant 3–10 dm. tall; tap-root 1–3 mm. in diameter, pale; stem often branched, hispidulous, often purple lined, finally glabrate below; petioles 3–15 cm. long; blades 5–12 cm. long, cordate or broadly ovate, dentate, often shallowly lobed, hispidulous, beneath paler green; fruit with the body 1.3–1.8 cm. long, ellipsoid, brown, glandular hispid; beaks 5–7 mm. long, stout, hispid, commonly incurved and hooked; prickles 3–7 mm. long, abundant, slender acicular, hispid to the middle, hooked at tip. *X. saccharatum* sensu Widder, in part; *X. californicum* sensu Widder, in part, not Greene; *X. affine* sensu Piper. Roadsides and stream banks, *U. Son., A. T.*

Xanthium oviforme Wallr. Plant 3–7 dm. tall; tap-root 2–5 mm. in diameter, brown; stem simple or branched above, rough hispidulous, below glabrate; cauline leaves with petioles 4–17 cm. long, the blades 5–15 cm. long, ovate to deltoid-ovate or cordate, doubly dentate and often somewhat lobed, the base from cuneate to subcordate, resinous atomiferous and scabrous hispidulous on both sides, beneath paler green; fruits brown with the body 20–25 mm. long, oblong-ovoid, glandular hispid; beaks 7–10 mm. long, heavy subulate, incurved, glandular hispid; prickles 7–12 mm. long, strong, arcuate, glandular hispid to above the middle, hooked at tip. Abundant, shores of Snake River, and tributaries, *U. Son.*

Xanthium pensylvanicum Wallr. Plant 3–9 dm. tall; tap-root 1–3 mm. in diameter, slender, pale; stem simple or branched, hispidulous, finally glabrate below; petioles 3–12 mm. long; blades 5–13 cm. long, deltoid-ovate or cordate, dentate often with shallow lobes, scabrous or hispidulous, beneath paler green; fruit with the body 1–2 cm. long, ellipsoid, glabrous or glandular puberulent; beaks 4–6 mm. long, somewhat incurved, the apex hooked, the stout base glandular puberulent; prickles 3–7 mm. long, acicular, numerous, the stout base glandular puberulent, the apex hooked. Stream banks, *A. T.*

Xanthium spinosum L. *Spiny Clotbur; Chinese Thistle.* Plant 3–12 dm. tall; tap-root and laterals becoming stout, brown; stem erect, usually branched, hirsutulous, finally glabrate below; cauline leaves with petioles 1–15 mm. long, the blades 4–10 cm. long, lanceolate or ovate-lanceolate, 2–4-lobed or the upper entire, acute, the base cuneate, beneath hispidulous canescent, above green but sparsely appressed hispidulous, densely so on the veins; nodes armed with yellow, 3-forked spines 10–32 mm. long; fruit with the body 9–12 mm. long, oblong-ellipsoid, viscid pilosulous, yellow; the beaks 2–3 mm. long, straight, subulate; prickles 2–2.5 mm. long, remote, hooked at tip. *Acanthoxanthium spinosum* (L.) Fourr. Weed, introduced from South America. Roadsides and fields, infrequent, *U. Son., A. T.*

Xanthium oligacanthum Piper was described from collections from Bolles, and Waitsburg. The fruits had a narrow, oblong body with very few (15–25) remote prickles. Millspaugh & Sherff made this a synonym of the earlier described similar species, *X. Wootoni* Cockerell. Widder kept them separate, calling ours *X. californicum* Greene, var. *oligacanthum* (Piper) Widder. J. L. Symons, in Bot. Gaz. **81:** 133, 1926, showed experimentally that the few prickled fruits occur solely on one plant or intermingled with many prickled fruits. Some of them breed true, some give many prickled offspring. Sparsely armed fruits have now been found associated with several species. They seem to be mutants. Ours, then, is probably a mutant from *X. italicum* Mor. Similar material later found at Hooper, *St. John et al.* 7,169.

GLOSSARY

Acaulescent. Stemless or apparently so, or with the stem underground.
Accrescent. Growing larger after flowering.
Accumbent. Lying against, as the cotyledons with their edges against the radicle.
Acerose. Needle-shaped; with a sharp rigid point.
Achene. A dry, 1-celled, indehiscent fruit, the 1 seed basal.
Acicular. Needle-shaped.
Acrid. Sharp and harsh to the taste.
Acuminate. Gradually tapering to a point.
Acute. Sharp pointed.
Adnate. United, especially when different organs are joined.
Aestival. Produced in summer.
Alternate. Leaves when attached one at a node. Stamens are alternate with the petals when they stand over the intervals between them.
Alveolate. Resembling a honeycomb.
Ament. A catkin; a scaly spike, usually unisexual.
Amphitropous (ovule or seed). Half inverted, straight, but with a lateral hilum.
Ampliate. Abruptly expanded.
Anatropous (ovule or seed). Inverted and straight, with the micropyle next to the hilum.
Androgynous. An inflorescence with the staminate flowers above the pistillate.
Angulate. More or less angular.
Annual. Within one year; lasting not more than a year.
Annulus. A ring; especially the ring of the thick walled cells on a fern sporangium.
Anterior. On the front side; away from the axis.
Anthesis. The period of, or the act of expansion of a flower.
Apices. The plural of apex, a tip or point.
Apiculate. Ending in a short, pointed tip.
Appressed. Lying close and flat against.
Arcuate. Bent or curved like a bow.
Areola. A small space marked out; a small cavity.
Aristate. Tipped with a stiff, short bristle.
Articulate. Jointed; with joints where it separates.
Asexual. Without sex.
Assurgent. Ascending.
Asymmetric. Irregular in outline or shape; as a flower which cannot be divided by a vertical plane into two similar halves.
Atomiferous. Bearing minute granules.
Attenuate. Narrowed; tapered.
Auricle. An ear-shaped appendage.
Autophytic. Independent; not dependent on humus.
Awn. A bristle-like appendage.
Axial. Relating to the axis.
Axillary. Occurring in an axil.
Axil. The angle on the upper side between a leaf and the stem.

Barbellate. Minutely barbed.
Basifixed. Attached by its base.
Beaked. Ending in a prolonged, narrow tip.
Bi-. Two; or twice.
Biennial. Of two years duration; germinating one year, flowering and dying the next.
Bifid. Two-cleft.
Bilabiate. Two-lipped.
Bimaculate. Two-spotted.
Biternate. Twice-ternate.
Blade. The expanded portion of a leaf.
Bract. A modified reduced leaf; especially those of an inflorescence.
Bracteate. Having bracts.
Bracteolate. Having small bracts.
Bracteole. A small bract.
Bulbiferous. Bearing or producing bulbs.

Caducous. Dropping off very early.
Caespitose. Growing in turf-like tufts or patches.
Callosity. A hardened thickening.
Callous. Hard and thick in texture.

Callus. A hard protuberance. In grasses, the tough swelling below the base of the lemma.
Calyculate. Having bracts around the flower which imitate a calyx.
Calyx. The outer set of floral envelopes or leaves.
Campanulate. Bell-shaped.
Canescent. Gray or hoary; covered with fine gray or whitish hairs.
Capillary. Hair-like.
Capitate. Shaped like a head; collected into a head.
Capsule. A dry, dehiscent, compound seed vessel.
Carinate. Keeled; with a sharp ridge on the lower side.
Carpel. A simple pistil; one element of a compound pistil.
Carpophore. A slender prolongation of the axis between the carpels.
Cartilaginous. Firm and tough; cartilage-like.
Caruncle. A protuberance on the hilum of a seed.
Caryopsis. A grain; seed-like fruit with the thin pericarp adherent to the single seed (*Gramineae*).
Castaneous. Chestnut-colored.
Catkin. Ament, a scaly spike, usually unisexual.
Caudate. Tailed; with a slender, tapering tip.
Caudex. An upright or short rootstock.
Caudicula. The thread-like or strap-shaped stalk of a pollinium.
Caulescent. Becoming stalked; with an obvious stem.
Caulicle. The stem-like part of an embryo; a hypocotyl.
Cauline. Pertaining to the stem; on the stem.
Cell. The locule or cavity, of an ovary or carpel.
Cernuous. Nodding; slightly drooping.
Chaff. Small dry bracts or scales; especially in the *Compositae* the scales subtending flowers on the receptacle.
Chartaceous. Papery.
Chlorophyll. The green coloring matter of plants.
Choripetalous. With sepatate petals; polypetalous.
Ciliate. The margin with a fringe of hairs.
Ciliolate. Minutely ciliate.
Cinereous. Ashy gray.
Circinate. Rolled inwards from the top.
Circumscissile. The upper part splitting off as a lid.
Clathrate. Like a lattice.
Clavate. Club-shaped; gradually thickened upwards.
Cleft. Cut half-way down.
Cleistogamous. Pollinated in the bud; the flowers not opening.
Coalescent. Growing together.
Cochleate. Coiled or shaped like a snail shell.
Coenosorus. Several sori touching and joined.
Coma. The tuft of hairs at the end of a seed.
Commissure. The surface on which two carpels join (*Umbelliferae*).
Compound. Of a leaf, when once or more times divided.
Compressed. Flattened, especially laterally.
Conduplicate. Folded together.
Confluent. Blended together; coherent.
Coniferous. Cone-bearing, as many *Gymnospermae*.
Connate. United; grown together from the first.
Connective. Of the anther, the sterile tissue between the anther sacs.
Connivent. Converging; brought close together.
Convolute. Rolled up lengthwise in the bud.
Coralloid. Resembling coral.
Coriaceous. Leathery.
Corm. A fleshy, enlarged stem base, resembling a bulb, but solid.
Corniculate. With a small horn or spur.
Corymb. A flat or convex, indeterminate inflorescence.
Corymbose. Arranged in corymbs.
Cotyledon. The first leaves of the embryo plant; the seed leaves.
Crenate. The margin with rounded, forward directed teeth.
Crenulate. Finely crenate.
Crested. With an upraised, crest-like appendage.

Cruciform. Cross-shaped.
Crustaceous. Hard and brittle.
Cucullate. Hooded.
Culm. The usually hollow stem of the grasses (*Gramineae*).
Cuneate. Wedge-shaped.
Cupulate. With or subtended by a little cup.
Cupuliform. Shaped like a little cup.
Cuspidate. Tipped with a cusp or sharp rigid point.
Cyme. A broad or flattened determinate inflorescence.
Cymose. Bearing cymes; cyme-like.
Cymule. A little cyme.
Cystolith. Mineral concretion, usually calcium oxalate, in special cells.

Deciduous. Falling off; of leaves which fall in the autumn; a plant naked of leaves in winter.
Declined. Bent downward.
Decompound. More than once compound.
Decumbent. Reclining at base, but the apex ascending.
Decurrent. Of leaves, prolonged down the stem below the point of insertion.
Dehiscence. The method of opening.
Dehiscent. Opening at maturity in a regular manner.
Deltoid. Triangular with the apex upward.
Dentate. Toothed.
Denticulate. Minutely toothed.
Di-. Two-; twice-.
Diadelphous. Stamens united in two groups.
Diandrous. With two stamens.
Dichotomous. Two-forked.
Dictyostelic. With the stele or vascular cylinder forming a network.
Didymous. Twinned.
Didynamous. Stamens in two pairs of unequal length.
Diffuse. Widely or loosely spreading.
Digitate. Like fingers. As leaflets or veins all springing from the apex of the petiole.
Dimorphic. Of two forms.
Dioecious. With the staminate and pistillate elements each on separate plants.
Discoid. Like a disk. In the *Compositae,* a discoid head contains only disk flowers, lacking ray flowers.
Disk. A ring-like expansion of the receptacle within the calyx or corolla; the face of any flat body; in the *Compositae* the central part of the head bearing only disk flowers.
Disk flowers. In *Compositae* the tubular, regular flowers, usually central.
Dissected. Divided into many fine segments.
Dissepiment. The partitions of a compound ovary or fruit.
Distal. The side remote from the place of attachment.
Diurnal. Occuring only in day time.
Divaricate. Widely divergent or spreading.
Divergent. Spreading away from each other.
Divided. Cut down to the midrib or base, but the divisions not quite distinct.
Dorsal. Relating to or attached to the back of an organ.
Drupaceous. Drupe-like.
Drupe. A stone-fruit, like a plum; a fleshy fruit, the mesocarp fleshy, the endocarp stony, containing one or more seeds.
Drupelet. Little drupe.

Ebracteate. Without bracts.
Ecological. Relating to ecology; in relation to the environment or habitat of the plant.
Elaters. A filament associated with the spores, hygroscopic and bending with the changes of humidity.
Emarginate. With a shallow notch at apex.
Embryo. The rudimentary plant contained in a seed.
Endocarp. The inner layer of a pericarp or carpel wall.
Endosperm. The albumen, the seed tissue stored with food, in the embryo sac, partly surrounding the embryo.
Entire. With smooth margins, not toothed or lobed.
Epi-. On; upon.

Epicotyl. The young stem above the cotyledons, in the seed.
Epidermis. The outer, cellular skin.
Epigynous. On top of the ovary; the calyx, corolla, or stamens adnate to and apparently growing from or near the top of the ovary.
Epipetalous. Borne on the petals or corolla.
Equitant. Astride, as when leaves are alternately folded over each other in two ranks.
Erose. Irregularly toothed on the margin, as if gnawed.
Evanescent. Soon disappearing or fading.
Ex-. Without; not.
Excavate. Hollowed out.
Excurrent. Running out, as a midrib projected beyond the blade apex; spire-shaped growth, resulting from the terminal bud always continuing the main stem.
Exocarp. The outer layer of the pericarp or carpel wall.
Exserted. Protruding out of, as the stamens from the corolla.

Falcate. Scythe-shaped.
Farinaceous. Containing starch; starch-like.
Fascicle. A close bundle or cluster.
Fasciculate. In close bundles.
Fastigiate. Parallel, clustered, and erect.
Fenestrate. Pierced with large holes like windows.
Fertile. Capable of producing fruits; of pollen in the case of anthers.
Fibrillose. Formed of small fibres.
Fibrous. Composed of or resembling fibres.
Filament. The stalk of a stamen; any thread-like body.
Filamentose. Formed of or bearing slender threads.
Filiform. Thread-like.
Fimbriate. Fringed.
Fimbrillate. Minutely fringed.
Fistulose. Hollow.
Flabellate. Fan-shaped.
Flabelliform. Fan-shaped.
Fleshy. Of firm pulp or flesh.
Flexuous. Zigzag; bending alternately in opposite directions.
Floccose. With tufts of soft hair or wool.
Floret. A small flower; one flower of a cluster.
Floricane. A biennial stem in its second year, bearing flowers and fruit (*Rubus*).
Floriferous. Flower-bearing.
Foliaceous. Leaf-like.
Foliage. The leafy covering.
Foliate. Provided with leaves.
Follicle. A dry fruit of one carpel, dehiscing along the ventral suture; where the seeds are attached.
Fornix (plural, fornices). Swelling or scale-like appendages in the throat of the corolla.
Foveate. Deeply pitted, like honey-comb.
Foveolate. With shallow pits, suggesting a honey-comb.
Frond. The leaf-like part of the ferns and related plants (*Pteridophyta*).
Fugacious. Fading or falling very early.
Funiculus. The stalk of an ovule.
Fuscous. Grayish-brown.
Fusiform. Spindle-shaped.

Galea. Shaped like a helmet, applied to a sepal or to the upper corolla lip.
Galeate. Helmet-shaped; having a galea.
Gamopetalous. Of united petals.
Geminate. Twin; in pairs.
Geniculate. Bent abruptly, like a knee.
Gibbous. With a protuberance or swelling on one side.
Glabrate. Becoming glabrous with age; almost glabrous.
Glabrous. Smooth; without hairs.
Gladiate. Sword-shaped.
Gland. A secreting organ.
Glandular. Furnished with glands; gland-like.
Glaucescent. Slightly glaucous; becoming glaucous.
Glaucous. Covered with a white, waxy bloom.
Glochid. A barbed hair or spine.
Glochidiate. Barbed; tipped with barbs.
Glomerate. In small, compact clusters.

Glomerule. A dense, head-like cluster.
Glumaceous. Glume-like.
Glumes. The two empty scales at the base of the grass spikelet (*Gramineae*).
Glutinous. Sticky; glue-like.
Grain. Caryopsis; a 1-celled, 1-seeded, indehiscent, seed-like fruit, the wall of the fruit adhering to the seed (*Gramineae*).
Granulate. Covered with small granules.
Granuliferous. Bearing granules.
Gynobase. A prolongation of the receptacle bearing the ovary, or in fruit the nutlets (*Boraginaceae*).
Gynophore. A stalk raising the pistil above the stamens.

Hastate. Halberd-shaped.
Haustoria. Root-like organs which parasites protrude into their hosts to absorb food.
Hemi-. Half.
Hemispheric. Half-spherical.
Herbaceous. Of the texture of an herb; not woody.
Hermaphrodite. Having both sexes; with both stamens and pistils.
Heterogamous. Bearing two kinds of flowers.
Hilum. The attachment scar on the seed.
Hirsute. With long, rather stiff hairs.
Hirsutulous. Minutely hirsute.
Hispid. With stiff hairs or bristles.
Hispidulous. Minutely hispid.
Hoary. Canescent; gray from fine pubescence.
Homogamous. With one kind of flowers.
Hyaline. Transparent or partly so.
Hydathode. A pore for the extrusion of water.
Hypanthium. The fused receptacle and base of calyx in perigynous or epigynous flowers.
Hypo-. Under; underneath.
Hypocotyl. Caulicle; the stem-like part of the embryo.
Hypogynous. Of perianth and stamens, attached to the receptacle below and free from the ovary.

Imbricate. Overlapping, like shingles on a roof.
Imperfect. Of flowers, when lacking one sex.
Inaequilateral. Unequal sided.
Incised. Cut sharply and deeply.
Incumbent. Leaning upon; of cotyledons when the back of one lies against the radicle; of anthers when turned inward.
Indehiscent. Not splitting open.
Induplicate. With the edges turned inwards.
Indurated. Hardened.
Indusium. An epidermal outgrowth or scale covering the sorus in the ferns (*Pteridophyta*).
Inferior. Growing from below or from a lower attachment than some other organ.
Inflorescence. The disposition of flowers on the axis; the flower cluster.
Infra-. Below.
Infrastipular. Attached below the stipules.
Inter-. Between.
Intercostal. Between the ribs or nerves of a leaf.
Internode. The part of the stem between two nodes.
Introrse. Turned inward; towards the axis.
Involucel. A secondary involucre, that subtending the umbellets (*Umbelliferae*).
Involucral. Belonging to an involucre.
Involucrate. Having an involucre.
Involucre. A circle or cluster of bracts surrounding a flower or flower cluster.
Involute. Rolled inwards from the edges.
Irregular. Asymmetric; as a flower which cannot be halved in any or not more than one plane.

Keel. A projecting central ridge; the two lower, united petals of a papilionaceous flower (*Leguminosae*).

Lacerate. Irregularly cleft, as if torn.
Laciniate. Cut into narrow, deep lobes.

Lamina. Blade of a leaf.
Lanate. Woolly; clothed with long, soft, entangled hairs.
Lanceolate. Lance-shaped.
Lanceoloid. A solid having a lance-shaped outline.
Lateral. Of the sides.
Leaflet. A division of a compound leaf.
Lemma. In the grass spikelet, the bract which bears a floret in its axil (*Gramineae*).
Lenticel. Raised, often light spots on the young bark, corresponding to epidermal stomata.
Lepidote. Covered with scurfy scales.
Lenticular. Lentil-like; doubly convex.
Ligulate. Strap-shaped; with strap-shaped corollas.
Ligule. A thin, often scarious projection from the tip of the leaf sheath in grasses (*Gramineae*); the limb of the ray flowers (*Compositae*).
Limb. The expanded part of a petal or sepal.
Linear. Long and narrow, the sides parallel.
Lobed. With the margin indented less than half way to the midrib.
Locule. One of the cavities (or cells) in an ovary.
Loculicidal. Dehiscing by the back of each locule, the clefts opening into the cavities or cells of the ovary.
Lodicule. The small scales or bodies beneath the stamens in the grass floret (*Gramineae*).
Lunate. Crescent-shaped; half-moon-shaped.
Lyrate. Pinnatifid with the terminal lobe large and rounded, the lower lobes small.

Macranthous. Large flowered,
Marcescent. Withering but not falling.
Megasporangium. The case in which the megaspores are produced.
Megaspore. The larger spores producing a female gametophyte (*Pteridophyta.*)
Membranaceous. Membrane-like; thin.
Mericarp. One carpel or one half of the fruit in the *Umbelliferae.*
Mesocarp. The middle part of the pericarp or ovary wall.
Microsporangium. The case in which the microspores are produced; the anther sac or cell of an anther.
Microspore. The small spores producing male gametophytes (*Pteridophyta*); a pollen grain.
Midrib. The middle or main vein of a leaf.
Monadelphous. Stamens all united in one cluster (*Leguminosae*).
Moniliform. Like a string of beads.
Monoecious. With the stamens and pistils in separate flowers on the same plant.
Mucronate. Tipped with a mucro or short, abrupt point.
Mucronulate. Slightly mucronate.
Multi-. Many.
Multicellular. Many celled.
Multicipital. With many heads.
Muricate. Roughened with short, hard, or prickly points.
Muriculate. Minutely muricate.
Mutant. Species or subdivision derived by an abrupt change from its parents.

Napiform. Turnip-shaped.
Narcotic. Numbing or sleep producing.
Nectariferous. Producing nectar.
Nectary. An organ or spot where nectar is secreted.
Nerve. A vein or rib of a leaf, especially when parallel and unbranched.
Neuter. Neutral; sexless.
Neutral. Sexless; without stamens or pistils.
Node. The joint of a stem, where leaves and buds are normally produced.
Nodose. Knotty; knobby.
Nodulose. With small knots or knobs.
Nut. A hard, indehiscent, 1-celled, 1-seeded fruit, though usually developing from a compound ovary.
Nutlet. A small nut; the small hard carpels separable at maturity, in the *Boraginaceae, Labiatae,* and *Verbenaceae.*

Ob-. Inverted.
Obcompressed. Flattened dorso-ventrally, or on anterior and posterior sides.
Obsolete. Not evident; wanting.
Obtuse. Blunt or rounded at the end.
Ochroleucous. Yellowish-white.
Ocrea. A sheath like or tubular stipule (*Polygonaceae*).
Ocreola (plural, ocreolae). A little ocrea; the sheathing stipules of the nodes of the inflorescence (*Polyganaceae*).
Olivaceous. Olive-green.
Operculum. A lid; the cap of a circumscissile capsule.
Opposite. Of the arrangement of leaves when the two from each node are attached at the interval of 180°. Stamens are opposite petals or corolla lobes when attached directly in front of their middle.
Oppositifolious. Attached opposite to a leaf.
Orthotropous. Ovule or seed, when erect, with the micropyle at the apex, oposite to the hilum.
Oval. Broadly elliptical.
Ovary. The enlarged basal part of the pistil containing the ovules.
Ovate. Egg-shaped, the broader end down.
Ovoid. A solid with the outline of an egg, broader end down.
Ovule. The rudiment in the ovary containing the egg, after fertilization becoming a seed.

Palea. The innermost bract enclosing the grass floret, facing the lemma; palet. (*Gramineae*).
Palmate. Lobed or veined so that the divisions spread from the tip of the petiole like fingers from the palm of a hand.
Panicle. A loose, elongate, indeterminate inflorescence, twice or more times branched, with pedicelled flowers.
Paniculate. Borne in panicles; resembling a panicle.
Papilionaceous. Butterfly-like. that is with the upper petal enlarged and spreading or reflexed, the two lateral petals small and oblique, the two lower ones connivent into a keel.
Papillate. With minute, pimple-like protuberances.
Papillose. Having minute papillae or nipple-like projections.
Pappus. A modified calyx-limb, of a ring of hairs, plumose hairs, bristles, awns, or scales (*Compositae*).
Parasitic. Depending on another living organism for its food.
Parietal. Of placentae, when attached to the walls of the ovary.
Parted. Separated or cleft into parts almost to the base or midrib.
Pectinate. Pinnatifid with narrow, close divisions like the teeth of a comb.
Pedicel. The stalk of a single flower of a cluster.
Peduncle. The common stalk of a flower cluster.
Pellucid. Clear; transparent.
Peltate. Shield-shaped; of a leaf of any shape when the petiole is attached at the middle or remote from the margin.
Penicillate. Tipped with a tuft of fine hairs.
Pentagonal. With five angles.
Pentamerous (5-merous). Five parted.
Penultimate. Next to the last.
Perennial. Lasting for several years.
Perfect. Of flowers when containing both sexes, or both stamens and pistils.
Perianth. The floral envelopes; calyx plus corolla.
Pericarp. The matured ovary wall; the outer layers of a fruit.
Perigynous. Around the ovary; of stamens or petals when adnate to the calyx or hypanthium cup, remote from the ovary.
Petaliferous. Petal bearing.
Petaloid. Petal-like; resembling petals in texture and color.
Petiole. The stalk of a leaf.
Petiolulate. Having a petiolule.
Petiolule. The stalk of a leaflet.
Pilose. With soft, slender hairs.
Pilosulous. Minutely pilose.
Pinna. A primary division of a pinnate leaf.
Pinnate. Of a leaf when compound, with the leaflets attached to the sides of a common rhachis; feather-like.
Pinnatifid. Pinnately cleft.

Pinnatisect. Pinnately parted.
Pinnule. The smaller division of a bipinnate or pinnately compound leaf.
Pistil. The female or seed-bearing organ of a flower, formed of ovary and stigma, and often with a style.
Placenta. The surface in an ovary to which the ovules are attached.
Plaited. Plicate; folded lengthwise.
Plicate. Plaited; folded lengthwise.
Plumose. Feather-like; with long, hair-like side branches.
Plumule. Epicotyl; the first shoot of a seedling above the cotyledons.
Pollinium. A mass of waxy pollen grains, of all the contents of an anther sac (*Orchidaceae*).
Poly-. Many.
Polygamo-dioecious. With perfect and staminate flowers on some plants, the others with perfect and pistillate flowers.
Polygamo-monoecious. Each plant bearing perfect, staminate, and pistillate flowers.
Polygamous. Having both perfect and unisexual flowers on the same plant.
Polymorphic. Of several or variable forms or shapes.
Pome. A fruit like an apple; a several-celled, inferior fruit, the principal fleshy parts formed by the fused calyx-tube and receptacle.
Primocane. A biennial stem in its first and vegetative year (*Rubus*).
Prismatic. Like a prism; angular with flat sides.
Procumbent. Trailing on the ground.
Protandrous. When the stamens discharge their pollen before the pistils are receptive.
Proterogynous. When the pistils are receptive before the anthers mature.
Prothallium (plural, prothallia). A thin, flat gametophyte (*Pteridophyta.*)
Proximal. The part nearest the axis.
Pruinose. As if frosted; covered with a whitish powder.
Puberulent. Covered with fine, short hairs.
Puberulous. Minutely puberulent.
Pubescence. The hairy coating of plants.
Pubescent. Hairy.
Pulvinate. Cushioned; like a cushion.
Pulvinus. A cushion-like swelling.
Puncate. Dotted.
Pungent. Terminated in a rigid, sharp point; penetrating.
Pustulate. As though blistered.
Pyramidal. Pyramid-like.
Pyriform. Pear-shaped.

Quadrifoliolate. With four leaflets.

Raceme. A simple inflorescence of pedicelled flowers on a common more or less elongate axis.
Racemiform. Shaped like a raceme.
Racemose. Bearing racemes; raceme-like.
Radiate. Arranged radially from a center; having ray flowers (*Compositae*).
Radical. Pertaining to the root; occurring at the base of the stem.
Raphe. On an anatropous seed, the ridge formed by the adherent funiculus.
Ray. A branch of an umbel; a ray flower.
Ray flower. A ligulate flower; a flower with a laterally prolonged, asymmetric, strap-shaped corolla limb (*Compositae*).
Receptacle. The modified apex of the stem to which the flower parts are attached.
Reflexed. Bent backwards or outwards.
Regular. Symmetrical in shape or structure; of a flower when all the parts of each series are alike in shape, size, and position.
Reniform. Kidney-shaped.
Repand. With a wavy or sinuous margin.
Resiniferous. Producing resin.
Reticulated. In the form of a net; net-veined.
Retrorse. Directed backward or downward.
Retuse. With a shallow notch in the rounded apex.
Revolute. Rolled backwards or under, as the margins of a leaf.

Rhachis. The main axis of an inflorescence or compound leaf or frond.
Rhachilla. A secondary axis of an inflorescence, leaf, or frond.
Rhizome. A rootstock; a root-like, underground stem.
Rosette. A circular (often basal) cluster of leaves, or other organs.
Rosulate. Concerning or collected into a rosette.
Rotate. Wheel-shaped; circular, flat, and spreading.
Rudiment. A partially developed, functionless organ.
Rufous. Reddish brown.
Rugose. Wrinkled.
Rugulose. Minutely wrinkled.
Runcinate. Coarsely saw-toothed or incised, the teeth retrorse.

Saccate. Sac-shaped.
Sagittate. Arrow-shaped.
Salient. Prominent.
Salverform. With a slender tube, abruptly expanded to a rotate limb.
Saprophytic. Depending on dead organic matter for its food.
Scaberulous. Minutely scabrous.
Scabrid. Minutely scabrous.
Scabrous. Rough or harsh to the touch.
Scalariform. With cross bands suggesting the steps of a ladder.
Scape. A peduncle arising from the ground.
Scapose. Bearing a, or resembling a scape.
Scarious. Thin, dry, membranous, not green.
Scorpiod. Like the tail of a scorpion; the form of a half or incomplete cyme, with the buds at the coiled tip, uncoiling with anthesis.
Scurfy. Covered with minute scales.
Secund. One-sided; when the flowers all turn to one side.
Semi-. Half.
Sepal. One division of the calyx.
Septicidal. Dehiscing through the dissepiments or partitions between the cells (locules).
Septifragal. Dehiscing so that the valves break away from the partitions (dissepiments).
Septum. A partition.
Sericeous. Silky; clothed with silky hairs.
Serrate. Saw-toothed; with sharp, forward pointed teeth.
Serratures. Serrate teeth.
Serrulate. Minutely serrate.
Sessile. Attached directly, without a stalk.
Seta (plural, setae). Bristle.
Setaceous. Bristle-like.
Setose. With bristles.
Setulose. With minute bristles.
Sheath. A tubular or enrolled part of an organ, as the petiole of grasses (*Gramineae*).
Sheathing. Wrapped around.
Sigmoid. Curved like the letter S.
Silicle. A short, 2-celled pod, not more than thrice as long as wide (*Cruciferae*).
Silique. An elongate, 2-celled pod, more than thrice as long as wide (*Cruciferae*).
Sinuate. With the outline strongly wavy.
Sinus. The recess or indentation between two lobes.
Solenostelic. With the stele or vascular cylinder a hollow cylinder.
Sordid. Dirty white; dirty or dull.
Sorus (plural, sori). A cluster of sporangia in ferns.
Spadix. A thick, fleshy spike, often with imperfect flowers.
Spathaceous. Spathe-like.
Spathe. A large, petal-like bract, enclosing or closely subtending an inflorescence.
Spathulate. Spatulate.
Spatulate. Shaped like a spatula; rounded at apex and tapering gradually to a narrow base.
Spicate. In a spike; resembling a spike.
Spiculate. Covered with fine points.
Spike. A simple inflorescence with the flowers sessile on a more or less elongate common axis.
Spikelet. A small spike; especially the specialized ultimate flower cluster in the grasses (*Gramineae*).
Spinescent. Becoming spiny.
Spinulose. With small spines.
Spiricles. Coiled gelatinous threads that emerge when wetted.

Sporangium (plural, sporangia). A spore case.
Spore. A single cell, becoming free, and capable of developing a new plant.
Sporocarp. A pod-like structure enclosing the sporangia.
Sporophyll. A leaf which bears the spores.
Spur. A hollow sac-like or tubular projection of the calyx or corolla, usually nectar bearing.
Stamen. The male organ of the flower, formed of filament and anther.
Staminate. Furnished with stamens.
Staminodium (plural, staminodia). An abortive stamen.
Standard. The upper petal of a papilionaceous flower (*Leguminosae*).
Stellate. Star-like; with branches like the arms of a star.
Sterile. Incapable of reproducing, as a flower without pistils or a stamen without pollen.
Stigma. The terminal, usually rough or sticky part of a pistil, on which pollen is received.
Stipe. A stalk, as that of a fern frond, of a pistil, etc.
Stipel. A stipule-like expansion at the base of a leaflet.
Stipitate. With a stalk.
Stipule. The leafy appendages often present on either side of the base of a leaf at its point of attachment.
Stolon. A runner or basal, usually prostrate branch that will strike root.
Stoloniferous. Bearing stolons.
Stoma (plural, stomata.) The breathing pore in the epidermis.
Striate. Marked with longitudinal lines or ridges.
Strigillose. With small sharp points.
Strigose. Covered with short, stiff, and appressed hairs.
Strophiole. Caruncle; an appendage near the hilum.
Style. A stalk between the ovary and the stigma.
Stylopodium. A conical or discoid expansion at the base of the style (*Umbelliferae*).
Sub-. Nearly; somewhat; almost.
Subulate. Awl-shaped.
Succulent. Juicy; fleshy.
Suffrutescent. Slightly shrubby or woody at base.
Superior. Growing or placed above; as to the ovary, when all the remaining flower parts are attached directly to the receptacle, at a lower level.
Supra.- Above.
Supra-axillary. Produced from above the axil.
Suture. The line of junction or of dehiscence.
Syn.- With; together with.
Syncarp. Of several carpels consolidated.
Sygenesious. With the anthers cohering in a ring, but the filaments separate.

Tap-root. The primary descending root, especially when stout and tapering, and when the secondary roots are small or inconspicuous.
Tawny. Dull yellowish brown.
Terete. Cylindric in cross-section.
Ternate. In threes.
Testa. The outer seed coat.
Tetra-. Four.
Tetradynamous. With four long and two short stamens.
Tetragonous. Four-angled.
Tetrahedral. Having four sides, like a tetrahedron.
Thallose. Resembling a thallus.
Thallus. A plant body not differentiated into stem and leaf.
Thyrse. A mixed inflorescence, a contracted panicle of cymes or cymules; thyrsus.
Thysoid. Resembling a thyrse.
Thyrsus. Thyrse.
Tomentose. Densely clothed with soft, matted hairs.
Tomentum. A coating of soft, matted hairs.
Tomentulose. Minutely tomentose.
Torulose. Cylindric, with swellings at intervals.
Trapeziform. With four unequal sides.
Tri-. Three.
Tricarpellary. Of three carpels.
Tridentate. With three teeth.
Trifoliate. Three leaved.
Trifoliolate. With three leaflets.

Trigonous. Three-angled.

Truncate. Ending abruptly, as if cut off.

Tuber. A short, thickened subterranean branch, with eyes or buds.

Tubercle. A small tuber or excrescence; a small tuber on the roots due to symbiotic growth with bacteria.

Tuberiferous. Bearing tubers.

Turbinate. Top-shaped.

Turgid. Swollen.

Turion. A strong, scaly sucker from the ground.

Umbel. An inflorescence with the pedicels or rays all from the summit of the peduncle.

Umbellate. In umbels, or resembling an umbel.

Umbellets. A small umbel; a secondary umbel.

Umbilicate. With a navel; depressed in the center.

Umbo. A low, rounded projection; a boss.

Uncinate. Hook-shaped; hooked at the end.

Uni-. One.

Unisexual. With only one sex; with stamens only; with pistils only.

Urceolate. Urn-shaped.

Utricle. A bladder-like, thin walled, 1-seeded fruit.

Valvate. Opening by valves; in buds, when the parts meet exactly, edge to edge but do not overlap.

Vascular. Having woody bundles or vessels.

Vein. The small nerves of a leaf or fibro-vascular bundles.

Velum. A membrane partly covering a sporangium (*Isoetes*).

Ventral. Belonging to the anterior or inner side of an organ; the opposite of dorsal.

Versatile. Attached loosely at one point and free to swing.

Verticillate. In a whorl.

Vespertine. Appearing or expanding at evening.

Vessel. A conducting duct.

Verrucose. Warty.

Villosulous. Minutely villous.

Villous. Shaggy with long, soft hairs.

Virgate. Wand-shaped; slender straight, erect.

Viscid. Glutinous; sticky.

Whorl. Leaves or other parts arranged in a circle of three or more at the same node.

Wing. Any thin, membranous expansion; a lateral petal of a papilionaceous flower (Leguminosae).

Zygomorphic. Bilaterally symmetrical; an irregular flower which can be bisected into similar halves in only one plane.

EXPLANATION OF AUTHORS' NAMES

A. A. Eaton. Eaton, Alvah Augustus, 1865-1908, American.
Abbe. Abbe, Ernest, 1905-, American.
A. Br. Braun, Alexander Carl Heinrich, 1805–1877, German.
Abrams. Abrams, Leroy, 1874–1956, American.
Achey. Achey, Daisy Bird, 1906–, American.
Adams. Adams, Johannes Michael Friedrich, 1780–183–, Russian.
A. DC. Candolle, Alphonse Louis Pierre Pyramus de, 1806–1893, Swiss.
A. Eaton. Eaton, Amos, 1776–1842, American.
Aellen. Aellen, Paul, 1896–, Swiss.
A. H. Moore. Moore, Albert Hanford, 1883–, American.
Ait. Aiton, William, 1731–1793, English.
Ambrosi. Ambrosi, Francesco, 1821–1897, Italian.
Ames. Ames, Oakes, 1874–1950, American.
A. M. Johnson. Johnson, Arthur Monrad, 1878–1943, American.
Anderss. Andersson, Nils Johan, 1821–1880, Swedish.
Andrz. Andrzejowski, Anton Lukianowicz, 1784–1868, Russian.
A. Nels. Nelson, Aven, 1859–1952, American.
Applegate. Applegate Elmer Ivan, 1867–1949, American.
Arrhen. Arrhenius, Johan Petter, 1811–1889, Swedish.
Asch. & Graebn. Ascherson, Paul Friedrich August, 1834–1913; Graebner, Karl Otto Robert Peter Paul, 1871–1933, German.
Aschers. Ascherson, Paul Friedrich August, 1834–1914, German.
Aschers. & Magn. Ascherson, Paul Friedrich August, 1834–1913; Magnus, Paul Wilhelm, 1844–1914, German.

Babc. Babcock, Ernest Brown, 1877–1954, American.
Babc. & Stebb. Babcock, Ernest Brown, 1877–1954, American; Stebbins, George Ledyard, Jr., 1906–, American.
Bailey. Bailey, Liberty Hyde, 1858–1954, American.
Baker. Baker, John Gilbert, 1834–1920, English.
Balf. Balfour, John Hutton, 1808–1884, English.
Ball. Ball, Carleton Roy, 1873–1958, American.
Banks. Banks, Sir Joseph, 1743–1820, English.
Barbey. Barbey, William, 1842–1914, Swiss.
Barneby. Barneby, Rupert Charles, 1911–, American.
Barnh. Barnhart, John Hendley, 1871–1949, American.
Barratt. Barratt, Joseph, 1796–1882, American.
Barton. Barton, William Paul Crillon, 1786–1856, American.
Batch. Batchelder, Frederick William, 1838–1911, American.
Beal. Beal, William James, 1833–1924, American.
Beauv. Palisot de Beauvois, Ambroise Marie François Joseph, 1752–1820, French.
Beck, G. Beck von Mannagetta, Guenther, 1856–1931, Austrian.
Beetle. Beetle, Alan Ackerman, 1913–, American.
Beissn. Beissner, Ludwig, 1843–1927, German.
Benke. Benke, Hermann Conrad, 1870–1947, American.
Benth. Bentham, George, 1800–1884, English.
Berger. Berger, Alwin, 1871–1931, German.
Bernh. Bernhardi, Johann Jacob, 1774–1850, German.
Bess. Besser, Wilibald Swibert Joseph Gottlieb, 1784–1842, Russian.
Bessey. Bessey, Charles Edwin, 1845–1915, American.
Betcke. Betcke, Ernst Friedrich, –1865, German.

B. & H. Bentham, George, 1800–1884; Hooker, Sir Joseph Dalton, 1817–1911, English.
Bickn. Bicknell, Eugene Pintard, 1859–1925, American.
Bieb. Bieberstein, Friedrich August, Marschall von, 1768–1826, Russian.
Biehler. Biehler, Johann Friedrich Theodor, German.
Bigel. Bigelow, Jacob, 1787–1879, American.
Bisch. Bischoff, Gottlieb Wilhelm, 1797–1854, German.
Biv. Bivona-Bernardi, Antonino, 1774–1837, Italian.
Blake. Blake, Sidney Fay, 1892–1959, American.
Blank. Blankinship, Joseph William, 1862–1938, American.
Blume. Blume, Carl Ludwig von, 1796–1862, Dutch.
Blytt. Blytt, Mathias Numsen, 1789–1862, Norwegian.
Boeckl. Boeckler, Johann Otto, 1803–1899, German.
Bogenhard. Bogenhard, Carl, 1811–, German.
Boiss. Boissier, Pierre Edmond, 1810–1885, Swiss.
Boivin. Boivin, Joseph Robert Bernard, 1916–, Canadian.
Boland. Bolander, Henry Nicholas, 1831–1897, American.
Bong. Bongard, Heinrich Gustav, 1786–1839, German-Russian.
Boott. Boott, Francis, 1792–1863, American-English.
Boott, W. Boott, William, 1805–1887, American.
Borbás. Borbás, Vinczé von, 1844–1905, Hungarian.
Boreau. Boreau, Alexandre, 1803–1875, French.
Br., A. Braun, Alexander Carl Heinrich, 1805–1877, German.
Bracelin. Bracelin, N. Floy (Mrs. H. P. Bracelin), 1890–, American.
Brack. Brackenridge, William Dunlop, 1810–1893, American.
Bradshaw. Bradshaw, Robert Vernon, 1896–, American.
Brain. Brainerd, Ezra, 1844–1924, American.
Brand. Brand, August, 1863–1931, German.
Brewer. Brewer, William Henry, 1828–1910, American.
Briq. Briquet, John Isaac, 1870–1931, Swiss.
Britt. & Rendle. Britten, James, 1846–1924; Rendle, Alfred Barton, 1865–1938, English.
Britton. Britton, Nathaniel Lord, 1859–1934, American.
Br., R. Brown, Robert, 1773–1858, English.
Broun. Broun, Maurice, 1906–, American.
Brown. Brown, Spencer Wharton, 1918–, American.
B. S. P. Britton, Nathaniel Lord, 1859–1934; Sterns, Emerson Ellick, 1846–1926; Poggenburg, Justus Ferdinand, 1840–1893; American.
Buch. Buchenau, Franz Georg Philipp, 1831–1906, German.
Buckl. Buckley, Samuel Botsford, 1809–1884, American.
Bunge. Bunge, Alexander von, 1803–1890, Russian.
Burm. f. Burman, Nicolaas Laurens, 1734–1793, Dutch.
Butters. Butters, Frederic King, 1878–1945, American.
Butters & Abbe. Butters, Frederic King, 1878–1945, American. Abbe, Ernest, 1905–, American.

Cambess. Cambessedes, Jacques, 1799–1863, French.
C. A. Mey. Meyer, Carl Anton von, 1795–1855, Russian.
C. & R. Coulter, John Merle, 1851–1928; Rose, Joseph Nelson, 1862–1928, American.
C. & S. Chamisso, Ludolf Adalbert von, 1781–1838; Schlechtendal, Diederich Franz Leonhard von, 1794–1866, German.
Camp. Camp, Wendell Holmes, 1904–1962, American.
Campbell. Campbell, Gloria Rae (later Day) 1934–, American.
Canby. Canby, William Marriott, 1831–1904, American.
Card. Card, Hamilton Hye, 1877–, American.
Carr. Carrière, Elie Abel, 1818–1896, French.
Cassidy. Cassidy, James, 1844?–1889, American.

Cav. Cavanilles, Antonio José, 1745–1804, Spanish.
Cavillier. Cavillier, François Georges, 1868–1953, Swiss.
Chabert. Chabert, Alfred, 1836–1916, Swiss.
Chaix. Chaix, Dominique, 1731–1800, French.
Cham. Chamisso, Ludolf Adalbert von, 1781–1838, German.
Chase. Chase, Mary Agnes (Merrill), 1869–, American.
Chat. Chatelain, Jean Jacques, 17—–17—, Swiss.
Ching. Ching, Ren-Chang, 1899–, Chinese.
Choisy. Choisy, Jacques Denis, 1799–1859, Swiss.
Church. Church, George Lyle, 1903–, American.
Clausen. Clausen, Robert Theodore, 1911–, American.
Clem. & Clem. Clements, Frederic Edward, 1874–1945, American; Clements, Edith Gertrude (Schwartz), 1877–, American.
C. L. Hitchc. Hitchcock, Charles Leo, 1902–, American.
Cockerell. Cockerell, Theodore Dru Alison, 1866–1948, American.
Coleman. Coleman, Nathan, 1825–1887, American.
Constance. Constance, Lincoln, 1909–, American.
Constance & Rollins. Constance, Lincoln, 1909–; Rollins, Reed Clarke, 1911–, American.
Coult. Coulter, John Merle, 1851–1928, American.
Cov. Coville, Frederick Vernon, 1867–1937, American.
Cov. & Britt. Coville, Frederick Vernon, 1867–1937; Britton, Nathaniel Lord, 1859–1934; American.
C. P. Sm. Smith, Charles Piper, 1877–1955, American.
C. P. Sm. & St. John. Smith, Charles Piper, 1877–1955; St. John, Harold, 1892–, American.
Crépin. Crépin, François, 1830–1903, Belgian.
Critchfield. Critchfield, William Burke, 1923–, American.
Croizat. Croizat, Léon Camille Marius, 1894–, Czechoslovakian-American.
Cronq. Cronquist, Arthur John, 1919–, American.
Curran. Curran, Mary Katherine (Layne) (later Brandegee), 1844–1920, American.
Cyrill. Cyrillo, Domenico Maria Leone, 1739–1799, Italian.

Dalla Torre. Dalla Torre, Karl Wilhelm von, 1850–1928, Austrian.
Daniels. Daniels, Francis Potter, 1869–1947, American.
Danser. Danser, Benedictus Hubertus, 1891–1943, Dutch.
Davidson. Davidson, John Fraser, 1911–, American.
Davis. Davis, Kary Cadmus, 1867–1936, American.
Davis, R. J. Davis, Ray Joseph, 1895–, American.
Davy. Davy, Joseph Burtt, 1870–1940, English.
DC. Candolle, Augustin Pyramus de, 1778–1841, Swiss.
DC., A. Candolle, Alphonse Louis Pierre Pyramus de, 1806–1893, Swiss.
D. C. Eaton. Eaton, Daniel Cady, 1834–1895, American.
Dcne. & Planch. Decaisne, Joseph, 1807–1882; Planchon, Jules Emile, 1823–1888; French.
D. Don. Don, David, 1799–1841, English.
Descourtilz. Descourtilz, Michel Etienne, 1775–1836, French.
Desf. Desfontaines, Réné Louiche, 1750–1833, French.
Desr. Desrousseaux, Louis Auguste Joseph, 1753–1838, French.
Desv. Desvaux, Nicaise Auguste, 1784–1856, French.
Dewey. Dewey, Chester, 1784–1867, American.
Dickson. Dickson, James, 1738–1822, English.
Dietr. Dietrich, Frederich Gottlieb, 1768–1850, German.
Dippel. Dippel, Leopold, 1827–1914, German.
Dode. Dode, Louis Albert, 1875–, French.
Doell. Döll, Johann Christoph, 1808–1885, German.
Don, D. Don, David, 1799–1841, English.

Don, G. Don, George, 1798–1856, English.
Donn. Donn, James, 1758–1813, English.
Dougl. Douglas, David, 1798–1834, British.
Drew, E. R. Drew, Elmer Reginald, 1865–1930, American.
Drew, W. B. Drew, William Brooks, 1908–, American.
Druce. Druce, George Claridge, 1851–1932, English.
Duby. Duby, Jean Etienne, 1798–1885, Swiss.
Dulac. Dulac, Joseph, 18— –189–, French.
Dumort. Dumortier, Barthélemy Charles Joseph, 1797–1878, Belgian.
Dunal. Dunal, Michel Félix, 1789–1856, French.
Dundas. Dundas, Frederick Winn, 1911–, American.
Dur. Durand, Elias Magloire, 1794–1873, American.
Dur. & Jacks. Durand, Theophile Alexis, 1855–1912, Belgian; Jackson, Benjamin Dayton, 1846–1927, English.
d'Urv. Dumont d'Urville, Jules Sébastian César, 1790 – 1842, French.
Dusén. Dusén, Per Karl Hjalmar, 1855–1926, Swedish.
Dyal. Dyal, Sarah Creecie (later Mrs. Etlar Nielsen), 1907–, American.

Eaton, A. Eaton, Amos, 1776–1842, American.
Eaton & Wright. Eaton, Amos, 1776–1842; Wright, John, 1811–1846, American.
Eaton, A. A. Eaton, Alvah Augustus, 1865–1908, American.
Eaton, D. C. Eaton, Daniel Cady, 1834–1895, American.
Eggleston. Eggleston, Willard Webster, 1863–1935, American.
Ehrh. Ehrhart, Friedrich, 1742–1795, Swiss-German.
Eichw. Eichwald, Carl Eduard von, 1794–1876, Russian.
Ell. Elliott, Stephen, 1771–1830, American.
Elmer. Elmer, Adolph Daniel Edward, 1870–, American.
E. Mey. Meyer, Ernst Heinrich Friedrich, 1791–1858, German.
Endl. Endlicher Stephan Ladislaus, 1804–1849, Austrian.
E. Nelson. Nelson, Elias Emanuel, 1876–, American.
Engelm. Engelmann, George, 1809–1884, American.
Engelm. & Gray. Engelmann, George, 1809–1884; Gray, Asa, 1810–1888, American.
Engl. & Irmsch. Engler, Heinrich Gustav Adolf, 1844–1930; Irmscher, Edgar, 1887–, German.
English. English, Carl Schurz, Jr., 1904–, American.
Epling. Epling, Carl Clawson, 1894–, American.
E. R. Drew. Drew E. R., ——, American.
Esch. Eschscholtz, Johann Friedrich Gustav von, 1793–1831, Russian.
Ewan, Joseph Andorfer, 1909–, American.

Farr. Farr, Edith May, 1863–American.
Farw. Farwell, Oliver Atkins, 1867–1942, American.
Fassett. Fassett, Norman Carter, 1900–1954, American.
Fedde. Fedde, Friedrich Karl Georg, 1873–1942, German.
Fée. Fée, Antoine Laurent Apollinaire, 1798–1874, French.
Fenzl. Fenzl, Eduard, 1808–1879, Austrian.
Fern. Fernald, Merritt Lyndon, 1873–1950, American.
Fern. & Griscom. Fernald, Merritt Lyndon, 1873–1950; Griscom, Ludlow, 1890–1959, American.
Fern. & Macbr. Fernald, Merritt Lyndon, 1873–1950; Macbride, J. Francis, 1892–, American.
Fern. & Weath. Fernald, Merritt Lyndon, 1873–1950; Weatherby, Charles Alfred, 1875–1949, American.
Fern. & Wieg. Fernald, Merritt Lyndon, 1873 – 1950; Wiegand, Karl McKay, 1873–1942, American.
Fiori. Fiori, Adriano, 1865–1950, Italian.

Fisch. Fischer, Friedrich Ernst Ludwig von, 1782–1854, Russian.
Fisch. & Mey. Fischer, Friedrich Ernst Ludwig von, 1782–1854; Meyer, Carl Anton von, 1795–1855, Russian.
Fisch. & Trautv. Fischer, Friedrich Ernst Ludwig von, 1782–1854, Russian; Trautvetter, Ernst Rudolph von, 1809–1889, Russian.
Fisch., G. Fischer, Gustav, 1889–, German.
Flous. Flous, Fernande, French.
Focke. Focke, Wilhelm Olbers, 1834–1922, German.
Forbes. Forbes, James, 1773–1861, English.
Formánek. Formánek, Edward, 1845–1900, Austrian.
Fosb. Fosberg, F(rancis) Raymond, 1908–, American.
Fourn. Fournier, Eugène Pierre Nicolas, 1834–1884, French.
Fourr. Fourreau, Pierre Jules, 1844–1871, French.
Franch. Franchet, Adrien, 1834–1900, French.
Franco. Franco, João Manuel Antonio Paes do Amaral, 1921– Portugese.
Frém. Frémont, John Charles, 1813–1890, American.
Fresen. Fresenius, Johann Baptist Georg Wolfgang, 1808–1866, German.
Fries. Fries, Elias Magnus, 1794–1878, Swedish.
Friesner. Friesner, Ray Clarence, 1894–, American.

Gaertn. Gaertner, Joseph, 1732–1791, German.
Garcke. Garcke, Friedrich August, 1819–1904, German.
Gaud. Gaudichaud-Beaupré, Charles, 1789–1854, French.
Gay, J. Gay, Jacques Etienne, 1786–1864, French.
G. Beck. Beck von Mannagetta, Guenther, 1856–1931, Austrian.
Gdgr. Gandoger, Michel, 1850–1926, French.
G. Don. Don, George, 1798–1856, English.
Geyer. Geyer, Carl Andreas, 1809–1853, German.
G. Fischer. Fischer, Gustav, 1889–, German.
G. F. W. Meyer. Meyer, Georg Friedrich Wilhelm, 1782–1856, German.
Gilbert. Gilbert, Benjamin Davis, 1835–1907, American.
Gilib. Gilibert, Jean Emmanuel, 1741–1814, French.
Gill. Gillies, John, ——, British.
Gill, L. S. Gill, Lake Shore, 1900–, American.
Glück. Glück, Christian Maximilian Hugo, 1868–1940, German.
G. M. & S. Gaertner, Philipp Gottfried, 1754–1825; Meyer, Bernhard, 1767–1836; Scherbius, Johannes, 1769–1813, German.
Gmel., J. F. Gmelin, Johaan Friedrich, 1748–1804, German.
Gmel. Gmelin, Johann Georg, 1709–1755, German.
G. N. Jones Jones, George Neville, 1904–, American.
Goldie. Goldie, John, 1793–1886, Canadian.
Goodd. Goodding Leslie Newton, 1880–, American.
Gould. Gould, Frank Walton, 1913–. American.
Graebn. Graebner, Karl Otto Robert Peter Paul, 1871–1893, German.
Graham. Graham, Robert, 1786–1845, British.
Grant. Grant, Adele Gerard (Lewis), 1881–, American.
Grant, V. Grant, Verne Edwin, 1917–, American.
Gray. Gray, Asa, 1810–1888, American.
Gray, S. F. Gray, Samuel Frederick, 1766–1836, English.
Greene. Greene, Edward Lee, 1842–1915, American.
Greenm. Greenman, Jesse More, 1867–1951, American.
Gren. & Godr. Grenier, Jean Charles Marie, 1808–1875; Godron, Dominique Alexandre, 1807–1880, French.
Griseb. Grisebach, August Heinrich Rudolf, 1814–1879, German.
G. S. Torr. Torrey, George Safford, 1891–, American.
Guss. Gussone, Giovanni, 1787–1866, Italian.

H. & A. Hooker, Sir William Jackson, 1785–1865; Arnott, George Arnott Walker, 1799–1868; English.
Haberer. Haberer, Joseph Valentine, 1855–1925, American.
Hack. Hackel, Eduard, 1850–1916, Austrian.
Hall. f. Haller, Albrecht von, 1758–1823, Swiss.
Hall, H. M. Hall, Harvey Monroe, 1874–1932, American.
Hamiliton. Hamilton, William, 1783–1856, English.
Hanson. Hanson, Herbert Christian, 1890–, American.
Hara. Hara, Hiroshi, 1911–, Japanese.
Hartm. Hartman, Carl Johan, 1790–1849, Swedish.
Haussk. Haussknecht, Heinrich Carl, 1838–1903, German.
Haw. Haworth, Adrian Hardy, 1768–1833, English.
Hayne. Hayne, Friedrich Gottlob, 1763–1832, German.
H. B. K. Humboldt, Friedrich Heinrich Alexander von, 1769–1859, German; Bonpland, Aimé Jacques Alexandre, 1773–1858, French; Kunth, Carl Sigismund, 1788–1850, German.
H. & C. Hall, Harvey Monroe, 1874–1932; Clements, Frederic Edward, 1874–1945, American.
Heller. Heller, Amos Arthur, 1867–1944, American.
Henders. Henderson, Louis Forniquet, 1853–1942, American.
Hieron. Hieronymus, Georg Hans Emmo Wolfgang, 1846–1921, German.
Hill. Hill, John, 1716–1775, English.
Hitchc. Hitchcock, Albert Spear, 1865–1935, American.
Hitchc, C. L. Hitchcock, Charles Leo, 1902–, American.
H. M. Hall. Hall, Harvey Monroe, 1874–1932, American.
Hoffm. Hoffmann, Georg Franz, 1761–1826, German.
Holm. Holm, Herman Theodor, 1854–1932, American.
Holz. Holzinger, John Michael, 1853–1929, American.
Hook. Hooker, Sir William Jackson, 1785–1865, English.
Hook. & Grev. Hooker, Sir William Jackson, 1785–1865; Greville, Robert Kaye, 1794–1866, English.
Hoover. Hoover, Robert Francis, 1913–, American.
Hoppe. Hoppe, David Heinrich, 1760–1846, German.
Hornem. Hornemann, Jens Wilken, 1770–1841, Danish.
Host. Host, Nicolaus Thomas, 1761–1834, Austrian.
House. House, Homer Doliver, 1878–, American.
Howell. Howell, Thomas, 1842–1912, American.
Howell, J. T. Howell, John Thomas, 1903–, American.
Hu. Hu, Shiu-ying, 1910–, Chinese-American.
Hubb. Hubbard, Frederick Tracy, 1875–1962, American.
Huds. Hudson, William, 1730–1793, English.
Hultén & St. John. Hultén, Oskar Eric Gunnar, 1894–, Swedish; St. John, Harold, 1892–, American.

Iljin. Iljin (or Il'in), Modest Mikhaĭlovich, 1889–, Russian.

Jacq. Jacquin, Nickolaus Josef von, 1727–1817, Austrian.
James. James, Edwin, 1797–1861, American.
J. & C. Presl. Presl, Jan Swatopluk, 1791–1849; Presl, Karel Bořiwog, 1794–1852, Czech.
Jepson. Jepson, Willis Linn, 1867–1946, American.
J. F. Gmel. Gmelin, Johann Friedrich, 1748–1804, German.
J. Gay. Gay, Jacques Etienne, 1786–1864, French.
J. G. Smith. Smith, Jared Gage, 1866–1957, American.
Johnson, A. M. Johnson, Arthur Monrad, 1878–1943, American.
Johnston. Johnston, Ivan Murray, 1898–1960, American.
Jones, G. N. Jones, George Neville, 1904–, American.
Jones, M. E. Jones, Marcus Eugene, 1852–1934, American.

Jones, Q. Jones, Quentin, 1920–, American.

Jordal. Jordal, Louis Henrik, 1919–1951, Norwegian-American.

J. T. Howell. Howell, John Thomas, 1903–, American.

Juss. Jussieu, Antoine Laurent de, 1748–1836, French.

Karsch. Karsch, Anton, 1822–1892, German.

Karst. Karsten, Gustav Karl Wilhelm Hermann, 1817–1908, Austrian.

Kaulf. Kaulfuss, Georg Friedrich, 1786–1830, German.

Kearney. Kearney, Thomas Henry, 1874–, American.

Keck. Keck, David Daniels, 1903–, American.

Keller. Keller, Robert, 1854–1939, Swiss.

Kellogg. Kellogg, Albert, 1813–1887, American.

Kirchner. Kirchner, Emil Otto Oskar von, 1851–1925, German.

Kit. Kitaibel, Paul, 1757–1817, Hungarian.

Kl. & Garcke. Klotzsch, Johann Friedrich, 1805–1860; Garcke, Friedrich August, 1819–1904, German.

Klett. & Richt. Klett, Gustav Theodor, —— –1827; Richter, Hermann Eberhard Friedrich, 1808–1876, German.

Koch. Koch, Wilhelm Daniel Joseph, 1771–1849, German.

Koehne. Koehne, Bernhard Adalbert Emil, 1848–1918, German.

Koeler. Koeler, Georg Ludwig, 1764–1807, German.

Kükenth. Kükenthal, Georg, 1864–1955, German.

Kuhn. Kuhn, Maximilian Friedrich Adalbert, 1842–1894, German.

Kunth. Kunth, Carl Sigismund, 1788–1850, German.

Kuntze. Kuntze, Carl Ernst Otto, 1843–1907, German.

L. Linnaeus, Carolus, 1707–1778, Swedish.

Lag. & Rodr. Lagasca y Segura, Mariano, 1776–1839; Rodriguez, José Demetrio, 1780–1846, Spanish.

Lakela. Lakela, Olga, 1890–, Finnish-American.

Lam. Lamarck, Jean Baptiste Antoine Pierre Monnet de, 1744–1829, French.

Lam. & DC. Lamarck, Jean Baptiste Antoine Pierre Monnet de, 1744–1829, French; Candolle, Augustin Pyramus de, 1778–1841, Swiss.

Lamb. Lambert, Aylmer Bourke, 1761–1842, English.

Lange. Lange, Johan Martin Christian, 1818–1898, Danish.

Lawson, Sir Charles, 1794–1873, Scotch.

Lawson. Lawson, George, 1827–1895, Canadian.

Lawson, P. & C. Lawson, Peter, –d. 1820, Scotch; Lawson, Sir Charles, 1794–1873, Scotch.

Lecoq & Lamotte. Lecoq, Henri, 1802–1871; Lamotte, Martial, 1820–1883, French.

Ledeb. Ledebour, Carl Friedrich von, 1785–1851, Russian.

Lehm. Lehmann, Johann Georg Christian, 1792–1860, German.

Leiberg. Leiberg, John Bernhard, 1853–1913, American.

Lejeune. Lejeune, Alexander Louis Simon, 1779–1858, Belgian.

Lemmon. Lemmon, John Gill, 1832–1908, American.

Lepage. Lepagc, (Abbé) Ernest, 1905–, Canadian.

Less. Lessing, Christian Friedrich, 1809–1862, German.

Leyss. Leysser, Friedrich Wilhelm von, 1731–1815, German.

L. f. Linnaeus, Carolus, the son, 1741–1783, Swedish.

L'Hér. L'Héritier de Brutelle, Charles Louis, 1746–1800, French.

Lindl. Lindley, John, 1799–1865, English.

Link. Link, Johann Heinrich Friedrich, 1767–1851, German.

Lodd. Loddiges, Conrad, 1738–1826, English.

Longyear. Longyear, Burton Orrange, 1868–, American.

Loud. Loudon, John Claudius, 1783–1843, English.

L. S. Gill. Gill, Lake Shore, 1900–, American.

Lyons. Lyons, Israel, 1739–1775, English.

Macbr. Macbride, J. Francis, 1892–, American.
McDermott. McDermott, Laura Frances, 18— –19—, American.
McGregor. McGregor, Ernst Alexander, 1880–, American.
Mack. Mackenzie, Kenneth Kent, 1877–1934, American.
Mack. & Bush. Mackenzie, Kenneth Kent, 1877–1934; Bush, Benjamin Franklin, 1858–1937, American.
MacM. MacMillan, Conway, 1867–1929, American.
Macoun. Macoun John 1832–1920, Canadian.
McVaugh. McVaugh, Rogers, 1909–, American.
Maguire. Maguire, Bassett, 1904–, American.
Malte. Malte, Malte Oskar, 1880–1933, Canadian.
Marsh. Marshall, Humphry, 1722–1801, American.
Mart. Martius, Karl Friedrich Philipp von, 1794–1868, German.
Mart. & Gal. Martens, Martin, 1797–1863; Galeoti, Henri Guillaume, 1814–1858, Belgian.
Martens. Martens, Martin, 1797–1863, Belgian.
Mason. Mason, Herbert Louis, 1896–, American.
Mathias. Mathias, Mildred Esther, 1906–, American.
Mattf. Mattfield, Joannes, 1895–1960, German.
Maxim. Maximowicz, Karl Johann, 1827–1891, Russian.
Maxon. Maxon, William Ralph, 1877–1948, American.
Medic. Medicus, Friedrich Casimir, 1736–1808, German.
Meinsh. Meinshausen, Karl Friedrich, 1819–1899, German.
Meisn. Meisner or Meissner, Carl Friedrich, 1800–1874, Swiss.
M. E. Jones. Jones, Marcus Eugene, 1852–1934, American.
Merr. & Davy. Merrill, Elmer Drew, 1876–, American; Davy, Joseph Burtt, 1870–1940, English.
Merrill. Merrill, Elmer Drew, 1876–, American.
Mert. & Koch. Mertens, Franz Carl, 1764–1831; Koch, Wilhelm Daniel Joseph, 1771–1849, German.
Mey, C. A. Meyer, Carl Anton von. 1795–1855, Russian.
Mey, E. Meyer, Ernst Heinrich Friedrich, 1791–1858, German.
Meyer, G. F. W. Meyer, Georg Friedrich Wilhelm, 1782–1856, German.
Michx. Michaux, André, 1746–1802, French.
Mill. Miller, Phillip, 1691–1771, English.
Miller, W. Miller, William Tyler, 1869–1938, American.
Milliken. Milliken, Jessie (later Mrs. Warner Brown), 1877–1951, American.
Millsp. Millspaugh, Charles Frederick, 1854–1923, American.
Miq. Miquel, Friedrich Anton Wilhelm, 1811–1871, Dutch.
Mirb. Mirbel, Charles Francois Brisseau de, 1776–1854, French.
Moench. Moench, Conrad, 1744–1805, German.
Molina. Molina, Juan Ignacio, 1737–1829, Spanish.
Moore. Moore, Thomas, 1821–1887, English.
Moore, A. H. Moore, Albert Hanford, 1883–, American.
Moq. Moquin-Tandon, Christian Horace Bénédict Alfred, 1804–1863, French.
Mor. Moretti, Giuseppi, 1782–1853, Italian.
Morong. Morong, Thomas, 1827–1894, American.
Morton. Morton, Conrad Vernon, 1905–, American.
Muell., O. F. Mueller, Otto Fridrich, 1730–1784, Danish.
Muhl. Mulhenberg, Gotthilf Henry Ernest, 1753–1815, American.
Mulford. Mulford, A. Isabel, 18— –, American.
Mulligan. Mulligan, Gerald Albert, 1928–, Canadian.
Munro. Munro, William, 1818–1880, English.
Munz. Munz, Philip Alexander, 1892–, American.
Murr. Murr, Josef, ——, German.
Murray. Murray, Johann Andreas, 1740–1791, German.

Nash. Nash, George Valentine, 1864–1921, American.
Nees. Nees von Esenbeck, Christian Gottfried Daniel, 1776–1858, German.
Nees & Mey. Nees von Esenbeck, Christian Gottfried Daniel, 1776–1858; Meyen, Franz Julius Ferdinand, 1804–1840, German.
Nels., A. Nelson, Aven, 1859–1952, American.
Nels. & Macbr. Nelson, Aven, 1859–1952; Macbride, J. Francis, 1892–, American.
Nelson, E. Nelson, Elias Emanuel, 1876–, American.
Nevskii. Nevskii, Sergeĭ Arsent'-evich, 1908–1938, Russian.
Newman. Newman, Edward, 1801–1876, English.
Nicholson. Nicholson, George, 1874–1958, English.
Nieuwl. Nieuwland, Julius Aloysius Arthur, 1878–1936, American.
Norton. Norton, John Bitting Smith, 1872–, American.
Nutt. Nuttall, Thomas, 1786–1859, American-English.

Oakes. Oakes, William, 1799–1848, American.
Oeder. Oeder, Georg Christian von, 1728–1791, Danish.
O. E. Schulz. Schulz, Otto Eugen, 1874–1936, German.
O. F. Muell. Mueller, Otto Fridrich, 1730–1784, Danish.
Olney. Olney, Stephen Thayer, 1812–1878, American.
Onno. Onno, Max, 1903–, Austrian.
Ottley. Ottley, Alice Maria, 1882–, American.

Pall. Pallas, Peter Simon, 1741–1811, German.
Parish. Parish, Samuel Bonsall, 1838–1928, American.
Parodi. Parodi, Lorenzo Raimundo, 1895–, Argentinian.
Parry. Parry, Charles Christopher, 1823–1890, American.
Pax. Pax, Ferdinand Albin, 1858–1942, German.
Pays. Payson, Edwin Blake, 1893–1927, American.
Payson & St. John. Payson, Edwin Blake, 1893–1927; St. John, Harold, 1892–, American.
P. & C. Lawson. Lawson, Peter, –d. 1920, Scotch; Lawson, Sir Charles, 1794–1873, Scotch.
Pease & Moore. Pease, Arthur Stanley, 1881–; Moore, Albert Hanford, 1883–, American.
Peck. Peck, Morten Eaton, 1871–1959, American.
Pennell. Pennell, Francis Whittier, 1886–1952, American.
Pers. Persoon, Christiaan Hendrik, 1761–1836, Dutch.
Peterm. Petermann, Wilhelm Ludwig, 1806–1855, German.
Pfeiffer. Pfeiffer, Norma Etta, 1889–, American.
Phil. Philippi, Rudolf Amandus, 1808–1904, Chilean.
Philipps, Lyle Llewellyn, 1923–, American.
Piper. Piper, Charles Vancouver, 1867–1926, American.
Piper & Beattie. Piper, Charles Vancouver, 1867–1926; Beattie, Rolla Kent, 1875–, American.
Piper & Brodie. Piper, Charles Vancouver, 1867–1926; Brodie, David Arthur, 1868–, American.
Piper & Hitchc. Piper, Charles Vancouver, 1867–1926; Hitchcock, Albert Spear, 1865–1935, American.
Planch. Planchon, Jules Emile, 1823–1888, French.
Podpera. Podpera, Josef, 1878–, Czech.
Poe. Poe, Ione, 1899–, American.
Poellnitz. Poellnitz, Karl von, 1896–1945, German.
Poepp. Poeppig, Eduard Friedrich, 1798–1868, German.
Poir. Poiret, Jean Louis Marie, 1755–1834, French.
Poll. Pollich, Johann Anton, 1740–1780, German.
Porter. Porter, Thomas Conrad, 1822–1901, American.
Porter & Coult. Porter, Thomas Conrad, 1822–1901, American; Coulter, John Merle, 1851–1928, American.
Presl. Presl, Karel Bořiwog, 1794–1852, Czech.

Presl. J. & C. Presl, Jan Swatopluk, 1791–1849; Presl, Karel Bořiwog, 1794–1852, Czech.
Pritzel. Pritzel, Georg August, 1815–1874, German.
Pursh. Pursh, Frederick Traugott, 1774–1820, German.

Rabenh. Rabenhorst, Gottlob Ludwig, 1806–1881, German.
Raf. Rafinesque, Constantine Samuel, 1783–1840, American.
Raup. Raup, Hugh Miller, 1901–, American.
R. Br. Brown, Robert, 1773–1858, English.
Rech. f. Rechinger, Karl Heinz, 1906–, Austrian.
Regel. Regel, Eduard August von, 1815–1892, Russian.
Rehder. Rehder, Alfred, 1863–1949, American.
Reichenb. Reichenbach, Heinrich Gottlieb Ludwig, 1793–1879, German.
Richards. Richardson, John, 1787–1865, English.
Richt. Richter, Karl, 1855–1891, German.
Rickett. Rickett, Harold William, 1896–, American.
Robins. Robinson Benjamin Lincoln, 1864–1935, American.
Robins. & Fern. Robinson, Benjamin Lincoln, 1864–1935; Fernald, Merritt Lyndon, 1873–1950, American.
Roemer. Roemer, Johann Jacob, 1763–1819, Swiss.
Rogers. Rogers, William Moyle, 1835–1920, English.
Rollins. Rollins, Reed Clark, 1911–, American.
Rose. Rose, Joseph Nelson, 1862–1928, American.
Rosend. Rosendahl, Carl Otto, 1875–1956, American.
Rostk. & Schmidt. Rostkovius, Friedrich Wilhelm Theophil, 1770–1848; Schmidt, Wilhelm Ludwig Ewald, 1804–1843, German.
Roth. Roth, Albrecht Wilhelm, 1757–1834, German.
Rottb. Rottböll, Christen Friis, 1727–1797, Danish.
Rouleau. Rouleau, Joseph Albert Ernest, 1916–, Canadian.
Rowland. Rowland, Verner Hawsbrook, 1883–, American.
R. & S. Roemer, Johann Jacob, 1763–1819, Swiss; Schultes, Joseph August, 1773–1831, Austrian.
Rupr. Ruprecht, Franz Josef, 1814–1870, Russian.
Rydb. Rydberg, Per Axel, 1860–1931, American.

St. John. St. John, Harold, 1892–, American.
St. John & Constance. St. John Harold, 1892–; Constance, Lincoln, 1909–, American.
St. John & Jones, G. N. St. John, Harold, 1892–; Jones, George Neville, 1904–, American.
St. John & Warren. St. John, Harold, 1892–; Warren, Fred Adelbert, 1902–1951, American.
St. John & Weitm. St. John, Harold, 1892–; Weitman, Gladys, (later Mrs. Will W. deNeff), American.
St.-Yves. Saint-Yves, Alfred, 1855–1933, French.
Salisb. Salisbury, Richard Anthony, 1761–1829, English.
Sam. Samuelsson, Gunnar, 1885–1944, Swedish.
Sanson. Sanson, M.
Sarg. Sargent, Charles Sprague, 1841–1927, American.
Schaffner. Schaffner, John Henry, 1866–1939, American.
Scheutz. Scheutz, Nils Johan Wilhelm, 1836–1889, Swedish.
Schinz & Keller. Schinz, Hans, 1858–1941, Swiss; Keller, Robert, 1854–1939, Swiss.
Schk. Schkuhr, Christian, 1741–1811, German.
Schlecht. Schlechtendal, Diederich Franz Leonhard von, 1794–1866, German.
Schleich. Schleicher, Johann Christoph, 1768–1834, Swiss.
Schleid. Schleiden, Matthias Jacob, 1804–1881, German.
Schmidt. Schmidt, Franz Wilibald, 1764–1796, Czech.
Schneid. Schneider, Camillo Karl, 1876–1951, German.

Schott. Schott, Heinrich Wilhelm, 1794–1865, Austrian.
Schrad. Schrader, Heinrich Adolph, 1767–1836, German.
Schrank. Schrank, Franz von Paula von, 1747–1835, German.
Schreb. Schreber, Johann Christian Daniel von, 1739–1810, German.
Schult. Schultes, Joseph August, 1773–1831, Austrian.
Schultz-Bip. Schultz, Carl Heinrich, 1805–1867, German.
Schulz, O. E. Schulz, Otto Eugen, 1874–1936, German.
Schweigg., Schweigger, Friedrich August, 1783–1821, German.
Schwein. Schweinitz, Lewis David de, 1780–1834, American.
Schwer. Schwerin, Fritz Kurt Alexander von, 1856–1934, German.
Scop. Scopoli, Johann Anton, 1723–1788, Italian.
Scribn. Scribner, Frank Lamson, 1851–1938, American.
Scribn. & Smith. Scribner, Frank Lamson, 1851–1938; Smith, Jared Gage, 1866–1957, American.
Scribn. & Tweedy. Scribner, Frank Lamson, 1851–1938; Tweedy, Frank, 1854–, American.
Scribn. & Williams. Scribner, Frank Lamson, 1851–1938; Williams, Thomas Albert, 1865–1900, American.
Sendt. Sendtner, Otto, 1813–1859, Montgasque.
Seub. Seubert, Moritz, 1818–1878, German.
S. F. Gray. Gray, Samuel Frederick, 1766–1836, English.
Sharp. Sharp, Seymour Sereno, 1893–, American.
Shear. Shear, Cornelius Lott, 1865–1956, American.
Sheldon. Sheldon, Edmund Perry, 1869–, American.
Sibth. Sibthorp, John, 1758–1796, English.
Sims. Sims, John, 1749–1831, English.
Slosson. Slosson, Margaret, 187– –, American.
Sm. Smith, James Edward, 1759–1828, English.
Sm., C. P. Smith, Charles Piper, 1877–1955, American.
Sm., C. P. & St. John. Smith, Charles Piper, 1877–1955; St. John, Harold, 1892–, American.
Small. Small, John Kunkel, 1869—1938, American.
Smiley. Smiley, Frank Jason, 1880–, American.
Smith, J. G. Smith, Jared Gage, 1866–1957, American.
Smyth. Smyth, Bernard Bryan, 1843–1913, American.
Sobol. Sobolewski, Gregor, 1741–1807, Russian.
Sonder. Sonder, Otto Wilhelm, 1812–1881, German.
Spach. Spach, Edouard, 1801–1879, French.
Spenner. Spenner, Fridolin Karl Leopold, 1798–1841, German.
Spreng. Sprengel, Curt Polycarp Joachim, 1766–1833, German.
Stanf. Stanford, Ernest Elwood, 1888–, American.
Stebbins. Stebbins, George Ledyard, 1906–, American.
Steud. Steudel, Ernst Gottlieb, 1783–1856, German.
Steyerm. Steyermark, Julian Alfred, 1909–, American.
Stockwell. Stockwell, William Palmer, 1898–1950, American.
Stokes. Stokes, Jonathan, 1755–1831, English.
Stuntz. Stuntz, Stephen Conrad, 1875–1918, American.
Stur. Stur, Dionys, 1827–1893, Austrian.
Sudw. Sudworth, George Bishop, 1864–1927, American.
Suksd. Suksdorf, Wilhelm Nikolaus, 1850–1932, American.
Svens. Svenson, Henry Knute, 1897–, American.
Sw. Swartz, Olaf Peter, 1760–1818, Swedish.
Swallen. Swallen, Jason Richard, 1903–, American.
Sweet. Sweet, Robert, 1783–1835, English.
Swingle. Swingle, Walter Tennuyson, 1871–1952, American.

Taylor, T. M. C. Taylor, Thomas Mayne Cunningham, 1904–, South African-Canadian.
Tenore. Tenore, Michele, 1780–1861, Italian.

T. & G. Torrey, John, 1796–1873; Gray, Asa, 1810–1888, American.
Thell. Thellung, Albert, 1881–1928, Swiss.
Thompson. Thompson, Henry Joseph, 1921–, American.
Thurb. Thurber, George, 1821–1890, American.
Th. Wolf. Wolf, (Franz) Theodor, 1841–1924, German.
Tidestr. Tidestrom, Ivar, 1865–1956, American.
T. M. C. Taylor. Taylor, Thomas Mayne Cunningham, 1904–, South African-Canadian.
Torr. Torrey, John, 1796–1873, American.
Torr., G. S. Torrey, George Safford, 1891–, American.
Torr. & Hook. Torrey, John, 1796–1873, American; Hooker, Sir William Jackson, 1785–1865, English.
Trautv. Trautvetter, Ernst Rudolf von, 1809–1889, Russian.
Trel. Trelease, William, 1857–1945, American.
Trev. Treviranus, Ludolf Christian, 1779–1864, German.
Trin. Trinius, Karl Bernhard von, 1778–1884, German-Russian.
Trin. & Rupr. Trinius, Karl Bernhard von, 1778–1884, German-Russian; Ruprecht, Franz Josef, 1814–1870, Russian.
Tryon. Tryon, Rolla Milton 1916–, American.
Tuckerm. Tuckerman, Edward, 1817–1886, American.
Turcz. Turczaninow, Nicolaus, 1796–1864, Russian.
Tweedy. Tweedy, Frank, 1854–1937, American.

Underw. Underwood, Lucian Marcus, 1853–1907, American.
Urb. & Gilg. Urban, Ignatz, 1848–1931; Gilg, Ernst Friedrich, 1867–1933, German.

Vahl. Vahl, Martin Hendriksen, 1749–1804, Danish.
Vail. Vail, Anna Murray, 1863–, American.
Van Hall. Hall, Herman Christiaan van, 1801–1874, Dutch.
Vasey. Vasey, George, 1822–1893, American.
Vasey & Scribn. Vasey, George, 1822–1893; Scribner, Frank Lamson, 1851–1938, American.
Vell. Vellozo, José Marianno da Conçeicão, 1742–1811, Brazilian.
V. Grant. Grant, Verne Edwin, 1917–, American.
Viviani. Viviani, Domenico, 1772–1840, Italian.
Voss. Voss, Andreas, 1857–1924, German.

Wahl. Wahlenberg, Göran, 1780–1851, Swedish.
Waldst. & Kit. Waldstein, Franz de Paula Adam von, 1759–1823; Kitaibel, Paul, 1757–1817, Hungarian.
Wallr. Wallroth, Carl Friedrich Wilhelm, 1792–1857, German.
Walp. Walpers, Wilhelm Gerhard, 1816–1853, German.
Walt. Walter, Thomas, 1740–1789, English-American.
Wangerin. Wangerin, Walther, 1884–, German.
Warren. Warren, Fred Adelbert, 1902–1951, American.
Wats. Watson, Sereno, 1826–1892, American.
W. A. Weber. Weber, William Alfred, 1918–, American.
W. B. Drew. Drew, William Brooks, 1908–, American.
W. Boott. Boott, William, 1805–1887, American.
Webb. Webb, Philipp Barker, 1793–1854, English.
Webb. & Moq. Webb, Philip Barker, 1793–1854, English; Moquin-Tandon, Christian Horace Bénédict Alfred, 1804–1863, French.
Weber. Weber, Georg Heinrich, 1752–1828, German.
Weber, W. A. Weber, William Alfred, 1918–, American.
Weigel. Weigel, Johann Adam Valentin, 1740–1806, German.
Weihe. Weihe, Carl Ernst August, 1779–1834, German.
Weihe & Nees. Weihe, Carl Ernst August, 1779–1834; Nees von Esenbeck, Christian Gottfried Daniel, 1776–1858, German.
Wheeler. Wheeler, Louis Cutter, 1910–, American.
Wheelock. Wheelock, William Efner, 1852–1927, American.

White. White, Theodore Greely, 1872–1901, American.

Widder. Widder, Felix Joseph, 1892–, Austrian.

Wieg. Wiegand, Karl McKay, 1873–1942, American.

Wiggers. Wiggers, Fredericus Henricus, –1799, German.

Wight. Wight, William Franklin, 1874–, American.

Wiks. Wikström, Johann Emanuel, 1789–1856, Swedish.

Willd. Willdenow, Carl Ludwig, 1765–1812, German.

Williams. Williams, Louis Otho, 1908–, American.

Wimm. Wimmer, Christian Friedrich Heinrich, 1803–1868, German.

Witasek. Witasek, Johanna, 1865–1910, Austrian.

With. Withering, William, 1741–1799, English.

Wittrock. Wittrock, Gustave Ludwig, 1895–, American.

W. Miller. Miller, Wilhelm, 1869–1938, American.

Th. Wolf. (Franz) Theodor, 1841–1924, German.

Woods. Woodson, Robert Everard, Jr. 1904–, American.

Woodville & Wood. Woodville, William, 1752–1805; Wood, William, 1745–1808, English.

Woot. & Standl. Wooton, Elmer Ottis, 1865–1945; Standley, Paul Carpenter, 1884–1963, American.

Woynar. Woynar, Heinrich, Karl, 1865–1917, Austrian.

Yuncker. Yuncker, Truman George, 1891–, American.

Zahn. Zahn, Karl Hermann, 1865–1940, German.

Zobel. Zobel, August, 1861–1934, German.

ADDITIONS AND CORRECTIONS

To insert on p. 262:

LEGUMINOSAE

Trifolium procumbens L. Annual, 20–50 cm. tall; stems appressed pilosulous, later glabrate, decumbent or ascending; leaflets 15–20 (–30) mm. long, oblong to obovate, blunt or obtuse, serrate; stipules ovate to lance-ovate, the base expanded and rounded; peduncles exceeding the leaves; heads 9–12 mm. long, elongating to 20–50 mm., 7–10 mm. broad; pedicels ½ length of calyx-tube; lower calyx teeth longer than the upper or than the tube; corolla 4–5 mm. long, yellow, turning light brown; standard obovate, expanded; wings and keel enfolded; style ¼ as long as the 2–2.5 mm. pod which is usually 1-seeded; seed ovoid, yellow, rather shiny. Introduced from Eurasia or Africa, to dry and sandy places.

To insert on p. 2:

SELAGINELLA

Dr. G. N. Jones of the University of Illinois kindly called my attention to a collection of *Selaginella Douglasii* made within our geographic limits. This report reached me while this book was in press. At my request he sent me a key and description of the species. Unlike my previous judgement, he accepts also *S. scopulorum* in the area. His manuscript arrived when this text was in page proof, too late to be incorporated on page 2. Since it is wholly his taxonomy and his writing, it is printed below just as he wrote it.

Key and Descriptions of Species of Selaginella
By G. Neville Jones

Leaves 4-ranked, the lateral ones larger, obtuse. *S. douglasii.*
Leaves arranged radially in several ranks, lanceolate,
  Leaves on the underside of the stem slightly longer that the others and tending to curve upward; stems densely tufted, 3–6 cm. long. *S. scopulorum.*
  Leaves uniform on all sides of the stem; stems loosely tufted, 5–15 cm. long. *S. wallacei.*

Selaginella douglasii (Hook. & Grev.) Spring. Stems prostrate, 15–30 cm. long, rooting throughout; branches alternate, distant, with about 3–6 divisions, these 1–3-divided, leafy throughout; lateral leaves complanate, yellowish green, oval, 2–3 mm. long, 1–2 mm. broad, obtuse, ciliate at the unequally auriculate base; leaves of the upper row smaller, narrowly rhombic-ovate, apiculate; spikes erect, numerous, quadrangular, 5–15 mm. long; sporophylls imbricate, deltoid-ovate, acuminate, membranous, sharply carinate. Moist wooded slopes along Clearwater River, five miles east of Spalding, Nez Perce Ce., April 2, 1950, *W. H. Baker* 6459.

Selaginella scopulorum Maxon. Plants densely tufted; stems prostrate, 3–6 cm. long, the branches numerous, ascending, 0.5–1.5 cm. long, with short, divaricate, rigid secondary branches; leaves linear-lanceolate, obtusish, those of the larger branches 2.5–3.5 mm. long, less than 0.5 mm. broad, tipped with a whitish-hyaline, scabrous awn, 0.3–0.6 mm. long, and with 4–8 cilia on each margin; spikes numerous, 1–3 cm. long, slender, curved; sporophylls ovate-deltoid, abruptly awned; megaspores orange, 0.3–0.4 mm. in diameter, reticulate; microspores about 0.004 mm. in diameter. On rocky ledges or talus, Mica Peak, Spokane Co., *Suksdorf* 8834, and other collections.

Selaginella wallacei Hieron. Plants loosely tufted; stems prostrate, 5–15 cm. long, the branches numerous, ascending, 1–4 cm. long, the divisons short, divaricate, rigid; leaves linear, obtusish, those of the larger branches 2.5–3.5 mm. long, less than 0.5 mm. broad, tipped with a whitish-hyaline, scabrous awn 0.1–0.3 mm. long and with 8–14 cilia on each margin; spikes numerous, 1–3 cm. long, slender, curved; sporophylls ovate-deltoid, abruptly awned; megaspores yellowish, 0.4 mm. in diameter, tuberculate-reticulate; microspores about 0.04 mm. in diameter. Abundant on rocky ledges.

By the regulations of the International Code of Botanical Nomenclature, 1961, it is now required that when a species has an accepted subspecific taxon, as subspecies, variety, etc., that there is automatically created a corresponding subspecific taxon for the species itself. This new subspecific taxon must bear the same epithet as the species and be cited without any author. For instance, the existence of *Polygonum bistortoides* Pursh, var. *oblongifolium* (Meisn.) St. John, causes us to refer to the species as *P. bistortoides* Pursh, var. *bistortoides.* The present author has complied with this regulation and applied it generally in this third edition of his book. However, in his reading of the last page proof, he discovered that in several instances the species had not been given a subspecific taxon with the identical epithet or name, to be parallel with the accepted other subspecific taxa. The failure to include this named subspecific taxon for the alpha minor taxon was inadvertent. Users of this book are advised to supply the needed name in the few cases where is was omitted.

LIST OF NEW SPECIES, VARIETIES, AND FORMS

INDEX TO SCIENTIFIC AND COMMON NAMES

Accepted names in Roman type; synonyms and excluded plants in Italics.